ENGINEERING DESIGN

FIFTH EDITION

George E. Dieter
University of Maryland

Linda C. Schmidt
University of Maryland

The McGraw-Hill Companies

ENGINEERING DESIGN, FIFTH EDITION

7 8 9 LHN 21 20 19 18

ISBN 978-0-07-339814-3
MHID 0-07-339814-4

Vice President & Editor-in-Chief: *Marty Lange*
Vice President of Specialized Publishing: *Janice M. Roerig-Blong*
Editorial Director: *Michael Lange*
Global Publisher: *Raghothaman Srinivasan*
Senior Sponsoring Editor: *Bill Stenquist*
Marketing Manager: *Curt Reynolds*
Developmental Editor: *Lorraine K. Buczek*
Project Manager: *Melissa M. Leick*
Design Coordinator: *Brenda A. Rolwes*
Cover Designer: *Studio Montage, St. Louis, Missouri*
Cover Image: © *Getty Images RF*
Buyer: *Laura Fuller*
Media Project Manager: *Balaji Sundararaman*
Compositor: *Cenveo Publisher Services*
Typeface: *10.5/12 Times Roman*
Printer: *LSC Communications*

Library of Congress Cataloging-in-Publication Data

Dieter, George Ellwood.
 Engineering design. — 5th ed. / George E . Dieter, Linda C . Schmidt.
  p. cm.
 ISBN 978-0-07-339814-3 (acid-free paper)
 1. Engineering design. I. Schmidt, Linda C. II. Title.
 TA174.D495 2013
 620'.0042—dc23

                                                          2011048958

www.mhhe.com

ABOUT THE AUTHORS

GEORGE E. DIETER is Glenn L. Martin Institute Professor of Engineering at the University of Maryland. The author received his B.S. Met.E. degree from Drexel University and his D.Sc. degree from Carnegie Mellon University. After a stint in industry with the DuPont Engineering Research Laboratory, he became head of the Metallurgical Engineering Department at Drexel University, where he later became Dean of Engineering. Professor Dieter later joined the faculty of Carnegie Mellon University as Professor of Engineering and Director of the Processing Research Institute. He moved to the University of Maryland in 1977 as professor of Mechanical Engineering and Dean of Engineering, serving as dean until 1994.

Professor Dieter is a fellow of ASM International, TMS, AAAS, and ASEE. He has received the education award from ASM, TMS, and SME, as well as the Lamme Medal, the highest award of ASEE. He has been chair of the Engineering Deans Council, and president of ASEE. He is a member of the National Academy of Engineering. He also is the author of *Mechanical Metallurgy,* published by McGraw-Hill, now in its third edition.

LINDA C. SCHMIDT is an Associate Professor in the Department of Mechanical Engineering at the University of Maryland. Dr. Schmidt's general research interests and publications are in the areas of mechanical design theory and methodology, design generation systems for use during conceptual design, design rationale capture, and effective student learning on engineering project design teams.

Dr. Schmidt completed her doctorate in Mechanical Engineering at Carnegie Mellon University with research in grammar-based generative design. She holds B.S. and M.S. degrees from Iowa State University for work in Industrial Engineering. Dr. Schmidt is a recipient of the 1998 U.S. National Science Foundation *Faculty Early Career Award* for generative conceptual design. She co-founded RISE, a summer research experience that won the 2003 Exemplary Program Award from the American College Personnel Association's Commission for Academic Support in Higher Education. Dr. Schmidt was awarded the American Society of Engineering Education's 2008 Merryfield Design Award.

Dr. Schmidt is active in engineering design theory research and teaching engineering design to third- and fourth-year undergraduates and graduate students in mechanical engineering. She has coauthored a text on engineering decision-making, two editions of a text on product development, and a team-training curriculum for faculty using engineering student project teams. Dr. Schmidt was the guest editor of the *Journal of Engineering Valuation & Cost Analysis* and has served as an Associate Editor of the ASME *Journal of Mechanical Design*. Dr. Schmidt is a member of ASME, SME, and ASEE.

BRIEF CONTENTS

DETAILED CONTENTS

PREFACE TO FIFTH EDITION

THE FIFTH EDITION of *Engineering Design* continues the reorganization and expansion of topics introduced in the fourth edition. Major reorganization of topics to improve flow of information and increase learning have been made in Chapter 3, Problem and Need Identification; Chapter 6, Concept Generation; and Chapter 7, Decision Making and Concept Selection. A new, progressive example has been introduced and is continued through these three chapters. A new Chapter 10, Design for Sustainability and the Environment, has been added. The book continues its tradition of being more oriented to material selection, design for manufacturing, and design for quality than other broad-based design texts.

The text is intended to be used in either a junior or senior engineering design course with an integrated hands-on design project. At the University of Maryland we present the design process material, Chapters 1 through 9, to junior students in a course introducing the design process. The whole text is used in the senior capstone design course that includes a complete design project, starting from selecting a market to creating a working prototype. Students move quickly through the first nine chapters and emphasize Chapters 10 through 17 in making embodiment design decisions.

The authors hope that students will consider this book to be a valuable part of their professional library. Toward this end we have continued and expanded the practice of giving key literature references and referrals to useful websites. Many new references have been added and all websites have been verified as of June 2011. References to many of the design handbooks and design monographs available at knovel.com have been added to this edition. We have also used the extensive series of *ASM Handbooks* to extend topics in Chapters 11, 12, 13, 14, and 15. These are also available at knovel.com.

New to This Edition

- Reorganization and new material in Chapters 3, 6, and 7, including a progressive example throughout these chapters
- New Chapter 10, Design for Sustainability and the Environment

- Chapter 16, Economic Decision Making, brought into the book from text website
- Section on Cost of Quality added to Chapter 17, Cost Evaluation
- Many additional connections to useful design information on the Internet
- Updated and new references including links to handbooks available through knovel.com
- PowerPoint lecture slides available to instructors through McGraw-Hill Higher Education

We want to acknowledge the willingness of students from our senior design course for permission to use material from their report in some of our examples. The JSR Design Team members are: Josiah Davis, Jamil Decker, James Maresco, Seth McBee, Stephen Phillips, and Ryan Quinn.

Special thanks to Peter Sandborn, Chandra Thamire, and Guangming Zhang, our colleagues in the Mechanical Engineering Department, University of Maryland, for their willingness to share their knowledge with us. Thanks also to Greg Moores of the DeWalt Division of Stanley Black and Decker, Inc. for his willingness to share his industrial viewpoint on several topics. We also thank the following reviewers for their helpful comments and suggestions: Bruce Floersheim, United States Military Academy; Mark A. Johnson, Michigan Tech University; Jesa Kreiner, California State University at Fullerton; David N. Kunz, University of Wisconsin, Platteville; Marybeth Lima, Louisiana State University; Bahram Nassersharif, University of Rhode Island; Ibrahim Nisanci, University of Arkansas at Little Rock; Keith E. Rouch, University of Kentucky; Paul Steranka, West Virginia University Institute of Technology; M. A. Wahab, Louisiana State University, John-David Yoder, Ohio Northern University; D. A. Zumbrunnen, Clemson University.

George E. Dieter and Linda C. Schmidt
College Park, MD
2012

1

ENGINEERING DESIGN

1.1
INTRODUCTION

What is design? If you search the literature for an answer to that question, you will find about as many definitions as there are designs. Perhaps the reason is that the process of design is such a common human experience. Webster's dictionary says that to design is "to fashion after a plan," but that leaves out the essential fact that to design is to create something that has never been. Certainly an engineering designer practices design by that definition, but so does an artist, a sculptor, a composer, a playwright, or any another creative member of our society.

Thus, although engineers are not the only people who design things, it is true that the professional practice of engineering is largely concerned with design; it is often said that design is the essence of engineering. *To design is to pull together something new or to arrange existing things in a new way to satisfy a recognized need of society.* An elegant word for "pulling together" is *synthesis.* We shall adopt the following formal definition of design: "Design establishes and defines solutions to and pertinent structures for problems not solved before, or new solutions to problems which have previously been solved in a different way."[1] The ability to design is both a science and an art. The science can be learned through techniques and methods to be covered in this text, but the art is best learned by doing design. It is for this reason that your design experience must involve some realistic project experience.

The emphasis that we have given to the creation of new things in our introduction to design should not unduly alarm you. To become proficient in design is a perfectly attainable goal for an engineering student, but its attainment requires the guided experience that we intend this text to provide. Design should not be confused with discovery. Discovery is getting the first sight of, or the first knowledge of something, as

---

1. J. F. Blumrich, *Science,* vol. 168, pp. 1551–1554, 1970.

when Columbus discovered America or Jack Kilby made the first microprocessor. We can discover what has already existed but has not been known before, but a design is the product of planning and work. We will present a structured design process to assist you in doing design in Sec. 1.5.

We should note that a design may or may not involve *invention*. To obtain a legal patent on an invention requires that the design be a step beyond the limits of the existing knowledge (beyond the state of the art). Some designs are truly inventive, but most are not.

Look up the word *design* in a dictionary and you will find that it can be either a noun or a verb. One noun definition is "the form, parts, or details of something according to a plan," as in the use of the word *design* in "My new design is ready for review." A common definition of the word *design* as a verb is "to conceive or to form a plan for," as in "I have to design three new models of the product for three different overseas markets." Note that the verb form of *design* is also written as "designing." Often the phrase "design process" is used to emphasize the use of the verb form of *design*. It is important to understand these differences and to use the word appropriately.

Good design requires both analysis and synthesis. Typically we approach complex problems like design by *decomposing* the problem into manageable parts. Because we need to understand how the part will perform in service, we must be able to calculate as much about the part's expected behavior as possible before it exists in physical form by using the appropriate disciplines of science and engineering science and the necessary computational tools. This is called *analysis*. It usually involves the simplification of the real world through models. *Synthesis* involves the identification of the design elements that will comprise the product, its decomposition into parts, and the combination of the part solutions into a total workable system.

At your current stage in your engineering education you may be much more familiar and comfortable with analysis. You have dealt with courses that were essentially disciplinary. For example, you were not expected to use thermodynamics and fluid mechanics in a course in mechanics of materials. The problems you worked in the course were selected to illustrate and reinforce the principles. If you could construct the appropriate model, you usually could solve the problem. Most of the input data and properties were given, and there usually was a correct answer to the problem. However, real-world problems rarely are that neat and circumscribed. The real problem that your design is expected to solve may not be readily apparent. You may need to draw on many technical disciplines (solid mechanics, fluid mechanics, electro magnetic theory, etc.) for the solution and usually on nonengineering disciplines as well (economics, finance, law, etc.). The input data may be fragmentary at best, and the scope of the project may be so huge that no individual can follow it all. If that is not difficult enough, usually the design must proceed under severe constraints of time and/or money. There may be major societal constraints imposed by environmental or energy regulations. Finally, in the typical design you rarely have a way of knowing the correct answer. Hopefully, your design works, but is it the best, most efficient design that could have been achieved under the conditions? Only time will tell.

We hope that this has given you some idea of the design process and the environment in which it occurs. One way to summarize the challenges presented by the design environment is to think of the *four C's of design*. One thing that should be clear

---

The Four C's of Design

Creativity
- Requires creation of something that has not existed before or has not existed in the designer's mind before

Complexity
- Requires decisions on many variables and parameters

Choice
- Requires making choices between many possible solutions at all levels, from basic concepts to the smallest detail of shape

Compromise
- Requires balancing multiple and sometimes conflicting requirements

---

by now is how engineering design extends well beyond the boundaries of science. The expanded boundaries and responsibilities of engineering create almost unlimited opportunities for you. In your professional career you may have the opportunity to create dozens of designs and have the satisfaction of seeing them become working realities. "A scientist will be lucky if he makes one creative addition to human knowledge in his whole life, and many never do. A scientist can discover a new star but he cannot make one. He would have to ask an engineer to do it for him."[1]

1.2
ENGINEERING DESIGN PROCESS

The engineering design process can be used to achieve several different outcomes. One is the design of products, whether they be consumer goods such as refrigerators, power tools, or DVD players, or highly complex products such as a missile system or a jet transport plane. Another is a complex engineered system such as an electrical power generating station or a petrochemical plant, while yet another is the design of a building or a bridge. However, the emphasis in this text is on product design because it is an area in which many engineers will apply their design skills. Moreover, examples taken from this area of design are easier to grasp without extensive specialized knowledge. This chapter presents the engineering design process from three perspectives. In Section 1.3 the design method is contrasted with the scientific method, and design is presented as a five-step problem-solving methodology. Section 1.4 takes the role of design beyond that of meeting technical performance requirements and introduces the idea that design must meet the needs of society at large. Section 1.5 lays out a cradle-to-the-grave road map of the design process, showing that the responsibility of the engineering designer extends from the creation of a design until its embodiment is

---

1. G.L. Glegg, *The Design of Design*, Cambridge University Press, New York, 1969.

disposed of in an environmentally safe way. Chapter 2 extends the engineering design process to the broader issue of product development by introducing more business-oriented issues such as product positioning and marketing.

1.2.1 Importance of the Engineering Design Process

In the 1980s when companies in the United States first began to seriously feel the impact of quality products from overseas, it was natural for them to place an emphasis on reducing their manufacturing costs through automation and moving plants to lower-labor-cost regions. However, it was not until the publication of a major study of the National Research Council (NRC)[1] that companies came to realize that the real key to world-competitive products lies in high-quality product design. This has stimulated a rash of experimentation and sharing of results about better ways to do product design. What was once a fairly cut-and-dried engineering process has become one of the cutting edges of engineering progress. This text aims at providing you with insight into the current best practices for doing engineering design.

The importance of design is nicely summed up in Fig. 1.1. This shows that only a small fraction of the cost to produce a product (≈ 5 percent) is involved with the design process, while the other 95 percent of cost is consumed by the materials, capital, and labor to manufacture the product. However, the design process consists of the accumulation of many decisions that result in design commitments that affect about 70 to 80 percent of the manufactured cost of the product. In other words, the decisions made beyond the design phase can influence only about 25 percent of the total cost. If the design proves to be faulty just before the product goes to market, it will cost a great deal of money to correct the problem. To summarize: *Decisions made in the design process cost very little in terms of the overall product cost but have a major effect on the cost of the product.*

The second major impact of design is on product quality. The old concept of product quality was that it was achieved by inspecting the product as it came off the production line. Today we realize that true quality is designed into the product. Achieving quality through product design will be a theme that pervades this book. For now we point out that one aspect of quality is to incorporate within the product the performance and features that are truly desired by the customer who purchases the product. In addition, the design must be carried out so that the product can be made without defect at a competitive cost. To summarize: *You cannot compensate in manufacturing for defects introduced in the design phase.*

The third area where engineering design determines product competitiveness is product cycle time. Cycle time refers to the development time required to bring a new product to market. In many consumer areas the product with the latest "bells and whistles" captures the customers' fancy. The use of new organizational methods, the widespread use of computer-aided engineering, and rapid prototyping methods are contributing to reducing product cycle time. Not only does reduced cycle time

---

1. "Improving Engineering Design," National Academy Press, Washington, D.C., 1991.

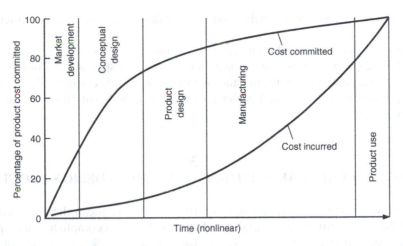

FIGURE 1.1
Product cost commitment during phases of the design process. (*After Ullman.*)

increase the marketability of a product, but it reduces the cost of product development. Furthermore, the longer a product is available for sale the more sales and profits there will be. To summarize: *The design process should be conducted so as to develop quality, cost-competitive products in the shortest time possible.*

1.2.2 Types of Designs

Engineering design can be undertaken for many different reasons, and it may take different forms.

- *Original design*, also called *innovative design*. This form of design is at the top of the hierarchy. It employs an original, innovative concept to achieve a need. Sometimes, but rarely, the need itself may be original. A truly original design involves invention. Successful original designs occur rarely, but when they do occur they usually disrupt existing markets because they have in them the seeds of new technology of far-reaching consequences. The design of the microprocessor was one such original design.
- *Adaptive design*. This form of design occurs when the design team adapts a known solution to satisfy a different need to produce a *novel application*. For example, adapting the ink-jet printing concept to spray binder to hold particles in place in a rapid prototyping machine.
- *Redesign*. Much more frequently, engineering design is employed to improve an existing design. The task may be to redesign a component in a product that is failing in service, or to redesign a component so as to reduce its cost of manufacture. Often redesign is accomplished without any change in the working principle or concept of the original design. For example, the shape may be changed to reduce a stress concentration, or a new material substituted to reduce weight or cost. When

redesign is achieved by changing some of the design parameters, it is often called *variant design.*

- *Selection design.* Most designs employ standard components such as bearings, small motors, or pumps that are supplied by vendors specializing in their manufacture and sale. Therefore, in this case the design task consists of selecting the components with the needed performance, quality, and cost from the catalogs of potential vendors.

1.3
WAYS TO THINK ABOUT THE ENGINEERING DESIGN PROCESS

We often talk about "designing a system." By a system we mean the entire combination of hardware, information, and people necessary to accomplish some specified task. A system may be an electric power distribution network for a region of the nation, a complex piece of machinery like an aircraft jet engine, or a combination of production steps to produce automobile parts. A large system usually is divided into *subsystems,* which in turn are made up of *components* or *parts.*

1.3.1 A Simplified Iteration Model

There is no single universally acclaimed sequence of steps that leads to a workable design. Different writers or designers have outlined the design process in as few as five steps or as many as 25. One of the first to write introspectively about design was Morris Asimow.[1] He viewed the heart of the design process as consisting of the elements shown in Fig. 1.2. As portrayed there, design is a sequential process consisting of many design operations. Examples of the operations might be (1) exploring the alternative concepts that could satisfy the specified need, (2) formulating a mathematical model of the best system concept, (3) specifying specific parts to construct a subsystem, and (4) selecting a material from which to manufacture a part. Each operation requires information, some of it general technical and business information that is expected of the trained professional and some of it very specific information that is needed to produce a successful outcome. Examples of the latter kind of information might be (1) a manufacturer's catalog on miniature bearings, (2) handbook data on the properties of polymer composites, or (3) personal experience gained from a trip to observe a new manufacturing process. Acquisition of information is a vital and often very difficult step in the design process, but fortunately it is a step that usually becomes easier with time. (We call this process *experience.*)[2] The importance of sources of information is considered more fully in Chap. 5.

Once armed with the necessary information, the design team (or design engineer if the task is rather limited) carries out the design operation by using the

---

1. M. Asimow, *Introduction to Design,* Prentice-Hall, Englewood Cliffs, NJ, 1962.
2. Experience has been defined, perhaps a bit lightheartedly, as just a sequence of nonfatal events.

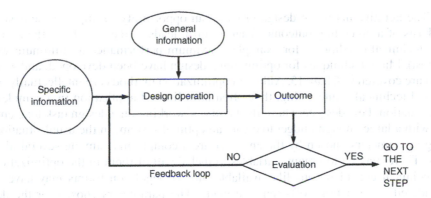

FIGURE 1.2
Basic module in the design process. (*After Asimow.*)

appropriate technical knowledge and computational and/or experimental tools. At this stage it may be necessary to construct a mathematical model and conduct a simulation of the component's performance on a computer. Or it may be necessary to construct a full-size prototype model and test it to destruction at a proving ground. Whatever it is, the operation produces one or more alternatives that, again, may take many forms. It can be 30 megabytes of data on a memory stick, a rough sketch with critical dimensions, or a 3-D CAD model. At this stage the design outcome must be evaluated, often by a team of impartial experts, to decide whether it is adequate to meet the need. If so, the designer may go on to the next step. If the evaluation uncovers deficiencies, then the design operation must be repeated. The information from the first design is fed back as input, together with new information that has been developed as a result of questions raised at the evaluation step. We call this *iteration.*

The final result of the chain of design modules, each like Fig. 1.2, is a new working object (often referred to as a prototype) or a collection of objects that is a new system. However, the goal of many design projects is not the creation of new hardware or systems. Instead, the goal may be the development of new information that can be used elsewhere in the organization. It should be realized that not all system designs are carried through to completion; they are stopped because it has become clear that the objectives of the project are not technically and/or economically feasible. Regardless, the system design process creates new information which, if stored in retrievable form, has future value, since it represents experience.

The simple model shown in Fig. 1.2 illustrates a number of important aspects of the design process. First, design of even the most complex system can be broken down into a sequence of design processes. Each outcome requires evaluation, and it is common for design to involve repeated trials or iterations. Of course, the more knowledge we have and can apply to the problem the faster we can arrive at an acceptable solution. This iterative aspect of design may take some getting used to. You will have to acquire a high tolerance for failure and the tenacity and determination to persevere and work the problem out one way or the other.

The iterative nature of design provides an opportunity to improve the design on the basis of a preceding outcome. That, in turn, leads to the search for the best possible technical condition—for example, maximum performance at minimum weight (or cost). Many techniques for optimizing a design have been developed, and some of them are covered in Chap. 15. Although optimization methods are intellectually pleasing and technically interesting, they often have limited application in a complex design situation. Few designers have the luxury of working on a design task long enough and with a large enough budget to create an optimal system. In the usual situation the design parameters chosen by the engineer are a compromise among several alternatives. There may be too many variables to include all of them in the optimization, or nontechnical considerations like available time or legal constraints may have to be considered, so that trade-offs must be made. The parameters chosen for the design are then close to but not at optimum values. We usually refer to them as *near-optimal values,* the best that can be achieved within the total constraints of the system.

1.3.2 Design Method Versus Scientific Method

In your scientific and engineering education you may have heard reference to the scientific method, a logical progression of events that leads to the solution of scientific problems. Percy Hill[1] has diagramed the comparison between the scientific method and the design method (Fig. 1.3). The scientific method starts with a body of existing knowledge based on observed natural phenomena. Scientists have curiosity that causes them to question these laws of science; and as a result of their questioning, they eventually formulate a hypothesis. The hypothesis is subjected to logical analysis that either confirms or denies it. Often the analysis reveals flaws or inconsistencies, so the hypothesis must be changed in an iterative process.

Finally, when the new idea is confirmed to the satisfaction of its originator, it must be accepted as proof by fellow scientists. Once accepted, it is communicated to the community of scientists and it enlarges the body of existing knowledge. The knowledge loop is completed.

The design method is very similar to the scientific method if we allow for differences in viewpoint and philosophy. The design method starts with knowledge of the state of the art. That includes scientific knowledge, but it also includes devices, components, materials, manufacturing methods, and market and economic conditions. Rather than scientific curiosity, it is really the needs of society (usually expressed through economic factors) that provide the impetus. When a need is identified, it must be conceptualized as some kind of model. The purpose of the model is to help us predict the behavior of a design once it is converted to physical form. The outcomes of the model, whether it is a mathematical or a physical model, must be subjected to a feasibility analysis, almost always with iteration, until an acceptable product is produced or the project is abandoned. When the design enters the production phase, it begins to compete in the world of technology. The design loop is closed when the

---

1. P. H. Hill, *The Science of Engineering Design,* Holt, Rinehart and Winston, New York, 1970.

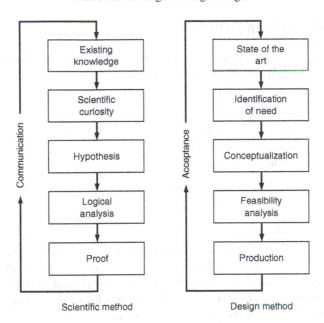

FIGURE 1.3
Comparison between the scientific method and the design method. (*After Percy Hill.*)

product is accepted as part of the current technology and thereby advances the state of the art of the particular area of technology.

A more philosophical differentiation between science and design has been advanced by the Nobel Prize–winning economist Herbert Simon.[1] He points out that science is concerned with creating knowledge about naturally occurring phenomena and objects, while design is concerned with creating knowledge about phenomena and *objects of the artificial.* Artificial objects are those made by humans rather than nature. Thus, science is based on studies of the observed, while design is based on artificial concepts characterized in terms of functions, goals, and adaptation.

In the preceding brief outline of the design method, the identification of a need requires further elaboration. Needs are identified at many points in a business or organization. Most organizations have research or development departments whose job it is to create ideas that are relevant to the goals of the organization. A very important avenue for learning about needs is the customers for the product or services that the company sells. Managing this input is usually the job of the marketing organization of the company. Other needs are generated by government agencies, trade associations, or the attitudes or decisions of the general public. Needs usually arise from dissatisfaction with the existing situation. The need drivers may be to reduce cost, increase reliability or performance, or just change because the public has become bored with the product.

---

1. H. A. Simon, *The Sciences of the Artificial,* 3rd ed., The MIT Press, Cambridge, MA, 1996.

1.3.3 A Problem-Solving Methodology

Designing can be approached as a problem to be solved. A problem-solving methodology that is useful in design consists of the following steps.[1]

- Definition of the problem
- Gathering of information
- Generation of alternative solutions
- Evaluation of alternatives and decision making
- Communication of the results

This problem-solving method can be used at any point in the design process, whether at the conception of a product or the design of a component.

Definition of the Problem

The most critical step in the solution of a problem is the *problem definition* or formulation. The true problem is not always what it seems at first glance. Because this step seemingly requires such a small part of the total time to reach a solution, its importance is often overlooked. Figure 1.4 illustrates how the final design can differ greatly depending upon how the problem is defined.

The formulation of the problem should start by writing down a problem statement. This document should express as specifically as possible what the problem is. It should include objectives and goals, the current state of affairs and the desired state, any constraints placed on solution of the problem, and the definition of any special technical terms. The problem-definition step in a design project is covered in detail in Chap. 3.

Problem definition often is called *needs analysis*. While it is important to identify the needs clearly at the beginning of a design process, it should be understood that this is difficult to do for all but the most routine design. It is the nature of the design process that new needs are established as the design process proceeds because new problems arise as the design evolves. At this point, the analogy of design as problem solving is less fitting. Design is problem solving only when all needs and potential issues with alternatives are known. Of course, if these additional needs require reworking those parts of the design that have been completed, then penalties are incurred in terms of cost and project schedule. Experience is one of the best remedies for this aspect of designing, but modern computer-based design tools help ameliorate the effects of inexperience.

Gathering Information

Perhaps the greatest frustration you will encounter when you embark on your first design project will be either the dearth or the plethora of information. Your assigned problem may be in a technical area in which you have no previous

---

1. A similar process called the guided iteration methodology has been proposed by J. R. Dixon; see J. R. Dixon and C. Poli, *Engineering Design and Design for Manufacturing,* Field Stone Publishers, Conway, MA, 1995. A different but very similar problem-solving approach using TQM tools is given in Sec. 4.6.

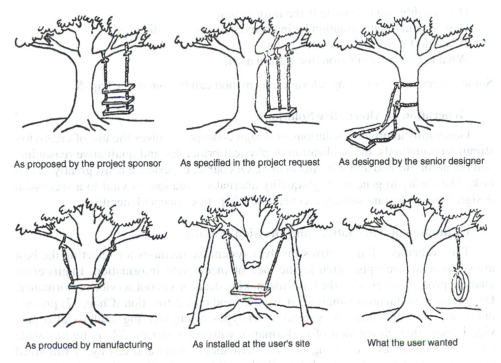

As proposed by the project sponsor   As specified in the project request   As designed by the senior designer

As produced by manufacturing   As installed at the user's site   What the user wanted

FIGURE 1.4
Note how the design depends on the viewpoint of the individual who defines the problem.

background, and you may not have even a single basic reference on the subject. At the other extreme you may be presented with a mountain of reports of previous work, and your task will be to keep from drowning in paper. Whatever the situation, the immediate task is to identify the needed pieces of information and find or develop that information.

An important point to realize is that the information needed in design is different from that usually associated with an academic course. Textbooks and articles published in the scholarly technical journals usually are of lesser importance. The need often is for more specific and current information than is provided by those sources. Technical reports published as a result of government-sponsored R&D, company reports, trade journals, patents, catalogs, and handbooks and literature published by vendors and suppliers of material and equipment are important sources of information. The Internet is a very useful resource. Often the missing piece of information can be supplied by an Internet search, or by a telephone call or an e-mail to a key supplier. Discussions with in-house experts (often in the corporate R&D center) and outside consultants may prove helpful.

The following are some of the questions concerned with obtaining information:

What do I need to find out?
Where can I find it and how can I get it?

How credible and accurate is the information?
How should the information be interpreted for my specific need?
When do I have enough information?
What decisions result from the information?

Some suggestions for finding relevant information can be found in Chap. 5.

Generation of Alternative Solutions

Generating alternative solutions or design concepts involves the use of creativity-stimulation methods, the application of physical principles and qualitative reasoning, and the ability to find and use information. Of course, experience helps greatly in this task. The ability to generate high-quality alternative solutions is vital to a successful design. This important subject is covered in Chap. 6, Concept Generation.

Evaluation of Alternatives and Decision Making

The evaluation of alternatives involves systematic methods for selecting the best among several concepts, often in the face of incomplete information. Engineering analysis procedures provide the basis for making decisions about service performance. Design for manufacturing analyses (Chap. 13) and cost estimation (Chap. 17) provide other important information. Various other types of engineering analysis also provide information. Simulation of performance with computer models is finding wide usage. Simulated service testing of an experimental model and testing of full-sized prototypes often provide critical data. Without this quantitative information it is not possible to make valid evaluations. Several methods for evaluating design concepts, or any other problem solution, are given in Chap. 7.

An important activity at every step in the design process, but especially as the design nears completion, is *checking*. In general, there are two types of checks that can be made: mathematical checks and engineering-sense checks. Mathematical checks are concerned with checking the arithmetic and the equations for errors in the conversion of units used in the analytical model. Incidentally, the frequency of careless math errors is a good reason why you should adopt the practice of making all your design calculations in a bound notebook. In that way you won't be missing a vital calculation when you are forced by an error to go back and check things out. Just draw a line through the section in error and continue. It is of special importance to ensure that every equation is dimensionally consistent.

Engineering-sense checks have to do with whether the answers "seem right." Even though the reliability of your intuition increases with experience, you can now develop the habit of staring at your answer for a full minute, rather than rushing on to do the next calculation. If the calculated stress is 10^6 psi, you know something went wrong! Limit checks are a good form of engineering-sense check. Let a critical parameter in your design approach some limit (zero, infinity, etc.), and observe whether the equation behaves properly.

We have stressed the *iterative* nature of design. An optimization technique aimed at producing a *robust design* that is resistant to environmental influences (water vapor, temperature, vibration, etc.) most likely will be employed to select the best values of key design parameters (see Chap. 15).

Communication of the Results

It must always be kept in mind that the purpose of the design is to satisfy the needs of a customer or client. Therefore, the finalized design must be properly communicated, or it may lose much of its impact or significance. The communication is usually by oral presentation to the sponsor as well as by a written design report. Surveys typically show that design engineers spend 60 percent of their time in discussing designs and preparing written documentation of designs, while only 40 percent of the time is spent in analyzing and testing designs and doing the designing. Detailed engineering drawings, computer programs, 3-D computer models, and working models are frequently among the "deliverables" to the customer.

It hardly needs to be emphasized that communication is not a one-time occurrence to be carried out at the end of the project. In a well-run design project there is continual oral and written dialog between the project manager and the customer.

Note that the problem-solving methodology does not necessarily proceed in the order just listed. While it is important to define the problem early on, the understanding of the problem improves as the team moves into solution generation and evaluation. In fact, design is characterized by its iterative nature, moving back and forth between partial solutions and problem definition. This is in marked contrast with engineering analysis, which usually moves in a steady progression from problem setup to solution.

There is a paradox inherent in the design process between the accumulation of problem (domain) knowledge and freedom to improve the design. When one is creating an original design, very little is known about its solution. As the design team proceeds with its work, it acquires more knowledge about the technologies involved and the possible solutions (Fig. 1.5). The team has moved up the learning curve. However, as the design process proceeds, the design team is forced to make many decisions about design

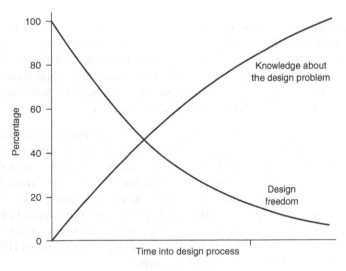

FIGURE 1.5
The design paradox between design knowledge and design freedom.

details, technology approaches, perhaps to let contracts for long-lead-time equipment, and so on. Thus, as Fig. 1.5 shows, the freedom of the team to go back and start over with their newly gained knowledge (experience) decreases greatly as their knowledge about the design problem grows. At the beginning the designer has the freedom to make changes without great cost penalty, but may not know what to do to make the design better. The paradox comes from the fact that when the design team finally masters the problem, their design is essentially frozen because of the great penalties involved with a change. The solution is for the design team to learn as much about the problem as early in the design process as it possibly can. This also places high priority on the team members learning to work independently toward a common goal (Chap. 4), being skilled in gathering information (Chap. 5), and being good at communicating relevant knowledge to their teammates. Design team members must become stewards of the knowledge they acquire. Figure 1.5 also shows why it is important to document in detail what has been done, so that the experience can be used by subsequent teams in future projects.

1.4
DESCRIPTION OF DESIGN PROCESS

Morris Asimow[1] was among the first to give a detailed description of the complete design process in what he called the morphology of design. Figure 1.6 shows the various activities that make up the first three phases of design: conceptual design, embodiment design, and detail design. The purpose of this graphic is to remind you of the logical sequence of activities that leads from problem definition to the detail design.

1.4.1 Phase I. Conceptual Design

Conceptual design is the process by which the design is initiated, carried to the point of creating a number of possible solutions, and narrowed down to a single best concept. It is sometimes called the feasibility study. Conceptual design is the phase that requires the greatest creativity, involves the most uncertainty, and requires coordination among many functions in the business organization. The following are the discrete activities that we consider under conceptual design.

- *Identification of customer needs*: The goal of this activity is to completely understand the customers' needs and to communicate them to the design team.
- *Problem definition*: The goal of this activity is to create a statement that describes what has to be accomplished to satisfy the needs of the customer. This involves analysis of competitive products, the establishment of target specifications, and the listing of constraints and trade-offs. Quality function deployment (QFD) is a valuable tool for linking customer needs with design requirements. A detailed listing of the product requirements is called a product design specification (PDS). Problem definition, in its full scope, is treated in Chap. 3.

---

1. I. M. Asimow, *Introduction to Design*, Prentice-Hall, Englewood Cliffs, NJ, 1962.

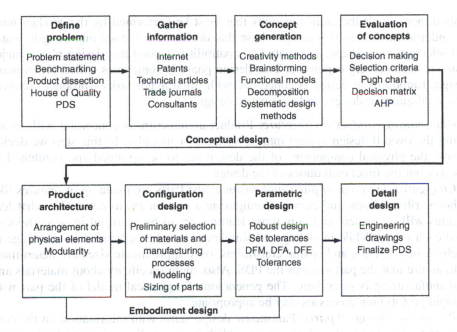

FIGURE 1.6
The design activities that make up the first three phases of the engineering design process.

- *Gathering information*: Engineering design presents special requirements over engineering research in the need to acquire a broad spectrum of information. This subject is covered in Chap. 5.
- *Conceptualization*: Concept generation involves creating a broad set of concepts that potentially satisfy the problem statement. Team-based creativity methods, combined with efficient information gathering, are the key activities. This subject is covered in Chap. 6.
- *Concept selection*: Evaluation of the design concepts, modifying and evolving into a single preferred concept, are the activities in this step. The process usually requires several iterations. This is covered in Chap. 7.
- *Refinement of the PDS*: The product design specification is revisited after the concept has been selected. The design team must commit to achieving certain critical values of design parameters, usually called critical-to-quality (CTQ) parameters, and to living with trade-offs between cost and performance.
- *Design review*: Before committing funds to move to the next design phase, a design review will be held. The design review will assure that the design is physically realizable and that it is economically worthwhile. It will also look at a detailed product-development schedule. This is needed to devise a strategy to minimize product cycle time and to identify the resources in people, equipment, and money needed to complete the project.

1.4.2 Phase II. Embodiment Design

Structured development of the design concept occurs in this engineering design phase. It is the place where flesh is placed on the skeleton of the design concept. An embodiment of all the main functions that must be performed by the product must be undertaken. It is in this design phase that decisions are made on strength, material selection, size, shape, and spatial compatibility. Beyond this design phase, major changes become very expensive. This design phase is sometimes called preliminary design. Embodiment design is concerned with three major tasks—product architecture, configuration design, and parametric design.

- *Determining product architecture*: Product architecture is concerned with dividing the overall design system into subsystems or modules. In this step we decide how the physical components of the design are to be arranged and combined to carry out the functional duties of the design.
- *Configuration design of parts and components*: Parts are made up of features like holes, ribs, splines, and curves. Configuring a part means to determine what features will be present and how those features are to be arranged in space relative to each other. While modeling and simulation may be performed in this stage to check out function and spatial constraints, only approximate sizes are determined to assure that the part satisfies the PDS. Also, more specificity about materials and manufacturing is given here. The generation of a physical model of the part with rapid prototyping processes may be appropriate.
- *Parametric design of parts:* Parametric design starts with information on the configuration of the part and aims to establish its exact dimensions and tolerances. Final decisions on the material and manufacturing processes are also established if this has not been done previously. An important aspect of parametric design is to examine the part, assembly, and system for design robustness. *Robustness* refers to how consistently a component performs under variable conditions in its service environment. The methods developed by Dr. Genichi Taguchi for achieving robustness and establishing the optimum tolerance are discussed in Chap. 15. Parametric design also deals with determining the aspects of the design that could lead to failure (see Chap. 14). Another important consideration in parametric design is to design in such a way that manufacturability is enhanced (see Chap. 13).

1.4.3 Phase III. Detail Design

In this phase the design is brought to the stage of a complete engineering description of a tested and producible product. Missing information is added on the arrangement, form, dimensions, tolerances, surface properties, materials, and manufacturing processes of each part. This results in a specification for each *special-purpose part* and for each *standard part* to be purchased from suppliers. In the detail design phase the following activities are completed and documents are prepared:

- Detailed engineering drawings suitable for manufacturing. Routinely these are computer-generated drawings, and they often include three-dimensional CAD models.
- Verification testing of prototypes is successfully completed and verification data is submitted. All critical-to-quality parameters are confirmed to be under control. Usually the building and testing of several preproduction versions of the product will be accomplished.

- Assembly drawings and assembly instructions also will be completed. The bill of materials for all assemblies will be completed.
- A detailed product specification, updated with all the changes made since the conceptual design phase, will be prepared.
- Decisions on whether to make each part internally or to buy from an external supplier will be made.
- With the preceding information, a detailed cost estimate for the product will be carried out.
- Finally, detail design concludes with a design review before the decision is made to pass the design information on to manufacturing.

Phases I, II, and III take the design from the realm of possibility to the real world of practicality. However, the design process is not finished with the delivery of a set of engineering drawings and specifications to the manufacturing organization. Many other technical and business decisions must be made to bring the design to the point where it can be delivered to the customer. Chief among these, as discussed in Sec. 9.5, are detailed plans for manufacturing the product, for planning its launch into the marketplace, and for disposing of it in an environmentally safe way after it has completed its useful life.

1.5
CONSIDERATIONS OF A GOOD DESIGN

Design is a multifaceted process. To gain a broader understanding of engineering design, we group various considerations of good design into three categories: (1) achievement of performance requirements, (2) life-cycle issues, and (3) social and regulatory issues.

1.5.1 Achievement of Performance Requirements

It is obvious that to be feasible the design must demonstrate the required performance. Performance measures both the function and the behavior of the design, that is, how well the device does what it is designed to do. Performance requirements can be divided into primary performance requirements and complementary performance requirements. A major characteristic of a design is its *function*. The function of a design is how it is expected to behave. For example, the design may be required to grasp an object of a certain mass and move it 50 feet in one minute. Functional requirements are usually expressed in capacity measures such as forces, strength, deflection, or energy or power output or consumption. Complementary performance requirements are concerns such as the useful life of the design, its robustness to factors occurring in the service environment (see Chap. 15), its reliability (see Chap. 14), and ease, economy, and safety of maintenance. Issues such as built-in safety features and the noise level in operation must be considered. Finally, the design must conform to all legal requirements and design codes.

A product[1] is usually made up of a collection of parts, sometimes called piece-parts. A *part* is a single piece requiring no assembly. When two or more parts are joined it is called an *assembly*. Often large assemblies are composed of a collection of smaller assemblies called *subassemblies*. A similar term for part is *component*. The two terms are used interchangeably in this book, but in the design literature the word *component* sometimes is used to describe a subassembly with a small number of parts. Consider an ordinary ball bearing. It consists of an outer ring, inner ring, 10 or more balls depending on size, and a retainer to keep the balls from rubbing together. A ball bearing is often called a component, even though it consists of a number of parts.

Closely related to the function of a component in a design is its form. *Form* is what the component looks like, and encompasses its shape, size, and surface finish. These, in turn, depend upon the material it is made from and the manufacturing processes that are used to make it.

A variety of analysis techniques must be employed in arriving at the features of a component in the design. By *feature* we mean specific physical attributes, such as the fine details of geometry, dimensions, and tolerances on the dimensions.[2] Typical geometrical features would be fillets, holes, walls, and ribs. The computer has had a major impact in this area by providing powerful analytical tools based on finite-element analysis. Calculations of stress, temperature, and other field-dependent variables can be made rather handily for complex geometry and loading conditions. When these analytical methods are coupled with interactive computer graphics, we have the exciting capability known as computer-aided engineering (CAE); see Sec. 1.6. Note that with this enhanced capability for analysis comes greater responsibility for providing better understanding of product performance at early stages of the design process.

Environmental requirements for performance deal with two separate aspects. The first concerns the service conditions under which the product must operate. The extremes of temperature, humidity, corrosive conditions, dirt, vibration, and noise, must be predicted and allowed for in the design. The second aspect of environmental requirements pertains to how the product will behave with regard to maintaining a safe and clean environment, that is, green design. Often governmental regulations force these considerations in design, but over time they become standard design practice. Among these issues is the disposal of the product when it reaches its useful life. Design for the Environment (DFE) is discussed in detail in Chap.10.

Aesthetic requirements refer to "the sense of the beautiful." They are concerned with how the product is perceived by a customer because of its shape, color, surface texture, and also such factors as balance, unity, and interest. This aspect of design usually is the responsibility of the industrial designer, as opposed to the engineering designer. The industrial designer is in part an applied artist. Decisions about the appearance of the product should be an integral part of the initial design concept.

---

1. Another term for product is *device,* something devised or constructed for a particular purpose, like a machine.

2. In product development the term *feature* has an entirely different meaning as "an aspect or characteristic of the product." For example, a product feature for a power drill could be a laser beam attachment for alignment of the drill when drilling a hole.

An important design consideration is adequate attention to *human factors engineering,* which uses the sciences of biomechanics, ergonomics, and engineering psychology to assure that the design can be operated efficiently by humans. It applies physiological and anthropometric data to such design features as visual and auditory display of instruments and control systems. It is also concerned with human muscle power and response times. The industrial designer often is responsible for considering the human factors. For further information, see Sec. 8.9.

Manufacturing technology must be closely integrated with product design. There may be restrictions on the manufacturing processes that can be used, because of either selection of material or availability of equipment within the company.

The final major design requirement is cost. Every design has requirements of an economic nature. These include such issues as product development cost, initial product cost, life cycle product cost, tooling cost, and return on investment. In many cases cost is the most important design requirement. If preliminary estimates of product cost look unfavorable, the design project may never be initiated. Cost enters into every aspect of the design process.

1.5.2 Total Life Cycle

The total life cycle of a part starts with the conception of a need and ends with the retirement and disposal of the product.

Material selection is a key element in shaping the total life cycle (see Chap. 11). In selecting materials for a given application, the first step is evaluation of the service conditions. Next, the properties of materials that relate most directly to the service requirements must be determined. Except in almost trivial conditions, there is never a simple relation between service performance and material properties. The design may start with the consideration of static yield strength, but properties that are more difficult to evaluate, such as fatigue, creep, toughness, ductility, and corrosion resistance may have to be considered. We need to know whether the material is stable under the environmental conditions. Does the microstructure change with temperature and therefore change the properties? Does the material corrode slowly or wear at an unacceptable rate?

Material selection cannot be separated from *manufacturability* (see Chap. 13). There is an inherent connection between design and material selection and the manufacturing processes. The objective in this area is a trade-off between the opposing factors of minimum cost and maximum durability. *Durability* is increased by designing so as to minimize material deterioration by corrosion, wear, or fracture. It is a general property of the product measured by months or years of successful service, and is closely related to reliability, a technical term that is measured by the probability of achieving a specified service life. Current societal issues of energy conservation, material conservation, and protection of the environment result in new pressures in the selection of materials and manufacturing processes. Energy costs, once nearly ignored in design, are now among the most prominent design considerations. Design for materials recycling also is becoming an important design consideration.

The life cycle of production and consumption that is characteristic of all products is illustrated by the materials cycle shown in Fig. 1.7. This starts with the mining of a

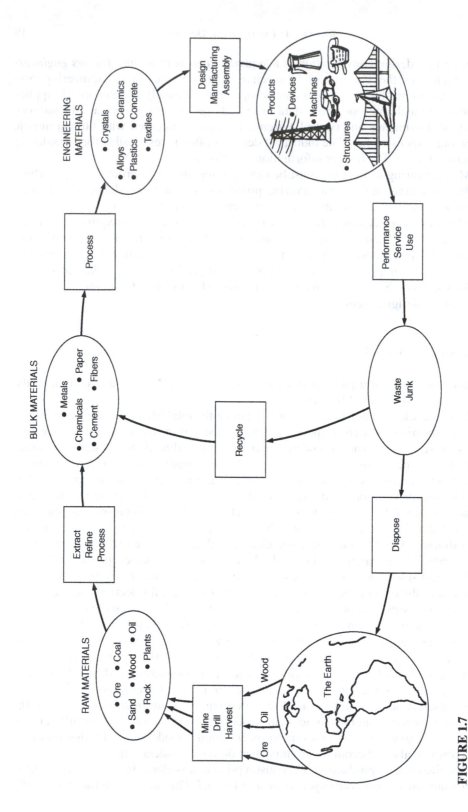

FIGURE 1.7
The total materials cycle. (*Reproduced from "Materials and Man's Needs," National Academy of Sciences, Washington, D.C., 1974.*)

mineral or the drilling for oil or the harvesting of an agricultural fiber such as cotton. These raw materials must be processed to extract or refine a bulk material (e.g., an aluminum ingot) that is further processed into a finished engineering material (e.g., an aluminum sheet). At this stage an engineer designs a product that is manufactured from the material, and the part is put into service. Eventually the part wears out or becomes obsolete because a better product comes on the market. At this stage, one option is to junk the part and dispose of it in some way that eventually returns the material to the earth. However, society is becoming increasingly concerned with the depletion of natural resources and the haphazard disposal of solid materials. Thus, we look for economical ways to recycle waste materials (e.g., aluminum beverage cans).

1.5.3 Regulatory and Social Issues

Specifications and standards have an important influence on design practice. The standards produced by such societies as ASTM and ASME represent voluntary agreement among many elements (users and producers) of industry. As such, they often represent minimum or least-common-denominator standards. When good design requires more than that, it may be necessary to develop your own company or agency standards. On the other hand, because of the general nature of most standards, a standard sometimes requires a producer to meet a requirement that is not essential to the particular function of the design.

The codes of ethics of all professional engineering societies require the engineer to protect public health and safety. Increasingly, legislation has been passed to require federal agencies to regulate many aspects of safety and health. The requirements of the Occupational Safety and Health Administration (OSHA), the Consumer Product Safety Commission (CPSC), the Environmental Protection Agency (EPA), and the Department of Homeland Security (DHS) place direct constraints on the designer in the interests of protecting health, safety, and security. Several aspects of the CPSC regulations have far-reaching influence on product design. Although the intended purpose of a product normally is quite clear, the unintended uses of that product are not always obvious. Under the CPSC regulations, the designer has the obligation to foresee as many unintended uses as possible, then develop the design in such a way as to prevent hazardous use of the product in an unintended but foreseeable manner. When unintended use cannot be prevented by functional design, clear, complete, unambiguous warnings must be permanently attached to the product. In addition, the designer must be cognizant of all advertising material, owner's manuals, and operating instructions that relate to the product to ensure that the contents of the material are consistent with safe operating procedures and do not promise performance characteristics that are beyond the capability of the design.

An important design consideration is adequate attention to human factors engineering, which uses the sciences of biomechanics, ergonomics, and engineering psychology to assure that the design can be operated efficiently and safely by humans. It applies physiological and anthropometric data to such design features as visual and auditory display of instruments and control systems. It is also concerned with human muscle power and response times. For further information, see Sec. 8.8.

1.6
COMPUTER-AIDED ENGINEERING

The advent of plentiful computing has produced a major change in the way engineering design is practiced. While engineers were among the first professional groups to adapt the computer to their needs, the early applications chiefly were computationally intensive ones, using a high-level language like FORTRAN. The first computer applications were conducted in batch mode, with the code prepared on punch cards. Overnight turnaround was the norm. Later, remote access to computer mainframes through terminals became common, and the engineer could engage in interactive (if still slow) computation. The development of the microprocessor and the proliferation of personal computers and engineering workstations with computational power equivalent to that of a mainframe has created a revolution in the way an engineer approaches and carries out problem solving and design.

The greatest impact of computer-aided engineering has been in engineering drawing. The automation of drafting in two dimensions has become commonplace. The ready ability to make changes and to use parts of old designs in new drawings offers a great saving in time. Three-dimensional modeling has become prevalent as it has become available on desktop computers. Three-dimensional solid modeling provides a complete geometric and mathematical description of the part geometry. Solid models can be sectioned to reveal interior details, or they can be readily converted into conventional two-dimensional engineering drawings. Such a model is very rich in intrinsic information so that it can be used not only for physical design but also for analysis, design optimization, simulation, rapid prototyping, and manufacturing. For example, geometric three-dimensional modeling ties in nicely with the extensive use of finite-element modeling (FEM) and makes possible interactive simulations in such problems as stress analysis, fluid flow, the kinematics of mechanical linkages, and numerically controlled tool-path generation for machining operations. The ultimate computer simulation is *virtual reality,* where the viewer feels like a part of the graphical simulation on the computer screen.

The computer extends the designer's capabilities in several ways. First, by organizing and handling time-consuming and repetitive operations, it frees the designer to concentrate on more complex design tasks. Second, it allows the designer to analyze complex problems faster and more completely. Both of these factors make it possible to carry out more iterations of design. Finally, through a computer-based information system the designer can share more information sooner with people in the company, like manufacturing engineers, process planners, tool and die designers, and purchasing agents. The link between computer-aided design (CAD) and computer-aided manufacturing (CAM) is particularly important. Moreover, by using the Internet and satellite telecommunication, these persons can be on different continents 10 time zones away.

Concurrent engineering is greatly facilitated by the use of computer-aided engineering. *Concurrent engineering* is a team-based approach in which all aspects of the product development process are represented on a closely communicating team. Team members perform their jobs in an overlapping and concurrent manner so as to minimize the time for product development (see Sec 2.4.4). A computer

---

Boeing 777

The boldest example of the use of CAD is with the Boeing 777 long-range transport. Started in fall 1990 and completed in April 1994, this was the world's first completely paperless transport design. Employing the CATIA 3-D CAD system, it linked all of Boeing's design and manufacturing groups in Washington, as well as suppliers of systems and components worldwide. At its peak, the CAD system served some 7000 workstations spread over 17 time zones.

As many as 238 design teams worked on the project at a single time. Had they been using conventional paper design, they might have experienced many interferences among hardware systems, requiring costly design changes and revised drawings. This is a major cost factor in designing a complex system. The advantage of being able to see what everyone else was doing, through an integrated solid model and digital data system, saved in excess of 50 percent of the change orders and rework expected for a design of this magnitude.

The Boeing 777 has more than 130,000 unique engineered parts, and when rivets and other fasteners are counted, there are more than 3 million individual parts. The ability of the CAD system to identify interferences eliminated the need to build a physical model (mockup) of the airplane. Nevertheless, those experienced with transport design and construction reported that the parts of the 777 fit better the first time than those of any earlier commercial airliner.

---

database in the form of a solid model that can be accessed by all members of the design team, as in the Boeing 777 example, is a vital tool for this communication. More and more the Internet, with appropriate security, is being used to transmit 3-D CAD models to tool designers, part vendors, and numerical-control programmers for manufacturing development in a highly networked global design and manufacturing system.

Computer-aided engineering became a reality when the power of the PC workstation, and later the laptop PC, became great enough at an acceptable cost to free the design engineer from the limitations of the mainframe computer. Bringing the computing power of the mainframe computer to the desktop of the design engineer has created great opportunities for more creative, reliable, and cost-effective designs.

CAE developed in two major domains: computer graphics and modeling, and mathematical analysis and simulation of design problems. The ability to do 3-D modeling is within the capability of every engineering student. The most common computer modeling software packages at the undergraduate level are AutoCAD, ProE, and SolidWorks. CAE analysis tools run the gamut from spreadsheet calculations to complex finite-element models involving stress, heat transfer, and fluid flow.

Spreadsheet applications may seem quaint to engineering students, but spreadsheet programs are useful because of their ability to quickly make multiple calculations without requiring the user to reenter all of the data. Each combination of row and column in the spreadsheet matrix is called a cell. The quantity in each cell can represent either a number entered as input or a number that the spreadsheet program calculates according to a prescribed equation.[1] The power of the spreadsheet is based on its ability to automatically recalculate results when new inputs have been entered in some cells. This can serve as a simple optimization tool as the values of one or two variables are changed and the impact on the output is readily observed. The usefulness of a spreadsheet in cost evaluations is self-evident. Most spreadsheet software programs contain built-in mathematical functions that permit engineering and statistical calculations. It is also possible to use them to solve problems in numerical analysis.

The solution of an equation with a spreadsheet requires that the equation be set up so that the unknown term is on one side of the equal sign. In working with equations it often is useful to be able to solve for any variable. Therefore, a class of equation-solving programs has been developed for small computations on the personal computer. The best-known examples are TK Solver, MathCAD, and EES (Engineering Equation Solver). Another important set of computational tools are the symbolic languages that manipulate the symbols representing the equation. Most common are Mathematica, Maple, and MATLAB. MATLAB[2] has found a special niche in many engineering departments because of its user-friendly computer interface, its ability to be programmable (and thus replace Fortran, Basic, and Pascal as programming languages), its excellent graphics features, excellent ability to solve differential equations, and the availability of more than 20 "toolboxes" in various applications areas.

Specialized application programs to support engineering design are appearing at a rapid rate. These include software for finite-element modeling, QFD, creativity enhancement, decision making, manufacturing process modeling and statistical modeling. Useful software packages of this type will be mentioned as these topics are introduced throughout the text.

1.7
DESIGNING TO CODES AND STANDARDS

While we have often talked about design being a creative process, the fact is that much of design is not very different from what has been done in the past. There are obvious benefits in cost and time saved if the best practices are captured and made available for all to use. Designing with codes and standards has two chief aspects: (1) it makes the best practice available to everyone, thereby ensuring efficiency and safety, and (2) it promotes interchangeability and compatibility. With respect to the second point, anyone who has traveled widely in other countries will understand the compatibility problems with connecting plugs and electrical voltage and frequency when trying to use small appliances.

A *code* is a collection of laws and rules that assists a government agency in meeting its obligation to protect the general welfare by preventing damage to property or injury or loss of life to persons. A *standard* is a generally agreed-upon set

1. B.S. Gottfried, *Spreadsheet Tools for Engineers*, McGraw-Hill, New York, 1996; S.C. Bloch, *EXCEL for Engineers and Scientists*, John Wiley & Sons, New York, 2000.
2. W.J. Palm III, *Introduction to MATLAB 7 for Engineers*, 2d ed., McGraw-Hill, New York, 2005; E.B. Magrab, et al, *An Engineer's Guide to MATLAB*, 2d ed., Prentice-Hall, Upper Saddle River, NJ. 2005.

of procedures, criteria, dimensions, materials, or parts. Engineering standards may describe the dimensions and sizes of small parts like screws and bearings, the minimum properties of materials, or an agreed-upon procedure to measure a property like fracture toughness.

The terms standards and specifications are sometimes used interchangeably. The distinction is that standards refer to generalized situations, while specifications refer to specialized situations. Codes tell the engineer what to do and when and under what circumstances to do it. Codes usually are legal requirements, as in the building code or the fire code. Standards tell the engineer how to do it and are usually regarded as recommendations that do not have the force of law. Codes often incorporate national standards into them by reference, and in this way standards become legally enforceable.

There are two broad forms of codes: performance codes and prescriptive codes. *Performance codes* are stated in terms of the specific requirement that is expected to be achieved. The method to achieve the result is not specified. *Prescriptive* or specification codes state the requirements in terms of specific details and leave no discretion to the designer. A form of code is government regulations. These are issued by agencies (federal or state) to spell out the details for the implementation of vaguely written laws. An example is the OSHA regulations developed by the U.S. Department of Labor to implement the Occupational Safety and Health Act (OSHA).

Design standards fall into three categories: performance, test methods, and codes of practice. There are published performance standards for many products such as seat belts, lumber, and auto crash safety. Test method standards set forth methods for measuring properties such as yield strength, thermal conductivity, or electrical resistivity. Most of these are developed for and published by the American Society for Testing and Materials (ASTM). Another important set of testing standards for products are developed by the Underwriters Laboratories (UL). Codes of practice give detailed design methods for repetitive technical problems such as the design of piping, heat exchangers, and pressure vessels. Many of these are developed by the American Society of Mechanical Engineers (ASME Boiler and Pressure Vessel Code), the American Nuclear Society, and the Society of Automotive Engineers.

Standards are often prepared by individual companies for their own proprietary use. They address such things as dimensions, tolerances, forms, manufacturing processes, and finishes. In-house standards are often used by the company purchasing department when outsourcing. The next level of standard preparation involves groups of companies in the same industry arriving at industry consensus standards. Often these are sponsored through an industry trade association, such as the American Institute of Steel Construction (AISC) or the Door and Hardware Institute. Industry standards of this type are usually submitted to the American National Standards Institute (ANSI) for a formal review process, approval, and publication. A similar function is played by the International Organization for Standardization (ISO) in Geneva, Switzerland. Another important set of standards are government (federal, state, and local) specification standards. Because the government is such a large purchaser of goods and services, it is important for the engineer to have access to these standards. Engineers working in high-tech defense areas must be conversant with MIL standards and handbooks of the Department of Defense.

In addition to protecting the public, standards play an important role in reducing the cost of design and of products. The use of standard components and materials leads to cost reduction in many ways. The use of design standards saves the designer, when involved in original design work, from spending time on finding solutions to a multitude of recurring identical problems. Moreover, designs based on standards provide a firm basis for negotiation and better understanding between the buyer and seller of a product. Failure to incorporate up-to-date standards in a design may lead to difficulties with product liability (see Chap. 18).

The engineering design process is concerned with balancing four goals: proper function, optimum performance, adequate reliability, and low cost. The greatest cost saving comes from reusing existing parts in design. The main savings come from eliminating the need for new tooling in production and from a significant reduction in the parts that must be stocked to provide service over the lifetime of the product. In much of new product design only 20 percent of the parts are new, about 40 percent are existing parts used with minor modification, while the other 40 percent are existing parts reused without modification.

An important aspect of standardization in CAD-CAM is in interfacing and communicating information between various computer devices and manufacturing machines. The National Institute of Standards and Technology (NIST) has been instrumental in developing and promulgating the IGES code, and more recently the Product Data Exchange Specification (PDES). Both of these standards represent a neutral data format for transferring geometric data between equipment from different vendors of CAD systems. This is an excellent example of the role of, and need for, a national standards organization.

1.8
DESIGN REVIEW

The design review is a vital aspect of the design process. It provides an opportunity for specialists from different disciplines to interact with generalists to ask critical questions and exchange vital information. A *design review* is a retrospective study of the design up to that point in time. It provides a systematic method for identifying problems with the design, determining future courses of action, and initiating action to correct any problem areas.

To accomplish these objectives, the review team should consist of representatives from design, manufacturing, marketing, purchasing, quality control, reliability engineering, and field service. The chairman of the review team is normally a chief engineer or project manager with a broad technical background and broad knowledge of the company's products. In order to ensure freedom from bias, the chairman of the design review team should not have direct responsibility for the design under review.

Depending on the size and complexity of the product, design reviews should be held from three to six times in the life of the project. The minimum review schedule consists of conceptual, interim, and final reviews. The conceptual review occurs once the conceptual design (Chap. 7) has been established. This review has the greatest impact on the design, since many of the design details are still fluid and changes can be made at this stage with least cost. The interim review occurs when the embodiment design is finalized and the product architecture, subsystems, performance characteristics, and critical design parameters are established. It looks critically at the interfaces between the subsystems. The final review takes place at completion of the detail design and establishes whether the design is ready for transfer to manufacturing.

Each review looks at two main aspects. The first is concerned with the technical elements of the design, while the second is concerned with the business aspects of the product (see Chap. 2). The essence of the technical review of the design is to compare the findings against the detailed product design specification (PDS) that is formulated at the problem definition phase of the project. The PDS is a detailed document that describes what the design must be in terms of performance requirements, the environment in which it must operate, the product life, quality, reliability, cost, and a host of other design requirements. The PDS is the basic reference document for both the product design and the design review. The business aspect of the review is concerned with tracking the costs incurred in the project, projecting how the design will affect the expected marketing and sales of the product, and maintaining the time schedule. An important outcome of the review is to determine what changes in resources, people, and money are required to produce the appropriate business outcome. It must be realized that a possible outcome of any review is to withdraw the resources and terminate the project.

A formal design review process requires a commitment to good documentation of what has been done, and a willingness to communicate this to all parties involved in the project. The minutes of the review meeting should clearly state what decisions were made and should include a list of "action items" for future work. Since the PDS is the basic control document, care must be taken to keep it always updated.

1.8.1 Redesign

A common situation is redesign. There are two categories of redesigns: *fixes* and *updates*. A fix is a design modification that is required due to less than acceptable performance once the product has been introduced into the marketplace. On the other hand, updates are usually planned as part of the product's life cycle before the product is introduced to the market. An update may add new features and improve performance to the product or improve its appearance to keep it competitive.

The most common situation in redesign is the modification of an existing product to meet new requirements. For example, the banning of the use of fluorinated hydrocarbon refrigerants because of the "ozone-hole problem" required the extensive redesign of refrigeration systems. Often redesign results from failure of the product in service. A much simpler situation is the case where one or two dimensions of a component must be changed to match some change made by the customer for that part. Yet another situation is the continuous evolution of a design to improve performance. An extreme example of this is shown in Fig. 1.8. The steel railroad wheel had been in its present design for nearly 150 years. In spite of improvements in metallurgy and the understanding of stresses, the wheels still failed at the rate of about 200 per year,

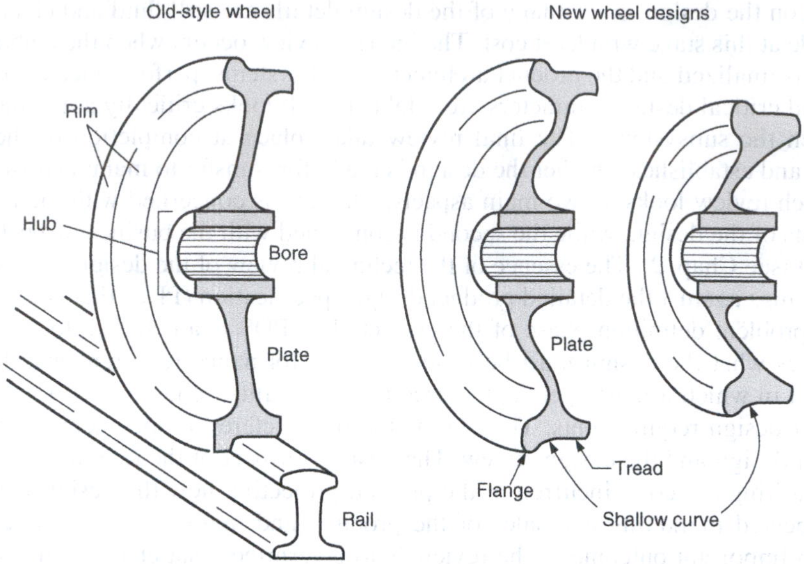

FIGURE 1.8
An example of a design update. Old design of railcar wheel versus improved design.

often causing disastrous derailments. The chief cause of failure was thermal buildup caused by failure of a railcar's braking system. Long-term research by the Association of American Railroads has resulted in the improved, current design. The chief design change is that the flat plate, the web between the bore and the rim, has been replaced by an S-shaped plate. The curved shape allows the plate to act like a spring, flexing when overheated, avoiding the buildup of stresses that are transmitted through the rigid flat plates. The wheel's tread has also been redesigned to extend the rolling life of the wheel. Car wheels last for about 200,000 miles. Traditionally, when a new wheel was placed in service it lost from 30 to 40 percent of its tread and flange while it wore away to a new shape during the first 25,000 miles of service. After that the accelerated wear stopped and normal wear ensued. In the new design the curve between the flange and the tread has been made less concave, more like the profile of a "worn" wheel. The new wheels last for many thousands of miles longer, and the rolling resistance is lower, saving on fuel cost.

1.9
SOCIETAL CONSIDERATIONS IN ENGINEERING DESIGN

The first fundamental canon of the ABET Code of Ethics states that "engineers shall hold paramount the safety, health, and welfare of the public in the performance of their profession." A similar statement has been in engineering codes of ethics since the early 1920s, yet there is no question that what society perceives to be proper treatment by the profession has changed greatly in the intervening time. Today's 24-hour news cycle and Internet make the general public, in a matter of hours, aware of events taking place anywhere in the world. That, coupled with a generally much higher standard of education and standard of living, has led to the development of a society that has high expectations, reacts to achieve change, and organizes to protest perceived wrongs. At the same time, technology has had major effects on the everyday life of the average citizen. Whether we like it or not, all of us are intertwined in complex technological systems: an electric power grid, a national network of air traffic controllers, and a gasoline and natural gas distribution network. Much of what we use to provide the creature comforts in everyday life has become too technologically complex or too physically large for the average citizen to comprehend. Moreover, our educational system does little to educate students to understand the technology within which they are immersed.

Thus, in response to real or imagined ills, society has developed mechanisms for countering some of the ills and/or slowing down the rate of social change. The major social forces that have had an important impact on engineering design are occupational safety and health, consumer rights, environmental protection, the antinuclear movement, and the freedom of information and public disclosure movement. The result of these social forces has been a great increase in federal regulations (in the interest of protecting the public) over many aspects of commerce and business and/or a drastic change in the economic payoff for new technologically oriented ventures. Those new factors have had a profound effect on the practice of engineering and the rate of innovation.

The following are some general ways in which increased societal awareness of technology, and subsequent regulation, have influenced the practice of engineering design:

- Greater influence of lawyers on engineering decisions, often leading to product liability actions
- More time spent in planning and predicting the future effects of engineering projects
- Increased emphasis on "defensive research and development," which is designed to protect the corporation against possible litigation
- Increased effort expended in research, development, and engineering in environmental control and safety

Clearly, these societal pressures have placed much greater constraints on how engineers can carry out their designs. Moreover, the increasing litigiousness of U.S. society requires a greater awareness of legal and ethical issues on the part of each engineer (see Chap. 18).

One of the most prevalent societal pressures at the present time is the environmental movement. Originally, governmental regulation was used to clean up rivers and streams, to ameliorate smog conditions, and to reduce the volume of solid waste that is sent to landfills. Today, there is a growing realization that placing environmental issues at a high priority (not doing them because the government demands it) represents smart business. Several major oil producers publicly take seriously the link between carbon dioxide emissions and rising global temperatures and have embarked on a major effort to become the leaders in renewable energy sources like solar power and fuel from biomass. A major chemical company has placed great emphasis on developing environmentally friendly products. Its biodegradable herbicides allow for a hundredfold

TABLE 1.1
Characteristics of an Environmentally Responsible Design

- Easy to dissassemble
- Able to be recycled
- Contains recycled materials
- Uses identifiable and recyclable plastics
- Reduces use of energy and natural materials in its manufacture
- Manufactured without producing hazardous waste
- Avoids use of hazardous materials
- Reduces product chemical emissions
- Reduces product energy consumption

reduction in the herbicide that must be applied per acre, greatly reducing toxic runoff into streams. This reorientation of business thinking toward environmental issues is often called *sustainable development*, businesses built on renewable materials and fuels.

The change in thinking, from fixing environmental problems at the discharge end of the pipe or smokestack to sustainable development, places engineering design at the heart of the issue. Environmental issues are given higher priority in design. Products must be designed to make them easier to reuse, recycle, or incinerate—a concept often called green design.[1] Green design also involves the detailed understanding of the environmental impact of products and processes over their entire life cycle. For example, life-cycle analysis would be used to determine whether paper or plastic grocery bags are more environmentally benign. Table 1.1 gives the chief aspects of an environmentally responsible design.

It seems clear that the future is likely to involve more technology, not less, so that engineers will face demands for innovation and design of technical systems of unprecedented complexity. While many of these challenges will arise from the requirement to translate new scientific knowledge into hardware, others will stem from the need to solve problems in "socialware." By socialware we mean the patterns of organization and management instructions needed for the hardware to function effectively.[2] Such designs will have to deal not only with the limits of hardware, but also with the vulnerability of any system to human ignorance, human error, avarice, and hubris. A good example of this point is the delivery system for civilian air transportation. While the engineer might think of the modern jet transport, with all of its complexity and high technology, as the main focus of concern, such a marvelous piece of hardware only satisfies the needs of society when embedded in an intricate system that includes airports, maintenance facilities, traffic controllers, navigation aids, baggage handling, fuel supply, meal service, bomb detection, air crew training, and weather monitoring. It is important to realize that almost all of these socialware functions are

1. Office of Technology Assessment, "Green Products by Design: Choices for a Cleaner Environment," OTA-E-541, Government Printing Office, Washington, DC, 1992.
2. E. Wenk, Jr., *Engineering Education,* November 1988, pp. 99–102.

TABLE 1.2
Future Trends in Interaction of Engineering with Society

- The future will entail more technology, not less.
- Because all technologies generate side effects, designers of technological systems will be challenged to prevent, or at least mitigate, adverse consequences.
- The capacity to innovate, manage information, and nourish knowledge as a resource will dominate economic growth as natural resources, capital, and labor once did. This places a high premium on the talent to design not simply hardware, but complex technology-based systems.
- Globalization will proceed at a rapid pace, with low-tech manufacturing moving to countries with low wage rates.
- Globalization will force developed countries to place great emphasis on creative, innovative design of products.
- Distribution of benefits in society will not be uniform, so disparity will grow between the "haves" and the "have nots."
- Conflicts between winners and losers will become more strenuous as we enter an age of scarcity, global economic competition, higher energy costs, increasing populations, associated political instabilities, and larger-scale threats to human health and the environment.
- Because of technology, we may be moving to "one world," with people, capital, commodities, information, culture, and pollution freely crossing borders. But as economic, social, cultural, and environmental boundaries dissolve, political boundaries will be stubbornly defended. The United States will sense major economic and geopolitical challenges to its position of world leadership in technology.
- Complexity of technological systems will increase, as will interdependencies, requiring management with a capacity for both systems planning and trouble-free, safe operations.
- Decision making will become more complex because of increases in the number and diversity of interconnected organizations and their separate motivations, disruptions in historical behavior, and the unpredictability of human institutions.
- Mass media will play an ever more significant role in highlighting controversy and publicizing technological dilemmas, especially where loss of life may be involved. Since only the mass media can keep everyone in the system informed, a special responsibility falls on them for both objective and courageous inquiry and reporting.
- Amidst this complexity and the apparent domination of decision making by experts and the commercial or political elite, the general public is likely to feel more vulnerable. Public interest lobbies will demand to know what is being planned that may affect people's lives and environment, to have estimates of a wide range of impacts, to weigh alternatives, and to have the opportunity to intervene through legitimate processes.
- Given the critical choices ahead, greater emphasis will be placed on moral vision and the exercise of ethical standards in delivering technology to produce socially satisfactory results. Accountability will be demanded more zealously.

Adapted from E. Wenk, Jr., "Tradeoffs," Johns Hopkins University Press, 1986.

driven by federal or local rules and regulations. Thus, it should be clear that the engineering profession is required to deal with much more than technology. Techniques for dealing with the complexity of large systems have been developed in the discipline of *systems engineering*.[1]

---

1. A. P. Sage, *Systems Enginering,* John Wiley & Sons, New York, 1992; B. S. Blanchard and W. K. Fabrycky, *Systems Engineering and Analysis,* Prentice Hall, Upper Saddle River, NJ. 1998.

Another area where the interaction between technical and human networks is becoming stronger is in consideration of risk, reliability, and safety (see Chap. 14). No longer can safety factors simply be looked up in codes or standards. Engineers must recognize that design requirements depend on public policy as much as industry performance requirements. This is an area of design where government influence has become much stronger.

There are five key roles of government in interacting with technology:

- As a stimulus to free enterprise through manipulation of the tax system
- By influencing interest rates and the supply of venture capital through changes in fiscal policy to control the growth of the economy
- As a major customer for high technology, chiefly in military systems
- As a funding source (patron) for research and development
- As a regulator of technology

Table 1.2 lists some major trends concerning the interaction of technology with a changing world society.

Engineering is concerned with problems whose solution is needed and/or desired by society. The purpose of this section is to reinforce that point, and hopefully to show the engineering student how important a broad knowledge of economics and social science is to modern engineering practice.

1.10
SUMMARY

Engineering design is a challenging activity because it deals with largely unstructured problems that are important to the needs of society. An engineering design process creates something that did not exist before, requires choices between many variables and parameters, and often requires balancing multiple and sometimes conflicting requirements. Product design has been identified as the real key to world-competitive business. The steps in the design process are:

Phase I. Conceptual design
- Recognition of a need
- Definition of the problem
- Gathering of information
- Developing a design concept
- Choosing between competing concepts (evaluation)

Phase II: Embodiment design
- Determining product architecture—arrangement of the physical functions
- Configuration design—preliminary selection of materials, modeling and sizing of parts
- Parametric design—creating a robust design, and selection of final dimensions and tolerances

Phase III: Detail design—finalizing all details of design. Creation of final drawings and specifications.

While many consider that the engineering design process ends with detail design, there are many issues that must be resolved before a product can be shipped to the customer. These additional phases of design are often folded into what is called the product development process; see Chap. 2.

Engineering design must consider many factors, which are documented in the product design specification (PDS). Among the most important of these factors are required functions with associated performance characteristics, environment in which it must operate, target product cost, service life, provisions for maintenance and logistics, aesthetics, expected market and quantity to be produced, man-machine interface requirements (ergonomics), quality and reliability, safety and environmental concerns, and provision for testing.

NEW TERMS AND CONCEPTS

Analysis	Form	Robust design
Code	Function	Specification
Component	Green design	Standard
Computer-aided engineering	Human factors engineering	Subsystem
Configuration design	Iterative	Synthesis
Critical to quality	Needs analysis	System
Design feature	Product design specification	Total life cycle
Detail design	Problem definition	Useful life
Embodiment design	Product architecture	

BIBLIOGRAPHY

Dym, C. I. and P. Little, *Engineering Design: A Project-Based Introduction,* 2d ed., John Wiley & Sons, New York, 2004.

Eggert, R. J., *Engineering Design,* Pearson Prentice Hall, Upper Saddle River, NJ, 2005.

Magrab, E. B. S. K. Gupta, F.P. McCluskey and P.A. Sandborn, *Integrated Product and Process Design and Development,* 2d ed., CRC Press, Boca Raton, FL, 1997, 2010.

Pahl, G. and W. Beitz, *Engineering Design,* 3d ed., Springer-Verlag, New York, 2006.

Stoll, H. W., *Product Design Methods and Practices,* Marcel Dekker, Inc., New York, 1999.

Ullman, D. G., *The Mechanical Design Process,* 4th ed., McGraw-Hill, New York, 2010.

PROBLEMS AND EXERCISES

1.1. A major manufacturer of snowmobiles needs to find new products in order to keep the workforce employed all year round. Starting with what you know or can find out about snowmobiles, make reasonable assumptions about the capabilities of the company. Then develop a needs analysis that leads to some suggestions for new products that the company could make and sell. Give the strengths and weaknesses of your suggestions.

1.2. Take a problem from one of your engineering science classes, and add and subtract those things that would frame it more as an engineering design problem.

1.3. There is a need in underdeveloped countries for building materials. One approach is to make building blocks (4 by 6 by 12 in.) from highly compacted soil. Your assignment is to design a block-making machine with the capacity for producing 600 blocks per day at a capital cost of less than $300. Develop a needs analysis, a definitive problem statement, and a plan for the information that will be needed to complete the design.

1.4. The steel wheel for a freight car has three basic functions: (1) to act as a brake drum, (2) to support the weight of the car and its cargo, and (3) to guide the freight car on the rails. Freight car wheels are produced by either casting or rotary forging. They are subjected to complex conditions of dynamic thermal and mechanical stresses. Safety is of great importance, since derailment can cause loss of life and property. Develop a broad systems approach to the design of an improved cast-steel car wheel.

1.5. The need for material conservation and reduced cost has increased the desirability of corrosion-resistant coatings on steel. Develop several design concepts for producing 12-in.-wide low-carbon-steel sheet that is coated on one side with a thin layer, e.g., 0.001 in., of nickel.

1.6. The support of thin steel strip on a cushion of air introduces exciting prospects for the processing and handling of coated steel strip. Develop a feasibility analysis for the concept.

1.7. Consider the design of aluminum bicycle frames. A prototype model failed in fatigue after 1600 km of riding, whereas most steel frames can be ridden for over 60,000 km. Describe a design program that will solve this problem.

1.8. (a) Discuss the societal impact of a major national program to develop synthetic fuel (liquid and gaseous) from coal. (It has been estimated that to reach the level of supply equal to the imports from OPEC countries would require over 50 installations, each costing several billion dollars.)

(b) Do you feel there is a basic difference in the perception by society of the impact of a synthetic fuel program compared with the impact of nuclear energy? Why?

1.9. You are a design engineer working for a natural gas transmission company. You are assigned to a design team that is charged with preparing the proposal to the state Public Utility Commission to build a plant to receive liquefied natural gas from ocean-going tankers and unload it into your company's gas transmission system. What technical issues and societal issues will your team have to deal with?

1.10. You are a senior design engineer at the design center of a major U.S. manufacturer of power tools. Over the past five years your company has outsourced component manufacturing and assembly to plants in Mexico and China. While your company still has a few plants operating in the United States, most production is overseas. Think about how your job as the leader of a product development team has changed since your company made this change, and suggest how it will evolve in the future.

1.11 The oil spill from BP well Deepwater Horizon is one of the world's greatest environmental disasters. Nearly 5 million barrels of crude oil spewed into the Gulf of Mexico for three months. As a team, do research on the following issues: (a) the technology of drilling for oil in water deeper than 1000 feet; (b) the causes of the well blowout; (c) the short-term damage to the U.S. economy; (d) the long-term effects on the U.S.; and (e) the impact on the owner of the well, BP Global.

1.12. Brazil is rapidly developing into one of the world's most dynamic economies. As a country it is blessed with great mineral resources, abundant unused farm land, and network of free flowing rivers. But, to achieve its potential, it will need a much expanded transportation network and considerably expanded electric generation capacity. Much of this expansion will occur in the undeveloped center of the country including the Amazon region. Define the major obstacles to large-scale development and suggest the best technologies to achieve it.

2

PRODUCT DEVELOPMENT PROCESS

2.1
INTRODUCTION

This text emphasizes the design of consumer and engineered products. Having defined the engineering design process in considerable detail in Chap. 1, we now turn to the consideration of the product development process. The engineering design of a product is a vital part of this process, but product development involves much more than design. The development of a product is undertaken by a company to make a profit for its stakeholders. There are many business issues, desired outcomes, and strategies that influence the structure of the product development process (PDP). The influence of business considerations, in addition to engineering performance, is seen in the structure of the PDP.

This chapter lays out a product development process that is more encompassing than the engineering design process described in Chap. 1. This chapter presents organizational structures for the design and product development functions and discusses markets and the vital function of marketing in detail. Since the most successful products are often innovative products, we conclude the chapter with some ideas about technological innovation.

2.2
PRODUCT DEVELOPMENT PROCESS

A generally accepted model of the product development process is shown in Fig. 2.1. The six phases shown in this diagram generally agree with those proposed by Asimow for the design process (see Sec.1.4) with the addition of the Phase 0, Planning.

Note that each phase in Fig. 2.1 narrows down to a point. This symbolizes the *"gate"* or review that the project must successfully pass through before moving on

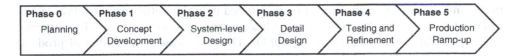

FIGURE 2.1
The product development process in stage-gate format.

to the next stage or phase of the process. This stage-gate product development process is used by many companies in order to encourage rapid product development and to cull out the least promising projects before large sums of money are committed. The amount of money to develop a project increases exponentially from Phase 0 to Phase 5. However, the money spent in product development is small compared to what it would cost in sunk capital and lost brand reputation if a defective product has to be recalled from the market. Thus, an important reason for using the *stage-gate process* is to quickly "get it right."

Phase 0 is the planning that should be done before the approval of the product development project. Product planning is usually done in two steps. The first step is a quick investigation and scoping of the project to determine the possible markets and whether the product is in alignment with the corporate strategic plan. It also involves a preliminary engineering assessment to determine technical and manufacturing feasibility. This preliminary assessment usually is completed in a month. If things look promising after this quick examination, the planning operation goes into a detailed investigation to build the *business case* for the project. This could take several months to complete and involves personnel from marketing, design, manufacturing, finance, and possibly legal. In making the business case, marketing completes a detailed marketing analysis that involves market segmentation to identify the target market, the product positioning, and the product benefits. Design digs more deeply to evaluate the technical capability, possibly including some proof-of-concept analysis or testing to validate some very preliminary design concepts, while manufacturing identifies possible production constraints, costs, and thinks about a supply chain strategy. A critical part of the business case is the financial analysis, which uses sales and cost projections from marketing to predict the profitability of the project. Typically this involves a discounted cash flow analysis (see Chap. 16) with a sensitivity analysis to project the effects of possible risks. The gate at the end of Phase 0 is crucial, and the decision of whether to proceed is made in a formal and deliberate manner, for costs will become considerable once the project advances to Phase 1. The review board makes sure that the corporate policies have been followed and that all of the necessary criteria have been met or exceeded. High among these is exceeding a corporate goal for return on investment (ROI). If the decision is to proceed, then a multifunctional team with a designated leader is established. The product design project is formally on its way.

Phase 1, Concept Development, considers the different ways the product and each subsystem can be designed. The development team takes what is known about the potential customers from Phase 0, adds its own knowledge base and fashions this into a carefully crafted *product design specification (PDS)*. This process of determining the needs and wants of the customer is more detailed than the initial market survey done in Phase 0. It is aided by using tools such as surveys and focus groups, benchmarking, and quality function deployment (QFD). The generation of a number of product concepts follows. The designers' creative instincts must be stimulated, but again tools are used to assist in the development of promising concepts. Now, having arrived at a small set of feasible concepts, the one best suited for development into a product must be determined using selection methods. Conceptual design is the heart of the product development process, for without an excellent concept you cannot have a highly successful product. These aspects of conceptual design are covered in Chapters 3, 6, and 7.

Phase 2, System-Level Design, is where the functions of the product are examined, leading to the division of the product into various subsystems. In addition, alternative ways of arranging the subsystems into a *product architecture* are studied. The interfaces between subsystems are identified and studied. Successful operation of the entire system relies on careful understanding of the interface between each subsystem. Phase 2 is where the form and features of the product begin to take shape, and for this reason it is often called *embodiment design*.[1] Selections are made for materials and manufacturing processes, and the configuration and dimensions of parts are established. Those parts whose function is *critical to quality* are identified and given special analysis to ensure *design robustness*.[2] Careful consideration is given to the product-human interface (ergonomics), and changes to form are made if needed. Likewise, final touches will be made to the styling introduced by the industrial designers. In addition to a complete computer-based geometrical model of the product, critical parts may be built with rapid protyping methods and physically tested. At this stage of development, marketing will most likely have enough information to set a price target for the product. Manufacturing will begin to place contracts for long-delivery tooling and will begin to define the assembly process. By this time legal will have identified and worked out any patent licensing issues.

Phase 3, Detail Design, is the phase where the design is brought to the state of a complete engineering description of a tested and producible product. Missing information is added on the arrangement, form, dimensions, tolerances, surface properties, materials, and manufacturing of each part in the product. These result in a specification for each special-purpose part to be manufactured, and the decision whether it will be made in the factory of the corporation or outsourced to a supplier. At the same time the design engineers are wrapping up all of these details, the manufacturing engineers are finalizing a process plan for each part, as well as designing the tooling to make these parts. They also work with design engineers to finalize any issue of product robustness and define the quality assurance processes that will be used to achieve a quality product. The output of the detail design phase is the *control documentation* for the product. This takes the form of CAD files for the product assembly and for each part and its tooling. It also involves detailed plans for production and quality assurance, as well as many legal documents in the form of contracts and documents

---

1. Embodiment means to give a perceptible shape to a concept.
2. Robustness in a design context does not mean strong or tough. It means a design whose performance is insensitive to the variations introduced in manufacturing, or by the environment in which the product operates.

protecting intellectual property. At the end of Phase 3, a major review is held to determine whether it is appropriate to let contracts for building the production tooling, although contracts for long lead-time items such as polymer injection molding dies are most likely let before this date.

Phase 4, Testing and Refinement, is concerned with making and testing many preproduction versions of the product. The first (alpha) prototypes are usually made with *production-intent parts*. These are working models of the product made from parts with the same dimensions and using the same materials as the production version of the product but not necessarily made with the actual processes and tooling that will be used with the production version. This is done for speed in getting parts and to minimize the cost of product development. The purpose of the alpha test is to determine whether the product will actually work as designed and whether it will satisfy the most important customer needs. The beta tests are conducted on products assembled from parts made by the actual production processes and tooling. They are extensively tested in-house and by selected customers in their own use environments. The purpose of these tests is to satisfy any doubts about the performance and reliability of the product, and to make the necessary engineering changes before the product is released to the general market. Only in the case of a completely "botched design" would a product fail at this stage gate, but it might be delayed for a serious fix that could delay the product launch. During Phase 4 the marketing people work on developing promotional materials for the product launch, and the manufacturing people fine-tune the fabrication and assembly processes and train the workforce that will make the product. Finally, the sales force puts the finishing touches on the sales plan.

At the end of Phase 4 a major review is carried out to determine whether the work has been done in a quality way and whether the developed product is consistent with the original intent. Because large monetary sums must be committed beyond this point, a careful update is made of the financial estimates and the market prospects before funds are committed for production.

At *Phase 5*, Production Ramp-Up, the manufacturing operation begins to make and assemble the product using the intended production system. Most likely they will go through a *learning curve* as they work out any production yield and quality problems. Early products produced during ramp-up often are supplied to preferred customers and studied carefully to find any defects. Production usually increases gradually until full production is reached and the product is *launched* and made available for general distribution. For major products there will certainly be a public announcement, and often special advertising and customer inducements. Some 6 to 12 months after product launch there will be a final major review. The latest financial information on sales, costs, profits, development cost, and time to launch will be reviewed, but the main focus of the review is to determine what were the strengths and weaknesses of the product development process. The emphasis is on *lessons learned* so that the next product development team can do even better.

The stage-gate development process is successful because it introduces schedule and approval to what is often an *ad hoc* process.[1] The process is relatively simple, and

1. R. G. Cooper, *Winning at New Products*, 3d ed., Perseus Books, Cambridge, MA, 2001.

the requirements at each gate are readily understood by managers and engineers. It is not intended to be a rigid system. Most companies modify it to suit their own circumstances. Neither is it intended to be a strictly serial process, although Fig. 2.1 gives that impression. Since the PDP teams are multifunctional, the activities as much as possible are carried out concurrently. Thus, marketing will be going on at the same time that the designers are working on their tasks, and manufacturing does their thing. However, as the team progresses through the stages, the level of design work decreases and manufacturing activities increase.

2.2.1 Factors for Success

In commercial markets the cost to purchase a product is of paramount importance. It is important to understand what the product cost implies and how it relates to the product price. More details about costing can be found in Chap. 17. Cost and price are distinctly different concepts. The product cost includes the cost of materials, components, manufacturing, and assembly. The accountants also include other less obvious costs such as the prorated costs of capital equipment (the plant and its machinery), tooling cost, development cost, inventory costs, and likely warranty costs, in determining the total cost of producing a unit of product. The price is the amount of money that a customer is willing to pay to buy the product. The difference between the price and the cost is the profit per unit of product sold.

$$\text{Profit} = \text{Product Price} - \text{Product Cost} \qquad (2.1)$$

This equation is the most important equation in engineering and in the operation of any business. If a corporation cannot make a profit, it soon is forced into bankruptcy, its employees lose their positions, and the owner or stockholders lose their investment. Everyone employed by a corporation seeks to maximize this profit while maintaining the strength and vitality of the product lines. The same statement can be made for a business that provides services instead of products. The price paid by the customer for a specified service must be more than the cost to provide that service if the business is to make a profit and prosper.

There are four key factors that determine the success of a product in the marketplace.

- The quality, performance, and price of the product.
- The cost to manufacture the product over its life cycle.
- The cost of product development.
- The time needed to bring the product to the market.

Let's discuss the product first. Is it attractive and easy to use? Is it reliable? Does it meet the needs of the customer? Is it better than the products now available in the marketplace? If the answer to all of these questions is an unqualified Yes, the customer may want to buy the product, but only if the price is right.

Equation (2.1) offers only two ways to increase profit on an existing product line with a mature market base. We can increase the product's price, justified by adding new features or improving quality, or we can reduce the product's cost, through improvements in the production process. In the highly competitive world market for consumer products the latter is more likely than the former.

Developing a product involves many people with talents in different disciplines. It takes time, and it costs a lot of money. Thus, if we can reduce the product development cost, the profit will be increased. First, consider development time. Development time, also known as the time to market, is the time interval from the start of the product development process (the kickoff) to the time that the product is available for purchase (the product release date). The product release date is a very important target for a development team because many significant benefits follow from being first to market. There are at least three competitive advantages for a company that has development teams that can develop products quickly. First, the product's life is extended. For each month cut from the development schedule, a month is added to the life of the product in the marketplace, generating an additional month of revenues from sales, and profit. We show the revenue benefits of being first to market in Fig. 2.2. The shaded region between the two curves showing time of market entry is the enhanced revenue due to the extra sales.

A second benefit of early product release is increased market share. The first product to market has 100 percent of the market share in the absence of a competing product. For existing products with periodic development of new models it is generally recognized that the earlier a product is introduced to compete with older models, without sacrificing quality, reliability, or performance and price, the better chance it has for acquiring and retaining a large share of the market. The effect of gaining a larger market share on sales revenue is illustrated in Fig. 2.2. The crosshatched region between the two curves at the top of the graph shows the enhanced sales revenue due to increased market share.

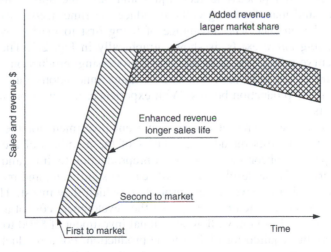

FIGURE 2.2
Increased sales revenue due to extended product life and larger market share.

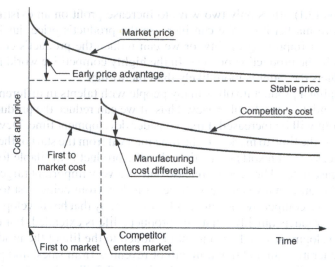

FIGURE 2.3

The team that brings the product first to market enjoys an initial price advantage and subsequent cost advantages from manufacturing efficiencies.

A third advantage of a short development cycle is higher *profit margins*. Profit margin is the net profit divided by the sales. If a new product is introduced before competing products are available, the corporation can command a higher price for the product, which enhances the profit. With time, competitive products will enter the market and force prices down. However, in many instances, relatively large profit margins can be maintained because the company that is first to market has more time than the competitor to learn methods for reducing manufacturing costs. They also learn better processing techniques and have the opportunity to modify assembly lines and manufacturing cells to reduce the time needed to manufacture and assemble the product. The advantage of being first to market, when a manufacturing *learning curve* exists, is shown graphically in Fig. 2.3. The manufacturing learning curve reflects the reduced cost of processing, production, and assembly with time. These cost reductions are due to many innovations introduced by the workers after mass production begins. With experience, it is possible to drive down production costs.

Development costs represent a very important investment for the company involved. Development costs include the salaries of the members of the development team, money paid to subcontractors, costs of preproduction tooling, and costs of supplies and materials. These development costs can be significant, and most companies must limit the number of development projects in which they invest. The size of the investment can be appreciated by noting that the development cost of a new automobile is an estimated $1 billion, with an additional investment of $500 to $700 million for the new tooling required for high-volume production. For a product like a power tool, the development cost can be one to several million dollars, depending on the features to be introduced with the new product.

2.2.2 Static Versus Dynamic Products

Some product designs are static, in that the changes in their design take place over long time periods through incremental changes occurring at the subsystem and component levels. Examples of static products are automobiles and most consumer appliances like refrigerators and dishwashers. Dynamic products like wireless mobile phones, digital video recorders and players, and software change the basic design concept as often as the underlying technology changes.

Static products exist in a market where the customer is not eager to change, technology is stable, and fashion or styling play little role. These are markets characterized by a stable number of producers with high price competition and little product research. There is a mature, stable technology, with competing products similar to each other. The users are generally familiar with the technology and do not demand significant improvement. Industry standards may even restrict change, and parts of the product are assembled from components made by others. Because of the importance of cost, emphasis is more on manufacturing research than on product design research.

With dynamic products, customers are willing to, and may even demand, change. The market is characterized by many small producers, doing active market research and seeking to reduce product cycle time. Companies actively seek new products employing rapidly advancing technology. There is high product differentiation and low industry standardization. More emphasis is placed on product research than on manufacturing research.

A number of factors serve to protect a product from competition. A product that requires high capital investment to manufacture or requires complex manufacturing processes tends to be resistant to competition. At the other end of the product chain, the need for an extensive distribution system may be a barrier to entry.[1] A strong patent position may keep out competition, as may strong brand identification and loyalty on the part of the customer.

2.2.3 Variations on the Generic Product Development Process

The product development process (PDP) described at the beginning of Sec. 2.2 was based on the assumption that the product is being developed in response to an identified market need, a *market pull* situation. This is a common situation in product development, but there are other situations that need to be recognized.[2]

The opposite of market pull is *technology push*. This is the situation where the company starts with a new proprietary technology and looks for a market in which to apply this technology. Often successful technology push products involve basic materials or basic process technologies, because these can be deployed in thousands of applications, and the probability of finding successful applications is therefore high.

1. The Internet has made it easier to set up direct marketing systems for products.
2. K. T. Ulrich and S. D. Eppinger, *Product Design and Development*, 3d ed., pp. 18–21, McGraw-Hill, New York, 2004.

The discovery of nylon by the DuPont Company and its successful incorporation into thousands of new products is a classic example. The development of a technology push product begins with the assumption that the new technology will be employed. This can entail risk, because unless the new technology offers a clear competitive advantage to the customer the product is not likely to succeed.

A *platform product* is built around a preexisting technological subsystem. Examples of such a platform are the Apple Macintosh operating system or the Black & Decker doubly insulated universal motor. A platform product is similar to a technology push product in that there is an *a priori* assumption concerning the technology to be employed. However, it differs in that the technology has already been demonstrated in the marketplace to be useful to a customer, so that the risk for future products is less. Often when a company plans to utilize a new technology in their products they plan to do it as a series of platform products. Obviously, such a strategy helps justify the high cost of developing a new technology.

For certain products the manufacturing process places strict constraints on the properties of the product, so product design cannot be separated from the design of the production process. Examples of *process-intensive products* are automotive sheet steel, food products, semiconductors, chemicals, and paper. Process-intensive products typically are made in high volume, often with continuous flow processes as opposed to discrete goods manufacturing. With such a product, it might be more typical to start with a given process and design the product within the constraints of the process.

Customized products are those in which variations in configuration and content are created in response to a specific order of a customer. Often the customization is with regard to color or choice of materials but more frequently it is with respect to content, as when a person orders a personal computer by phone, or the accessories with a new car. Customization requires the use of modular design and depends heavily on information technology to convey the customer's wishes to the production line. In a highly competitive world marketplace, *mass customization* appears to be one of the future trends.

2.3
PRODUCT AND PROCESS CYCLES

Every product goes through a cycle from birth, into an initial growth stage, into a relatively stable period, and finally into a declining state that eventually ends in the useful life of the product (Fig. 2.4). Since there are challenges and uncertainties any time a new product is brought to market, it is useful to understand these cycles.

2.3.1 Stages of Development of a Product

In the introductory stage the product is new and consumer acceptance is low, so sales are low. In this early stage of the product life cycle the rate of product change is rapid as management tries to maximize performance or product uniqueness in an attempt to enhance customer acceptance. When the product has entered the growth stage, knowledge of the product and its capabilities has reached an increasing number of customers,

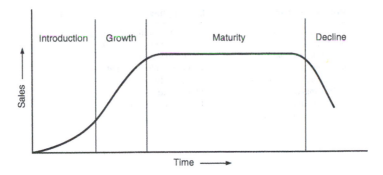

FIGURE 2.4
Product life cycle.

and sales growth accelerates. There may be an emphasis on custom tailoring the product by making accessories for slightly different customer needs. At the maturity stage the product is widely accepted and sales are stable and are growing at the same rate as the economy as a whole. When the product reaches this stage, attempts should be made to rejuvenate it by the addition of new features or the development of still new applications. Products in the maturity stage usually experience considerable competition. Thus, there is great emphasis on reducing the cost of a mature product. At some point the product enters the decline stage. Sales decrease because a new and better product has entered the market to fulfill the same societal need.

During the product introduction phase, where the volume of production is modest, expensive to operate but flexible manufacturing processes are used and product cost is high. As we move into the period of product market growth, more automated, higher-volume manufacturing processes can be justified to reduce the unit cost. In the product maturity stage, emphasis is on prolonging the life of the product by modest product improvement and significant reduction in unit cost. This might result in outsourcing to a lower-labor-cost location.

If we look more closely at the product life cycle, we will see that the cycle is made up of many individual processes (Fig. 2.5). In this case the cycle has been divided into the premarket and market phases. The former extends back to the product concept and includes the research and development and marketing studies needed to bring the product to the market phase. This is essentially the product development phases shown in Fig. 2.1. The investment (negative profits) needed to create the product is shown along with the profit. The numbers along the profit versus time curve correspond to the processes in the product life cycle. Note that if the product development process is terminated prior to entering the market, the company must absorb the PDP costs.

2.3.2 Technology Development and Insertion Cycle

The development of a new technology follows an S-shaped growth curve (Fig. 2.6a) similar to that for the growth of sales of a product. In its early stage, progress in technology tends to be limited by the lack of ideas. A single good idea can make

Premarket phase	Market phase
1. Idea generation	9. Product introduction
2. Idea evaluation	10. Market development
3. Feasibility analysis	11. Rapid growth
4. Technical R&D	12. Competitive market
5. Product (market) R&D	13. Maturity
6. Preliminary production	14. Decline
7. Market testing	15. Abandonment
8. Commercial production	

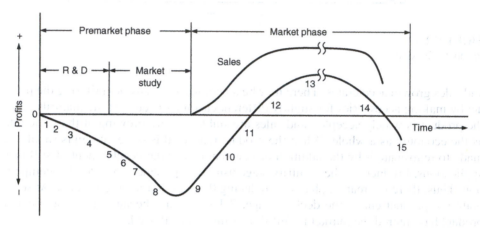

FIGURE 2.5
Expanded view of product development cycle.

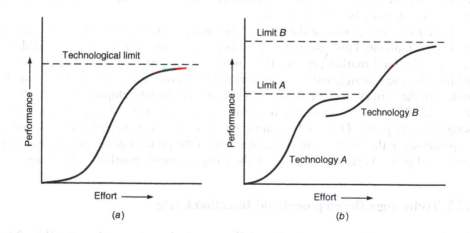

FIGURE 2.6
(a) Simplified technology development cycle. (b) Transferring from one technology growth curve (A) to another developing technology (B).

several other good ideas possible, and the rate of progress becomes exponential as indicated by a steep rise in performance that creates the lower steeply rising curve of the S. During this period a single individual or a small group of individuals can have a pronounced effect on the direction of the technology. Gradually the growth becomes more nearly linear when the fundamental ideas are in place, and technical progress is concerned with filling in the gaps between the key ideas. This is the period when commercial exploitation flourishes. Specific designs, market applications, and manufacturing occur rapidly in a field that has not yet matured. Smaller entrepreneurial firms can have a large impact and capture a dominant share of the market. However, with time the technology begins to run dry, and improvements come with greater difficulty. Now the market tends to become stabilized, manufacturing methods become fixed in place, and more capital is expended to reduce the cost of manufacturing. The business becomes capital-intensive; the emphasis is on production know-how and financial expertise rather than scientific and technological expertise. The maturing technology grows slowly, and it approaches a limit asymptotically. The limit may be set by a social consideration, such as the fact that the legal speed of automobiles is set by safety and fuel economy considerations, or it may be a true technological limit, such as the fact that the speed of sound defines an upper limit for the speed of a propeller-driven aircraft.

The success of a technology-based company lies in recognizing when the core technology on which the company's products are based is beginning to mature and, through an active R&D program, transferring to another technology growth curve that offers greater possibilities (Fig. 2.6b). To do so, the company must manage across a *technological discontinuity* (the gap between the two S-curves in Fig. 2.6b), and a new technology must replace the existing one (*technology insertion*). Past examples of technological discontinuity are the change from vacuum tubes to transistors and from the three- to the two-piece metal can. Changing from one technology to another may be difficult because it requires different kinds of technical skills, as in the change from vacuum tubes to transistors.

A word of caution. Technology usually begins to mature before profits top out, so there is often is a management reluctance to switch to a new technology, with its associated costs and risks, when business is doing so well. Farsighted companies are always on the lookout for the possibility for technology insertion because it can give them a big advantage over the competition.

2.3.3 Process Development Cycle

Most of the emphasis in this text is on developing new products or existing products. However, the development process shown in Fig. 2.1 can just as well be used to describe the development of a process rather than a product. Similarly, the design process described in Sec. 1.5 pertains to process design as well as product design. One should be aware that there may be differences in terminology when dealing with processes instead of products. For example in product development we talk about the *prototype* to refer to the early physical embodiment of the product, while in process design one is more likely to call this the *pilot plant* or *semi works*.

Process development is most important in the materials, chemicals, or food processing industries. In such businesses the product that is sold may be a coil of aluminum to be made into beverage cans or a silicon microchip containing hundreds of thousands of transistors and other circuit elements. The processes that produced this product create most of its value.

We also need to recognize that process development often is an enabler of new products. Typically, the role of process development is to reduce cost so that a product becomes more competitive in the market. However, revolutionary processes can lead to remarkable products. An outstanding example is the creation of microelectromechanical systems (MEMS) by adapting the fabrication methods from integrated circuits.

2.4
ORGANIZATION FOR DESIGN AND PRODUCT DEVELOPMENT

The organization of a business enterprise can have a major influence on how effectively design and product development are carried out. There are two fundamental ways for organizing a business: with regard to *function* or with respect to *projects*.

A brief listing of the functions that encompass engineering practice is given in Fig. 2.7. At the top of this ladder is research, which is closest to the academic experience, and as we progress downward we find that more emphasis in the job function is given to financial and administrative matters and less emphasis is given to strictly technical matters. Many engineering graduates find that with time their careers follow the progression from heavy emphasis on technical matters to more emphasis on administrative and management issues.

A *project* is a grouping of activities aimed at accomplishing a defined objective, like introducing a particular product into the marketplace. It requires certain activities:

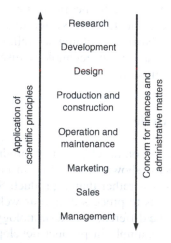

FIGURE 2.7
Spectrum of engineering functions.

identifying customer needs, creating product concepts, building prototypes, designing for manufacture, and so on. These tasks require people with different functional specialties. As we shall see, the two organizational arrangements, by function or by project, represent two disparate views of how the specialty talents of people should be organized.

An important aspect of how an enterprise should be organized is concerned with the links between individuals. These links have to do with:

- Reporting relationships: A subordinate is concerned about who his or her supervisor is, since the supervisor influences evaluations, salary increases, promotions, and work assignments.
- Financial arrangements: Another type of link is budgetary. The source of funds to advance the project, and who controls these funds, is a vital consideration.
- Physical arrangement: Studies have shown that communication between individuals is enhanced if their offices are within 50 feet of each other. Thus, physical layout, whether individuals share the same office, floor, or building, or are even in the same country, can have a major impact on the spontaneous encounters that occur and hence the quality of the communication. The ability to communicate effectively is most important to the success of a product development project. The use of video teleconferencing using the Internet has greatly reduced the need for travel, but it does not replace the importance of face-to-face discussion at critical times in a project.

We now discuss the most common types of organizations for carrying out product development activities. As each is presented, examine it with regard to the links between people.

2.4.1 A Typical Functional Organization

Figure 2.8 shows an organization chart of a typical manufacturing company of modest size organized along conventional functional reporting lines. All research and engineering report to a single vice president; all manufacturing activity is the responsibility of another vice president. Take the time to read the many functions under each vice president that are needed even in a manufacturing enterprise that is modest in size. Note that each function is a column in the organizational chart. These reporting chain columns are often called "silos" or "stove pipes" because they can represent barriers to communication between functions. A chief characteristic of a functional organization is that each individual has only one boss. By concentrating activities in units of common professional background, there are economies of scale, opportunities to develop deep expertise, and clear career paths for specialists. Generally, people gain satisfaction from working with colleagues who share similar professional interests. Since the organizational links are primarily among those who perform similar functions, formal interaction between different functional units, as between engineering and manufacturing, is forced to the level of the unit manager or higher.

Concentrating technical talent in a single organization produces economies of scale and opportunities to develop in-depth technical knowledge. This creates an efficient organization for delivering technical solutions, but because of communication problems inherent in this structure it may not be the optimum organization for effective product

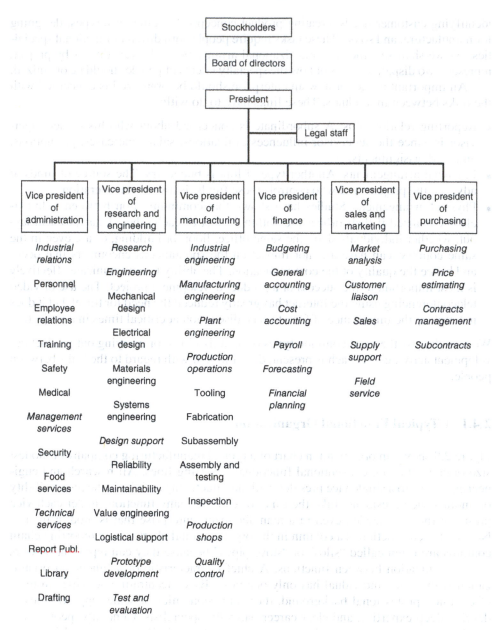

FIGURE 2.8
Example of a functional organization.

development. It may be acceptable for a business with a narrow and slowly changing set of product lines, but the inevitable slow and bureaucratic decision making that this type of structure imposes can be a problem in a dynamic product situation. Unless effective communication can be maintained between engineering and manufacturing and marketing, it will not produce the most cost-effective and customer-oriented designs.

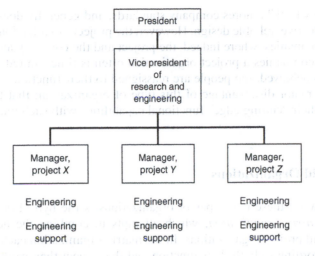

FIGURE 2.9
A simplified project organization.

2.4.2 Organization by Projects

The other extreme in organizational structure is the *project organization,* where people with the various functional abilities needed for the product development are grouped together to focus on the development of a specific product or product line (Fig. 2.9). These people often come on special assignment from the functional units of the company. Each development team reports to a project manager, who has full authority and responsibility for the success of the project. Thus the project teams are autonomous units, charged with creating a specific product. The chief advantage of a project organization is that it focuses the needed talents exclusively on the project goal, and it eliminates issues with communication between functional units by creating teams of different functional specialists. Thus, decision-making delays are minimized. Another advantage of the project organization is that members of a project team are usually willing to work outside of their specialty area to get the work done when bottlenecks arise in completing the many tasks required to complete a design. They do not have to wait for some functional specialist to finish her current assignment to work on their project. Therefore, working in a project team develops technical breadth and management skills.

A product created by a project organization is not as economical in its utilization of scarce technical expertise as the functional organization. While an autonomous project team will create a product much more quickly than the functional team, it often is not as good a design as would be produced by the functional design organization.[1] The problem arises when the project team really believes that it is an independent unit and ignores the existing knowledge base of the organization. It tends to

---

1. D.G. Reinertsen, *Managing the Design Factory,* The Free Press, New York, 1997, pp. 102–5.

"reinvent the wheel," ignores company standards, and generally does not produce the most cost-effective, reliable design. However, the project organization is very common in start-up companies, where indeed, the project and the company are synonymous.

In large companies a project organization often is time limited; once the goal of the project is achieved, the people are reassigned to their functional units. This helps to address a major disadvantage of this type of organization: that technical experts tend to lose their "cutting edge" functional capabilities with such intense focus on the project goal.

2.4.3 Hybrid Organizations

Midway between these two types of organizations is the hybrid organization, often called the *matrix organization,* which attempts to combine the advantages of the functional and project organizations. In the matrix organization each person is linked to others according to both their function and the project they work on. As a consequence, each individual has two supervisors, one a functional manager and the other a project manager. While this may be true in theory, in practice either the functional manager or the project manager predominates.[1] In the *lightweight project organization* the functional links are stronger than the project links (Fig. 2.10a). In this matrix the functional specialties are shown along the y-axis and the various project teams along the x-axis. The project managers assign their personnel as required by the project teams. While the project managers are responsible for scheduling, coordination, and arranging meetings, the functional managers are responsible for budgets, personnel matters, and performance evaluations. Although an energetic project manager can move the product development along faster than with a strict functional organization because there is one person who is dedicated and responsible for this task, in fact he or she does not have the authority to match the responsibility. A lightweight matrix organization may be the worst of all possible product development organizations because the top management may be deluded into thinking that they have adopted a modern project management approach when in effect they have added one layer of bureaucracy to the traditional functional approach.[2]

In the *heavyweight matrix organization* the project manager has complete budgetary authority, makes most of the resource allocation decisions, and plays a strong role in evaluating personnel (Fig. 2.10b). Although each participant belongs to a functional unit,[3] the functional manager has little authority and control over project decisions. However, he continues to write his people's reviews, and they return to his organization at the end of the project. The functional organization or the lightweight project organization works well in a stable business environment, especially one where the product predominates in its market because of technical excellence. A heavyweight

1. R.H. Hayes, S.C. Wheelwright, and K.B. Clark, *Dynamic Manufacturing: Creating the Learning Organization,* The Free Press, New York, 1988, pp. 319–23.

2. P.G. Smith and D.G. Reinertsen, *Developing Products in Half the Time,* Van Nostrand Reinhold, New York, 1991, pp. 134–45.

3. Sometimes a functional specialist may be working on different product teams at the same time.

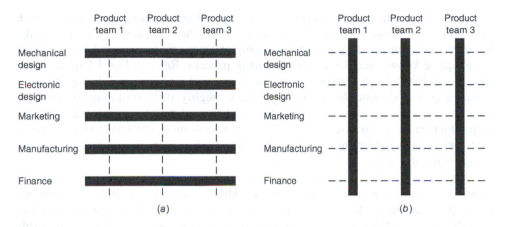

FIGURE 2.10
(a) A lightweight project organization; (b) a heavyweight project organization.

project organization has advantages in introducing radically new products, especially where speed is important. Some companies have adopted the project form of organization where the project team is an organizationally separate unit in the company. Often this is done when they plan to enter an entirely new product area that does not fit within the existing product areas.

We have mentioned the concern that an empowered product development team may get carried away with its freedom and ignore the corporate knowledge base to create a fast-to-market product that is less than optimum in some aspects such as cost or reliability. To prevent this from occurring, the product team must clearly understand the boundaries on its authority. For example, the team may be given a limit on the cost of tooling, which if exceeded requires approval from an executive outside the team. Or, they may be given an approved parts list, test requirements, or vendors from which to make their selections, and any exceptions require higher approval.[1] It is important to define the boundaries on team authority early in the life of the team so that it has a clear understanding of what it can and cannot do. Moreover, the stage-gate review process should provide a deterrent to project teams ignoring important company procedures and policy.[2]

2.4.4 Concurrent Engineering Teams

The conventional way of doing product design has been to carry out all of the steps serially. Thus, product concept, product design, and product testing have been done prior to process planning, manufacturing system design, and production. Commonly

---

1. D. G. Reinertsen, *op. cit.*, pp. 106–8.
2. For current examples of successful use of matrix organizations see J. R. Galbraith, *Designing Matrix Organizations That Actually Work,* Jossey-Bass, San Franciso, CA, 2009.

these serial functions have been carried out in distinct and separate organizations with little interaction between them. Thus, it is easy to see how the design team will make decisions, many of which can be changed only at great cost in time and money, without adequate knowledge of the manufacturing process. Refer to Fig. 1.1 to reinforce the concept that a large percentage of a product's cost is committed during the conceptual and embodiment phases of design. Very roughly, if the cost to make a change at the product concept stage is $1, the cost is $10 at the detail design stage and $100 at the production stage. The use of a serial design process means that as changes become necessary there is a doubling back to pick up the work, and the actual process is more in the nature of a spiral.

Starting in the 1980s, as companies met increasing competitive pressure, a new approach to integrated product design evolved, which is called *concurrent engineering*. The impetus came chiefly from the desire to shorten product development time, but other drivers were the improvement of quality and the reduction of product life-cycle costs. Concurrent engineering is a systematic approach to the integrated concurrent design of products and their related processes, including manufacture and support. With this approach, product developers, from the outset, consider all aspects of the product life cycle, from concept to disposal, including quality, cost, schedule, and user requirements. A main objective is to bring many viewpoints and talents to bear in the design process so that these decisions will be valid for downstream parts of the product development cycle like manufacturing and field service. Toward this end, computer-aided engineering (CAE) tools have been very useful (see Sec. 1.6). Concurrent engineering has three main elements: cross-functional teams, parallel design, and vendor partnering.

Of the various organizational structures for design that were discussed previously, the heavyweight project organization, usually called just a *cross-functional design team* or an *integrated product and process product development (IPPD)* team, is used most frequently with concurrent engineering. Having the skills from the functional areas embedded in the team provides for quick and easy decision making, and aids in communication with the functional units. For cross-functional teams to work, their leader must be empowered by the managers of the functional units with decision-making authority. It is important that the team leader engender the loyalty of the team members toward the product and away from the functional units from which they came. Functional units and cross-functional teams must build mutual respect and understanding for each other's needs and responsibilities. The importance of teams in current design practice is such that Chap. 4 is devoted to an in-depth look at team behavior.

Parallel design, sometimes called simultaneous engineering, refers to each functional area implementing their aspect of the design at the earliest possible time, roughly in parallel. For example, the manufacturing process development group starts its work as soon as the shape and materials for the product are established, and the tooling development group starts its work once the manufacturing process has been selected. These groups have had input into the development of the product design specification and into the early stages of design. Of course, nearly continuous communication between the functional units and the design team is necessary in order to know what the other functional units are doing. This is decidedly different from the old practice of completely finishing a design package of drawings and specifications before transmitting it to the manufacturing department.

Vendor partnering is a form of parallel engineering in which the technical expertise of the vendor or supplier for certain components is employed as an integral member of the cross-functional design team. Traditionally, vendors have been selected by a bidding process after the design has been finalized. In the concurrent engineering approach, key suppliers known for proficient technology, reliable delivery, and reasonable cost are selected early in the design process before the parts have been designed. Generally, these companies are called *suppliers,* rather than vendors, to emphasize the changed nature of the relationship. A strategic partnership is developed in which the supplier becomes responsible for both the design and production of components, in return for a major portion of the business. Rather than simply supplying standard components, a supplier can partner with a company to create customized components for a new product. Supplier partnering has several advantages. It reduces the amount of component design that must be done in-house, it integrates the supplier's manufacturing expertise into the design, and it ensures a degree of allegiance and cooperation that should minimize the time for receipt of components.

2.5
MARKETS AND MARKETING

Marketing is concerned with the interaction between the corporation and the customer. Customers are the people or organizations that purchase products. However, we need to differentiate between the customer and the user of the product. The corporate purchasing agent is the customer in so far as the steel supplier is concerned, for she negotiates price and contract terms, but the design engineer who developed the specification for a highly weldable grade of steel is the end user (indirect customer), as is the production supervisor of the assembly department. Note that the customer of a consulting engineer or lawyer is usually called a client. Methods for identifying customer needs and wants are considered in Sec. 3.2.

2.5.1 Markets

The market is an economic construct to identify those persons or organizations that have an interest in purchasing or selling a particular product, and to create an arena for their transactions. We generally think of the stock market as the prototypical market.

A quick review of the evolution of consumer products is a good way to better understand markets. At the beginning of the Industrial Revolution, markets were mainly local and consisted of close-knit communities of consumers and workers in manufacturing companies. Because the manufacturing enterprise was locally based, there was a close link between the manufacturers and the users of their product, so direct feedback from customers was easily acquired. With the advent of railroads and telephone communication, markets expanded across the country and very soon became national markets. This created considerable *economy of scale,* but it required new ways of making products available to the customer. Many companies created a national

2

distribution system to sell their products through local stores. Others depended on retailers who offered products from many manufacturers, including direct competitors. Franchising evolved as an alternative way of creating local ownership while retaining a nationally recognized name and product. Strong *brand names* evolved as a way of building customer recognition and loyalty.

As the capability to produce products continued to grow, the markets for those products expanded beyond the borders of one country. Companies then began to think of ways to market their products in other countries. The Ford Motor Company was one of the first U.S. companies to expand into overseas markets. Ford took the approach of developing a wholly owned subsidiary in the other country that was essentially self-contained. The subsidiary designed, developed, manufactured, and marketed products for the local national market. The consumer in that country barely recognized that the parent company was based in the United States. This was the beginning of *multinational companies*. The chief advantage of this approach was the profits that the company was able to bring back to the United States. However, the jobs and physical assets remained overseas.

Another approach to multinational business was developed by the Japanese automakers. These companies designed, developed, and manufactured the product in the home nation and marketed the product in many locations around the world. This became possible with a product like automobiles when roll-on/roll-off ships made low-cost transportation a reality. Such an approach to marketing gives the maximum benefit to the home nation, but with time a backlash developed because of the lost jobs in the customer countries. Also, developing a product at a long distance from the market makes it more difficult to satisfy customer needs when there is a physical separation in cultural backgrounds between the development team and the customers. More recently, Japanese companies have established design centers and production facilities in their major overseas markets.

It is very clear that we are now dealing with a *world market*. Improved manufacturing capabilities in countries such as China and India, coupled with low-cost transportation using container ships, and instant worldwide communication with the Internet, have enabled an increasing fraction of consumer products to be manufactured overseas. In 2010, manufacturing jobs in the United States accounted for only one in eleven jobs, down from one in three in 1950. This is not a new trend. The United States became a net importer of manufactured goods in 1981, but in recent years the negative balance of trade has grown to possibly unsustainable proportions. The reduction in the percentage of the U.S. engineering workforce engaged in manufacturing places greater incentive and emphasis on knowledge-based activities such as innovative product design.

2.5.2 Market Segmentation

Although the customers for a product are called a "market" as though they were a homogeneous unit, this generally is not the case. In developing a product, it is important to have a clear understanding of which segments of the total market the product is intended to serve. There are many ways to segment a market. Table 2.1 lists the broad types of markets that engineers typically address in their design and product development activities.

TABLE 2.1
Markers for Engineered Products, Broadly Defined

Type of Product Market	Examples	Degree of Engineering Involvement with Customer
Large one-off design	Petrochemical plant; skyscraper; automated production line	Heavy: close consultation with customer. Job sold on basis of past experience and reputation
Small batch	Typically 10–100 items per batch. Machine tools; specialized control systems	Moderate: based mostly on specifications developed with customer
Raw materials	Ores, oil, agricultural products	Low: buyer sets standards
Processed materials	Steel, polymer resins, Si crystal	Low: buyer's engineers set specifications
High-volume engineered products	Motors, microprocessors, bearings, pumps, springs, shock absorbers, instruments	Low: vendor's engineers design parts for general customer
Custom-made parts	Made for specific design to perform function in product	Moderate: buyer's engineers design and specify; vendors bid on manufacture
High-volume consumer products	Automobiles, computers, electronic products, food, clothing	Heavy in best of companies
Luxury consumer goods	Rolex watch; Harley Davidson	Heavy, depending on product
Maintenance and repair	Replacement parts	Moderate, Depending on product
Engineering services	Specialized consultant firms	Heavy: Engineers sell as well as do technical work

One-of-a-kind installations, such as a large office building or a chemical plant, are expensive, complex design projects. With these types of projects the design and the construction are usually separate contracts. Generally these types of projects are sold on the basis of a prior successful record of designing similar installations, and a reputation for quality, on-time work. Typically there is frequent one-on-one interaction between the design team and the customer to make sure the user's needs are met.

For small-batch engineered products, the degree of interaction with the customer depends on the nature of the product. For a product like railcars the design specification would be the result of extensive direct negotiation between the user's engineers and the vendor. For more standard products like a CNC lathe, the product would be considered an "off-the-shelf" item available for sale by regional distributors or direct from catalog sales.

Raw materials, such as iron ore, crushed rock, grain and oil, are *commodities* whose characteristics are well understood. Thus, there is little interaction between the buyer's engineers and the seller, other than to specify the quality level (grade) of the commodity. Most commodity products are sold chiefly on the basis of price.

When raw materials are converted into processed materials, such as sheet steel or a silicon wafer, the purchase is made with agreed-upon industry standards of quality, or in extreme cases with specially engineered specifications. There is little interaction of the buyer's and seller's engineers. Purchase is highly influenced by cost and quality.

Most technical products contain standard components or subassemblies (COTS, or commercial off-the-shelf products) that are made in high volumes and purchased from distributors or directly from the manufacturer. Companies that supply these parts are called *vendors* or *suppliers,* and the companies that use these parts in their products are called *original equipment manufacturers* (OEM). Usually, the buyer's engineers depend on the specifications provided by the vendor and their record for reliability, so their interaction with the vendor is low. However, it will be high when dealing with a new supplier, or a supplier that has developed quality issues with its product.

All products contain parts that are custom designed to perform one or more functions required by the product. Depending on the product, the production run may vary from several thousand to a few million piece parts. Typically these parts will be made as castings, metal stampings, or plastic injection moldings. These parts will be made in either the factory of the product producer or the factory of independent parts-producing companies. Generally these companies specialize in a specific manufacturing process, like precision forging, and increasingly they may be located worldwide. This calls for considerable interaction by the buyer's engineers to decide, with the assistance of purchasing agents, where to place the order to achieve reliable delivery of high-quality parts at lowest cost.

Luxury consumer products are a special case. Generally, styling and quality materials and workmanship play a major role in creating the brand image. In the case of a high-end sports car, engineering interaction with the customer to ensure quality may be high, but in most products of this type styling and salesmanship play a major role.

After-sale maintenance and service can be a very profitable market for a product producer. The manufacturers of inkjet printers make most of their profit from the sale of replacement cartridges. The maintenance of highly engineered products like elevators and gas turbine engines increasingly is being done by the same companies that produced them. The profits over time for this kind of engineering work can easily exceed the initial cost of the product.

The corporate downsizings of their staff specialists that occurred in the 1990s resulted in many engineers organizing specialist consulting groups. Now, rather than using their expertise exclusively for a single organization, they make this talent available to whoever has the need and ability to pay for it. The marketing of engineering services is more difficult than the marketing of products. It depends to a considerable degree on developing a track record of delivering competent, on-time results, and in maintaining these competencies and contacts. Often these firms gain reputations for creative product design, or for being able to tackle the most difficult computer modeling and analysis problems. An important area of engineering specialist service is *systems integration.* Systems integration involves taking a system of separately produced subsystems or components and making them operate as an interconnected and interdependent engineering system.

Having looked at the different types of markets for engineering products, we now look at the way any one of these markets can be segmented. Market segmentation recognizes that markets are not homogeneous, but rather consist of people buying things, no two of whom are exactly alike in their purchasing patterns. *Market segmentation* is the attempt to divide the market into groups so that there is relative homogeneity within each group and distinct differences between groups. Cooper[1] suggests that four broad categories of variables are useful in segmenting a market.

- **State of Being**
  - a. Sociological factors—age, gender, income, occupation
  - b. For industrial products—company size, industry classification (NAICS code), nature of the buying organization
  - c. Location—urban, suburban, rural; regions of the country or world

- **State of Mind**—This category attempts to describe the attitudes, values, and life-styles of potential customers.

- **Product Usage**—looks at how the product is bought or sold

  - a. Heavy user; light user; nonuser
  - b. Loyalty: to your brand; to competitor's brand; indifferent

- **Benefit Segmentation**—attempts to identify the benefits people perceive in buying the product. This is particularly important when introducing a new product. When the target market is identified with benefits in mind, it allows the product develop-ers to add features that will provide those benefits. Methods for doing this are given in Chap. 3.

For more details on methods for segmenting markets see the text by Urban and Hauser.[2]

2.5.3 Functions of a Marketing Department

The marketing department in a company creates and manages the company's relation-ship with its customers. It is the company's window on the world with its customers. It translates customer needs into requirements for products and influences the creation of services that support the product and the customer. It is about understanding how people make buying decisions and using this information in the design, building, and selling of products. Marketing does not make sales; that is the responsibility of the sales department.

The marketing department can be expected to do a number of tasks. First is a preliminary marketing assessment, a quick scoping of the potential sales, competition, and market share at the very early stages of the product planning. Then they will do a detailed market study. This involves face-to-face interviews with potential customers to determine their needs, wants, preferences, likes, and dislikes. This will be done be-fore detailed product development is carried out. Often this involves meeting with the end user in the location where the product is used, usually with the active participation of the design engineer. Another common method for doing this is the focus group. In this method a group of people with a prescribed knowledge about a product or service is gathered around a table and asked their feelings and attitudes about the product

---

1. R.G. Cooper, *Winning at New Products,* 3d ed., Perseus Books, Cambridge, MA, 2001.
2. G.L. Urban and J.R. Hauser, *Design and Marketing of New Products,* 2d ed., Prentice Hall, Englewood Cliffs, NJ, 1993.

under study. If the group is well selected and the facilitator of the focus group is experienced, the sponsor can expect to receive a wealth of opinions and attitudes that can be used to determine important attributes of a potential product.

The marketing department also plays a vital role in assisting with the introduction of the product into the marketplace. They perform such functions as undertaking customer tests or field trials (beta test) of the product, planning for test marketing (sales) in restricted regions, advising on product packaging and warning labels, preparing user instruction manuals and documentation, arranging for user instruction, and advising on advertising. Marketing may also be responsible for providing for a product support system of spare parts, service representatives, and a warranty system.

2.5.4 Elements of a Marketing Plan

The marketing plan starts with the identification of the target market based on market segmentation. The other main input of the marketing plan is the *product strategy,* which is defined by product positioning and the benefits provided to the customer by the product. A key to developing the product strategy is the ability to define in one or two sentences the *product positioning,* that is, how the product will be perceived by potential customers. Of equal importance is to be able to express the *product benefits.* A product benefit is not a product feature, although the two concepts are closely related. A product benefit is a brief description of the main advantage to using the product as seen through the eyes of the customer. The chief features of the product should derive from the product benefit.

> **EXAMPLE 2.1**
> A manufacturer of garden tools might decide to develop a power lawnmower targeted at the elderly population. Demographics show that this segment of the market is growing rapidly, and that they have above-average disposable income. The product will be positioned for the upper end of the elderly with ample disposable income. The chief benefit would be ease of use by elderly people. The chief features to accomplish this goal would be power steering, an automatic safety shutoff while clearing debris from the blade, an easy-to-use device for raising the mower deck to get at the blade, and a clutchless transmission.

A marketing plan should contain the follow information:

- Evaluation of market segments, with clear explanation of reasons for choosing the target market
- Identification of competitive products
- Identification of early product adopters
- Clear understanding of benefits of product to customers
- Estimation of the market size in terms of dollars and units sold, and market share
- Determination of the breadth of the product line, and number of product variants
- Estimation of product life
- Determination of the product volume and price relationships
- Complete financial plan including time to market, 10-year projection of costs and income

2.6
TECHNOLOGICAL INNOVATION

Many of the products that engineers are developing today are the result of new technology. Much of the technology explosion started with the invention of the digital computer and transistor in the 1940s and their subsequent development through the 1950s and 1960s. The transistor evolved into micro-integrated circuits, which allowed the computer to shrink in size and cost, becoming the desktop computer we know today. Combining the computer with communications systems and protocols like optical fiber communications gave us the Internet and cheap, dependable worldwide communications. At no other time in history have several breakthrough technologies combined to so substantially change the world we live in. Yet, if the pace of technology development continues to accelerate, the future will see even greater change.

2.6.1 Invention, Innovation, and Diffusion

Generally, the advancement of technology occurs in three stages:

- Invention: The creative act whereby an idea is conceived, articulated, and recorded.
- Innovation: The process by which an invention or idea is brought into successful practice and is utilized by the economy.
- Diffusion: The successive and widespread implementation and adoption of successful innovations.

Without question, innovation is the most critical and most difficult of the three stages. Developing an idea into a product that people will buy requires hard work and skill at identifying market needs. Diffusion of technology throughout society is necessary to preserve the pace of innovation. As technologically advanced products are put into service, the technological sophistication of consumers increases. This ongoing education of the customer base paves the way for the adoption of even more sophisticated products. A familiar example is the proliferation of bar codes and bar code scanners.

Many studies have shown that the ability to introduce and manage technological innovation is a major factor in a country's leadership in world markets and also a major factor in raising its standard of living. Science-based innovation in the United States has spawned such key industries as jet aircraft, computers, plastics, and wireless communication. Relative to other nations, however, the importance of the United States' role in innovation appears to be decreasing. If the trend continues, it will affect our well-being.

Likewise, the nature of innovation has changed over time. Opportunities for the lone inventor have become relatively more limited. As one indication, independent investigators obtained 82 percent of all U.S. patents in 1901, while by 1937 this number had decreased to 50 percent, indicating the rise of corporate research laboratories. Today the number is about 25 percent, but it is on the rise as small companies started by entrepreneurs become more prevalent. This trend is attributable to the venture capital industry, which stands ready to lend money to promising innovators, and to various federal programs to support small technological companies.

FIGURE 2.11
A market-pull model for technological innovation.

Figure 2.11 shows the generally accepted model for a technologically inspired product. This model differs from one that would have been drawn in the 1960s, which would have started with basic research at the head of the innovation chain. The idea then was that basic research results would lead to research ideas that in turn would lead directly to commercial development. Although strong basic research obviously is needed to maintain the storehouse of new knowledge and ideas, it has been well established that innovation in response to a market need has greater probability of success than innovation in response to a technological research opportunity. Market pull is far stronger than technology push when it comes to innovation.

The introduction of new products into the marketplace is like a horse race. The odds of picking a winner at the inception of an idea are about 5 or 10 to 1. The failure rate of new products that actually enter the marketplace is around 35 to 50 percent. Most of the products that fail stumble over market obstacles, such as not appreciating the time it takes for customers to accept a new product.[1] The next most common cause of new product failure is management problems, while technical problems comprise the smallest category for failure.

The digital imaging example illustrates how a basic technological development created for one purpose can have greater potential in another product area. However, its initial market acceptance is limited by issues of performance and manufacturing cost. Then a new market develops where the need is so compelling that large development funding is forthcoming to overcome the technical barriers, and the innovation becomes wildly successful in the mass consumer market. In the case of digital imaging, the innovation period from invention to widespread market acceptance was about thirty-five years.

2.6.2 Business Strategies Related to Innovation and Product Development

A common and colorful terminology for describing business strategy dealing with innovation and investment was advanced by the Boston Consulting Group in the 1970s. Most established companies have a portfolio of businesses, usually called business units. According to the BCG scheme, these business units can be placed into one of four categories, depending on their prospects for sales growth and gain in market share.

- *Star businesses*: High sales growth potential, high market share potential
- *Wildcat businesses:* High sales growth potential, low market share
- *Cash-cow businesses:* Low growth potential, high market share
- *Dog businesses:* Low growth potential, low market share

---

1. R. G. Cooper, *Research Technology Management* July–August, 1994, pp. 40–50.

The Innovation of Digital Imaging

It is instructive to trace the history of events that led to the innovation of digital imaging, the technology at the heart of the digital camera.

In the late 1960s Willard Boyle worked in the division of Bell Laboratories concerned with electronic devices. The VP in charge of this division was enamored with *magnetic bubbles,* a new solid-state technology for storing digital data. Boyle's boss was continually asking him what Boyle was contributing toward this activity.

In late 1969, in order to appease his boss, Boyle and his collaborator George Smith sat down and in a one-hour brainstorming session came up with the basic design for a new memory chip they called a *charge-coupled device* or CCD. The CCD worked well for storing digital data, but it soon became apparent that it had outstanding potential for capturing and storing digital images, a need that had not yet been satisfied by technology in the rapidly developing semiconductor industry. Boyle and Smith built a proof-of-concept model containing only six pixels, patented their invention, and went on to other exciting research discoveries.

While the CCD was a good digital storage device, it never became a practical storage device because it was expensive to manufacture and was soon supplanted by various kinds of disks coated with fine magnetic particles, and finally the hard drive went on to capture the digital storage market.

In the meantime, two space-related applications created the *market pull* to develop the CCD array to a point where it was a practical device for digital photography. The critical issues were decreasing the size and the cost of a CCD array that captures the image.

Astronomers had never been really happy about capturing the stars on chemical-based film, which lacks the sensitivity to record events occurring far out into space. The early CCD arrays, although heavy, bulky, and costly, had much greater inherent sensitivity. By the late 1980s they became standard equipment at the world's astronomical observatories.

An even bigger challenge came with the advent of military satellites. The photographs taken from space were recorded on film, which was ejected from space and picked out of the air by airplanes or fished out of the ocean, both rather problematic operations. When further development reduced the size and weight of CCD arrays and increased their sensitivity, it became possible to digitally transmit images from space, and we saw the rings of Saturn and the landscape of Mars in graphic detail. The technology advances achieved in these application areas made it possible for digital still and video cameras to become a commercial success roughly thirty years after the invention of the CCD.

In 2006 Willard Boyle and George Smith received the Draper Prize of the National Academy of Engineering, the highest award for technological innovation in the United States, and shared the Nobel prize for physics in 2009.

Excerpted from G. Gugliotta, "One-Hour Brainstorming Gave Birth to Digital Imaging," *Wall Street Journal,* February 20, 2006, p. A09.

In this classification scheme, the break between high and low market share is the point at which a company's share equals that of its largest competitor. For a cash-cow business, cash flow should be maximized but investment in R&D and new plant costs should be kept to a minimum. The cash these businesses generate should be used in star and wildcat businesses, or for new technological opportunities. Heavy investment is required in star businesses so they can increase their market share. By pursuing this strategy, a star becomes a cash-cow business over time, and eventually a dog business. Wildcat businesses require generous funding to move into the star category. That only a limited number of wildcats can be funded will result in the survival of the fittest. Dog businesses receive no investment and are sold or abandoned as soon as possible. This whole approach is artificial and highly stylized, but it is a good characterization of corporate reasoning concerning business investment with respect to available product areas or business units. Obviously, the innovative engineer should avoid becoming associated with the dogs and cash cows, for there will be little incentive for creative work.

There are other business strategies that can have a major influence on the role engineers play in engineering design. A company that follows a *first in the field* strategy is usually a high-tech innovator. Some companies may prefer to let others pioneer and develop the market. This is the strategy of being a *fast follower* that is content to have a lower market share at the avoidance of the heavy R&D expense of the pioneer. Other companies may emphasize process development with the goal of becoming the high-volume, low-cost producer. Still other companies adopt the strategy of being the key supplier to a few major customers that market the product to the public.

A company with an active research program usually has more potential products than the resources required to develop them. To be considered for development, a product should fill a need that is presently not adequately served, or serve a current market for which the demand exceeds the supply, or has a differential advantage over an existing product (such as better performance, improved features, or lower price).

2.6.3 Characteristics of Innovative People

Studies of the innovation process by Roberts[1] have identified five behavioral types of people who are needed in a product team devoted to technological innovation.

- Idea generator: The creative individual
- Entrepreneur: The person who "carries the ball" and takes the risks
- Gatekeepers: People who provide technical communication from outside to inside the product development organization
- Program manager: The person who manages without inhibiting creativity
- Sponsor: The person who provides financial and moral support, often senior management or a venture capital company

1. E.B. Roberts and H.A. Wainer, *IEEE Trans. Eng. Mgt.*, vol. EM-18, no. 3, pp. 100–9, 1971; E.B. Roberts (ed.), *Generation of Technological Innovation*, Oxford University Press, New York, 1987.

Roughly 70 to 80 percent of the people in a technical organization are routine problem solvers and are not involved in innovation. Therefore, it is important to be able to identify and nurture the small number who show promise of becoming technical innovators.

Innovators tend to be the people in a technical organization who are the most familiar with current technology and who have well-developed contacts with technical people outside the organization.[1] These innovators receive information directly and then diffuse it to other technical employees. Innovators tend to be predisposed to "do things differently" as contrasted with focusing on "doing things better." Innovators are early adopters of new ideas. They can deal with unclear or ambiguous situations without feeling uncomfortable. That is because they tend to have a high degree of self-reliance and self-esteem. Age is not a determinant or barrier to becoming an innovator, nor is experience in an organization, so long as it is sufficient to establish credibility and social relationships. It is important for an organization to identify the true innovators and provide a management structure that helps them develop. Innovators respond well to the challenge of diverse projects and the opportunity to communicate with people of different backgrounds.

A successful innovator is a person who has a coherent picture of what needs to be done, although not necessarily a detailed picture. Innovators emphasize goals, not methods of achieving the goal. They can move forward in the face of uncertainty because they do not fear failure. Many times the innovator is a person who has failed in a previous venture and knows why. The innovator is a person who identifies what he or she needs in the way of information and resources and gets them. The innovator aggressively overcomes obstacles by breaking them down, or hurdling over them, or running around them. Frequently the innovator works the elements of the problem in parallel, not serially.

2.6.4 Types of Technology Innovation

We have seen in Fig. 2.6 that a natural evolution of a technology-based business is for a new technology to substitute for the old. There are two basic ways for the new technology to arise.

- *Need-driven innovation,* where the development team seeks to fill an identified gap in performance or product cost (technology pull)
- *Radical innovation,* which leads to widespread change and a whole new technology, and arises from basic research (technology push)

Most product development is of the need-driven type. It consists of small, almost imperceptible improvements, which when made over a long time add up to major progress. These innovations are most valuable if they lead to patent protection for the existing product line. Typically these improvements come about by redesign of

---

1. R. T. Keller, *Chem. Eng.,* Mar. 10, 1980, pp. 155–58.

products for easier manufacture or the addition of new features, or the substitution of less expensive components for those used in the earlier design. Also important are changes in the manufacturing processes to improve quality and decrease cost. A methodology for conducting *continuous product improvement* is presented in Sec. 4.6.

Radical innovation is based on a *breakthrough idea*[1] that is outside the scope of conventional thinking. It is an invention that is surprising and discontinuous from previous thought. Such a creative leap usually requires a completely new perspective of the problem (a shift to a new location in the design space). Breakthrough ideas create something new or satisfy a previously undiscovered need, and when converted to a radical innovation they can create new industries or product lines. An extreme example is the transistor that replaced the vacuum tube and finally made possible the digital revolution in computing and communication.

2.7
SUMMARY

Product development encompasses much more than conceiving and designing a product. It involves the preliminary assessment of the market for the product, the alignment of the product with the existing product lines of the company, and an estimate of the projected sales, cost of development, and profits. These activities take place before permission is given to proceed with concept development, and they occur throughout the product development process as better estimates are obtained for the cost of development and estimated sales.

The keys to creating a winning product are:

- Designing a quality product with the features and performance desired by its customers at a price they are willing to pay
- Reducing the cost to manufacture the product over its life cycle
- Minimizing the cost to develop the product
- Quickly bringing the product to market

The organization of a product development team can have a major influence on how effectively product development is carried out. For minimizing the time to market, some kind of project team is required. Generally, a heavyweight matrix organization with appropriate management controls works best.

Marketing is a key function in product development. Marketing managers must understand market segmentation, the wants and needs of customers, and how to advertise and distribute the product so it can be purchased by the customer. Products can be classified with respect to markets in several ways:

- A product developed in response to market pull or technology push
- A platform product that fits into an existing product line and uses its core technology

---

1. M. Stefik and B. Stefik, *Breakthrough: Stories and Strategies of Radical Innovation,* MIT Press, Cambridge, MA, 2004.

- A process-intensive product whose chief attributes are due to the processing
- A customized product whose configuration and content are created in response to a specific customer order

Many products today are based on new and rapidly developing technologies. A technology evolves in three stages:

- Invention—the creative act by which a novel idea is conceived
- Innovation—the process by which an invention is brought into successful practice and is utilized by the economy
- Diffusion—the widespread knowledge of the capabilities of the innovation

Of these three stages, innovation is the most difficult, most time consuming, and most important. While technological innovation used to be the purview of a relatively small number of developed nations, in the 21st century it is occurring worldwide at a rapid pace.

NEW TERMS AND CONCEPTS

Brand name	Lightweight matrix organization	Platform product
Concurrent engineering team	Market	Profit margins
Control document	Market pull	Project organizations
Economy of scale	Marketing	Product positioning
Functional organization	Matrix organization	PDS
Learning curve	OEM supplier	Supply chain
Lessons learned	Product development cycle	Systems integration

BIBLIOGRAPHY

Cooper, R. G., *Winning at New Products,* 3d ed., Perseus Books, Reading, MA, 2001.
Otto, K. and K. Wood, *Product Design: Techniques in Reverse Engineering and New Product Development,* Prentice Hall, Upper Saddle River, NJ, 2001.
Reinertsen, D. G., *Managing the Design Factory,* The Free Press, New York, 1997.
Smith, P. G. and D. G. Reinertsen, *Developing Products in Half the Time: New Rules, New Tools,* 2d ed., John Wiley & Sons, New York, 1996.
Ulrich, K. T. and S. D. Eppinger, *Product Design and Development,* 5th ed., McGraw-Hill, New York, 2011.

PROBLEMS AND EXERCISES

2.1. Consider the following products: (a) a power screwdriver for use in the home; (b) a desktop inkjet printer; (c) an electric car. Working in a team, make your team estimate of the following factors needed for the development project to launch each of the products: (i) annual units sold, (ii) sales price, (iii) development time, years, (iv) size of development team, (v) development cost.

2.2. List three products that are made from a single component.

2.3. Discuss the spectrum of engineering job functions shown in Fig. 2.7 with regard to such factors as (a) need for advanced education, (b) intellectual challenge and satisfaction, (c) financial reward, (d) opportunity for career advancement, and (e) people versus "thing" orientation.

2.4. Strong performance in your engineering discipline ordinarily is one necessary condition for becoming a successful engineering manager. What other conditions are there?

2.5. Discuss the pros and cons of continuing your education for an MS in an engineering discipline or an MBA on your projected career progression.

2.6. Discuss in some detail the relative roles of the project manager and the functional manager in the matrix type of organization.

2.7. List the factors that are important in developing a new technologically oriented product.

2.8. In Sec. 2.6.2 we briefly presented the four basic strategies suggested by the Boston Consulting Group for growing a business. This is often called the BCG growth-share matrix. Plot the matrix on coordinates of market growth potential versus market share, and discuss how a company uses this model to grow its overall business.

2.9. List the key steps in the technology transfer (diffusion) process. What are some of the factors that make technology transfer difficult? What are the forms in which information can be transferred?

2.10. John Jones is an absolute whiz in computer modeling and finite-element analysis. These skills are badly needed on your product development team. However, Jones is also the absolute loner who prefers to work from 4 P.M. to midnight, and when asked to serve on a product development team he turns the offer down. If ordered to work on a team he generally fails to turn up for team meetings. As team leader, what would you do to capture and effectively utilize John Jones's strong expertise?

2.11. An important issue in most product development projects is making sure that the project schedule can take advantage of the "window of opportunity." Use Fig. 2.6b to help explain what is meant by this concept.

2.12. The development of the steel shipping container that can be transferred from a ship to a truck or train has had a huge impact on world economies. Explain how such a simple engineering development could have such far-reaching consequences.

2.13. Explain the physics behind the charge-coupled device (CCD) discussed in Section 2.6.1, and explain why this was the invention that made digital photography practical.

2.14. What other technological developments besides the steel shipping container were required to produce the global marketplace that we have today? Explain how each contributed to the global marketplace.

2.15. The demand for most edible fish exceeds the supply. While fish can be raised in ponds on land or in ocean enclosures close to shore, there are limitations of scale. The next step is mariculture—fish farming in the open sea. Develop a new product business development plan for such a venture.

2.16. Conventional thinking in product development has been that innovation starts in advanced developed countries like the United States and Japan. Products marketed in countries where the average income is much lower often are older models of U.S. products or used but still serviceable equipment. Several U.S. multinational companies have established R&D labs in India and China. Originally this was to take advantage of the large number of well-educated engineers who could be employed at salaries much lower than the going U.S. rate, but soon it was found that these engineers were adept at developing products for sale to the mass markets in these local countries. Typically these are products with somewhat reduced functionality, but they still are useful quality products. Now these U.S. companies are beginning to market these products in the United States as a low-cost product line that is attractive to a new low-end market segment.

Search the business literature for examples of this new approach to *trickle-up* product innovation. Discuss advantages of this new approach to product development and discuss possible risks.

3

PROBLEM DEFINITION AND NEED IDENTIFICATION

3.1
INTRODUCTION

The engineering design process has been depicted as a stream of potential designs for a new product that will fit the needs of a targeted group of consumers. The stream is channeled through a pipeline of narrowing diameter with filters at key junctions that screen out less valuable candidate designs. At the end of the pipeline, a nearly ideal single design (or a very small set of designs) emerges. The filters represent key decision points in the design evaluation process where candidate designs are evaluated by a panel of reviewers overseeing product development for the business unit. Candidate designs are rejected when they fail to meet one or more of the engineering or business objectives of the unit.

Candidate design concepts are not stored and waiting to be released like ice cubes dispensed one glassful at a time from a port in a refrigerator door. Design is a much more complex activity that requires intense focus at the very beginning to determine the full and complete description of what the final product will do for a particular customer base with a set of specific needs. The design process only proceeds into concept generation once the product is so well described that it has met with the approval of groups of technical and business discipline specialists and managers. These review groups include the R&D division of the corporation and may also include employees anywhere in the company, as well as customers and key suppliers. New product ideas must be checked for their fit with the technology and product market strategies of the company, and their requirement for resources. A senior management team will review competing new product development plans championed by different product managers to select those in which to invest resources. The issues involved in planning for the design of a new product are discussed in various sections of Chap. 2 namely: Product and Process Cycles; Markets and Marketing, and Technological Innovation. Certain decisions about the PDP are made even before the engineering design process begins. Chap. 2 sections point out certain types of development work and decision making that must be completed before the design problem definition starts.

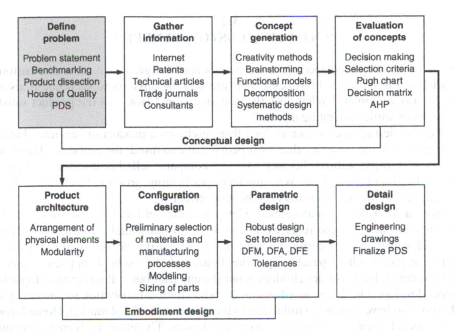

FIGURE 3.1
The engineering design process showing problem definition as the start of the conceptual design process.

Product development begins by determining what the needs are that a product must meet. Problem definition is the most important of the steps in the PDP (Fig. 3.1). Understanding any problem thoroughly is crucial to reaching an outstanding solution. This axiom holds for all kinds of problem solving, whether it be math problems, production problems, or design problems. In product design the ultimate test of a solution is meeting management's goal in the marketplace, so it is vital to work hard to understand and provide what it is that the customer wants.

This chapter emphasizes the customer satisfaction aspect of problem definition, an approach not always taken in engineering design. This view turns the design problem definition process into the identification of the outcome the customer or end user of the product wants to achieve. *Therefore, in product development, the problem definition process is mainly the need identification step.* The need identification methods in this chapter draw heavily on processes introduced and proven effective by the total quality management (TQM) movement. TQM emphasizes customer satisfaction. The TQM tool of *quality function deployment* (QFD) will be introduced. QFD is a process devised to identify the *voice of the customer* and channel it through the entire product development process. The most popular step of QFD, producing the House of Quality (HOQ), is presented here in detail. The chapter ends by proposing an outline of the *product design specification* (PDS), which serves as the governing document for the product design. A design team must generate a starting PDS at this point in the design process to guide its design generation. However, the PDS is an evolving document that will not be finalized until the detail design phase of the PDP process.

3.2
IDENTIFYING CUSTOMER NEEDS

Increasing worldwide competitiveness creates a need for greater focus on the customer's wishes. Engineers and businesspeople are seeking answers to such questions as: Who are my customers? What does the customer want? How can the product satisfy the customer while generating a profit?

Webster defines a customer as "one that purchases a product or service." This is the definition of the customer that most people have in mind, the *end user*. These are the people or organizations that buy what the company sells because they are going to be using the product. However, engineers performing product development must broaden their definition of *customer* to be most effective.

From a total quality management viewpoint, the definition of *customer* can be broadened to "anyone who receives or uses what an individual or organization provides." However, not all customers who make purchasing decisions are end users. Clearly the parent who is purchasing action figures, clothes, school supplies, and even breakfast cereal for his or her children is not the end user but still has critical input for product development. Large retail customers who control distribution to a majority of end users also have increasing influence. In the do-it-yourself tool market, Home Depot and Lowes act as customers but they are not end users. Therefore, both customers *and* those who influence them must be consulted to identify needs the new product must satisfy.

The needs of customers outside of the company are important to the development of the product design specifications for new or improved products. A second set of critical constituents are the internal customers, such as a company's own corporate management, manufacturing personnel, the sales staff, and field service personnel whose needs must be considered. For example, the design engineer who requires information on the properties of three potential materials for his or her design is an *internal customer* of the company's materials specialist.

The product under development defines the range of customers that a design team must consider. Remember that the term *customer* implies that the person is engaging in more than just a one-time transaction. Every great company strives to convert each new buyer into a customer for life by delivering quality products and services. A customer base is not necessarily captured by a fixed demographic range. Marketing professionals are attuned to changes in customer bases that will lead to new definitions of markets for existing product improvements and new target markets for product innovations.

3.2.1 Preliminary Research on Customers Needs

In a large company, the research on customer needs for a particular product or for the development of a new product is done using a number of formal methods and by different business units. The initial work may be done by a marketing department specialist or a team made up of marketing and design professionals (see Sec. 2.5). The

3

natural focus of marketing specialists is the buyer of the product and similar products. Designers focus on needs that are unmet in the marketplace, products that are similar to the proposed product, historical ways of meeting the need and technological approaches to engineering similar products of the type under consideration. Clearly, information gathering is critical for this stage of design. Chapter 5 outlines sources and search strategies for finding published information on existing designs. Design teams will also need to gather information directly from potential customers.

The Shot-Buddy: A Product Developed by a Team of Engineering Students

A great basketball player has the ability to make shots from a variety of distances and at a variety of angles measured from the basketball hoop. Michael Jordan may be known for his great leaping ability, but it was his game winning shots that allowed the Chicago Bulls to win seven NBA titles. In order to develop a great jump or set shot, an athlete must practice for hours, taking hundreds or thousands of shots. For amateur players most of the practice time is spent retrieving the basketball after it goes careening off the rim or backboard or after it falls through the basket. As a result, there is a need to allow players to maximize shooting time by minimizing the time spent retrieving basketballs.

A senior design course team, JSR Design, is developing a product called the *Shot-Buddy,* a system that returns a thrown basketball to the place of the shooter without manual rotation of the shooting return device. There are products on the market for rotationally adjustable ball returns, but all of them require manual adjustment and will not change automatically as the shooter moves around the court.

Driving ranges are popular because they allow the golfer to hit hundreds of golf balls, one after the other, without ever having to chase down or locate a golf ball. This allows the golfer to focus the entire practice time on technique.

In contrast, a young basketball player practicing on his or her jump shot will usually have only one basketball with which to shoot. This means that a large portion of practice time involves not only shooting the basketball, but retrieving both made and missed shots. Depending on the distance the shooter is from the basket, errant shots can rebound in almost any direction, with nearly the same velocity with which the basketball was shot. Coaches and experts estimate that nearly 70 percent of shots taken from the wings (or sides of the basket) will rebound to the weak (or opposite) side from which the ball was shot.[1] Figure 3.2 illustrates this point. Even in the case where the shooter is successful in making a basket, the ball still needs to be retrieved from underneath the hoop, which can be as far as 24 feet away. More time is spent running after the ball than actually shooting. The Shot-Buddy will allow basketball players to spend more time practicing their ball shooting skills.

---

1. "Basketball Zone Defense—Rebounding out of the Zone", The Coach's Clipboard, n.d., 8/15/2010 <www.coachesclipboard.net/ZoneRebounding.html>.

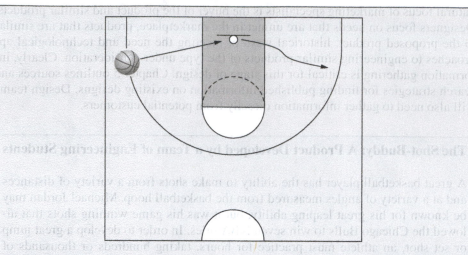

FIGURE 3.2
Shot from the "left wing" on basketball court.

Adapted from Josiah Davis, Jamil Decker, James Maresco, Seth McBee, Stephen Phillips, and Ryan Quinn, "JSR Design Final Report: Shot-Buddy," unpublished, ENME 472, University of Maryland, May 2010.

EXAMPLE 3.1 Determining the Market
The JSR Design team must begin the Shot-Buddy product development process by determining their target end-user.

The market for the Shot-Buddy will be focused on, but not limited to, the parents of basketball players between the ages of 10 and 18 years old. The reason that 10 years old was chosen as the lower limit is that JSR Design members feel it is at this age when a person usually has developed the necessary strength and motor skills required to begin training for basketball team play. Younger athletes, who have not yet developed the upper body strength to shoot from long range, are not concerned with the unpredictable rebounds that result from longer range shooting. Children under age 10 are also not usually as competitive and serious regarding their athletics, which means they will have less of a need for individual practice time.

At age 18, the upper limit, many young adults are transitioning into a time when their need for a product such as this diminishes, as new life changes become more of a priority. At this age students either enter college athletics or become more focused on their careers and academics. If they become involved in college athletics, improved facilities and increased coaching staff make the need for this product obsolete. Nevertheless, the Shot-Buddy would still be a useful practice tool for young adults who continue to play basketball for recreation and have a hoop at their homes.

One way to begin to understand needs of the targeted customers is for the development team to use their own experience. The team can begin to identify the needs that current products in their area of interest do not meet and those that an ideal new product should meet. In fact, there's no better group of people to start articulating unmet needs than members of a product development team who also happen to be end users of what they are designing. Thus members of JSR Design are well-suited to start describing performance and features of a basketball return system.

Brainstorming is a natural idea generation tool that can be used at this point in the process. Brainstorming will be covered in more detail in Chap. 4. It is such a familiar process that a brief example of how brainstorming can be carried out to provide insight into customer needs is given here.

EXAMPLE 3.2 Brainstorming Product Performance and Features
JSR Design team members play basketball for recreation. As a group they can use the guidelines of brainstorming to begin to determine the performance the Shot-Buddy must provide. JSR Design developed the following problem statement for their brainstorming session:

Problem Statement: *Design a basketball return device for players from about age 10 to age 18 that will automatically return the ball to the shooting player.*

The following list is a subset of the team's ideas for the Shot-Buddy.

1. Return missed shots near the hoop
2. Return missed shots even when they aren't hitting the hoop or the backstop
3. Track where the shooter is on the court
4. Return the ball to the position of the shooter
5. Return the ball quickly
6. Do not block the shooter's access to the basket
7. Fit any kind of hoop that a young player might have (e.g., a height adjustable hoop)
8. Be easily set up on a hoop and court
9. Fit hoops that are set up on home courts (e.g., free standing systems and those mounted on a garage or home wall)
10. Be able to be stored in small space
11. Withstand the elements if left attached to a hoop for an extended period of time
12. Return shots taken from the wings of the baskets (not just in front of the basket)
13. Return balls with enough energy to reach a shooter standing as far away as the three-point line
14. Return the ball accurately—so the shooter doesn't have to move to get the ball

Next, the ideas for improvement were grouped into common areas by using an *affinity diagram* (see Chap. 4). A good way to achieve this is to write each of the ideas on a Post-it note and place them randomly on a wall. The team then examines the ideas and arranges them into columns of logical groups. After grouping, the team determines a heading for the column and places that heading at the top of the column. The team created an affinity diagram for their improvement ideas, and it is shown in Table 3.1.

TABLE 3.1
Affinity Diagram Created from Brainstormed
Shot-Buddy Features

Ball Catch Area	Return Direction	Return Characteristics	Size and Shape	Other
1	3	4	6	11
2		5	7	
6		13	8	
12		14	9	
			10	

The five product improvement categories appearing in Table 3.1 emerged from the within-team brainstorming session. This information helps to focus the team's design scope. It also aids the team in determining areas of particular interest for more research from direct interaction with customers and from the team's own testing processes.

3.2.2 Gathering Information from Customers

It is the customer's desires that ordinarily drive the development of the product, not the engineer's vision of what the customer should want. (An exception to this rule is the case of technology driving innovative products that customers have never seen before, Sec. 2.6.4.) Information on the customer's needs is obtained through a variety of channels: [1]

- *Interviews with customers:* Active marketing and sales forces should be continuously meeting with current and potential customers. Some corporations have account teams whose responsibility is to visit key customer accounts to probe for problem areas and to cultivate and maintain friendly contact. They report information on current product strengths and weaknesses that will be helpful in product upgrades. An even better approach is for the design team to interview customers in the service environment where the product will be used. Key questions to ask are: What do you like or dislike about this product? What factors do you consider when purchasing this product? What improvements would you make to this product?
- *Focus groups:* A focus group is a moderated discussion with 6 to 12 customers or targeted customers of a product. The moderator is a facilitator who uses prepared questions to guide the discussion about the merits and disadvantages of the product. A trained moderator will follow up on any surprise answers in an attempt to uncover implicit needs and latent needs of which the customer is not consciously aware.
- *Customer complaints:* A sure way to learn about needs for product improvement is from customer complaints. These may be recorded by communications (by telephone, letter, or email) to a customer information department, service center or warranty department, or a return center at a larger retail outlet. Third-party Internet websites can be another source of customer input on customer satisfaction with a product. Purchase sites often include customer rating information. Savvy marketing departments monitor these sites for information on their products and competing products.
- *Warranty data:* Product service centers and warranty departments are a rich and important source of data on the quality of an existing product. Statistics on warranty claims can pinpoint design defects. However, gross return numbers can be misleading. Some merchandise is returned with no apparent defect. This reflects customer dissatisfaction with paying for things, not with the product.
- *Customer surveys:* A written questionnaire is best used for gaining opinions about the redesign of existing products or new products that are well understood by the public. Other common reasons for conducting a survey are to identify or prioritize problems and to assess whether an implemented solution to a problem was successful. A survey can be done by mail, e-mail, telephone, or in person.

The creation of customer surveys is now presented in more detail.

---

1. K. T. Ulrich and S. D. Eppinger, *Product Design and Development,* 4th ed., McGraw-Hill, New York, 2007.

Constructing a Survey Instrument

Regardless of the method used to gain information from customers, considerable thought needs to go into developing the survey instrument.[1] Creating an effective survey requires the following steps. The steps described here will be illustrated by imagining how members of the JSR Design team would follow this process to collect information from their targeted market for the Shot-Buddy.

1. Determine the survey purpose. Write a short paragraph stating the purpose of the survey, what will be done with the results, and by whom.
2. Identify what specific information is needed. Each question should have a clear purpose for eliciting responses to inform on specific issues. You should have no more questions than the absolute minimum. The sample survey for the "Shot Buddy" (Fig. 3.3) contains two sections of questions. The first set, questions 1 to 4, determine if the respondent is in the target market for a basketball return.
3. Design the questions. Each question should be unbiased, unambiguous, clear, and brief. There are three categories of questions: (1) attitude questions—how the customers feel or think about something; (2) knowledge questions—questions asked to determine whether the customer knows the specifics about a product or service; and (3) behavior questions—usually contain phrases like "how often," "how much," or "when." Some general rules to follow in writing questions are:

- Do not use jargon or sophisticated vocabulary.
- Every question should focus directly on one specific topic.
- Use simple sentences. Two or more simple sentences are preferable to one compound sentence.
- Do not lead the customer toward the answer you want.
- Avoid questions with double negatives because they may create misunderstanding.
- In any list of options given to the respondents, include the choice of "Other" with a space for a write-in answer.
- Always include one open-ended question. Open-ended questions can reveal insights and nuances and tell you things you would never think to ask.
- The number of questions should be such that they can be answered in about 15 (but no more than 30) minutes.
- Design the survey form so that tabulating and analyzing data will be easy.
- Include instructions for completing and returning it.

Questions can have the following types of answers:

- Yes—no—don't know
- A Likert-type rating scale made up of an odd number of rating responses, e.g., strongly disagree—mildly disagree—neutral—mildly agree—strongly agree. On a 1–5 scale such as this, always set up the numerical scale so that a high number means a good answer. The question must be posed so that the rating scale makes sense.
- Rank order—list in descending order of preference
- Unordered choices—choose (b) over (d) or (b) from a, b, c, d, e.

---

1. P. Slanat and D. A. Dillman, *How to Conduct Your Own Survey,* Wiley, New York, 1994 and "Survey Design," Creative Research Systems, n.d., www.surveysystem.com/sdesign.htm, accessed August 15, 2010.

Basketball Return Device
Product Design Survey

Students from a senior capstone design course are designing an improved basketball return device for players from about ages 10 to 18. These survey answers will be used to guide the design process. Please take 10 minutes to complete this survey.

For this set of questions circle the number from 1 to 5 that most accurately reflects your answer.	Response from Participant				
	Strongly Disagree (or Never)		Neutral		Strongly Agree (or Always)
1. Members of my family play basketball at my home	1	2	3	4	5
2. Members of my family practice their shooting skills alone	1	2	3	4	5
3. Members of my family have more than one basketball	1	2	3	4	5
4. I believe that basketball practice is important to my family members	1	2	3	4	5

If all your answers to the previous questions are "1" you may return the survey to the administrator without further answers. Otherwise, please continue with your answers on a scale from 1 to 5.

5. Members of my family are on basketball teams	1	2	3	4	5
6. Members of my family wish to improve their shooting skills	1	2	3	4	5
7. Members of my family should practice basketball more than they do at home	1	2	3	4	5
8. I like to assist my family members by practicing with them	1	2	3	4	5
9. Members of my family frequently give and get sports gifts	1	2	3	4	5

For the next set of answers, check the box for "Yes" or "No"	Yes	No
10. My family has a basketball hoop attached to a building	❑	❑
11. My family has a free standing basketball hoop	❑	❑
12. My family has an adjustable height basketball hoop	❑	❑

Additional

How much would you pay for an automatic basketball rebounding system that will attach to any standard basketball hoop installation and allow the ball to return to the shooter's location (circle one price range)?

I would pay $50 < $100 $100 < $150 $150 < $200 $200<$250 $250 and over

What features of a basketball return device would be most important to you as a potential buyer?

_____

_____

_____

Voluntary Demographic Information: Age: _____ Gender: _____

FIGURE 3.3
Customer survey for the Shot-Buddy.

Select the type of answer option that will elicit responses in the most revealing format without overtaxing the respondent.

4. Arrange the order of questions so that they provide context to what you are trying to learn from the customer. Group the questions by topic, and start with easy ones. The second set of questions in the Shot-Buddy survey (Fig. 3.3) seek to learn the attitude of the respondent to basketball practice at the home. The third set of questions investigates the type of hoop the respondents have.
5. Pilot the survey. Before distributing the survey to the customer, always pilot it on a smaller sample group and review the reported information. This will tell you whether any of the questions are poorly worded and sometimes misunderstood, whether the rating scales are adequate, and whether the survey is too long.

6. Administer the survey. Key issues in administering the survey are assuring that the people surveyed constitute a representative sample for fulfilling the purpose of the survey, and determining size sample must be used to achieve statistically significant results. Answering these questions requires special expertise and experience. Consultants in the area of marketing should be used for really important situations.

Evaluating Customer Surveys

Evaluating a survey question depends on the type of question and the kind of information sought. To evaluate the customer responses, we could calculate the average score for each question, using a 1–5 scale. Those questions scoring highest would represent aspects of the product ranked highest in the minds of the customers. Alternatively, we can take the number of times a feature or attribute of a design is mentioned in the survey and divide by the total number of customers surveyed. For the survey shown in Fig. 3.3, we might use the number of responses to each question rating a feature as either a 4 or a 5.

It is worth noting that a response to a questionnaire of this type really measures the need obviousness as opposed to need importance. To get at true need importance, it is necessary to conduct face-to-face interviews or focus groups and to record the actual words used by the persons interviewed. These responses need to be studied in depth. It is important to realize that often respondents will omit talking about factors that are very important to them because they seem so obvious. Safety and durability are good examples. It is also possible for an end user to forget to mention a feature of a product that has become standard. This is addressed in Sec. 3.3.2.

The relative frequency of responses from a survey can be displayed in a bar graph or a Pareto chart. In a bar graph the frequency of responses to each of the questions is plotted in order of the question number. Figure 3.4 displays a bar graph of simulated responses (no actual survey was done) to one group of questions to the "Basketball

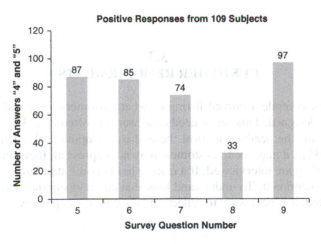

FIGURE 3.4
Chart of simulated answers to Shot-Buddy Survey questions 5–9.

Return Device" survey. A similar plot would be made for questions 1–4 and 10–12. In a Pareto chart the frequency of responses is arranged in decreasing order with the item of highest frequency at the left-hand side of the plot. This plot clearly identifies the most important customer requirements—the vital few. The responses indicate that a basketball return would be viewed as a good gift idea for members of the family and that the families in this response group are in the target market. Most importantly, answers to question 8 indicate that respondents want a return system that frees them from practicing with their basketball player. A more precise question could ask how often the respondent rebounds for a practice session.

Ethnographic Studies

Surveys can be a powerful means of collecting answers to known questions. However, finding out the complete story about how customers interact with a product is often more difficult than asking for answers to a brief survey. Customers are inventive, and much can be discovered from them. A method called *ethnographic investigation* is valuable to learning about the way people behave in their regular environments.[1]

Ethnography is the process of investigation and documentation of the behavior of a specific group of people under particular conditions. Ethnography entails close observation, even to the point of immersion, in the group being studied while they are experiencing the conditions of interest. This way the observer can get a comprehensive and integrated understanding of the scenario under investigation. It is not unusual for a company to support this type of study by setting up situations that enable members of a product development team to observe end users in their natural work settings.

The design team can employ this method to determine how a customer uses (or misuses) a product. Ethnographic study of products involves observing actual end users interacting with the product under typical use conditions. Team members collect photographs, sketches, videos, and interview data during an ethnographic study. The team can further explore product use by playing the roles of typical end users. (A detailed interview with a few end users is more useful than a survey of students acting as end users.)

3.3
CUSTOMER REQUIREMENTS

Designers must compile a ranked listing of what customers need and want from the product being designed. This set of needs and wants is often called *customer requirements*. These are the needs that form the end user's opinion about the quality of a product. As odd as it may seem, customers may not express all their requirements of a product when they are interviewed. If a feature has become standard on a product they may forget to mention it. To understand how that can happen and how the omissions can be mitigated, it is necessary to reflect on how customers perceive "needs."

---

1. H. Mariampolski, *Ethnography for Marketers: A Guide to Consumer Immersion,* Sage Publications, 2005.

From a global viewpoint, we should recognize that there is a hierarchy of human needs that motivate individuals in general.[1]

Rank of 1. *Physiological needs* such as thirst, hunger, sex, sleep, shelter, and exercise. These constitute the basic needs of the body, and until they are satisfied, they remain the prime influence on the individual's behavior.

Rank of 2. *Safety and security needs,* which include protection against danger, deprivation, and threat. When the bodily needs are satisfied, the safety and security needs become dominant.

Rank of 3. *Social needs* for love and esteem by others. These needs include belonging to groups, group identity, and social acceptance.

Rank of 4. *Psychological needs* for self-esteem and self-respect and for accomplishment and recognition.

Rank of 5. *Self-fulfillment needs* for the realization of one's full potential through self-development, creativity, and self-expression.

As each need in this ranking is satisfied, a person becomes driven by the next highest ranking need. The first concerns of a parent of a 10-year-old child are food, shelter and clothing. No parent would be interested in the child's sports equipment unless basic needs were met. Our design problem should be related to the basic human needs, some of which may be so obvious that in our modern technological society they are taken for granted. For example, in the brainstorming listing of performance and features of the Shot-Buddy there is no mention of *safety* in the performance of the device. Nevertheless, safety is always a requirement for engineering designers.

3.3.1 Differing Views of Customer Requirements

From a design team point of view, the customer requirements fit into a broader picture of the PDP requirements, which include product performance, time to market, cost, and quality.

- *Performance* deals with what the design should do when it is completed and in operation. Design teams do not blindly adopt the customer requirements set determined thus far. However, that set is the foundation for design team actions. Other factors may include requirements by internal customers (e.g., manufacturing) or large retail distributors.
- The *time* dimension includes all time aspects of the design. Currently, much effort is being given to reducing the PDP cycle time, also known as the time to market, for new products.[2] For many consumer products, the first to market with a great product captures the market (Fig. 2.2).
- *Cost* pertains to all monetary aspects of the design. It is a paramount consideration of the design team. When all other customer requirements are roughly equal, cost determines most customers' buying decisions. From the design team's point

---

1. A. H. Maslow, *Psych. Rev.,* vol. 50, pp. 370–396, 1943.
2. G. Stalk, Jr., and T. M. Hout, *Competing against Time,* The Free Press, New York, 1990.

TABLE 3.2
Garvin's Eight Dimensions of Quality

Dimension	Description
Performance	The primary operating characteristics of a product. This dimension can be expressed in measurable quantities and ranked objectively
Features	Characteristics that supplement a product's basic functions. Features customize or personalize a product to the customer's needs or taste.
Reliability	The probability of a product failing or malfunctioning within a specified time period. See Chap. 14.
Durability	A measure of the amount of use one gets from a product before it breaks down and replacement is preferable to continued repair. Durability is a measure of product life. Durability is not the same as reliability.
Serviceability	Ease of repair and time to repair after breakdown. Other issues are courtesy and competence of repair personnel and cost and ease of repair.
Conformance	The degree to which a product meets both customer expectations and established standards. These standards include industry standards, government regulations, and safety and environmental standards.
Aesthetics	How a product looks, feels, sounds, tastes, and smells. Customer response is a matter of personal judgment and individual preference.
Perceived Quality	Customers' judgment of the product prior to purchase. This dimension is associated with past experience with similar products or the same manufacturer's products. Advertising seeks to influence this perception.

of view, cost is a result of many design decisions and must often be used to make trade-offs among features and deadlines.

- *Quality* is a complex characteristic with many aspects and definitions. A good definition of quality for the design team is the totality of features and characteristics of a product or service that bear on its ability to satisfy stated or implied needs.

A more inclusive customer requirement than the four listed above is *value*. Value is the worth of a product or service. It can be expressed by the function provided divided by the cost, or the quality provided divided by the cost. Studies of large, successful companies have shown that the return on investment is correlated with high market share and high quality.

Garvin[1] identified the *eight basic dimensions of quality* (Table 3.2) for a manufactured product. These have become a standard list that design teams use as a guide for completeness of customer requirement data gathered in the PDP. Not all dimensions of quality are equally important to each product, so not all are critical customer requirements. Some dimensions highlight the need for a multidisciplinary product development team. Aesthetics in design falls into the domain of the *industrial designer,* who is part artist. An important technical issue that affects aesthetics is *ergonomics,* how well the design fits the human user. Ergonomics falls into the skill set of the *industrial engineer.*

---

1. D. A. Garvin, *Harvard Business Review,* November–December 1987, pp. 101–9.

The challenge for the design team is to combine all the information gathered about customers' needs for a product and interpret it. The customer data must be filtered into a manageable set of requirements that drive the generation of design concepts. The design team must clearly identify preference levels among the customer requirements before adding in considerations like time to market or the requirements of the company's internal customers.

3.3.2 Classifying Customer Requirements

Not all customer requirements are equal. This essentially means that customer requirements have different values for different people. The design team must identify those requirements that are most important to the success of the product in its target market and must ensure that those requirements are satisfied by the product.

This is a difficult distinction for some design team members to make because the pure engineering viewpoint is to deliver the best possible performance in all product aspects. A Kano diagram is a good tool to visually partition customer requirements into categories that will allow for their prioritization. Kano recognized that there are four levels of customer requirements: (1) expecters, (2) spokens, (3) unspokens, and (4) exciters.[1]

- *Expecters:* These are the basic attributes that one would expect to see in the product, i.e., standard features. Expecters are frequently easy to measure and are used often in benchmarking.
- *Spokens:* These are the specific features that customers say they want in the product. Because the customer defines the product in terms of these attributes, the designer must be willing to provide them to satisfy the customer.
- *Unspokens:* These are product attributes the customer does not generally talk about, but they remain important to him or her. They cannot be ignored. They may be attributes the customer simply forgot to mention or was unwilling to talk about or simply does not realize he or she wants. It takes great skill on the part of the design team to identify the unspoken requirements.
- *Exciters:* Often called *delighters,* these are product features that make the product unique and distinguish it from the competition. Note that the absence of an exciter will not make customers unhappy, since they do not know what is missing.

A Kano diagram depicts how expected customer satisfaction (shown on y-axis) can vary with the success of the execution (shown on x-axis) of customer requirements. The success of execution can also be interpreted as product performance. The adequate level of performance is at the zero point on the x-axis. Performance to the right of the y-axis indicates higher quality than required. Performance to the left represents decreasing quality to the point where there is no performance on a requirement.

Figure 3.5 depicts three types of relationships between product performance and customer requirements. Curve 2 is the 45° line that begins in the region of "absent"

---

1. L. Cohen, *Quality Function Deployment: How to Make QFD Work for You,* Addison-Wesley, *Publishing Company,* New York, 1995.

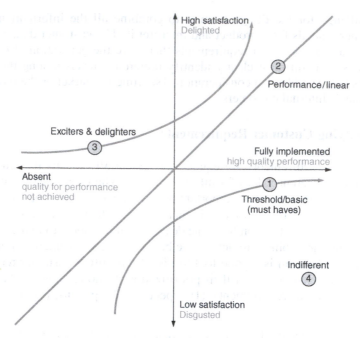

FIGURE 3.5
Kano diagram.

performance on a requirement and lowest customer satisfaction or "disgust" and progresses to the point of high quality performance and customer delight. Since it is a straight line, it represents customer requirements that are basic to the intended function of the product and will, eventually, result in delight. Customer Requirements (CRs) in the Expecter category are represented on Curve 2. Most Spoken CRs also follow Curve 2.

Curve 1 on Fig. 3.5 begins in the region of existing but less than adequately implemented performance and rises asymptotically to the positive x-axis. Customer requirements that follow Curve 1 will never contribute to positive customer satisfaction because the requirements are expected in the product's performance. However, failing to meet the expected performance will disproportionately decrease quality perceptions. Expecter CRs follow Curve 1. Unspoken CRs will also follow Curve 1.

On Curve 3 any product performance that helps to satisfy these CRs will increase the customer's impression of quality. The improvement in quality rating will increase dramatically as product performance increases. These are the CRs in the Exciter category. The Kano diagram of Fig. 3.5 shows that a design team must be aware of the nature of each CR so that they know which ones are the most important to meet. This understanding of the nature of CRs is necessary for prioritizing design team efforts and making decisions on performance trade-offs.

Considering all the information on customer requirements that has been presented up to this point, the design team can now create a more accurately prioritized list of customer requirements. This set is comprised of

- Basic CRs that are discovered by studying competitor products during benchmarking
- Unspoken CRs that are observed by ethnographic observation
- High-ranking customer requirements (CRs) found from the surveys
- Exciter or Delighter CRs that the company is planning to address with new technology.

The highest-ranked CRs are called *critical to quality customer requirements* (CTQ CRs). The designation of CTQ CRs means that these customer requirements will be the focus of design team efforts because they will lead to the biggest payoff in customer satisfaction.

EXAMPLE 3.3 Shot-Buddy Customer Requirements
The JSR Design team has been researching information on their market and end-user groups for the Shot-Buddy. Following is their set of customer requirements.[1]

1. Weatherproof—System is not vulnerable to rusting from being exposed to rain and snow to give the option of leaving it in its in-use position for long periods of time.
2. Accurate Shot Return—An effective ball return system must be able to return the ball to the place of the shooter at the time when the ball leaves the shot return system.
3. Tool-less Installation—System does not require any tools to be used in order to assemble, disassemble, or install; this includes hand tools or power tools. This CR stems from a desire to save customer time and energy.
4. Five-year Lifetime—This includes the ability to handle environmental factors as well as dropping hazards from heights up to the maximum usage height of the product (12 feet).
5. Quick Return—The Shot-Buddy must return balls quickly, even if they are missed shots. In practice, a shooter can get into a rhythm which helps with building and maintaining a particular shooting "touch."
6. Ability to Store in Garage—System should fit it in a small portion of owner's garage or a shed without having to significantly adjust the placement of other belongings.
7. Compatibility with Most Basket Configurations—Basketball return system must be compatible to attach to any brand of basketball hoop.
8. Does not Jam—The Shot-Buddy must return shots that are coming from all angles and at different velocities without letting the ball get stuck in the system and fail to return.
9. Ability to Catch Most Shots (Missed and Made)—The Shot-Buddy must work with a wide range of shots, both falling into the basket and missing the basket.
10. Non-obtrusive—The Shot-Buddy cannot limit the number of shots that can be taken by having components that block a shooter's access to the basket on the floor or in the air.

---

1. Adapted from: Josiah Davis, Jamil Decker, James Maresco, Seth McBee, Stephen Phillips, and Ryan Quinn, "JSR Design Final Report: Shot-Buddy," unpublished, ENME 472, University of Maryland, May 2010.

The team knows that not all customer requirements (CRs) have the same weight in determining customers' attitudes about the product. The Shot-Buddy's ability to automatically return the ball to where the shooter is standing (2 in the list on the previous page) is the innovation. It is an Exciter CR. High-ranking CRs include Does Not Jam (8), Ability to Catch Most Shots (9), and Compatibility (7). CRs in these two categories would be considered CTQ CRs. The remaining CRs include items that improve the quality of the product (e.g., Quick Return) and those items that are unspoken (e.g., Tool-less Installation).

3.4
GATHERING INFORMATION ON EXISTING PRODUCTS

Exploring and understanding performance is a crucial process in the earliest stages of product development. Gathering information on a product can be done by conducting firsthand observation, reading product and technical literature, and applying the principals of physics and engineering sciences to the task.

3.4.1 Product Dissection

Observing a product during its use is one of the most natural ways to gather information about it and is suggested in Sec. 3.3 on ethnographic studies. The next logical step in product investigation is to take the object apart to see how it works. This process is known as both *product dissection* and *reverse engineering.*

Product dissection is the dismantling of a product to determine the selection and arrangement of component parts and to gain insight about how the product is made. It is carried out to learn about a product from the physical *artifact*[1] itself. Product dissection should be an important part of the engineering design learning process. The information collected during dissection can lead to an understanding of the design decisions made by the producers of the artifact.

The product dissection process includes four activities. Listed with each activity are important questions to be answered during that step in the dissection process.

1. *Discover the operational requirements of the product.* How does the product operate? What conditions are necessary for proper functioning of the product?
2. *Examine how the product performs its functions.* What mechanical, electrical, control systems or other devices are used in the product to generate the desired functions? What are the energy and force flows through the product? What are the spatial constraints for subassemblies and components? Is clearance required for proper functioning? If a clearance is present, why is it present?
3. *Determine the relationships between parts of the product.* What are the major subassemblies? What are the key part interfaces?

---

1. An artifact is a man-made object.

4. *Determine the manufacturing and assembly processes used to produce the product.* Of what material and by what process does it appear that is each part is made? What are the joining methods used on the key components? What kinds of fasteners are used and where are they located on the product?

Discovering the operational requirements of the product is the only step that proceeds with the product fully assembled. Disassembling the product is necessary to complete the other activities. If an assembly drawing is not available with the product, it is a good idea to sketch one as the product is disassembled for the first time. In addition to creating an assembly drawing, creating thorough documentation during this phase is critical. This may include a detailed list of disassembly steps and a listing of each component.

The term *reverse engineering* is typically used for the product dissection process when the goal is to learn about a competitor's products. Engineers do reverse engineering to discover information that *they cannot access any other way.* Reverse engineering is an unsavory process when done for the sole purpose of copying a design for profit. Reverse engineering can show a design team what the competition has done, but it will not explain why the choices were made. Designers doing reverse engineering should be careful not to assume that they are seeing the best design of their competition. Factors other than creating the best performance influence all design processes and are not captured in the physical description of the product.

3.4.2 Product and Technical Literature

Most products purchased by customers come with information on their packaging or labels. Both might include a version of use instructions, warnings, performance ratings, certifications, and producer's contact information. Simple products may have this information included on a label affixed directly on the product. Others have information printed on their exteriors, as is the case with recycling codes on plastics. Other products include the information on their packaging and in data sheets or manuals that accompany the product.

There are labeling requirements included in federal regulations for some products. There are standards organizations and government agencies that regulate the label contents for certain products. The Federal Trade Commission's publication Title 16, Part 305 of Title 16 Commercial Practices, includes a section (Part 305) titled, "Rule Concerning Disclosures Regarding Energy Consumption and Water Use of Certain Home Appliances and Other Products Required Under the Energy Policy and Conservation Act". This act is also known as the "Appliance Labeling Rule." Section 305.13 provides a description of the content, size, font, and placement on fan for the label.[1] This regulation assures that anyone can find out a ceiling fan's high-speed air flow (in cubic feet per minute), electricity usage (in watts) at high speed and an efficiency rating determined by a standard procedure. This is an excellent example of the information that one can find on the product's exterior, packaging, or in literature

---

1. 73 FR 63068, Oct. 23, 2008.

accompanying the product as purchased. Researching product regulations is described in Chap. 5, Gathering Information.

Producers may choose to provide buyers with more information than can be included on a label. Many products, like electronics, come with instruction manuals. Often the product will come with a "Quick Start Guide" for users who do not read instruction manuals. Many larger manufacturers maintain websites with product manuals available for download to product owners and those researching similar products.

Consumer Product Literature

There are private nonprofit organizations dedicated to informing consumers about products. One is the Consumers Union, the publisher of the popular magazine *Consumer Reports*. This periodical is published by the Consumers Union group, an organization that conducts its own testing and research providing results through publications and online content to subscribers at ConsumerReports.org.[1] The combination of independence and Internet dissemination makes this one of the first places buyers should look for product information. Consumer Reports provides information on a broad base of products ranging from cars and appliances to electronics and pet accessories.

There are a number of print and online consumer publications that focus on one particular segment of products. Edmunds, Inc. is a corporation that was founded in 1966 to provide information to automobile buyers.[2] The company established on online presence in 1994 and was the first online information outlet for automobiles. Edmunds' offerings now cover new and used cars as well as loans and insurance for automobiles. This is an example of a consumer information database targeted on one product area.

Another subset of product literature is safety information. The United States Consumer Product Safety Commission (CPSC) is responsible for the identification of unsafe consumer goods. The CPSC focuses on products that tend to be used in or around a household and products that children may be able to use or abuse. The CPSC maintains a listing of regulated products[3] and links to other government agencies for regulated products not under the CPSC's jurisdiction. Safety actions taken by the CPSC in July of 2010 include a recall on specific types of mini bikes and go carts due to the possibility of fire hazard. In the same month the CPSC also proposed new rules[4] for cribs to improve their safety.

Internet Shopping Sites

Internet sites exist to compile information for specialty products. A specialty site is Competitive Edge Products, Inc.[5] That site provides information on a suite of

---

1. ConsumerReports, http://www.consumerreports.org, accessed September 12, 2010.

2. "Welcome to Edmunds.com," Edmunds, http://www.edmunds.com, accessed September 12, 2010.

3. "Regulated Products," Consumer Product Safety Commission, http://www.cpsc.gov/businfo/regl. html, accessed September 12, 2010.

4. "CPSC Proposes New Rules for Full-Size and Non-Full-Size Cribs," *NEWS from CPSC,* Release # 10-301, U.S. Consumer Product Safety Commission, http://www.cpsc.gov/cpscpub/prerel/prhtml10/10301.html, accessed September 12, 2010.

5. "Lifetime Basketball Systems, Hoops, Goals, Backboards and Sports Accessories from Competitive Edge Products," http://www.competitiveedgeproducts.com/basketballsystems.aspx, accessed July 14, 2010.

basketball products ranging from rim and backboard set ups (in-ground and pool-side) to accessories like backboard shatter guards, pole padding, and ball return systems. Available products are displayed with photographs, labeling information, and specifications. On some sites one can find customer reviews input by purchasers. Users of a specialty marketing website must keep in mind that the information provided is not necessarily unbiased.

Technical Literature

In addition to information from special interest publications, there are scholarly journals that publish research quality information. These journals are peer-reviewed and provide material that is deemed worthy of publication to increase the body of knowledge in a topic area. Journal articles can provide important information on a technology that is new to the marketplace. Journal articles can also provide technical analysis that is pertinent to existing products. Using research procedures outlined in Chap. 5, Gathering Information, anyone can search academic journals for pertinent literature. For example, the team developing the Shot-Buddy needs to be able to predict the behavior of a basketball that is thrown at the net in a regulation court. Here are three articles of particular interest to the team:

1. H. Okubo and Hubbard, M. (2006), "Dynamics of the basketball shot with application to the free throw," *Journal of Sports Sciences,* 24:12, 1303–1314.
2. Tran, C. M. and Silverberg, L. M. (2008), "Optimal release conditions for the free throw in men's basketball," *Journal of Sports Sciences,* 26:11, 1147–1155.
3. H. Okubo and Hubbard, M. (2004), "Dynamics of basketball-rim interactions," *Sports Engineering,* 7:1, 15–29.

The Patent Literature

Not all products are patented, but patent literature does include inventions that have become successful products. Patents are a certification by the Patent and Trademark Office of the United States to the inventor of a novel and useful device. A discussion of the U.S. Patent System is included in Chap. 5 along with sections on searching for patents by a variety of classification tags. Patent information is easy to retrieve if the patent number is known. The patent system is also organized by application category so once the proper classification is found, information on inventions proposed (but not necessarily built) can be uncovered.

EXAMPLE 3.4 Finding Patents for Products Like the Proposed Shot-Buddy
U.S. Patent 5540428[1] is an example of a hybrid basketball retrieval apparatus. It is shown in Fig. 3.6. The device works by utilizing a large net (78) set underneath and around the rim to funnel both missed and made shots into a channel (82), at the base of the device. This channel eventually returns the basketball, to the user via gravity and the momentum of the basketball. The net used to funnel the basketballs is sufficiently large to catch the majority of balls that will rebound off of the rim (36) or backboard (10). The net itself is attached to the rim as well as the support pole (74) of the backboard.

---

1. John. G. Joseph, "Basketball Retrieval and Return Apparatus," Patent 5540428, July 30, 1996.

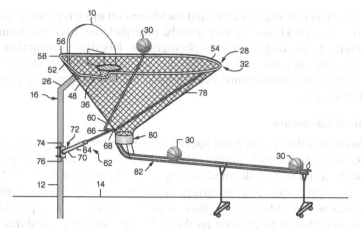

FIGURE 3.6
Basketball retrieval and return device figure from Patent 5540428.

The two major advantages of this design are the ability to retrieve a wide range of missed shots and to consistently return the ball to a position at the end of the ball channel. This design has disadvantages in that it is large, has some fixed supports necessary for use (74, 76), and only returns the ball to the one location regardless of where the shooter is on the court. Finally, this device is designed to be used on the pole-supported baskets normally found in playgrounds or household driveways. While this covers the majority of applications, it still leaves out those baskets found in gymnasiums and recreational centers, which are usually supported, in a more complicated fashion.

Another interesting basketball return device is shown in Fig. 3.7.[1]

3.4.3 Physics of the Product or System

Engineering courses teach first principles in subjects like statics, dynamics, mechanics of materials, electric circuits, controls, fluids, and thermodynamics. Word problems are given describing a physical system and its immediate environment, and students learn to solve these problems using a variety of analytical, logical, mathematical, and empirical methods. In their engineering science courses students are typically given all the detail necessary to translate a description of a product, device, or system into a problem evaluating its performance. This process amounts to setting up *models* and using them for evaluation purposes.

Engineering Models

Efficient analysis of products and systems requires descriptions of each design or system option which is *just detailed enough* that performance measures of interest can be accurately calculated. This description required for analysis is called a *model*.

1. Harold F. Krings, "Automatic Basketball Return Apparatus," Patent 5681230, Oct 28, 1997.

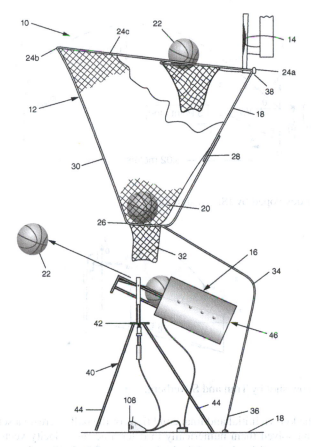

FIGURE 3.7
A different basketball return device figure from Patent 5681230.

The model can include a representation of the physical aspects of the product or system (i.e., a sketch or geometric model), constraints on the design detail to be modeled, physical laws that govern its behavior, and mathematical equations that describe its behavior. (See Sec. 7.4 for more information on developing models.)

The practice of building a model to describe the behavior of a system to be designed is shown here. For the Shot-Buddy to work effectively, it must be capable of enduring certain forces that will be applied to it when in use.

EXAMPLE 3.5 Estimating Forces in Use

Determine the variables needed to estimate the maximum force a basketball shot will have as it hits any kind of ball return device. To estimate the forces a ball return system must withstand, the JSR Design must determine the speed and direction with which a basketball could hit it. Figure 3.8 is the team's diagram representing the motion of a basketball through the air, when released from the three-point shot line (6.02 meters from the basket), which is likely to be the shot that will have the maximum force. JSR Design assumes a shooter's height is the average of an eighth-grade male and that the ball is thrown at head height. The model neglects drag effects of the air on the ball. The design

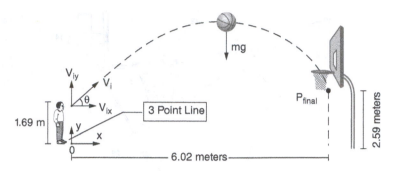

FIGURE 3.8
Model of shooter developed by JSR Design Team.

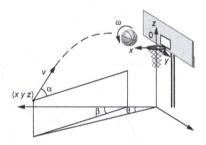

FIGURE 3.9
Model of free throw shot by Tran and Silverberg.

team used the known initial and final conditions of the ball to create a set of simultaneous equations and solved them numerically to determine the velocity vector components at the start and finish of the shot and the highest point of the ball's trajectory and its distance from the shooter to the basket. The calculations were done for a variety of different ball release angles. JSR Design's estimate for initial velocity towards the basket (v_x) at a 45° release angle is 6.5 m/s, with a final velocity when hitting a point just below the basket at 8.6 m/s. At this point, JSR Design estimates the contact time of the basketball and the ball return system to be 0.1 second (Δt) and uses the relationship for momentum ($p = mv$) to estimate a force. According to JSR Design, the estimated force is 55 N and occurs at a release angle of 30°. (Note that if JSR Design members had found the technical literature published by Tran and Silverberg[1] they would have had a reference for estimates of forces and other variables and would not have needed to do so much analysis.)

To verify their modeling, JSR Design can review the technical literature to find a shooting model (see Fig. 3.9) developed by researchers Tran and Silverberg,[1] for their study of basketball free-throws. Tran and Silverberg don't show a height variable as they use 6 feet 6 inches as the average. They also report that a typical free-throw shot is released at about 6 inches above the head of the shooter. The formal model and JSR Design's model include a velocity vector with a release angle, and both sets of

1. C. M. Tran and L. M. Silverberg (2008), "Optimal release conditions for the free throw in men's basketball," *Journal of Sports Sciences,* 26:11, 1147–1155.

researchers recognize that the angle will change. The formal model includes the back-spin on the ball (ω) and two additional angles. Angle β is the side angle of the velocity vector and θ indicates the angle the sagittal plane of the shooter's body makes with the normal line from the backboard. These angles are relevant when the shooter is not facing straight at the backboard during a shot.

The technical paper's model includes a lot of detail that is not necessary during conceptual design. For example, professionals always put a backspin on the basketball to increase the chances of the ball rebounding downward towards the basket if it hits the backboard. Research literature cited in the previous section places the best back-spin in the range of 3 to 4 Hz. This is a detail that is safely omitted in the JSR model used to determine a force level for the ball return device to withstand.

Free Body Diagrams

Free body diagrams are tools to explore the physical nature (existence in form) and operation of the product as it is used. Engineers are taught to create a model to describe the forces and moments that act on physical objects in a defined environment. This type of model is called a *free body diagram*.[1] The object being modeled is sketched with all forces acting upon it. The modeled object must be at rest, so all forces and moments must be balanced. Any unbalanced energy forces and moments result in moving the object in direction of the resultant force.

A free body diagram consists of a declared coordinate system indicated by a set of x-, y-, and z-axes; a sketch of an object oriented in the defined space; labeled arrows that indicate all forces acting on the object oriented appropriately in the coordinate space (e.g., the object's weight will normally be indicated by an arrow labeled "mg" to represent the force due to gravity, where "m" is the object's mass and "g" is the acceleration due to gravity); and rotational forces (torques), indicated by labeled arcs, acting to rotate the object about the declared x-, y-, and z-directions. This model of an object is called "free" because it is represented alone in the sketch without any other object or person acting on it. They are replaced by the equivalent forces. In a correct free body diagram the set of force equations will sum all forces to zero, and the set of moment equations will sum all moments to zero.[2]

The free body diagram is a model that can be used to determine the forces that will be acting on a product during its operation. Free body diagrams can also be created for components within a product to understand and estimate the forces acting upon them as a device operates.

EXAMPLE 3.6 Free Body Diagram of Basketball Goal
The ball return has not been designed yet, but JSR Design needs to understand how the shots on the goal will transmit forces. The simplest way to model this is with a free body diagram. Figure 3.10 is a free body diagram used to estimate the forces on the basketball rim when hit by a shot on the front of the rim. The basketball rim is treated as a simple beam fixed at one end. It will experience forces and a moment at the point it connects with the post.

---

1. An excellent set of notes on constructing free body diagrams, "Some Notes on Free-Body Diagrams" by Professor William Hallett, Dept. of Mechanical Engineering, University of Ottawa, can be found at www.mhhe.com/dieter.
2. Ibid.

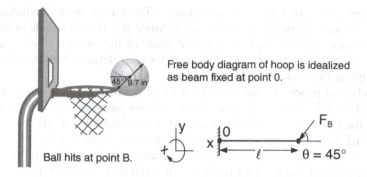

Ball hits at point B.

Free body diagram of hoop is idealized as beam fixed at point 0.

$$0 = F_{xo} - F_B \sin \theta$$
$$0 = F_{yo} - F_B \cos \theta$$
$$0 = M_o + F_B \sin \theta$$

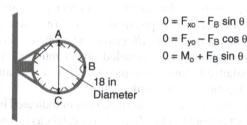

18 in Diameter

FIGURE 3.10
Free body diagram for Example 3.6.

3.5
ESTABLISHING THE ENGINEERING CHARACTERISTICS

Establishing the engineering characteristics is a critical step toward writing the product design specification (Sec. 3.7). The process of identifying the needs that a product must fill is a complicated undertaking. Earlier sections of this chapter focused on gathering and understanding the total picture of what the customer wants from a product. A major challenge of this step is to hear and record the fullness of customer ideas without applying assumptions. For example, if a customer is talking about carry-on luggage they may say, "I want it to be easy to carry." An engineer might interpret that phrase to mean "make it lightweight," and set weight as a design parameter that should be minimized. However, the customer may really want a carry-on case that is easy to fit into the overhead luggage compartment of a plane. The carrying task is already easy due to the design innovation of wheeled luggage.

Just knowing what a customer or end user wants from a product is not sufficient for generating designs. Recall that the design process only proceeds into concept generation once the product is so well-described that it meets with the approval of groups of technical and business discipline specialists and managers. This description is comprised of solution-neutral specifications, meaning that the specification *at this time* should not be so complete as to suggest a single concept or class of concepts.

This description is a set of engineering characteristics that are defined as follows:

- *Design Parameters.* Parameters are a set of physical properties whose values determine the form and behavior of a design. Parameters include the features of a design that can be set by designers *and* the values used to describe the performance of a design. Note: It must be clear that designers make choices in an attempt to *achieve* a particular product performance level, but they cannot *guarantee* they will succeed until embodiment design activities are finalized.
- *Design Variable.* A design variable is a parameter over which the design team has a choice. For example, the gear ratio for the RPM reduction from the rotating spindle of an electric motor can be a variable.
- *Constraints.* A design parameter whose value has been fixed becomes a constraint during the design process. Constraints are limits on design freedom. They can take the form of a selection from a particular color scheme, or the use of a standard fastener, or a specific size limit determined by factors beyond the control of both the design team and the customers.[1] Constraints may be limits on the maximum or minimum value of a design variable or a performance parameter. Constraints can take the form of a range of values.

The product description that a design team must present for approval before getting authorization to continue the PDP process is a set of solution-neutral specifications made up of engineering characteristics. These will include parameters that have been set prior to the design process, design variables, and their constraints. These are the framework for the final set of product design specifications, but they are not the final specifications.

Customers cannot describe the product they want in engineering characteristics because they lack the knowledge base and expertise. Engineering and design professionals are able to describe products in solution-neutral form because they can imagine the physical parts and components that create specific behaviors. Engineers can use a common product development activity called *benchmarking* to expand and refresh their understanding of products of similar type to what they must design.

3.5.1 General and Competitive Performance Benchmarking

Benchmarking is a process for measuring a company's operations against the best practices of companies both inside and outside of their industry.[2] It takes its name

---

1. A good example of this kind of constraint is the size limitation on luggage that may be carried onto a commercial airplane.

2. R. C. Camp, *Benchmarking,* 2d ed., Quality Press, American Society for Quality, Milwaukee, 1995; M. J. Spendolini, *The Benchmarking Book,* Amacom, New York, 1992; M. Zairi, *Effective Benchmarking: Learning from the Best,* Chapman & Hall, New York, 1996 (many case studies).

from the surveyor's benchmark or reference point from which elevations are measured. Benchmarking can be applied to all aspects of a business. It is a way to learn from other businesses through an exchange of information.

Benchmarking operates most effectively on a quid pro quo basis—as an exchange of information between companies that are not direct competitors but can learn from each other's business operations. Other sources for discovering best practices include business partners (e.g., a major supplier to your company), businesses in the same supply chain (e.g., automobile manufacturing suppliers), companies in collaborative and cooperative groups, or industry consultants. Sometimes trade or professional associations can facilitate benchmarking exchanges. More often, it requires good contacts and offering information from your own company that may seem useful to the companies you benchmark.

A company can look for benchmarks in many different places, including within its own organizational structure. Identifying intra-company best practices (or gaps in performance of similar business units) is one of the most efficient ways to improve overall company performance through benchmarking.

Even in enlightened organizations, resistance to new ideas may develop. Benchmarking is usually introduced by a manager who has studied it after learning about success experienced by other companies using the process. Since not all personnel involved in the process have the same education or comfort level with benchmarking, an implementation team can encounter resistance. The more common sources of resistance to benchmarking are as follows:

- Fear of being perceived as copiers.
- Fear of yielding competitive advantages if information is traded or shared.
- Arrogance. A company may feel that there is nothing useful to be learned by looking outside of the organization, or it may feel that it is the benchmark.
- Impatience. Companies that engage in an improvement program often want to begin making changes immediately. Benchmarking provides the first step in a program of change—an assessment of a company's relative position at the current point in time.

To overcome barriers to benchmarking, project leaders must clearly communicate to all concerned the project's purpose, scope, procedure, and expected benefits. All benchmarking exercises begin with the same two steps, regardless of the focus of the benchmarking effort.

- Select the product, process, or functional area of the company that is to be benchmarked. That will influence the selection of key performance *metrics* that will be measured and used for comparison. From a business viewpoint, metrics might be fraction of sales to repeat customers, percent of returned product, or return on investment.
- Identify the *best-in-class companies* for each process to be benchmarked. A best-in-class company is one that performs the process at the lowest cost with the highest degree of customer satisfaction, or has the largest market share.

Finally, it is important to realize that benchmarking is not a one-time effort. Competitors will also be working hard to improve their operations. Benchmarking should be viewed as the first step in a process of continuous improvement if an organization intends to maintain operational advantages.

Competitive performance benchmarking involves testing a company's product against the best-in-class that can be found in the current marketplace. It is an important step for making comparisons in the design and manufacturing of products. Benchmarking is used to develop performance data needed to set functional expectations for new products and to classify competition in the marketplace.

The design engineer's competitive-performance benchmarking procedure is summarized in the following eight steps: [1,2]

1. Determine features, functions, and any other factors that are the most important to end user satisfaction.
2. Determine features and functions that are important to the technical success of the product.
3. Determine the functions that markedly increase the costs of the product.
4. Determine the features and functions that differentiate the product from its competitors.
5. Determine which functions have the greatest potential for improvement.
6. Establish metrics by which the most important functions or features can be quantified and evaluated.
7. Evaluate the product and its competing products using performance testing.
8. Generate a benchmarking report summarizing all information learned about the product, data collected, and conclusions about competitors.

3.5.2 Determining Engineering Characteristics

There is a need to translate the customer requirements into language that expresses the parameters of interest in the language of engineering characteristics. Defining any conceptual design requires that the design team or its approving authority set the level of detail that is necessary to uniquely define every design alternative. This is the set of engineering characteristics (EC) that will include the parameters, design variables, and constraints the design team has begun to collect through research, including benchmarking and reverse engineering activities. The team may have some idea of what the most important engineering characteristics are, but this cannot be determined until the next activity is completed, and that is creating the House of Quality.

EXAMPLE 3.7 Shot-Buddy Engineering Characteristics
The JSR Design team has been researching ball return devices that exist in the market place and comparing them to their customer requirements to develop a set of engineering

1. B. B. Anderson and P. G. Peterson, *The Benchmarking Handbook: Step-by-Step Instructions*, Chapman & Hall, New York, 1996.
2. C. C. Wilson, M. E. Kennedy, and C. J. Trammell, *Superior Product Development, Managing the Process for Innovative Products*, Blackwell Business, Cambridge, MA, 1996.

characteristics that cover the key parameters of the Shot-Buddy as it is imagined. JSR Design had to make certain highlevel design decisions prior to the making a list of possible design characteristics. To make the ball return practical, it is necessary to designate lanes for returning the ball (as shown in Fig. 3.11). The return lane is the one in which the shooter is standing at the time the ball return is actuated. Designs can vary in the number of lanes created.

It is not necessary to have a design defined in order to create a set of engineering characteristics. Team members must understand the problem well enough to create a list of parameters that describe the behavior of the system to be created. The design team will revise their list of ECs throughout the design process. The list presented below is the product of several iterations by the JSR Design Team.

The set of parameters is as follows:

1. **Catch area**—the volume around the basket that indicates the zone in which any basketball thrown will be returned to the shooter
2. **Probability of jamming**—the configuration (mouth size, length, number of turns) of the ball return guide will determine the likelihood of a basketball getting stuck
3. **Accuracy of ball return**—% of time the ball returns to the lane of the shooter
4. **Average time of ball return**—length of time from shot passing the height of the basket to when it is returned to the shooter
5. **Sensing position of shooter**—a key functionality of the Shot-Buddy is to determine where the shooter is on the court in order to accurately aim the ball's return
6. **Lane change time**—time it takes the ball return aiming device to rotate through a lane
7. **Lane span**—degrees in radians that the lane traces out in rotation centered on the basket
8. **Energy or torque to rotate ball return subsystem**—the Shot-Buddy must include a moving system to aim the ball to the lane of the shooter
9. **Weight**
10. **Time to install system**—length of time it takes for a homeowner to assemble and mount the system and get it working

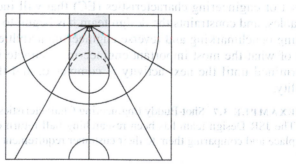

FIGURE 3.11
Ball return lanes for Shot-Buddy (six lane design shown).

Adapted from Josiah Davis, Jamil Decker, James Maresco, Seth McBee, Stephen Phillips, and Ryan Quinn, "JSR Design Final Report: Shot-Buddy," unpublished, ENME 472, University of Maryland, May 2010.

11. **Material rigidity**—any part of the system that is vulnerable to impact by the basketball must be able to withstand a deflection without displaying permanent deformation
12. **Material toughness at attachment areas**—The Shot-Buddy will be attached to some part of the existing basketball hoop installation or supporting structure and all parts of the attachment must be able to withstand a hard hit imparted by a basketball
13. **Weather resistance**—the Shot-Buddy is designed to be installed on outdoor basketball hoops, meaning it must withstand the elements for a period of five years

The ECs listed here are mix of physical and performance characteristics. Some ECs like number 5, sensing the position of the shooter, describe a key functionality of the system. It is likely that many different methods of sensing can be proposed for the Shot-Buddy; each would describe a different design.

The list of engineering characteristics (ECs)—developed in Ex. 3.7 represents aspects of the Shot-Buddy's performance or physical characteristics that are variables to be determined by the design team. Each EC will contribute to determining the overall performance of the Shot-Buddy, but some ECs will be more critical to satisfying the customer requirements than others. The QFD method introduced in Sec. 3.6 will aid design teams in determining the most critical ECs.

3.6
QUALITY FUNCTION DEPLOYMENT

Quality function deployment (QFD) is a planning and team problem-solving tool that has been adopted by a wide variety of companies as the tool of choice for focusing a design team's attention on satisfying customer needs throughout the product development process. The term *deployment* in QFD refers to the fact that this method determines the important set of requirements for each phase of PDP planning and uses them to identify the set of technical characteristics of each phase that most contribute to satisfying the requirements. QFD is a largely graphical method that aids a design team in systematically identifying all of the elements that go into the product development process and creating relationship matrices between key parameters at each step of the process. Gathering the information required for the QFD process forces the design team to answer questions that might be glossed over in a less rigorous methodology and to learn what it does not know about the problem. Because it is a group decision-making activity, it creates a high level of buy-in and group understanding of the problem. QFD, like brainstorming, is a tool for multiple stages of the design process. In fact, it is a complete process that provides input to guide the design team.

The complete QFD process is diagrammed in Fig. 3.12.[1] Three aspects of the QFD process are depicted here. It is clear why the phases of QFD, especially the

---

1. S. Pugh, *Total Design*, Chap. 3, Addison-Wesley, Reading, MA, 1990.

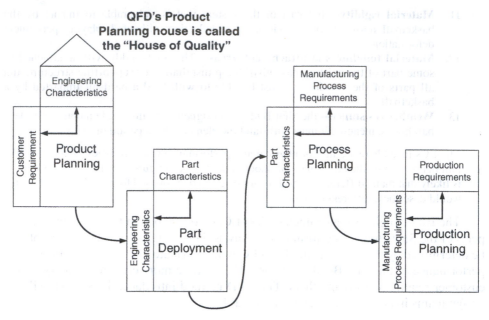

FIGURE 3.12
Diagram showing the four houses of the complete QFD process.

first, product planning, are called houses. Second, the QFD process is made up of four phases that proceed in sequence and are connected as a chain with the output from each phase becoming the input to the next. The product planning phase of QFD, called the *House of Quality,* feeds results into the design of individual parts, giving inputs into the process planning design stage, which become inputs into the production planning phase of QFD. For example, the important engineering characteristics determined by the House of Quality become the input for the part design house. Third, the QFD process is created to transform or map input requirements to each house into the characteristics output from the house. Since QFD is a linked, sequential, and transformational process, the first set of inputs strongly influences all subsequent transformations. Thus, the QFD process is known as a methodology for infusing *the voice of the customer* into every aspect of the design process.

The implementation of the QFD method in U.S. companies is often reduced to the use of only its first house, the House of Quality. The House of Quality develops the relationships between what the customer wants from a product and which of the product's features and overall performance parameters are most critical to fulfilling those wants. The House of Quality translates *customer requirements*[1] into generally quantifiable design variables, called *engineering characteristics*. This mapping of customer wants to engineering characteristics informs the remainder of the design

---

1. It is usual to refer to the set of desirable characteristics of a product as *customer requirements* even though the more grammatically correct term is *customers' requirements.*

> ## The Development of QFD in Brief
>
> QFD was developed in Japan in the early 1970s, with its first large-scale application in the Kobe Shipyard of Mitsubishi Heavy Industries. It was rapidly adopted by the Japanese automobile industry. By the mid-1980s many U.S. auto, defense, and electronics companies were using QFD. A recent survey of 150 U.S. companies showed that 71 percent of these have adopted QFD since 1990. These companies reported that 83 percent believed that using QFD had increased customer satisfaction with their products, and 76 percent felt it facilitated rational design decisions. It is important to remember these statistics because using QFD requires a considerable commitment of time and effort. Most users of QFD report that the time spent in QFD saves time later in design, especially in minimizing changes caused by poorly defining the original design problem.

process. When the HOQ is constructed in its most comprehensive configuration, the process will identify a set of essential features and product performance measures that will be the target values to be achieved by the design team.

The House of Quality can also be used to determine which engineering characteristics should be treated as constraints for the design process and which should become decision criteria for selecting the best design concept. This function of the HOQ is explained in Sec. 3.6.3. Therefore, creating QFD's House of Quality is a natural precursor to establishing the product design specification (see Sec. 3.7).

3.6.1 The House of Quality Configurations

Engineers today can find many different versions of QFD's House of Quality. As with many TQM methods, there are hundreds of consultants specializing in training people in the use of QFD. A quick Internet search will identify scores of websites that describe QFD in general and the House of Quality in particular. Some use the same texts on QFD that we cite in this section. Others develop and copyright their own materials. These sites include consulting firms, private consultants, academics, professional societies, and even students who have developed HOQ software packages and templates. These applications range from simple Excel spreadsheet macros to sophisticated, multi-versioned families of software.[1] Naturally, each creator of HOQ software uses a slightly different configuration of the HOQ diagram and slightly different terminology. The HOQ configuration used in this text is a compilation of a variety of different HOQ terminologies that is presented in a format for the product development team. It is important to understand the basics of the HOQ so that you

---

1. Three packages are QFD/Capture, International Techne Group, 5303 DuPont Circle, Milford, OH, 45150; QFD Scope, Integrated Quality Dynamics; and QFD Designer from American Supplier Institute.

can easily recognize how different versions of HOQ software are oriented. The main purpose of the HOQ will remain the same.

The HOQ takes information developed by the design team and guides the team into translating it into a format that is more useful for new product generation. This text uses an eight-room version of the House of Quality as shown in Fig. 3.13. As in all HOQ layouts, the relationship matrix (Room 4 in Fig. 3.13) is central to the goal of relating the CRs to the ECs. The CRs are processed through the HOQ in such a

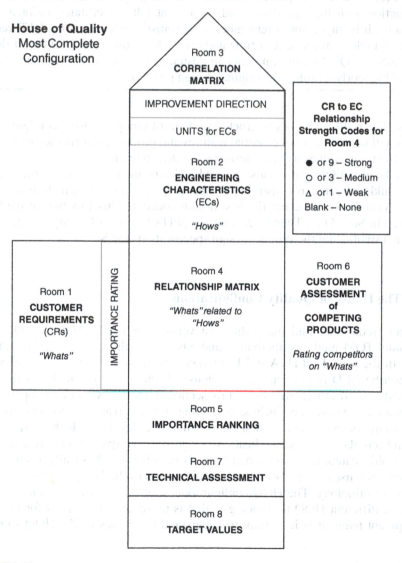

FIGURE 3.13

The House of Quality translates the voice of the customer, input as CRs in Room 1, into target values for ECs in Room 8.

way that their influence is embedded throughout the design process. The Critical to Quality ECs are determined by the simple calculations done in Room 5. Additional data gathered through examination of competitor products, benchmarking, and customer survey results are recorded in Rooms 6 and 7, the assessments of competing products.

The visual nature of the House of Quality should be apparent. Notice that all the rooms of the HOQ that are arranged horizontally pertain to customer requirements (CRs). Information compiled from identifying the needs of the customer and end user is inserted in Room 1 in the form of customer requirements and their importance ratings. Clearly, the initial work to obtain customer preferences, or "Whats," is driving the HOQ analysis. Similarly, the HOQ rooms aligned vertically are organized according to engineering characteristics (ECs), the "Hows." The nature of the ECs and how they are arrived at were described in Sec. 3.5.2. The ECs that you have already identified as constraints can be included in Room 2. They can also be omitted if you do not think that they are major aspects of what the customer will perceive as quality. An example of a constraint like this is 110V AC current for a household appliance.

The end result of the HOQ is the set of target values for ECs that flow through the HOQ and exit at the bottom of the house in Room 8. This set of target values guides the selection and evaluation of potential design concepts. Note that the overall purpose of the HOQ process is broader than establishing target values. Creating the HOQ requires that the design team collects, relates, and considers many aspects of the product, competitors, customers, and more. Thus, by creating the HOQ the team has developed a strong understanding of the issues of the design.

You can see that the House of Quality summarizes a great deal of information in a single diagram. The determination of the "Whats" in Room 1 drives the HOQ analysis. The results of the HOQ, target values for "Hows" in Room 8, drives the design team forward into the concept evaluation and selection processes (topics addressed in Chap. 7). Thus, the HOQ will become one of the most important reference documents created during the design process. Like most design documents, the HOQ should be updated as more information is developed about the design.

3.6.2 Steps for Building a House of Quality

Not all design projects will call for the construction of a House of Quality in its full configuration (Rooms 1 through 8) as shown in Fig. 3.13.

The Streamlined House of Quality

The basic translation of CRs into ECs can be accomplished with an HOQ consisting of Rooms 1, 2, 4, and 5. This streamlined configuration of the House of Quality is shown in Fig. 3.14. Additional detail is given to the three parts of Room 5, the Importance Ranking of ECs. This section describes the construction of the streamlined HOQ in a step-by-step process, followed by a sample HOQ built for the Shot-Buddy design project introduced in Example 3.1.

> Room 1: *Customer requirements* are listed by rows in Room 1. The CRs and their importance ratings are gathered by the team as discussed in Sec. 3.3. It is

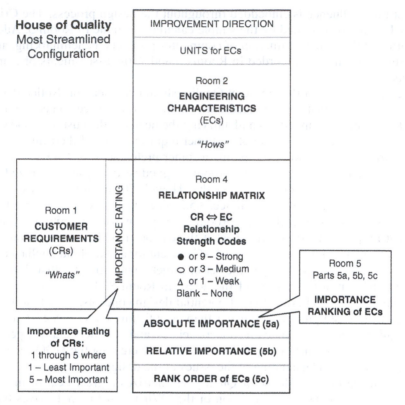

FIGURE 3.14
The Minimal HOQ Template includes Rooms 1, 2, 4, and 5.

common to group these requirements into related categories as identified by an affinity diagram. Also included in this room is a column with an importance rating for each CR. The ratings range from 1 to 5. These inputs to the HOQ are the set of CRs that *includes but is not limited to the CTQ CRs*. The CTQ CRs will be those with importance ratings of 4 and 5.

Room 2: *Engineering characteristics* are listed by columns in Room 2. ECs are product performance measures and features that have been identified as the means to satisfy the CRs. Sec. 3.5.2 discusses how the ECs are identified. One basic way is to look at a particular CR and answer the question, "What can I control that allows me to meet my customer's needs?" Typical ECs include weight, force, velocity, power consumption, and key part reliability. ECs are usually measurable values (unlike the CRs) and their units are placed near the top of Room 2. Symbols indicating the preferred improvement direction of each EC are placed at the top of Room 2. Thus a ↑ symbol indicates that a higher value of this EC is better, and a ↓ symbol indicates that a lower value is better. It is also possible that an EC will not have an improvement direction.

Room 4: The *Relationship matrix* is at the center of an HOQ. It is created by the intersection of the rows of CRs with the columns of ECs. Each cell in the matrix is marked with a symbol that indicates the strength of the causal association between the EC of its column and the CR of its row. The coding scheme for each cell is given as a set of symbols[1] that represent an exponential range of numbers (e.g., 9, 3, 1, and 0). To complete the Relationship Matrix systematically, take each EC in turn, and move down the column cells row by row, asking whether the EC will contribute to fulfilling the CR in the cell's row significantly (9), moderately (3), or slightly (1). The cell is left blank if the EC has no impact on the CR.

Room 5: *Importance Ranking of ECs.* The main contribution of the HOQ is to determine which ECs are of critical importance to satisfying the CRs listed in Room 1. Those ECs with the highest rating are given special consideration, for these are the ones that have the greatest effect upon customer satisfaction.

- *Absolute importance* (Room 5a) of each EC is calculated in two steps. First multiply the numerical value in each of the cells of the Relationship Matrix by the associated CR's importance rating. Then, sum the results for each column, placing the total in Room 5a. These totals show the absolute importance of each engineering characteristic in meeting the customer requirements.
- *Relative importance* (Room 5b) is the absolute importance of each EC, normalized on a scale from 1 to 0 and expressed as a percentage of 100. To arrive at this, total the values of absolute importance. Then, take each value of absolute importance, divide it by the total, and multiply by 100.
- *Rank order of ECs* (Room 5c) is a row that ranks the ECs' Relative Importance from 1 (highest % in Room 5b) to *n*, where *n* is the number of ECs in the HOQ. This ranking allows viewers of the HOQ to quickly focus on ECs in order from most to least relevant to satisfying the customer requirements.

The HOQ's Relationship Matrix (Room 4) must be reviewed to determine the sets of ECs and CRs before accepting the EC Importance rankings of Room 5. The following are interpretations of patterns[2] that can appear in Room 4:

- An empty row signals that no ECs exist to meet the CR.
- An empty EC column signals that the characteristic is not pertinent to customers.
- A row without a "strong relationship" to any of the ECs highlights a CR that will be difficult to achieve.
- An EC column with too many relationships signals that it is really a cost, reliability, or safety item that must be always considered, regardless of its ranking in the HOQ. This EC could be considered a constraint.

---

1. In the first HOQ applications in Japan, the teams liked to use the relationship coding symbols • for Strong, o for Medium, and Δ for Weak. These were taken from the racing form symbols for *win, place,* and *show.*
2. Adapted from S. Nakui, "Comprehensive QFD," *Transactions of the Third Symposium on QFD,* GOAL/QPC, June 1991.

- Two EC columns with nearly the same relationships may indicate that the ECs are similar and need to be combined.
- An HOQ displaying a diagonal matrix (1:1 correspondence of CRs to ECs) signals that the ECs may not yet be expressed in the proper terms (rarely is a quality requirement the result of a single technical characteristic).

If one or more of the patterns is present in Room 4, the CRs and ECs involved should be reviewed and altered if appropriate.

Construction of this HOQ requires inputs from the design team in the form of CRs and ECs. The processing of the HOQ inputs enables the design team to convert the set of CRs into a set of ECs and to determine which ECs are the most important to the design of a successful product. The output of this HOQ is found in Room 5. This information allows a design team to allocate design resources to the product performance aspects or features (ECs) that are most critical to the success of the product. These can be called critical to quality engineering characteristics or CTQ ECs.

EXAMPLE 3.8 Streamlined House of Quality

A streamlined House of Quality is constructed (Fig. 3.15) for the Shot-Buddy in accordance with the instructions for Room 4. The CRs listed in Room 1 are from the list developed in Example 3.3. The Importance Weight factors are determined by the JSR Design team through their research. Room 2, Engineering Characteristics, names the ECs that were developed by completing the activities described in Example 3.7. The cells of the Relationship Matrix in Room 4 hold the rating that describes how much the execution of the EC in the column's heading contributes to satisfying the CR of that row.

The HOQ in Fig. 3.15 shows that the most important engineering characteristics to the design of the Shot-Buddy are the catch area, low jamming probability, weather resistance and sensing the position of the shooter. These are the most important basic parameters of the Shot-Buddy and are defined as CTQ ECs. It may seem odd that the weather resistance of the system is one of the CTQ ECs of the Shot-Buddy. Further consideration of the CRs indicates how important it is to make a ball return system that works for basketball hoops that are usually installed outside the home and remain in place for several years. The HOQ analysis shows that the weather resistance of the system is of critical importance and one EC that JSR Design might have overlooked. This illustrates the value of the HOQ to draw attention to engineering characteristics of real value to the customer.

The least important ECs are lane change time, weight, energy or torque to rotate, accuracy of ball return, and average time to return ball. It is interesting to note that most of these characteristics concern the functions of returning the ball to the proper lane, so one would think that they would be of major importance. The team may decide that two or more ECs should be combined into a more meaningful performance measure. For example, if we combined "accuracy of ball return" with the "average time to return the ball" we would create an EC called "effectiveness of ball return" with a relative weight of 11.6 percent, raising it into the top three ECs. This is a change that the team could make after a critical review of the HOQ.

The results of the HOQ are dependent on the members of the design team who are following the process. Another group working on the same design task may have

Customer Requirements	Importance Weight Factor	Catch Area	Jamming Probability	Accuracy of Ball Return	Average Time to Return Ball	Sensing Position of Shooter	Lane Change Time	Lane Span	Energy or Torque to Rotate	Weight	Time to Install System	Material Rigidity	Material Toughness at Attachment Areas	Weather Resistance
Engineering Characteristics														
Improvement Direction		↑	↓	↑	↓		↓	↓	↓	↓	↓	↑	↑	↑
Units		m²	%	m	sec	n/a	sec	rad	N	kg	min	MPa	MPa√m	n/a
Weatherproof	4											1	3	9
Accurate Ball Return	4	3	9	9		9			9			3		
Tool-less Installation	2	3								3	9	1		
Five-Year Lifetime	4	1										3	9	9
Quick Return	3		9	1	9	9	3	3	9					
Store in Garage	3	9									3			
Compatible with All Hoop Installations	4	1									9			
Does Not Jam	5	3	9		3							3		
Catch Most Shots	5	9												
Non-Obtrusive	2	9												
Raw Score (698)		131	108	39	42	63	9	45	27	15	54	45	48	72
Relative Weight %		18.8	15.5	5.6	6.0	9.0	1.3	6.4	3.9	2.1	7.7	6.4	6.9	10.3
Rank Order		1	2	9	8	4	13	6	11	12	10	6	5	3

FIGURE 3.15
HOQ example of streamlined configuration for the Shot-Buddy.

Adapted from Josiah Davis, Jamil Decker, James Maresco, Seth McBee, Stephen Phillips, and Ryan Quinn, "JSR Design Final Report: Shot-Buddy," unpublished, ENME 472, University of Maryland, May 2010.

different outcomes. However, as the knowledge of the design teams and their experience become more similar their HOQ's will too.

The Correlation Matrix or Roof of the House of Quality

A *correlation matrix* (Room 3) can be built for the House of Quality for the Shot-Buddy design example. The correlation matrix is shown in Fig. 3.16. The correlation matrix, Room 3, records possible interactions between ECs for future trade-off decisions.

> Room 3: The *Correlation matrix* shows the degree of dependence among the engineering characteristics in the roof of the HOQ. It is best to recognize these correlated relationships early so that appropriate trade-offs can be made during embodiment design. The correlation matrix in Fig. 3.16 shows that there are four strong positive correlations (indicated by "++") among EC pairs. One is the correlation between the catch area and the average time to return a ball.

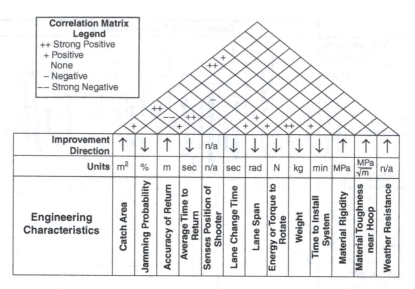

FIGURE 3.16

Shot-Buddy Design House of Quality Rooms 2 and 3.

Adapted from Josiah Davis, Jamil Decker, James Maresco, Seth McBee, Stephen Phillips, and Ryan Quinn, "JSR Design Final Report: Shot-Buddy," unpublished, ENME 472, University of Maryland, May 2010.

This is logical because as the catch area expands the distance that a trapped basketball might travel from a missed shot back to the shooter increases. This signals the design team to remember that if they increase the overall catch area they must be wary of the increase in the average time to return the basketball. Another correlation in shown is the negative correlation (indicated by "−") between the lane span and the accuracy of ball return. Clearly, as the span of the lanes (arc width in radians) increases the less likely it is that the ball return will be absolutely aligned with the shooter when the ball is released. Other correlations are indicated in the matrix.

Determining the strength of the correlations between ECs requires knowledge of the use of the product being designed and engineering experience. It is not necessary to have exact correlation data at this point. The rating serves as a visual reminder for the design team for use in future phases of the design process, like embodiment design (Chap. 8).

Assessment of Competitors' Products in House of Quality

The data available from the HOQ can be augmented by adding the results of any benchmarking activities conducted for the product. The results are shown in two different places.

Room 6: *Competitive assessment* is a table that displays how the top competitive products rank with respect to the customer requirements listed across the HOQ in Room 2. This information comes from direct customer surveys, industry consultants, and

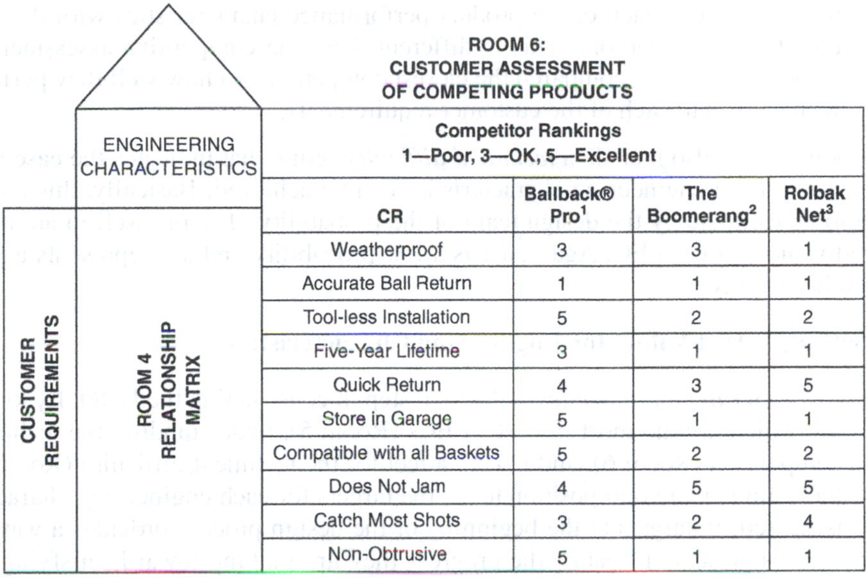

| | | ROOM 6: CUSTOMER ASSESSMENT OF COMPETING PRODUCTS | | |
| | | | | |

ENGINEERING CHARACTERISTICS

CUSTOMER REQUIREMENTS

ROOM 4 RELATIONSHIP MATRIX

	Competitor Rankings 1—Poor, 3—OK, 5—Excellent		
CR	Ballback® Pro[1]	The Boomerang[2]	Rolbak Net[3]
Weatherproof	3	3	1
Accurate Ball Return	1	1	1
Tool-less Installation	5	2	2
Five-Year Lifetime	3	1	1
Quick Return	4	3	5
Store in Garage	5	1	1
Compatible with all Baskets	5	2	2
Does Not Jam	4	5	5
Catch Most Shots	3	2	4
Non-Obtrusive	5	1	1

FIGURE 3.17

HOQ example of streamlined configuration for the Shot-Buddy.

[1]"Ballback® Pro Basketball Return System," Sports Authority, http://www.sportsauthority.com/product/index.jsp?productId=2923, accessed October 27, 2010.

[2]"The Boomerang," www.boomerangbasketball.net/boomerang.html, accessed October 27, 2010.

[3]"Rolbak Net," http://www.basketballgoalstore.com/goalrilla-accessories/rolbak-net.aspx, accessed October 27, 2010.

Adapted from Josiah Davis, Jamil Decker, James Maresco, Seth McBee, Stephen Phillips, and Ryan Quinn, "JSR Design Final Report: Shot-Buddy," unpublished, ENME 472, University of Maryland, May 2010.

marketing departments. In Fig. 3.17 it appears that all competitors meet the requirement of not jamming. This means the Shot-Buddy cannot jam and still be competitive. The Shot-Buddy will be able to improve on "Accurate Ball Return" with its ability to return the ball to the shooter's position, even when the shooter moves. Note: it is not unusual to have sparse data on some of the competitors and very detailed data on another. Certain competitors are targets for new products and, therefore, are studied more closely than others.

Room 7 (refer to the complete HOQ back in Fig. 3.13) in the lower levels of the House of Quality provides another area for the comparison to competing products. Room 7, *Technical Assessment,* is located under the Relationship Matrix. Technical Assessment data can be located above or below the Importance Ranking sections of Room 5. (Recall that there are many different configurations of the House of Quality.)

Room 7: *Technical assessment,* indicates how your competing products score on achieving the suggested levels of each of the engineering characteristics listed in the column headings atop the Relationship Matrix. Generally a scale of 1 to 5 (best) is used. Often this information is obtained by getting examples of the competitor's product and testing them. Note that the data in this room compares each of the product performance characteristics with those of the closest competitors. This is different from the competitive assessment in Room 6, where we compared the closest competitors on how well they perform with respect to each of the customer requirements.

Room 7 may also include a *technical difficulty* rating that indicates the ease with which each of the engineering characteristics can be achieved. Basically, this comes down to an estimate by the design team of the probability of doing well in attaining desired values for each EC. Again, a 1 is a low probability and a 5 represents a high probability of success.

Setting Target Values for Engineering Characteristics

Room 8: *Setting target values* is the final step in constructing the HOQ. By knowing which are the most important ECs (Room 5), understanding the technical competition (Room 6), and having a feel for the technical difficulty (Room 7), the team is in a good position to set the targets for each engineering characteristic. Setting targets at the beginning of the design process provides a way for the design team to gauge the progress they are making toward satisfying the customer's requirements as the design proceeds.

3.6.3 Interpreting Results of HOQ

The design team has collected a great deal of information about the design and processed it into the completed House of Quality. The creation of the HOQ required consideration of the connections between what the customers expect of the product, CRs, and the parameters that are set by the design team. The set of parameters make up the solution-neutral specifications for the product and were defined in Sec. 3.4. Some of the parameters of the design of the product are *already defined*. They may be defined as the result of a decision by the approving authority that initiated the design process, they may be defined by the physics applied to the product while it is in use, or they may be defined by regulations set up by a standards organization or other regulatory bodies. The design variables that are already defined as constraints or that have already been given values do not need to appear in the HOQ.

The highest-ranking ECs from the HOQ are either constraints or design variables whose values can be used as decision-making criteria for evaluating candidate designs (see Chap. 7). If a high-ranking EC has only a few possible candidate values then it may be appropriate to treat that EC as a constraint. There are certain design parameters that can only take a few discrete values. If so, the design team should review the possible values of the EC, determine which is best at meeting correlated EC targets of the design, and then use only the selected value of the EC in generating conceptual designs.

If a high-ranking EC is a design variable that can take many values, like weight, or power output, it is good to use that EC as a metric by which you compare conceptual designs. Thus, your highest-ranking ECs may become your design selection criteria. The results from the HOQ act as a guide to assist the team in determining the selection criterion for evaluating designs.

The lowest-ranking ECs of the HOQ are not as critical to the success of the design. These ECs allow freedom during the design process because their values can be set according to priorities of the designer or approving authority. Values for the low-ranking ECs can be determined by whatever means is most conducive to achieving a good design outcome. They can be set in such a way as to reduce cost or to preserve some other objective of the design team. As long as low-ranking ECs are independent of the CTQ ECs, they can be set expeditiously and not require a great deal of design team effort. Once EC values are set, they are documented in the PDS.

3.7
PRODUCT DESIGN SPECIFICATION

The goal of design process planning is to identify, search, and assemble enough information to decide whether the product development venture is a good investment for the company, and to decide what time to market and level of resources are required. The resulting documentation is typically called a *new product marketing report*. This report can range in size and scope from a one-page memorandum describing a simple product change to a business plan of several hundred pages. The marketing report includes details on such things as the business objectives, a product description and available technology base, the competition, expected volume of sales, marketing strategy, capital requirements, development cost and time, expected profit over time, and return to the shareholders.

In the product development process, the results of the design planning process that governs the engineering design tasks are compiled in the form of a set of product design specifications (PDS). The PDS is the basic control and reference document for the design and manufacture of the product. The PDS is a document that contains all of the facts related to the outcome of the product development. It should avoid forcing the design direction toward a particular concept and predicting the outcome, but it should also contain the realistic constraints.

Creating the PDS finalizes the process of establishing the customer needs and wants, prioritizing them, and beginning to cast them into a technical framework so that design concepts can be established. The process of group thinking and prioritizing that developed the HOQ provides excellent input for writing the PDS. However, it must be understood that the PDS will change as the design process proceeds. Nevertheless, at the end of the process the PDS will describe in writing the product that is intended to be manufactured and marketed.

Table 3.3 is a typical listing of elements that are included in a product design specification. The elements are grouped by categories, and some categories include questions that should be answered by the design team and replaced with their decisions. Not every product will require consideration of every item in this list, but many will. The list demonstrates the complexity of product design. The Shot-Buddy design example used throughout this chapter is again the example in the PDS of Table 3.4.

At the beginning of the concept generation process, the PDS should be as complete as possible about what the design should do. However, it should say as little as possible about *how the requirements are to be met*. Whenever possible the specifications

TABLE 3.3
Template for Product Design Specification

Product Design Specification

Product Identification

- Product name (# of models or different versions, related in-house product families)
- Basic functions of the product
- Special features of the product
- Key performance targets (power output, efficiency, accuracy)
- Service environment (use conditions, storage, transportation, use and predictable misuse)
- User training required

Key Project Deadlines

- Time to complete project
- Fixed project deadlines (e.g., review dates)

Physical Description

What is known (or has already been decided) about the physical requirements for the new product?

- Design variable values that are known or fixed prior to the conceptual design process (e.g., external dimensions)
- Constraints that determine known boundaries on some design variables (e.g., upper limit on acceptable weight)

Financial Requirements

What are the assumptions of the firm about the economics of the product and its development?

What are the corporate criteria on profitability?

- Pricing policy over life cycle (target manufacturing cost, price, estimated retail price, discounts)
- Warranty policy
- Expected financial performance or rate of return on investment
- Level of capital investment required

Life Cycle Targets

What targets should be set for the performance of the product over time? (This will relate to the product's competition.)

What are the most up-to-date recycling policies of the corporation and how can this product's design reflect those policies?

- Useful life and shelf life
- Cost of installation and operation (energy costs, crew size, etc.)
- Maintenance schedule and location (user-performed or service centered)
- Reliability (mean time to failure): Identify critical parts and special reliability targets for them
- End-of-life strategy (% and type of recyclable components, remanufacture of the product, company take back, upgrade policy)

Market Identification

- Description of target market and its size
- Anticipated market demand (units per year)
- Competing products
- Branding strategy (trademark, logo, brand name)

What is the need for a new (or redesigned) product? How much competition exists for the new product? What are the relationships to existing products?

TABLE 3.3 *(continued)*

Product Design Specification

Social, Political, and Legal Requirements

Are there government agencies, societies, or regulation boards that control the markets in which this product is to be launched?
Are there opportunities to patent the product or some of its subsystems?

- Safety and environmental regulations. Applicable government regulations for all intended markets.
- Standards. Pertinent product standards that may be applicable (Underwriters Laboratories, OSHA).
- Safety and product liability. Predictable unintended uses for the product, safety label guidelines, applicable company safety standards.
- Intellectual property. Patents related to product. Licensing strategy for critical pieces of technology.

Manufacturing Specifications

Which parts or systems will be manufactured in-house?

- Manufacturing requirements. Processes and capacity necessary to manufacture final product.
- Suppliers. Identify key suppliers and procurement strategy for purchased parts.

---

should be expressed in quantitative terms and include all known ranges (or limits) within which acceptable performance lies. For example: The power output of the engine should be 5 hp, plus or minus 0.25 hp. Remember that the PDS is a dynamic document. While it is important to make it as complete as possible at the outset of design, do not hesitate to change it as you learn more as the design evolves. The PDS is a document that should always be up to date and reflect the current design.

3.8
SUMMARY

Problem definition in the engineering design process takes the form of identifying the needs of the customer that a product will satisfy. If the needs are not properly defined, then the design effort may be futile. This is especially true in product design, where considerable time and effort is invested in listening to and analyzing the "voice of the customer."

Collecting customer opinions on what they need from a product is done in many ways. For example, a marketing department research plan can include interviewing existing and target customers, implementing customer surveys, and analyzing warranty data on existing products. The design team recognizes that there are many classes of customer needs, and research data must be studied intently to determine which needs will motivate customers to select a new product. Some customer needs are identified as critical to quality and take on added priority for the design team.

Design teams describe products in terms of engineering characteristics: parameters, design variables, and constraints that communicate how the customer needs will be satisfied. More than one engineering characteristic will contribute to satisfying a single customer need. Engineering characteristics are discovered through benchmarking competing products, performing reverse engineering on similar products,

TABLE 3.4
PDS for Shot-Buddy Device after the Problem Description
and Need Identification Steps Are Complete

Product Design Specification: Shot-Buddy

Product Identification

- Basketball return that automatically directs ball to the shooter enabling effective practice shooting
- Fits all structure-mounted and free-standing, standard-size hoops
- User installation

Special Features

- Shooter wears sensor that enables return targeting
- Targeting works up to 3-point arc

Key Performance Targets

- Returns all made shots and missed shots falling within 8-inches of the hoop
- Returns basketball accurately and quickly to the user at any location on the court
- Powered by rechargeable batteries

User Training Required: NONE

Service Environment:

- Outdoor: −20 to 120°F
- Indoor: 50 to 80°F
- Up to 100% humidity

Key Project Deadlines

- Six month to finalize design
- Target advertising for holiday season

Physical Description:

- External Dimensions:
  - Catch area approximately 6 feet by 4 feet
  - Control housing approximately 2 feet wide, 2 feet long, and 10 inches tall
  - Return device approximately 2 feet by 2 feet
- Material: To be determined (TBD)
- Weight Targets:
  - Ball catching device <15 pounds
  - Base component <15 pounds

Manufacturing Specifications

- All framing and support components will be manufactured in house. Others will be COTS
- Suppliers: TBD

Market Identification

- The target market for this product will be Middle School and High School age users
- Initial Launch: Baltimore—DC metro area
- Initial production run 2500 units
- Year 2–3: based on market acceptance expand to nationwide market in 4th year
- Competing products:
  - Current products can only return a basketball to a very limited range of the court
  - No products involve the sensor technology
- Brand Name: Shot-Buddy

Financial Requirements:

- Pricing policy over life cycle:
  - Target manufacturing cost: $250
  - Estimated Retail Price: $500
- Warranty Policy: 1 year complete warranty
- Expected financial performance or rate of return on investment: TBD
- Level of capital investment required: TBD

Life Cycle Targets:

- Useful life 5 years and beyond
- Maintenance schedule: No maintenance required if sensors and control equipment are stored properly
- Reliability (mean time to failure): 5 years
- End of life strategy: Shot-Buddy will be recyclable with batteries requiring special handling

Social Political, and Legal Requirements

- Safety and environmental regulations will be followed
- Standards: Research federal regulations on sports equipment
- Safety and product liability: The only safety aspect of the Shot-Buddy is the installation process where a ladder might be involved to hang the device from the rim/backboard
- Intellectual property: Will investigate patent potential

and technical research. The TQM tool called Quality Function Deployment (QFD) is a well-defined process that will lead a design team in translating the important customer needs into critical-to-quality engineering characteristics. This enables the product development team to focus design effort on the right aspects of the product.

The House of Quality (HOQ) is the first step in QFD and is the most used in the product development process. The HOQ has a number of different configurations. There is a minimum number of "rooms" of the HOQ that must be completed to gain the benefits of the method. The HOQ will provide relative weight information for the engineering characteristics. Using this data the design team can determine which ECs are critical to quality (CTQ) and which should be set as constraints for concept generation. Other rooms of the HOQ can be used to identify EC correlations (Room 3) and assess competing products (Room 6).

The product design process results in a document called the Product Design Specification (PDS). The PDS is a living document that will be refined at each step of the PDP. The PDS is the single most important document in the design process as it describes the product and the market it is intended to satisfy.

NEW TERMS AND CONCEPTS

Affinity diagram	Engineering characteristics	Quality function deployment
Benchmarking	Ethnographic study	Reverse engineering
Constraint	Focus group	Survey instrument
Customer requirement	House of quality (HOQ)	TQM
Design parameter	Kano diagram	Value
Design variable	Pareto chart	Voice of the customer

BIBLIOGRAPHY

Customer Needs and Product Alignment

Mariampolski, H.: *Ethnography for Marketers: A Guide to Consumer Immersion,* Sage Publications, 2005.
Meyer, M. H., and A. P. Lehnerd: *The Power of Product Platforms,* The Free Press, New York, 1997.
Smith, P. G., and D. G. Reinertsen: *Developing Products in Half the Time: New Rules, New Tools,* 2d ed., Wiley, New York, 1996.
Ulrich, K. T., and S. D. Eppinger: *Product Design and Development,* 4th ed., McGraw-Hill, New York, 2007, Chap. 4.
Urban, G. L., and J. R. Hauser: *Design and Marketing of New Products,* 2nd ed., Prentice Hall, Englewood Cliffs, NJ, 1993.

Quality Function Deployment

Bickell, B. A., and K. D. Bickell: *The Road Map to Repeatable Success: Using QFD to Implement Change,* CRC Press, Boca Raton, FL, 1995.

Clausing, D.: *Total Quality Development,* ASME Press, New York, 1995.
Cohen, L.: *Quality Function Deployment,* Addison-Wesley, Reading, MA, 1995.
Day, R. G.: *Quality Function Deployment,* ASQC Quality Press, Milwaukee, WI, 1993.
Guinta, L. R., and N. C. Praizler: *The QFD Book,* Amacom, New York, 1993.
King, B.: *Better Designs in Half the Time,* 3d ed., GOAL/QPC, Methuen, MA, 1989.

Customer Requirements and PDS

Pugh S.: *Total Design,* Addison-Wesley, Reading, MA, 1990.
Ullman, D. G.: *The Mechanical Design Process,* 4th ed., McGraw-Hill, New York, 2009.

PROBLEMS AND EXERCISES

3.1 Select 10 products from a department store's online catalog for a supplier of household items (not clothing) and decide which needs in Maslow's hierarchy of human needs they satisfy. Then, identify the particular product features that make the products attractive to you. Divide your customer needs into the four categories described by Kano.

3.2 The transistor, followed by the microprocessor, is one of the most far-reaching products ever developed. Make a list of the major products and services that have been impacted by these inventions.

3.3 Take 10 minutes and individually write down small things in your life, or aspects of products that you use, that bother you. You can just name the product, or better yet, give an attribute of the product that "bugs you." Be as specific as you can. You are really creating a needs list. Combine this with other lists prepared by members of your design team. Perhaps you have created an idea for an invention.

3.4 Write a survey to determine the customers' wants for a microwave oven.

3.5 List a complete set of customer needs for cross-country skis to allow skiing on dirt or grass. Divide the list of customer needs into "must haves" and "wants."

3.6 Suppose you are the inventor of a new device called the helicopter. By describing the functional characteristics of the machine, list some of the societal needs that it is expected to satisfy. Which of these have come to fruition, and which have not?

3.7 Assume that a focus group of college students was convened to show them an innovative thumb drive memory unit and to ask what characteristics they wanted it to have. The comments were as follows:

- It needs to have enough memory to meet student needs.
- It should interface with any computer a student would encounter.
- It must have a reliability of near 100%.
- It should have some way to signal that it is working.

Translate these customer requirements into engineering characteristics of the product.

3.8 Complete the streamlined configuration of the House of Quality (i.e., Rooms 1, 2, 4, and 5) for a heating and air-conditioning design project. The customer requirements are lower operating costs; improved cash flow; managed energy use; increased occupant comfort; and easy to maintain. The engineering characteristics are an energy efficiency ratio of 10; zonal controls; programmable energy management system; payback 1 year; and 2-hour spare parts delivery.

3.9 A product design team is designing an improved flip-lid trash can such as that which would be found in a family kitchen. The problem statement is as follows:

> Design a user-friendly, durable, flip-lid trash can that opens and closes reliably. The trash can must be lightweight yet tip-resistant. It must combat odor, fit standard kitchen trash bags, and be safe for all users in a family environment.

With this information, and a little research and imagination where needed, construct a House of Quality (HOQ) for this design project.

3.10 Write a product design specification for the flip-lid trash can described in Prob. 3.9.

4

TEAM BEHAVIOR AND TOOLS

4.1
INTRODUCTION

Engineering design is really a "team sport." Certainly in the context of being an engineering student, there is so much to learn for your design project and so little time to do everything required for a successful design that being a member of a smoothly functioning team is clearly a major benefit. Also, as discussed in the next paragraph, the ability to work effectively in teams is highly prized in the business world. A team provides two major benefits: (1) a diversity of teammates with different educations and life experiences results in a knowledge base that is broader and often more creative than a single individual, and (2) by team members taking on different tasks and responsibilities, the work gets finished more quickly. Therefore, this chapter has three objectives:

- To provide time-tested tips and advice for becoming an effective team member.
- To introduce you to a set of problem-solving tools that you will find useful in carrying out your design project, as well as being useful in your everyday life.
- To emphasize the importance of project planning to success in design, and to provide you with some ideas of how to increase your skill in this activity.

A recent column in *The Wall Street Journal* was titled "Engineering Is Reengineered into a Team Sport." The article went on to say, "These firms want people who are comfortable operating in teams and communicating with earthlings who know nothing about circuit-board design or quantum mechanics." This is to emphasize that when industry leaders are asked what they would like to see changed in engineering curricula they invariably respond, "Teach your students to work effectively in teams." A more near-term reason for devoting this chapter to team behavior is that the engineering design courses for which this text is intended are mostly focused around team-based projects. All too often we instructors thrust you students into a team situation without providing proper understanding of what it takes to achieve a smoothly functioning team. Most often things work out just fine, but at a cost of extra hours of trial and error to find the best way to function as a team. Indeed, the greatest complaint that students have about project design courses is "it takes too much time." This chapter is designed to give you an understanding of the team-building process and to introduce you to some tools that people have found helpful in getting results through teams.

A team is a small number of people with complementary skills who are committed to a common purpose, performance goals, and approach for which they hold themselves mutually accountable.[1] There are two general types of teams: teams that do real work, like design teams, and teams that make recommendations. Both are important, but we focus here on the former. Most people have worked in groups, but a working group is not necessarily a team. A team is a high order of group activity. Many groups do not reach this level, but it is a goal truly worth achieving.

4.2
WHAT IT MEANS TO BE AN EFFECTIVE TEAM MEMBER

There is a set of attitudes and work habits that you need to adopt to be a good team member. First and foremost, you need *to take responsibility for the success of the team*. Without this commitment, the team is weakened by your presence. Without this commitment, you shouldn't be on the team.

Next, you need to *be a person who delivers on commitments*. This means that you consider membership on the team as something worthwhile and that you are willing to rearrange your job and personal responsibilities to satisfy the needs of the team. On occasions when you cannot complete an assignment, always notify the team leader as soon as possible so other arrangements can be made.

Much of the team activity takes place in meetings where members share their ideas. Learn to *be a contributor to discussions*. Some of the ways that you can contribute are by asking for explanations to opinions, guiding the discussion back on track, and pulling together and summarizing ideas.

Listening is an art that not all of us have learned to practice. Learn to *give your full attention to whomever is speaking and demonstrate this by asking helpful questions*. To help focus on the speaker, take notes and never do distracting things like reading unrelated material, using your mobile phone, walking around, or interrupting the speaker.

Develop techniques for getting your message across to the team. This means thinking things through briefly in your own mind before you speak. Always speak in a loud, clear voice. Have a positive message, and avoid "put-downs" and sarcasm. Keep focused on the point you are making. Avoid rambling discussion.

Learn to give and receive useful feedback. The point of a team meeting is to benefit from the collective knowledge and experience of the team to achieve an agreed-upon goal. Feedback is of two types. One is a natural part of the team discussion. The other involves corrective action for improper behavior by a member of the team that is best done after the meeting.

---

1. J. R. Katzenbach and D. K. Smith, *The Wisdom of Teams,* HarperCollins, New York, 1994.

The following are characteristics of an effective team:

- Team goals are as important as individual goals.
- The team understands the goals and is committed to achieving them.
- Trust replaces fear, and people feel comfortable taking risks.
- Respect, collaboration, and open-mindedness are prevalent.
- Team members communicate readily; diversity of opinions is encouraged.
- Decisions are made by consensus and have the acceptance and support of the members of the team.

We hope you will want to learn how to become an effective team member. Much of this chapter is devoted to helping you do that. Being a good team member is not a demeaning thing at all. Rather, it is a high form of group leadership. Being recognized as an effective team member is a highly marketable skill. Corporate recruiters say that the traits they are looking for in new engineers are communication skills, team skills, and problem-solving ability.

4.3
TEAM LEADERSHIP ROLES

We have just discussed the behavior that is expected of a good team member. Within a team, members assume different roles in addition to being active team members. The discussion that follows is oriented toward how teamwork is practiced in business and industry. However, student design teams differ in several important respects from a team in the business world: (1) the team members are all close to the same age and level of formal education, (2) they are peers and no one has authority over the other team members, and as a result, (3) they often prefer to work without a designated leader in a shared leadership environment.

An important role that is external to the team but vital to its performance is the *team sponsor*. The team sponsor is the manager who has the need for the output of the team. He or she selects the team leader, negotiates the participation of team members, provides any special resources needed by the team, and formally commissions the team.

The *team leader* convenes and chairs the team meetings using effective meeting management practices (see Sec. 4.5). He or she guides and manages the day-to-day activity of the team by tracking the team's accomplishment toward stated goals, helping team members to develop their skills, communicating with the sponsor about progress, trying to remove barriers toward progress, and helping to resolve conflict within the team. In general, there are three styles of team leadership: the traditional or autocratic leader, the passive leader, and the facilitative leader. Table 4.1 lists some major characteristics of these types of leaders. Clearly, the facilitative leader is the modern type of leader who we want to have leading teams.

Many teams in industry include a *facilitator*, a person trained in group dynamics who assists the leader and the team in achieving its objectives by coaching them in team skills and problem-solving tools, and assisting in data-collection activities. While the facilitator functions as a team member in most respects, she or he must

TABLE 4.1
Characteristics of Three Leadership Types

Traditional Leader	Passive Leader	Facilitative Leader
Directive and controlling	Hands off	Creates open environment
No questions—just do it	Too much freedom	Encourages suggestions
Retains all decision-making authority	Lack of guidance and direction	Provides guidance
Nontrusting	Extreme empowerment	Embraces creativity
Ignores input	Uninvolved	Considers all ideas
Autocratic	A figurehead	Maintains focus; weighs goals vs. criteria

remain neutral in team discussions and stand ready to provide interventions to attain high team productivity and improved participation by team members or, in extreme situations, to resolve team disputes. A key role of the facilitator is to keep the group focused on its task. When a facilitator is not available the team leader must take on these responsibilities.

Suggestions on organization of student design teams and the duties that need to be shared by the team members can be found in the document Team Organization and Duties at www.mhhe.com/dieter.

4.4
TEAM DYNAMICS

Students of team behavior have observed that most teams go through five stages of team development.[1]

1. *Orientation* (*forming*): The members are new to the team. They are probably both anxious and excited, yet unclear about what is expected of them and the task they are to accomplish. This is a period of tentative interactions and polite discourse, as the team members undergo orientation and acquire and exchange information.

2. *Dissatisfaction* (*storming*): Now the challenges of forming a cohesive team become real. Differences in personalities, working and learning styles, cultural backgrounds, and available resources (time to meet, access to and agreement on the meeting place, access to transportation, etc.) begin to make themselves known. Disagreement, even conflict, may break out in meetings. Meetings may be characterized by criticism, interruptions, poor attendance, or even hostility.

3. *Resolution* (*norming*): The dissatisfaction abates when team members establish group norms, either spoken or unspoken, to guide the process, resolve conflicts,

---

1. R. B. Lacoursiere, *The Life Cycle of Groups,* Human Service Press, New York, 1980; B. Tuckman, "Developmental Sequence in Small Groups," *Psychological Bulletin,* no. 63, pp. 384–99, 1965.

and focus on common goals. The norms are given by rules of procedure and the establishment of comfortable roles and relationships among team members. The arrival of the resolution stage is characterized by greater consensus seeking,[1] and stronger commitment to help and support each other.

4. *Production* (*performing*): This is the stage of team development we have worked for. The team is working cooperatively with few disruptions. People are excited and have pride in their accomplishments, and team activities are fun. There is high orientation toward the task, and demonstrable performance and productivity.

5. *Termination* (*adjourning*): When the task is completed, the team prepares to disband. This is the time for joint reflection on how well the team accomplished its task, and reflection on the functioning of the team. In addition to a report to the team sponsor on results and recommendations of the team, another report on team history and dynamics may be written to capture the "lessons learned" to benefit future team leaders.

It is important for teams to realize that the dissatisfaction stage is perfectly normal and that they can look forward to its passing. Many teams experience only a brief stage 2 and pass through without any serious consequences. However, if there are serious problems with the behavior of team members, they should be addressed quickly.

One way or another, a team must address the following set of team challenges:

- *Safety*: Are the members of the team safe from destructive personal attacks? Can team members freely speak and act without feeling threatened?
- *Inclusion*: Team members need to be allowed equal opportunities to participate. Rank is not important inside the team. Make special efforts to include new, quiet members in the discussion.
- *Appropriate level of interdependence*: Is there an appropriate balance between the individuals' needs and the team needs? Is there a proper balance between individual self-esteem and team allegiance?
- *Cohesiveness*: Is there appropriate bonding between members of the team?
- *Trust*: Do team members trust each other and the leader?
- *Conflict resolution*: Does the team have a way to resolve conflict?
- *Influence*: Do team members or the team as a whole have influence over members? If not, there is no way to reward, punish, or work effectively.
- *Accomplishment*: Can the team perform tasks and achieve goals? If not, frustration will build up and lead to conflict.

It is important for the team to establish some guidelines for working together. Team guidelines will serve to ameliorate the dissatisfaction stage and are a necessary condition for the resolution stage. The team should begin to develop these guidelines early in the orientation stage. Team guidelines are often given in a Team Charter, which the team develops and then agrees to with their signatures. An example is given on the text website www.mhhe.com/dieter under Chapter 4.

---

1. Consensus means general agreement or accord. Consensus does not require 100 percent agreement of the group. Neither is 51 percent agreement a consensus.

TABLE 4.2
Different Behavioral Roles Found in Groups

Helping Roles		Hindering Roles
Task Roles	**Maintenance Roles**	
Initiating: proposing tasks; defining problem	Encouraging	Dominating: asserting authority or superiority
Information or opinion seeking	Harmonizing: attempting to reconcile disagreement	Withdrawing: not talking or contributing
Information or opinion giving	Expressing group feeling	Avoiding: changing the topic; frequently absent
Clarifying	Gate keeping: helping to keep communication channels open	Degrading: putting down others' ideas; joking in barbed way
Summarizing	Compromising	Uncooperative: Side conversations: whispering and private conversations across the table
Consensus testing	Standard setting and testing: checking whether group is satisfied with procedures	

People play various roles during a group activity like a team meeting. It should be helpful in your role as team leader or team member to recognize some of the behavior listed briefly in Table 4.2. It is the task of the team leader and facilitator to try to change the hindering behavior and to encourage team members in their various helping roles.

4.5
EFFECTIVE TEAM MEETINGS

Much of the work of teams is accomplished in team meetings. It is in these meetings that the collective talent of the team members is brought to bear on the problem, and in the process, all members of the team "buy in" to the problem and together develop a solution. Students who complain about design projects taking too much time often are really expressing their inability to organize their meetings and manage their time effectively.

At the outset it is important to understand that an effective meeting requires planning. This is the responsibility of the person who will lead the meeting. Meetings should begin on time and last for about 90 minutes, the optimum time to retain all members' concentration. A meeting should have a written agenda, with the name of the designated person to present each topic and an allotted time for discussion of the topic. If the time allocated to a topic proves to be insufficient, it can be extended by the consent of the group, or the topic may be given to a small task group to study further and report back at the next meeting of the team. In setting the agenda, items of greatest urgency should be placed first on the agenda.

The team leader directs but does not control discussion. As each item comes up for discussion on the agenda, the person responsible for that item makes a clear statement of the issue or problem. Discussion begins only when it is clear that every participant understands what is intended to be accomplished regarding that item. One reason for keeping teams small is that every member has an opportunity to contribute to the discussion. Often it is useful to go around the table in a round-robin fashion, asking each person for ideas or solutions, while listing them on a flip chart or blackboard. No criticism or evaluation should be given here, only questions for clarification. Then the ideas are discussed by the group, and a decision is reached. It is important that this be a group process and that an idea become disassociated from the individual who first proposed it.

Decisions made by the team should be consensus decisions. When there is a consensus, people don't just go along with the decision, they invest in it. Arriving at consensus requires that all participants feel that they have had their full say. Try to help team members to avoid the natural tendency to see new ideas in a negative light. However, if there is a sincere and persuasive negative objector, try to understand their real objections. Often they have important substance, but they are not expressed in a way that they can be easily understood. It is the responsibility of the leader to keep summing up for the group the areas of agreement. As discussion advances, the area of agreement should widen. Eventually you come to a point where problems and disagreement seem to melt away, and people begin to realize that they are approaching a decision that is acceptable to all.

4.5.1 Helpful Rules for Meeting Success

1. Pick a regular meeting location and try not to change it.
2. Pick a meeting location that: (a) is agreeable, accessible to all, and conducive to work, (b) has breathing room when there is full attendance, (c) has a pad and easel in the room, (d) isn't too hot, too cold, or too close to noisy distractions.
3. Regular meeting times are not as important as confirming the time of meetings. Once a meeting time has been selected, confirm it immediately by e-mail. Remain flexible on selecting meeting length and frequency. Shape the time that the team spends together around the needs of the work to be accomplished. This being said, it is important for every student design team to have a two-hour block of time when they can meet weekly without interference from class or work schedules.
4. Send an e-mail reminder to team members just before the first of several meetings.
5. If you send materials out in advance of a meeting, bring extra copies just in case people forget to bring theirs, or they did not arrive.
6. Start on time, or no later than 5 to 7 minutes from the stated starting time.
7. Pass out an agenda at the beginning of the meeting and get the team's concurrence with the agenda. Start every meeting with "what are we trying to accomplish today?"
8. Rotate the responsibility for writing summaries of each meeting. The summaries should document: (a) when did the team meet and who attended, (b) what were the issues discussed (in outline form), (c) decisions, agreements, or apparent consensus on issues, (d) next meeting date and time, (e) action items, with assignment to team members for completion by the next meeting. In general, meeting summaries should not exceed one page, unless you are attaching results from

group brainstorming, lists of issues, ideas, and so on. Meeting summaries should be distributed by the assigned recorder within 48 hours of the meeting.

9. Notice members who come late, leave early, or miss meetings. Ask if the meeting time is inconvenient or if competing demands are keeping them from meetings.

10. Observe team members who are not speaking. Near the end of the discussion, ask them directly for their opinion on an issue. Consult them after the meeting to be sure that they are comfortable with the team and discussion.

11. Occasionally use meeting evaluations (perhaps every second or third meeting) to gather anonymous feedback on how the group is working together. Meeting evaluations should be turned in to the facilitator, who should summarize the results, distribute a copy of those results to everyone, and lead a brief discussion at the next meeting on reactions to the meeting evaluations and any proposed changes in the meeting format.

12. Do not bring guests or staff support or add team members without seeking the permission of the team.

13. Avoid canceling meetings. If the team leader cannot attend, an interim discussion leader should be designated.

14. End every meeting with an "action check": (a) What did we accomplish/agree upon today? (b) What will we do at the next meeting? (c) What is everyone's "homework," if any, before the next meeting?

15. Follow up with any person who does not attend, especially people who did not give advance notice. Call to update them about the meeting and send them any materials that were passed out at the meeting. Be sure they understand what will take place at the next meeting.

For smooth team operation, it is important to:

- Create a team roster. Ask team members to verify mailing addresses, e-mail addresses, names, and phone numbers or a team website. Include information about the team sponsor. Use e-mail addresses to set up a distribution list for your team.

- Organize important material in team binders or a team website. Include the team roster, team charter, essential background information, data, critical articles, and so on.

A well-functioning team achieves its objectives quickly and efficiently in an environment that induces energy and enthusiasm. However, it would be naive to think that everything will always go well with teams. Suggestions for dealing with people problems in teams can be found in the text website www.mhhe.com/dieter.

4.6
PROBLEM-SOLVING TOOLS

In this section we present some problem-solving tools that are useful in any problem situation, whether as part of your design project or in any other business situation—as in trying to identify new sources of income for the student ASME chapter. These tools are especially well suited for problem solving by teams. They have a strong element of common sense and do not require sophisticated mathematics, so they can be learned

TABLE 4.3
Problem-Solving Tools

Problem Definition	Cause Finding	Solution Finding and Implementation
Brainstorming (see Sec. 6.3.1)	*Gathering data* Interviews (see Sec. 3.2.2)	*Solution finding* Brainstorming (see Sec. 6.3.1)
Affinity diagram	Focus groups (see Sec. 3.2.2)	How-how diagram
Pareto chart	Surveys (see Sec. 3.2.2)	Concept selection (see Chap. 7)
	Analyzing data Checksheet	*Implementation* Force field analysis
	Histogram	Written implementation plan
	Flowchart	
	Pareto chart	
	Search for root causes Cause-and-effect diagram	
	Why-why diagram	
	Interrelationship digraph	

and practiced by any group of educated people. They are easy to learn, but a bit tricky to learn to use with real expertise. These tools have been codified within the discipline called *total quality management*.[1]

Many strategies for problem solving have been proposed. The one that we have used and found effective is a simple three-phase process.[2]

- Problem definition
- Cause finding
- Solution finding and implementation

Table 4.3 lists the tools that are most applicable in each phase of the problem-solving process. Most are described below in examples that illustrate their use. A few are found in other sections of this text.

We view problem definition as the critical phase in any problem situation. A problem can be defined as the difference between a current state and a more desirable state. Often the problem is posed by management or the team sponsor, but until the team redefines it for itself, the problem has not been defined. The problem should be based on data, which may reside in the reports of previous studies, or in surveys or tests that the team undertakes to define the problem. In working toward an acceptable problem definition, the team uses *brainstorming* and the *affinity diagram*.

The objective of the cause-finding stage is to identify all of the possible causes of the problem and to narrow them down to the most probable *root causes*. This phase

1. J. W. Wesner, J. M. Hiatt, and D. C. Trimble, *Winning with Quality: Applying Quality Principles in Product Development,* Addison-Wesley, Reading, MA, 1995; C. C. Pegels, *Total Quality Management,* Boyd & Fraser, Danvers, MA, 1995; W. J. Kolarik, *Creating Quality,* McGraw-Hill, New York, 1995; S. Shiba, A. Graham, and D. Walden, *A New American TQM,* Productivity Press, Portland, OR, 1993.
2. Ralph Barra, *Tips and Techniques for Team Effectiveness,* Barra International, PO Box 325, New Oxford, PA.

starts with the gathering of data and analyzing the data with simple statistical tools. The first step in data analysis is the creation of a *checksheet* in which data is recorded by classifications. Numeric data may lend itself to the construction of a histogram, while a *Pareto chart* or simple bar chart may suffice for other situations. Run charts may show correlation with time, and scatter diagrams show correlation with critical parameters. Once the problem is understood with data, the *cause-and-effect diagram* and the *why-why diagram* are effective tools for identifying possible causes of the problem. The *interrelationship digraph* is a useful tool to help identify root causes.

With the root causes identified, the objective of the solution-finding phase is to generate as many ideas as possible as to how to eliminate the root causes. Brainstorming clearly plays a role, but this is organized with a *how-how diagram*. A concept selection method such as the Pugh chart (Sec. 7.5) is used to select among the various solutions that evolve. With the best solutions identified, the pros and cons of a strategy for implementing them is identified with the help of *force field analysis*. Finally, the specific steps required to implement the solution are identified and written into an *implementation plan*. Then, as a last step, the implementation plan is presented to the team sponsor.

We have outlined briefly a problem-solving strategy that utilizes a number of tools that are often associated with total quality management (TQM).[1] They are useful for finding solutions to problems of a technical, business, organization, or personal nature. We present these problem-solving tools in the order that they would usually be used for a problem that might be found in an organizational setting. Then, in Example 4.1, we use them again to solve a technical problem.

Step 1. Problem Definition

Problem Statement. A group of engineering honors students[2] was concerned that more engineering seniors were not availing themselves of the opportunity to do a senior research project. All engineering departments listed this as a course option, but only about 5 percent of the students chose this option. To properly define the problem, the team brainstormed about the question, "Why do so few senior engineering students choose to do a research project?"

Brainstorming. Brainstorming is a group technique for generating ideas in a nonthreatening atmosphere. It is a group activity in which the collective creativity of the group is tapped and enhanced. The objective of brainstorming is to generate the *greatest number of alternative ideas* from the uninhibited responses of the group. Brainstorming is most effective when it is applied to specific rather than general problems. It is frequently used in the problem definition phase and solution-finding phase of problem solving.

---

1. M. Brassard and D. Ritter, *The Memory Jogger™ II, A Pocket Guide of Tools for Continuous Improvement,* GOAL/QCP, Methuen, MA, 1994; N. R. Tague, *The Quality Toolbox,* ASQC Quality Press, Milwaukee, WI, 1995.

2. The team of students making this study was Brian Gearing, Judy Goldman, Gebran Krikor, and Charnchai Pluempitiwiriyawej. The results of the team's study have been modified appreciably by the authors.

There are four fundamental brainstorming principles.

1. *Criticism is not allowed.* Any attempt to analyze, reject, or evaluate ideas is postponed until after the brainstorming session. The idea is to create a supportive environment for free-flowing ideas.
2. *Ideas brought forth should be picked up and built upon by the other members of the team.* Individuals should focus only on the positive aspects of ideas presented by others. The group should attempt to create chains of mutual associations that result in a final idea that no one has generated alone. All output of a brainstorming session is to be considered a group result.
3. *Participants should divulge all ideas entering their minds without any constraint.* All members of the group should agree at the outset that a seemingly wild and unrealistic idea may contain an essential element of the ultimate solution.
4. *A key objective is to provide as many ideas as possible within a relatively short time.* It is not unusual for a group to generate 20 to 30 ideas in a half hour of brainstorming. Obviously, to achieve that output the ideas are described only roughly and without details.

It is helpful for a brainstorming session to have a facilitator to control the flow of ideas and to record the ideas. Start with a clear, specific written statement of the problem. Allow a few minutes for members to collect their thoughts, and then begin. Go around the group, in turn, asking for ideas. Anyone may pass, but all should be encouraged to contribute. Build on (piggyback on) the ideas of others. Encourage creative, wild, or seemingly silly notions. There is no questioning, discussion, or criticism of ideas. Generally the ideas build slowly, reach a point where they flow faster than they can be written down, and then fall off. When the group has exhausted all ideas, stop. A good format for brainstorming is to write ideas on large sticky notes and place them on the wall where the entire team can view them and hopefully will build upon them. This procedure also facilitates performing the next step in problem definition, the affinity diagram. A variety of brainstorming in which ideas are written down without public disclosure is called *brainwriting* or 6-3-5 brainstorming (www.ifm.eng.cam.ac.uk/dng/tools/project/brainwrite.html).

When the student team brainstormed, they obtained the following results.

Problem: Why do so few engineering seniors do a research project?
    Students are too busy.
    Professors do not talk up research opportunity.
    Students are thinking about getting a job.
    Students are thinking about getting married.
    They are interviewing for jobs.
    They don't know how to select a research topic.
    I'm not interested in research. I want to work in manufacturing.
    I don't know what research the professors are interested in.
    The department does not encourage students to do research.
    I am not sure what research entails.

It is hard to make contact with professors.
I have to work part-time.
Pay me and I'll do research.
I think research is boring.
Lab space is hard to find.
Faculty just use undergraduates as a pair of hands.
I don't know any students doing research.
I haven't seen any notices about research opportunities.
Will working in research help me get into grad school?
I would do it if it was required.

Affinity Diagram. The affinity diagram identifies the inherent similarity between items. It is used to organize ideas, facts, and opinions into natural groupings. If you have used sticky memo notes, a good way to start building the affinity diagram is to put all of the brainstorming responses on the wall in no particular order. Each idea is "scrubbed," that is each person explains what they wrote on each note so that each team member understands it the same way. This often identifies more than one note with the same thought, or reveals cards that have more than one idea on them. If this happens, additional notes are made up. Then the notes or cards are sorted into columns of loosely related groupings. As the nature of a grouping becomes clear, place a header card at the top of each column to denote its content. Also, add a column header "Other" to catch the outliers. If an idea keeps being moved between two groups because of disagreement as to where it belongs, make a duplicate and put the same idea in both groups.

Unlike brainstorming, building the affinity diagram is a time for plenty of discussion so that everyone understands what is being proposed. Team members may be called upon to defend their idea or where it has been placed in the diagram. The creation of affinity groups serves several purposes. First, it breaks a problem down into its major issues; subdivision of a problem is an important step toward solution. Second, the act of the team assembling the affinity diagram stimulates a clear understanding of the ideas that were put forth hurriedly in the brainstorming session, and often leads to new ideas through clarification or combination. It also provides an opportunity to abandon obviously poor or frivolous ideas.

The team arranged their brainstorming ideas into the following *affinity diagram*. Note that in the discussion a few of the ideas were judged to be not worthy of further consideration, but rather than drop them from the list, they have been placed in brackets to indicate they have been removed from active consideration. In this way, none of the ideas proposed in brainstorming have been lost.

Time constraints
　　Students are too busy.
　　Students are interviewing for jobs.
　　I have to work part-time.
Faculty issues
　　Professors don't talk up research opportunities.
　　The department does not encourage students to do research.

It is hard to make contact with professors.

Faculty just use undergraduates as a pair of hands.

Lack of interest

Students are thinking about getting a job.

[They are thinking about getting married.]

I'm not interested in research. I want to work in manufacturing.

[Pay me and I'll do research.]

I think research is boring.

I would do it if it was required. (2)

Lack of information

They don't know how to select a research topic.

I don't know what research the professors are interested in.

I'm not sure what research entails.

I don't know any students doing research.

I haven't seen any notices about research opportunities.

Will working in research help me get into graduate school?

Other

Lab space is hard to find.

To focus more clearly on the problem definition, the team took the results of their brainstorming, as represented by the affinity diagram, and narrowed the problem down to the seven issues (A through G) shown in the following table. Note that four main subheadings from the affinity diagram are represented in this list, along with three issues from within the subheadings that the team thought were worthy of further consideration.

Issues		Brian	Judy	Gebran	Charn	Total
A.	Lack of readily available information about research topics	3	3	4	4	14
B.	Lack of understanding of what it means to do research	2	5		1	8
C.	Time constraints			5		5
D.	Lack of a strong tradition for undergraduate research					0
E.	Lack of a mandatory research course	2				2
F.	Lack of student interest				3	3
G.	Lack of incentives	3	2	1	2	8

The team then practiced *list reduction* using a method called *multivoting*. Each team member received 10 votes that they could distribute any way they wished among the seven issues. Note that Gebran felt strongly that time constraint was the main issue and placed half of his votes on this topic. The other team members distributed their votes more widely. From this multivoting three issues stood out—A, B, and G.

A second round of list reduction was conducted by simple *ranking*. Each team member was asked to pick which of the three issues they favored (3), which was lowest in importance (1), and which was intermediate in importance (2). The results were as follows:

Ideas		Brian	Judy	Gebran	Charn	Total
A.	Lack of readily available information about research topics	2	1	1	2	6
B.	Lack of understanding of what it means to do research	3	3	3	3	12
C.	Lack of incentives	1	2	2	1	6

As a result of a second round of ranking, the team of four students formed the tentative impression that a lack of understanding on the part of undergraduates about what it means to do research is the strongest contributor to the low participation by students in research projects. This is at variance with their earlier ranking. However, the two issues, lack of understanding of what it means to do research and lack of information about possible research topics are really part of a large topic of lack of information concerning research. Therefore, the problem statement was formulated as follows:

Revised Problem Statement

The lack of information among undergraduate engineering students about what it means to do research, including the lack of information on specific research opportunities with faculty, is responsible for the low participation of students in elective research courses.

However, the team realized that they were but four students, whose ideas might be different from a wider group of engineering students. They realized that a larger database was needed as they went into the cause-finding stage of problem solving.

Step 2. Cause Finding

Survey. One hundred surveys were distributed to senior engineering students. The questions were based on the A to G issues listed previously, with issue D omitted. The students were asked to rank the importance of each issue on a 1–7 Likert scale, and they were asked whether they were interested doing a research project. Of the 75 surveys received from undergraduate students, a surprising 93 percent said they were interested in doing a research project, while 79 percent felt there was a lack of undergraduate involvement in research. A very similar survey was given to faculty.

Pareto Chart. The results of the survey are best displayed by a *Pareto chart*. This is a bar chart used to prioritize causes or issues, in which the cause with the highest frequency of occurrence is placed at the left, followed by the cause with the next frequency of occurrence, and so on. It is based on the Pareto principle, which states that a few causes account for most of the problem, while many other causes are relatively unimportant. This is often stated as the 80/20 rule, that roughly 80 percent of the problem is caused by only 20 percent of the causes, or 80 percent of the sales come from

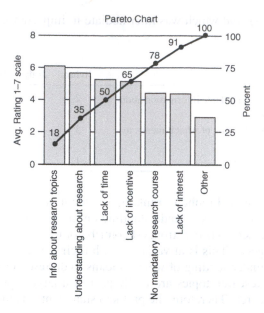

FIGURE 4.1
Pareto chart for average rating of reasons why undergraduate students do not do research projects. Based on responses from 75 students.

20 percent of the customers, or 80 percent of the tax income comes from 20 percent of the taxpayers, etc. A Pareto chart is a way of analyzing the data that identifies the *vital few in contrast to the trivial many.*

The Pareto chart for the student ranking of the causes why they do not do research is shown in Fig. 4.1. Lack of understanding of what it means to do research has moved to second place, to be replaced in first place by "lack of information about research topics." However, if one thinks about these results one would conclude that "no mandatory research course" is really a subset of "lack of understanding about research," so that this remains the number one cause of the problem. It is interesting that the Pareto chart for the faculty surveys showed lack of facilities and funding, and lack of incentives, in the one/two position. Otherwise the order of causes of the problem was about the same. Referring again to Fig. 4.1, note that this contains another piece of information in addition to relative importance. Plotted along the right axis is the cumulative percent of responses. We note that the first five categories (first four when the above correction is made) contain 80 percent of the responses.

Cause-and-Effect Diagram. The cause-and-effect diagram, also called the fishbone diagram (after its appearance), or the Ishikawa diagram (after its originator), is a powerful graphical way of identifying the factors that cause a problem. It is used after the team has collected data about possible causes of the problem. It is often used in conjunction with brainstorming to collect and organize all possible causes and converge on the most probable root causes of the problem.

Constructing a cause-and-effect diagram starts with writing a clear statement of the problem (effect) and placing it in a box to the right of the diagram. Then the backbone of the "fish" is drawn horizontally out from this box. The main categories of causes, "ribs of the fish," are drawn at an angle to the backbone, and labeled at the ends. Usually these end labels are categories specific to the problem that come mainly from the headers of the affinity diagram. Sometimes more generic categories such as methods, machines (equipment), materials, and people for a problem dealing with a production process, and policies, procedures, plant (equipment and space), and people for a service-related or organizational problem. Ask the team, "What causes this?" and record the cause, not the symptom, along one of the ribs. Dig deeper, and ask what causes the cause you just recorded, so the branches develop subbranches and the whole chart begins to look like the bones of a fish. A good fishbone diagram should go down three levels. In recording ideas from the brainstorming session, be succinct but use problem-oriented statements to convey the sense of the problem. As the diagram builds up, look for possible root causes. One way to identify root causes is to look for causes that appear frequently within or across main categories.

Figure 4.2 shows the cause-and-effect diagram generated by the students to understand the causes for the low student involvement in research. We note that time pressures caused by heavy course loads and necessity to work part-time are one possible root cause, while others center around the lack of understanding of students about what it means to do research and the lack of appreciation by faculty of student interest in doing research.

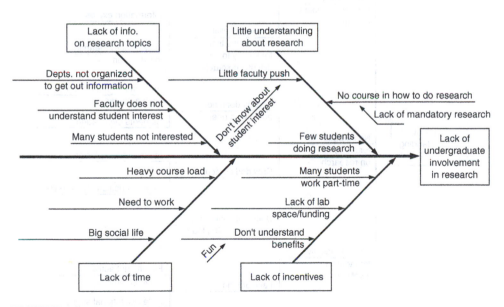

FIGURE 4.2
Cause-and-effect diagram for lack of undergraduate student involvement in research.
Note: Most third level causes are omitted in this diagram for clarity.

Why-Why Diagram. To delve deeper into root causes, we turn to the why-why diagram. The why-why diagram can be used as an alternative to the cause-and effect diagram, but more commonly it is used to dig deeper about one of the more likely root causes. This is a tree diagram, which starts with the problem and asks, "Why does this problem exist?" in order to develop a tree with a few main branches and several smaller branches. The team continues to grow the tree by repeatedly asking "why" until patterns begin to show up. Root causes are identified by causes that begin to repeat themselves on several branches of the why-why tree. The why-why diagram should extend to four levels, counting the problem statement as the first level.

The Pareto chart, when reinterpreted, shows that student lack of understanding about research was the most important cause of low student participation in research. The cause-and-effect diagram also shows this as a possible root cause. To dig deeper we construct the why-why diagram shown in Fig. 4.3. This begins with the clear statement of the problem. The lack of understanding about research on the part of the undergraduates is two-sided: the faculty doesn't communicate with the students about opportunities, and the students don't show initiative to find out about it. The team, in asking why, came up with three substantial reasons. Again, they asked why, about each of these three causes, and asking why yet a third time builds up a tree of causes.

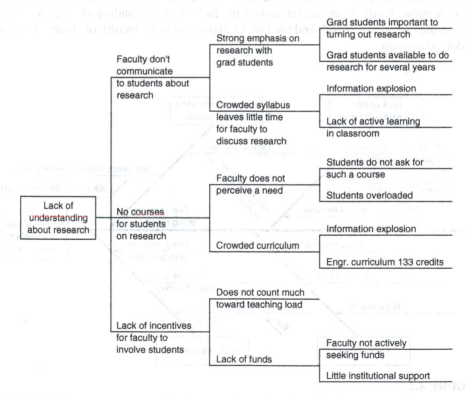

FIGURE 4.3
Why-why diagram for lack of student understanding about research.

At this stage we begin to see patterns of causes appearing in different branches of the tree—a sign that these are possible root causes. These are:

- Students and curriculum are overloaded.
- The information explosion is a major cause of the above.
- The faculty doesn't perceive a need to provide information about research.
- The faculty perceive a low student interest in doing research.
- A lack of resources, funding, and space limits faculty involvement in under-graduate research.

Narrowing down this set of causes to find the root cause is the job of the next tool.

Interrelationship Digraph.   This is a tool that explores the cause-and-effect relationships among issues and identifies the root causes. Start with a clear statement of the problem. The causes that you examine with the IR digraph will be suggested by common issues appearing in the fishbone or why-why diagram, or that are clearly defined by the team as being important. Generally try to limit the possible root causes to six. The possible root causes are laid out in a large circular pattern (Fig. 4.4). The cause and influence relationships are identified by the team between each cause or factor in turn. Starting with A we ask whether a causal relationship exists between A and B, and if so, whether the direction is stronger from A to B or B to A. If the causal relationship is stronger from B to A, then we draw an arrow in that direction. Next we explore the relationship between A and C, A and D, etc., in turn, until causal relationships have been explored between all of the factors. Note that there will not be a causal relationship between all factors. For each cause or factor, the number of arrows going in and coming out should be recorded. A high number of outgoing arrows indicates the cause or factor is a root cause or driver. A factor with a high number of incoming arrows indicates that it is a key indicator and should be monitored as a measure of improvement. To aid in making good decisions about relationships, write a defining sentence or statement about each possible root cause. Usually short one- or two-word statements are not specific enough and lead to fuzzy decisions as to whether a relationship exists between a pair of causes.

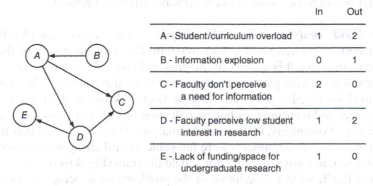

	In	Out
A - Student/curriculum overload	1	2
B - Information explosion	0	1
C - Faculty don't perceive a need for information	2	0
D - Faculty perceive low student interest in research	1	2
E - Lack of funding/space for undergraduate research	1	0

FIGURE 4.4
Interrelationship digraph to identify root causes from why-why diagram (Fig. 4.3).

In Fig. 4.4, the root causes are the overloaded students and curriculum, and the fact that the faculty perceive that there is a low undergraduate student interest in doing research. The key input is that the faculty do not perceive a need to supply information on research to the undergraduates. Solutions to the problem should then focus on ways of reducing student overload and developing a better understanding of the student interest in doing research. It was decided that reducing student overload had to precede any efforts to change faculty minds that students are not interested in doing research.

Step 3. Solution Planning and Implementation

While this is the third of three phases in the problem-solving process, it does not consume one-third of the time in the problem-solving process. This is because, having identified the true problem and the root causes, we now are most of the way home to a solution. The objective of solution finding is to generate as many ideas as possible on "how" to eliminate the root causes and to converge on the best solution. To do this we first employ brainstorming and then use multivoting or other evaluation methods to arrive at the best solution. The concept-selection method and other evaluation methods are discussed in Chap. 7.

How-How Diagram. A useful technique for suggesting solutions is the how-how diagram. Like the why-why diagram, the how-how diagram is a tree diagram, but it starts with a proposed solution and asks the question, "How do we do that?" The how-how diagram is best used after brainstorming has generated a set of solutions and an evaluation method has narrowed them to a small set.

A how-how diagram is constructed for the question, "How can we reduce the overload on students?" Brainstorming and multivoting had shown the main issues to be:

- Curriculum reform
- Student time management
- Student and faculty financial issues

Specific solutions that would lead to improvements in each of these areas are recorded in Fig. 4.5. Study of the first level of solutions—curriculum reform, helping students improve time management skills, and financial issues—showed that the only broad solution that would reduce student overload was curriculum reform.

Force Field Analysis. Force field analysis is a technique that identifies those forces that both help (drive) and hinder (restrain) the implementation of the solution of a problem. In effect, it is a chart of the pros and cons of a solution, and as such, it helps in developing strategies for implementation of the solution. This forces team members to think together about all the aspects of making the desired change a permanent change, and it encourages honest reflection on the root causes of the problem and its solution. Fortunately, the force field analysis, Fig. 4.6, showed that the college and higher education environments were favorable toward changing the curriculum.

The first step in constructing the force field diagram (Fig. 4.6) is to draw a large T. At the top of the T, write a description of the problem that is being addressed. To the far right of the T, write a description of the ideal solution that we would like to achieve. Participants then list forces (internal and external) that are driving the organization

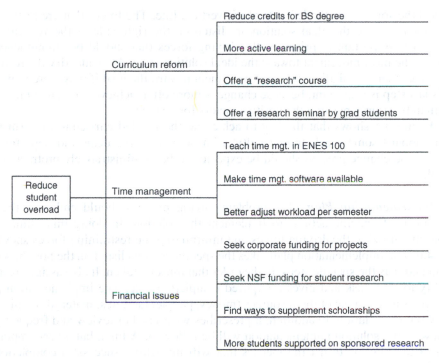

FIGURE 4.5
How-how diagram for problem of reducing student overload, so more students will be able to engage in research projects.

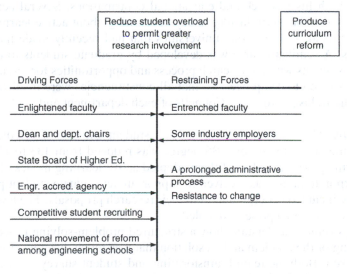

FIGURE 4.6
Force field diagram for implementing solutions to reducing student overload.

toward the solution on the left side of the vertical line. The forces that are restraining movement toward the ideal solution are listed on the right side of the vertical line. Often it is important to prioritize the driving forces that should be strengthened to achieve the most movement toward the ideal solution state. Also, identify the restraining forces that would allow the most movement toward the goal if they were removed. This last step is important, because change is more often achieved by removing barriers than by simply pushing the positive factors for change.

Figure 4.6 shows that the key to achieving the needed curriculum reform is to bring aboard some recalcitrant faculty, with help from the dean and departmental chairs. The change process should be expected to be administratively protracted, but doable.

Implementation Plan. The problem-solving process should end with the development of specific actions to implement the solution. In doing this, think hard about maximizing the driving forces and minimizing the restraining forces shown in Fig. 4.6. The implementation plan takes the specific actions listed on the how-how diagram and lists the specific steps, in the order that must be taken. It also assigns responsibility to each task and gives a required completion date. The implementation plan also gives an estimate of the resources (money, people, facilities, material) required to carry out the solution. In addition, it prescribes what level of review and frequency of review of the solution implementation will be followed. A final, but a very important part of the plan, is to list the metrics that will measure a successful completion of the plan.

The implementation plan for reducing the overload on the students by introducing a new curriculum is shown in Fig. 4.7. A Curriculum Action Team was established by the dean, with representation from both faculty and undergraduate students. The team leader was a distinguished faculty member who was recognized widely throughout the College for both his research and educational contributions. Several activities were created to involve the entire faculty: a day of learning about active learning methods and seminar speakers from other universities that had recently made major curriculum changes. A seminar course was developed by graduate students to acquaint undergraduate students with the research process and opportunities for research. Careful attention was given to due process so that all constituencies were involved. One such group was the Industry Advisory Councils of each department and the College.

Epilogue. This was not just an isolated student exercise. Over the next three years the number of credits for a BS degree was reduced from 133 to 122 credits in all engineering programs. Most of them adopted active learning modes of instruction. A major corporate grant was received to support undergraduate student projects, and many faculty included undergraduates in their research proposals. The level of student participation in research projects doubled.

It is important to understand how a structured problem-solving process led to an understanding of the problem and its solution that was much different from that originally perceived. Both the team brainstorming and student surveys viewed the cause of the problem as lack of information about the process of doing research and about actual areas in which research could be conducted. Yet the root cause analysis pointed

IMPLEMENTATION PLAN

Date:8/10/00

PROBLEM STATEMENT: Increase the undergraduate student participation in research.

PROPOSED SOLUTION: Create an action team of faculty and students within the college to produce major curriculum reform, to include reduction of credits for the BS degree from 133 to 123 credits, more teaching by active learning, and more opportunity for undergraduate students to do research.

SPECIFIC STEPS:

	Responsibility	Completion date
1. Create curriculum reform action team	Dean	9/30/00
2. Discuss issues with Faculty Council/Dept. Chairs	Dean	10/30/00
3. Hold discussion with dept. faculty	Team	11/15/00
4. Discuss with College Industrial Advisory Council	Dean/Team	11/26/00
5. Discuss with Student Council	Team	11/30/00
6. Day of learning about active learning	Team	1/15/01
7. Dept. curriculum committees begin work	Dept. Chairs	1/30/01
8. Teach "research course" as honors seminar	Team	5/15/01
9. Organize "research seminar ," taught by grad students	Team	5/15/01
10. Preliminary reports by dept. curriculum committees	Dean/Team	6/2/01
11. Fine-tuning of curriculum changes	Curric. Com.	9/15/01
12. Faculty votes on curriculum	Dept. Chairs	10/15/01
13. Submittal of curriculum to Univ. Senate	Dean	11/15/01
14. Vote on curriculum by Univ. Senate		2/20/02
15. Implementation of new curriculum	Dean/Chairs	9/1/02

RESOURCES REQUIRED
Budget: $15,000. Speakers for Day of Learning
People: None additional; redirection of priorities is needed.
Facilities: Reserve Dean's Conference Room, each month, 1st and 3rd Wed, 3-5 pm.
Materials: Covered in budget above.

REVIEWS REQUIRED
Monthly meeting between team leader and Dean.

MEASURES OF SUCCESSFUL PROJECT ACHIEVEMENT
Reduction in credits for BS degree from 133 to 123 credits.
Increase in number of undergraduates doing research project from 8% to 20%.
Increase in number of engineering students graduating in 4 years.
Increase in number of undergraduates going to graduate school.

FIGURE 4.7
Implementation plan for creating curriculum reform.

to the underlying cause being the crowded curriculum with students too overloaded to think about becoming involved in a research project. The ultimate solution was a completely new curriculum that reduced the number of total credits and required courses, introduced more opportunities for elective courses, and emphasized active learning by providing "studio hours" for most required courses.

4.6.1 Applying the Problem-Solving Tools in Design

The problem-solving tools described previously are very useful in design, but they are applied in a somewhat different way. Customer interviews and surveys are important in both business and design environments. In engineering design the problem definition step is often much more tightly prescribed and less open-ended, but achieving full understanding of the problem requires using some specific tools like Quality Function Deployment (QFD), as described in Chap. 3. In design the full suite of problem-solving tools are rarely used from problem definition to problem solution. Brainstorming is used extensively in developing design concepts (Chap. 6), but the affinity diagram could be used to more advantage than it normally is. The cause-finding tools are becoming more frequently used to improve the quality of products by seeking out the root causes of defects (Chap. 14). This is shown in Example 4.1.

EXAMPLE 4.1

Early prototype testing of a new game box with a selected group of energetic 10-year-olds revealed that in 20 out of 100 units the indicator light failed to function after three weeks of active use.

Problem Definition: The indicator light on the SKX-7 game box does not have the required durability to perform its function.

The nature of the failures could be characterized as either a poorly made solder joint, a break in the wiring to the bulb, a loose socket, or excessive current passing through the filament. These results are displayed in Fig. 4.8 as a Pareto chart.

Cause Finding

The Pareto chart points to faulty solder joints as the chief cause of failure. There is a high degree of confidence that the issue of excessive current will be readily fixed when the electronic circuits are redesigned. This is scheduled for next week.

The indicator light is but one of many components included on a printed circuit board (PCB), also called a card, that is the heart of the game box. If the simple light circuit is failing then there is concern that more critical circuits may fail with time due to solder defects. This calls for a detailed root cause investigation of the process by which the PCBs are made.

A printed circuit board (PCB) is a reinforced plastic board laminated with copper. Electronic components such as integrated circuit (IC) chips, resistors, and capacitors are placed at specified positions on the board and connected with a pathway of copper. The

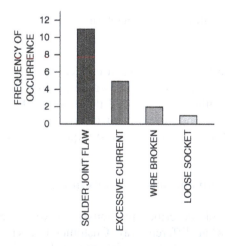

FIGURE 4.8
Pareto chart for the general issues with the failure of the indicator light to function.

circuit path is produced by silk screen printing a layer of acid-resistant ink where the wires are to go, and then acid etching away the rest of the copper layer. The components are connected to the copper circuit by soldering.

Soldering is a process by which two metals are joined using a low-melting-point alloy. Traditionally lead-tin alloys have been used for soldering copper wires, but because lead is toxic it is being replaced by tin-silver and tin-bismuth alloys. Solder is applied as a paste consisting of particles of metallic solder held together in a plastic binder. The solder paste also contains fluxing and wetting agents. The flux acts to remove any oxide or grease on the metal surfaces to be joined and the wetting agent lowers the surface tension so the molten solder spreads out over the surface to be joined. The solder paste is applied to the desired locations on the PCB by forcing it through a screen with a squeegee action. To control the height of the solder pad or ball, the distance between the screen and the PCB surface (standoff) must be accurately controlled.

Flowchart. A flowchart is a map of all of the steps involved in a process or a particular segment of a process. Flowcharting is an important tool to use in the early steps of cause finding because the chart quickly allows the team to understand all of the steps that can influence the causes of the problem. A flowchart for the reflow soldering process is shown in Fig. 4.9.

The symbols in the flowchart have particular meaning. The input and output to the process are placed inside the ovals. A rectangle is used to show a task or activity performed in the process. Decision points are shown by diamonds. Typically these are points where a yes or no decision must be made. The direction of flow in the process is shown with arrows.

The flowchart shows that after the solder and components have been placed the PCB is put in an oven and carefully heated. The first step is to drive off any solvents and to

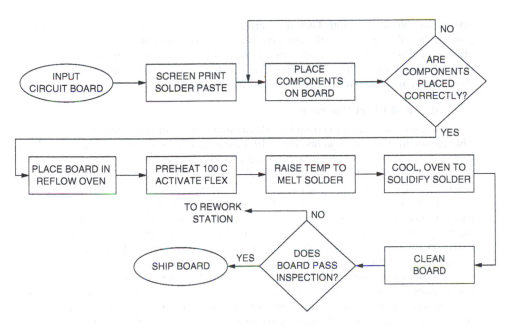

FIGURE 4.9
A simplified flowchart for the reflow soldering process.

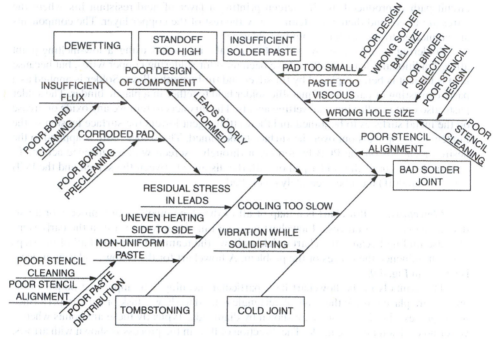

FIGURE 4.10
Cause-and-effect diagram for the production of flawed solder joints.

activate the fluxing reaction. Then the temperature is increased to just above the melting point of the solder where it melts and wets the leads of the components. Finally the assembly is cooled slowly to room temperature to prevent generating stresses due to differential thermal contraction of the components. The last step is to carefully clean the PCB of any flux residue, and the board is inspected visually for defects.

Cause-and-Effect Diagram:

The cause-and-effect diagram provides a visual way to organize and display the possible causes for bad solder joints, Fig. 4.10. Five generic causes for bad solder joints are shown in the rectangles, and more detailed reasons for these causes of defective joints are given by the lines feeding into these major "bones." We now look at this level of the diagram to identify possible root causes. Not providing enough solder paste to the joint is a broad generic cause that involves such possible root causes as using the wrong grade of paste or old paste that is approaching the end of its shelf life. Other issues have to do with the design or application of the screen (stencil) through which the paste gets on the PCB. Failure of the solder to adequately wet the leads of the component (dewetting) is a failure of the flux and wetting agent to perform their function, which ties in again with using the wrong solder paste. Tombstoning is a defect in PCB manufacture where instead of a component lying flat on the board it moves upright when going through the soldering process. As shown in Fig. 4.10, this is caused by lack of uniformity of temperature or stress or issues with the alignment of the stencil. Tombstoning is apparent on final inspection of the PCB. Since it was not observed, it was not considered further as a possible root cause. A cold joint occurs when the solder does not make good contact with the component lead or the solder pad on the PCB. This can occur when movement occurs before the solder

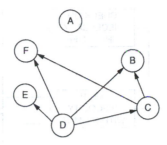

FIGURE 4.11
Interrelationship digraph used to reduce the possible root causes to a single root cause.

is completely cooled or when vibration occurs. Improper maintenance of the soldering machine can cause vibrations.

Interrelationship Digraph. The interrelationship digraph, Fig. 4.11, is helpful in reducing the number of possible root causes. By examining Fig. 4.10, the following list of possible root causes was developed. In developing this list, be as explicit as you can in writing each possible cause, so that there is no misunderstanding among team members as to what is intended.

Possible Root Causes

		Arrows In	Arrows Out	
A	Poor design of component leads, or errors in fabrication of leads	0	0	
B	Improper board cleaning	2	0	
C	Solder paste used beyond its shelf life	1	2	
D	Incorrect selection of paste (solder/binder/flux mixture)	0	3	Root cause
E	Poor operation or maintenance of reflow soldering machine	1	0	
F	Design or maintenance of stencil	2	0	

As described earlier, each combination of possible causes is examined to asking the question, "Is there a relationship between the two causes, and if so, which cause is the driver of the problem?" In this way, Fig. 4.11 was completed. The root cause is the possible cause with the greatest number of arrows directed out from it, that is, it is driving the greatest number of other causes. The root cause was found to be incorrect selection of the solder paste. This is not a surprising result given that new technology with nonleaded solder was being used.

Solution Finding and Implementation

Finding a solution in this case does not depend on brainstorming so much as on careful engineering application of well-known practices. The how-how diagram is useful in organizing the information needed to achieve a good solution.

How-How Diagram. The how-how diagram is a tree diagram that starts with the problem requiring a solution. The how-how diagram is filled out by repeatedly asking the

4

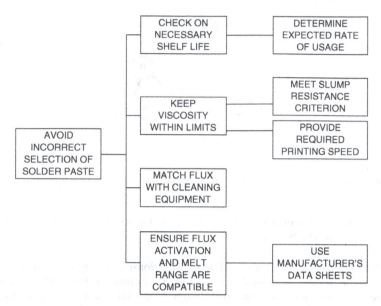

FIGURE 4.12
The development of the solution steps with the how-how diagram.

question, "How can we achieve this?" Figure 4.12 shows the how-how diagram. It serves as a visual checklist for proper selection of solder paste for a given application.

The two remaining tools in the problem-solving suite of tools, force field analysis and implementation planning, could be developed in the way described earlier in this section. In a design or manufacturing environment, often the process stops with finding a good workable solution. The busy engineer is off to solve another problem.

4.7
TIME MANAGEMENT

Time is an invaluable and irreplaceable commodity. You will never recover the hour you squandered last Tuesday. All surveys of young engineers making an adjustment to the world of work point to personal time management as an area that requires increased attention. The chief difference between time management in college and as a practicing engineer is that time management in the world of work is less repetitive and predictable than when you are in college. For instance, you are not always doing the same thing at the same time of the day as you do when you are taking classes as a college student. If you have not done so, you need to develop a personal time management system that is compatible with the more erratic time dimension of professional practice. Remember, effectiveness is doing the right things, but efficiency is doing those things the right way, in the shortest possible time.

An effective time management system is vital to help you focus on your long-term and short-term goals. It helps you distinguish urgent tasks from important tasks. It is the only means of gaining free time for yourself. Each of you will have to work out a time management system for yourself. The following are some time-tested points to achieve it: [1]

Find a place for everything. This means you should have a place for the tools of your profession (books, reports, data files, research papers, software manuals, etc.). It means that you need to develop a filing system and to have the perseverance to use it. It does not mean that you need to keep every piece of paper that passes through your hands.

Schedule your work. You do not need to have an elaborate computerized scheduling system, but you need a scheduling system. Professor David Goldberg suggests you need three things: (1) a monthly calendar to keep track of day-to-day and future appointments and commitments; (2) a diary to keep track of who you talked with and what you did (this can often be combined with a lab notebook), and (3) a to-do list. His system for this is as simple as an 8½ × 11-inch lined pad of paper. All tasks are rated as either To-Do (needed in the next two weeks) or Pending (those tasks two weeks out or "would be nice to do").

It works like this: Every morning create a list of activities for the day. It may contain meetings or classes you must attend, e-mails you need to send, and people you need to talk with. When you complete a task, celebrate silently and cross it off the list. The next morning review the previous day and make a new list of the current day's activities. At the beginning of each week, make a new sheet updating the to-do and pending lists.

Stay current with the little stuff. Learn to quickly decide between the big items and the small stuff. Be cognizant of the 80/20 rule that 80 percent of your positive results will come from the vital 20 percent of your activities, the urgent and important. Big items, such as reports or design reviews, go on the pending list and time is set aside to give these major tasks the thoughtful preparation they require. With the small stuff that is too important to throw away or ignore but is not really major, learn to deal with it as soon as it gets to you. If you don't let the small stuff pile up, it allows you to keep a clearer calendar for when the big, important jobs need your undivided attention.

Learn to say no. This takes some experience to accomplish, especially for the new employee who does not want to get a reputation of being uncooperative. However, there is no reason you should volunteer for every guided tour or charity drive, spend time with every salesperson who cold calls on you, or interview every potential hire unless they are in your area of specialization. And—be ruthless with junk mail.

Find the sweet spot and use it. Identify your best time of day, in terms of energy level and creative activity, and try to schedule your most challenging tasks for that time period. Conversely, group more routine tasks like returning phone calls or writing simple memos into periods of common activity for more efficient performance. Occasionally make appointments with yourself to reflect on your work habits and think creatively about your future.

---

1. Adapted from D. E. Goldberg, *Life Skills and Leadership for Engineers,* McGraw-Hill, New York, 1995.

4.8
PLANNING AND SCHEDULING

It is an old business axiom that time is money. Therefore, planning future events and scheduling them so they are accomplished with a minimum of delay is an important part of the engineering design process. For large construction and manufacturing projects, detailed planning and scheduling is a must. Computer-based methods for handling the large volume of information that accompanies such projects have become commonplace. However, engineering design projects of all magnitudes can benefit greatly from the simple planning and scheduling techniques discussed in this chapter.

One of the most common criticisms leveled at young graduate engineers is that they overemphasize the technical perfection of the design and show too little concern for completing the design on time and below the estimated cost. Therefore, the planning and scheduling tools presented in this chapter are decidedly worth your attention.

In the context of engineering design, *planning* consists of identifying the key activities in a project and ordering them in the sequence in which they should be performed. *Scheduling* consists of putting the plan into the time frame of the calendar. The major decisions that are made over the life cycle of a project fall into four areas: performance, time, cost, and risk.

- *Performance*: The design must possess an acceptable level of operational capability or the resources expended on it will be wasted. The design process must generate satisfactory specifications to test the performance of prototypes and production units.
- *Time*: In the early phases of a project the emphasis is on accurately estimating the length of time required to accomplish the various tasks and scheduling to ensure that sufficient time is available to complete those tasks. In the production phase the time parameter becomes focused on setting and meeting production rates, and in the operational phase it focuses on reliability, maintenance, and resupply.
- *Cost*: The importance of cost in determining what is feasible in an engineering design has been emphasized in earlier chapters. Keeping costs and resources within approved limits is one of the chief functions of the project manager.
- *Risk*: Risks are inherent in anything new. Acceptable levels of risk must be established for the parameters of performance, time, and cost, and they must be monitored throughout the project. The subject of risk is considered in Chap. 14.

4.8.1 Work Breakdown Structure

A *work breakdown structure* (WBS) is a tool used to divide a project into manageable segments to ensure that the complete scope of work is understood. The WBS lists the tasks that need to be done. Preferably, these are expressed as *outcomes* (deliverables) instead of planned *actions*. Outcomes are used instead of actions because they are easier to predict accurately at the beginning of a project. Also, specifying outcomes rather than actions leaves room for ingenuity in delivering results. Table 4.4 shows the WBS for a project to develop a small home appliance.

TABLE 4.4
Work Breakdown Structure for the Development of a Small Appliance

1.0 Development Process for Appliance	Time (Person Weeks)
1.1 Product specification	
1.1.1 Identify customer needs (market surveys, QFD)	4
1.1.2 Conduct benchmarking	2
1.1.3 Establish and approve product design specifications (PDS)	2
1.2 Concept generation	
1.2.1 Develop alternative concepts	8
1.2.2 Select most suitable concept	2
1.3 Embodiment design	
1.3.1 Determine product architecture	2
1.3.2 Complete part configurations	5
1.3.3 Select materials. Analyze for design for manufacture & assembly	2
1.3.4 Design for robustness for CTQ requirements	4
1.3.5 Analyze for reliability and failure with FMEA and root cause analysis	2
1.4 Detail design	
1.4.1 Integration check of subsystems; tolerance analysis	4
1.4.2 Finish detail drawings and bill of materials	6
1.4.3 Prototype test results	8
1.4.4 Correct product deficiencies	4
1.5 Production	
1.5.1 Design production system	15
1.5.2 Design tooling	20
1.5.3 Procure tooling	18
1.5.4 Make final adjustments to tooling	6
1.5.5 Make pilot manufacturing run	2
1.5.6 Complete distribution strategy	8
1.5.7 Ramp-up to full production	16
1.5.8 Ongoing product production	20
1.6 Life cycle tracking	Ongoing
TOTAL TIME (if done sequentially)	160

This work breakdown structure has been developed at three levels: (1) the overall project objective, (2) the design project phases, and (3) the expected outcomes in each design phase. For large, complicated projects the work breakdown may be taken to one or two more levels of detail. When taken to this extreme level of detail the document, called a *scope of work,* will be a thick report with a narrative paragraph

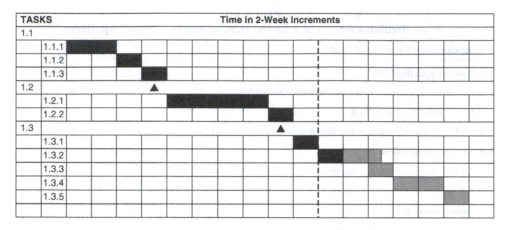

FIGURE 4.13
Gantt chart for the first three phases of the work breakdown structure in Table 4.4.

describing the work to be done. Note that the estimated time for achieving each outcome is given in terms of person weeks. Two persons working for an elapsed time of two weeks equals four person weeks.

4.8.2 Gantt Chart

The simplest and most widely used scheduling tool is the *Gantt chart,* Fig. 4.13. The tasks needed to complete the project are listed sequentially in the vertical axis and the estimated time to accomplish the task are shown along the horizontal axis. The time estimates are made by the development team using their collective experience. In some areas like construction and manufacturing there are databases that can be accessed through handbooks or scheduling and cost estimation software.

The horizontal bars represent the estimated time to complete the task and produce the required deliverable. The left end of the bar represents the time when the task is scheduled to start; the right end of the bar represents the expected date of completion. The vertical dashed line at the beginning of week 20 indicates the current date. Tasks that have been completed are shown in black. Those yet to be completed are in gray. The black cell for task 1.3.2 indicates that the team is ahead of schedule and already working on designing part configurations. Most of the schedule is sequential, showing that there is not much use of concurrent engineering principles in this design team. However, the tasks of selecting materials and performing design for manufacturing activities are started before task 1.3.2 is scheduled for completion. The symbol ▲ indicates *milestone events.* These are design reviews, scheduled to take place when the product design specification (PDS) and conceptual design are finished.

A deficiency of the Gantt chart is that succeeding tasks are not readily related to preceding tasks. For example, it is not apparent what effects a delay in a preceding task will have on the succeeding tasks and the overall project completion date.

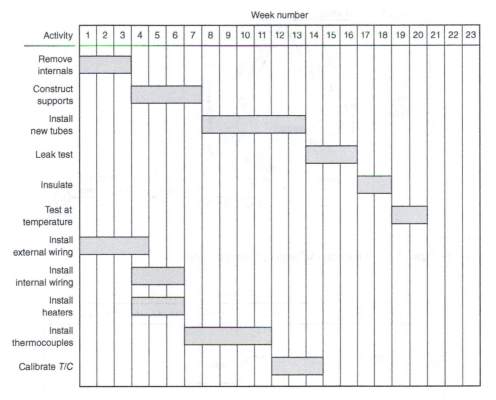

FIGURE 4.14
Gantt chart for prototype testing a heat exchanger.

EXAMPLE 4.2

The *project objective* of a development team is to install a prototype of a new design of heat transfer tubes in an existing shell and determine the performance of the new tube bundle design. The Gantt chart is shown in Fig. 4.14. Note that the process proceeds along two paths: (1) remove the internals from the shell and install the new tubes, and (2) install the wiring and instrumentation.

The dependence of one task on another can be shown by a network logic diagram like Fig. 4.15. This diagram clearly shows the precedence relationships, but it loses the strong correspondence with time that the Gantt chart displays.

The longest path through the project from start to end of testing can be found from inspection. This is called the *network critical path*. From Fig. 4.15 it is the 20 weeks required to traverse the path *a-b-c-d-e-f-g*. The critical path is shown on the modified Gantt chart, at the bottom, Fig. 4.16. On this figure the parts of the schedule that have slack time are shown dashed. *Slack* is the amount of time by which an activity can exceed its estimated duration before failure to complete the activity becomes critical. For example, for the activities of installing heaters, there is a seven-week slack before the activities must be completed to proceed with the leak testing. Thus, the identification of the longest path focuses attention on the activities that must be given special management attention, for any delay in those activities would critically lengthen the project. Conversely, identification of activities with slack indicates the activities in which some natural slippage can occur without serious consequences. This, of course, is not license to ignore the activities with slack.

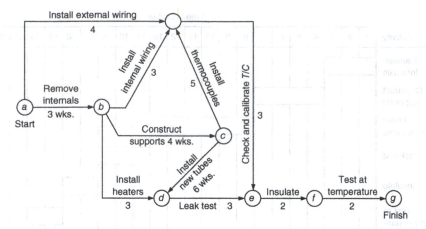

FIGURE 4.15
Network logic diagram for heat exchanger prototyping tests.

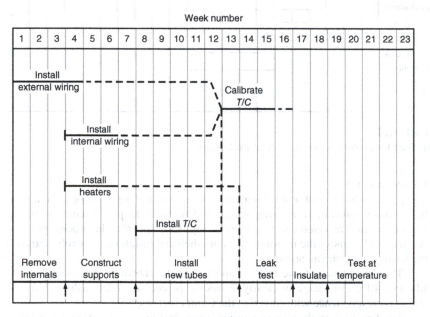

FIGURE 4.16
Modified Gantt chart for heat exchanger prototype tests.

4.8.3 Critical Path Method

The *critical path method* (CPM) is a graphical network diagram that focuses on identifying the potential bottlenecks in a project schedule. While it was relatively easy to identify the critical path in a simple network like Fig. 4.15, most construction or product development projects are very complex and require a systematic method of analysis like CPM. The basic tool of CPM is an arrow network diagram similar to Fig. 4.15. The chief definitions and rules for constructing this diagram are:

- An *activity* is a time-consuming effort that is required to perform part of a project. An activity is shown on an arrow diagram by a directed line segment with an arrowhead pointing in the direction of progress in completion of the project.
- An *event* is the end of one activity and the beginning of another. An event is a point of accomplishment and/or decision. However, an event is assumed to consume no time. A circle is used to designate an event. Every activity in a CPM diagram is separated by two events.

There are several logic restrictions to constructing the network diagram.

1. An activity cannot be started until its tail event is reached. Thus, if $\underset{A}{\longrightarrow}\bigcirc\overset{B}{\longrightarrow}$ activity B cannot begin until activity A has been completed. Similarly, if
$\underset{C}{\longrightarrow}\bigcirc\underset{E}{\overset{D}{\diagdown}}$ activities D and E cannot begin until activity C has been completed.

2. An event cannot be reached until all activities leading to it are complete. If

$\overset{F}{\searrow}\bigcirc\overset{H}{\longrightarrow}$ activities F and G must precede H.
$\underset{G}{\nearrow}$

3. Sometimes an event is dependent on another even preceding it, even though the two events are not linked together by an activity. In CPM we record that situation by introducing a dummy activity, denoted ---▶. A *dummy activity* requires zero time and has zero cost. Consider two examples:

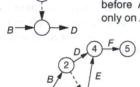

Activities *A* and *B* must both be completed before Activity *D*, but Activity *C* depends only on *A* and is independent of Activity *B*.

Activities *A* must precede both *B* and *C*.
B must precede *D* and *E*.
C must precede *E*.
D and *E* must precede *F*.

To develop a methodology for finding the longest path or paths through the network (the critical path) requires defining some additional parameters.

- *Duration* (D): The *duration of an activity* is the estimated time to complete the activity.
- *Earliest start* (ES): The earliest start of an activity is the earliest time when the activity can start. To find ES trace a path from the start event of the network to the tail of the selected activity. If multiple paths are possible, use the one with the longest duration.
- *Latest start* (LS): The latest time an activity can be initiated without delaying the minimum completion time for the project. To find LS take a backward pass (from head to tail of each activity) from the last event of the project to the tail of the activity in question. If multiple paths are possible use the path with the largest duration.

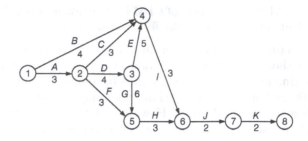

FIGURE 4.17
CPM network based on Example 4.3, prototype testing of new heat exchanger design.

- *Earliest finish time* (EF): EF = ES + D, where D is the duration of each activity.
- *Latest finish time* (LF): LF = LS + D
- *Total float* (TF): The slack between the earliest and latest start times. TF = LS − ES. An activity on the critical path has zero total float.

EXAMPLE 4.3

The network diagram in Fig. 4.15 has been redrawn as a CPM network in Fig. 4.17. The activities are labeled with capital letters, and their duration is given below each line in weeks. To facilitate solution with computer methods, the events that occur at the nodes have been numbered serially. The node number at the tail of each activity must be less than that at the head. The ES times are determined by starting at the first node and making a forward pass through the network while adding each activity duration in turn to the ES of the preceding activity. The details are shown in Table 4.5.

The LS times are calculated by a reverse procedure. Starting with the last event, a backward pass is made through the network while subtracting the activity duration from the LS at each event. The calculations are given in Table 4.6. Note that for calculating LS, each activity starting from a common event can have a different late start time, whereas all activities starting from the same event had the same early start time.

TABLE 4.5
Calculation of Early Start Times Based on Fig. 4.17

Event	Activity	ES	Comment
1	A, B	0	Conventional to use ES = 0 for the initial event
2	C, D, F	3	$ES_2 = ES_1 + D = 0 + 3 = 3$
3	E, G	7	$ES_3 = ES_2 + D = 7$
4	I	12	At a merge like 4 the largest ES + D of the merging activities is used
5	H	13	$ES_5 = ES_3 + 6 = 13$
6	J	16	$ES_6 = ES_5 + 3 = 16$
7	K	18	
8	—	20	

TABLE 4.6
Calculation of Late Start Times Based on Fig. 4.17

Event	Activity	LS	Event	Activity	LS
8	—	20	5–2	F	10
8–7	K	18	4–3	E	8
7–6	J	16	4–2	C	10
6–5	H	13	4–1	B	9
6–4	I	13	3–2	D	3
5–3	G	7	2–1	A	0

A summary of the results is given in Table 4.7. The total float (TF) was determined from the difference between LS and ES. The total float for an activity indicates how much the activity can be delayed while still allowing the complete project to be finished on time. When TF = 0 it means that the activity is on the critical path. From Table 4.7 the critical path consists of activities A-D-G-H-J-K.

In CPM the estimate of the duration of each activity is based on the most likely estimate of time to complete the activity. All time durations should be expressed in the same units, whether they be hours, days, or weeks. The sources of time estimates are records of similar projects, calculations involving personnel and equipment needs, legal restrictions, and technical considerations.

PERT (program evaluation and review technique) is a popular scheduling method that uses the same ideas as CPM. However, instead of using the most likely estimate of time duration, it uses a probabilistic estimate of the time for completion of an activity. Details about PERT will be found in the Bibliography to this chapter.

TABLE 4.7
Summary of Scheduling Parameters for Prototype Testing Project

Activity	Description	D, weeks	ES	LS	TF
A	Remove internals	3	0	0	0
B	Install external wiring	4	0	9	9
C	Install internal wiring	3	3	10	7
D	Construct supports	4	3	3	0
E	Install thermocouples	5	7	8	1
F	Install heaters	3	3	10	7
G	Install new tubes	6	7	7	0
H	Leak test	3	13	13	0
I	Check thermocouples	3	12	13	1
J	Insulate	2	16	16	0
K	Test prototype at temperature	2	18	18	0

4.9
SUMMARY

This chapter considered methods for making you a more productive engineer. Some of the ideas, time management and scheduling, are aimed at the individual, but most of this chapter deals with helping you work more effectively in teams. Most of what is covered here falls into two categories: attitudes and techniques.

Under attitudes we stress:

- The importance of delivering on your commitments and of being on time
- The importance of preparation—for a meeting, for benchmarking tests, and so on
- The importance of giving and learning from feedback
- The importance of using a structured problem-solving methodology
- The importance of managing your time

With regard to techniques, we have presented information on the following:

Team processes:
- Team guidelines (rules of the road for teams)
- Rules for successful meetings

Problem-solving tools (TQM):
- Brainstorming
- Affinity diagram
- Multivoting
- Pareto chart
- Cause-and-effect diagram
- Why-why diagram
- Interrelationship digraph
- How-how diagram
- Force field analysis
- Implementation plan

Scheduling tools:
- Gantt chart
- Critical path method (CPM)

Further information on these tools can be found in the references listed in the Bibliography. Also given there are names of software packages for applying some of these tools.

NEW TERMS AND CONCEPTS

Consensus	Gantt chart	PERT
Critical path method (CPM)	How-how diagram	Total quality
Facilitator	Interrelationship digraph	management (TQM)
Float (in CPM)	Milestone event	Work breakdown
Flowchart	Multivoting	structure
Force field analysis	Network logic diagram	

BIBLIOGRAPHY

Team Methods

Cleland, D. I.: *Strategic Management of Teams,* Wiley, New York, 1996.
Harrington-Mackin, D.: *The Team Building Tool Kit,* American Management Association, New York, 1994.
Katzenbach, J. R., and D. K. Smith: *The Wisdom of Teams,* HarperBusiness, New York, 1993.
Scholtes, P. R., et al.: *The Team Handbook,* 3d ed., Joiner Associates, Madison, WI, 2003.
West, M. A.: *Effective Teamwork: Practical Lessons from Organizational Research,* 2d ed., BPS Blackwell, Malden, MA 2004.

Problem-Solving Tools

Barra, R.: *Tips and Techniques for Team Effectiveness,* Barra International, New Oxford, PA, 1987.
Brassard, M., and D. Ritter: *The Memory Jogger™ II,* GOAL/QPC, Methuen, MA, 1994.
Folger, H. S., and S. E. LeBlanc: *Strategies for Creative Problem Solving,* Prentice Hall, Englewood Cliffs, NJ, 1995.
Tague, N. R.: *The Quality Toolbox,* ASQC Quality Press, Milwaukee, WI, 1995.

Planning and Scheduling

Gido, J. and J. D. Clements, Successful Project Management, Southwestern, Mason, OH, 2009.
Mantel, S. J. and J. R. Meredeth, S. M. Shafer, M. M. Sutton, *Project Management in Practice,* John Wiley & Sons, Hoboken, NJ, 2008.
Rosenau, M. D. and G. D. Githens: *Successful Project Management,* 4th ed., Wiley, New York, 1998.
Shtub, A., J. F. Bard, and S. Globerson: *Project Management: Process, Methodologies, and Economics,* 2d ed., Prentice Hall, Upper Saddle River, NJ, 2005.

Scheduling Software

Microsoft Project 2010 is the most widely used midrange scheduling software for making Gantt charts and determining the critical path. It is also capable of assigning resources to tasks and managing budgets. The software is compatible with Microsoft Office tools.

Oracle Primavera Project Portfolio Management offers a suite of planning and scheduling software tools that can be used on very large construction and development projects, e.g., 100,000 activities. Depending on the choice of software it can be used to define project scope, schedule, and cost. The software can be integrated with a corporate enterprise resource planning (ERP) system.

PROBLEMS AND EXERCISES

4.1 For your first meeting as a team, do some team-building activities to help you get acquainted. (a) Ask a series of questions, with each person giving an answer in turn. Start with the first question and go completely around the team, then the next, etc. Typical questions might be: (1) What is your name? (2) What is your major and class? (3) Where did you grow up or go to school? (4) What do you like best about school? (5) What do you like least about school? (6) What is your hobby? (7) What special skills do you feel you bring to the team? (8) What do you want to get out of the course? (9) What do you want to do upon graduation?

(b) Do a brainstorming exercise to come up with a team name and a team logo.

4.2 Brainstorm about uses for old newspapers.

4.3 Teams often find it helpful to create a team charter between the team sponsor and the team. What topics should be covered in the team charter?

4.4 To learn to use the TQM tools described in Sec. 4.7, spend about four hours total of team time to arrive at a solution for some problem that is familiar to the students and that they feel needs improvement. Look at some aspect of an administrative process in the department or campus. Be alert to how you can use the TQM tools in your design project.

4.5 The *nominal group technique* is a variation on using brainstorming and the affinity diagram as a way to generate and organize ideas for the definition of a problem. Do research about NGT, and use it as an alternative to the methods discussed in this chapter.

4.6 There are certain short statements (killer phrases) that unthinking persons often say during brainstorming sessions that destroy the free flow of ideas. The team should make a list of 10 or 12 killer phrases as a reminder of what not to do when brainstorming.

4.7 After about two weeks of team meetings, invite a disinterested and knowledgeable person to attend a team meeting as an observer. Ask this person to give a critique of what he or she found. Then invite this person back in two weeks to see if you have improved your meeting performance.

4.8 Develop a rating system for the effectiveness of team meetings.

4.9 Keep a record of how you spend your time over the next week. Break it down by 30-minute intervals. What does this tell you about your time management skills?

4.10 The following restrictions exist in a scheduling network. Determine whether the network is correct, and if it is not, draw the correct network.

(a) A precedes C
     B precedes E
     C precedes D and E

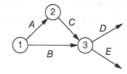

(b) A precedes D and E
     B precedes E and F
     C precedes F

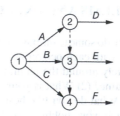

4.11 The development of an electronic widget is expected to follow steps.

Activity	Description	Time est., weeks	Preceded by
A	Define customer needs	4	
B	Evaluate competitor's product	3	
C	Define the market	3	
D	Prepare product specs	2	B
E	Produce sales forecast	2	B
F	Survey competitor's marketing methods	1	B
G	Evaluate product vs. customer needs	3	A,D
H	Design and test the product	5	A,B,D
I	Plan marketing activity	4	C,F
J	Gather information on competitor's pricing	2	B,E,G
K	Conduct advertising campaign	2	I
L	Send sales literature to distributors	4	E,G
M	Establish product pricing	3	H,J

Determine the arrow network diagram and determine the critical path by using the CPM technique.

5

GATHERING INFORMATION

5.1
THE INFORMATION CHALLENGE

The need for information can be crucial at many steps in a design project. You will need to find these bits of information quickly, and validate them as to their reliability. For example, you might need to find suppliers and costs of fractional-horsepower motors with a certain torque and speed. At a lower level of detail, you would need to know the geometry of the mounting brackets for the motor selected for the design. At a totally different level, the design team might need to know whether the new trade name they created for a new product infringes on any existing trade names, and further, whether it will cause any cultural problems when pronounced in Spanish, Japanese, and Mandarin Chinese. Clearly, the information needed for an engineering design is more diverse and less readily available than that needed for conducting a research project, for which the published technical literature is the main source of information. We choose to emphasize the importance of the information-gathering step in design by placing this chapter early in this text (Fig. 5.1).

Figure 5.1 requires some explanation. The need for information permeates the entire engineering design or process design process. By placing the Gathering Information step between the Problem Definition and Concept Generation steps, we are emphasizing the critical need for information to achieve a creative concept solution. Moreover, we think that the suggestions described in this chapter for finding information, and suggestions for sources of information, will be equally useful in the embodiment and detail design phases. You will find that as you progress into these phases of design the information required becomes increasingly technical. Of course, there is information, mostly marketing information, that was needed to accomplish the problem definition.

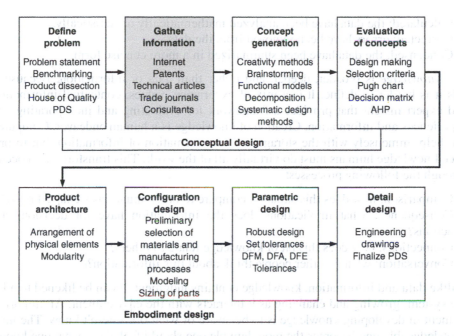

FIGURE 5.1
Steps in the design process, showing early placement of the gathering information step.

5.1.1 Data, Information, and Knowledge

We are told that the future prosperity of the United States and other developed countries will depend on the ability of their *knowledge workers*, such as engineers, scientists, artists, and other innovators, to develop new products and services as the manufacturing of these things is sent offshore to less developed countries with lower wage rates.[1] Thus, it behooves us to learn something about this elusive thing called knowledge, and how it is different from just plain facts.

Data is a set of discrete, objective facts about events. These data may be experimental observations about the testing of a new product, or data on sales that are part of a marketing study. *Information* is data that has been treated in some way that it conveys a message. For example, the sales data may have been analyzed statistically so as to identify potential markets by customer income level, and the product test data may have been compared with competitive products. Information is meant to change the way the receiver of the message perceives something, i.e., to have an impact on his or her judgment and behavior. The word *inform* originally meant "to give shape to." Information is meant to shape the person who gets it and to make some difference in his outlook or insight.

Data becomes information when its creator adds meaning. This can be done in the following ways.[2]

- Contextualized: we know for what purpose the data was gathered.
- Categorized: we know the units of analysis or key components of the data.

1. T. L. Friedman, *The World Is Flat*, Farrar, Strauss and Giroux, New York, 2005.
2. T. H. Davenport and L. Prusak, *Working Knowledge*, Harvard Business School Press, Boston, 1998.

- Calculated: the data have been analyzed mathematically or statistically.
- Corrected: errors have been removed from the data.
- Condensed: the data have been summarized in a more concise form.

Knowledge is broader, deeper, and richer than data or information. Because of this it is harder to define. It is a mix of experience, values, contextual information, and expert insight that provides a framework for evaluating and incorporating new experiences and information. Creation of knowledge is a human endeavor. Computers can help immensely with the storage and transformation of information, but to produce knowledge humans must do virtually all of the work. This transformation occurs through the following processes:

- Comparison: how does this situation compare to other situations we have known?
- Consequence: what implications does the information have for decisions and actions?
- Connections: how does this bit of knowledge relate to others?
- Conversation: what do other people think about this information?

Unlike data and information, knowledge contains judgment. It can be likened to a living system, growing and changing as it interacts with the environment. An important element in developing knowledge is to be aware of what one doesn't know. The more knowledgeable one becomes the more humble one should feel about what one knows. Much knowledge, especially design knowledge, is applied through "rules of thumb." These are guides to action that have been developed through trial and error over long periods of observation and serve as shortcuts to the solution of new problems that are similar to problems previously solved by experienced workers. Rules of thumb occur frequently in areas where detailed knowledge is needed, as in decisions concerning design for manufacture (DFM).

Under this schema a component, a specification, or a material data sheet is *data*. A catalog containing the dimensions and performance data of bearings made by a certain manufacturer is *information*. An article about how to calculate the failure life of bearings published in an engineering technical journal is *knowledge*. The output of a design review session is information, but the output of a more in-depth review of lessons learned upon completing a major design project is most likely knowledge. Since it is not easy to decide whether something is information or knowledge without having a deep understanding of the context in which it exists, in this text we shall generally call most things information unless it is quite clearly knowledge.

5.2
TYPES OF DESIGN INFORMATION

The information needed to do engineering design is of many types and occurs in many forms other than the written word. Some examples are CAD files, computer data files, models, and prototypes. Table 5.1 shows the broad spectrum of information needed in design.

TABLE 5.1
Types of Design Information

Customer
　　Surveys and feedback
　　Marketing data
Related designs
　　Specs and drawings for previous versions of the product
　　Similar designs of competitors (reverse engineering)
Analysis methods
　　Technical reports
　　Specialized computer programs, for example, finite element analysis
Materials
　　Performance in past designs (failure analysis)
　　Properties
Manufacturing
　　Capability of processes
　　Capacity analysis
　　Manufacturing sources
　　Assembly methods
Cost
　　Cost history
　　Current material and manufacturing costs
Standard components
　　Availability and quality of vendors
　　Size and technical data
Technical standards
　　ISO
　　ASTM
　　Company specific
Governmental regulations
　　Performance based
　　Safety
Life cycle issues
　　Maintenance/service feedback
　　Reliability/quality data
　　Warranty data

5.3
SOURCES OF DESIGN INFORMATION

Just as design requires a variety of types of information, so there is a variety of sources in which to find this information; see Table 5.2.

When college students today are asked to find some information, invariably their first inclination is to "Google" the topic. There is no question that the extreme growth

TABLE 5.2
Sources of Information Pertinent to Engineering Design

Libraries
  Dictionaries and encyclopedias
  Engineering handbooks
  Texts and monographs
  Periodicals (technical journals and magazines, and newspapers)
Internet
  A massive depository of information. See Sec. 5.6 for more detail.
Government
  Technical reports
  Databases
  Search engines
  Laws and regulations
Engineering professional societies and trade associations
  Technical journals and news magazines
  Technical conference proceedings
  Codes and standards, in some cases
Intellectual property
  Patents, both national and international
  Copyrights
  Trademarks
Personal activities
  Buildup of knowledge through work experience and study
  Contacts with colleagues
  Personal network of professionals
  Contacts with suppliers and vendors
  Contacts with consultants
  Attendance at conferences, trade shows, exhibitions
  Visits to other companies
Customers
  Direct involvement
  Surveys
  Feedback from warranty payments and returned products

of the World Wide Web through the *Internet* has been an exhilarating, liberating movement. It provides much entertainment and near-universal access in developed countries. Business has found many ways to use the Web to speed communication and increase productivity. However, it is important to realize that much of the information retrieved from the Internet is raw information in the sense that it has not been reviewed for correctness by peers or an editor. Thus, articles retrieved from the Internet do not generally have the same credibility as articles published in a reputable technical or business journal. Also, there is a tendency to think that everything on the Web is current material, but that may not be so. Much material gets posted and is never updated. Another problem is the volatility of Web pages. Web pages disappear when their webmaster changes job or loses interest. With increasing use of advertisement on the Internet there is a growing concern about the objectivity of the information that is posted there. All of these are points that the intelligent reader must consider when enjoying and utilizing this fast-growing information resource.

Table 5.2 lists the complete range of sources of design information. In subsequent sections we will briefly discuss each of these types and sources of information so that you can judge for yourself whether one of these is applicable to your problem. In most cases we will also give a few carefully chosen reference materials and websites.

In reviewing this list, you can divide the sources of information into (1) people who are paid to assist you, for example, the company librarian or consultant, (2) people who have a financial interest in helping you, for example, a potential supplier of equipment for your project, (3) people who help you out of professional responsibility or friendship, and (4) customers.

All suppliers of materials and equipment provide sales brochures, catalogs, and technical manuals, that describe the features and operation of their products. Usually this information can be obtained at no cost by checking the reader service card that is enclosed in most technical magazines. Much of this information is available on the Internet. Practicing engineers commonly build up a file of such information. Generally a supplier who has reason to expect a significant order based on your design will most likely provide any technical information about the product that is needed for you to complete your design.

It is only natural to concentrate on searching the published technical literature for the information you need, but don't overlook the resources available among your colleagues. The professional files or notebooks of engineers more experienced than you can be a gold mine of information if you take the trouble to communicate your problem in a proper way. Remember, however, that the flow of information should be two-way. Be willing to share what you know, and above all, return the information promptly to the person who lent it to you.

It is important to remember that information costs time and money. It is actually possible to acquire too much information in a particular area, far more than is needed to make an intelligent design decision. Also, as noted in Chap. 6, this could actually inhibit your ability in coming up with creative design concepts. However, do not underestimate the importance of information gathering or the effort required in searching for information. Many engineers feel that this isn't real engineering, yet surveys of how design engineers use their time show that they spend up to 30 percent of their time searching for information.[1] There is a marked difference in the information profiles of engineers engaged in concept design and those involved in detail design. The former group use large volumes of information, rely heavily on their own personal collections, and search widely for information. The detail designers use much less information,

---

1. A. Lowe, C. McMahon, T. Shah, and S. Culley, "A Method for the Study of Information Use Profiles for Design Engineers," *Proc. 1999 ASME Design Engineering Technical Conference.* DETC99DTM-8753

rely heavily on company design guidelines and information sources, and apply this information most frequently in working with engineering drawings and CAD models.

5.4
LIBRARY SOURCES OF INFORMATION

In Sec. 5.3 we considered the broad spectrum of information sources and focused mostly on the information that can be obtained in a corporate design organization. In this section we shall deal with the type of information that can be obtained from libraries. The library is the most important resource for students and young engineers who wish to develop professional expertise quickly.

When you are looking for information in a library you will find a hierarchy of information sources. Where you enter the hierarchy depends on your own state of knowledge about the subject and the nature of the information you want to obtain. If you are a complete novice, it may be necessary to use a technical dictionary and read an encyclopedia article to get a good overview of the subject. If you are quite familiar with the subject, then you may simply want to use an index or abstract service to find pertinent technical articles.

Starting with a limited information base, you should consult technical encyclopedias and the library's electronic catalog to search out broad introductory texts. As you become expert in the subject, you should move to more detailed monographs and/or use abstracts and indexes to find pertinent articles in the technical literature. Reading these articles will suggest other articles (cross references) that should be consulted. Another route to important design information is the patent literature (Sec. 5.9).

The task of translating your own search needs into the terminology that appears in the library catalog is often difficult. Library catalogs have been developed for more traditional scholarly and research activities than for the information needs of engineering design. The kinds of questions raised in the context of engineering design, where graphical information and data on suppliers may be more valuable than scholarly knowledge, suggest that a quick search with an Internet browser such as Google may be a useful step early in your search. Also, pay close attention to the lists of *keywords* found in all abstracts and many technical articles. These give you alternative places to search for information.

5.4.1 Dictionaries and Encyclopedias

At the outset of a project dealing with a new technical area, there may be a need to acquire a broad overview of the subject. You may find that the technical terms are unfamiliar and that you need to consult a technical dictionary. Two good sources are:

Davis, J. R. (ed.): *ASM Materials Engineering Dictionary,* ASM International, Materials Park, OH, 1992.

Nayler, G. H. F.: *Dictionary of Mechanical Engineering,* 4th ed., Butterworth-Heinemann, Boston, 1996.

Technical encyclopedias are written for the technically trained person who is just beginning to learn about a new subject. Thus, encyclopedias are a good place to start out if you are only slightly familiar with a subject because they give a broad overview rather quickly. In using an encyclopedia, spend some time checking the index for the entire set of volumes to discover subjects you would not have looked up by instinct. Some useful technical encyclopedias are:

McGraw-Hill Encyclopedia of Environmental Science and Engineering, 3d ed., McGraw-Hill, New York, 1993.
McGraw-Hill Encyclopedia of Physics, 2d ed., McGraw-Hill, New York, 1993.
McGraw-Hill Encyclopedia of Science and Engineering, 8th ed., 20 vols., McGraw-Hill, New York, 1997. Also available on CD-ROM.

5.4.2 Handbooks

Undoubtedly, at some point in your engineering education a professor has admonished you to reason out a problem from "first principles" and not be a "handbook engineer." That is sound advice, but it may put handbooks in a poor perspective that is undeserved. Handbooks are compendia of useful technical information and data. They are usually compiled by an expert in a field who decides on the organization of the chapters and then assembles a group of experts to write the individual chapters. Many handbooks provide a description of theory and its application, while others concentrate more on detailed technical data. You will find that an appropriately selected collection of handbooks will be a vital part of your professional library.

There are hundreds of scientific and engineering handbooks, far more than we can possibly list. A good way to find out what is available in your library is to visit its reference section and spend time looking at the books on the shelf. To get a list of those handbooks in your library, go to the electronic catalog and enter handbook of ___. When we did this we got the following small sampling of the many handbooks that were there:

- Handbook of engineering fundamentals
- Handbook of mechanical engineering
- Handbook of mechanical engineering calculations
- Handbook of engineering design
- Handbook of design, manufacturing, and automation
- Handbook of elasticity solutions
- Handbook of formulas for stress and strain[1]
- Handbook of bolts and bolted joints
- Handbook of fatigue tests

The point of this is that you can find an engineering handbook on practically any topic, from fundamental engineering science to very specific engineering details and

---

1. A very useful reference source is *Roark's Formulas for Stress and Strain,* 7th ed., McGraw-Hill. This is available online at knovel.com.

data. Many handbooks are becoming available online for a modest subscription fee. This greatly extends the search capability of the engineer's laptop computer. Many engineering libraries subscribe to knovel.com, an online reference service with a special emphasis on engineering design. Knovel provides online access to handbooks and specialized monographs along with information retrieval capabilities.

5.4.3 Textbooks and Monographs

New technical books are continually being published. Monographs are books with a narrower and more specialized content than the books you used as texts. A good way to keep up to date is to scan the books-in-print column of your professional society's monthly magazine, or to belong to a technical book club.

5.4.4 Finding Periodicals

Periodicals are publications that are issued periodically, every month, every three months, or daily (as a newspaper). The main periodicals that you will be interested in are *technical journals*, which describe the results of research in a particular field, like engineering design or applied mechanics, and *trade magazines*, which are less technical and more oriented to current practice in a particular industry.

Indexing and abstracting services provide current information on periodical literature, and more importantly they also provide a way to retrieve articles published in the past. An indexing service cites the article by title, author, and bibliographic data. An abstracting service also provides a summary of the contents of the article. Although indexing and abstracting services primarily are concerned with articles from periodicals, many often include books and conference proceedings, and some list technical reports and patents. Until the digital age, abstracts and indexes were contained in thick books in the reference section. Now they can be accessed from your computer by tying into the *reference port* of your library. Table 5.3 lists the most common *abstract databases* for engineering and science.

Conducting a search in the published literature is like putting together a complex puzzle. One has to select a starting place, but some starts are better than others. A good strategy[1] is to start with the most recent subject indexes and abstracts and try to find a current review article or general technical paper. The references cited in it will be helpful in searching back along the "ancestor references" to find the research that led to the current state of knowledge. However, this search path will miss many references that were overlooked or ignored by the original researchers. Therefore, the next step should involve *citation searching* to find the "descendant references" using *Science Citation Index*. Once you have a reference of interest, you can use Citation Index to find all other references published in a given year that cited the key reference. Because the index is online, such searches can be done quickly and precisely.

---

1. L. G. Ackerson, *Reference Quarterly* (RQ), vol. 36, pp. 248–60, 1996.

TABLE 5.3
Common Databases for Electronic Access to Engineering Abstracts and Indexes

Name	Description
Academic Search Premier	Abstracts and indexing for over 7000 journals. Many full text.
Aerospace Database	Indexes journals, conferences, reports by AIAA, IEEE, ASME.
Applied Science & Technology	Includes buyers guides, conf. proceedings. Most applied of group.
ASCE Database	All American Society of Civil Engineers documents.
Compendex	Electronic replacement for Engineering Index.
Engineered Materials	Covers polymers, ceramics, composites.
General Science Abstracts	Coverage of 265 leading journals in U.S. and UK.
INSPEC	Covers 4000 journals in physics, EE, computing and info. techn.
Mechanical Engineering	Covers 730 journals and magazines.
METADEX	Covers metallurgy and materials science.
Safety Science and Risk	Abstracts from 1579 periodicals.
Science Citation Index (Web of Science)	Covers 5700 journals in 164 science and technology disciplines.
Science Direct	Coverage of 1800 journals; full text for 800.

These two search strategies will uncover as many references as possible about the topic. The next step is to identify the key documents. One way to do this is to identify the references with the greatest number of citations, or those that other experts in the field cite as particularly important. You must remember that it takes 6 to 12 months for a reference to be included in an index or abstract service, so current research will not be picked up using this strategy. Current awareness can be achieved by searching *Current Contents* on a regular basis using keywords, subject headings, journal titles, and authors already identified from your literature search. One must also be aware that much information needed in engineering design cannot be accessed through this strategy because it is never listed in scientific and technical abstract services. For this information, the Internet is an important resource (see Sec. 5.6).

5.4.5 Catalogs, Brochures, and Business Information

An important category of design information is catalogs, brochures, and manuals giving information on materials and components that can be purchased from outside suppliers. Most engineers build up a collection of this trade literature, often using the reply cards in trade magazines as a way of obtaining new information. Visits to trade shows are an excellent way to become acquainted quickly with the products offered by many vendors. When faced with the problem of where to turn to find information about an unfamiliar new component or material, start with the *Thomas Register of American Manufacturers* (www.thomasnet.com). This is the most comprehensive resource for finding information on suppliers of industrial products and services in North America.

Most technical libraries also contain certain types of business or commercial information that is important in design. Information on the consumption or sales of commodities and manufactured goods by year and state is collected by the federal government and is available in the U.S. Department of Commerce Census of Manufacturers and the Bureau of the Census Statistical Abstract of the United States. This type of statistical information, important for marketing studies, is also sold by commercial vendors. This data is arranged by industry according to the *North American Industry Classification System* (NAICS) code. The NAICS is the replacement for the former *Standard Industrial Classification* (SIC) code. Businesses that engage in the same type of commerce will have the same NAICS code regardless of size. Therefore, the NAICS code is often needed when searching in government databases. See Sec. 5.6.3 for useful websites for finding business information.

5.5
GOVERNMENT SOURCES OF INFORMATION

The federal government either conducts or pays for about 35 percent of the research and development performed in this country. That generates an enormous amount of information, mostly in the form of technical reports. This R&D enterprise is concentrated in defense, space, environmental, medical, and energy-related areas. It is an important source of information, but all surveys indicate that it is not utilized nearly as much as it ought to be.

Government-sponsored reports are only one segment of what is known among information specialists as the *gray literature*. Other components of the gray literature are trade literature, preprints, conference proceedings, and academic theses. This is called gray literature because it is known to exist but it is difficult to locate and retrieve. The organizations producing the reports control their distribution. Concerns over intellectual property rights and competition result in corporate organizations being less willing to make reports generally available than governmental and academic organizations.

The Government Printing Office (GPO) is the federal agency with the responsibility for reproducing and distributing federal documents. Although it is not the sole source of government publications, it is a good place to start, particularly for documents dealing with federal regulations and economic statistics (www.gpoaccess.gov). Some GPO publications also can be purchased as e-books from Google.

Reports prepared under contract by industrial and university R&D organizations ordinarily are not available from the GPO. These reports may be obtained from the *National Technical Information Service* (NTIS), a branch of the Department of Commerce. NTIS, a self-supporting agency through the sale of information, is the nation's central clearinghouse for U.S. and foreign technical reports, federal databases, and software. Searches can be made online at www.ntis.gov.

In searching for government sources of information, the GPO covers a broader spectrum of information, while NTIS will focus you on the technical report literature. However, even the vast collection at NTIS does not have all federally sponsored technical reports. The Office of Scientific and Technical Information, sponsored by DOE, provides access to reports from DOE, EPA, NIST, and other agencies at www.osti.gov.

While not government publications, academic theses to a large extent are dependent for their existence on government support to the authors who did the research. The *Dissertation Abstracts* database gives abstracts to over 1.5 million doctoral dissertations and masters' theses awarded in the United States and Canada. Copies of the theses can also be purchased from this source.

5.6
INFORMATION FROM THE INTERNET

The fastest-growing communication medium is the Internet.[1] Not only is this becoming the preferred form of personal and business communication via e-mail, but it is rapidly becoming a major source for information retrieval and a channel of commerce. For a discussion of how the Internet developed and functions see www.isoc .org/history/brief.shtml.

5.6.1 Searching with Google

There are over 150 search engines that search the World Wide Web, but the most commonly used general-purpose search engine by far is *Google* (www.google.com). Like many search engines, it builds up its search index by using robot crawlers to traverse the Web and add the URLs to its index. Obviously, the search engine can only find what has been indexed. However, since the indexing of pages is performed automatically, a tremendous number of Web pages are indexed. Also, because any page is added by the crawler without any human judgment, the number of documents returned for a query can be very large. Google ranks the order in which search results appear primarily by how many other sites link to each Web page. This is a kind of popularity vote based on the assumption that other pages would create a link to the best or most useful pages.

The issue with Google is not getting "hits" for your keywords; rather, it is limiting the responses to a manageable number. The following simple rules help achieve this:

- Suppose we want to find responses on the topic **proportional control.** When entered into the search box of Google this results in 11,800,000 responses.
- Obviously, some of these were achieved because Google found the word **proportional** in some web pages, **control** in others, and **proportional control** in still others. To search for the exact phrase **proportional control**, place the phrase within quotation marks in the search field, i.e, **"proportional control."** This reduced the number of responses to 134,000.
- The search can be restricted further by excluding a term from the search. Suppose we wanted to exclude from the search any references that pertain to temperature control. We could do this by typing a minus sign before the word temperature.

---

1. An excellent tutorial for finding information on the Internet is http://www.lib.berkeley.edu/TeachingLib/ Guides/Internet/Find.

Thus, we would type **"proportional control"−temperature**, and the responses are reduced to 76,300.

- There certainly is not a paucity of responses for the search in this example, but if there were, and we were trying to increase the responses, we could purposely tell the search engine to search for either term by using an OR search. We would enter **proportional OR control**, and the responses would rise to 1.05 billion.

These operations are performed more easily with the Advanced Search options found in Google. Advanced Search also allows you to restrict the search to only Web sites written in the English language, or any other language, and to pages for which the search term appeared in the title of the page. The latter restriction indicates greater relevance. When the restrictions of English language and appearing in the title of the page were used in our example search, the number of responses was reduced to 363. If this was further restricted to pages updated within the past year, the number came down to 244 responses. You also can restrict the search to documents having a certain format, for example documents formatted in PDF, Word, or PowerPoint.

Google contains a large number of features not usually associated with a search engine.[1] Just above the search field on the Google Home Page are the frequently used features **Image, Video, Maps, News,** and **Gmail.** The last feature in this lineup is **More.**

One of the more useful categories under **More** is **Directory.** This allows you to browse the Web by topic. Click on **Science → Technology → Mechanical Engineering** to find 17 categories listed, everything from academic mechanical engineering departments worldwide, to design, to tribology. A single click on an entry takes you to a website. Generally, these tend to be more general than the Web pages turned up by the regular Google search, but they often are good places to start a search because they may open up new ideas for keywords or topics. Many of the URLs listed in the next section were found in the Google Directory.

An important category under **More** is **Google Scholar.** It serves the same purpose as the online abstract services. The sources searched include peer-reviewed papers in journals, academic theses, books, and articles from professional societies, universities, and other scholarly organizations. The results are ordered by relevance, which considers the author, the publication in which the article appeared, and how often the article has been cited in the published scholarly literature. When "proportional controller"−temperature was entered in Google Scholar it received 1720 responses, much fewer than was found with the main search but presumably all of higher quality.

Other useful categories under **More** are **Patents** (access to the full text of over seven million U.S. patents), **Finance** (business information and news, including stock charts) and **Product Search,** a quick way to find if your idea for a design project is original.

---

1. Google changes the features listed over the search field from time to time as new features become available. If you do not find one of the features discussed in this paragraph at that location, click on **More.**

Not many people realize that Google has a built-in calculator. It is also very adept at conversion of units. The following calculations and conversions were performed by typing the appropriate expression on the left of the equal sign in the search box.

- $(2 \wedge 4) + 25 = 41$
- sqrt $125 = 11.1802291$
- 45 degrees in radians = 0.785398
- $\tan(0.785398) = 0.999967$ Google does trig functions using radians, not degrees.
- 50,000 psi in MPa = 344.73 megapascals

Another important feature of Google is its ability to translate from one language to another. You can find the Translate tab under **More**. Simply type or paste the text to be translated into the Translate Text box, select the original and translation languages, and click the translate button. References to articles that are not in English have a Translate tab built-in.

Yet another very useful function for Google is to get a quick definition of a word or term that you do not understand. Just type the word define into the Search Space followed by the word or phrase you want defined. One or more definitions will pop up at the head of the search results. Try this with the words *reverse engineering* and *harbinger.*

Yahoo (www.yahoo.com) is much older than Google. It has evolved into more of a general-purpose directory that covers a very wide spectrum of information than a general-purpose search engine. It is particularly strong in the business and finance areas. Search engines are continually being introduced. One that has achieved a growing following is Bing from Microsoft.

5.6.2 Some Helpful URLs for Design

Listed in this section are some websites that have been found to be useful in providing technical information for design projects. This section deals chiefly with references to mechanical engineering technical information. Similar information will be found in Chap. 11 for materials and in Chap. 12 for manufacturing processes. Section 5.6.3 gives references of a more business-oriented nature that are useful in design.

Directories

Directories are collections of websites on specific topics, like mechanical engineering or manufacturing engineering. Using directories narrows down the huge number of hits you get when using a search engine. They direct you to more specific sites of information. Also, the information specialists who build directories are more likely to screen the directory content for the quality of the information. Following are some directories to information in mechanical engineering. The reader should be able to use these URLs to find similar directories in other areas of science, engineering, and technology.

> WWW Virtual Library: Gives a comprehensive set of websites for most U.S. mechanical engineering departments and many commercial vendors. http://vlib .org/Engineering

Intute: Science, Engineering and Technology. A large catalog of Internet sources in the three broad areas. Focus is on UK sources. www.intute.ac.uk/sciences/engineering/

Yahoo Directory: http://dir.yahoo.com/Science/Engineering/Mechanical_Engineering

Google Directory: http:directory.google.com/Top/Science/Technology/Mechanical_Engineering/

NEEDS: A digital library with links to online learning materials in engineering. www.needs.org/needs/

Wikipedia.com is the popular online encyclopedia. Articles are submitted by readers with little editorial review. Thus they can contain errors or biases. For technical topics this is a good place to get a quick overview of a new subject, but it should be read with caution for political or economic topics where prejudices often run high.

Technical Information

Knovel (http://knovel.com) is a web-based engineering information service that is available through many engineering libraries. It offers direct access to thousands of engineering handbooks and design-oriented monographs that are search optimized for engineering. Although it is a subscription service, there is free access to a limited number of handbooks and databases.

An online text on Design of Machine Elements, with a good discussion of design creativity, review of mechanics of materials, and design of components. Excellent problems with answers. http://www.mech.uwa.edu.au/DANotes

Browse back issues of *Mechanical Engineering* and *Machine Design* magazines. http://www.memagazine.org/index.html; http://www.machinedesign.com

ESDU Engineering Data Service, http://www.esdu.com, began as a unit of the Royal Aeronautical Society in the UK, and now is part of IHS Inc, a large U.S. engineering information products company. On a subscription basis, it provides well-researched reports on design data and procedures for topics ranging from aerodynamics to fatigue to heat transfer to wind engineering.

How Stuff Works: Simple but very useful descriptions, with good illustrations and some animations, of how technical machines and systems work. http://www.howstuffworks.com. For common engineering devices click on Science → Engineering.

Working models of common mechanical mechanisms. www.brockeng.com/mechanism. Simple but very graphic models of mechanical mechanisms. www.flying-pig.co.uk/mechanisms. A world-famous collection of kinematic models. http://kmoddl.library.cornell.edu

eFunda, for Engineering Fundamentals, bills itself as the ultimate online reference for engineers. http://www.efunda.com. The main sections are materials, design data, unit conversions, mathematics, and engineering formulas. Most equations from engineering science courses are given with brief discussion, along with nitty-gritty design data like screw thread standards and geometric dimensioning and tolerancing. It is basically a free site, but some sections require a subscription fee for entry.

Engineers Edge is similar to eFunda but with more emphasis on machine design calculations and details. Also, there is good coverage of design for manufacture for most metal and plastic manufacturing processes. www.engineersedge.com.

For useful websites on materials properties go to Chap. 11; for manufacturing process go to Chap. 13.

Access to Supplier Information

When searching for suppliers of materials or equipment with which to build prototypes for your design project, it is important to contact your local purchasing agent. He or she may know of local vendors who can provide quick delivery at good prices. For more specialized items, you may need to shop on the Web. Three supply houses that have a national network of warehouses and good online catalogs are:

McMaster-Carr Supply Co. http://www.mcmaster.com
Grainger Industrial Supply. http://www.grainger.com
MSC Industrial Supply Co. http://www1.mscdirect.com

A good place to start a search of vendors is the website section of Google. Introducing a product or equipment name in the Search box will turn up several names of suppliers, with direct links to their websites. For many years, the very large books of *Thomas Register of American Manufacturers* was a standard fixture in design rooms. This important source of information can now be found on the Web at http://www.thomasnet.com. One of its features is PartSpec®, over one million predrawn mechanical and electrical parts and their specifications that can be downloaded into your CAD system. Directories of suppliers can be found in eFunda, the website for *Machine Design* magazine, www.industrylink.com, and www.engnetglobal.com. Be advised that the companies that will turn up in these directories are basically paid advertisers to these directories.

5.6.3 Business-Related URLs for Design and Product Development

We have made the point many times, and it will be repeated many times elsewhere in this text, that design is much more than an academic exercise. Engineering design does not have real meaning unless it is aimed at making a profit, or at least reducing cost. Hence, we have assembled a group of references to the WWW that are pertinent to the business side of the product development process. These all are subscription services, so it is best to enter them through your university or company website.

General Websites

LexisNexis, http://web.lexis-nexis.com, is the world's largest collection of news, public records, legal, and business information. The major divisions show the scope of its contents: News, Business, Legal Research, Medical, Reference.

General Business File ASAP provides references to general business articles dating from 1980 to the present.

Business Source Premier gives full text for 7800 academic and trade magazines in a spectrum of business fields. It also gives profiles of 10,000 of the world's largest companies.

Marketing

North American Industry Classification System (NAICS) can be found at http://www.census.gov/epcd/www/naics.html. Knowledge of the NAICS code often is useful when working with the following marketing databases:

Hoovers is the place to go to get detailed background on companies. It provides key statistics on sales, profits, the top management, the product line, and the major competitors.

Standard and Poors Net Advantage provides financial surveys by industry sector and projections for the near future.

IBIS World provides world market industry reports on 700 U.S. industries and over 8000 companies.

RDS Business & Industry is a broad-based business information database that focuses on market information about companies, industries, products, and markets. It covers all industries and is international in scope. It is a product of the Gale Group of the Thomson Corporation.

Dialog (www.dialog.com) is a major online business and technical information system.

Statistics

A large amount of U.S business, trade, and economic statistics is available from federal government agencies. Some of the most commonly used sources are discussed below. For a guide to even more U.S. government departments and bureaus see http://guides.ucf.edu/statusa.

Bureau of Economic Analysis, Department of Commerce. http://www.bea.gov. This is the place to find information on the overview of the U.S. economy and detailed data on such things as gross domestic product (GDP), personal income, corporate profits and fixed assets, and the balance of trade.

Bureau of Census, Department of Commerce. http://census.gov/. This is the place to find population figures and population projections by age, location, and other factors.

Bureau of Labor Statistics, Department of Labor. http://bls.gov. This is the place to find data on the consumer price index, producer price index, wage rates, productivity factors, and demographics of the labor force.

Federal Reserve Bank of St. Louis. http://www.stls.frb.org. If you really are into economic data, this website contains a huge depository of historical economic data, as well as full text of many federal publications.

5.7
PROFESSIONAL SOCIETIES AND TRADE ASSOCIATIONS

Professional societies are organized to advance a particular profession and to honor those in the profession for outstanding accomplishments. Engineering societies advance the profession chiefly by disseminating knowledge through sponsoring annual

meetings, conferences and expositions, local chapter meetings, by publishing technical journals (archival journals), magazines, books, and handbooks, and sponsoring
short courses for continuing education. Unlike some other professions, engineering
societies rarely lobby for specific legislation that will benefit their membership. Some
engineering societies develop codes and standards; see Sec. 5.8.

The first U.S. engineering professional society was the American Society of
Civil Engineers (ASCE), followed by the American Society of Mining, Metallurgi-
cal and Petroleum Engineers (AIME), the American Society of Mechanical Engineers
(ASME), the Institute of Electrical and Electronic Engineers (IEEE), and the American
Institute of Chemical Engineers (AIChE). These five societies are called the Five
Founder Societies, and were all established in the latter part of the 19th century and
early 1900s. As technology advanced rapidly, new groups were formed, such as the
Institute of Aeronautics and Astronautics, Institute of Industrial Engineers, American
Nuclear Society, and such specialty societies as the American Society of Heating,
Refrigerating, and Air-conditioning Engineers (ASHRAE), the International Society
for Optical Engineering (SPIE), the Biomedical Engineering Society and the American
Society for Engineering Education (ASEE). One count of engineering societies comes
to 30,[1] while another totals 85. These references should serve as an entrée to the websites of most engineering societies.

The lack of a central society focus for engineering, such as exists in medicine
with the American Medical Association, has hampered the engineering profession
in promoting the public image of engineering, and in representing the profession
in discussions with the federal government. The American Association of Engineering Societies (AAES) serves as the "umbrella organization" for engineering
representation in Washington, although a number of the larger societies also have
a Washington office. The current membership in the AAES is 13 societies, including the five founder societies. The National Academy of Engineering (NAE) is the
engineering counterpart to the National Academy of Science. It exists to honor distinguished engineers and to advise the government on technical issues that affect
the nation.

Trade associations represent the interests of the companies engaged in a particular sector of industry. All trade associations collect industrywide business statistics
and publish a directory of members. Most lobby on behalf of their members in such
things as import controls and special tax regulations. Some, such as the American
Iron and Steel Institute (AISI) and the Electric Power Research Institute (EPRI),
sponsor research programs to advance their industries. A trade association like the
National Association of Manufacturers is a multi-industry association with a heavy
educational program aimed at Congress and the general public. Others like the Steel
Tank Institute are much more focused and issue such things as *Standards for Inspection of Above Ground Storage Tanks*. Yahoo gives a good listing of trade associations
of all kinds at Business→Trade Associations.

---

1. http://www.englib.cornell.edu/erg/soc.php

5.8
CODES AND STANDARDS

The importance of codes and standards in design is discussed in Sec. 1.7 and Sec. 13.7. A *code* is a set of rules for performing some task, as in the local city building code or fire code. A *standard* is less prescriptive and can be defined as a set of technical definitions and guidelines. It establishes a basis for comparison. Many standards describe a best way to perform some test so that the data obtained can be reliably compared with data obtained by other persons. A *specification* describes how a system should work, and is usually is much more specific and detailed than a standard, but sometimes it is difficult to differentiate between documents that are called standards and those called specifications.[1]

The United States is the only industrialized country in which the national standards body is not a part of or supported by the national government. The American National Standards Institute (ANSI) is the coordinating organization for the voluntary standards system of the United States (www.ansi.org). Codes and standards are developed by professional societies or trade associations with committees made up mostly of industry experts, with representation from university professors and the general public. The standards may then be published by the technical organizations themselves, but most are also submitted to ANSI. This body certifies that the standards-making process was carried out properly and publishes the document also as an ANSI standard. ANSI may also initiate new standards-making projects, and it has the important responsibility of representing the United States on the International Standards Committees of the International Organization for Standardization (ISO). The standard development process in the United States does not involve substantial support from the federal government, but it does represent a substantial commitment of time from volunteer industry and academic representatives, and cost to their sponsoring organizations for salary and travel expenses. Because the cost of publishing and administering the ANSI and other standards systems must be covered, the cost for purchasing standards is relatively high, and they are not generally available free on the World Wide Web.

The standards responsibility of the U.S. government is carried out by the *National Institute for Standards and Technology* (NIST), a division of the Department of Commerce. The Standards Services Division (SSD) of NIST (http://www.nist.gov/ts/ssd/index.cfm) is the focal point for standards in the federal government that coordinates activities among federal agencies and with the private sector. Since standards can serve as substantial barriers to foreign trade, SSD maintains an active program of monitoring standards globally and supporting the work of the U.S. International Trade Administration. SSD also manages the national program by which testing laboratories become nationally accredited. NIST, going back to its origins as the National Bureau of Standards, houses the U.S. copies of the international standards for weights and measures, such as the standard kilogram and meter, and maintains a program for

---

1. S. M. Spivak and F. C. Brenner, *Standardization Essentials: Principles and Practice*, Marcel Dekker, New York, 2001.

calibrating other laboratories' instruments against these and other physical standards. The extensive laboratories of NIST are also used, when necessary, to conduct research to develop and improve standards.

ASTM International is the major organization that prepares standards in the field of materials and product systems. It is the source of more than half of the existing ANSI standards. Most technical libraries will have a set of the Annual Book of ASTM Standards (http://astm.org).

The ASME prepares the well-known Boiler and Pressure Vessel Code that is incorporated into the laws of most states. The ASME Codes and Standards Division also publishes performance test codes for turbines, combustion engines, and other large mechanical equipment (http://asme.org/Codes/). For a long list of standard developing organizations go to http://engineers.ihs.com/products/standards and click on standards to find the list.

ANSI provides an educational website about standards (http://standardslearn.org). It lists many standards developing organizations (SDO), a broad tutorial about standards, and case studies showing where standards can be critical in design.

The Department of Defense (DOD) is the most active federal agency in developing specifications and standards. DOD has developed a large number of standards, generally by the three services, Army, Navy, and Air Force. Defense contractors must be familiar with and work to these standards. In an effort to reduce costs through common standards, DOD has established a Defense Standardization Program (DSP) Office (www.dsp.dla.mil). One aim is to lower costs through the use of standardized parts as a result of reduced inventories. The other major goal is to achieve improved readiness through shortened logistics chains and improved interoperability of joint forces. Other important federal agencies that write standards are:

- Department of Energy (DOE)
- Occupational Safety and Health Administration (OSHA)
- Consumer Product Safety Commission (CPSC)

The General Services Agency (GSA) is the federal government's landlord charged with providing office space and facilities of all kinds, and procuring common items of business like floor coverings, automobiles, and light bulbs. Thus it has issued over 700 standards for common everyday items. A listing of these specifications can be found at http://apps.fss.gsa.gov/pub/fedspecs/. A quick scan of the index found standards for the abrasion resistance of cloth, the identification of asbestos, and turbine engine lubricants. These standards are not downloadable and must be purchased. Links to all of these federal sources of standards can be found at http://www.uky.edu/Subject/standards.html.

Because of the growing importance of world trade, foreign standards are becoming more important. Some helpful websites are:

- International Organization for Standardization (ISO); http://www.iso.org
- British Standards Institution (BSI); http://www.bsigroup.com
- DIN (Deutsches Institut fur Normung), the German standards organization. Copies of all DIN standards that have been translated into English can be purchased from ANSI at http://webstore.ansi.org
- Another website from which to purchase foreign standards is World Standards Services Network, http://www.wssn.net

An important website to use to search for standards is the National Standards System Network http://www.nssn.org. NSSN was established by ANSI to search for standards in its database of over 250,000 references. For example, a search for standards dealing with nuclear waste found 50 records, including standards written by ASTM, ISO, ASME, DIN and the American Nuclear Society (ANS).

5.9
PATENTS AND OTHER INTELLECTUAL PROPERTY

Creative and original ideas can be protected with patents, copyrights, and trademarks. These legal documents fall within the broad area of property law. Thus, they can be sold or leased just like other forms of property such as real estate and plant equipment. There are several different kinds of intellectual property. A *patent,* granted by a government, gives its owner the right to prevent others from making, using, or selling the patented invention. We give major attention to patents and the patent literature in this section because of their importance in present-day technology. A *copyright* gives its owner the exclusive right to publish and sell a written or artistic work. It therefore gives its owner the right to prevent the unauthorized copying by another of that work. A *trademark* is any name, word, symbol, or device that is used by a company to identify its goods or services and distinguish them from those made or sold by others. The right to use trademarks is obtained by registration and extends indefinitely so long as the trademark continues to be used. A *trade secret* is any formula, pattern, device, or compilation of information that is used in a business to create an opportunity over competitors who do not have this information. Sometimes trade secrets are information that could be patented but for which the corporation chooses not to obtain a patent because it expects that defense against patent infringement will be difficult. Since a trade secret has no legal protection, it is essential to maintain the information in secret.

5.9.1 Intellectual Property

Intellectual property has received increasing attention in the high-tech world. *The Economist* states that as much as three-quarters of the value of publicly traded companies in the United States comes from intangible assets, chiefly intellectual property.[1] The revenue from licensing technology-based intellectual property in the United States is estimated at $45 billion annually, and around $100 billion worldwide. At the same time, it has been estimated that only about 1 percent of patents earn significant royalties, and only about 10 percent of all patents issued are actually used in products. The majority of patents are obtained for defensive purposes, to prevent the competition from using your idea in their product.

---

1. "A Market for Ideas," *The Economist,* Oct. 20, 2005.

How Intellectual Property Can Pay Off Big!

An excellent example of a company whose profitability depends on intellectual property is Qualcomm. A pioneer in mobile telephones, Qualcomm early on developed a communication protocol called CDMA. Most other cell phone makers went a different way, but Qualcomm persisted, and now CDMA is generally accepted and forms the basis of third-generation wireless networks.

Initially Qualcomm made handsets, but it sold this business in 1999 to focus on developing CDMA and the semiconductor chips that make it possible. Today it spends 19 percent of sales on R&D and has over 1800 patents, with 2200 being processed through the patent system. Sixty percent of its profits come from royalties on other companies' cell phones that use CDMA. Qualcomm is not a "patent troll," but it is a good example of how if you get the technology right and pursue its intellectual property you can make a very nice business.

The new emphasis on patents in the high-tech industries has been driven by several broad industry trends.

- The technology in the information technology and telecommunication businesses has become so complex that there is a greater willingness to accept the innovations of other companies. The industry has changed from vertically integrated firms dealing with every aspect of the product or service to a large number of specialist companies that focus on narrow sectors of the technology. These companies must protect their intellectual property before licensing it to other companies.
- Since the technology is moving so fast, there is a tendency for cutting-edge technology to quickly change to a commodity-type business. When this happens, profit margins are reduced, and licensing of the technology is one way to improve the profit situation.
- Customers are demanding common standards and interoperability between systems. This means that companies must work together, which often requires pooling of patents or cross-licensing agreements.
- For start-up companies, patents are important because they represent assets that can be sold in case the company is not successful and goes out of business. In the high-tech world, large companies often buy out small start-ups to get their intellectual property, and their talented workforce.

It seems clear that a major force behind the great increase in patent development is that everyone seems to be doing it. IBM has around 40,000 patents, and that number is increasing by 3000 every year. Nokia, a company with a rather narrow product line, has over 12,000 patents worldwide. Hewlett-Packard, a company that refrained from intensive patenting of its technology in its early days because founder David Packard felt it would help the industry innovate, recently created an intellectual property team of 50 lawyers and engineers, and in three years increased its annual licensing revenue from $50 million to over $200 million.

Some observers of the scene say that this reminds them of the mutually assured destruction scenario that existed during the Cold War. "You build up your patent portfolio, I'll build up mine." The question is whether this proliferation of intellectual property is damping down innovation. It certainly means that it becomes more difficult to build new products without accidentally infringing on a patent owned by another company. The worst of this situation is exemplified by the "patent trolls," small companies of lawyers who write patents without much reduction to practice to back them up, or who buy patents for critical bits of a technology, and then shop for settlements from companies who they claim are infringing on their patents.

5.9.2 The Patent System

Article 1, Section 8 of the Constitution of the United States declares that Congress shall have the power to promote progress in science and the useful arts by securing for limited times to inventors the exclusive right to their discoveries. A patent granted by the U.S. government gives the patentee the right to prevent others from making, using, or selling the patented invention for a set period of time. Any patent application filed since 1995 has a term of protection that begins on the date of the grant of the patent and ends on a date 20 years after the filing date of the application. The 20-year term from the date of filing brings the United States into harmony with most other countries in the world in this respect. The most common type of patent, the *utility patent*, may be issued for a new and useful machine, process, article of manufacture, or composition of matter. In addition, *design patents* are issued for new ornamental designs and *plant patents* are granted on new varieties of plants. Computer software, previously protected by copyright, became eligible for patenting in 1981. In 1998 a U.S. court allowed *business practices* to be patented. In addition, new uses for an invention in one of the above classes are patentable.

Laws of nature and physical phenomena cannot be patented. Neither can mathematical equations and methods of solving them. In general, abstract ideas cannot be patented. A 2010 Supreme Court decision ruled that a business method for hedging energy purchases was too abstract to qualify for a patent. At the same time the court rejected a lower court's reasoning that only machines and physical transformations could be patented. Some experts hailed the decision as a move to broaden the scope of patent eligible inventions to be more aligned with the information age. Patents cannot be granted merely for changing the size or shape of a machine part, or for substituting a better material for an inferior one. Artistic, dramatic, literary, and musical works are protected by copyright, not by patents. Prior to 20 years ago, computer software was protected by copyrights. Today, this form of intellectual property is protected by patents.

There are three general criteria for awarding a patent:

- The invention must be new or novel.
- The invention must be useful.
- It must not be obvious to a person skilled in the art covered by the patent.

A key requirement is novelty. Thus, if you are not the first person to propose the idea you cannot expect to obtain a patent. If the invention was made in another country, but it was known or used in the United States before the date of the invention in the United States, it would not meet the test of novelty. Finally, if the invention was published anywhere in the world before the date of invention but was not known to the inventor, it would violate the requirement of novelty. The requirement for usefulness is rather straightforward. For example, the discovery of a new chemical compound (composition of matter) which has no useful application is not eligible for a patent. The final requirement, that the invention be unobvious, can be subject to considerable debate. A determination must be made as to whether the invention would have been the next logical step based on the state of the art at the time the discovery was made. If it was, then there is no patentable discovery. Note that if two people worked on the invention they both must be listed as inventors, even if the work of one person resulted in only a single claim in the patent. The names of financial backers cannot be on the patent if they did not do any of the work. Since most inventors today work for a company, their patent by virtue of their employment contract will be assigned to their company. Hopefully the company will suitably reward its inventors for their creative work.

The requirement for novelty places a major restriction on disclosure prior to filing a patent application. In the United States the printed publication or public presentation at a conference of the description of the invention anywhere in the world more than one year before the filing of a patent application results in automatic rejection by the Patent Office. It should be noted that to be grounds for rejection the publication must give a description detailed enough so that a person with ordinary skill in the subject area could understand and make the invention. Also, public use of the invention or its sale in the United States one year or more before patent application results in automatic rejection. The patent law also requires diligence in *reduction to practice*. If development work is suspended for a significant period of time, even though the invention may have been complete at that time, the invention may be considered to be abandoned. Therefore, a patent application should be filed as soon as it is practical to do so.

In the case of competition for awarding a patent for a particular invention, the patent is awarded to the inventor who can prove the earliest date of conception of the idea and can demonstrate reasonable diligence in reducing the idea to practice.[1] The date of invention can best be proved in a court of law if the invention has been recorded in a bound laboratory notebook with numbered pages and if the invention has been witnessed by a person competent to understand the idea. For legal purposes, corroboration of an invention must be proved by people who can testify to what the inventor did and the date when it occurred. Therefore, having the invention disclosure notarized is of little value since a notary public usually is not in a position to understand a highly technical disclosure. Similarly, sending a registered letter to oneself is of little value.

---

1. A major difference between U.S. patent law and almost every other country's laws is that in the United States a patent is awarded to the first person to invent the subject matter, while in other countries the patent is awarded to the first inventor to file a patent application. There is a bill in Congress at the time of this writing to change the U.S. patent law so that it conforms to the rest of the world. Another difference is that in any country but the United States public disclosure of the invention before filing the applications results in loss of patent rights on grounds of lack of novelty. The U.S. patent system provides for a filing of a Provisional Patent Application, which sets the filing date and gives the inventor one year to decide whether to file a regular and more expensive patent.

For details about how to apply, draw up, and pursue a patent application the reader is referred to the literature on this subject.[1]

5.9.3 Technology Licensing

The right to exclusive use of technology that is granted by a patent may be transferred to another party through a licensing agreement. A license may be either an exclusive license, in which it is agreed not to grant any further licenses, or a nonexclusive license. The licensing agreement may also contain details as to geographic scope, for example, one party gets rights in Europe, another gets rights in South America. Sometimes the license will involve less than the full scope of the technology. Frequently consulting services are provided by the licensor for an agreed-upon period.

Several forms of financial payment are common. One form is a paid-up license, which involves a lump sum payment. Frequently the licensee will agree to pay the licensor a percentage of the sales of the products (typically 2 to 5 percent) that utilize the new technology, or a fee based on the extent of use of the licensed process. Before entering into an agreement to license technology, it is important to make sure that the arrangement is consistent with U.S. antitrust laws or that permission has been obtained from appropriate government agencies in the foreign country. Note that some defense-related technology is subject to export control laws.

5.9.4 The Patent Literature

The U.S. patent system is the largest body of information about technology in the world. At present there are over 7 million U.S. patents, and the number is increasing by about 190,000 each year. Old patents can be very useful for tracing the development of ideas in an engineering field, while new patents describe what is happening at the frontiers of a field. Patents can be a rich source of ideas. Since only about 20 percent of the technology that is contained in U.S. patents can be found elsewhere in the published literature,[2] the design engineer who ignores the patent literature is aware of only the tip of the iceberg of information.

The U.S. Patent and Trademark Office (USPTO) has been highly computerized. Its official website at www.uspto.gov contains a great deal of searchable information about specific patents and trademarks, information about patent laws and regulations, and news about patents. Typical reasons for making a patent search are:

1. You have been asked to comment on a patent used by a competitor.
2. You are looking for ideas to improve your design concept.

---

1. W. G. Konold, *What Every Engineer Should Know about Patents,* 2d ed., Marcel Dekker, New York, 1989; M. A. Lechter (ed.), *Successful Patents and Patenting for Engineers and Scientists,* IEEE Press, New York, 1995; D. A. Burge, *Patent and Trademark Tactics and Practice,* 3d ed., John Wiley & Sons, New York, 1999; H. J. Knight, *Patent Strategy,* John Wiley & Sons, New York, 2001; "A Guide to Filing a Non-Provisional (utility) Patent Application, U.S. Patent and Trademark Office (available electronically).
2. P. J. Terrago, *IEEE Trans. Prof. Comm.,* vol. PC-22, no. 2, pp. 101–4, 1974.

3. You have come up with a really cool design concept, and you want to determine if the idea is novel enough to warrant the expense of preparing a patent submission.
4. You want to continue to update yourself on a particular technology of interest.

Let us see how you could approach the first task listed above. At www.uspto.gov, decide whether you wish to search for Patents or Trademarks and click on **Search.** Under Search Full Text patents, since you know the patent number, click on **Patent Number Search.** This gives the name of the patent holder, the date of issue, the owner of the patent, the filing date, an abstract, references to other pertinent patents, the claims, and the patent classes and subclasses that it is filed under. A copy of the full patent, with drawings, can be obtained by clicking on **Images** at the top of the first page. These full-page images are in tag image file format (TIF).[1]

The second task is a bit harder. Without knowing the patent number you will need to use the Quick Search feature. You enter keywords, as in searching the abstracts and indexes for technical literature, and get back lists of patent numbers and titles in descending date of issue. By clicking **Advanced Search,** you can find things like all the patents owned by a certain company or all patents issued in a certain person's name. Be sure to read the Help page before using this search tool.

The third task is more complicated than the others. To gain confidence that your idea is new, you will need to do a thorough search using the Patent Classification System. U.S. patents have been organized into about 450 classes, and each class is subdivided into many subclasses. All told, there are 150,000 classes/subclasses listed in *The Manual of Classification.* This classification system helps us to find patents between closely related topics. The use of this classification system is a first step in making a serious patent search.[2] If you have already found some pertinent patents, for example, by using Google ® Patent Search, these will suggest typical classes and subclasses for the subject. The Manual of Classification can be found at http://www.uspto.gov/go/ classification. The issue is that the Quick Search is not guaranteed to produce all relevant patents because they are often filed under categories that seem strange to the inventor but perfectly logical to the patent examiner. Patent searching is more of an art than a science, even though information science has been brought to bear on the problem.

Once the classes and subclasses for appropriate patents have been obtained by clicking on **Tools to Help Searching by Patent Classification** (under **Links** on the left at bottom), you can enter the class/subclass in the appropriate boxes. This will give a list of patents in the classification.

To stay up-to-date on an area of technology with the patented literature, you can read the weekly issues of the *Official Gazette for Patents.* An electronic version is available from the USPTO home page. Starting on page 2 of the USPO website, click on **Patent Office Gazette.** You can browse by classification, name of inventor or assignee, and state in which the inventor resides. The last 52 weeks of issues can be

---

1. Another important source of patent information is worldwide.espacenet.com. This website for the European Patent Office has over 59 million patents from 72 countries.
2. An excellent online tutorial on the use of the patent classification system is available from the McKinney Engineering Library, University of Texas, Austin. http://www.lib.utexas.edu/engin/patent-tutorial/index.htm.

read online. After that they are available in the *Annual Index of Patents,* available on DVD-ROM, or in the printed index available in many libraries. The Patent Office has established a nationwide system of Patent Depository Libraries where patents can be examined and copied. Many of these are at university libraries.

Many people experience difficulty printing the figures from patents viewed on the USPTO website. An alternative site that is more user-friendly is the Patents Search application in Google. Another website from which clear copies of patents can be downloaded is www.pat2pdf.org. While Google Patents Search is a user-friendly site, it does not have the capability for finding patents when the topic is spread over multiple classes and subclasses, which is common, and therefore it cannot guarantee a complete search.

> **EXAMPLE 5.1** **Objective** Find patents on the making of parts by powder forging.
>
> Starting with www.uspto.gov/go/classification, we click on **Index to the U.S. Patent Classification System** at the top of the page. Next click on **U.S. Patent Classification (USPC) Index.** Under P, find **Powder/metallurgy/Sintering then working** because forging is a metalworking process. This gives classification 419/28. Clicking on 28 produces a list of 696 patents and titles. Clicking on any one will give the details of the patent.
>
> Using Quick Search with the key words "powder forging" gave 94 patents.

5.9.5 Reading a Patent

Because a patent is a legal document, it is organized and written in a style much different from the style of the usual technical paper. Patents must stand on their own and contain sufficient disclosure to permit the public to practice the invention after the patent expires. Therefore, each patent is a complete exposition on the problem, the solution to the problem, and the applications for the invention in practical use.

Figure 5.2 shows the first page of a patent for a basketball return apparatus. This page carries bibliographic information, information about the examination process, an abstract, and a general drawing of the invention. At the very top we find the inventor, the patent number, and the date of issuance. Below the line on the left we find the title of the invention, the inventor(s) and address(es), the date the patent application was filed, and the application number. Next are listed the class and subclass for both the U.S. patent system and the international classification system and the U.S. classes in which the examiner searched for prior art. The references are the patents that the examiner cited as showing the most prior art at the time of the invention. The rest of the page is taken up with a detailed abstract and a key drawing of the invention. Additional pages of drawings follow, each keyed to the description of the invention.

The body of the patent starts with a section on the Background of the Invention followed by the Summary of the Invention and a Brief Description of the Drawings. Most of the patent is taken up by the description of the Preferred Embodiment. This comprises a detailed description and explanation of the invention, often in legal terms and phrases that are strange-sounding to the engineer. The examples cited show as broadly as possible how to practice the invention, how to use the products, and how the invention is superior to prior art. Not all examples describe experiments that were actually run, but they do provide the inventor's teaching of how they should best be run.

US005540428A

United States Patent [19]

Joseph

[11] **Patent Number:** 5,540,428

[45] **Date of Patent:** Jul. 30, 1996

[54] **BASKETBALL RETRIEVAL AND RETURN APPARATUS**

[76] Inventor: **John G. Joseph,** 3305 C.H. 47, Upper Sandusky, Ohio 43351

[21] Appl. No.: **393,351**

[22] Filed: **Feb. 23, 1995**

[51] **Int. Cl.**[6] **A63B 69/00**

[52] **U.S. Cl.** .... **273/1.5 A**

[58] **Field of Search** .... 273/1.5 A, 396, 273/397

[56]   **References Cited**

U.S. PATENT DOCUMENTS

4,786,731	11/1988	Postol	273/1.5 A
4,936,577	6/1990	Kington et al.	273/1.5 A
5,333,853	8/1994	Hektor	273/1.5 A
5,393,049	2/1995	Nelson	273/1.5 A

Primary Examiner—Paul E. Shapiro
Attorney, Agent, or Firm—George C. Atwell

[57]   **ABSTRACT**

A basketball retrieval and return apparatus is used in com-
bination with a pole-supported basketball backboard to collect shot basketballs that either ricochet off the backboard or fall through a rim attached to the backboard for returning the basketballs to the practicing player or players. The apparatus includes a bracket removably mountable to the lowest portion of the backboard, an elongated support bar pivotally mounted to the bracket, and a U-shaped ring bar attached to the support bar and which extends outwardly from and perpendicular to the backboard when the ring bar is pivoted from a non-use to a use position. In order to maintain the ring bar in its use position, a U-shaped support member is attached to the ring bar and pivots downward toward the level surface or ground so that a pole brace attached to the support member can have one end mounted to a pole bracket which is secured generally to the mid-point of the pole. When the U-shaped ring bar is disposed in its use position, a flexible, collapsibly-extensible netting attached to the ring bar encompasses the rim and net, collects the thrown basketballs, and directs the basketballs to a ball return structure located at the lowest part of the netting whereupon the basketballs can be retrieved by the players.

8 Claims, 5 Drawing Sheets

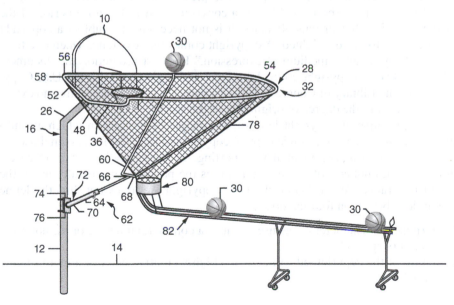

FIGURE 5.2

The first page of a U.S. patent for a basketball return apparatus.

The last part of the patent comprises the claims of the invention. These are the legal description of the rights of invention. The broadest claims are usually placed first, with more specific claims toward the end of the list. The strategy in writing a patent is to aim at getting the broadest possible claims. The broadest claims are often disallowed first, so it is necessary to write narrower and narrower claims so that not all claims are disallowed.

There is a very important difference between a patent and a technical paper. In writing a patent, inventors and their attorneys purposely broaden the scope to include all materials, conditions, and procedures that are believed to be equally likely to be operative as the conditions that were actually tested and observed. The purpose is to develop the broadest possible claims. This is a perfectly legitimate legal practice, but it has the risk that some of the ways of practicing the invention that are described in the embodiments might not actually work. If that happens, then the way is left open to declare the patent to be invalid if the patent is ever contested in court.

Another major difference between patents and technical papers is that patents usually avoid any detailed discussion of theory or why the invention works. Those subjects are avoided to minimize any limitations to the claims of the patent that could arise through the argument that the discovery would have been obvious from an understanding of the theory.

5.9.6 Copyrights

A copyright is the exclusive legal right to publish a tangible expression of literary, scientific, or artistic work, whether it appears in digital, print, audio, or visual form. It gives a right to the owner of the copyright to prevent the unauthorized copying by another of that work. In the United States a copyright is awarded for a period of the life of the copyright holder plus 50 years. It is not necessary to publish a copyright notice for a work to be copyrighted. A copyright comes into existence when one fixes the work in "any tangible medium of expression." For best protection the document should be marked © copyright 2006, John Doe, and registered with the U.S. Copyright Office of the Library of Congress. Unlike a patent, a copyright requires no extensive search to ensure the degree of originality of the work.

A basic principle of copyright law is the principle of *fair use* in which an individual has the right to make a single copy of copyrighted material for personal use for the purpose of criticism, comment, news reporting, teaching, scholarship, or research. Copying that does not constitute fair use requires the payment of a royalty fee to the Copyright Clearance Center. While the U.S. Copyright Act does not directly define fair use, it does base it on four factors:[1]

- The purpose and character of the use—is it of a commercial nature or for nonprofit educational purposes?
- The nature of the copyrighted work—is it a highly creative work or a more routine document?

---

1. D. V. Radack, *JOM*, February 1996.

- The amount of the work used in relation to the copyrighted work as a whole.
- The effect of the use on the potential market value of the copyrighted work. Usually this is the most important of the factors.

5.10
COMPANY-CENTERED INFORMATION

We started this chapter with an attempt to alert you to the magnitude of the problem with gathering information for design. Then we introduced you to each of the major sources of engineering information in the library and on the Internet, as well as giving you many trusted places to get started in your "information treasure hunt." This last section deals more specifically with company-based information and alerts you to the importance of gaining information by networking with colleagues at work and within professional organizations.

We can differentiate between *formal* (explict) sources of information and *informal* (tacit) sources. The sources of information considered in this chapter have been of the formal type. Examples are technical articles and patents. Informal sources are chiefly those in which information transfers on a personal level. For example, a colleague may remember that Sam Smith worked on a similar project five years ago and suggests that you check the library or file room to find his notebooks and any reports that he may have written.

The degree to which individual engineers pursue one or the other approaches to finding information depends on several factors:

- The nature of the project. Is it closer to an academic thesis, or is it a "firefighting" project that needs to be done almost immediately?
- The personality and temperament of the individual. Is he a loner who likes to puzzle things out on his own, or a gregarious type who has a wide circle of friends willing to share their experience at any time?
- Conversations are sometimes crucial to the solution of a problem. In this environment, knowledge sharing can form a community of understanding in which new ideas are created.
- The corporate culture concerning knowledge generation and management. Has the organization emphasized the importance of sharing information and developed methods to retain the expertise of senior engineers in ways that it can be easily accessed?
- Perhaps the necessary information is known to exist but it is classified, available only to those with a need to know. This requires action by higher management to gain you access to the information.

Clearly, the motivated and experienced engineer will learn to utilize both kinds of information sources, but each person will favor either explicit or tacit information sources.

In the busy world of the design engineer, relevance is valued above all else. Information that supplies just the needed answer to a particular stress analysis problem is more prized than a source that shows how to work a class of stress problems and contains the nugget of information that can be applied to the actual problem. Books are generally considered to be highly reliable, but out-of-date. Periodicals can provide the timeliness that is required, but there is a tendency to be overwhelmed by sheer numbers. In deciding which article to sit down and read, many engineers quickly read the abstract, followed by a scan of the graphs, tables, and conclusions.

The amount of design information that can be obtained from within the company is quite considerable and of many varieties. Examples are:

- Product specifications
- Concept designs for previous products
- Test data on previous products
- Bill of materials on previous products
- Cost data on previous projects
- Reports on previous design projects
- Marketing data on previous products
- Sales data on previous products
- Warranty reports on previous products
- Manufacturing data
- Design guides prepared for new employees
- Company standards

Ideally this information will be concentrated in a central engineering library. It may even be neatly packaged, product by product, but most likely much of the information will be dispersed between a number of offices in the organization. Often it will need to be pried out individual by individual. Here is where the development of a good network among your colleagues pays big dividends.

5.11
SUMMARY

The gathering of design information is not a trivial task. It requires knowledge of a wide spectrum of information sources. These sources are, in increasing order of specificity:

- The World Wide Web, and its access to digital databases
- Business catalogs and other trade literature
- Government technical reports and business data
- Published technical literature, including trade magazines
- Network of professional friends, aided by e-mail
- Network of professional colleagues at work
- Corporate consultants

At the outset it is a smart move to make friends with a knowledgeable librarian or information specialist in your company or at a local library who will help you become familiar with the information sources and their availability. Also, devise a plan to develop your own information resources of handbooks, texts, tearsheets from magazines, computer software, websites, and a digital portfolio of your own work products.

NEW TERMS AND CONCEPTS

Citation searching	Keyword	TCP/IP
Copyright	Monograph	Technical journal
Gray literature	Patent	Trade magazine
HTML	Periodical	Trademark
Intellectual property	Reference port	URL
Internet	Search engine	World Wide Web

BIBLIOGRAPHY

Anthony, L. J.: *Information Sources in Engineering,* Butterworth, Boston, 1985. *Guide to Materials Engineering Data and Information,* ASM International, Materials Park, OH, 1986.

Lord, C. R.: *Guide to Information Sources in Engineering,* Libraries Unlimited, Englewood, CO, 2000 (emphasis on U.S. engineering literature and sources).

MacLeod, R. A.: *Information Sources in Engineering,* 4th ed., K. G. Saur, Munich, 2005 (emphasis on British engineering literature and sources).

Osif, B. A.: *Using the Engineering Literature,* CRC Press, Boca Raton, FL, 2006.

Wall, R. A. (ed.): *Finding and Using Product Information,* Gower, London, 1986.

PROBLEMS AND EXERCISES

5.1 Prepare in writing a personal plan for combating technological obsolescence. Be specific about the things you intend to do and read.

5.2 Select a technical topic of interest to you.

   (a) Compare the information that is available on this subject in a general encyclopedia and a technical encyclopedia.

   (b) Look for more specific information on the topic in a handbook.

   (c) Find five current texts or monographs on the subject.

5.3 Use the indexing and abstracting services to obtain at least 20 current references on a technical topic of interest to you. Use appropriate indexes to find 10 government reports related to your topic.

5.4 Search for:

   (a) U.S. government publications dealing with the disposal of nuclear waste;

   (b) metal matrix composites.

5.5 Where would you find the following information?

   (a) The services of a taxidermist.

   (b) A consultant on carbon-fiber-reinforced composite materials.

(c)   The price of an X3427 semiconductor chip.

(d)   The melting point of osmium.

(e)   The proper hardening treatment for AISI 4320 steel.

5.6 Find and read a technical standard on the air flow performance characteristics of vacuum cleaners in the ASTM Standards. List some other standards concerning vacuum cleaners. Write a brief report about the kind of information covered in a standard.

5.7 Find a U.S. patent on a favorite topic. Print it out and identify each element of the patent as described in Sec. 5.9.5.

5.8 Discuss how priority is established in patent litigation.

5.9 Find out more information on the U.S. Provisional Patent. Discuss its advantages and disadvantages.

5.10 Find out about the history of Jerome H. Lemelson, who holds over 500 U.S. patents and who endowed the Lemelson prize for innovation at M.I.T.

5.11 In the discussion of copyrights in Section 5.9.6,

(a)   What is the meaning of the term "tangible expression"?

(b)   Explain the concept of fair use of creative works that have copyright protection.

(c)   How does ready electronic access affect the fair use guidelines that were established in 1978 before personal computers and the Internet?

(d)   What are some proposals that have been made to modify the copyright law to better address the needs of the digital age?

6

CONCEPT GENERATION

The most innovative products are the result of not only remembering useful design concepts but also recognizing promising concepts that arise in other disciplines. The best engineers will use creative thinking methods and design processes that assist in the synthesis of new concepts not previously imagined. Practical methods for enhancing creativity like brainstorming and Synectics, developed in the 20th century, are now adapted and adopted as methods for generating design concepts.

Creative idea generation is an intuitive way to proceed to a feasible design solution. However, being able to find one or two good concept ideas from a creative idea making session is not the same as generating a feasible conceptual design in engineering. Engineering systems are typically very complex, and their design requires structured problem solving at many points in the process. This means that all of the creativity available to an engineer or designer is called on several times in the design process and is used to arrive at alternative concepts for a small portion of an overall design task. Thus, all the creativity-enhancing methods are valuable to engineering designers during the conceptual design process (see Fig. 6.1).

Creative thinking is highly valued across many fields of endeavor, especially those that deal with problem solving. Naturally then, creativity-enhancing methods are offered in workplace seminars, and recruiters of new talent are including creativity as a high-value characteristic in job applicants. This chapter opens with a short section on how the human brain is able to perform creatively, and how successful problem solving is seen as a demonstration of creative skill. Methods for thinking in ways that increase creative results in problem-solving contexts have been codified by specialists in several fields and are presented here.

No engineering activity requires more creativity than design. The ability to identify concepts that will achieve particular functions required by a product is a creative task. Section 6.3 shows how creativity methods and creative problem-solving

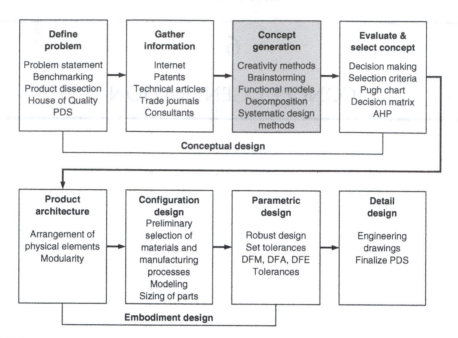

FIGURE 6.1
Product development process diagram displaying where creativity methods fit into the conceptual design process.

techniques are fundamental skills of engineering designers. It follows then that some methods for concept generation in the product development process blend engineering science and creative thinking techniques. The remainder of the chapter introduces three of the most common engineering design methods: Functional Decomposition and Synthesis in Sec. 6.5; Morphological Analysis in Sec. 6.6; and, the Theory of Inventive Problem Solving, TRIZ, in Sec. 6.7. The basics of each method are presented with examples illustrating the method's core ideas. Each section includes many excellent references for the reader wishing to study the design methods in more detail.

6.1
INTRODUCTION TO CREATIVE THINKING

During past periods of growth in the United States, manufacturing managers believed that a product development organization could be successful with only a small number of creative people and the majority of the professionals being detail-oriented doers. Today's fierce worldwide competition for markets, new products, and engineering dominance is changing that mindset. Current business strategists believe that only organizations that create the most innovative and advanced products and processes will survive, let alone thrive. Thus, each engineer has a strong incentive to improve his or her own creative abilities and put them to work in engineering tasks.

Society's view of creativity has changed over time. During the 19th century, creativity was seen as a romantic and mysterious characteristic. Scholars believed creativity to be an unexplainable personal talent present at the birth of an artist. It was thought that creativity was unable to be taught, copied, or mimicked. Individual creativity was a kind of genius that was nurtured and developed in those with the natural gift. The rising popularity of the scientific approach in the 20th century changed the perception of creativity. Creativity was measurable and, therefore, controllable. That perspective grew into the progressive notion that creativity is a teachable skill for individuals and groups. Today's managers recognize that the same kind of psychological and physiologically based cognitive processes that produce artistic creativity are used in the deliberate reasoning about and development of solutions.

6.1.1 Models of the Brain and Creativity

The science of thinking and the more narrow science of design are classified as sciences of the artificial.[1] Exploring natural sciences is based on investigating phenomena that can be observed by the scientist. Unfortunately, it is not possible to observe and examine the steps that a creative person's brain follows while solving a problem or imagining a potential design. One can only study the results of the process (e.g., a problem solution or a design) and any commentary on how they developed as stated or recorded by the producer.[2]

Advances in medicine and technology have expanded the boundaries of the activities of the brain that are observable and can be studied in real time. Modern neuroscience uses sophisticated tools such as functional MRI and positron emission tomography to observe the brain in action. The field is making great strides in revealing how the brain works by identifying which parts of the brain are responsible for particular actions. While technology is helping scientists to investigate the physical workings of the brain, cognitive scientists are still at work on investigating the workings of the human mind so that the best thinking skills and methods of thought can be learned and taught for the benefit of all.

Understanding thinking is the realm of cognitive scientists and psychologists.[3] In general terms, cognition is the act of human thinking. Thinking is the execution of cognitive processes like the activities of collecting, organizing, finding, and using knowledge. Cognitive psychology is the more specialized study of the acquisition and use of knowledge by humans in their activities. The psychological aspects of human behavior must be considered in helping us to understand a person's thinking because cognitive processes are naturally influenced by an individual's perceptions and representations of knowledge. Skills for developing creative thinking come from sciences that study human thinking, actions, and behavior.

---

1. H. A. Simon, *Sciences of the Artificial,* 3d ed., MIT Publishing, Cambridge, MA, 1996.
2. Thinking about one's own thought process as applied to a particular task is called metacognition.
3. M. M. Smyth, A. F. Collins, P. E. Morris, and P. Levy, *Cognition in Action*, 2d ed. Psychology Press, East Sussex, UK, 1994.

Freud's Model of Levels of the Mind

Psychologists have developed several models of how the brain processes information and creates thoughts. Sigmund Freud developed a topographical model of the mind consisting of three levels:

- *Conscious mind*: the part of the mind where our current thinking and objects of attention take place. You can verbalize about your conscious experience, and you can think about it in a logical fashion. The conscious mind has relatively small capacity for storage of information in its memory. This memory can be categorized as immediate memory, lasting only milliseconds, and working memory lasting about a minute.
- *Preconscious mind*: the long-term memory, lasting anywhere from about an hour to several years. This is a vast storehouse of information, ideas, and relationships based on past experience and education. While things stored here are not in the conscious, they can be readily brought into the conscious mind.
- *Subconscious mind*: the content of this mind level is out of reach of the conscious mind. Thus, the subconscious acts independently of the conscious mind. It may distort the relation of the conscious and preconscious through its control of symbols and the generation of bias.

Freud developed his model to explain personality types and their behaviors based on his own training, experience, and beliefs about cognition. Freud's work led to the important conclusion that much behavior is driven directly from the subconscious mind, and these actions cannot be controlled by the conscious mind. One needs to be clear that Freud's levels of the mind are not necessarily physical locations in the brain. They are a model of the brain that helps to explain the ways that the brain appears to work when judged only by observing the actions of its owner.

Freud's levels of consciousness are used to help explain the process by which problems are solved in a creative fashion. The actions of the conscious mind are used to collect relevant information about a task while the pre- and subconscious levels of the mind are suspected of working on that information over time and then passing a solution to the conscious level of the brain in a flash of insight.

Brain-Dominance Theory

A second important model of the brain is the brain-dominance theory. Nobel Prize winner Roger Sperry studied the relationships between the brain's right and left hemispheres. He found that the left side of the brain tends to function by processing information in an analytical, rational, logical, sequential way. The right half of the brain tends to function by recognizing relationships, integrating and synthesizing information, and arriving at intuitive insights. Thinking that utilizes the left hemisphere of the brain is called critical or convergent thinking. Other terms for left-brained thinking are analytic or vertical thinking. It is generally associated with persons educated in the technical disciplines. Thinking that utilizes the right hemisphere of the brain is called creative or divergent thinking. Other terms for right-brained thinking are associative or lateral thinking. It is found most often with persons educated in the arts or social sciences. Examples of these two classifications of thinking operations are given in Table 6.1.

TABLE 6.1
Comparison of Left-Brained and Right-Brained Thinking

Critical Thinking (Left Brain)	Creative Thinking (Right Brain)
Logical, analytic, judgmental process	Generative, suspended judgment
Linear	Associative
Leads to only one solution	Creates many possible solutions
Considers only relevant information	Considers broad range of information
Movement is made in a sequential, rule-based manner	Movement is made in a more random pattern
Embodies scientific principles	Heavily influenced by symbols and imagery
Classifications and labels are rigid	Reclassifies objects to generate ideas
Vertical	Lateral
Convergent	Divergent

The understanding of the physiology of the brain is useful in research on cognition. Study of the brain physiology has revealed that there are connections between the two hemispheres of the brain and within the same hemisphere.[1] The number of connections between the hemispheres varies. In general, women appear to have a higher number of these cross connections, and this difference is noted as one explanation for women's higher capacity for multitasking. Connections found within the same hemisphere of the brain allow closer connections between the specialized areas of thought.

Researchers like Herrmann have developed a means of characterizing how individuals think according to the preference with which they seem to access different areas of the brain. Herrmann's instrument is a standardized test, the Herrmann Brain Dominance Instrument (HBDI™). It is similar in nature to the Kolb Learning Style Inventory or the Myers-Briggs Type Indicator (MBTI) personality classification instruments. Using tests like these, engineers often test as being as left-brained, preferring to think in logical, linear, and convergent ways. This skill set is ideal for analysis and deductive problem solving, but is not ideal for creative activities.

The brain-dominance model of thinking seems to fall short of giving concrete steps that one can follow to think up a creative idea when it is needed. However with study and practice, there is no reason that you cannot become adept at using both sides of your brain. Many training methods exist to encourage the use of the right side of the brain in problem solving such as that proposed by Buzan.[2] This model also provides support for having a team of members with diverse thinking styles working on problems requiring creativity and invention.

---

1. E. Lumsdain, M. Lumsdain, and J. W. Shelnutt, *Creative Problem Solving and Engineering Design*, McGraw-Hill, New York, 1999.
2. T. Buzan, *Use Both Sides of Your Brain*, Penguin Group, USA, New York, 1991.

6.1.2 Thinking Processes That Lead to Creative Ideas

Researchers have discovered that, generally speaking, the thought processes or mental operations used to develop a creative idea are the same processes that are routinely used. The good news about this view of creativity is that these strategies for achieving creative thinking can be accomplished by deliberate use of particular techniques, methods, or in the case of computational tools, software programs.

The study of creativity usually focuses on both the creator and the created object.[1] The first step is to study people who are considered to be creative and to study the development of inventions that display creativity. The assumption is that studying the thinking processes of the creative people will lead to a set of steps or procedures that can improve the creativity of the output of anyone's thinking. Similarly, studying the development of a creative artifact should reveal a key decision or defining moment that accounts for the outcome. This is a promising path if the processes used in each case have been adequately documented.

The first research strategy will lead us to creativity process techniques like those introduced in Sec. 6.2.1 and 6.3. The second strategy of studying creative objects to discover the winning characteristic has led to the development of techniques that use a previous set of successful designs to find inspiration for new ones. Analogy-based methods fall into this category, as do methods that generalize principles for future use, like TRIZ (see Sec. 6.7).

6.2

CREATIVITY AND PROBLEM SOLVING

Creative thinkers are distinguished by their ability to synthesize new combinations of ideas and concepts into meaningful and useful forms. A creative engineer is one who produces a lot of useful ideas. These can be completely original ideas inspired by a discovery. More often, creative ideas result from putting existing ideas together in novel ways. A creative person is adept at breaking a problem-solving task down to take a fresh look at its parts, or in making connections between the current problem and seemingly unrelated observations or facts.

We would all like to be called "creative," yet many of us feel that creativity is reserved for only the gifted few. There is the popular myth that creative ideas arrive with flash-like spontaneity—the flash of lightning and clap of thunder routine. However, researchers of the creative process assure us that most ideas occur by a slow, deliberate process that can be cultivated and enhanced with study and practice.

A characteristic of the creative process is that initially the idea is only imperfectly understood. Usually the creative person senses the total structure of the idea but initially perceives only a limited number of its details. There ensues a slow process of clarification and exploration as the entire idea takes shape. The creative process can be

---

1. K. S. Bowers, P. Farvolden, and L. Mermigis, "Intuitive Antecedents of Insight," in *The Creative Cognition Approach,* Steven Smith, Thomas Ward, and Ronald Finke (eds.), The MIT Press, Cambridge, MA, 1995.

viewed as moving from a vague idea to a well-structured idea, from the chaotic to the organized, from the implicit to the explicit. Engineers, by nature and training, usually value order and explicit detail and abhor chaos and vague generality. Thus, we need to train ourselves to be open to these aspects of the creative process. Recognizing that the flow of creative ideas cannot be turned on upon command, we need to identify the conditions that are most conducive to creative thought. Recognizing that creative ideas are elusive, we need to be alert to capture and record our creative thoughts.

6.2.1 Supports to Creative Thinking

A group of researchers in the sciences named the successful use of thought processes and existing knowledge to produce creative ideas *creative cognition*.[1] Creative cognition is the use of regular cognitive operations to solve problems in novel ways. One way to increase the likelihood of positive outcomes is to apply methods found to be useful for others. Following are steps you can take to enhance your creative thinking.

1. *Develop a creative attitude*: To be creative it is essential to develop confidence that you can provide a creative solution to a problem. Although you may not visualize the complete path through to the final solution at the time you first tackle a problem, you must have self-confidence; you must believe that a solution will develop before you are finished.
2. *Unlock your imagination*: You must rekindle the vivid imagination you had as a child. One way to do so is to begin to question again. Ask "why" and "what if," even at the risk of displaying a bit of naïveté. Scholars of the creative process have developed thought games that are designed to provide practice in unlocking your imagination and sharpening creative ability.
3. *Be persistent*: Creativity often requires hard work. Most problems will not succumb to the first attack. They must be pursued with persistence. After all, Edison tested over 6000 materials before he discovered the species of bamboo that acted as a successful filament for the incandescent light bulb. It was also Edison who made the famous comment, "Invention is 95 percent perspiration and 5 percent inspiration."
4. *Develop an open mind*: Having an open mind means being receptive to ideas from any and all sources. The solutions to problems are not the property of a particular discipline, nor is there any rule that solutions can come only from persons with college degrees. Ideally, problem solutions should not be concerned with company politics. Because of the NIH factor (not invented here), many creative ideas are not picked up and followed through.
5. *Suspend your judgment*: Nothing inhibits the creative process more than critical judgment of an emerging idea. Engineers, by nature, tend toward critical attitudes, so special forbearance is required to avoid judgment at an early stage of conceptual design.

---

1. Steven Smith, Thomas Ward, and Ronald Finke (eds.), *The Creative Cognition Approach,* The MIT Press, Cambridge, MA, 1995.

6. *Set problem boundaries*: We place great emphasis on proper problem definition as a step toward problem solution. Establishing the boundaries of the problem is an essential part of problem definition. Experience shows that setting problem boundaries appropriately, not too tight or not too open, is critical to achieving a creative solution.

Some psychologists describe the creative thinking process and problem solving in terms of a simple four-stage model.[1]

- Preparation (stage 1): The elements of the problem are examined and their interrelations are studied.
- Incubation (stage 2): You "sleep on the problem." Sleep disengages your conscious mind, allowing the unconscious mind to work on a problem freely.
- Inspiration (stage 3): A solution or a path toward the solution emerges.
- Verification (stage 4): The inspired solution is checked against the desired result.

The preparation stage should not be slighted. The design problem is clarified and defined. Information is gathered, assimilated, and discussed among the team. Generally, more than one session will be required to complete this phase. Between team meetings the subconscious mind works on the problem to provide new approaches and ideas. The incubation period then follows. A creative experience often occurs when the individual is not expecting it and after a period when they have been thinking about something else. Observing this relationship between fixation and incubation led Smith to conclude that incubation time is a necessary pause in the process. Incubation time allows fixation to lessen so that thinking can continue.[2] Other theorists suggest that this time allows for the activation of thought patterns and searches to fade, allowing new ones to emerge when thinking about the problem is resumed.[3]

One prescription for improving creativity is to fill the mind and imagination with the context of the problem and then relax and think of something else. As you read or play a game there is a release of mental energy that your preconscious can use to work on the problem. Frequently there will be a creative "Ah-ha" experience in which the preconscious will hand up into your conscious mind a picture of what the solution might be.

Insight is the name science gives to the sudden realization of a solution. There are many explanations of how insight moments occur. Consultants in creativity train people to encourage the insight process, even though it is not a well-understood one. Insight can occur when the mind has restructured a problem in such a way that the previous impediments to solutions are eliminated, and unfulfilled constraints are suddenly satisfied.

---

1. S. Smith, "Fixation, Incubation, and Insight in Memory and Creative Thinking," in *The Creative Cognition Approach,* Steven Smith, Thomas Ward, and Ronald Finke (eds.), The MIT Press, Cambridge, MA, 1995.
2. Ibid.
3. J. W. Schooler and J. Melcher, "The Ineffability of Insight," in *The Creative Cognition Approach,* Steven Smith, Thomas Ward, and Ronald Finke (eds.), The MIT Press, Cambridge, MA, 1995.

Since the preconscious has no vocabulary, the communication between the conscious and preconscious will be by pictures or symbols. This is why it is important for engineers to be able to communicate effectively through sketches. If the inspiration stage does not occur in the dramatic manner just described, then the prepared minds of the team members achieve the creative concept through a more extended series of meetings using the methods considered in the balance of this chapter. Finally, the ideas generated must be validated against the problem specification using the evaluation methods discussed in Chap. 7.

To achieve a truly creative solution to a problem, one must utilize two thinking styles: convergent thinking and divergent thinking. Convergent thinking is the type of analytical thought process reinforced by most engineering courses where one moves forward in sequential steps after a positive decision has been made about the idea. If a negative decision is made at any point in the process, you must retrace your steps along the analysis trail until the original concept statement is reached. In lateral thinking your mind moves in many different directions, combining different pieces of information into new patterns (synthesis) until several solution concepts appear.

6.2.2 Barriers to Creative Thinking

It is important for you to recognize how *mental blocks* interfere with creative thinking.[1] A mental block is a mental wall that prevents the problem solver from correctly perceiving a problem or conceiving its solution. A mental block is an event that inhibits the successful use of normal cognitive processes to come to a solution. There are many different types of mental blocks.

Perceptual Blocks

Perceptual blocks have to do with not properly defining the problem and not recognizing the information needed to solve it.

- *Stereotyping*: Thinking conventionally or in a formulaic way about an event, person, or way of doing something. The brain classifies and stores information in labeled groups. When new information is taken in, it is compared with established categories and assigned to the appropriate group. This leads to stereotyping of ideas since it imposes preconceptions on mental images. As a result, it is difficult to combine apparently unrelated images into an entirely new creative solution for the design.
- *Information overload*: You become so overloaded with minute details that you are unable to sort out the critical aspects of the problem. This scenario is termed "not being able to see the forest for the trees." Cognitively this is a situation of engaging all the available short-term memory so that there is no time for related searches in long-term memory.
- *Limiting the problem unnecessarily*: Broad statements of the problem help keep the mind open to a wider range of ideas.

---

1. J. L. Adams, *Conceptual Blockbusting*, 3d ed., Addison-Wesley, Reading, MA, 1986.

- *Fixation*:[1] People's thinking can be influenced so greatly by their previous experience or some other bias that they are not able to sufficiently recognize alternative ideas. Since divergent thinking is critical to generating broad sets of ideas, fixation must be recognized and dealt with. A kind of fixation called memory blocking is discussed in the section on intellectual blocks.
- *Priming or provision of cues*: If the thinking process is started by giving examples or solution cues, it is possible for thinking to stay within the realm of solutions suggested by those initial starting points. This is known as the conformity effect. Some capstone design instructors have noted this commenting that once students find a relevant patent for solving a design problem, many of their new concepts follow the same solution principle.

Emotional Blocks

These are obstacles that are concerned with the psychological safety of the individual. They reduce the freedom with which you can explore and manipulate ideas. They also interfere with your ability to conceptualize readily.

- *Fear of risk taking*: This is the fear of proposing an idea that is ultimately found to be faulty. This is inbred in us by the educational process. Truly creative people must be comfortable with taking risks.
- *Unease with chaos*: People in general, and many engineers in particular, are uncomfortable with highly unstructured situations.
- *Inability or unwillingness to incubate new ideas*: In our busy lives, we often don't take the time to let ideas lie dormant so they can incubate properly. It is important to allow enough time for ideas to incubate before evaluation of the ideas takes place. Studies of creative problem-solving strategies suggest that creative solutions usually emerge as a result of a series of small ideas rather than from a "home run" idea.
- *Motivation*: People differ considerably in their motivation to seek creative solutions to challenging problems. Highly creative individuals do this more for personal satisfaction than personal reward. However, studies show that people are more creative when told to generate many ideas, so it shows that the motivation is not all self-generated.

Intellectual Blocks

Intellectual blocks arise from a poor choice of the problem-solving strategy or having inadequate background and knowledge.

- *Poor choice of problem-solving language or problem representation*: It is important to make a conscious decision concerning the "language" for your creative problem solving. Problems can be solved in either a mathematical, verbal, or a visual mode. Often a problem that is not yielding to solution using, for example, a verbal

1. S. Smith, "Fixation, Incubation, and Insight in Memory and Creative Thinking," in *The Creative Cognition Approach,* Steven Smith, Thomas Ward, and Ronald Finke (eds.), The MIT Press, Cambridge, MA, 1995.

mode can be readily solved by switching to another mode such as the visual mode. Changing the representation of a problem from the original one to a new one (presumably more useful for finding a solution) is recognized as fostering creativity.[1]

- *Memory block*: Memory holds strategies and tactics for finding solutions as well as solutions themselves. Therefore, blocking in memory searches is doubly problematic to creative thinking. A common form of blocking is maintaining a particular search path through memory because of the false belief that it will lead to a solution. This belief may arise from a false hint, reliance on incorrect experience, or any other reason that interrupts or distracts the mind's regular problem-solving processes.

- *Insufficient knowledge base*: Usually, ideas are generated from a person's education and experience. Thus, an electrical engineer is more likely to suggest an electronics-based idea, when a cheaper and simpler mechanical design would be better. This is a strong reason for working in interdisciplinary design teams. A good approach to gathering information is to do enough to get a good feel for the problem and then use this knowledge base to try to generate creative concepts. After that it is important to go back and exhaustively develop an information base to use in evaluating the creative ideas.

- *Incorrect information*: It is obvious that using incorrect information can lead to poor results. One form of the creative process is the combining of previously unrelated elements or ideas (information); if part of the information is wrong then the result of creative combination will be flawed. For example, if you are configuring five elements of information to achieve some result, and the ordering of the elements is critical to the quality of the result, you have 120 different orderings. If one of the elements is wrong, all 120 alternative orderings are wrong. If you only need to take two (2) of the five (5) elements, then there are 20 possible combinations. Of these 20, four will lead to wrong results because they will contain the incorrect element. The higher the number of elements that are combined, the more difficult it will be to sort out the correct combinations from those that are flawed.

Environmental Blocks

These are blocks that are imposed by the immediate physical or social environment.

- *Physical environment*: This is a very personal factor in its effects on creativity. Some people can work creatively with all kinds of distractions; others require strict quiet and isolation. It is important for each person to determine their optimum conditions for creative work, and to try to achieve this in the workplace. Also, many people have a time of day in which they are most creative. Try to arrange your work schedule to take advantage of this.

- *Criticism*: Nonsupportive remarks about your ideas can be personally hurtful and harmful to your creativity. It is common for students in a design class to be hesitant

---

1. R. L. Dominowski, "Productive Problem Solving," in *The Creative Cognition Approach*, Steven Smith, Thomas Ward, and Ronald Finke (eds.), The MIT Press, Cambridge, MA, 1995.

to expose their ideas, even to their team, for fear of criticism. This lack of confidence comes from the fact that you have no basis of comparison as to whether the idea is good. As you gain experience you should gain confidence, and be able to subject your ideas to friendly but critical evaluations. Therefore, it is very important for the team to maintain an atmosphere of support and trust, especially during the concept design phase.

6.3
CREATIVE THINKING METHODS

Improving creativity is a popular endeavor. A search of Google under Creative Methods yielded over 137 million hits, many of them books or courses on creativity improvement. Over 150 creativity improvement methods have been cataloged.[1] These methods are aimed at improving the following characteristics of the problem solver:

- *Sensitivity*: The ability to recognize that a problem exists
- *Fluency*: The ability to produce a large number of alternative solutions to a problem
- *Flexibility*: The ability to develop a wide range of approaches to a problem
- *Originality*: The ability to produce original solutions to a problem

Following are descriptions of some of the most commonly used creativity methods. Many of these creativity improvement methods directly eliminate the most common mental blocks to creativity.

6.3.1 Brainstorming

Brainstorming is the most common method used by design teams for generating ideas. This method was developed by Alex Osborn[2] to stimulate creative magazine advertisements, but it has been widely adopted in other areas such as design. The word *brainstorming* has come into general usage in the language to denote any kind of idea generation.

Brainstorming is a carefully orchestrated process. It makes use of the broad experience and knowledge of groups of individuals. The brainstorming process is structured to overcome many of the mental blocks that curb individual creativity in team members who are left to generate ideas on their own. Active participation of different individuals in the idea generation process overcomes most perceptual, intellectual, and cultural mental blocks. It is likely that one person's mental block will be different from another's, so that by acting together, the team's combined idea generation process flows well.

A well-done brainstorming session is an enthusiastic session of rapid, free-flowing ideas. The brainstorming process was first described in Sec. 4.7. Please review

---

1. www.mycoted.com.
2. A. Osborn, *Applied Imagination,* Charles Scribner & Sons, New York, 1953.

this section before proceeding further. To achieve a good brainstorming session, it is important to carefully define the problem at the start. Time spent here can help to avoid wasting time generating solutions to the wrong problem. It is also necessary to allow a short period for individuals to think through the problem quietly and on their own before starting the group process.

Participants in brainstorming sessions react to ideas they hear from others by recalling their own thoughts about the same concepts. This action of redirecting a stream of thought uncovers new possibilities in the affected team member. Some new ideas may come to mind by adding detail to a recently voiced idea or taking it in different, but related, directions. This building upon others' ideas is known as piggy-backing or scaffolding, and it is an indicator of a well-functioning brainstorming session. It has been found that the first 10 or so ideas will not be the most fresh and creative, so it is critical to get at least 30 to 40 ideas from your session. An important attribute of this method is that brainstorming creates a large number of ideas, some of which will be creative.

The evaluation of your ideas should be done at a meeting on a day soon after the brainstorming session. This removes any fear that criticism or evaluation is coming soon and keeps the brainstorming meeting looser. Also, making the evaluation on the day after the idea generation session allows incubation time for more ideas to generate and time for reflection on what was proposed. The evaluation meeting should begin by adding to the original list any new ideas realized by the team members after the incubation period. Then the team evaluates each of the ideas. Hopefully, some of the wild ideas can be converted to realistic solutions. Chapter 7 will discuss methods of evaluation.

One way to help the brainstorming process is to break up the normal thought pattern by using a *checklist* to help develop new ideas. The originator of brainstorming proposed such a list, which Eberle[1] modified into the acrostic SCAMPER (Table 6.2). Generally, the SCAMPER checklist is used as a stimulant when the flow of ideas begins to fall off during the brainstorming activity. The questions in the SCAMPER checklist are applied to the problem in the following way:[2]

- Read aloud the first SCAMPER question.
- Write down ideas or sketch ideas that are stimulated by the question.
- Rephrase the question and apply it to the other aspects of the problem.
- Continue applying the questions until the ideas cease to flow.

Because the SCAMPER questions are generalized, they sometimes will not apply to a specific technical problem. Therefore, if a question fails to evoke ideas, move on quickly to the next question. A group that will be doing product development over time in a particular area should attempt to develop their own checklist questions tailored to the situation.

Brainstorming has benefits and is an appropriate activity for idea generation in a team setting. However, brainstorming does not surmount many emotional and

1. R. Eberle, *SCAMPER: Games for Imagination Development*, D.O.K. Press, Buffalo, NY, 1990.
2. B. L. Tuttle, "Creative Concept Development," *ASM Handbook*, vol. 20, pp. 19–48, ASM International, Materials Park, OH, 1997.

TABLE 6.2
SCAMPER Checklist to Aid in Brainstorming

Proposed Change	Description
Substitute	What if used in a different material, process, person, power source, place, or approach?
Combine	Could I combine units, purposes, or ideas?
Adapt	What else is like this? What other idea does it suggest? Does the past offer a parallel? What can I copy?
Modify, magnify, minify	Could I add a new twist? Could I change the meaning, color, motion, form, or shape? Could I add something? Make stronger, higher, longer, thicker? Could I subtract something?
Put to other uses	Are there new ways to use this as is? If I modify it, does it have other uses?
Eliminate	Can I remove a part, function, person without affecting outcome?
Rearrange, reverse	Could I interchange components? Could I use a different layout or sequence? What if I transpose cause and effect? Could I transpose positive and negative? What if I turn it backward, upside down, or inside out?

environmental mental blocks. In fact, the process can intensify some of the mental blocks in some team members (e.g., unease with chaos, fear of criticism, and perpetuation of incorrect assumptions). To mitigate these effects that dampen creativity, a team can conduct a *brainwriting*[1] exercise prior to the formal brainstorming session.

6.3.2 Refinement and Evaluation of Ideas

The objective of creative idea evaluation is not to winnow down the set of ideas into a single or very small number of solutions. The primary purpose of the refinement and evaluation step in concept generation is the identification of creative, feasible, yet still practical ideas. (Convergent thinking dominates this process.)

The type of thinking used in refining the set of creative ideas is more focused than the divergent type of thinking that was used in generating creative ideas. (Recall that teams often use techniques that purposely encourage divergent thinking (e.g., SCAMPER). Here we use convergent thinking to clarify concepts and arrive at ideas that are physically realizable.

The first step is to sort the ideas into feasibility categories following the method of the affinity diagram as discussed in Sec. 4.7. A quick way to do this is to group the ideas into three categories based on the judgment of the team as to their feasibility.

- Ideas that are feasible as they stand. (You would be happy to show them to your boss.)
- Ideas that may have potential after more thought or research are applied. (These ideas you would not want to show your boss.)

---

1. CreatingMinds, http://creatingminds.org/tools/brainwriting.htm, accessed February 16, 2007.

- Ideas that are very unfeasible and have no chance of becoming good solutions. Before discarding an idea, ask, "What about this idea makes it not feasible?" and "What would have to change for this idea to become feasible? " This type of examination of wacky ideas can lead to new insights into the design task.

Checking concept ideas for feasibility is a critical step in the design process. Time is a valuable and limited resource the team cannot spend on developing design solutions with a low probability of success.

It is difficult to choose the right time to eliminate early design concepts. If the time is too early, team members may not yet have enough information to determine the level of feasibility of some concepts. The more ambitious the design task, the more likely this is to be true. A valuable strategy used by successful teams is to document ideas and the rationale made for choosing to pursue them or not. When documentation is thorough, a team can take some risks in moving rapidly because they can retrace their steps through the documented design rationale.

An alternate strategy for classifying concepts is to group the ideas according to common engineering characteristics, as in the Shot-Buddy consideration of different ideas for a basketball ball return. It would make sense to use critical-to-quality engineering characteristics. There will always be a category for wild ideas.

Next, the team examines each category of designs, one at a time. The team discusses the concepts within the class with the objective of seeing how they can be combined or rearranged into more completely developed solutions.

Unlike the original brainstorming session, where emphasis was on quantity of ideas and discussion was minimized, here discussion and critical thought are encouraged.

Team members can elaborate on ideas, piggyback on other ideas, or force-fit and combine ideas to create a new idea. This is shown in Fig. 6.2 by representing each idea with a different symbol. First ideas are grouped into categories (Task 1). Then concepts are synthesized by combining ideas from the different categories (Task 2). Notice that the ideas that are combined to form a concept may come from any of the previous categories. Sometimes *force-fitting* results in further consolidation of the ideas (Task 3). The overall objective is to come out of this session with several well-developed design concepts.

The above example is idealized. It uses only visual design elements to represent ideas, but mechanical design is more complex because functionality is the prime consideration in the generation of concepts. Also, aspects of form must be accommodated by the design concept.

Please realize that this evaluation session is as important as the original meeting in which ideas were first generated. It should not be rushed. Typically it will take two or three times as long as the first brainstorming session, but it is worth it.

6.3.3 Idea Generating Techniques Beyond Brainstorming

Brainstorming is commonly used as the first tool in generating creative ideas. There are many other tools and methods that are also effective. This section presents simple

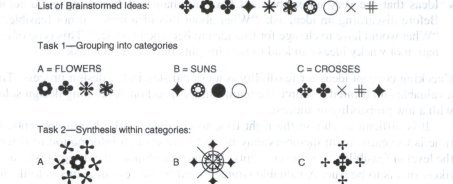

Task 3—Force-fitting between categories:

FIGURE 6.2
Schematic diagram of the creative idea evaluation process. (From E. Lumsdaine and M. Lumsdaine, *Creative Problem Solving,* McGraw-Hill, New York, 1995, p. 226.)

methods that support creative thinking.[1] These methods consist of prompting new thinking or blocked thinking by providing questions that lead team members to considered new perspectives on a problem or creative task. You will note that the SCAMPER questions listed in Table 6.2 have the same intent as the methods listed in this section.

Six Key Questions

Journalism students are taught to ask six simple questions to ensure that they have covered the entire story. These same questions can be used to help you approach the problem from different angles.

- Who? Who uses it, wants it, will benefit by it?
- What? What happens if X occurs? What resulted in success? What resulted in failure?
- When? Can it be speeded up or slowed down? Is sooner better than later?
- Where? Where will X occur? Where else is possible?
- Why? Why is this done? Why is that particular rule, action, solution, problem, failure involved?
- How? How could it be done, should it be done, prevented, improved, changed, made?

---

1. R. Harris, Creative Thinking Techniques, http://www.virtualsalt.com/crebook2.htm

Five Whys

The Five Whys technique is used to get to the root of a problem. It is based on the premise that it is not enough to just ask why one time. For example:

- Why has the machine stopped? A fuse blew because of fan overload.
- Why was there an overload? There was inadequate lubrication for the bearings.
- Why wasn't there enough lubrication? The lube pump wasn't working.
- Why wasn't the pump working? The pump shaft was vibrating because it had worn due to abrasion.
- Why was there abrasion? There was no filter on the lube pump, allowing debris into the pump.

Checklists

Checklists of various types often are used to help stimulate creative thoughts. Osborn was the first to suggest this method. Table 6.3 is a modification of his original checklist of actions to take to stimulate thought in brainstorming. Please note that checklists are used often in design in a completely different way. They are used in a way to remember important functions or tasks in a complex operation. See, for example, the checklist for a final design review in Chap. 9. Table 6.3 is an example of a checklist devised for a specific technical problem.

Fantasy or Wishful Thinking

A strong block to creativity is the mind's tenacious grip on reality. One way to stimulate creativity is to entice the mind to think in a flight of fancy, in the hope of bringing out really creative ideas. This can be done by posing questions in an "invitational way" so as to encourage an upbeat, positive climate for idea generation. Typical questions would be:

- Wouldn't be nice if . . . . ?
- What I really want to do is . . . .
- If I did not have to consider cost, ...
- I wish . . .

TABLE 6.3
A Checklist for Technological Stretching
(G. Thompson and M. London)

What happens if we push the conditions to the limit?

Temperature, up or down?

Pressure, up or down?

Concentration, up or down?

Impurities up or down?

G. Thompson and M. London, "A Review of Creativity Principles Applied to Engineering Design," *Proc. Instn. Mech. Engrs.,* vol. 213, part E, pp. 17–31, 1999.

The use of an invitational turn of phrase is critical to the success of this approach. For example, rather than stating, "this design is too heavy," it would be much better to say "how can we make the design lighter?" The first phrase implies criticism, the latter suggests improvement for use.

6.3.4 Random Input Technique

Edward de Bono is a long-time developer of creativity methods.[1] He stresses the importance of thought patterns, and he coined the term *lateral thinking* for the act of cutting across thought patterns. One of the key tenets of lateral thinking is the concept that an act of provocation is needed to make the brain switch from one pattern of thought to another. The provocative event interrupts the current thinking process by introducing a new problem representation, providing a new probe for a memory search, or leading to a restructuring of the solution plan.

Suppose you are thinking about a problem and you have a need for a new idea. In order to force the brain to introduce a new thought, all you have to do is to introduce a new random word. The word can be found by turning at random to a page in a dictionary, arbitrarily deciding to take the ninth word on the page, or turning randomly to a page in any book and at random selecting a word. Now, the provocation is to find how the chosen word is related to the problem under consideration.

As an example,[2] consider a group of students who were working on the problem of how the rules of basketball could be changed to make shorter players (under 5' 9") competitive. The word *humbug* was chosen, which led to the word *scrooge*, which led to *mean,* which led to *rough*, which led to the idea of *more relaxed foul rules for short players.* De Bono points out that this forced relationship from a random word works because the brain is a self-organizing patterning system that is very good at making connections even when the random word is very remote from the problem subject. He says, "It has never happened to me that the random word is too remote. On the contrary, what happens quite often is that the random word is so closely connected to the focus that there is very little provocative effect." It is also worth noting that the random input technique does not apply only to random words. It also works with objects or pictures. Ideas can be stimulated by reading technical journals in fields other than your own, or by attending technical meetings and trade shows in fields far from your own. The overarching principle is the willingness to look for unconventional inputs and use these to open up new lines of thinking.

6.3.5 Synectics: An Inventive Method Based on Analogy

In design, like in everyday life, many problems are solved by analogy. The designer recognizes the similarity between the design under study and a previously solved

1. E. de Bono, *Lateral Thinking*, Harper & Row, New York, 1970; *Serious Creativity*, Harper Collins, New York, 1993.
2. S. S. Folger and S. E. LeBlanc, *Strategies for Creative Problem Solving*, Prentice Hall, Englewood Cliffs, NJ, 1995.

problem. Whether it is a creative solution depends on the degree to which the analogy leads to a new and different design. One type of solution based on analogy recognizes the similarities between an existing product and its design specification and the design specification of the product under study. This most likely will not be a creative design, and it may not even be a legal design, depending on the patent situation of the older product.

Synectics (from the Greek word *synektiktein,* meaning joining together of different things into unified connection) is a methodology for creativity based on reasoning by analogy that was first described in the book by Gordon.[1] It assumes that the psychological components of the creative processes are more important in generating new and inventive ideas than the intellectual processes. This notion is counterintuitive to engineering students, who are traditionally very well trained in the analysis aspects of design.

Synectics is a formalized process led by a highly trained facilitator that proceeds in stages. The first stage of Synectics is to understand the problem. The problem is examined from all angles with the goal of *"making the strange familiar."* However, examining all aspects of the problem to the extent that is done in Synectics is likely to have blocked one's capacity for creative solution of the problem. Therefore, the second phase searches for creative solutions drawing heavily on the four types of analogies discussed in this section. The objective is to distance your mind from the problem using analogies, and then to couple them with the problem in the last phase of Synectics. This is done by *force-fitting* the ideas generated by analogy into the various aspects of the problem definition. We have already seen an example of force fitting in the Random Input Technique discussed in Sec. 6.3.4.

Synectics is used in creative problem solving because of the power of the use of analogies. Knowing how to use the four different types of analogies differentiated in Synectics is valuable for anyone wishing to generate ideas about an existing problem. Synectics recognizes four types of analogy: (1) direct analogy, (2) fantasy analogy, (3) personal analogy, and (4) symbolic analogy.

- *Direct analogy*: The designer searches for the closest physical analogy to the situation at hand. In describing the motion of electrons about the nucleus of an atom it is common to use the analogy of the moon's rotation about Earth or Earth's rotation about our sun. The analogy is direct because in each system there are matched physical objects behaving the same way—rotating about a central object. A direct analogy may take the form of a similarity in physical behavior (as in the previous example), similarity in geometrical configuration, or in function. Analogies are not necessarily the result of complex mental model restructuring of ideas if they are from the same domain. Novices are likely to find analogies based on physical similarities. It takes special training (like that provided by formal methods) to recognize analogies based on more abstract characteristics like functional similarity. Bio-inspired design is a specific type of analogy under increased research in the past decade. Bio-inspired design is based on the similarity

1. W. J. J. Gordon, *Synectics: The Development of Creative Capacity,* Harper & Brothers, New York, 1961.

between biological systems and engineering systems. This topic is discussed further in this section.

- *Fantasy analogy*: The designer disregards all problem limitations and laws of nature, physics, or reason. Instead, the designer imagines or wishes for the perfect solution to a problem. For example, suppose you enter a large parking lot on a cold, windy, and rainy day, only to discover that you have forgotten where your car is parked. In a perfect world, you could wish your car to materialize in front of you or to turn itself on and drive to where you are standing when you call it. These are far-fetched ideas but they contain potential. Many cars now have a chip in their key ring that flashes the car lights when activated to send you a locator signal. Perhaps the design team used some aspect of the fantasy analogy to solve the lost car problem.

- *Personal analogy*: The designer imagines that he or she is the device being designed, associating his or her body with the device or the process under consideration. For example, in designing a high-quality industrial vacuum cleaner, we could imagine ourselves as the cleaner. We can suck up dirt through a hose like drinking through a straw. We can pick up dirt and debris by running our hands across a smooth surface or by combing our fingers through a thick and fibrous material. We could also lick the surface clean using moisture, friction, and an absorbent material like we do when we lick frosting off a cupcake.

- *Symbolic analogy*: This is perhaps the least intuitive of the approaches. Using symbolic analogy the designer replaces the specifics of the problem with symbols and then uses manipulation of the symbols to discover solutions to the original problem. For example, there are some mathematical problems that are converted (mapped) from one symbolic domain to another to allow for easier processing. LaPlace transforms are an example of this type of symbolic analogy. There is a method for the structural synthesis of mechanisms that requires drawing a graph representing the joints and linkages of the mechanism and then converting the graph into a set of equations for solution.[1]

6.3.6 Biomimetic Design

A particularly intriguing source of direct analogies is those that are inspired by biological systems. This subject is called *biomimetics,* the mimicking of biological systems. A well-known example of biomimetics is the invention of the Velcro fastener. Its inventor, George de Mestral, conceived the idea when he wondered why cockleburs stuck to his trousers after a walk in the woods. Mestral was trained as an engineer. Under the microscope he found that the hook-shaped projections on the burs adhered to the small loops on his wool trousers. After a long search he found that nylon tape could be shaped into a hook tape with small, stiff hooks and a loop tape with small loops. Velcro tape was born. This example also illustrates the principle of *serendipitous discovery*—discovery by accident. It also shows that discovery of this

1. L. W. Tsai, *Mechanism Design: Enumeration of Kinetic Structures According to Function*, CRC Press, Boca Raton, FL, 1997.

type also requires a curious mind, often called *the prepared mind*. In most cases of serendipitous discovery, the idea comes quickly, but as in the case of Velcro, a long period of hard work is required to develop the innovation.

Biomimetic design combines the principles of using design by direct analogy with the knowledge of biological phenomena. Mechanical design is based on satisfying what a product or system must do (i.e., the function for which the device is created). Therefore, the effective use of biological analogies is based on identifying how biological systems manage to produce behaviors designers seek in physical systems. The challenge for designers is twofold: (1) engineering designers are not trained in a wide variety of biological systems and (2) the words engineers use to express behavior do not always match words used to describe biological systems. The value of identifying biological analogies to mechanical systems has created a rich research literature for design. One example of a tool that engineers can use is the website Ask Nature,[1] an open source website created to facilitate the use of biomimicry. A growing body of literature includes many other examples of biological analogies.[2]

6.3.7 Concept Map

A very useful tool for the generation of ideas by association, and for organizing information in preparation for writing a report, is the *concept map*,[3] and its close relation the *mind map*.[4] A concept map is good for generating and recording ideas during brainstorming. Because it is a visual method instead of a verbal one, it encourages left-brained thinking. Because it requires the mapping of associations between ideas it stimulates creative thought. Thus, it also can be very useful in generating solution concepts.

A concept map is made on a large sheet of paper. A concise label for the problem or issue is placed at the center of the sheet. Then the team is asked to think about what concepts, ideas, or factors are related to the problem.

- Write down team-generated thoughts surrounding the central problem label.
- Underline or circle them and connect them to the central focus.
- Use an arrow to show which issue drives what.
- Create new major branches of concepts to represent major subtopics.
- If the process develops a secondary or separate map, label it and connect it to the rest of the map.

---

1. "Ask Nature,"The Biomimicry Institute, www.asknature.org, accessed July 19, 2011.

2. T. W. D'Arcy, *Of Growth and Form*, Cambridge Univ. Press, 1961; S. A. Wainwright et al., *Mechanical Design in Organisms*, Arnold, London, 1976; M. J. French, *Invention and Evolution: Design in Nature and Engineering*, Cambridge Univ. Press, 1994; S. Vogel, *Cat's Paws and Catapults: Mechanical Worlds of Nature and People*, W. W. Norton & Co., New York, 1998; Y. Bar-Cohen, *Biomimetics: Biologically Inspired Technologies*, Taylor & Francis, Inc., 2006; A. von Gleich, U. Petschow, C. Pade, E. Pissarskoi, *Potentials and Trends in Biomimetics*, Springer-Verlag, New York, 2010; J. M. Benyus, *Biomimcry: Innovation Inspired by Nature*, Harper Collins, New York, 2002; P. Forbes, *The Gecko's Foot: Bio-inspiration: Engineering New Materials from Nature*, W. W. Norton & Co., New York, 2005.

3. J. D. Novak and D. B. Gowan, *Learning How to Learn*, Cambridge Univ. Press, New York, 1984.

4. T. Buzan, *The MindMap Book*, 2d ed., BBC Books, London, 1995.

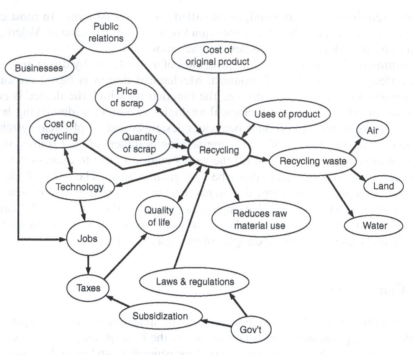

FIGURE 6.3
Concept map for the recycling of a metal like steel or aluminum.

The process of creating a concept map builds a network of associations around a central problem or topic. The requirement to fit these into a coherent, logical map stimulates new ideas. Note that such a process can quickly produce a messy and hard to read map. One way to avoid this is to first write your ideas on file cards or "sticky notes," and arrange them on an appropriate surface before committing to a written map. Color coding may be helpful in improving the clarity of the map. Figure 6.3 shows a concept map developed for a project on the recycling of steel and aluminum scrap.[1]

6.4
CREATIVE METHODS FOR DESIGN

The motivation for applying any creativity technique to a design task is to generate as many ideas as possible. Quantity counts above quality, and wild ideas are encouraged at the early stages of the design work. Once an initial pool of concepts for alternative designs exists, these alternatives can be reviewed more critically. Then the goal

---

1. I. Nair, "Decision Making in the Engineering Classroom," *J. Engr. Education,* vol. 86, no. 4, pp. 349–56, 1997.

becomes sorting out infeasible ideas. The team is identifying a smaller subset of ideas that can be developed into practical solutions.

6.4.1 Generating Design Concepts

Systematic methods for generating engineering designs exist. The task of the designer is to find the best of all possible candidate solutions to a design task. *Generative design* is a theoretical construct of a process that creates many feasible alternatives to a given product design specification (PDS). The set of all possible and feasible designs created in response to the articulation of a design task is pictured as a problem space or a design space that consists of states as shown in Fig. 6.4. Each state is a different conceptual design. The space has a boundary that encloses only the feasible designs, many of which are unknown to the designer.

The set of all possible designs is an *n*-dimensional hyperspace called a *design space*. The space is more than three dimensions because there are so many characteristics that can categorize a design (e.g., cost, performance, weight, size, etc.). A stationary solar system is a useful analogy for a design space. Each planet or star in the system is different from the others. Each known body in the space is a potential solution to the design task. There are also a number of undiscovered planets and stars. These represent designs that no one has discovered. The vastness of outer space is also a good analogy for a design space. There are many, many, many, different solutions for any design problem. The number of potential solutions can be as high as the order of *n*! where *n* is equal to the number of different engineering characteristics it takes to fully describe the design.

Allen Newell and Herbert Simon popularized this view of a set of problem solutions while working together at Carnegie Mellon University. The design space of solutions is the dominant model of problem solving in both the artificial intelligence

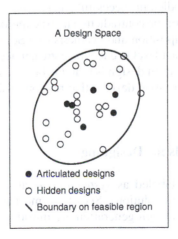

FIGURE 6.4
Schematic of an *n*-dimensional design space.

and cognitive psychology fields.[1] It is also a well-recognized model for a given set of designs to many engineering design researchers.

The design space is discrete, meaning that there are distinct and distinguishable differences between design alternatives. It is the job of the designer to find the best of all available designs. In the context of a design space that defines all feasible solutions, design becomes a search of the space to find the best available state that represents a solution to the task.

Searching a design space is a job complicated by the fact that the feasible designs differ in many ways (i.e., the values assigned to the engineering characteristics). There is no common metric to pinpoint the coordinates of any single design. It is reasonable to assume that once one feasible design is found, another feasible design that is close to the first one will be similar in all but one or a very few engineering characteristics. Once a designer finds a feasible solution to a design problem, she searches the nearby design space by making small changes to one or more of the design's engineering characteristics. This is good if the first design is close to the best design, but this will not help the designers sample different parts of the design space to find a set of very different designs. Creative idea generation methods can help a design team find designs in different areas of the space but are not as reliable as engineering design requires.

Systematic design methods help the design team consider the broadest possible set of feasible conceptual designs for a given task. Many of these methods are easier to understand when they are explained using the design space model. Some methods make the search through the design space more efficient. Others focus on narrowing into the area of the design space that is most likely where the best solution exists. Still other systematic design methods provide operations that allow a designer to travel from one design in the space to the next closest design.

Just as some of the creativity improving methods are intended to directly overcome barriers to creativity, some of the conceptual design generation methods are created to directly apply strategies of the past that were found useful in generating alternative design solutions. For example, the method called TRIZ (see Sec. 6.7) uses the concepts of inventive solution principles embodied in successful patents and equivalent databases in other countries as the foundation for the contradiction matrix approach to inventive design. The method of functional decomposition and synthesis (see Sec. 6.5) relies on restructuring a design task to a more abstract level to encourage greater access to potential solutions.

The key idea to remember in design is that it is beneficial in almost every situation to develop a number of alternative designs that rely on different means to accomplish a desired behavior.

6.4.2 Systematic Methods for Designing

Some design methods are labeled as *systematic* because they involve a structured process for generating design solutions. Six of the most popular systematic methods for mechanical, conceptual design generation are introduced in this section. The first

---

1. J. R. Anderson, *Cognitive Psychology and Its Implications,* W. H. Freeman and Company, New York, 1980.

three methods will be presented in much greater detail in subsequent sections of this chapter. We mention them briefly here for the sake of completeness.

Functional Decomposition and Synthesis (Sec. 6.5): Functional analysis is a logical approach for describing the transformation between the initial and final states of a system or device. The ability to describe devices in terms of physical behavior or actions, rather than components, allows for a logical breakdown of a product in the most general way, which often leads to creative concepts of how to achieve the function.

Morphological Analysis (Sec. 6.6): The morphological chart approach to design generates alternatives from an understanding of the structure of necessary component parts. Entries from an atlas, directory, or one or more catalogs of components can then be identified and ordered in the prescribed configuration. The goal of the method is to achieve a nearly complete enumeration of all feasible solutions to a design problem. Often, the morphological method is used in conjunction with other generative methods like the functional decomposition and synthesis method (Sec. 6.5.3).

Theory of Inventive Problem Solving (Sec. 6.7): TRIZ, the better-known Russian acronym for this method, is a creative problem-solving methodology especially tailored for scientific and engineering problems. Genrich Altshuller and coworkers in Russia started developing the method around 1940. From a study of over 1.5 million Russian patents they were able to deduce general characteristics of technical problems and recurring inventive principles.

Axiomatic Design:[1] Design models that claim legitimacy from the context of "first principles" include Suh's Axiomatic Design that articulates and explicates Design Independence and Information Axioms (i.e., maintain functional independence and minimize information content).[2] Suh's methods provide a means to translate a design task into functional requirements (the engineering equivalent of what the customer wants) and use those to identify design parameters (the physical components of the design). Suh's principles lead to theorems and corollaries that help designers diagnose a candidate solution now represented as a matrix equation with function requirements and design parameters.

Design Optimization (discussed in Chap. 15): Many of the strongest and currently recognized design methods are actually searches of a design space using optimization strategies. These algorithms predict a design engineering performance once the design specifications have been set. This method is treating design as an engineering science problem and is effective at analyzing potential designs. There are many valid and verified optimization approaches to design. They range from single-objective and single-variable models to multi-objective, multi-variable models that are solved using different decompositions and sequences. Methods are deterministic, stochastic, and combinations of the two.

Decision-Based Design (DBD) is an advanced way of thinking about design.[3,4] The DBD perspective on design differs from past design models that focus on problem solving in two major ways. The first is the incorporation of the customers' requirements

---

1. A section on Axiomatic Design can be found online at www.mhhe.com/Dieter.
2. Nam P. Suh, *Axiomatic Design,* Oxford University Press, New York, 2001; Nam P. Suh, *The Principles of Design,* Oxford University Press, New York, 1990.
3. G. Hazelrigg, *Systems Engineering: An Approach to Information-Based Design,* Upper Saddle River, NJ, 1986.
4. K. E. Lewis, W. Chen, L. C. Schmidt (eds), *Decision Making in Engineering Design,* ASME Press, 2006.

as the driver of the process. The second is using the design outcomes (e.g., maximum profit, market share capture, or high-quality image) as the ultimate assessment of good designs.

6.5
FUNCTIONAL DECOMPOSITION AND SYNTHESIS

A common strategy for solving any complex task or describing any complex system is to decompose it into smaller units that are easier to manage. Decomposing must result in units that meaningfully represent the original entity. The units of the decomposition must also be obvious to the decomposer. Standard decomposition schemes reflect natural groupings of the units that comprise an entity or are mutually agreed upon by users. This text decomposes the product development process into three major design phases and eight specific steps. The decompositions are useful for understanding the design task and allocating resources to it. The decomposition defined in this section is the breaking up of the product itself, not the process of design. Mechanical design is *recursive*. That means the same design process applied to the overall product applies to the units of the product and can be repeated until a successful outcome is achieved.

The product development process includes methods that use product decomposition. For example, QFD's House of Quality decomposes an emerging product into engineering characteristics that contribute to customers' perceptions of quality. There are other ways to decompose a product for ease of design. For example, an automobile decomposition is major subsystems of engine, drive train, suspension system, steering system, and body. This is an example of physical decomposition and is discussed in Sec. 6.5.1.

Functional decomposition is the second type of representational strategy common in early stages of concept generation. Here the emphasis is on identifying the functions and subfunctions necessary to achieve the overall behavior required by the end user. Functional decomposition is a top-down strategy where a general description of a device is refined into more specific arrangements of functions and subfunctions. The decomposed function diagram is a map of focused design problems. Functional decomposition can be done with a standardized representation system that models a device very generally. Functional decomposition does not initially impose a design, allowing more leeway for creativity and generates a wide variety of alternative solutions. This feature of the functional decomposition method is called *solution-neutrality*.

6.5.1 Physical Decomposition

To understand a device, most engineers instinctively begin with physical decomposition. Sketching the parts of a system, a subassembly, or a physical part is a way to represent the product and begin accessing all the relevant knowledge about the product. Sketching some kind of assembly drawing or schematic is a way to contemplate the design without thinking explicitly about the functions each component performs.

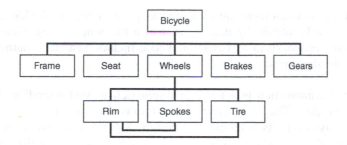

FIGURE 6.5
Physical decomposition of a bicycle with two levels of decomposition detail on the wheel subassembly.

Physical decomposition means separating the product or subassembly directly into its subsidiary subassemblies and components and accurately describing how these parts are joined together to create the behavior of the product. The result is a schematic diagram that holds some of the connectivity information found by doing reverse engineering. Figure 6.5 displays a partial physical decomposition of a standard bicycle.

Decomposition is a recursive process. This is shown in Fig. 6.5, where the entity "wheels" is further decomposed on the lower level in the hierarchy. The recursion continues until the entity is an individual part that is still essential for the overall functioning of the product. The steps to create a physical decomposition tree diagram as shown in Fig. 6.5 are:

1. Define the physical system in total and draw it as the root block of a tree diagram.[1]
2. Identify and define the first major subassembly of the system described by the root block and draw it as a new block below the root.
3. Identify and draw in the physical connections between the subassembly represented by the newly drawn block and all other blocks in the next higher level of the hierarchy in the decomposition diagram. There must be at least one connection to a block on the next higher level or the new subassembly block is misplaced.
4. Identify and draw in the physical connections between the subassembly and any other subassemblies on the same hierarchical level of the diagram's structure.
5. Examine the first subassembly block in the now complete level of the diagram. If it can be decomposed into more than one distinct and significant component, treat it as the root block and return to Step 2 in this list. If the block under examination cannot be decomposed in a meaningful way, move on to check the other blocks at the same level of the diagram hierarchy.

---

1. The physical decomposition diagram is not a true tree diagram because there may be connections between blocks on the same level of the hierarchy. There also may be connections to more than one higher-level block in the diagram. This is analogous to having a leaf grow from two different branches at the same time.

6. End the process when there are no more blocks anywhere in the hierarchical diagram that can be physically decomposed in a meaningful way. Some parts of a product are secondary to its behavior. Those include fasteners, nameplate, bearings, and similar types.

Physical decomposition is a top-down approach to understanding the physical nature of the product. The decomposition diagram is not solution-neutral because it is based on the physical parts of an existing design. A physical decomposition will lead designers to think about alternatives to parts already called out in the product. That will limit the number of alternative designs generated in the design space surrounding the existing solution.

Functional decomposition results in a solution-neutral representation of a product called a *function structure*. This type of representation is useful for generating a wide variety of design solutions. Functional decomposition is the focus of the rest of this section.

6.5.2 Functional Representation

Systematic design is a highly structured design method developed in Germany starting in the 1920s. The method was formalized by two engineers named Gerhard Pahl and Wolfgang Beitz. The stated goal of Pahl and Beitz was to "set out a comprehensive design methodology for all phases of the product planning, design, and development process for technical systems."[1] The first English translation of their text was published in 1976 as the result of enormous effort by Ken Wallace, University of Cambridge. The work's popularity continues with the publication of the third English edition in 2007.[2]

Systematic design represents all technical systems as *transducers* interacting with the world around them. The system interacts with its users and use environment by exchanging flows of energy, material, and signal with them. The technical system is modeled as a transducer because it is built to respond in a known way to flows from the use environment.

A kitchen faucet can be modeled as a transducer that alters the amount and temperature of water flowing into a kitchen sink. A person controls the amount and temperature of the water by manually moving one or more handles. If someone is at the sink to fill a drinking glass with cold water, they may hold their hand in the water flow to determine when it is cold enough to drink. Then they watch as they position the glass in the flow of water and wait for it to fill. When the glass is full, the user moves it out of the water flow and adjusts the faucet handle to stop the flow. This happens during a short time interval. The user operates the system by applying human energy

---

1. G. Pahl and W. Beitz, *Engineering Design: A Systematic Approach*, K. Wallace (translator), Springer-Verlag, New York, 1996.
2. G. Pahl, W. Beitz, J. Feldhusen, and K. H. Grote, *Engineering Design: A Systematic Approach,* 3d ed., K. Wallace (ed.), K. Wallace and L. Blessing and F. Bauert (translators), Springer-Verlag, New York, 2007.

TABLE 6.4
Standard Flow Classes and Member Flow Types

Flow Classes		
Energy	**Material**	**Signal**
Human	Human	Status
Hydraulic	Solid	• Auditory
Pneumatic	Gas	• Olfactory
Mechanical	Liquid	• Tactile
• translational	Plasma	• Taste
• rotational	Mixture	• Visual
Electrical		Control
Acoustic		• Analog
Thermal		• Discrete
Electromagnetic		
Chemical		
Biological		

R. E. Stone, "Functional Basis", *Design Engineering Lab Webpage,* designengineeringlab.org/FunctionCAD/FB.htm, accessed November 10, 2011.

to move the faucet control handle and the glass. The user collects information about the operation through his or her senses throughout the entire operation. The same system can be designed to operate automatically with other sources of energy and a control system. In either case, the kitchen faucet is modeled by describing interactions of flows of energy, material (water), and information signals with the user.

A focused research effort to standardize a function language began in 1997.[1] The work was motivated by the vision of developing a broad design repository of thousands of devices all represented from the function transformation view of mechanical design. This work resulted in the establishment of a *function basis*.[2] The expanded list of flow types is given in Table 6.4 and the function listing is given in Table 6.5. Naturally, Pahl and Beitz's function description scheme was prominent among the work consulted to develop the basis.

The standardized flow types and function block names are organized as general classes divided by more specific basic types. This allows designers to represent components and systems at different levels of abstraction. Using the most general level of function representation, function class names, allows the reader to re-represent the design problem in the broadest possible terms. This abstraction encourages diverse thinking required in conceptual design.

1. A. Little, K. Wood, and D. McAdams, "Functional Analysis," *Proceedings of the 1997 ASME Design Theory and Methodology Conference,* ASME, New York, 1997.
2. J. Hirtz, R. Stone, D. McAdams, S. Szykman, and K. Wood, "A Functional Basis for Engineering Design: Reconciling and Evolving Previous Efforts," *Research in Engineering Design,* vol. 13, 65–82, 2002.

TABLE 6.5
Standardized Function Names

Function Class	Basic Function Names	Alternate Wording of Basic Functions
Branch	Separate	Detach, disassemble, disconnect, divide, disconnect, subtract
	Remove	Cut, polish, punch, drill, lathe
	Distribute	Absorb, dampen, diffuse, dispel dispense, disperse, empty, resist, scatter
	Refine	Clear, filter, strain, purify
Channel	Import	Allow, capture, input, receive
	Export	Eject, dispose, output, remove
	Transfer	
	Transport	Lift, move
	Transmit	Conduct, convey
	Guide	Direct, straighten, steer
	Translate	
	Rotate	Spin, turn
	Allow DOF	Constrain, unlock
Connect	Couple	Assemble, attach, join
	Mix	Add, blend, coalesce, combine, pack
Control Magnitude	Actuate	Initiate, start
	Regulate	Allow, control, enable, limit, prevent
	Change	Adjust, amplify, decrease, increase, magnify, multiply, normalize, rectify, reduce, scale
	Form	Compact, compress, crush, pierce, shape
	Condition	Prepare, adapt, treat
	Stop	Inhibit, end, halt, pause, interrupt, restrain, protect, shield
	Inhibit	Shield, insulate, protect, resist
Convert	Convert	Condense, differentiate, evaporate, integrate, liquefy, process, solidify, transform
Provision	Store	Contain, collect, reserve, capture
	Supply (extract)	Expose, fill, provide, replenish
Signal	Sense	Discern, locate, perceive, recognize
	Indicate	Mark
	Display	
	Process	Calculate, compare, check
Support	Stabilize	Steady
	Secure	Attach, fasten, hold, lock, mount
	Position	Align, locate, orient

R. E. Stone, "Functional Basis", *Design Engineering Lab Webpage,* designengineeringlab.org/FunctionCAD/
FunctionCAD/FB.html, accessed November 10, 2011.

TABLE 6.6
Components Abstracted into Function Blocks

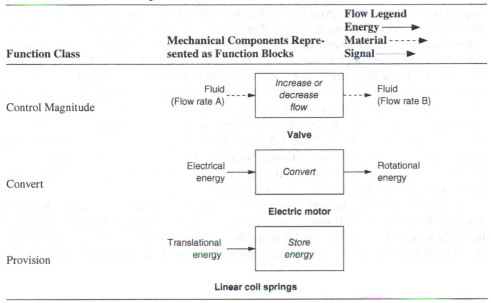

Function Class	Mechanical Components Represented as Function Blocks	Flow Legend Energy ——▶ Material - - - -▶ Signal ·········▶
Control Magnitude	Fluid (Flow rate A) - - - -▶ [Increase or decrease flow] - - - -▶ Fluid (Flow rate B) **Valve**	
Convert	Electrical energy ——▶ [Convert] ——▶ Rotational energy **Electric motor**	
Provision	Translational energy ——▶ [Store energy] **Linear coil springs**	

Systematic design represents mechanical components abstractly by a labeled *function block* and its interacting flow lines. Three standard mechanical components are listed in Table 6.6. The function flows and class names are expressed in the most general possible terms.

Systematic design provides a way to describe an entire device or system in a general way. A device can be modeled as a single component entity that transforms inputs of energy, material, and signal into desired outputs. An abstract model of a basketball return modeled as a single function block is presented in Fig. 6.6.

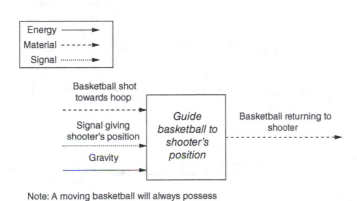

Energy ——▶
Material - - - -▶
Signal ·········▶

Basketball shot towards hoop - - - - - - ▶ [Guide basketball to shooter's position] Basketball returning to shooter - - - - - - ▶

Signal giving shooter's position ·········▶

Gravity ——▶

Note: A moving basketball will always possess direction and kinetic energy.

FIGURE 6.6
Function structure black box for a basketball ball return.

6.5.3 Performing Functional Decomposition

Functional decomposition produces a diagram called a *function structure*. A function structure is a block diagram depicting flows of energy, material, and signal as labeled arrows taking paths between function blocks, like those in Table 6.6. The function structure represents mechanical devices by the arrangement of function blocks and flow arrows. Flow lines are drawn with arrows to indicate direction and labels to define the flow connecting the function blocks (see Fig. 6.7). Designers use function blocks in the diagram to represent the transformations done by the system, assembly, or component, and label each block by selecting function names from a predefined set of transformational verbs in Table 6.5. The function structure is very different from the physical decomposition of a product because a function is the combined behavior of mechanical components and their physical arrangement. There is no one-to-one correspondence of function block to component.

The most general function structure is a single function block description of a device, like the basketball return model of Fig. 6.6. This type of function structure (a single function block) is called a *black box* representation of a device. It must list the overall function of the device and supply all appropriate input and output flows.

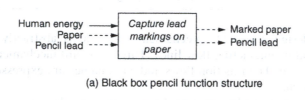

(a) Black box pencil function structure

(b) Standard function blocks to describe pencil behavior

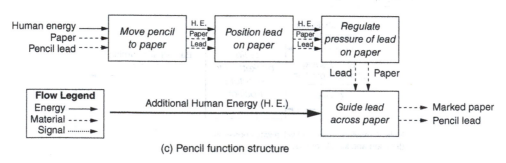

(c) Pencil function structure

FIGURE 6.7
Function structure for a mechanical pencil.

In the case of designing a new device, the black box representation is the most logical place to begin the process.

A simplified method for creating a function structure is described in the following steps. The example used is that of a lead pencil.

1. Identify the overall function that needs to be accomplished using function basis terms. Identify the energy, material, and signal flows that will be input to the device. Identify the energy, material, and signal flows that will be output from the device once the transformations are complete. Use the standard flow classes defined in Table 6.4. Common practice is to use different line styles for arrows to represent each general flow type (i.e., energy, material, and signal). Label each arrow with the name of the specific flow. This "black box" model of the product (Fig. 6.7a for the pencil) shows the input and output flows for the primary high-level function of the design task.
2. Using everyday language, write a description of the individual functions that are required to accomplish the overall task described in the black box model of the pencil in Fig. 6.7a. The most abstract function of a pencil is to capture lead markings on paper. The input flows of material include both lead and paper. Since a human user is needed to operate the pencil, the energy flow type is human. For example, in everyday language the general functions to be accomplished by the pencil and its user are:

    - Movement of pencil lead to the appropriate area of the paper
    - Applying the sufficient but not overwhelming force to the lead while moving it through specific motions to create markings on the paper
    - Raising and lowering the lead to contact the paper at appropriate times

    The list describes the use of the pencil in a conventional way with everyday language. This list is not unique. There are different ways to describe the behavior of writing with a pencil.

3. Having thought about the details of accomplishing the pencil's function described in the black box, identify more precise functions (from Table 6.5) necessary to fulfill the more detailed description of the pencil's function in solution-neutral language. This process creates function blocks for a more detailed description of the pencil. One set of function blocks for the pencil is shown in Fig. 6.7b.
4. Arrange the function blocks in the order that they must take place for the desired functions. The arrangement depicts the precedence required by the functions. This means that function block arrangements will include blocks in parallel, in series, and in all combinations possible. Sticky notes are a great tool to use in this process, especially when decisions are made by team consensus. Rearrangement is often necessary.
5. Add the energy, material, and signal flows between the function blocks. Preserve the input and output flows from the black box representation of the device. Not all flows will travel through each function block. Remember that the function structure is a visual representation, not an analytical model. For example, flows in a function structure do not adhere to the conservation laws used to model systems for thermodynamic analysis. An example of this different behavior is the representation of a coil spring in Table 6.6. It accepts translational energy without discharging any energy. The preliminary function structure for the pencil is depicted in Fig. 6.7c.

6. Examine each block in the function structure to determine if additional energy, material, or signal flows are necessary to perform the function. In the pencil function structure, an additional human energy flow is input to the "Guide lead" function block to reinforce the idea that there is a second type of activity that the user must perform.

7. Review each function block again to see if additional refinement is necessary. The objective is to refine the function blocks as much as possible. Refinement stops when a function block can be fulfilled by a single solution that is an object or action, and the level of detail is sufficient to address the customer needs.

Designers make unstated assumptions that are revealed by examining the pencil function structure. The function structure built here presumes that a user can hold and manipulate a piece of pencil lead directly. We know that is not the case. Thin lead requires a casing.

Function structures are not necessarily unique. Another designer or design team can create a slightly different set of descriptive function blocks for a lead pencil. This demonstrates the creative potential of design by functional decomposition and synthesis. A designer can look at a portion of a function structure and replace it with a new set of function blocks as long as the functional outcome is preserved.

Figure 6.8 displays a function structure for a basketball return device. This function structure was created from the blackbox representation given in Fig 6.6. This is one possible version of a function structure for the Shot-Buddy. Some designers may use different combinations of function blocks in the diagram. For example, the initial

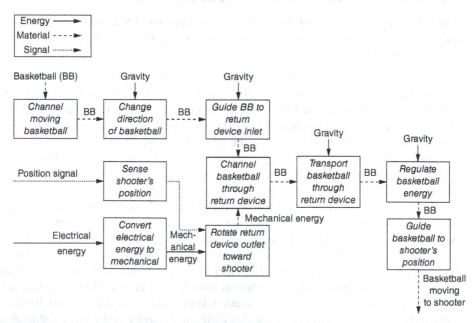

FIGURE 6.8
Function structure for a basketball return device.

functionality of the Shot-Buddy is to provide a means of catching a basketball shot into or near the net. Different functions in the Function Class of *channel* would be appropriate. Figure 6.8 shows several instances where energy in the form of gravity is designated. This indicates that the designers are focusing on the natural downward forces on a basketball and are probably thinking of using that energy in the design.

Functional decomposition is not easy to implement in all situations. It is well-suited for mechanical systems that include components in relative motion with one another. It is a poor method for representing load-bearing devices that exist to resist other forces. An example is a desk.

6.5.4 Strengths and Weaknesses of Functional Synthesis

The modeling of a mechanical product in a form-independent and solution-neutral way will allow for more abstract thinking about the problem and enhance the possibility of more creative solutions. The function structure's model of flows and functions may provide cues for making decisions on how to segment the device into systems and subsystems. This is known as determining the product architecture. By creating function structures, flows separate, begin, end, and transform as they pass through the device. It may be advantageous to combine functions that act on the same flow into subsystems or physical modules. Flow descriptions provide a way to plan for measuring the effectiveness of a system, subsystem, or function because a flow is measurable.

The advantages of functional decomposition and synthesis follow from two key elements of the method.

- First, creating function structures forces re-representation into a language that is useful for the manipulation of mechanical design problems.
- Second, using a function structure to represent a design lends functional labels to potential solution components, and these labels serve as hints for new memory searches.

Again, we see that the methods use strategies suggested to improve creativity. The great advantage of functional decomposition is that the method facilitates the examination of options that most likely would not have been considered if the designer moved quickly to selecting specific physical principles or, even worse, selecting specific hardware.

There are several weaknesses to the functional decomposition method. Briefly:

- Some products are better suited to representation and design by functional decomposition and synthesis than are others. Products that consist of function-specific modules arranged in a way that all the material flowing through the product follows the same path are the best candidates for this method. Examples include a copying machine, a factory, or a peppermill. Any product that acts sequentially on some kind of material flowing through it is well suited for description by a function structure.
- The function structure is a flow diagram where flows are connecting different functions performed by the product the structure represents. Each function applied to a

flow is articulated separately by a function block in the function structure, even if the action is at essentially the same time. Thus, the ordering of the function structure boxes seems to imply a sequence in time that may or may not be accurately depicting the device's action.

- There are weaknesses in using functional structures during conceptual design. A function structure is not a complete conceptual design. Even after developing a function structure, you still need to select devices, mechanisms, or structural forms to fulfill the function. There are no comprehensive catalogs of solution embodiments like those available in the German technical literature.

- Functional decomposition can lead to excess parts and subsystems if the designer does not stop to integrate common function blocks and flows. Employing function sharing or taking advantage of emergent behavior is difficult when the method is so focused on the parts instead of the whole.

- A final criticism of this method is that the results are not necessarily unique. This can bother researchers who want a repeatable process. Ironically, many students who are trained in this method find it too constrained because of the requirements of expressing functions in predefined categories.

6.6
MORPHOLOGICAL METHODS

Morphological analysis is a method for representing and exploring all the relationships in multidimensional problems. The word *morphology* means the study of shape and form. Morphological analysis is a way of creating new forms. Morphological methods have been recorded in science as a way to enumerate and investigate solution alternatives as far back as the 1700s. The process was developed into a technique for generating design solutions by Zwicky.[1] Zwicky formalized the process of applying morphological methods to design in the mid-1960s with the publication of a text that was translated into English in 1969.

Generating product design concepts from a given set of components is one such problem. There are many different combinations of components that can satisfy the same functionality required in a new product. Examining every candidate design is a combinatorially explosive problem. Yet, one wonders how many great designs are missed because the designer or team ran out of time for exploring alternative solutions. Morphological methods for design are built on a strategy that helps designers uncover novel and unconventional combinations of components that might not ordinarily be generated. Success with morphological methods requires broad knowledge of a wide variety of components and their uses, and the time to examine them. It's unlikely that any design team will have enough resources (time and knowledge) to completely search a design space for any given design problem. This makes a method

1. F. Zwicky, *The Morphological Method of Analysis and Construction,* Courant Anniversary Volume, pp. 461–70, Interscience Publishers, New York, 1948.

like morphological analysis of great interest to design teams. It is a method that is especially useful when merged with other generative methods.

The function structure of a design, discussed in Sec. 6.5, is a template for generating design options by examining combinations of known devices to achieve the behavior described by each function block. Morphological analysis is very effective for solution synthesis when paired with functional decomposition. The treatment provided here assumes that the team has first used systematic design to create an accurate function structure for the product to be designed and now seeks to generate a set of feasible concepts for further consideration.

6.6.1 Morphological Method for Design

Morphological methods help structure the problem for the synthesis of different components to fulfill the same required functionality. This process is made easier by access to a component catalog. Yet it does not replace the interaction of designers on a team. Teams are vital for refining concepts, communication, and building consensus. The best procedure is for each team member to spend several hours working as an individual on some subset of the problem, such as how to satisfy the need described by an identified function. Morphological analysis assists a team in compiling individual research results into one structure to allow the full team to process the information.

The general morphological approach to design is summarized in the following three steps.

1. Divide the overall design problem into simpler subproblems.
2. Generate solution concepts for each subproblem.
3. Systematically combine subproblem solutions into different complete solutions and evaluate all combinations.

The morphological approach to mechanical design begins with the functional decomposition of the design problem into a detailed function structure. We will use the redesign of a basketball return device as an illustrative example. The function structure is, in itself, a depiction of a number of smaller design problems or subproblems. Each consists of finding a solution to replace the function block in the larger function structure. If each subproblem is correctly solved, then any combination of subproblem solutions comprises a feasible solution to the total design problem. The morphological chart is the tool used to organize the subproblem solutions.

The designer or team can continue with morphological analysis once they have an accurate decomposition of the problem. The process proceeds with completing a morphological chart (Table 6.7). The chart is a table organizing the subproblem solutions. The chart's column headings are the names of the sub problems identified in the decomposition step. The rows hold solutions to the subproblem. Descriptive words or very simple sketches depict the subproblem solution in every chart cell. Some columns in the morphological chart may hold only a single solution concept. There are two possible explanations. The design team may have made a fundamental assumption that limits the subproblem solution choices. Another reason could be that a satisfactory physical embodiment is given, or it could be that the design team is weak on ideas. We call this limited domain knowledge.

TABLE 6.7
Morphological Chart for Shot-Buddy Basketball Return System

Subproblem Solution Concepts				
Channel Moving Basketball	**Change Direction of Basketball**	**Sense Shooter's Position**	**Guide Basketball to Return Device Inlet**	**Rotate Return Device Outlet Toward Shooter**
Catch net	Sheet of flexible material	RFID tag worn by shooter	Funnel(net or solid material)	Ratchet device
Plastic sheeting with wire ribs	Solid deflector panels	Motion sensor	Set of rails	None (rely on ball's direction and gravity)
Finger-like converging structure	Shaped foam	Optical sensor	Tube of netting	Cam mechanism
Tubing (partially open or closed)	Paddle arms	Acoustic sensor	Metal guide (moving or stationary)	Geared shaft

6.6.2 Generating Concepts from a Morphological Chart

The next step in morphological design is to generate all designs by synthesizing possible combinations of alternatives for each subfunction solution identified in Table 6.7. One possible design concept to consider is combining the component alternatives appearing in the first row under each subfunction. Another potential design is comprised of the random selection of one subproblem solution from each column. Designs generated from the chart must be checked for feasibility and may not represent a viable overall design alternative. The advantage of creating a morphological chart is that it allows a systematic exploration of many possible design solutions.

One possible basketball return concept for the Shot-Buddy is shown as rough sketch in Fig. 6.9. It is made from the first subproblem solution listed under each heading in Table 6.7. It is easy to understand how this concept could be changed by substituting some other type of system to catch a basketball shot at the net. The advantages of the morphological approach become clear when illustrated with an example such as this.

Table 6.7 only includes five of the ten function blocks given in the basketball return systems function structure. Still the set of possible combinations is quite large. For the five function blocks example given here there are $4 \times 4 \times 4 \times 4 \times 4 = 1024$ combinations, clearly too many to follow up in detail. Some may be clearly infeasible or impractical. Care should be taken not to make this judgment too hurriedly. Also, realize that some concepts will satisfy more than one subproblem. Likewise, some subproblems are coupled, not independent. This means that their solutions can be evaluated only in conjunction with the solutions to other subproblems.

Do not rush into evaluation of design concepts. Outstanding designs often evolve out of several iterations of combining concept fragments from the morphological chart and working them into an integrated solution. This is a place where a smoothly functioning team pays off.

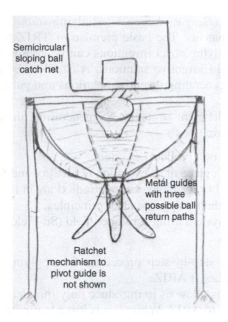

Semicircular
sloping ball
catch net

Metal guides
with three
possible ball
return paths

Ratchet
mechanism to
pivot guide is
not shown

FIGURE 6.9
Sketch of Shot-Buddy concept.

Adapted from Josiah Davis, Jamil Decker, James Maresco, Seth McBee, Stephen Phillips, and Ryan Quinn, "JSR Design Final Report: Shot-Buddy," unpublished, ENME 472, University of Maryland, May 2010.

Although design concepts are quite abstract at this stage, it often is very helpful to utilize rough sketches. Sketches help us associate function with form, and they aid with our short-term memory as we work to assemble the pieces of a design. Moreover, sketches in a design notebook are an excellent way of documenting the development of a product for patent purposes.

6.7
TRIZ: THE THEORY OF INVENTIVE PROBLEM SOLVING

The Theory of Inventive Problem Solving, known by the acronym "TRIZ,"[1] is a problem-solving methodology tailored to provide innovative solutions for scientific and engineering problems. Genrich Altshuller, a Russian inventor, developed TRIZ in the late 1940s and 1950s. After World War II, Altshuller worked on design problems in the Soviet Navy. Altshuller was convinced that he could improve the creativity of design engineers. He began by looking into Synectics but was not impressed with the method. So in 1946 Altshuller started his work to create a new science of invention.[2]

---

1. TRIZ is an acronym for Teoriya Resheniya Izobreatatelskikh Zadatch.
2. K. Gadd, *TRIZ for Engineers: Enabling Inventive Problem Solving,* John Wiley & Sons, Incorporated, 2011. M. A. Orloff, *Inventive Thought through TRIZ,* 2d ed., Springer, New York, 2006; L. Shulyak, ed., *The Innovation Algorithm,* Technical Innovation Center, Inc., Worchester, MA, 2000.

Altshuller and a few colleagues began by studying *author certificates*, the Soviet Union's equivalent to patents. The basic premise of TRIZ is that the solution principles derived from studying novel inventions can be codified and applied to related design problems to yield inventive solutions. Altshuller and colleagues constructed their methodology for generating inventive solutions and published the first article on TRIZ in 1956.

TRIZ offers four different strategies for generating an innovative solution to a design problem. They are:

1. Increase the ideality of a product or system.
2. Identify the product's place in its evolution to ideality and force the next step.
3. Identify key physical or technological contradictions in the product and revise the design to overcome them using inventive principles.
4. Model a product or system using substance-field (Su-Field) analysis and apply candidate modifications.

Altshuller developed a step-by-step procedure for applying strategies of inventive problem solving and called it ARIZ.

Space considerations allow us to introduce only the idea of contradictions and to give a brief introduction to ARIZ. While this is just a beginning introduction to TRIZ, it can serve as a significant stimulation to creativity in design and to further study of the subject. Note that this section follows the TRIZ conventions in using the term *system* to mean the product, device, or artifact that is invented or improved.

6.7.1 Invention: Evolution to Increased Ideality

Altshuller's examination of inventions led to his observation that systems had a level of goodness he called *ideality* and that inventions result when changes were made to improve this attribute of a product or system. Altshuller modeled ideality as a mathematical construct defined as the ratio of the useful effects of a system to its harmful effects. Like any ratio, as the harmful effects decrease to approach a value of zero, the ideality grows to infinity.

Improving system ideality is one of the TRIZ inventive design strategies. Briefly, the six specific design suggestions to examine for improving the ideality of a system are as follows:

1. Exclude auxiliary functions (by combining them or eliminating the need for them).
2. Exclude elements (i.e., subsystems or components) in the existing system.
3. Identify self-service functions (i.e., exploit function sharing by identifying an existing element of a system that can be altered to satisfy another necessary function).
4. Replace elements or parts of the total system.
5. Change the system's basic principle of operation.
6. Utilize resources in system and its immediate surroundings.

The TRIZ strategy of improving ideality is more complex than simply following the six guidelines, but the scope of this text limits us to this introduction.

The patent research led Altshuller and his colleagues to a second strategy for invention. They observed that engineering systems are refined over time to achieve higher states of ideality. The history of systems displayed consistent patterns of design evolution that a system follows as it is reinvented. Again, this inventive strategy of forcing the next step in product evolution is complex. The redesign patterns identified in TRIZ are listed here.

- Development toward increased dynamism and controllability
- Develop first into complexity then combine into simpler systems
- Evolution with matching and mismatching components
- Evolution toward micro level and increasing use of fields (more functions)
- Evolution toward decreased human involvement

Altshuller believed that an inventor could use one of the suggestions to inspire inventive improvements in existing systems, giving the inventor a competitive advantage.

These strategies for producing inventive designs follow from the theory of innovation that Altshuller proposes with the TRIZ methodology. Notice that the guidelines developed from researching inventions are similar to suggestions or prompts in creativity-enhancing methods for general problem solving. Like many theories of design, TRIZ has been demonstrated but not proven. Nevertheless, the principles behind the theory are observable and lead to guidelines for producing inventive design solutions.

6.7.2 Innovation by Overcoming Contradictions

Developing a formal and systematic design method requires more than guidelines drawn from experience. Continuing with the examination of the inventions verified by author certificates, Altshuller's group noted differences in the type of change proposed by the inventor over the existing system design. The solutions fell into one of five very specific levels of innovation. The following list describes each innovation level and shows its relative frequency.

- Level 1: (32%) Conventional design solutions arrived at by methods well known in the technology area of the system.
- Level 2: (45%) Minor corrections made to an existing system by well-known methods at the expense of some compromise in behavior.
- Level 3: (18%) Substantial improvement in an existing system that resolves a basic behavior compromise by using the knowledge of the same technology area; the improvement typically involves adding a component or subsystem.
- Level 4: (4%) Solutions based on application of a new scientific principle to eliminate basic performance compromises. This type of invention will cause a paradigm shift in the technology sector.
- Level 5: (1% or less) Pioneering inventions based on a discovery outside of known science and known technology.

In 95 percent of the cases, inventors arrived at new designs by applying knowledge from the same technical field as the existing system. The more innovative design solutions improved a previously accepted performance compromise. In 4 percent of the inventions, the compromise was overcome by application of new knowledge to the field. These cases are called inventions outside of technology and often proceed to revolutionize an industry. One example is the development of the integrated circuit that replaced the transistor. Another is the digitizing technology used in audio recordings that led to the compact disc.

Diligent application of good engineering practice in the appropriate technical specialty already leads a designer to Level 1 and 2 inventions. Conversely, the pioneering scientific discoveries driving the inventions of Level 5 are serendipitous in nature and cannot be found by any formal method. Therefore, Altshuller focused his attention on analyzing innovations on Levels 3 and 4 in order to develop a design method for inventive solutions.

Altshuller had about 40,000 instances of Level 3 and 4 inventions within his initial sample of 200,000 Soviet author certificates. These inventions were improvements over systems containing a fundamental *technical contradiction*. This condition exists when a system contains two important attributes related such that an improvement in the first attribute degrades the other. For example, in aircraft design a technical contradiction is the inherent trade-off between improving an aircraft's crashworthiness by increasing the fuselage wall thickness and minimizing its weight. These technical contradictions create design problems within these systems that resist solution by good engineering practice alone. A compromise in performance is the best that can be obtained by ordinary design methods. The redesigns that inventors proposed for these problems were truly *inventive*, meaning that the solution surmounts a basic contradiction that occurs because of conventional application of known technology.

As seen with other design methods, it is useful to translate a design problem into general terms so that designers are not restricted in their search for solutions. TRIZ required a means to describe the contradictions in general terms. In TRIZ, the technical contradiction represents a key design problem in solution-neutral form by identifying the engineering parameters that are in conflict. TRIZ uses a list of 39 engineering parameters (see Table 6.8) to describe system contradictions.

The parameters in Table 6.8 are self-explanatory and the list is comprehensive. The terms seem general, but they can accurately describe design problems.[1] Consider the example of competing goals of the airplane, being both crashworthy and lightweight. Proposing an increase in the thickness of the fuselage material increases the strength of the fuselage but also negatively affects the weight. In TRIZ terms, this design scenario has the technical contradiction of improving strength (parameter 14) at the expense of the weight of a moving object (parameter 1).

---

1. An excellent description of each TRIZ parameter can be found online in Ellen Domb with Joe Miller, Ellen MacGran, and Michael Slocum, "The 39 Features of Altshuller's Contradiction Matrix," *The TRIZ Journal*, http://www.triz-journal.com, November, 1998.

TABLE 6.8
TRIZ List of 39 Engineering Parameters

Engineering Parameters Used to Represent Contradictions in TRIZ	
1. Weight of moving object	21. Power
2. Weight of nonmoving object	22. Waste of energy
3. Length of moving object	23. Waste of substance
4. Length of nonmoving object	24. Loss of information
5. Area of moving object	25. Waste of time
6. Area of nonmoving object	26. Amount of substance
7. Volume of moving object	27. Reliability
8. Volume of nonmoving object	28. Accuracy of measurement
9. Speed	29. Accuracy of manufacturing
10. Force	30. Harmful factors acting on object
11. Tension, pressure	31. Harmful side effects
12. Shape	32. Manufacturability
13. Stability of object	33. Convenience of use
14. Strength	34. Repairability
15. Durability of moving object	35. Adaptability
16. Durability of nonmoving object	36. Complexity of device
17. Temperature	37. Complexity of control
18. Brightness	38. Level of automation
19. Energy spent by moving object	39. Productivity
20. Energy spent by nonmoving object	

6.7.3 TRIZ Inventive Principles

TRIZ is based on the notion that inventors recognized technical contradictions in design problems and overcame them using a principle that represented a new way of thinking about the situation. Altshuller's group studied inventions that overcame technical contradictions, identified the solution principles used in each case, and distilled them into 40 unique solution ideas. These are the 40 Inventive Principles of TRIZ, and they are listed in Table 6.9.

Several elements in the list of Inventive Principles, like Combining (#5) and Asymmetry (#4), are similar to the prompts provided in some of the creativity-enhancing methods like SCAMPER and are self-explanatory. Some of the principles are very specific like numbers 29, 30, and 35. Others, like Spheroidality[1] (#14) require more explanation before they can be applied. Many of the inventive principles listed have special meaning introduced by Altshuller.

---

1. Principle 14, Spheroidality, means to replace straight-edged elements with curved ones, use rolling elements, and consider rotational motion and forces.

TABLE 6.9
The 40 Inventive Principles of TRIZ

Names of TRIZ Inventive Principles	
1. Segmentation	21. Rushing through
2. Extraction	22. Convert harm into benefit
3. Local quality	23. Feedback
4. Asymmetry	24. Mediator
5. Combining	25. Self-service
6. Universality	26. Copying
7. Nesting	27. An inexpensive short-lived object instead of an expensive durable one
8. Counterweight	28. Replacement of a mechanical system
9. Prior counteraction	29. Use of a pneumatic or hydraulic construction
10. Prior action	30. Flexible film or thin membranes
11. Cushion in advance	31. Use of porous material
12. Equipotentiality	32. Change the color
13. Inversion	33. Homogeneity
14. Spheroidality-Curvature	34. Rejecting and regenerating parts
15. Dynamicity	35. Transformation of physical and chemical states of an object
16. Partial or overdone action	36. Phase transition
17. Moving to a new dimension	37. Thermal expansion
18. Mechanical vibration	38. Use of strong oxidizers
19. Periodic action	39. Inert environment
20. Continuity of useful action	40. Composite materials

The five most frequently used Inventive Principles of TRIZ are listed here with more detail and examples.

Principle 1: Segmentation
   a. Divide an object into independent parts.
      o Replace mainframe computer with personal computers.
      o Replace a large truck with a truck and trailer.
      o Use a work breakdown structure for a large project.
   b. Make an object easy to disassemble.
   c. Increase the degree of fragmentation or segmentation.
      o Replace solid shades with Venetian blinds.
      o Use powdered welding metal instead of foil or rod to get better penetration of the joint.

Principle 2: Extraction—Separate an interfering part or property from an object, or single out the only necessary part (or property) of an object.
   a. Locate a noisy compressor outside the building where the air is used.
   b. Use the sound of a barking dog, without the dog, as a burglar alarm.

Principle 10: Prior action
   a. Perform the required change (fully or partially) before it is needed.
      o Prepasted wallpaper.
      o Sterilize all instruments needed for a surgical procedure on a sealed tray.
   b. Prearrange objects such that they can come into action from the most convenient place and without losing time for their delivery.
      o Kanban arrangements in a just-in-time factory.
      o Flexible manufacturing cell.

Principle 28: Replacement of mechanical system
   a. Replace a mechanical means with a sensory (optical, acoustic, taste or smell) means.
      o Replace a physical fence to confine a dog or cat with an acoustic "fence" (signal audible to the animal).
      o Use a bad-smelling compound in natural gas to alert users to leakage, instead of a mechanical or electrical sensor.
   b. Use electric, magnetic, and electromagnetic fields to interact with the object.
   c. Change from static to movable fields or from unstructured to structured.

Principle 35: Transformation of properties
   a. Change an object's physical state (e.g., to a gas, liquid, or solid).
      o Freeze the liquid centers of filled candies prior to coating them.
      o Transport oxygen or nitrogen or natural gas as a liquid, instead of a gas, to reduce volume.
   b. Change the concentration or consistency.
   c. Change the degree of flexibility.
   d. Change the temperature.

The 40 principles of TRIZ have a remarkably broad range of application. However, they do require considerable study to understand them fully. Complete listings of the 40 Inventive Principles are available in book form[1] and online through the TRIZ Journal website. There, the TRIZ principles are listed with explanations and examples.[2] The TRIZ Journal has also published listings of the principles interpreted for nonengineering application areas, including business, architecture, food technology, and microelectronics, to name a few.

6.7.4 The TRIZ Contradiction Matrix

TRIZ is a process of reframing a designing task so that the key contradictions are identified and appropriate inventive principles are applied. TRIZ leads designers to represent problems as separate technical contradictions within the system. Typical

---

1. Genrich Altshuller with Dana W. Clarke, Sr., Lev Shulyak, and Leonoid Lerner, "40 Principles Extended Edition," published by Technical Innovation Center, Worcester, MA, 2006. Or online at www.triz.org.
2. "TRIZ 40 Principles", www.TRIZ40.com, Solid Creativity, 2004. accessed November 10, 2011.

conflicts are reliability versus complexity, productivity versus accuracy, and strength versus ductility. TRIZ then provides one or more inventive principles that have been used to overcome this contradiction in the past, as found by searching documentation of prior inventions. The TRIZ Contradiction Matrix is the key tool for selecting the right inventive principles to use to find a creative way to overcome a contradiction. The matrix is square with 39 rows and columns. It includes about 1250 typical system contradictions, a low number given the diversity of engineering systems.

The TRIZ Contradiction Matrix guides designers to the most useful inventive principles. Recall that a technical contradiction occurs when an improvement in a desired engineering parameter of the system results in deterioration of the other parameter. Therefore, the first step to finding a design solution is to phrase the problem statement to reveal the contradiction. In this format, the parameters to be improved are identified, as are those parameters that are being degraded. The rows and columns of the Contradiction Matrix are numbered from 1 to 39, corresponding to the numbers of the engineering parameters. Naturally, the diagonal of the matrix is blank. To resolve a contradiction where parameter i is improved at the expense of parameter j, the designer locates the cell of the matrix in row i and column j. The cell includes the number of one or more inventive principles that other inventors used to overcome the contradiction.

The TRIZ Contradiction Matrix for parameters 1 through 10 is displayed in Table 6.10. An interactive TRIZ Contradiction Matrix is published online at http://triz40.com/ with thanks to Ellen Domb of PQR Group consulting and training firm (www.trizpqrgroup.com) and SolidCreativity.

EXAMPLE 6.1

A metal pipe pneumatically transports plastic pellets.[1] A change in the process requires that metal powder now be used with the pipe instead of plastic. The metal must also be delivered to the station at the end of the transport pipe at a higher rate of speed. Changes in the transport system must be done without requiring significant cost increases. The hard metal powder causes erosion of the inside of the pipe at the elbow where the metal particles turn 90° (Fig. 6.10).

Conventional solutions to this problem include: (1) reinforcing the inside of the elbow with abrasion-resistant, hard-facing alloy; (2) redesigning the path so that any compromised section of pipe could be easily replaced; and (3) redesigning the shape of the elbow to reduce or eliminate the instances of impact. However, all of these solutions require significant extra costs. TRIZ is employed to find a better and more creative solution.

Consider the function that the elbow serves. Its primary function is to change the direction of the flow of metal particles. However, we want to increase the speed at which the particles flow through the system and at the same time reduce the energy requirements. We must identify the engineering parameters involved in the design change in order to express this as a number of smaller design problems restated as TRIZ contradictions. There are two engineering parameters that must be improved upon: the speed of the metal powder through the system must be increased, and the energy used in the system must improve, requiring a decrease in energy use.

Consider the design objective of increasing the speed (parameter 9) of the metal powder. We must examine the system to determine the engineering parameters that will be

---

1. Example adapted from J. Terninko, A. Zusman, B. Zlotin, "Step-by-Step TRIZ", Nottingham, NH, 1997.

TABLE 6.10

Partial TRIZ Contradiction Matrix (Parameters 1 to 10)

TRIZ CONTRADICTION MATRIX FOR ENGINEERING PARAMETERS 1 THROUGH 10		DEGRADING ENGINEERING PARAMETER									
		1 Weight of moving object	2 Weight of stationary object	3 Length of moving object	4 Length of stationary object	5 Area of moving object	6 Area of stationary object	7 Volume of moving object	8 Volume of stationary object	9 Speed	10 Force (Intensity)
Improving Engineering Parameter	1 Weight of moving object	+	–	15, 8, 29, 34	–	29, 17, 38, 34	–	29, 2, 40, 28	–	2, 8, 15, 38	8, 10, 18, 37
	2 Weight of stationary object	–	+	–	10, 1, 29, 35	–	35, 30, 13, 2	–	5, 35, 14, 2	–	8, 10, 19, 35
	3 Length of moving object	8, 15, 29, 34	–	+	–	15, 17, 4	–	7, 17, 4, 35	–	13, 4, 8	17, 10, 4
	4 Length of stationary object	–	35, 28, 40, 29	–	+	–	17, 7, 10, 40	–	35, 8, 2, 14	–	28, 10
	5 Area of moving object	2, 17, 29, 4	–	14, 15, 18, 4	–	+	–	7, 14, 17, 4	–	29, 30, 4, 34	19, 30, 35, 2
	6 Area of stationary object	–	30, 2, 14, 18	–	26, 7, 9, 39	–	+	–		–	1, 18, 35, 36
	7 Volume of moving object	2, 26, 29, 40	–	1, 7, 4, 35	–	1, 7, 4, 17	–	+	–	29, 4, 38, 34	15, 35, 36, 37
	8 Volume of stationary object	–	35, 10, 19, 14	19, 14	35, 8, 2, 14	–	–	–	+	–	2, 18, 37
	9 Speed	2, 28, 13, 38	–	13, 14, 8	–	29, 30, 34	–	7, 29, 34	–	+	13, 28, 15, 19
	10 Force (Intensity)	8, 1, 37, 18	18, 13, 1, 28	17, 19, 9, 36	28, 10	19, 10, 15	1, 18, 36, 37	15, 9, 12, 37	2, 36, 18, 37	13, 28, 15, 12	+

"TRIZ 40 Principles", www.TRIZ40.com, Solid Creativity, 2004. accessed November 10, 2011.

FIGURE 6.10
Metal powder hitting bend in pipe.

degraded by the increase in speed. Then Inventive Principles are identified from querying the TRIZ contradiction matrix. If we think about increasing the speed of the particles, we can envision that other parameters of the system will be degraded, or affected in a negative way. For example, increasing the speed increases the force with which the particles strike the inside wall of the elbow, and erosion increases. This and other degraded parameters are listed in Table 6.11. Also included in the table are the inventive principles taken from a contradiction table for each pair of parameters. For example, to improve speed (9) without having an undesirable effect on force (10), the suggested inventive principles to apply are 13, 15, 19, and 28.

The most direct way to proceed is to look at each inventive principle and sample applications of the principle and attempt to use a similar design change on the system under study.

Solution Idea 1: Principle 13, inversion, requires the designer to look at the problem in reverse or the other way around. In this problem, we should look at the next step of the processing of the metal powder and see what kind of solution can come from bringing materials for the next step to the location of the metal powder. This eliminates the contradiction by removing the need to transport the powder through any kind of direction-changing flow.

Solution Idea 2: Principle 15, dynamicity or dynamics, suggests: (a) allowing the characteristics of an object to change to become more beneficial to the process; and (b) make a rigid or inflexible object moveable or adaptable. We could apply this principle by redesigning the elbow bend in the pipe to have a higher wall thickness through the bend so that the erosion of the inner surface will not compromise the structure of the bend. Another option might be to make the bend area elastic so that the metal particles would transmit some of their impact energy to deformation instead of erosion. Other interpretations are possible.

TABLE 6.11
**Technical Contradictions for Improving Speed of
Metal Powder and Principles to Eliminate Them**

Improved Speed (9) Degraded Parameter	Parameter Number	Principle to be Applied to Eliminate Contradiction
Force	10	13, 15, 19, 28
Durability	15	8, 3, 14, 26
Loss of matter	23	10, 13, 28, 38
Quantity of substance	26	10, 19, 29, 38

Another tactic for using TRIZ would be to determine which principles are most often suggested when looking across all degraded engineering parameters. A count of the frequency with which individual inventive principles were suggested shows that four inventive principles appear twice as suggested redesign tactics. They are: Principle 10—Prior action, 19—Periodic action, 28—Replacement of a mechanical system, and 38—Use strong oxidizers.

Solution Idea

The full description of Principle 28, Replacement of a mechanical system

    a. Replace a mechanical system with an optical, acoustical, or odor system.
    b. Use an electrical, magnetic, or electromagnetic field for interaction with the object.
    c. Replace fields. Example: (1) stationary field change to rotating fields; (2) fixed fields become fields that change in time; (3) random fields change to structured ones.
    d. Use a field in conjunction with ferromagnetic particles.

Principle 28(b) suggests the creative solution of placing a magnet at the elbow to attract and hold a thin layer of powder that will serve to absorb the energy of particles navigating the 90° bend, thereby preventing erosion of the inside wall of the elbow. This solution will only work if the metal particles are ferromagnetic so that they can be attracted to the pipe wall.

The example of improving the transport of metal powder through a pipe seems simple. Use of the TRIZ Contradiction Matrix yielded three diverse, alternative solutions that used unconventional principles to eliminate a couple of the technical contradictions identified in the problem statement. A practice problem is included at the end of the chapter that will allow you to continue the solution generation process. The power of TRIZ inventive principles and their organization should be evident now that the use of the Contradiction Matrix has been demonstrated.

The Contradiction Matrix is powerful, but it only makes use of one of the TRIZ creative solution generation strategies. ARIZ is the more complete, systematic procedure for developing inventive solutions. ARIZ is a Russian acronym and stands for Algorithm to Solve an Inventive Problem. Like Pahl and Beitz's systematic design, the ARIZ algorithm is multiphased, exceedingly prescriptive, precise in its instructions, and uses all the strategies of TRIZ. The interested reader can find more details on ARIZ in a number of texts—for example, see Altshuller.[1]

6.7.5 Strengths and Weaknesses of TRIZ

TRIZ presents a complete design methodology based on a theory of innovation, a process for describing a design problem, and several strategies for solving a design problem. Altshuller intended that TRIZ be systematic in guiding designers to a nearly ideal solution. He also intended that TRIZ be repeatable and reliable, unlike the tools for improving creativity in design (for example, brainstorming).

---

1. G. Altshuller, *The Innovation Algorithm,* L. Shulyak and S. Rodman (translators), Technical Innovation Center, Inc., Worcester, MA, 2000.

Strengths of TRIZ

The TRIZ design method has achieved popularity outside of academic circles un-matched by other methods for technical design. This is due in part to the connection between the application of TRIZ principles and patents.

- The principles at the heart of TRIZ are based on designs that are certified as inventive through the patent-type system of the country of the inventor.
- The developers of TRIZ continued to expand their database of inventive designs beyond the original 200,000.
- A dedicated TRIZ user community (including students of Altshuller) continues to expand the examples of inventive principles, keeping the TRIZ examples contemporary.

Weaknesses of TRIZ

TRIZ has weaknesses common to all design methods that rely on designer inter-pretation. These include:

- Inventive Principles are guidelines subject to designer interpretation.
- The principles are too general for application in a particular design domain, espe-cially in newly developed areas like nanotechnology.
- The designer must develop her own analogous design solution for the given problem, even with an example of an Inventive Principle in the same technical application domain. This calls into question the repeatability of TRIZ principle applications.
- There are differences in the interpretation of TRIZ concepts. For example, some treatments of TRIZ also describe a separate set of four separation principles that can be used to overcome strictly physical contradictions. Two of the separation principles direct the inventor to consider separating conflicting elements of the sys-tem in space or time. The other two are more vague. Some works on TRIZ con-clude that the separation principles are included in the inventive principles, so they are redundant and not mentioned.
- There are aspects of TRIZ that are less intuitive, less available in application ex-amples, and largely overlooked. TRIZ includes techniques for representing techni-cal systems graphically for additional insight and solution. This strategy is called Su-Field Analysis. Altshuller created 72 standard solutions, represented as transfor-mations of Su-Field graphs.

This section presents an introduction to the complex methodology of TRIZ and the philosophy supporting it. The TRIZ Contradiction Matrix and Inventive Principles represent a design methodology that has appeal within the engineering community and may continue to grow in prominence.

6.8
SUMMARY

Engineering design success requires the ability togenerate concepts that are broad in how they accomplish their functions but are also feasible. This requires that each design team member be trained and ready to use all the tools. In presenting this subject we have discussed both the attitudes with which you should approach these tasks and techniques for creativity.

Current research on creativity shows that all people naturally perform basic intellectual functions required to find creative solutions to problems, including design problems. Many methods have been developed that can lead one or more designers in finding creative solutions to any problem. Designers must only be open to using the methods that have been shown to work. There is a four-stage model proposed for creative thinking: preparation, incubation, inspiration, and verification. There are many barriers to creative thinking, including different types of blockages in normal thinking processes. There are also techniques to help people to push through the mental blocks. Some of these methods seem far-fetched, like using the SCAMPER technique, fantasy analogy, asking series of general questions, and incorporating random input into solution ideas. Nevertheless, these methods are useful and can be applied to increase the number of high-quality solution concepts and less formalized design ideas. The idea of a design space filled with alternative solutions is introduced as a meta-model for the conceptual design problem.

The chapter introduced several specific methods for generating conceptual design solutions. Each method includes steps that capitalize on some technique known to be effective in creative problem solving. For example, Synectics is a process of purposefully searching for a variety of analogies that can be used whenever a designer must provide optional solution principles.

Three formal methods for design are introduced in this chapter. Systematic design's functional decomposition process works on intended behavior like physical decomposition works on the form of an existing design. The function structures created with standard function and flow terms serve as templates for generating design solutions. Morphological analysis is a method that works well with a decomposed structure (like that provided in a function structure) to guide in the identification of subproblem solutions that can be combined into alternative design concepts. TRIZ is one of the most recognized and commercially successful design methods today. TRIZ is the method based on innovations extracted from patents and generalized into inventive principles by G. Altshuller. TRIZ's most popular tool for design innovation is the Contradiction Matrix.

NEW TERMS AND CONCEPTS

Axiomatic Design	Functional decomposition	Morphological analysis
Biomimetics	Function structure	Synectics
Concept maps	Generative design	Technical contradiction
Creative cognition	Intellectual blocks	TRIZ
Design fixation	Lateral thinking	
Design space	Mental blocks	

BIBLIOGRAPHY

Creativity

De Bono, E.: *Serious Creativity,* HarperCollins, New York, 1992.
Lumsdaine, E., and M. Lumsdaine: *Creative Problem Solving,* McGraw-Hill, New York, 1995.
Weisberg, R. W.: *Creativity: Beyond the Myth of Genius,* W. H. Freeman, New York, 1993.

Conceptual Design Methods

Cross, N: *Engineering Design Methods,* 3d ed., John & Sons Wiley, Hoboken, NJ, 2001.
French, M. J.: *Conceptual Design for Engineers,* Springer-Verlag, New York, 1985.
Orloff, M. A., *Inventive Thought through TRIZ,* 2d ed., Springer, New York, 2006.
Otto, K. N., and K. L. Wood: *Product Design: Techniques in Reverse Engineering and New Product Development,* Prentice Hall, Upper Saddle River, NJ, 2001.
Suh, N. P.: *The Principles of Design,* Oxford University Press, New York, 1990.
Ullman, D. G.: *The Mechanical Design Process,* 4th ed., McGraw-Hill, New York, 2010.
Ulrich, K. T., and S. D. Eppinger: *Product Design and Development,* 5th ed., McGraw-Hill, New York, 2011.

PROBLEMS AND EXERCISES

6.1 Go to an online catalog of personal use items. Randomly select two products from their inventory and combine them into a useful innovation. Describe the key functionality.

6.2 A technique for removing a blockage in the creative process is to apply transformation rules (often in the form of questions) to an existing but unsatisfactory solution. Apply the key question techniques to the following problem: As a city engineer, you are asked to suggest ways to eliminate puddles from forming on pedestrian walkways. Start with the current solution: waiting for the puddles to evaporate.

6.3 Create a concept map to track your progress through a team brainstorming exercise. Show your map to those present during the session and record their comments.

6.4 Central power plant operators consider converting their energy sources from existing fuels to coal only to discover that they lack the empty property near their facility to store massive piles of coal. Conduct a brainstorming session to propose new ways to store coal.

6.5 Dissect a small appliance and create a physical decomposition diagram. Write a narrative accompanying the diagram to explain how the product works.

6.6 Using the function basis terms provided in the chapter, create a valid function structure for the device chosen in Problem 6.5.

6.7 Create a function structure of a dishwasher.

6.8 Use the idea of a morphological box (a three-dimensional morphological chart) to develop a new concept for personal transportation. Use as the three main factors (the axes of the cube) power source, media in which the vehicle operates, and method of passenger support.

6.9 Sketch and label an exploded view of your favorite mechanical pencil. Create a function structure for it. Use the function structure to generate new designs.

6.10 Use the morphological chart of subproblem solution concepts in Table 6.7 to generate two new basketball return design concepts. Sketch and label your concepts.

6.11 Create a morphological chart for a mechanical pencil.

6.12 Research the personal history of Genrich Altshuller and write a short report on his life.

6.13 Return to Example 6.1, the metal powder transport through an elbow bend. The second engineering parameter to improve is 19. Use the TRIZ Contradiction Matrix to identify inventive principles and generate new solutions to the problem.

7
DECISION MAKING AND CONCEPT SELECTION

7.1
INTRODUCTION

Some writers have described the engineering design process as a series of decisions carried out with less than adequate information. Certainly, creativity, the ability to acquire information, and the ability to combine physical principles into working concepts are critically important in making wise design decisions. So, too, are an understanding of the psychological influences on the decision maker, the nature of the trade-offs embodied in the selection of different options, and the uncertainty inherent in the alternatives. The need to understand the principles behind good decision making is equally important to the business executive, the surgeon, or the military commander as it is to the engineering designer.

Figure 7.1 depicts the concept generation and selection processes as a succession of divergent and convergent steps. Initially we spread the net wide to capture all kinds of customer and industry information about a proposed design. This is then "boiled down" into a product design specification, PDS. Then, with efficient information gathering and creativity stimulation methods, assisted with systematic design methods like function structure analysis and TRIZ, we formulate a set of design concepts using divergent ways of thinking. Convergent thinking comes into play as the design concepts are evaluated at a high level. Often new concepts emerge as the team begins to think about new combinations and adaptations among the concepts—a divergent step. Once again there is an evaluation of concepts against obvious selection criteria that assess broad acceptability of the concepts. The steps of widening the pool of possible concepts and eliminating the clearly inferior ones can repeat until only a small set of concepts remain.

The rounds of successive concept generation and selection cycles modeled in Fig. 7.1 will result in a set of improving concepts if the cycles are controlled by the proper design specification criteria in both the generation and elimination of concepts. The creation of a set of proper design criteria was formalized in Chap. 3. The product or system's *design selection criteria* are the engineering characteristics that emerged

244

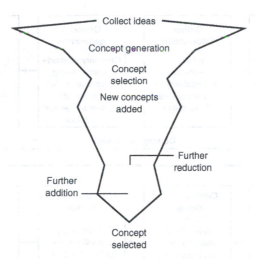

FIGURE 7.1
Concept generation and selection, viewed as alternating divergent and convergent processes.

from the House of Quality as the most important design variables whose values were not set by constraints. Additional design selection criteria may be discovered in consultation with design process sponsors as the process continues, as the result of changing regulations or the unceasing demands of a competitive marketplace.

During any stage of the design process selecting among design alternatives requires: (1) a set of design selection criteria; (2) a set of alternatives believed to satisfy the set of criteria; and (3) a means to evaluate the design alternatives with respect to each criterion. Earlier chapters presented strategies and methods to set design specifications and design criteria, and to generate design alternatives that are likely to meet the criteria. This chapter focuses on determining a decision strategy appropriate to both the design environment and the phase of the design process, creating models to assess design alternatives on decision criteria, and using evaluation processes to reduce a set of alternatives to a few or a single best alternative. Using these methods, a designer or team can decide on one design to carry forward into the embodiment design process after having produced a set of good design alternatives.

The evaluation, modeling, and decision methods described in this chapter are first used in selecting alternatives during conceptual design. The methods will also be useful in any phase of engineering design during which a selection must be made from a set of alternatives. What will differ is the amount of information required for the evaluation, the detail and accuracy of the performance models, and the detail of the design alternatives. The amount of detail increases as design teams move forward in their process.

7.2
DECISION MAKING

Theory for decision making is rooted in many different academic disciplines, including pure mathematics, economics (macro and micro), psychology (cognitive and behavioral), probability, and many others. For example, the discipline of *operations*

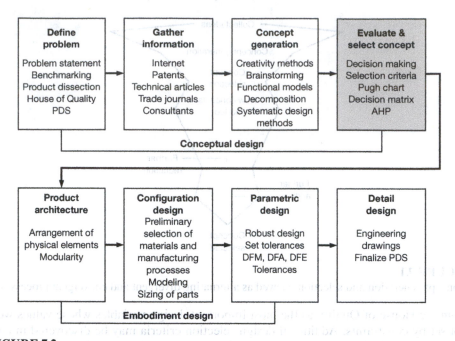

FIGURE 7.2
Steps in the design process, showing evaluation and selection of concepts as the completing step in conceptual design.

research contributed to decision theory. Operations research evolved from the work of a brilliant collection of British and American physicists, mathematicians, and engineers who used their technical talent to provide creative solutions to problems of military operations[1] in World War II. We discuss some of these ideas as they pertain to decision making in the first part of the chapter.

This is followed by a discussion of methods for evaluating and selecting between alternative concepts. As Fig. 7.2 shows, these steps complete the conceptual design phase of the design process.

7.2.1 Behavioral Aspects of Decision Making

Behavioral psychology provides an understanding of the influence of risk taking in individuals and teams.[2] Making a decision is a stressful situation for most people when there is no way to be certain about the information about the past or the predictions of the future. This psychological stress arises from at least two sources.[3] First,

1. A typical problem was how to arrange the ships in a convoy to best avoid being sunk by submarines.
2. R. L. Keeney, *Value-Focused Thinking,* Harvard University Press, Cambridge, MA, 1992.
3. I. L. Janis and L. Mann, *Am. Scientist,* November–December 1976, pp. 657–67.

decision makers are concerned about the material and social losses that will result from either course of action that is chosen. Second, they recognize that their reputations and self-esteem as competent decision makers are at stake. Severe psychological stress brought on by decisional conflict can be a major cause of errors in decision making. There are five basic patterns by which people cope with the challenge of decision making.

1. *Unconflicted adherence*: Decide to continue with current action and ignore information about risk of losses.
2. *Unconflicted change*: Uncritically adopt whichever course of action is most strongly recommended.
3. *Defensive avoidance*: Evade conflict by procrastinating, shifting responsibility to someone else, and remaining inattentive to corrective information.
4. *Hypervigilance*: Search frantically for an immediate problem solution.
5. *Vigilance*: Search painstakingly for relevant information that is assimilated in an unbiased manner and appraised carefully before a decision is made.

All of these patterns of decision making, except the last one, are defective.

A decision is made on the basis of available facts. Great effort should be made to evaluate possible bias and relevance of the facts. It is important to ask the right questions to pinpoint the problem. Emphasis should be on prevention of arriving at the right answer to the wrong question. When you are getting facts from subordinates, it is important to guard against the selective screening out of unfavorable results. Remember that the same set of facts may be open to more than one interpretation. Of course, the interpretation of qualified experts should be respected, but blind faith in expert opinion can lead to trouble.

Facts must be carefully weighed in an attempt to extract the real meaning: knowledge. In the absence of real knowledge, we must seek advice. It is good practice to check your opinions against the counsel of experienced associates. There is an old adage that there is no substitute for experience, but the experience does not have to be your own. You should try to benefit from the successes and failures of others. Unfortunately, failures rarely are recorded and reported widely. There is also a reluctance to properly record and document the experience base of people in a group.

Before a decision can be made, the facts, the knowledge, and the experience must be brought together and evaluated in the context of the problem. Previous experience will suggest how the present situation differs from other situations that required decisions, and thus precedent will provide guidance. If time does not permit an adequate analysis, then the decision will be made on the basis of intuition, an instinctive feeling as to what is probably right (an educated guess). An important help in the evaluation process is discussion of the problem with peers and associates.

The last and most important ingredient in the decision process is judgment. Good judgment cannot be described, but it is an integration of a person's basic mental processes and ethical standards. Judgment is a highly desirable quality, as evidenced by the fact that it is one of the factors usually included in personal evaluation ratings. Judgment is particularly important because most decisional situations are shades of gray rather than either black or white. An important aspect of good judgment is to understand clearly the realities of the situation.

A decision usually leads to an *action*. A situation requiring action can be thought of as having four aspects:[1] should, actual, must, and want. The *should* aspect identifies what ought to be done if there are no obstacles to the action. A should is the expected standard of performance if organizational objectives are to be obtained. The should is compared with the *actual*, the performance that is occurring at the present point in time. The *must* action draws the line between the acceptable and the unacceptable action. A must is a requirement that cannot be compromised. A *want* action is a requirement that is subject to bargaining and negotiation. Want actions are usually ranked and weighted to give an order of priority. They do not set absolute limits but instead express relative desirability.

To summarize this discussion of the behavioral aspects of decision making, we list the sequence of steps that are taken in making a good decision.

1. The objectives of a decision must be established first.
2. The objectives are classified as to importance. (Sort out the musts and the wants.)
3. Alternative actions are developed.
4. The alternatives are evaluated against the objectives.
5. The choice of the alternative that holds the best promise of achieving all of the objectives represents the tentative decision.
6. The tentative decision is explored for future possible adverse consequences.
7. The effects of the final decision are controlled by taking other actions to prevent possible adverse consequences from becoming problems and by making sure that the actions decided on are carried out.

7.2.2 Decision Theory

An important area of activity within the broad discipline of operations research has been the development of a mathematically based theory of decisions.[2] Decision theory is based on utility theory, which develops values, and probability theory, which assesses our stage of knowledge. Decision theory was first applied to business management situations and has now become an active area for research in engineering design.[3] The purpose of this section is to acquaint the reader with the basic concepts of decision theory and point out references for future study.

A decision-making model contains the following six basic elements:

1. *Alternative courses of action* can be denoted as $a_1, a_2, \ldots a_n$. As an example of alternative actions, the designer may wish to choose between the use of steel (a_1), aluminum (a_2), or fiber-reinforced polymer (a_3) in the design of an automotive fender.

---

1. C. H. Kepner and B. B. Tregoe, *The New Rational Manager: A Systematic Approach to Problem Solving and Decision Making,* Kepner-Tregoe, Inc., Skillman, NJ, 1997.
2. H. Raiffa, *Decision Analysis,* Addison-Wesley, Reading, MA, 1968; S. R. Watson and D. M. Buede, *Decision Synthesis: The Principles and Practice of Decision Analysis,* Cambridge University Press, Cambridge, 1987.
3. K. E. Lewis, W. Chen, and L. C. Schmidt, eds. *Decision Making in Engineering Design,* ASME Press, New York, 2006.

2. *States of nature* are the environment of the decision model. Usually, these conditions are out of the control of the decision maker. If the part being designed is to withstand salt corrosion, then the state of nature might be expressed by θ_1 = no salt, θ_2 = weak salt concentration, etc.
3. *Outcome* is the result of a combination of an action and a state of nature.
4. *Objective* is the statement of what the decision maker wants to achieve.
5. *Utility* is the measure of satisfaction that the decision maker associates with each outcome.
6. *States of knowledge* is the degree of certainty that can be associated with the states of nature. This is expressed in terms of probabilities.

Decision-making models usually are classified into four groups with respect to the state of knowledge.

- *Decision under certainty*: Each action results in a known outcome that will occur with a probability of 1.
- *Decision under uncertainty*: Each state of nature has an assigned probability of occurrence.
- *Decision under risk*: Each action can result in two or more outcomes, but the probabilities of the outcomes are unknown.
- *Decision under conflict*: The states of nature are replaced by courses of action determined by an opponent who is trying to maximize his or her objective function. This type of decision theory usually is called game theory.

In the situation of *decision under certainty,* the decision maker has all the information necessary to evaluate the outcome of her choices. She also has information about different conditions under which the decision must be made. Therefore, the decision maker need only recognize the situation in which the decision is occurring and look up the outcomes of all possible choices. The challenge here is having the information on the outcomes ready when needed. This decision strategy is illustrated with Example 7.1.

EXAMPLE 7.1 Decision Under Certainty
To select the best material to resist road salt corrosion in an automotive fender, we construct a table of the utilities for each outcome. The possible states of nature are the road conditions for driving. They are: θ_1: no salt; θ_2: mild salt; and, θ_3: heavy salt. A utility can be thought of as a generalized loss or gain, all factors of which (cost of material, cost of manufacturing, corrosion resistance) have been converted to a common scale. Assume that utility has been expressed on a scale of "losses." Table 7.1 shows the loss table for this material selection decision. Note that, alternatively, the utility could be expressed

TABLE 7.1
Loss Table for a Material Selection Decision

	State of Nature		
Course of Action	θ_1	θ_2	θ_3
a_1: steel	1	4	10
a_2: aluminum	3	2	4
a_3: FRP	5	4	3

in terms of gains, and then the table would be called the *payoff matrix*. Using a *decision under certainty condition,* we only have to look at the values of a column to determine the appropriate selection. Examination of Table 7.1 would lead us to conclude that a_1 (steel) is the material of choice (lowest loss) when there is no salt present, a_2 (aluminum) is the choice when mild salt is present in the environment, and a_3 (FRP) is the best material when heavy salt corrosion is present.

For *decision making under uncertainty,* the probability of occurrence for each of the states of nature must be able to be estimated. This allows us to determine the expected value for each of the alternative design parameters (courses of action).

EXAMPLE 7.2 Decision Under Uncertainty
The probability of occurrence of the states of nature are estimated as:

State of nature	θ_1	θ_2	θ_3
Probability of occurrence	0.1	0.5	0.4

The *expected value* of an action, a_1, is given by

$$\text{Expected value of } a_i = E(a_i) \sum_i P_i a_i \tag{7.1}$$

Thus, for the three materials in Table 7.1, the expected losses would be

Steel: $\quad E(a_1) = 0.1(1) + 0.5(4) + 0.4(10) = 6.1$

Aluminum: $\quad E(a_2) = 0.1(3) + 0.5(2) + 0.4(4) = 2.9$

FRP: $\quad E(a_3) = 0.1(5) + 0.5(4) + 0.4(3) = 3.7$

Therefore, we would select aluminum for the car fender since it has the lowest value of loss in utility.

The assumption in *decision making under risk* is that the probabilities associated with the possible outcomes are not known. The approach used in this situation is to form a matrix of outcomes, usually expressed in terms of utilities, and base the selection (decision) on various decision rules. Ex. 7.3 and 7.4 illustrate decision rules *Maximin* and *Maximax,* respectively.

EXAMPLE 7.3 Maximin Rule
The *maximin decision rule* states that the decision maker should choose the alternative that maximizes the minimum payoff that can be obtained. Since we are dealing with *losses* in utility, we should select the alternative that minimizes the maximum loss.

Looking at Table 7.1, we find the following maximum losses for each alternative:

$$a_1 : \theta_3 = 10 \qquad a_2 : \theta_3 = 4 \qquad a_3 : \theta_1 = 5$$

The maximin rule requires selection of aluminum, a_2, because it has the smallest of the maximum losses. The best outcome is to select the worst-case situation that results in the lowest loss.

EXAMPLE 7.4 Maximax Rule
An opposite extreme in decision rules is the *maximax decision rule*. This rule states that the decision maker should select the alternative that maximizes the maximum value of

the outcomes. This is an optimistic approach because it assumes the best of all possible worlds. For the loss table in Table 7.1 the alternative selected would be the one with the smallest possible loss.

$$a_1 : \theta_1 = 1 \qquad a_2 : \theta_2 = 2 \qquad a_3 : \theta_3 = 3$$

The decision based on a maximax criterion would be to select steel, a_1, because it has the smallest loss of the best outcome for each alternative.

The use of the maximin decision rule implies that the decision maker is very pessimistic. On the other hand, the decision maker who adopts the maximax approach is an optimist who places little utility on values below the maximum. Neither decision rule is particularly logical. Since the pessimist is too cautious and the optimist is too audacious, we would like to have an in-between decision rule. It can be obtained by combining the two rules. By using an index of optimism, α, the decision maker can weigh the relative amount of pessimistic and optimistic components of the combined decision rule.

EXAMPLE 7.5 Combined Criterion

We weight the decision criterion as three-tenths optimistic. Next we construct Table 7.2. Under the optimistic column, place the lowest loss for each alternative, while under the pessimistic column place the largest loss for each material. When each term is multiplied by α and $(1 - \alpha)$ and summed to total we obtain Table 7.2. After a quick read of the table, aluminum, a_2, is selected once again for use as the fender material.

TABLE 7.2
Revised Loss Estimates Combining States
of Nature Information (with $\alpha = 0.3$)

Alternative	Optimistic	Pessimistic	Total
Steel	0.3(1)	+0.7(10)	= 7.3
Aluminum	0.3(2)	+0.7(4)	= 3.4
FRP	0.3(3)	+0.7(5)	= 4.4

Decisions will be different if the conditions under which they are made vary. Table 7.1 shows that there is a state of nature that justifies the use of each material on the basis that it provides the best outcome. Knowing that the states of nature in which the car will be used can vary, the decision maker must determine a strategy for choosing fender material. Several examples in this section showed how different decision rules (maximin and maximax) have been developed to take into account the decision maker's comfort with uncertainty and risk.

7.2.3 Decision Trees

Decision trees are a graphical and mathematical method for decision making under uncertainty. The construction of a decision tree is a useful technique when a string of decisions must be made in succession and probabilities of each outcome are known or can be estimated. Figure 7.3 shows the decision tree concerned with deciding whether an electronics firm should carry out R&D in order to develop a new product. The firm is a

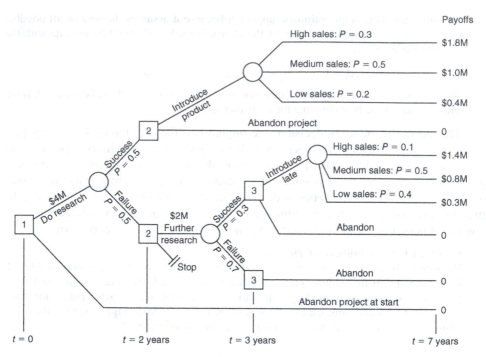

FIGURE 7.3
Decision tree for an R&D project.

large conglomerate that has had extensive experience in electronics manufacture but no direct experience with the product in question. With the preliminary research done so far, the director of research estimates that a $4 million ($4M) R&D program conducted over two years would provide the knowledge to introduce the product to the marketplace.

A decision point in the decision tree is indicated by a square, and circles designate chance events (states of nature) that are outside the control of the decision maker. The length of line between nodes in the decision tree is not scaled with time, although the tree does depict precedence relations.

The first decision point is whether to proceed with the $4M research program or abandon it before it starts. We assume that the project will be carried out. At the end of the two-year research effort the research director estimates there is a 50-50 chance of being ready to introduce the product. If the product is introduced to the market, it is estimated to have a life of five years. If the research is a failure, it is estimated that an investment of an additional $2M would permit the R&D team to complete the work in an additional year. The chances of successfully completing the R&D in a further year are assessed at 3 in 10. Management feels that the project should be abandoned if a successful product is not developed in three years because there will be too much competition. On the other hand, if the product is ready for the marketplace after three years, it is given only a 1 in 10 chance of producing high sales.

The payoffs expected at the end are given to the far right at the end of each branch. The dollar amounts should be discounted back to the present time by using techniques of the time value of money (Chap. 16). When selecting branches in a decision tree the decision rule used the largest expected value of the payoff. (It is possible to use other decision rules, such as maximin).

The best place to start in this problem is at the ends of the branches and work backward. The expected values for the chance events are:

$$E = 0.3(1.8) + 0.5(1.0) + 0.2(0.4) = \$1.12M \text{ for the on-time project}$$

$$E = 0.1(1.4) + 0.5(0.8) + 0.4(0.3) = \$0.66M \text{ for the delayed project at}$$
$$\text{decision point 3}$$

$$E = 0.3(0.66) + 0.7(0) - 2 = -\$1.8M \text{ for the delayed project at decision point 2}$$

Thus, carrying the analysis for the delayed project backward to decision point 2 shows that to continue the project beyond that point results in a large negative expected payoff. The proper decision, therefore, is to abandon the research project if it is not successful in the first two years. Further, the calculation of the expected payoff for the on-time project at point 1 is a large negative value.

$$E = 0.5(1.12) + 0.5(0) - 4.0 = -\$3.44M$$

Thus, either the expected payoff is too modest or the R&D costs are too great to be warranted by the payoff. Therefore, based on the estimates of payoff, probabilities, and costs, this R&D project should not have been undertaken.

7.2.4 Utility Theory

Maximax and maximin are strategies that incorporate attitude toward risk in decision problems. The examples presented in the previous section presuppose the ability to determine the utility of each outcome. A more direct method is to use Utility Theory in establishing the problem.

In Utility Theory, everyday words take on precise meanings that are not the same as in common usage. Definitions are required:

- *Value* is an attribute of an alternative that is implied by choice (e.g., if A is chosen over B, it is assumed that A has more value than B). Nowadays, money is the medium of exchange that is used to express value. A buyer will exchange an amount of money (B) for a material good (A) only if the buyer perceives A to be worth more than B at the time of the exchange.
- *Preference* is the statement of relative value in the eyes of the decision maker. Preference is a subjective quality that depends totally on the decision maker.
- *Utility* is a measure of preference order for a particular user. Utility is not necessarily equal to the value of exchange in the marketplace.
- *Marginal utility*: A key concept of utility theory is the understanding of the nature of what is gained by adding one more unit to the amount already possessed. Most decision makers have utility functions that are consistent with the Law of Diminishing Marginal Utility.[1]

---

1. It may seem intuitive that more is always better. However, consider servings of a favorite dessert. In 1738 Bernoulli established the fact that money has decreasing marginal utility. The more one has, the less value the next unit brings to the decision maker.

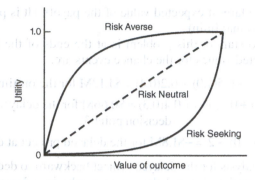

FIGURE 7.4
Utility functions implied by decision maker's risk tolerance.

Utility for a particular set of alternatives is often represented by a function, and that function is usually assumed to be continuous. When presented with a utility function, $U(x)$, you can draw some conclusions about the preferences of the person from whom it was constructed. First, you can determine a preference ordering of two different amounts of something. Second, you can determine some idea of the decision maker's attitude toward risk as shown in Fig. 7.4. The utility functions curves are for a risk-averse and risk-taking individual.

EXAMPLE 7.6
Table 7.3 lists the probabilities associated with various outcomes related to the acceptance of two contracts that have been offered to a small R&D laboratory. Using expected values only, a decision maker would choose Contract I because it has a greater expected value than Contract II.

TABLE 7.3
Probabilities and outcomes to illustrate utility

Contract I		Contract II	
Outcome	Probability	Outcome	Probability
+100,000	0.6	+60,000	0.5
+15,000	0.1	+30,000	0.3
−40,000	0.3	−10,000	0.2

$$E(\text{I}) = 0.6(100,000) + 0.1(15,000) + 0.3(-40,000) = \$62,700$$
$$E(\text{II}) = 0.5(60,000) + 0.3(30,000) + 0.2(-10,000) = \$37,000$$

The decision in Example 7.6 is straightforward when the probability of each outcome is known and the decision maker is going to act according to the expected value calculation. Complications arise when the probabilities are not known or the decision maker includes more than just expected value of the outcome in the decision process.

Reviewing Example 7.6, Contract I has a higher expected value ($62,700) than Contract II ($37,000). However, Contract I has a 30 percent chance of incurring a fairly large loss (–$40,000), whereas Contract II has only a 20 percent chance of a much smaller loss. If the decision maker decided to take the worst-case scenario into account and minimize the loss exposure of the company, Contract II would be selected. In this case, expected value analysis is inadequate because it does not include the value of minimizing loss to the decision maker.

What is needed is *expected utility analysis* so that the attitude of the decision maker toward risk becomes part of the decision process. Under expected utility theory, the decision maker always chooses the alternative that maximizes expected utility. The decision rule is: maximize expected utility.

To establish the utility function, we rank the outcomes in numerical order: +100,000, +60,000, +30,000, +15,000, 0, –10,000, –40,000. The value $0 is introduced to represent the situation in which we take neither contract. Because the scale of the utility function is wholly arbitrary, we set the upper and lower limits as

$$U(+100,000) = 1.00 \qquad U(-40,000) = 0 \tag{7.2}$$

Note that in the general case the utility function is not linear between these limits.

EXAMPLE 7.7
Determine the utility value of the outcome of earning $60,000 under a contract to the decision maker choosing contracts. To establish the utility associated with the outcome of +60,000, decision makers (DM) ask themselves a series of questions.

    Question 1: Which would I prefer?
      A: Gaining $60,000 for certain; or,
      B: Having a 75% chance of gaining $100,000 and a 25% chance of losing $40,000.

    DM Answer: I'd prefer option A because option B is too risky.

    Question 2: Changing the probabilities of option B, which would I now prefer?
      A: Gaining $60,000 for certain; or,
      B: Having a 95% chance of gaining $100,000 and a 5% chance of losing $40,000.

    DM Answer: I'd prefer option B with those probabilities.

    Question 3: Again changing the probabilities for option B, which would I prefer?
      A: Gaining $60,000 for certain; or,
      B: Having a 90% chance of gaining $100,000 and a 10% chance of losing $40,000?

    DM Answer: It would be a toss-up between A and B with those chances.

These answers tell us that this decision maker sees the utility of option A and has found the certainty equivalent to the chances given by option B. He's determined that the certain outcome of gaining $60,000 is equivalent to the uncertain outcome expressed by the lottery of option B.

    $U(+60,000) = 0.9U(+100,000) + 0.1U(-40,000)$,
    substituting in values from Eq. (7.2),
    $U(+60,000) = 0.9(1.0) + 0.1(0)$
    $U(+60,000) = 0.9$.

Example 7.7 shows us that a technique for finding utility values is to vary the odds on the choices until the decision maker is indifferent to the choice between A and B. The same procedure is repeated for each of the other values of outcomes to establish the utility for those points. A difficulty with this procedure is that many people have difficulty in distinguishing between small differences in probability at the extremes, for example, 0.80 and 0.90 or 0.05 and 0.01.

A critical concept about expected utility is that it is not the same as the expected value. This can be emphasized by reviewing the choices the decision maker gave in response to questions 1 and 2 in Example 7.7. The expected values of option B in questions 1 and 2 are +65,000 and +95,000, respectively. In Question 1, the decision maker rejected an option that had an expected value of a 65,000 gain in favor of a certain gain of 60,000. Here the decision maker wants to avoid risk, making him risk adverse. It takes the possibility of a huge increase in gain to convince the decision maker to accept risk. Question 2's option B has an expected value of +93,000. That's a differential of 33,000 over the certain option of a 60,000 gain.

Nonmonetary values of outcome can be converted to utility in various ways. Clearly, quantitative aspects of a design performance, such as speed, efficiency, or horsepower, can be treated as dollars were in Example 7.7. Qualitative performance indicators can be ranked on an ordinal scale, for example, 0 (worst) to 10 (best), and the desirability evaluated by a questioning procedure similar to the above.

Two common types of utility functions that are found for the design variables are shown in Fig. 7.5. The utility function shown in Fig. 7.5a is the most common. Above the design value the function shows diminishing marginal return for increasing the value of the outcome. The dependent variable (outcome) has a minimum design value set by specifications, and the utility drops sharply if the outcome falls below that value. The minimum pressure in a city water supply system and the rated life of a turbine engine are examples. For this type of utility function a reasonable design criterion would be to select the design with the maximum probability of exceeding the design value. The utility function sketched in Fig. 7.5b is typical of a high-performance situation. The variable under consideration is very dominant, and we are concerned with maximum performance. Although there is a minimum value below which the design is useless, the probability of going below the minimum value is considered to be very low.

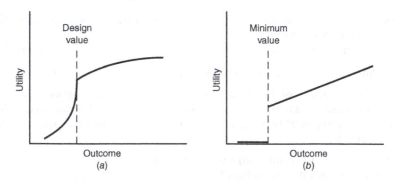

FIGURE 7.5
Common types of utility functions in engineering design.

In the typical engineering design problem more than one dependent variable is important to the design. This requires developing a multiattribute utility function.[1] These ideas, originally applied to problems in economics, have been developed into a design decision methodology called methodology for the evaluation of design alternatives (MEDA).[2] Using classical utility theory, MEDA extends the usual design evaluation methods to provide a better measure of the worth of the performance levels of the attributes to the designer and more accurately quantify attribute trade-offs. The price is a considerable increase in the resources required for evaluation analysis.

7.3
EVALUATION PROCESSES

We have seen that decision making is the process of identifying alternatives and the outcomes from each alternative and subjecting this information to a rational process of making a decision. *Evaluation* is a type of process in which alternatives are first appraised according to some standard. Their scores or rank as determined by that standard are compared to make the decision as to which is best.

Figure 7.6 reviews the main steps in concept generation (Chap. 6) and shows the steps that make up concept evaluation. Note that these evaluation steps are not limited

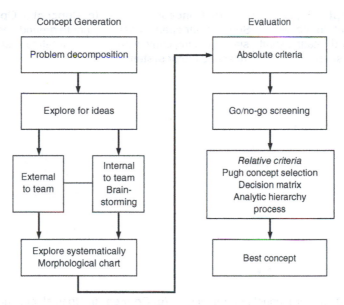

FIGURE 7.6
Steps that are involved in concept generation and its evaluation.

1. R. L. Keeney and H. Raiffa, *Decisions with Multiple Objectives,* Cambridge University Press, Cambridge, UK, 1993.
2. D. L. Thurston, *Research in Engineering Design,* vol. 3, pp. 105–22, 1991.

to the conceptual design phase of the design process. They are just as applicable, and should be used, in embodiment design when deciding which of several component designs is best or which of five possible fabrication materials should be chosen. Figure 7.7 displays a set of five concepts for automated basketball return devices that were generated by the JSR Design team.

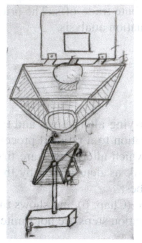

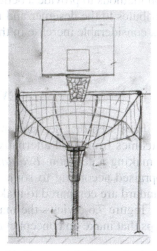

(a) **Concept 1**: Square-opening net, trampoline, ground-based rotating shaft system

(b) **Concept 2**: Semicircular opening net, single rotating chute with attached motor system

(c) **Concept 3**: Open sloping net to ground based shaft rotating system

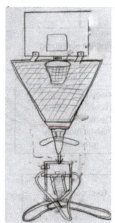

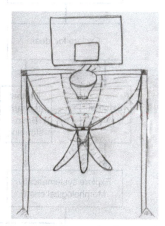

(d) **Concept 4**: Square funnel non-rotating director and multiple chutes

(e) **Concept 5**: Open sloping net to pivot a single guide to multiple positions

FIGURE 7.7
Shot-Buddy concepts generated by design team.[1]

---

1. Adapted from Josiah Davis, Jamil Decker, James Maresco, Seth McBee, Stephen Phillips, and Ryan Quinn, "JSR Design Final Report: Shot-Buddy," unpublished, ENME 472, University of Maryland, May 2010.

In an *absolute comparison* the concept is directly compared with a fixed and known set of requirements such as a PDS or design code. In a *relative comparison* the concepts are compared with each other on the basis of a metric. Checking to see if a design alternative would be under the weight limit specified in the PDS is an example of an absolute comparison. On the other hand, if the best design possible would be the lightest design, the design team would need to estimate the weight of each design alternative, and then compare the results. The most suitable alternative in terms of weight would be the one with the lowest estimate. This is a relative comparison.

7.3.1 Design Selection Based on Absolute Criteria

It makes no sense to subject several design concepts to a rigorous evaluation process if it is obvious, or soon becomes clear, that some aspect about the concept disqualifies it for selection. Therefore, it is good practice to begin the evaluation process by using a series of absolute filters.[1]

1. **Evaluation based on judgment of functional feasibility of the design:** The initial screening is based on the overall evaluation of the design team as to the feasibility of each concept. Concepts should be placed into one of three categories:
   (a) It is not feasible (it will never work). Before discarding an idea, ask "why is it not feasible?" The answer may provide new insight into the problem.
   (b) Feasibility is conditional—it might work if something else happens. The something else could be the development of a critical element of technology or the appearance in the market of a new microchip that enhances some function of the product.
   (c) It will work. This is a concept that seems worth developing further.

The reliability of these judgments is strongly dependent on the expertise of the design team. When making this judgment, err on the side of accepting a concept unless there is strong evidence that it will not work.

2. **Evaluation based on assessment of technology readiness:** Except in unusual circumstances, the technology used in a design must be mature enough that it can be used in the product design without additional research effort. *Product design is not the appropriate place to do R&D.* Some indicators of technology maturity are:
   (a) Can the technology be manufactured with known processes?
   (b) Are the critical parameters that control the function identified?
   (c) Are the safe operating latitude and sensitivity of the parameters known?
   (d) Have the failure modes been identified?
   (e) Does hardware exist that demonstrates positive answers to the above four questions?

---

1. D.G. Ullman, *The Mechanical Design Process,* 4th ed., McGraw-Hill, New York, 2010.

3. **Evaluation based on go/no-go screening of the constraints and threshold levels of engineering characteristics:** After a design concept has passed filters 1 and 2, the emphasis shifts to establishing whether it satisfies the constraints of the problem. The emphasis is not on a detailed examination (that comes below) but on eliminating any design concepts that clearly are not able to meet constraints or minimum acceptable levels of important engineering characteristics.

EXAMPLE 7.8

In Sec. 6.6.2, a morphological chart was used to generate a concept for the automated basketball return device designed by the JSR Design team and is shown in Fig. 6.9. This alternative is also shown in Fig. 7.7 as Concept 5. It consists of a roughly semicircular shot catch net supported on a frame connected to the court edge at the ground which fits under the basketball net. The catch net tapers down to the size of a basketball and terminates in a curved metal guide somewhat like a sloping ski jump ramp that the ball will follow as it continues its downward travel after passing through or near the basketball hoop. It is assumed that the ball's kinetic energy will provide enough force to allow it to ride the guide ramp back to the direction of the shooter. Figures 6.9 and 7.7 do not include any detail about the system that will be used to pivot the ball return guide between the three possible positions shown in the sketch. Nor does the sketch detail the ability of the pivoting mechanism to sense the location of the shooter to determine the acceptable guide position. This is the typical amount of detail that would be provided in an early concept.

Apply the functional feasibility screening criterion to this Shot-Buddy concept.

Question: Can this concept return a basketball to the shooter?

Answer: There are some missing subsystems as described above, but they could be specified and work to control the position of the guide.

Question: Assuming you augment the design, is it feasible as a concept?

Answer: This is *not* a feasible design.

- The catch net is only supported at the sides. Some means of extending the net out over the basketball court is needed, which would obstruct the play.
- The guide rail appears to be hanging from the catch net. This is not a rigid position, so the guide's ability to direct the motion of the basketball would be jeopardized.

Summary and Decision: This Shot-Buddy concept is not functionally feasible as represented in the sketch. (1) A catch net of the size required in the specification (Table 3.4) could not be supported as shown. (2) Adjustments to the design to provide the support for the physics of changing the basketball's motion would violate an implied but critical constraint of not interfering with the shooter's play. There may be value in the pivoting guide mechanism if it were supported from a fixed position.

Proceed in this way through all of the proposed concepts. Note that if a design concept shows mostly "go" responses, but it has a few no-go responses, it should not be summarily discarded. The weak areas in the concept may be able to be fixed by borrowing ideas from another concept. Or the process of doing this go/no-go analysis may trigger a new idea.

7.3.2 Measurement Scales

Rating a design parameter of several alternative designs is a measurement process. Therefore, we need to understand the various scales of measurement that can be used in this type of process.[1]

- *Nominal scale* is a named category or identifier like "thick or thin," "red or black," or "yes or no." The only comparison that can be made is whether the categories are the same or not. Variables that are measured on a nominal scale are called categorical variables.
- *Ordinal scale* is a measurement scale in which the items are placed in rank order, first, second, third, and so on. These numbers are called ordinals, and the variables are called ordinal or rank variables. Comparisons can be made as to whether two items are greater or less than each other, or whether they are equal, but addition or subtraction is not possible using this scale. The ordinal scale says nothing about how far apart the elements are from each other. However, the mode can be determined for data measured on this scale. (The Pugh concept selection method uses an ordinal scale).

Ranking on an ordinal scale calls for decisions based on subjective preferences. One method of ranking alternatives on an ordinal scale is to use *pairwise comparison*. Each design criterion is listed and is compared to every other criterion, two at a time. In making the comparison the objective that is considered the more important of the two is given a 1 and the less important objective is given a 0. The total number of possible comparisons is $N = n(n-1)/2$, where n is the number of criteria under consideration.

Consider the case where there are five design alternatives, A, B, C, D, and E. In comparing A to B we consider A to be more important, and give it a 1. (In building this matrix, a 1 indicates that the objective in the row is preferred to the objective in the column.) In comparing A to C we feel C ranks higher, and a 0 is recorded in the A line and a 1 on the C line. Thus, the table is completed. The rank order established is B, D, A, E, C. Note that we used head-to-head comparisons to break ties, as shown in the rows of the following table.

Design Criterion	A	B	C	D	E	Row Total
A	—	1	0	0	1	2
B	0	—	1	1	1	3
C	1	0	—	0	0	1
D	1	0	1	—	1	3
E	0	0	1	0	—	1
						10

1. K. H. Otto, "Measurement Methods for Product Evaluation," *Research in Engineering Design*, vol. 7, pp. 86–101, 1995.

Since the ratings are ordinal values, we cannot say that A has a weighting of 2/10 because division is not a possible arithmetic operation on an ordinal scale. In other words, it is mathematically incorrect to use the numerical values in the table as weighting factors.

- *Interval scale* is the type needed to determine how much worse A is compared with D. On an interval scale of measurement, differences between arbitrary pairs of values can be meaningfully compared, but the zero point on the scale is arbitrary. Addition and subtraction are possible, but not division and multiplication. Central tendency can be determined with the mean, median, or mode.

For example, we could distribute the results from the previous table along a 1 to 10 scale to create an interval scale. This can be done only if additional information is available to quantify the differences between the alternatives.

C	E			A		D	B
1	2	3	4	5	6	7	8	9	10

The most important alternative designs have been given a value of 10, and the others have been given values relative to this.

- *Ratio scale* is an interval scale in which a zero value is used to anchor the scale. Each data point is expressed in cardinal numbers (2, 2.5, etc.) and is ordered with respect to an absolute point. All arithmetic operations are allowed. A ratio scale is needed to establish meaningful weighting factors. Most engineering characteristics in engineering design, like weight, force, and velocity, are measured on a ratio scale.

7.4
USING MODELS IN EVALUATION

Analyzing performance is an important step in conceptual design. In evaluating competing concepts, it is necessary to analyze information obtained from models of various sorts. Models fall into three categories: iconic, analog, and symbolic.

An *iconic model* is a physical model that looks like the real thing but is a scaled representation. Generally the model scale is reduced from the real situation, as in a scale model of an aircraft for wind tunnel tests. An advantage of iconic models is that they tend to be smaller and simpler than the real object, so they can be built and tested more quickly and at lower cost. Iconic models are geometric representations. They may be two-dimensional, as in maps, photographs, or engineering drawings, or three-dimensional as in machined parts. Three-dimensional CAD models are commonly used with a computer to do analysis and simulate behavior.

Analog models are models that are based on an analogy, or similarity, between different physical phenomena. This approach allows the use of a solution based in one physical science discipline, for example, electric circuits, to solve a problem in a completely different field, for example, heat transfer. Analog models are often used to compare something that is unfamiliar with something that is very familiar.

An ordinary graph is really an analog model because distances represent the magnitudes of the physical quantities plotted on each axis. Since the graph describes the real functional relation that exists between those quantities, it is a model. Another common class of analog models is process flow charts.

Symbolic models are abstractions of the important quantifiable components of a physical system that use symbols to represent properties of the real system. A mathematical equation expressing the dependence of the system output parameter on the input parameters is a common symbolic or *mathematical model*. A symbol is a shorthand label for a class of objects, a specific object, a state of nature, or simply a number. Symbols are useful because they are convenient, assist in explaining complex concepts, and increase the generality of the situation. Symbolic models probably are the most important class of model because they provide the greatest generality in attacking a problem. The use of a symbolic model to solve a problem calls on our analytical, mathematical, and logical abilities. A symbolic model is also important because it leads to quantitative results. When a mathematical model is reduced to computer software, we can use the model to investigate design alternatives in a relatively inexpensive way.

In conceptual design we use both iconic and symbolic models. Simple mathematical models like free body diagrams and heat balances are used to help formalize a concept and to provide data, not just opinions, to use in decision evaluation tools. A *proof-of-concept prototype* is typically made by the end of conceptual design. Ideally, a succession of models, some physical, others rough sketches, are made to serve as learning tools until reaching the final proof-of-concept model. This is just the first of a succession of prototypes (physical models) that will be made until the product reaches the marketplace (see Sec. 8.11.1).

Choosing Appropriate Models

The type of model and its level of detail and accuracy changes depending upon the stage of the design process in which you are working.

- In **conceptual design,** the emphasis is on geometrical modeling using multiple hand sketches supplemented with quick physical prototypes made from wood, foam board, and so on. Simple mathematical models based on concepts learned from your engineering science courses are applied in concept evaluation using hand-calculation levels of precision. After concept selection is completed, it is usually capped off by the development of a geometrical computer-based model (CAD model). This serves as a *proof-of-concept prototype* that is frequently supplemented with a physical prototype, often made by a rapid prototyping process.
- In **embodiment design,** where major emphasis is given to establishing shape, dimensions, and tolerances, the level of detail in mathematical and physical models increases. It is usually helpful to use a computational tool such as Excel, MATLAB, or a specialized software program. Often a finite analysis program is used to determine stresses in a part with complex shape or critical to quality issues. This design phase ends with the testing of a *proof-of-product prototype* using full size parts made from the materials selected for the product.
- In **detail design,** more complex mathematical modeling may be conducted to optimize some product characteristic or to improve its robustness. A complete set of detail and

assembly drawings suitable to manufacture the product will be completed. A *proof-of-process prototype* will be tested using the exact materials and processes that will be used to manufacture the product. For more details on the sequence of prototypes used throughout the product design process see Sec. 8.11.1.

7.4.1 Aids to Mathematical Modeling

Engineering courses teach first principles in subjects like statics, dynamics, mechanics of materials, fluids, and thermodynamics by describing a physical system and its immediate environment in a complex word problem that you learned to solve using a variety of analytical, logical, mathematical, and empirical methods. The key to finding a solution is to understand the mathematical model appropriate for the problem. Engineering design courses provide the opportunity to use this knowledge in more applied ways.

Dimensional Analysis

A useful tool in model building is dimensional analysis. There are usually fewer dimensionless groups than there are physical quantities in the problem, so the groups become the real variables of the problem. You most likely learned about dimensional analysis in a course on fluid mechanics[1] or heat transfer. The importance of dimensional analysis is that it allows you to express a problem with a minimum number of design variables. Also, representing a complex phenomenon in a concise way can make difficult problems understandable. An important advantage to using dimensional analysis is that it significantly reduces the number of trials required when seeking to improve the robustness of a design or to optimize it for some property such as minimum weight.[2]

Scale Models

Scale models are often used in design because they can be made more quickly and at less cost. In using physical models, it is necessary to understand the conditions under which *similitude* prevails for both the model and the prototype.[3] By similitude we mean the condition of physical response is similar between the model and the prototype. There are several forms of similitude: geometric, kinematic (similar velocities), and dynamic (similar forces). Geometric similarity is the form most usually encountered in product design. The conditions for it are a three-dimensional equivalent of a photographic enlargement or reduction, that is, identity of shape, equality of corresponding angles or arcs, and a constant proportionality or scale factor relating corresponding linear dimensions.

---

1. B. R. Munson, D. F. Young, and T. H. Okiishi, *Fundamentals of Fluid Mechanics*, 5th ed., John Wiley & Sons, Hoboken, NJ, 2006, pp. 347–69. For an advanced treatment see T. Szirtes, *Applied Dimensional Analysis and Modeling*, 2d ed., Butterworth-Heinneman, Boston, 2007. See Wikipedia at *Dimensional quantities* for a large collection of dimensionless numbers.
2. D. Lacey and C. Steele, *Journal of Engineering Design*, vol. 17, no. 1, pp. 55–73, 2006.
3. D. J. Schuring, *Scale Models in Engineering*, Pergamon Press, New York, 1977; E. Szucs, *Similitude and Modeling*, Elsevier Scientific Publ. Co., New York, 1977.

To illustrate scale modeling, consider a bar loaded in tension. The stress in the bar due to axial loading is:

$$\sigma = P/A = P/(\pi D^2/4) \qquad (7.3)$$

where P is the axial load on the bar
A is the cross sectional area with a diameter D
If the left side of Eq. (7.3) is divided by the right side we obtain

$$\frac{\pi \sigma D^2}{4P} = 1 \qquad (7.4)$$

Equation (7.4) is dimensionless. This illustrates that for a relationship to be a valid indicator of similitude it must be dimensionless. If we designate the model with a subscript m and the prototype with subscript p, we can write one equation for m and another for p, and equate them because they each are equal to unity.

$$\sigma_p P_m D_p^2 = \sigma_m P_p D_m^2 \qquad (7.5)$$

We are testing the model and want to determine what this tells us about the performance of the prototype. Therefore, solving Eq. (7.5) for σ_p

$$\sigma_p = \left(\frac{D_m}{D_p}\right)^2 \left(\frac{P_p}{P_m}\right) \sigma_m \qquad (7.6)$$

Equation (7.6) tells us what to expect from the prototype for a measured stress on the model. The answer depends on two scale factors that emerge in Eq. (7.6). If we have a 1/10th scale model, it means that the *geometric scale factor* $\mathbf{S} = D_m/D_p$ is 1/10. The second scale factor is the *load scale factor* $\mathbf{L} = P_m/P_p$. Since the model is much smaller than the prototype, it cannot withstand the same loads as the prototype. For example, $\mathbf{L} = 1/3$ might be an appropriate load factor. Then Eq. (7.6) can be written

$$\sigma_p = (S^2/L)\sigma_m \qquad (7.7)$$

The form of the scaling relationship between the prototype and the model will change depending on the physical situation, but the approach will be as above. For example, if we wanted to model the displacement of the axially loaded bar, δ, based on the strength of materials relationship, $\delta = PL/AE$, the scaling equation would contain three terms, $\mathbf{S}$, $\mathbf{L}$, and $\mathbf{E}$, the last one being an elastic modulus scaling factor.

7.4.2 A Process for Mathematical Model Building

There are four distinct characteristics of mathematical models consisting of two classes each (1) steady-state or transient (dynamic), (2) continuous media or discrete events, (3) deterministic or probabilistic, and (4) lumped or distributed. A steady-state model is one in which the input variables and their properties do not change with time. In a dynamic (transient) model the parameters change with time. Models based on continuous media, such as solids or fluids, assume that the medium transmitting a stress or flow vector does not contain voids or holes, while a discrete model deals with individual entities, such as cars in a traffic model or digital packets in a wireless transmission.

The following is a listing of the general steps required to build a mathematical design model.

Step 1. Determine problem statement
Step 2. Define the boundaries of the model
Step 3. Determine which physical laws are pertinent to the problem and identify the data that is available to support building the model
Step 4. Identify assumptions
Step 5. Construct the model
Step 6. Perform computations and verify the model
Step 7. Validate the model

1. Problem Statement

Determine the purpose of the model, its inputs and desired outputs. For example, is the purpose of the model to decide between alternative shapes, to determine the value of a critical dimension, or to improve the efficiency of an entire system? Write out the questions that you expect the model to help you answer. An important task in this step is to determine the desired inputs and outputs of the model. The amount of resources spent on the model will depend on the importance of the decision that needs to be made.

2. Define the Boundaries of the Model

Closely related to the previous step is to define the model's boundaries. The boundary of the design problem distinguishes a part of the model from the model's environment. The boundaries of the model are often called the *control volume*. The control volume can be drawn either as a *finite* control volume, which defines the overall system behavior, or a differential control volume at some point in the system. The latter is the standard way to set up a model for something like the stress state at a point or the flow of heat in conduction.

3. Determine What Physical Laws Are Pertinent to the Problem and What Data Is Available to Support Building the Model

With all the thought that has gone into defining the problem, we should now know what physical knowledge domain(s) we will use to represent the physical situation. Assemble the necessary textbooks, handbooks, and class notes to review the theoretical basis for constructing the model.

4. Assumptions

In building a model we should be aware that the model is an abstraction of reality. Model building walks a fine line between simplification and authenticity. One way to achieve simplification is to minimize the number of physical quantities that must be considered in the model to make it easier to achieve a mathematical solution. We do this by making assumptions to neglect what we believe to be small effects. Thus, we may assume a structural member is completely rigid when its elastic deformation is considered of little consequence to the problem. One of the distinctions between an engineering design model and a scientific model is our willingness to make these kinds of assumptions so long as we can justify that they will not lead to wrong conclusions.

Modeling is often an iterative process, where we start with an order of magnitude model that aims to predict outputs to within a factor of 10. Then as we gain confidence that the variables have been properly identified and their behavior understood, we can remove some of the assumptions to gain the needed precision. Remember that design modeling is always a balance between the necessary resources and the required precision of outputs.

Some common modeling simplifications are: (1) neglecting changes in physical and mechanical properties with temperature, (2) starting with a two-dimensional model when it is really a 3-D problem, (3) replacing the distributed properties of a variable with "lumped" parameters, (4) assuming a linear model when most real-world behavior is nonlinear.

5. Construct the Model

A helpful first step in building the model is to make a careful sketch of the physical elements of the problem. Try to make the sketch approximately to scale, as this will help in visualization. Next, relate the various physical quantities to one another by the appropriate physical laws. These are modified in ways appropriate to the model to provide the governing equations that transform the input quantities into the desired output. Usually the analytical description of the model starts with either appropriate conservation laws, like the conservation of energy, or balance equations like the summation of the forces and moments equal zero.

6. Computation and Verification

With the model developed the next step is to try it out with a computational tool. For simple models hand calculators will suffice, but spreadsheet computation is often very helpful. The model needs to be tested to see that it contains no mathematical errors and gives reasonable answers. This is the process of *model verification*. Verification is checking to see that the model works as you intended. For more advanced models involving finite element analysis, the preparation and verification of the model is much more detailed and time consuming.

7. Validation of the Model

Validation[1] is checking to see if the model gives an accurate representation of the real world. A common way to validate a model is to vary the inputs over a wide range to see if the outputs of the model appear to be physically reasonable, especially at the limits of performance. Find how sensitive the outputs are to the inputs. If the impact of a particular variable is weak, then it may be possible to replace that variable in the model with a constant. Full validation of a model requires a set of critical physical tests to establish how well the model describes the model.

---

1. D. D. Frey and C. L. Dym, *Research in Engineering Design,* vol. 17, pp. 45–57, 2006.

Although the foundations of engineering design models are firmly based in physical principles, sometimes the problem is just too complex to create a mathematical model of sufficient precision with the available resources, and the design engineer must use experimental test data to create an *empirical model*. This is an acceptable approach, since the goal of a design model is not to advance scientific understanding but rather to predict actual system behavior with sufficient accuracy and resolution for decision making. Empirical data needs to be treated with *curve-fitting methods* that describe the design parameter as a high-order polynomial equation. It must be understood that an empirical model is only valid over the range of parameters for which the tests were conducted.

7.4.3 Shot-Buddy Example

The JSR Design team's Shot-Buddy is designed to catch a ball falling through or around the rim and direct it back toward the user. For the Shot-Buddy to work effectively, it must be capable of enduring certain forces that will be applied to it when in use. The Shot-Buddy must also provide a sufficient catch area for shots that miss the basketball hoop. Reviewing JSR Design's concepts for the Shot-Buddy (see Fig. 7.7) reveals that a key challenge will be to design a ball catch subsystem that is large enough to catch a majority of missed shots while being light enough to be supported by attachments to the backboard of the basketball hoop (and compact enough to be readily packaged). This creates a common engineering design trade-off of stiffness vs. weight. Example 3.5 provided an overview of the mathematical model JSR Design used to estimate the maximum force exerted by the intended end-user of the Shot-Buddy. A more formal model building will be shown in Example 7.9.[1]

EXAMPLE 7.9 Part 1

Create a model for the force exerted by a basketball on the Shot-Buddy. This model will be used to determine the maximum loading that the parts must withstand. This example follows the process described in Sec. 7.4.2.

Step 1: Determine problem statement.

The purpose of this model is to determine the maximum forces that will be applied to the Shot-Buddy catch area subsystem support frame during use.

The answers to this question will help address the following, more specific issues about Shot-Buddy concept performance, which will need to be addressed during embodiment design.

1. *Can a catch area subsystem be designed that will withstand the direct impact of a missed basketball shot from the three-point line?*
2. *What is the approximate weight of a catch area subsystem?*
3. *Can this weight be supported by simple attachments to a basketball backboard?*

Step 2: Define the boundaries of the model.

The boundary of the model is a rectangular volume of one-half of the basketball court extending to a line just behind the basketball hoop and 15 feet above the surface of the court.

---

1. Adapted from Josiah Davis, Jamil Decker, James Maresco, Seth McBee, Stephen Phillips, and Ryan Quinn, *"JSR Design Final Report: Shot-Buddy,"* unpublished, ENME 472, University of Maryland, May 2010.

Step 3: *Determine which physical laws are pertinent to the problem, and identify the data that is available to support building the model.*

The position and velocity of an object under constant accelerated motion is equal to:

$$\mathbf{p} = \mathbf{p}_0 + \mathbf{v}_0 t + \frac{1}{2}\mathbf{a}t^2$$

$$\mathbf{v}_t = \mathbf{v}_0 + \mathbf{a}t$$

where
t is the time during the object's motion
$\mathbf{p}$ is a position vector that has components in the x and y directions
$\mathbf{v}$ is a the velocity vector with components in the x and y directions
$\mathbf{a}$ is a constant acceleration vector applied to the object components in the x and y directions

Any projectile object acting under constant acceleration (like gravity) will follow a parabolic trajectory with a path determined by its initial velocity and angle (θ) that the velocity vector makes with respect to a horizontal axis.

The impulse momentum theorem leads to the result that an applied force can be calculated from the change in momentum of a body.

$$I = F\Delta t_c = m\Delta v_c$$

where
F is the average force (a vector) applied during contact
Δt_c is the duration of time over which the force is applied during contact
m is mass of the object
v_t is velocity (a vector) of the object at time of first contact

Step 4: *Identify assumptions.*

1. The shooter has a height of 1.69 meters. This height is based on the height of the average 8th grade male.
2. A regulation basketball ($m = 0.624$ kg) will be used by the shooters.
3. The basketball court has a regulation height hoop, 3.05 meters high.
4. The shooter releases the ball very near the top of his head.
5. The shooter releases the ball from a regulation 3-point line, 6.02 meters away from the basket in the x direction. The ball will move in the plane defined by the shooter and the center of the basketball hoop and mounted Shot-Buddy.
6. The Shot-Buddy catch area assembly is rectangular with dimensions of 6-feet wide by 4-feet deep. These dimensions are called out in the PDS.
7. The only force acting on the basketball once it is in the air is gravity. Any other forces (i.e., drag, wind, ball spin) are neglected.
8. The collision of the basketball and Shot-Buddy catch area causes the basketball to lose its velocity. This assumption is based on the fact that the system is meant to get the ball to stop its motion and roll back to the user. Some of the velocity remains as it slides down the system, but because we want to assume the maximum impulse, we assume the momentum of the ball reaches zero. This momentum is transferred into the Shot-Buddy.
9. For the purpose of these calculations, all objects will be treated as rigid bodies.

Step 5: *Construct the model.*

Known

- For any time t:

$$a_x = 0\frac{m}{s^2}, a_y = g \text{ and } g = -9.81\frac{m}{s^2}.$$

- For any time t: $v(t)_x = v(t)_0$
- Time $t = 0$, the ball leaves the shooter's hands at an angle θ,

$$x_0 = 0$$

$$y_0 = 1.69$$

- Adjusting the frame of reference so that ball is released at point $(0, 0)$ gives

$$y_0 = 0 \, m$$

$$y_f = 1.36 \, m$$

- Time $t = t_f$, the ball hits the front of the Shot-Buddy frame

$$x_f = 5.33 \, m$$

$$y_f = 1.36 \, m$$

- Unknowns: $\theta(t)$, $v(t)$, and t_f.

Find an expression for the velocity of the ball when it hits the Shot-Buddy frame, without knowing the initial velocity vector's magnitude or direction.

$$x_t = x_0 + (v_0)_x t \tag{a}$$

$$y_t = y_0 + (v_0)_y t + \frac{1}{2}gt^2 \tag{b}$$

$$(v_t)_y = (v_0)_y + gt \tag{c}$$

$$\tan\theta(t) = \frac{(v_t)_y}{(v_t)_x} \tag{d}$$

At this point in the design process only a good approximation is needed. In a general study on the science of basketball,[1] the minimum shooting velocity to make a basket at a distance of 6.10 m (20 ft) was reported as 8.138 m/s, and the recommended best launch angle from the horizontal was 47.9°. We also know that the recommended best launch angle for a free throw shot is 52°[2] and that shot is made from 15 feet (4.572 m) from the target point. To find the velocity of the ball as it might hit the Shot-Buddy frame from the 3-point line, we will try out the velocity vector of 8.138 m/s at 52°.

1. S. L. Blanding and J. J. Monteleone, *The Science of Sports,* Barnes and Noble Books, New York, 2003.
2. C. M. Tran and L. M. Silverberg, "Optimal release conditions for the free throw in men's basketball," *Journal of Sports Sciences,* 26:11, 1147–1155, 2008.

Finding the x and y components of the launch velocity:

$$(v_0)_x = \left(8.138\frac{m}{s}\right)\cos(52°) = 5.232\frac{m}{s}$$

$$(v_0)_y = \left(8.138\frac{m}{s}\right)\sin(52°) = 6.234\frac{m}{s}$$

Step 6: *Perform calculations and verify the model.*

A. Calculate the time, t_f, to reach the Shot-Buddy frame:

Use Eq. (a) to determine the time at which the basketball travels the 5.33 m to the frame.

$$t_f = \frac{x_f}{(v_0)_x} = \frac{5.33m}{5.232\frac{m}{s}} = 1.02s$$

B. Use this result to calculate the final velocities of the ball when it impacts the Shot-Buddy's frame.

Using Eq. (b), find $(v_f)_y$, knowing $t_f = 1.02$ seconds.

$$y_f = y_0 + (v_f)_y t_f + \frac{1}{2}gt_f^2$$

$$(v_f)_y = \frac{y_f - y_0 - \frac{1}{2}gt_f^2}{t_f} = \frac{1.36m - \frac{1}{2}\left(9.81\frac{m}{s^2}\right)(1.02s)^2}{1.02s} = -3.67\frac{m}{s}m$$

The negative sign indicates that the basketball is now moving downward in its trajectory toward the Shot-Buddy frame.

Find the magnitude and direction of the velocity vector at time $t_f = 1.02$ seconds.

$$|v_f| = \sqrt{(v_f)_x^2 + (v_f)_y^2} = 6.39\frac{m}{s}, \qquad \theta(t_f) = -35°$$

C. To verify that this model and assumptions work: use Eq. (b) to calculate the basketball's vertical position at time t_f.

$$y_f = y_0 + (v_f)_y t_f + \frac{1}{2}gt_f^2$$

$$y_f = 0 + \left(-3.67\frac{m}{s}\right)(1.02s) + \frac{1}{2}\left(9.81\frac{m}{s^2}\right)(1.02s)^2 = 1.36$$

The coordinates for the middle of the Shot-Buddy frame as shown in Fig. 7.8 are $x = 5.33$ m and $y = 1.36$ m. The model for finding the velocity is verified.

D. Use the impulse force model to determine the force on the Shot-Buddy frame.

$$I = F\Delta t_c = m\Delta v_c \qquad \text{(e)}$$

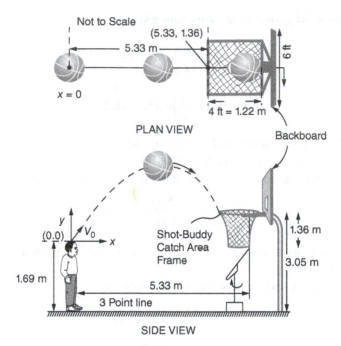

FIGURE 7.8
Sketch of key model building factors.

The velocity at the moment the basketball hits the Shot-Buddy frame was found in Step 6, part B. The standard basketball has a mass of 0.624 kg. Assume that the frame will stop the ball before it rebounds and that the duration of the contact, Δt_c, is 0.1 second.

$$F = \frac{m((v_f)_y - 0)}{\Delta t}$$

$$F = \frac{0.624 \text{ kg}\left(0 - 6.39\frac{m}{s}\right)}{0.1s} = 39.8 \text{ N}$$

Step 7: *Validate the model.*

It is difficult to validate the force finding model. The assumption of the time elapsed during collision with the Shot-Buddy catch area frame could be estimated using collisions with the basketball rim. Empirical data in the form of a video recording could be analyzed to estimate contact time, taking into account the deflection of the ball, and so on. This level of detail is not necessary during conceptual design, but will be for sizing Shot-Buddy components during parametric design.

Since each of the design concepts depends on having a large catch area to direct the ball into the return device, a major decision in conceptual design is to determine whether a large frame (6 ft × 4 ft) can be made light enough to be easily mounted and stiff enough to resist permanent deformation when struck at mid-span by a basketball.

Possible materials from which to make the frame are low-carbon steel, aluminum, and an extruded thermoplastic polymer. We first investigate a carbon steel since it has the highest elastic modulus and elastic limit of these materials. If the design cannot work with steel, the concept will have to be radically changed. It is desirable to minimize bracing so as to not obstruct the movement of the ball in the catch area; therefore deflection due to the weight of the frame will also be a consideration. We continue Example 7.9, again using the method for creating a mathematical model.

EXAMPLE 7.9 Part 2

Step 1: Determine problem statement.

Can a catch area subsystem frame be designed that will withstand the direct impact of a missed basketball shot from the three-point line?

Step 2: Define the boundaries of the model.

The boundary of the model is a rectangular volume enclosing the basketball and the front rod of the Shot-Buddy frame at the time of impact.

Step 3: Determine which physical laws are pertinent to the problem and identify the data that is available to support building the model.

$$\delta = \frac{PL^3}{48EI} \tag{f}$$

where
δ is the deflection (inches) at the tip of the beam where the load is applied
P is the load applied at the midpoint of a simply beam
L is the length of the beam
E is the elastic modulus of the material
I is the moment of inertia of the cross-sectional shape of the beam

Step 4: Identify assumptions.

1. The Shot-Buddy catch area frame will be constructed from 1-in. diameter steel rods with special corner connectors, so it can be assembled by the user. The front and back rods will be $L = 6$ ft, and the side rods will be 4 ft long.
2. The end connectors of the sides of the frame will behave as simple beam supports so we can use the standard equation for the deflection a beam loaded at mid-span by a point load.
3. The low-carbon steel selected for the frame has an elastic limit (yield strength), σ_y, of 50,000 psi and the elastic modulus, E, of 30×10^6 psi.
4. The moment of inertia for a circular cross-section is $I = \pi r^4/4$.

Step 5: Construct the model.

Known: The front rod of the Shot-Buddy is modeled as a simply supported beam.

- $P = 40$ N or 9 lb$_f$[1]
- $L = 6$ ft
- $E = 30 \times 10^6$ psi

---

1. lb$_f$ used as the abbreviation for pound-force.

Step 6: *Perform calculations and verify the model.*

A. We can use Hooke's law to find the strain at which elastic deformation gives way to permanent deformation: $e = \sigma_y/E = 50 \times 10^3/30 \times 10^6 = 0.00167$ in/in.

B. Using Eq. (f) find the deflection, δ, for a simply supported beam loaded at mid-point by a force $P = 40$ N or 9 lb$_f$:

$$\delta = \frac{PL^3}{48EI} = \frac{9(6\times12)^3}{48(30\times10^6)(0.0491)} = 0.0474 \text{ in.}$$

where I = moment of inertia = $\pi r^4/4 = 0.0491$ in^4

C. Using Eq. (g) below, determine if the strain induced by the load exceeds the elastic limit of the front rod. The strain in the fibers of an elastically bent beam is the ratio of the distance from its neutral axis, y, to the radius of curvature of the beam given by ρ. One over the curvature is given by M/EI, where M is the bending moment.[1]

$$e_x = \frac{y}{\rho} = \frac{My}{EI} = \frac{(PL/4)y}{EI} = \frac{9\times72\times0.5}{(30\times10^6)0.0491} = 0.00022 \text{ in.} \tag{g}$$

The strain in bending is well below the limiting elastic strain of 0.00167, indicating that the beam behaves elastically.

D. Verification: Structural *static loads* are applied slowly and continuously. The classic beam theory is inaccurate for the Shot-Buddy situation because the load is applied by an impact. When the basketball strikes the front of the frame, it hits with an estimated velocity of 6.39 m/s or 251.6 in/s. This *dynamic load* is assumed to be applied in less than 0.1 s.

A dynamic load magnifies the force for the short time it is applied, and the deflection is greater than the static deflection determined with Eq. (f). Equating the potential energy released by a falling mass with the elastic energy absorbed by the structure results in the relationship[2] between dynamic deflection δ_{dy} and static deflection δ_{st}.

$$\delta_{dy} = \delta_{st}\left(1+\sqrt{1+\frac{v^2}{g\delta_{st}}}\right) = \delta_{st}(\text{I.F.}) \tag{h}$$

where v is the impact velocity, g = acceleration of gravity, 32.2 ft/s^2 or 386.4 in/s.^{2} I.F. is called the *impact factor*. It can be applied to the deflection as shown above, or it can be applied to the force P. Calculating I.F. from Eq. (h) gives a value of 60.

Using an impact factor of 60 leads to a deflection of the front bar in the amount of 2.84 in. and a strain of 0.013. This value of the strain exceeds the elastic limit given by Eq. (g), which would mean that the steel rod would be plastically deformed by the impact of the basketball. Engineering reasoning leads to questioning this result. Further research is needed.

Step 7: *Validate the model.*

Equation (h) is derived on the assumption that all of the energy of the colliding body will go into the stationary object. This gives very conservative results and is not valid for a

1. F. P. Beer, E. R. Johnston. and J. T. DeWolf, *Mechanics of Materials*, 4th ed., McGraw-Hill, New York, 2006, p. 218.
2. R. C. Juvinall and K. M. Marshek, *Fundamentals of Machine Component Design,* 4th ed., John Wiley & Sons, Hoboken, NJ, 2006, pp. 269–279.

basketball, which deforms *viscoelastically*. Research shows that for a basketball colliding with the frame only about 15 percent of the energy goes into the frame.[1] A more realistic value would be I.F. = 0.15(60) = 9 and δ_{dy} = 9(0.047) = 0.426 in. Thus, calculations show that a frame made from 1-in. diameter rods should resist strikes from a basketball without damage. Further calculations can be done to show that deflection of the frame from its own weight (53 lbs) and from nylon netting would be negligible.

The analysis that went into the Shot-Buddy's conceptual design model revealed issues that need to be addressed in embodiment design. Among these are: (1) trade-offs between rod material, cross-section shape, and cost to make the frame lighter and easier to package, ship, and assemble, (2) the design of rugged brackets to attach the frame to the backboard, and (3) designs that will allow brackets for stiffness without obstructing ball travel so that aluminum or plastics could be used in the frame.

7.4.4 Geometric Modeling on the Computer

Geometric modeling on the computer was the fastest-changing area of engineering design in the late 20th century. When computer-aided design (CAD) was introduced in the late 1960s, it essentially provided an electronic drafting board for drawing in two dimensions. Through the 1970s CAD systems were improved to provide three-dimensional wireframe and surface models. By the mid-1980s nearly all CAD products had true solid modeling capabilities. In the beginning CAD required mainframe or minicomputers to support the software. Today, with the enhanced capabilities of personal computers, solid modeling software runs routinely on desktop machines.

An aspect of CAD modeling that has grown in importance is *data associativity,* the ability to share digital design data with other applications such as finite element analysis or numerical controlled machining without each application having to translate or transfer the data. An important aspect of associativity is that the database of the application is to be updated when a change is made in the basic CAD design data. In order to integrate digital design models from design to manufacturing, there must be a data format and transfer standard. First IGES, Initial Graphics Exchange Specification, and now Standard for the Exchange of Product model data (STEP) has been adopted by major CAD vendors. STEP has evolved into a complex system of interlocking standards and applications. (See Wikipedia at List of STEP (ISO 103-03) parts.) STEP also makes possible an open system of engineering information exchange using the World Wide Web or private networks based on the Internet (intranets).

Computer modeling software increasingly includes analysis tools for simulation of manufacturing processes (see Chap. 13). Solid modeling software can handle large assemblies with thousands of parts. It can deal with the associativity of the parts and manage the subsequent revisions to the parts. An increasing number of systems are providing top-down assembly modeling functions, where the basic assembly can be laid out and then populated later with parts.

---

1. L. C. Silverberg, C. Tran, and K. Adcock, "Numerical Analysis of the Basketball Shot," *Journal of Dynamic Systems, Measurement and Control,* December 2003, vol. 125, pp. 531–540.

For more details on computer generation of solids and creation of features in solid models, see Computer Modeling at www.mhhe.com/dieter.

7.4.5 Finite Element Analysis

Most classical models treat solids and fluids as continuous, homogeneous bodies so that properties such as stress or heat flux can only be predicted on an average basis. This is one of the modeling assumptions that is commonly negated by reality. It has been realized since the 1940s that if a continuum could be divided into small, well-defined finite elements, it would be possible to determine field-properties on a localized basis. Each element's behavior would be determined by its material and geometrical properties, interacting with all other elements in its vicinity. The theory was sound, but computational difficulty of solving thousands of simultaneous equations prevented much progress. With the advent of the digital computer applications of finite element analysis (FEA) grew steadily, but were mainly confined to large mainframe computers. It has only been in the past 20 years that FEA has become available for use on the design engineer's computer.

FEA applications that are available to the design engineer are almost endless: static and dynamic, linear and nonlinear, stress and deflection analysis; buckling analysis; free and forced vibrations; heat transfer; thermally induced stresses and deflections; fluid mechanics, acoustics, electrostatics, and magnetics. An important development is multiphysics software which allows ready interaction of models from multiple engineering sciences with excellent computer graphics capability.

In FEA, a continuum solid or fluid is divided into small elements. The behavior over each element is described by the value of the unknown variables evaluated at *nodes* and the physical laws for the behavior of the material (constitutive equations). All elements are then linked together taking care to ensure continuity at the boundaries between elements. Provided the boundary conditions are satisfied, a unique solution can be obtained for the large system of linear algebraic equations that result.

Since the elements can be arranged in virtually any fashion, they can be used to model very complex shapes. Thus, it is no longer necessary to find an analytical solution that treats a close "idealized" model and guess at how the deviation from the model affects the prototype. As the finite element method has developed, it has replaced a great deal of expensive preliminary cut-and-try experimentation with quicker and cheaper computer modeling. In contrast to the analytical methods that often require the use of higher-level mathematics, the finite element method is based on linear algebraic equations. For an elementary introduction to the mathematics behind FEA and a discussion of types of elements, see FEA Math and Elements at www.mhhe.com/dieter.

Phases in the FEA Process

Finite element modeling is divided into three phases: preprocessing, computation, and post processing. However, even before entering the first phase, a careful engineer will perform a preliminary analysis to define the problem. Is the physics of the problem known well enough? What is an approximate solution based on simple methods of analysis?

Preprocessing: In the preprocessing phase the following actions are taken.

- Import the geometry of the part from the CAD model. Because solid models contain great detail, they often must be simplified by deleting small nonstructural features and taking advantage of symmetry to reduce computation time.
- Determine the division of the geometry into elements, often called *meshing*. The issue with selecting a mesh is knowing which types of elements to use, linear, quadratic, or cubic interpolation functions, and building a mesh that will provide a solution with the needed accuracy and efficiency. Most FEA software provides a means for automatically meshing the geometry.
- Determine how the structure is loaded and supported, or in a thermal problem determine the initial conditions of temperature. Make sure you understand the boundary conditions. It is important to incorporate sufficient restraints to displacement so that rigid body motion of the structure is prevented.
- Select the constitutive equation for describing the material (linear, nonlinear, etc.) that relates displacement to strain and then to stress.

Computation: The operations in this phase are performed by the FEA software.

- The FEA program renumbers the nodes in the mesh to minimize computational resources.
- It generates a stiffness matrix for each element and assembles the elements together so that continuity is maintained to form the *global* matrix. Based on the load vector, the software generates the external loads and applies displacement boundary conditions.
- Then the computer solves the massive matrix equation for the displacement vector or whatever is the dependent variable in the problem. The constraint forces are also determined.

Post processing: These operations are also performed by the FEA software.

- In a stress analysis problem, post processing takes the displacement vector and converts it into strains, element by element, and then, with the appropriate constitutive equation, into a field of stress values.
- A finite element solution could easily contain thousands of field values. Therefore, post processing operations are needed to interpret the numbers efficiently. Typically the geometry of the part is shown over which contours of constant stress have been plotted. Mathematical operations may have to be performed on the data by the FEA software before it is displayed, such as determining the Von Mises effective stress.
- Increasingly, FEA software is being combined with an optimization package and used in iterative calculations to optimize a critical dimension or shape.

The key to practical utilization of finite element modeling is for the FEA software to be integrated with CAD so that FEA is executed without leaving the CAD program. This means the use of solid modeling, parametric, feature-based CAD software. In this way unimportant geometric features can be temporarily suppressed without permanently deleting them, and different design configurations can be easily examined using the parametric formulation of the CAD model. While in most cases the default choices in meshing and element selection are acceptable, the FEA software should provide the ability for custom settings.

To minimize cost, the model should contain the smallest number of elements to produce the needed accuracy. The best procedure is to use an iterative modeling strategy whereby coarse meshes with few elements are increasingly refined in critical areas of the model. Coarse models can be constructed with beam and plate structural models, ignoring details like holes and flanges. Once the overall structural characteristics have been found with the coarse model, a fine-mesh model is used, with many more elements constructed in regions where stress and deflection must be determined more accurately. Accuracy increases rapidly as a function of the number of degrees of freedom (DOF), defined as the product of the number of nodes times the number of unknowns per node. However, cost increases exponentially with DOF.

The application of FEA to the complex problem of a truck frame is illustrated in Figure 7.9. A "stick figure" or beam model of the frame is constructed first to find the deflections and locate the high-stress areas. Once the critical stresses are found, a fine-mesh model is constructed to get detailed analysis. The result is a computer generated drawing of the part with the stresses plotted as contours.

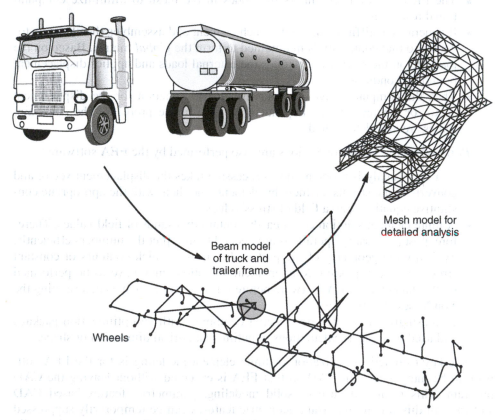

FIGURE 7.9
Example of use of FEA in design.

7.4.6 Simulation

Design models are created to imitate the behavior of a part or system under a particular set of conditions. When we exercise the model by inputting a series of values to determine the behavior of the proposed design under a stated set of conditions, we are performing a *simulation*. The purpose of the simulation is to explore the various outputs that might be obtained from the real system by subjecting the model to environments that represent the situations requiring additional understanding. Simulation models are built from individual models of parts of a larger system. The parts are modeled by logic rules that decide which of a set of predefined behaviors will occur and mathematical models to calculate the values of the behavior variables. The part models often rely on a probability distribution to select one of the predefined behaviors. It is the arrangement of the individual models that creates an overall system for the prediction of the behavior under study.

A simulation model can also be used to understand an existing system when data is not readily available. In this instance, a behavior model is usually made of logic rules, and mathematical descriptions of all potential outputs from the component are created for each component of the system. The model is streamlined so that it outputs only the characteristics required for the study. The model is verified by running it under historically accurate inputs and checking model outputs against prior data. Often a simulation model is created for an existing system so that design changes can be proposed or tested.

7.5
PUGH CHART

A particularly useful method for identifying the most promising design concepts among the alternatives generated is the *Pugh chart*.[1] Pugh's method compares each concept relative to a reference or datum concept and for each criterion determines whether the concept in question is better than, poorer than, or about the same as the reference concept. Thus, it is a relative comparison technique. The Pugh chart is created by the design team, usually in iterative rounds of examination and deliberation. The design concepts submitted for the Pugh method should all have passed the absolute filters discussed in Sec. 7.3.1. The steps in the concept selection method, as given by Clausing, are:

1. *Choose the criteria by which the concepts will be evaluated:* QFD's House of Quality is is the starting place from which to develop the criteria. If the concept is well worked out, then the criteria will be based on the engineering characteristics listed in the columns of the House of Quality.

---

1. S. Pugh, *Total Design,* Addison-Wesley, Reading, MA, 1991; S. Pugh, *Creating Innovative Products Using Total Design,* Addison-Wesley, Reading, MA, 1996; D. Clausing, *Total Quality Development,* ASME Press, New York, 1994. D. D. Frey, P. M. Herder, Y. Wijnia, E. Subrahamanian, K. Kastsikopoulos, and D. P. Clausing, "The Pugh Controlled Convergence Method: Model-Based Evaluation and Implications for Design Theory," *Research in Engineering Design,* (2009) 20, pp. 41–58.

In formulating the final list of criteria, it is important to consider the ability of each criterion to differentiate among concepts. A criterion may be very important, but if every design concept satisfies it well, it will not help you to select the final concept. Therefore, this criterion should be left out of the concept selection matrix. Also, some teams want to determine a relative weight for each criterion. This should be avoided at this point in the selection process, since it adds a degree of detail that is not justified at the concept level of information. Instead, list the criteria in approximate decreasing order of priority.

2. *Formulate the decision matrix:* The criteria are entered into the matrix as the row headings. The concepts are the column headings of the matrix. Again, it is important that concepts to be compared be the same level of abstraction. If a concept can be represented by a simple sketch, this should be used in the column heading. Otherwise, each concept is defined by a text description or a separate set of sketches, as shown in Fig. 7.7.

3. *Clarify the design concepts:* The goal of this step is to bring all members of the team to a common level of understanding about each concept. If done well, this will also develop team "ownership" in each concept. This is important, because if individual concepts remain associated with different team members the final team decision could be dominated by political negotiation. A good team discussion about the concepts often is a creative experience. New ideas often emerge and are used to improve concepts or to create entirely new concepts that are added to the list.

4. *Choose the datum concept:* One concept is selected by the team as a datum for the first round. This is the reference concept to which all other concepts are compared. In making this choice it is important to choose one of the better concepts. A poor choice of datum would cause all of the concepts to be positive and would unnecessarily delay arriving at a solution. It is good to choose the leading product in the market if one exists. For a redesign, the datum is the existing design reduced to the same level of abstraction as the other concepts. The column chosen as datum is marked accordingly, DATUM.

5. *Complete the matrix entries:* It is now time to do the comparative evaluation. Each concept is compared with the datum for each criterion. A three-level ordinal scale is used. At each comparison we ask the question, is this concept better (+), worse (−), or about the same (S) as the datum, and the appropriate symbol is placed in the cell of the matrix. Same (S) means that the concept is judged to be roughly the same as the datum with regard to the criterion for that row.

There should be brief constructive discussion when scoring each cell of the matrix. It will be necessary to conduct research or model the concepts to determine estimates of some performance criteria before completing the matrix. Divergent opinions lead to greater team insight about the design problem. Long, drawn-out discussion usually results from insufficient information and should be terminated with an assignment to someone on the team to generate the needed information.

Again, the team discussion often stimulates new ideas that lead to additional improved concepts. Someone will suddenly see that combining this idea from concept 3 solves a deficiency in concept 8, and a hybrid concept evolves. Another column is added for the new concept. A major advantage of the Pugh method is that it helps the team to develop better insights into the types of features that strongly satisfy the design requirements.

6. *Evaluate the ratings:* Once the comparison matrix is completed, the sum of the + and – ratings is determined for each concept. Do not become too quantitative with these ratings. Be careful about rejecting a concept with a high negative score without further examination. The few positive features in the concept may really be "gems" that could be picked up and used in another concept. For the highly rated concepts determine what their strengths are and what criteria they treat poorly. Look elsewhere in the set of concepts for ideas that may improve these low-rated criteria. Also, if most concepts get the same rating on a certain criterion, examine it to see whether it is stated clearly or not uniformly evaluated from concept to concept. If this is an important criterion, then you will need to spend effort to generate better concepts or to clarify the criterion.

7. *Establish a new datum and rerun the matrix:* The next step is to establish a new datum, usually the concept that received the highest rating in the first round, and run the matrix again. Eliminate the lowest rating concepts from this second round. The main intent of this round is not to verify that the selection in round 1 is valid but to gain added insight to inspire further creativity. The use of a different datum will give a different perspective at each comparison that will help clarify relative strengths and weaknesses of the concepts.

8. *Examine the selected concept for improvement opportunities:* Once the superior concept is identified, consider each criterion that performed worse than the datum. Keep asking questions about the factors detracting from the merits of an idea. New approaches emerge; negative scores can change to positive scores. Answers to your questions often lead to design modifications that eventually provide a superior concept.

Example 7.10 describes the use of the Pugh chart as applied to the Shot-Buddy concept selection task.

EXAMPLE 7.10 Pugh Concept Selection Process

The JSR Design team generated five concepts for the automated basketball return device using the tools and methods found in Chap. 6.[1] These early stage concepts are shown in Fig. 7.7. Apply the Pugh Concept Selection Process to the set of five concepts to reduce the group to the three best alternatives for future examination. Note: In Example 7.8 it was determined that Concept 5 in the set is not functionally feasible. We will include it here for purposes of demonstrating the Pugh Concept Selection method.

The decision criteria for the selection process are determined from the development and interpretation of the House of Quality for the design of the Shot-Buddy reported in Example 3.8. The critical to quality engineering characteristics (CTQ ECs) are listed below along with cost. To complete the list of decision criteria, it is necessary to review the product design specification, PDS (Table 3.4), for the Shot-Buddy for any threshold constraints to be used in this process. (A threshold constraint is an engineering characteristic that has a firm target level. However, if different concepts exceed the target level by various amounts that threshold constraint can be used as a valid selection criterion.) The PDS includes the requirement that the Shot-Buddy work on battery power, so the JSR Design team added a criterion of the power needed to operate the ball return device. The less power required by the device, the longer it can be used without recharging or replacing the batteries.

---

1 Adapted from Josiah Davis, Jamil Decker, James Maresco, Seth McBee, Stephen Phillips, and Ryan Quinn, *"JSR Design Final Report: Shot-Buddy,"* unpublished, ENME 472, University of Maryland, May 2010.

TABLE 7.4
Pugh Selection Chart 1 for Shot-Buddy Concepts shown in Fig. 7.7

Selection Criteria	RolBak Gold Pro	Concepts				
		1	2	3	4	5
Catch area		+	+	+	+	+
Probability of jamming		S	S	+	+	+
Weather resistance		−	−	−	−	−
Sensing position of shooter	DATUM	+	+	+	S	+
Effectiveness of ball return		+	+	+	+	+
Cost		−	−	−	S	S
Weight		−	−	−	−	−
Time to mount to hoop		−	−	+	+	−
Work required to rotate		−	−	−	S	−
Storage volume		−	−	−	−	−
# of Pluses		3	3	5	4	4
# of Minuses		6	6	5	4	5

The list of decision criteria for selecting a Shot-Buddy concept is as follows:

- Catch area configuration
- Low jamming probability
- Weather resistance
- Sensing the position of the shooter
- Effectiveness of ball return (i.e., a measure that includes accuracy and time)
- Cost
- Weight
- Time to mount to existing basketball hoop (if necessary)
- Work required to rotate ball return mechanism
- Storage volume required when not in use

There is no existing automatic basketball return device, so JSR Design decides to use a simple net return system called the RolBak™ Basketball Return Net System[1] as the datum design. The RolBak uses a 10-foot high net, mounted on the basketball backboard, that catches and returns balls that are in or near the rim. However, the net projects outward onto the court, obstructing any close shot that the user may want to practice, like a lay-up. The RolBak system is the simplest of the net systems on the market, and is priced at $189.90.

JSR Design completes the Pugh Concept Selection Matrix shown in Table 7.4. At first it seems apparent that none of the concepts is an outstanding improvement over the RolBak Gold Pro product. All proposed concepts offer improvements in the catch area and sensing the position of the shooter. All concepts fail to meet the same level of performance on weather resistance, price, weight, and storage volume.

---

1. "The Rolbak Basketball Protecto Net," http://www.jumpusa.com/rolbak.htm, accessed July 8, 2011.

TABLE 7.5
Pugh Selection Chart 2 for Shot-Buddy Concepts from in Fig. 7.7

			Concepts		
Selection Criteria		3	1	2	5
Catch area			S	S	S
Probability of jamming			+	S	S
Weather resistance			S	S	S
Sensing position of shooter			+	+	+
Effectiveness of ball return	DATUM		+	S	−
Cost			S	S	S
Weight			−	+	+
Time to mount to hoop			S	S	S
Work required to rotate			S	S	+
Storage volume			S	+	+
# of Pluses			3	3	4
# of Minuses			1	0	1

Concept 4 has the fewest minus ratings and matches three other concepts for plus ratings. The criteria that differentiate Concept 4 from the other proposed concepts must be examined. Concept 4 has a better rating on mounting to existing basketball hoops (because it stands on the court). Concept 4 is the only concept that does not sense the position of the shooter. It does not improve on the Datum design in this criterion row. This is a serious functional feasibility deficiency that could have been avoided if the team had checked the absolute criteria first! Thus the Rolbak design is not a great selection for a datum concept. Based on the results of the chart, Concept 4 can be eliminated. A new Pugh chart is created using Concept 3 as the datum (this concept has the highest number of pluses) and appears in Table 7.5.

The second Pugh Selection Chart (Table 7.5) indicates that there are good concepts in the set of those generated. The number of minus ratings is much lower than in the previous chart. Focusing again on the areas of difference between the ratings, Concept 5 is showing relative weakness in the effectiveness of the ball return. This deficiency is enough to overcome positive aspects in terms of work to rotate and storage volume. The team decides to eliminate Concept 5 and take Concepts 1, 2, and 3 forward for more modeling and development.

7.6
WEIGHTED DECISION MATRIX

A decision matrix is a method of evaluating competing concepts by ranking the design criteria with weighting factors and scoring the degree to which each design concept meets the criterion. To do this it is necessary to convert the values obtained for different design criteria into a consistent set of values. The simplest way of dealing with design criteria expressed in a variety of ways is to use a point scale. A 5-point

TABLE 7.6
Evaluation Scheme for Design Alternatives or Objectives

11-point Scale	Description	5-point Scale	Description
0	Totally useless solution	0	Inadequate
1	Very inadequate solution		
2	Weak solution	1	Weak
3	Poor solution		
4	Tolerable solution		
5	Satisfactory solution	2	Satisfactory
6	Good solution with a few drawbacks		
7	Good solution	3	Good
8	Very good solution		
9	Excellent (exceeds the requirement)	4	Excellent
10	Ideal solution		

scale is used when the knowledge about the criteria is not very detailed. An 11-point scale (0–10) is used when the information is more complete (Table 7.6). It is best if several knowledgeable people participate in this evaluation.

Determining weighting factors for criteria is an inexact process. Intuitively we recognize that a valid set of weighting factors should sum to 1. Therefore, when n is the number of evaluation criteria and w is the weighting factor,

$$\sum_{i=1}^{n} w_i = 1.0 \quad and \quad 0 \le w_i \le 1 \tag{7.8}$$

Systematic methods can be followed for determining weighting factors. Three are listed below.

- *Direct Assignment:* The team decides how to assign 100 points between the different criteria according to their importance. Dividing each criterion's score by 100 normalizes the weights. This method is *only recommended* for design teams where there are many years of experience designing the same product line.
- *Objective Tree:* Weighting factors can be determined by using a hierarchical objective tree as shown in Example 7.11. Better decisions regarding preferences will be made when the comparisons are made at the same level in the hierarchy, because you will be comparing "apples with apples and oranges with oranges". Again, this method relies on some experience with the importance of the criteria in the design process.
- *Analytic Hierarchy Process* (AHP): AHP is the least arbitrary method for determining weighting factors. This method is presented in detail in Sec. 7.7.

EXAMPLE 7.11
A heavy steel crane hook, for use in supporting ladles filled with molten steel as they are transported through the steel mill, is being designed. Two crane hooks are needed for each steel ladle. These large, heavy components are usually made to order in the steel mill machine shop when one is damaged and needs to be replaced.

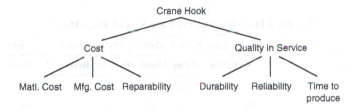

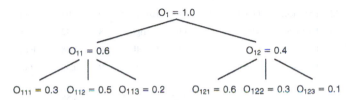

FIGURE 7.10
Objective tree for the design of a crane hook.

Three concepts have been proposed: (1) built up from flame-cut steel plates, welded together; (2) built up from flame-cut steel plates, riveted together; (3) a monolithic cast-steel hook.

The first step is to identify the design criteria by which the concepts will be evaluated. The product design specification is a prime source of this information. The design criteria are identified as (1) material cost, (2) manufacturing cost, (3) time to produce a replacement hook if one fails, (4) durability, (5) reliability, (6) reparability.

The next step is to determine the weighting factor for each of the design criteria. We do this by constructing a hierarchical objective tree (Fig. 7.10). We do this by direct assignment based on engineering judgment. This is easier to do using the objective tree because the problem is broken down into two levels. The weights of the individual categories at each level of the tree must add to 1.0. At the first level we decide to weight cost at 0.6 and quality at 0.4. Then at the next level it is easier to decide the weights between cost of material, cost of manufacturing, and cost to repair, than it would be if we were trying to assign weights to six design criteria at the same time. To get the weight of a factor on a lower level, multiply the weights as you go up the chain. Thus, the weighting factor for material cost, $O_{111} = 0.3 \times 0.6 \times 1.0 = 0.18$.

The decision matrix is given in Table 7.7. The weighting factors are determined from Fig. 7.10. Note that three of the design criteria in Table 7.7 are measured on an ordinal scale, and the other three are measured on a ratio scale. The score for each concept for each criterion is derived from Table 7.6 using the 11-point scale. When a criterion based on a ratio scale changes its magnitude from one design concept to another, this does not necessarily reflect a linear change in its score. The new score is based on the team assessment of suitability of the new design based on the descriptions in Table 7.7.

The rating for each concept at each design criterion is obtained by multiplying the score by the weighting factor. Thus, for the criterion of material cost in the welded-plate design concept, the rating is $0.18 \times 8 = 1.44$. The overall rating for each concept is the sum of these ratings.

The weighted decision matrix indicates that the best overall design concept would be a crane hook made from elements cut from steel plate and fastened together with rivets.

TABLE 7.7
Weighted Decision Matrix for a Steel Crane Hook

Design Criterion	Weight Factor	Units	Built-Up Plates Welded			Built-Up Plates Riveted			Cast Steel Hook		
			Magnitude	Score	Rating	Magnitude	Score	Rating	Magnitude	Score	Rating
Material cost	0.18	c/lb	60	8	1.44	60	8	1.44	50	9	1.62
Manufacturing cost	0.30	$	2500	7	2.10	2200	9	2.70	3000	4	1.20
Reparability	0.12	Experience	Good	7	0.84	Excellent	9	1.08	Fair	5	0.60
Durability	0.24	Experience	High	8	1.92	High	8	1.92	Good	6	1.44
Reliability	0.12	Experience	Good	7	0.84	Excellent	9	1.08	Fair	5	0.60
Time to produce	0.04	Hours	40	7	0.28	25	9	0.36	60	5	0.20
					7.42			8.58			5.66

The simplest procedure in comparing design alternatives is to add up the ratings for each concept and declare the concept with the highest rating the winner. A better way to use the decision matrix is to examine carefully the components that make up the rating to see what design factors influenced the result. This may suggest areas for further study or raise questions about the validity of the data or the quality of the individual decisions that went into the analysis. Pugh points out[1] that the outcome of a decision matrix depends heavily on the selection of the criteria. He worries that the method may instill an unfounded confidence in the user and that the designer will tend to treat the total ratings as being absolute.

7.7
ANALYTIC HIERARCHY PROCESS (AHP)

The Analytic Hierarchy Process (AHP) is a problem-solving methodology for making a choice from among a set of alternatives when the selection criteria represent multiple objectives, have a natural hierarchical structure, or consist of qualitative and quantitative measurements. AHP was developed by Saaty.[2] AHP builds upon the mathematical properties of matrices for making consistent pairwise comparisons. An important property of these matrices is that their principal eigenvector can generate legitimate weighting factors. Not only is AHP mathematically sound, but it is also intuitively correct.

AHP is a decision analysis tool that is used throughout a number of fields in which the selection criteria used for evaluating competing solutions that do not have exact, calculable outcomes. Operations research scholars Forman and Gass describe the AHP's key functions as structuring complexity, measurement, and synthesis.[3] Like

---

1. S. Pugh, op. cit., pp. 92–99.
2. T. L. Saaty, *The Analytic Hierarchy Process,* McGraw-Hill, New York, 1980; T. L. Saaty, *Decision Making for Leaders,* 3d ed., RWS Publications, Pittsburgh, PA, 1995.
3. E. H. Forman and S. I. Gass, "The Analytic Hierarchy Process—An Exposition," *Operations Research,* vol. 49, July–August 2001, pp. 469–86.

other mathematical methods, AHP is built on principles and axioms such as top-down decomposition and reciprocity of paired comparisons that enforces consistency throughout an entire set of alternative comparisons.

AHP is an appropriate tool for selecting among alternative engineering designs. AHP is relevant for choice problems in the following categories: comparing untested concepts; structuring a decision-making process for a new situation; evaluating non-commensurate trade-offs; performing and tracking group decision making; integrating results from different sources (e.g., analytical calculations, HOQ relative values, group consensus, and expert opinion); and performing strategic decision making. Many evaluation problems in engineering design are framed in a hierarchy or system of stratified levels, each consisting of many elements or factors.

AHP Process

AHP leads a design team through the calculation of weighting factors for decision criteria for one level of the hierarchy at a time. AHP also defines a pairwise, comparison-based method for determining relative ratings for the degree to which each of a set of options fulfills each of the criteria. AHP includes the calculation of an inconsistency measurement and threshold values that determine if the comparison process has remained consistent.

AHP's application to the engineering design selection task requires that the decision maker first create a hierarchy of the selection criteria. We will use the crane hook design problem of Ex. 7.11 to illustrate AHP's workings. We no longer need the intermediate level of the hierarchy since it's not necessary for setting the weights, and all the criteria are similar. The criteria all measure aspects of the product's design performance. We have six criteria as follows: (1) material cost, (2) manufacturing cost, (3) reparability, (4) durability, (5) reliability, and (6) time to produce.

Table 7.8 shows the rating system for the pairwise comparison of two criteria and gives explanations for each rating. The rating of pair A to pair B is the reciprocal of the

TABLE 7.8
AHP's Ratings for Pairwise Comparison of Selection Criteria

Rating Factor	Relative Rating of Importance of Two Selection Criteria A and B	Explanation of Rating
1	A and B have equal importance.	A and B both contribute equally to the product's overall success.
3	A is thought to be moderately more important than B.	A is slightly more important to product success than B.
5	A is thought to be strongly more important than B.	A is strongly more important to product success than B.
7	A is thought to be very much more important than B, or is demonstrated to be more important than B.	A's dominance over B has been demonstrated.
9	A is demonstrated to have much more importance than B.	There is the highest possible degree of evidence that proves A is more important to product success than B.

The ratings of even numbers 2, 4, 6, and 8 are used when the decision maker needs to compromise between two positions in the table.

TABLE 7.9
Development of Candidate Set of Criteria Weights {W} for Crane Hook

Criteria Comparison Matrix [C]

	Material Cost	Mfg Cost	Reparability	Durability	Reliability	Time Prod
Material Cost	1.00	0.33	0.20	0.11	0.14	3.00
Mfg Cost	3.00	1.00	0.33	0.14	0.33	3.00
Reparability	5.00	3.00	1.00	0.20	0.20	3.00
Durability	9.00	7.00	5.00	1.00	3.00	7.00
Reliability	7.00	3.00	5.00	0.33	1.00	9.00
Time Prod	0.33	0.33	0.33	0.14	0.11	1.00
Sum	25.33	14.67	11.87	1.93	4.79	26.00

Normalized Criteria Comparison Matrix [Norm C]

	Material Cost	Mfg Cost	Reparability	Durability	Reliability	Time Prod	Criteria Weights {W}
Material Cost	0.039	0.023	0.017	0.058	0.030	0.115	0.047
Mfg Cost	0.118	0.068	0.028	0.074	0.070	0.115	0.079
Reparability	0.197	0.205	0.084	0.104	0.042	0.115	0.124
Durability	0.355	0.477	0.421	0.518	0.627	0.269	0.445
Reliability	0.276	0.205	0.421	0.173	0.209	0.346	0.272
Time Prod	0.013	0.023	0.028	0.074	0.023	0.038	0.033
Sum	1.000	1.000	1.000	1.000	1.000	1.000	1.000

rating of pair B to A. That means if it is determined that A is strongly more important than B, the rating of A to B is set as 5. This makes the rating of B to A 1/5 or 0.20.

AHP Process for Determining Criteria Weights

We will now use the AHP rating system to create the initial comparison matrix [C] shown in Table. 7.9. Enter the data into Excel to do the simple mathematics and the matrix multiplication. The process is:

1. Complete criteria comparison matrix [C] using 1–9 ratings described in Table 7.8.
2. Normalize the matrix [C] to give [NormC].
3. Average row values. This is the Criteria Weights vector {W}.
4. Perform a consistency check on [C] as described in Table 7.10.

The matrix [C] is square with n rows and columns, n being the number of selection criteria. The matrix is constructed one pairwise comparison at a time. The diagonal entries are all 1 because comparing (A) with (A) means they are of equal importance. Once [C] is complete, the matrix entries are normalized by dividing each column cell by the column sum. The normalized matrix is called [NormC] in Table 7.9. Average each row to calculate a candidate set of criteria weights shown in vector {W} in Table 7.9.

Each pair of criteria are compared and assigned a value for the matrix entry. The first comparison of two different criteria in [C] is done between material cost (A) and manufacturing cost (B). The rating factor becomes the entry for the first row, second column of [C] (also referred to as entry $C_{i,j}$). Referring back to Table 7.8, we determine that material and manufacture costs are both important in determining the goodness of the crane hook design. Yet, material cost is slightly less critical than manufacturing cost to the design of a hook. Therefore the value of $C_{1,2}$ is set at 1/3. The corresponding value of $C_{2,1}$ is 3.

Now consider the rating factor comparing material cost (A) to reliability (B), to set the value of $C_{1,5}$. These are not easy criteria to compare. In product design, reliability is almost taken for granted. The materials of a product contribute to the overall reliability, but some are more critical to functionality than others are. The crane hook is designed to be a single component, so the material properties are of higher importance than if the hook were an assembly of five components. One of our design alternatives is a cast steel hook that has properties tied closely to the integrity of the casting, i.e., whether it is free of voids and porosity. This perspective can lead us to setting $C_{1,5}$ to a value between 3 and 7. Another factor to consider is the application of the crane. Since the hook is for use in a steel melting shop, failure could be catastrophic and would cause a work stoppage or even loss of life. The same is not true if the hook is to be fitted onto a small crane used by a roofer to lift shingles up to the roof of a one- or two-story home. We set $C_{1,5}$ to 1/7 because reliability is more critical to the operation than material cost. That means $C_{5,1}$ is 7, as shown in Table 7.9.

This process may seem as easy as the simple binary rating scheme used in an earlier section. However, creating a *consistent* set of rating factors is difficult. The pair rating factors for the crane design discussed in the last two paragraphs involve relationships among material cost, manufacturing cost, and reliability. The pair not yet discussed is manufacturing cost (A) and reliability (B) for $C_{2,5}$. It's tempting to use 1/7 again since the logic applied to material cost should be similar for manufacturing cost. However, earlier decisions set manufacturing cost as more important than material cost. This difference must carry through to the relationships manufacturing and material costs have to other criteria.

Consistency Check Process for AHP Comparison Matrix [C]

As the number of criteria increases, it is difficult to assure consistency. That is why the AHP process includes a consistency check on [C]. The process is as follows:

1. Calculate weighted sum vector, $\{Ws\} = [C] \times \{W\}$
2. Calculate consistency vector, $\{Cons\} = \{Ws\}/\{W\}$
3. Estimate λ as the average of values in $\{Cons\}$
4. Evaluate consistency index, $CI = (\lambda - n)/(n - 1)$
5. Calculate consistency ratio, $CR = CI/RI$. The random index (RI) values are the consistency index values for randomly generated versions of [C]. The values for RI are listed in Table 7.11. The rationale for this comparison is that the [C] matrix constructed by a knowledgeable decision maker will show much more consistency than a matrix randomly populated with values from 1 to 9.
6. If $CR < 0.1$ the $\{W\}$ is considered to be valid; *otherwise adjust [C] entries and repeat.*

The consistency check for the crane hook design problem's criteria weights is shown in Table 7.10. An Excel spreadsheet provides an interactive and updatable tool for setting up [C] and working through the consistency checking process.

TABLE 7.10
Consistency Check for {W} for Crane Hook

Consistency Check		
{Ws} = [C]{W}[1] **Weighted Sum Vector**	{W} Criteria **Weights**	{Cons} = {Ws}/{W} **Consistency Vector**
0.286	0.047	6.093
0.515	0.079	6.526
0.839	0.124	6.742
3.090	0.445	6.950
1.908	0.272	7.022
0.210	0.033	6.324
Average of {Cons} = λ		6.610
Consistency Index, CI = $(\lambda - n)/(n - 1)$		0.122
Consistency Ratio, CR = CI/RI		0.098[2]
Is Comparison Consistent: CR < 0.10		*Yes*

[1] The values in column are the matrix product of the [C] and {W} arrays. Excel has a function MMULT(array1, array2) that will easily calculate the matrix product. The number of columns in array1 must be equal to the number of rows in array2. The result of the matrix product is a single column matrix with the same number of rows as [C]. When using the Excel function MMULT, remember that the arrays must be entered as array formula by pressing Ctrl-Shift-Enter.
[2] If this value is equal to or greater than 0.10 the [C] matrix must be reset.

TABLE 7.11
RI Values for Consistency Check

# of Criteria	RI Value
3	0.52
4	0.89
5	1.11
6	1.25
7	1.35
8	1.40
9	1.45
10	1.49
11	1.51
12	1.54
13	1.56
14	1.57
15	1.58

TABLE 7.12
AHP's Ratings for Pairwise Comparison of Design Alternatives

Rating Factor	Relative Rating of the Performance of Alternative A Compared to Alternative B	Explanation of Rating
1	A = B	The two are the same with respect to the criterion in question.
3	A is thought to be moderately superior to B.	Decision maker slightly favors A over B.
5	A is thought to be strongly superior to B.	Decision maker strongly favors A over B.
7	A is demonstrated to be superior to B.	A's dominance over B has been demonstrated.
9	A is demonstrated to be absolutely superior to B.	There is the highest possible degree of evidence that proves A is superior to B under appropriate conditions.

The ratings of even numbers 2, 4, 6, and 8 are used when the decision maker needs to compromise between two positions in the table.

The AHP process does not stop with the criteria weights. It continues by providing a similar comparison method for rating the design alternatives. The mathematical benefits of AHP are only realized if you continue through the process.

Before proceeding to evaluate each of the alternative designs using AHP, review the weighting factors. Members of the design team may have insight into the expected ranking of the factors. They should apply their experience in this review process before accepting the weights. If there is one that is much less significant than the others, the design team could eliminate that criterion from further use in evaluation before rating the alternative designs against each criterion.

Determining Ratings for Design Alternatives with Respect to a Criterion

AHP's pairwise comparison step is different from the simple one introduced in Sec. 7.3.2 on measurement scales. In AHP's pairwise comparison the decision maker must judge which of two options (A and B) is superior to the other with respect to some criterion and then make a judgment about the number of times better the superior option is to the inferior one (the comparison is unit-less). AHP allows the decision maker to use a scale of 1 to 9 to describe the strength of the rating. In this way, AHP's rating factors are not interval values. They are ratios and can be added and divided for the evaluation of competing design alternatives.[1]

Table 7.12 shows the rating system for the pairwise comparison of two alternatives, A and B, with respect to *one specific engineering selection criterion*. The explanation of each rating is given in the third column. The scale is the same as that described in Table 7.8, but the explanations have been adjusted for comparing the performance of design alternatives. The differences in performance are likely to be fractional improvements, like a $0.10/lb lower cost.

The process of using AHP will ultimately give us a priority vector $\{P_i\}$ of the design alternatives with respect to their performance for each selection criterion. This

---

1. T. L. Saaty, *Journal of Multi-Criteria Decision Analysis,* vol. 6:324–35, 1997.

will be used in the same way as the ratings developed in Sec. 7.6. The process is summarized as:

1. Complete comparison matrix [C] using 1–9 ratings of Table 7.12 to evaluate pairs of competing design alternatives.
2. Normalize the matrix [NormC].
3. Average row values—This is the vector priority {P_i} of design alternative ratings.
4. Perform a consistency check on [C].

Notice that steps 2, 3, and 4 are the same as the steps to determine the criteria weight factors.

The design alternatives for the crane hook design example are: (1) built up plates with welding, (2) built up plates with rivets, and (3) a monolithic steel casting. Consider the material cost criterion. Design teams use their standard cost estimation practices and experience to determine estimates of the material costs of each of the design alternatives. These costs are embedded in Table 7.7 in Sec. 7.6. We know that the material costs for each design are 0.60 $/lb for both plate designs and 0.50 $/lb for cast steel. Since we are comparing three design alternatives, the comparison matrix [C] is 3 × 3 (see Table 7.13). All the diagonal elements are ratings of 1, and reciprocals will be used for the lower triangular matrix. That leaves only three comparisons to rate as follows:

- $C_{1,2}$ is the comparison of the welded plate design's material cost (A) to the riveted plate design's material cost (B). This rating is 1 since the costs are the same.
- $C_{1,3}$ is the comparison of the welded plate design's material cost (A) to the cast steel design's material cost (B). Alternative A is slighty more expensive than alternative B, so the rating is set to 1/3. (If the $0.10/lb cost differential is significant to the decision maker, the rating could be set lower as in 1/5, 1/6, . . . 1/9.)
- $C_{2,3}$ is the comparison of the riveted plate design's cost (A) to the cast steel design's material cost (B). Since the riveted plate's material cost is the same as the welded plate's cost, $C_{2,3}$ must be set the same as $C_{1,3}$ at 1/3. This is enforcing the consistency of the matrix.

The development of the matrix [C] and {P_i} for the alternative design's material costs are shown in Table 7.13. Notice that the consistency check is almost trivial in this case because the relationships were clear to us as we set the [C] values.

The process is repeated for each of the five other criteria until all the {P_i} of design alternative ratings are complete for each criterion. The {P_i} vectors will be used to determine the [FRating] decision matrix Table 7.14, as described next.

Determine Best of Design Alternatives

The process of using AHP to select the best design alternative can be done once all alternatives have been rated to produce a separate and consistent priority matrix for each criterion. The process is summarized below:

1. Compose Final Rating Matrix [FRating]. Each {P_i} is transposed to give the ith row of the [FRating] matrix. Table 7.14 is a 6 × 3 matrix describing the relative priority of each criterion for the three alternative designs.

TABLE 7.13
Design Alternative Ratings for Material Cost

Material Cost Comparison [C]

	Plates Weld	Plates Rivet	Cast Steel
Plates Weld	1.000	1.000	0.333
Plates Rivet	1.000	1.000	0.333
Cast Steel	3.000	3.000	1.000
Sum	5.000	5.000	1.667

Normalized Cost Comparison [NormC]

	Plates Weld	Plates Rivet	Cast Steel	Design Alternative Priorities $\{P_1\}$
Plates Weld	0.200	0.200	0.200	0.200
Plates Rivet	0.200	0.200	0.200	0.200
Cast Steel	0.600	0.600	0.600	0.600
	1.000	1.000	1.000	1.000

Consistency Check

$\{Ws\} = [C]\{P_1\}^1$ Weighted Sum Vector	$\{P_1\}$ Alternative Priorities	$\{Cons\} = \{Ws\}/\{P_1\}$ Consistency Vector
0.600	0.200	3.000
0.600	0.200	3.000
1.800	0.600	3.000
	Average of {Cons} =	3.000
	Consistency Index, CI =	0
	Consistency Ratio, CR =	0
	Is Comparison Consistent	*YES*

$n = 3$, RI = 0.52; λ Estimate; $(\lambda - n)/(n - 1)$; CI/RI; CR < 0.10
[1]The weighted sum vector {Ws} can be calculated in Excel using the function MMULT.

TABLE 7.14
Final Rating Matrix

Selection Criteria	[FRating] Welded Plates	Riveted Plates	Cast Steel
Material Cost	0.200	0.200	0.600
Manufacturing Cost	0.260	0.633	0.106
Reparability	0.292	0.615	0.093
Durability	0.429	0.429	0.143
Reliability	0.260	0.633	0.105
Time to Produce	0.260	0.633	0.106

2. Calculate $[FRating]^T\{W\}=\{$Alternative Value$\}$ by first taking the transpose of [FRating]. Now matrix multiplication is possible because we are multiplying a (3×6) times (6×1) matrix. This produces a column matrix, the Alternative Value. Weighting vector $\{W\}$ was calculated in Table 7.9.

	Alternative Value
Welded plate design	0.336
Riveted plate design	0.520
Monolithic casting	0.144

3. Select the alternative with the highest rating relative to others.

The design alternative with the highest alternative relative value is the riveted plates design.

Since this is the same conclusion as found using the weighted design matrix approach that is interpreted from Table 7.7, one might question the value of using the AHP method. The AHP advantage is that the criteria weights are determined in a more systematic fashion and have been judged to meet a standard of consistency. The design selection process template has been set up (assuming Excel is used), and different decision maker assumptions can be used to test the sensitivity of the selection.

This section used Excel to implement the AHP process. One reference for additional information on this topic is a text on decision models by J. H. Moore et al.[1] The popularity of AHP for decision making can be measured by searching for business consultants who provide AHP training and software for implementing AHP. For example, one commercially available software package for AHP is called Expert Choice (http://www.expertchoice.com).

7.8
SUMMARY

In all stages of the design process, decisions are made to select options from a set of alternatives. The decision making process involves understanding the nature of the decision to be made as well as the nature of the decision-maker. These topics are addressed by aspects of decision theory and utility theory. Decision in design requires identifying choices, predicting the expectations for the outcomes of each choice, determining a way to rate alternatives against a set of criteria, and performing the selection process in a mathematically valid and consistent way.

Modeling the physical behavior of design alternatives is a prerequisite for good engineering decision making. Section 7.4 addresses the kinds of models available to designers and provides a logical method for building models that can be used throughout all the engineering design stages. The example presented in the section is customized to match the model to the level of concept detail available at the conceptual design stage.

1. J. H. Moore (ed.), L. R. Weatherford (ed.), Eppen, Gould, and Schmidt, *Decision Modeling with Microsoft Excel,* 6th edition, Prentice-Hall, Upper Saddle River, NJ, 2001.

The first evaluation of alternative designs should be a screening process based on meeting absolute criteria (e.g., functional feasibility, technology readiness, constraint satisfaction). The chapter presented three frequently used design tools for decision making: the Pugh chart, weighted decision matrix, and AHP. Each tool uses comparisons of alternatives to make a selection.

The use of the Pugh chart deserves a special note. This evaluation tool is used frequently by engineering students. However, students often fail to realize that the numbers resulting from creating a Pugh chart are less important than the insights about the problem and solution concepts that are obtained from vigorous team participation in the process. Creating a Pugh chart should be an intensive team exercise from which improved concepts often result.

The reality of modern engineering is that mere analysis of engineering performance is not sufficient for making choices among design alternatives. Engineers are increasingly required to factor other outcomes (e.g., performance in the marketplace and risk to meet a product launch schedule) into their decision-making process as early as conceptual design.

NEW TERMS AND CONCEPTS

Absolute comparison	Expected value	Preference
Analytic hierarchy process (AHP)	Evaluation	Pugh concept selection chart
Decision Based Design	Marginal utility	Ratio scale
Decision tree	Maximin strategy	Relative comparison
Decision under certainty	Minimax strategy	Utility
Decision under risk	Objective tree	Value
Decision under uncertainty	Ordinal scale	Weighted decision matrix

BIBLIOGRAPHY

Clemen, R. T.: *Making Hard Decisions: An Introduction to Decision Analysis,* 2d ed., Wadsworth Publishing Co., Belmont, CA, 1996.

Cross, N.: *Engineering Design Methods,* 2d ed., John Wiley & Sons, New York, 1994.

Dym, C.I. and P. Little: *Engineering Design,* 3rd ed, Chap. 3, John Wiley & Sons, Hoboken, NJ, 2008.

Lewis, K. E., W. Chen, and L. C. Schmidt: *Decision Making in Engineering Design,* ASME Press, New York, 2006.

Pugh, S.: *Total Design,* Addison-Wesley, Reading, MA, 1990.

Starkey, C. V.: *Engineering Design Decisions,* Edward Arnold, London, 1992.

PROBLEMS AND EXERCISES

7.1   Construct a simple personal decision tree (without probabilities) for whether to take an umbrella when you go to work on a cloudy day.

7.2 You are the owner of a new company that is deciding to invest in the development and launch of a household product. You have learned that there are two other companies preparing to enter the same market that have products close to one of your models. Company 1, Acme, will market a basic version of the same household item. Company 2, Luxur, will market the item with several extra features. Some end users will not need all Luxur's extra features. There is also a possibility that both Acme and Luxur will have their products in the marketplace when you launch yours.

You have designed three different versions of the product. However, resources limit you to launching only one product model.

- Model a_1 is a basic functional model with no extra features. You have designed model a_1 to be of higher quality than Acme's proposed product, and it will also cost more.
- Model a_2 is your model with a set of controls allowing variable output. This functionality is not on Acme's product but is on Luxur's Model. a_2 will be priced between the two competitors' products.
- Model a_3 is the deluxe, top-of-the-line model with features exceeding those on the Luxur model. It will also be priced above the Luxur model.

Your best marketing team has developed the following table summarizing the anticipated market share that your company can expect under the different competition scenarios with Acme and Luxur products. However, no one knows which products will be on the market when you launch your new product.

**Predicted Market Share for Your New Product
When It Faces Competition**

Your Model To Be Launched	Competitors in Market When Product a_x Is Launched		
	Acme	Luxur	Acme & Luxur
a_1	45%	60%	25%
a_2	35%	40%	30%
a_3	50%	30%	20%

You must decide which product model to develop and launch, a_1, a_2 or a_3?

(a) Assume that you will know which competing products will be in the market. Choose the model you will launch under each of the three possible conditions.

(b) Assume that you have inside information about the likelihood of the competitors entering the market with their products. You are told that Acme will enter the market alone with a 32% probability; Luxur will enter the market alone with a 48% probability; and there is a 20% probability that both companies will enter the market together when you are ready to launch your product.

(c) Assume that you have no information on the actions of the competitors. You are told that you need to be very conservative in your decision so that you will capture the largest share of the market even if the competition is fierce.

7.3 This decision concerns whether to develop a microprocessor-controlled machine tool. The high-technology microprocessor-equipped machine costs $4 million to develop, and the low-technology machine costs $1.5 million to develop. The low-technology machine is less likely to receive wide customer acclaim ($P = 0.3$) versus $P = 0.8$ for the microprocessor-equipped machine. The expected payoffs (present worth of all future profits) are as follows:

	Strong Market Acceptance	Minor Market Acceptance
High technology	$P = 0.8$	$P = 0.2$
	PW = $16M	PW = $10M
Low technology	$P = 0.3$	$P = 0.7$
	PW = $12M	PW = 0

If the low-technology machine does not meet with strong market acceptance (there is a chance its low cost will be more attractive than its capability), it can be upgraded with microprocessor control at a cost of $3.2 million. It will then have an 80 percent chance of strong market acceptance and will bring in a total return of $10 million. The non-upgraded machine will have a net return of $3 million. Draw the decision tree and decide what you would do on the basis of (a) net expected value and (b) net opportunity loss. Opportunity loss is the difference between the payoff and the cost for each strategy.

7.4 The prototype of a tie rod is designed to be 10 feet long and have a rectangular cross section with width, $w = 2$ inches and breadth, $b = 1$ inch. The material will be a heat-treated steel with Young's modulus of 30×10^6 lb/in². The tie rod is intended to be loaded axially in tension. A model of the rod is to be made and tested from a soft, easy-to-machine, aluminum alloy with Young's modulus of 10×10^6 lb/in². The model must remain elastic during testing as must the prototype during service. The yield strength for the aluminum alloy is 20,000 psi (or lb/in²). Therefore, the model cannot be loaded as heavily as the prototype. It has been decided that every pound of load on the model will be equivalent to 10 pounds on the prototype. Now we need to determine the dimensions of the model based on scale relationships.

(a) Derive the scaling relationship between the predicted deflection of the prototype, δ_p, for the deflection of the model, δ_m.

(b) Determine the geometric, load, and elastic scale factors, and determine δ_p when the model is at its largest possible deflection.

7.5 In the search for more environmentally friendly design, paper cups have replaced Styrofoam cups in most fast-food restaurants. These cups are less effective insulators, and the paper cups often get too hot for the hand. A design team is in search of a better disposable coffee cup. The designs to be evaluated are: (a) a standard Styrofoam cup, (b) a rigid injection-molded cup with a handle, (c) a paper cup with a cardboard sleeve, (d) a paper cup with a pull-out handle, and (e) a paper cup with a cellular wall. These design concepts are to be evaluated with the Styrofoam cup as the datum.

The engineering characteristics on which the cups are evaluated are:

1. Temperature in the hand
2. Temperature of the outside of the cup
3. Material environmental impact
4. Indenting force of cup wall
5. Porosity of cup wall
6. Manufacturing complexity

    7. Ease of stacking the cups
    8. Ease of use by customer
    9. Temperature loss of coffee over time
   10. Estimated cost for manufacturing the cup in large quantities

Using your knowledge of fast-food coffee cups, use the Pugh concept selection method to select the most promising design.

7.6 Four concepts for improving the design of an on/off switch in a right-angle drill are sketched in the figure below. Determine a set of criteria for an on/off switch. Use this information to prepare a Pugh chart and select the best option from the given alternatives. Concept A is a modest change to the existing switch, and will be the DATUM. Concept B adds three buttons for on/off/ and reverse. Concept C is a track and slider design, and D is an add-on accessory to make it easier to operate the existing switch.

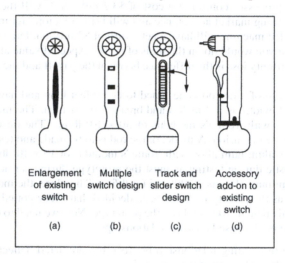

Enlargement of existing switch	Multiple switch design	Track and slider switch design	Accessory add-on to existing switch
(a)	(b)	(c)	(d)

7.7 Four preliminary designs for sport-utility vehicles had the characteristics listed in the following table. Using the weighted decision matrix, which design looks to be the most promising?

Characteristics	Parameter	Weight factor	Design A	Design B	Design C	Design D
Gas mileage	Miles per gal	0.175	20	16	15	20
Range	Miles	0.075	300	240	260	400
Ride comfort	Rating	0.40	Poor	Very good	Good	Fair
Ease to convert to 4-wheel drive	Rating	0.07	Very good	Good	Good	Poor
Load capacity	lb.	0.105	1000	700	1000	600
Cost of repair	Avg. of 5 parts	0.175	$700	$625	$600	$500

7.8 Repeat Prob. 7.7 using the AHP method. Determine your own weighting factors for the characteristics according the AHP method. Then continue applying AHP until you can recommend the best Design for a customer with your weight factors.

8
EMBODIMENT DESIGN

8.1
INTRODUCTION

Prior chapters have described the engineering design process to the point where a set of concepts has been generated and evaluated to produce a single concept or small set of concepts for further development. It may be that some of the major dimensions have been established roughly, and the major components and materials have been tentatively selected.

The next phase of the design process is often called *embodiment design*. It is the phase where the design concept is invested with physical form, where we "put meat on the bones." We have divided the embodiment phase of design into three activities (Fig. 8.1):

- *Product architecture*—setting the arrangement of the physical elements of the design into groupings, called modules
- *Configuration design*—designing special-purpose parts and the selection of standard components, like pumps or motors
- *Parametric design*—determining the exact values, dimensions, or tolerances of the components or component features that are deemed critical-to-quality

Also, in this chapter we consider such important issues as setting the dimensions on parts, designing to enhance the aesthetic values of the design, and achieving a design that is both user friendly and environmentally benign. These are but a small sample of the requirements that a good design needs to meet. Therefore, we conclude this chapter with a listing of the many other issues that must be considered in completing the design, and point the reader to where these subjects are discussed in detail in the text.

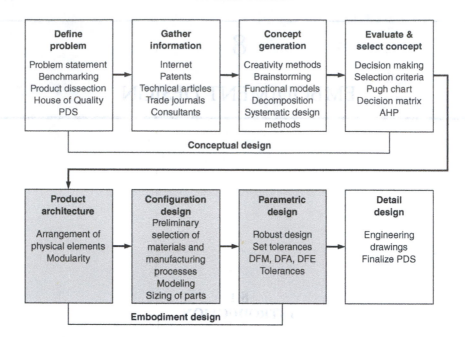

FIGURE 8.1
Steps in the design process showing that embodiment design consists of establishing the product architecture and carrying out the configuration and parametric design.

8.1.1 Comments on Nomenclature Concerning the Phases of the Design Process

It is important to understand that writers about engineering design do not use the same nomenclature to label the phases of the design process. Nearly everyone agrees that the first step in design is *problem definition* or needs analysis. Some writers consider problem definition to be the first phase of the design process, but in agreement with most designers we consider it to be the first step of the conceptual design phase, Fig. 8.1. The design phase that we consider in this chapter, which we call *embodiment design,* is also often called *preliminary design.* It has also been called *system-level design* in the description of the PDP given in Fig. 2.1. The term *embodiment design* comes from Pahl and Beitz[1] and has been adopted by most European and British writers about design. We continue the trend that adopts the terminology conceptual design, embodiment design, and detail design because these words seem to be more descriptive of what takes place in each of these design phases.

However, doing this raises the question of what is left in the design process for Phase 3, detail design. The last phase of design is uniformly called *detail design,* but the activities included in detail design vary. Prior to the 1980s it had been the design

---

1. G. Pahl and W. Beitz, *Engineering Design: A Systematic Approach,* First English edition, Springer-Verlag, Berlin, 1996.

phase where final dimensions and tolerances were established, and all information on the design was gathered into a set of "shop drawings" and bill of materials. However, moving the setting of dimensions and tolerances into embodiment design is in keeping with the adoption of computer-aided engineering methods to move the decision making forward as early as possible in the design process to shorten the product development cycle. Not only does this save time, but it saves cost of rework compared to when errors are caught in detail design at the very end of the design process. Most of the specifics of the design of components are set during parametric design, yet detail design is still required to provide information to describe the designed object fully and accurately in preparation for manufacturing. As will be shown in Chap. 9, detail design is becoming more integrated into information management than just detailed drafting.

Returning once more to the consideration of design nomenclature, it needs to be recognized that engineering disciplines other than mechanical often use different nomenclature to describe the phases of the design process. For example, one text on designing steel building and bridge structures uses the terms *conceptual design, design development,* and *construction documentation,* while another uses the descriptors *conceptual design, preliminary design,* and *final design.* One long-standing text in chemical process design, where the emphasis is on designing by assembling standard components like piping and evaporators into economical process systems uses the terminology *preliminary (quick-estimate) designs, detailed estimate designs,* and *firm process designs* for the three design phases we have been considering.

8.1.2 Idealization of the Design Process Model

It is important to realize that Fig. 8.1 does not capture the intricacies of the design process in at least two major respects. In this figure the design process is represented as being sequential, with clear boundaries between each phase. Engineering would be easy if the design process flowed in a nice serial fashion from problem to solution, but it does not. To be more realistic, Fig. 8.1 should show arrows looping back from every phase to those phases previous to it in the process. This would represent the fact that design changes may be needed as more information is uncovered. For example, increases in weight brought about by the addition of heavier components demanded by a failure modes and effects analysis would require going back and beefing up support members and bracing. Information gathering and processing is not a discrete event. It occurs in every phase of the process, and information obtained late in the process may necessitate changes to decisions made at an earlier phase of the process.

The second simplification is that Fig. 8.1 also implies that design is a linear process. For purposes of learning, we characterize design as a phased process in time sequence. But we learned in the discussion of concurrent engineering in Sec. 2.4.4 that performing some design activities in parallel is the key to shortening the product development cycle time. Thus, it is quite likely that one member of the design team is testing some subassembly that has been finished early, while other team members are still sizing the piping, and yet another member may be designing tooling to make another component. Different team members are often working on different design steps in parallel.

Not all engineering design is of the same type or level of difficulty.[1] Much of design is routine, where all possible solution types are known and often prescribed in codes and standards. Thus, in routine design the attributes that define the design and the strategies and methods for attaining them are well known. In adaptive design not all attributes of the design may be known beforehand, but the knowledge base for creating the design is known. While no new knowledge is added, the solutions are novel, and new strategies and methods for attaining a solution may be required. In original design neither the attributes of the design nor the precise strategies for achieving them are known ahead of time.

The conceptual design phase is most central to original design. At the opposite end of the spectrum is selection design, which is more central to routine design. Selection design involves choosing a standard component, like a bearing or a cooling fan, from a catalog listing similar items. While this may sound easy, it really can be quite complex owing to the presence of many different items with slightly different features and specifications. In this type of design the component is treated as a "black box" with specified properties, and the designer selects the item that will meet the requirements in the best way. In the case of selecting dynamic components (motors, gearboxes, clutches, etc.) its characteristic curve and transfer function must be carefully considered.[2]

8.2
PRODUCT ARCHITECTURE

Product architecture is the arrangement of the physical elements of a product to carry out its required functions. Product architecture begins to emerge in the conceptual design phase from such things as diagrams of functions, rough sketches of concepts, and perhaps a proof-of-concept model. However, it is in the embodiment design phase that the layout and architecture of the product must be established by defining the basic building blocks of the product and their interfaces. (Some organizations refer to this as system-level design.) Note that a product's architecture is related to its function structure, but it does not have to match it. In Chap. 6 function structure was presented as a way of generating design concepts. A product's architecture is selected to establish the best system for functional success once a design concept has been chosen.

The physical building blocks that the product is organized into are usually called modules. Other terms are subsystem, subassembly, cluster, or chunk. Each module is made up of a collection of components that carry out functions. The architecture of the product is given by the relationships among the components in the product and the functions the product performs. There are two entirely opposite styles of product architecture, *modular* and *integral*. Systems with modular architecture are most common; they usually are a mixture of standard modules and customized components.

Understanding the interfaces between modules is critical to successful product functioning. These are often the sites for corrosion and wear. Unless interfaces are

1. M. B. Waldron and K. J. Waldron (eds.), *Mechanical Design: Theory and Methodology,* Chap. 4, Springer-Verlag, Berlin, 1996.
2. J. F. Thorpe, *Mechanical System Components,* Allyn and Bacon, Boston, 1989.

designed properly, they can cause residual stresses, unplanned deflections, and vibration. Examples of interfaces are an IC engine piston and its chamber or the connection between a computer monitor and the CPU. Interfaces should be designed to be as simple and stable as possible (see Sec. 8.4.2). Standard interfaces, those that are well understood by designers and parts suppliers, should be used if possible. The personal computer is an outstanding example of the use of standard interfaces. PCs can be customized, module by module, from parts supplied by many different suppliers. A USB port can attach a variety of drives, printers, and PDAs to any computer.

8.2.1 Integral Architecture

In an *integral architecture* the implementation of functions is accomplished by only one or a few modules. In integral product architectures, components perform multiple functions. This reduces the number of components, generally decreasing cost unless the integral architecture is obtained at the expense of extreme part complexity. A simple example is the humble crowbar, where a single part provides both the functions of leverage and acting as a handle. A more complex example is found in the BMW model R1200S motorcycle where the transmission case serves as part of the structural frame, thereby saving both weight and cost. When a component provides more than one function it enables *function sharing*.

Integral product architecture is often adopted when constraints of weight, space, or cost make it difficult to achieve required performance. Another strong driver toward integration of components is the design for manufacturing and assembly (DFMA) strategy, which calls for minimizing the number of parts in a product (see Chap. 13). There is a natural trade-off between component integration to minimize costs and integral product architecture. Thus, product architecture has strong implications for manufacturing costs. DFM studies should begin early in design when the product architecture is being established to define these trade-offs. The trade-off is that with integral architecture design, parts tend to become more complex in shape and features because they serve multiple purposes, thus costing more to manufacture.

8.2.2 Modular Architecture

A modular architecture makes it easier to evolve a design over time. The product can be adapted to the needs of different customers by adding or deleting modules. Obsolescence can be dealt with by replenishing components as they wear out, and at the end of its useful life the product can be remanufactured. Modular design may even be carried to the point of using the same set of basic components in multiple products, creating a *product family*. This form of standardization allows the component to be manufactured in higher quantities than would otherwise be possible, achieving cost savings due to economy of scale. An excellent example of a modular component with many uses is the rechargeable battery pack used in many electrical hand tools.

In a modular architecture, each module implements only one or a few functions, and the interactions between modules are well defined. An example would be

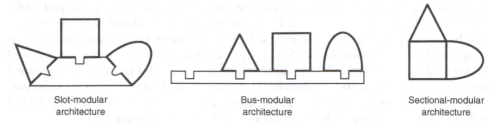

| Slot-modular architecture | Bus-modular architecture | Sectional-modular architecture |

FIGURE 8.2
Three types of modular architectures.

a personal computer where different functionality can be achieved with an external mass storage device or adding special-purpose software.

A modular architecture also tends to shorten the product development cycle because modules can be developed independently provided that interfaces are well laid out and understood. A module's design can be assigned to a single individual or small design team because the decisions regarding interactions and constraints are confined within that module. In this case, communication with other design groups is concerned primarily with the interfaces. However, if a function is implemented using two or more modules, the interaction problem becomes much more challenging. That explains why designs "farmed out" to an outside supplier or remote location within the corporation usually are subsystems of a highly modular design, for example, automotive seats.

There are three types of modular architectures defined by the type of interfaces used to assemble them to the body of the product: *slot, bus,* and *sectional.* Each of the modular types involves a one-on-one mapping from the functional elements to the physical product and well-defined interfaces. Differences in the types of modular architectures lie in the way the interfaces between the modules are laid out. Figure 8.2 illustrates these differences.

- *Slot-modular.* Each of the interfaces between modules is of a different type from the others. This is the most common situation for modular architecture since typically each module requires a different interface to perform its function with the product. For example, an automobile radio cannot be interchanged with the DVD player.
- *Bus-modular.* The modules can be assembled along a common interface, or *bus.* Therefore, interchange of modules can be done readily. The use of a power bus is common in electrical products, but it can also be found in such mechanical systems as shelving systems.
- *Sectional-modular.* All interfaces are of the common type, but there is no single element to which the other chunks attach. The design is built by connecting the chunks to each other through identical interfaces, as in a piping system.

8.2.3 Modularity and Mass Customization

Society has benefited from the exploitation of *mass production,* by which the unit price of most consumer goods has been reduced through large-scale production aimed at large, homogeneous consumer markets. However, current competitive conditions make it difficult to maintain this situation. Increasingly, customers look for products with variety and distinctiveness. Thus, there is growing interest in finding ways to produce products at a reasonable cost but also with enough variety and customization (*mass customization*) so that everyone can buy exactly what they want. Such products have *economy of scope* as well as *economy of scale*. Designing products with a modular architecture is one of the best ways to approach the goal of mass customization.

There are four distinct strategies for using modularity in product design and manufacturing.

- *Component-sharing modularity.* This type of modularity exists when a family of dissimilar products uses the same assembly or component. For example, an entire family of rechargeable battery-powered hand tools would be designed to use the same battery, thus achieving lower cost as a result of economy of scale in manufacture and providing a desirable marketing feature in that the user would need only a single recharging station for several different tools.
- *Component-swapping modularity.* This type of modularity exists in a product that is differentiated only by a single component or assembly. Automobiles are good examples of this type of modularity. Consumers buy a certain model car and select one or more options that differentiate their car from others. A purchaser may order or select a model with a power package that includes power windows, door locks, and seat adjustment controls. Once the car is in service, it is not a simple matter of exchanging modules to switch from power to manual door locks. The module selection must occur prior to final assembly. Another example of component swapping modularity occurs in some refrigerator lines that feature in-door water and ice dispensing options. The differentiation occurs in the manufacturing process by exploiting the modular design architecture.
- *Cut-to-fit modularity.* This is a customization strategy whereby a component's parameters or features can be adjusted within a given range to provide a variety of products. Tailored clothing is one example of cut-to-fit modularity. So are window blinds, shelving units, and housing siding.
- *Platform modularity.* This form of modularity describes products that consist of different combinations of modules assembled on the same basic structure, as in the bus modularity discussed above or automobiles. It is now common for an automaker to design different vehicles on the same frame. Design with common platforms is necessary in the auto business because of the huge investment in tooling required to manufacture frames and the relentless need to introduce new car models into the marketplace every year.

8.2.4 Budgeted Resources

In any design there is at least one scarce resource that needs to be carefully allocated or budgeted. While cost or performance/cost ratio comes first to mind, often other design variables fit into this category, for example, weight, cubic space to be installed in a fixed volume, temperature rise in a computer chip, battery life, and fuel consumption.

Establishing product architecture is the first place in the design process where resource budgeting can be accomplished. For effective resource budgeting, the design team needs to decide on the need for the budgeted resource. In addition, there should be one person responsible for allocating and tracking the resource. All team members must know what their allocation is and be informed regularly how close they are to their limit of the resource.

8.3
STEPS IN DEVELOPING PRODUCT ARCHITECTURE

Establishing the product architecture is the first task of embodiment design. Product subsystems, called modules or chunks, are defined and details of integration with each other are determined. To establish a product's architecture, a designer defines the geometric boundaries of the product and lays out the proposed elements of the design within its envelope. The design elements are both functional elements and physical elements. The functional elements are the functions that the product must perform to conform to its PDS. The physical elements are components, either standard parts or special-purpose parts, needed to achieve the functions. As will be seen below, at the time of developing the product architecture not all functions have been selected down to the part level, so the designer must leave room in the architecture for developing the physical realization of the function. A function element is inserted as a place-holder in the layout.

The process of developing the product architecture includes clustering the physical elements and the functional elements into groupings, often called chunks, to perform specific functions or sets of functions. The chunks are then placed in locations and orientations relative to each other within the overall physical constraints imposed on the product.

Ulrich and Eppinger[1] propose a four-step process for establishing product architecture.

- Create a schematic diagram of the product
- Cluster the elements of the schematic
- Create a rough geometric layout
- Identify the interactions between modules

8.3.1 Create a Schematic Diagram of the Design

The schematic diagram ensures that the team understands the basic elements of the product needed to produce an operating design. Some of these elements will be actual components that the team recognizes are required for the design, like the ball return trampoline. Other elements will still be in functional form because the team has not yet specified their embodiment, like the trampoline turning mechanism. Figure 8.3 shows the schematic diagram for the Shot-Buddy.

---

1. K. T. Ulrich and S. H. Eppinger, *Product Design and Development,* 4th ed., McGraw-Hill, New York, 2008.

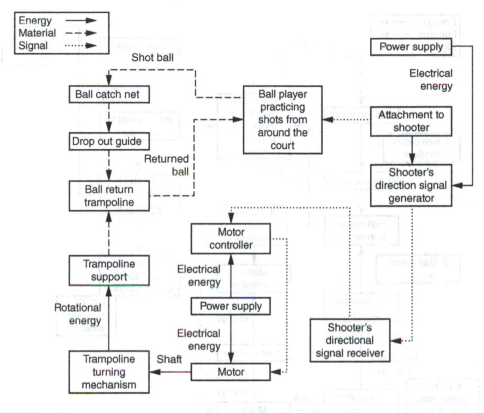

FIGURE 8.3
Schematic diagram of the Shot-Buddy showing flows between components.[1]

Development of the schematic diagram starts with the function structure, Fig. 6.8, and the concept sketch, Fig. 7.7. Note that the flows of energy, material, and signal that were used in functional analysis are traced through the schematic diagram.

Judgment should be used in deciding what level of detail to show on the schematic. Generally, no more than 30 elements should be used to establish the initial product architecture. Also, realize that the schematic is not unique. As with everything in design, the more options you investigate (i.e., the more you iterate in the process), the better the chance of arriving at a good solution.

8.3.2 Cluster the Elements of the Schematic Diagram

The second step of setting product architecture is to create groups of the elements in the schematic. The purpose of this step is to arrive at an arrangement of design elements (clusters) that will become modules. Looking at Fig. 8.4, we see that the

---

1. Adapted from Josiah Davis, Jamil Decker, James Maresco, Seth McBee, Stephen Phillips, and Ryan Quinn, *"JSR Design Final Report: Shot-Buddy,"* unpublished, ENME 472, University of Maryland, May 2010.

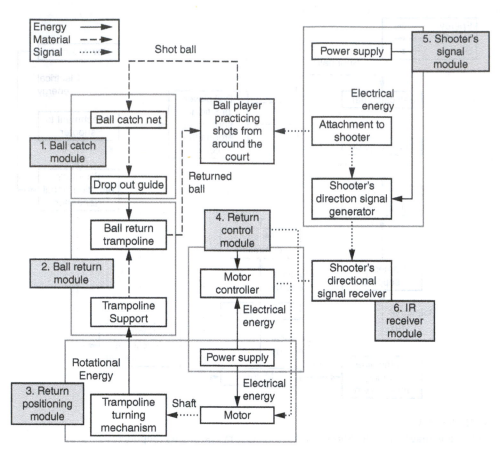

FIGURE 8.4
Schematic diagram of the Shot-Buddy showing components clustered into modules.[1]

following modules have been established: (1) ball catch module, (2) ball return module, (3) return positioning module, (4) return control module, (5) shooter's signal module, and (6) infrared (IR) receiver module. The Shot-Buddy has one module made up of a single component (the IR receiver module). Another interesting feature in Fig. 8.4 is that there are two modules (return positioning and return control) sharing a power supply. This is denoted by the overlap of modules 3 and 4. This reflects the practical nature of engineering. We could draw the schematic with two separate power supplies, but it is inevitable that designers will choose to use only one.

One way of deciding on the formation of modules is to start with the assumption that each design element will be an independent module and then cluster the elements to realize advantages, or commonalities. Some of the reasons for clustering elements include requiring close geometric relationship or precise location, elements that can share a function or an interface, the desire to outsource part of the design, and the portability

---

1. Adapted from Josiah Davis, Jamil Decker, James Maresco, Seth McBee, Stephen Phillips, and Ryan Quinn, *"JSR Design Final Report: Shot-Buddy,"* unpublished, ENME 472, University of Maryland, May 2010.

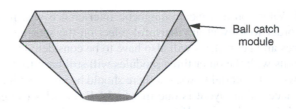

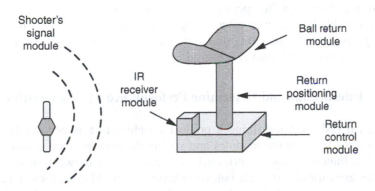

FIGURE 8.5
Geometric layout of the Shot-Buddy.[1]

of interfaces. For example, digital signals are much more portable and can be distributed more easily than mechanical motions. Clustering is natural for elements that have the same flows through them. Other issues that could affect clustering include the use of standard parts or modules, the ability to customize the product in the future (make a product family), or the allowance for improved technology in future versions of the product.

8.3.3 Create a Rough Geometric Layout

Making a geometric layout allows the designer to investigate whether there is likely to be geometrical, thermal, or electrical interference between elements and modules. A trial layout positions modules in a possible physical configuration. For some problems a two-dimensional drawing is adequate (Fig. 8.5), while for others a three-dimensional model (either physical or computer) is required.

The layout of the Shot-Buddy in Fig. 8.5 indicates no physical contact between the shooter's signal module and any other module in the product. The ball catch module doesn't connect to other components of the Shot-Buddy but is designed to be mounted on the basketball hoop and backboard (a fact not indicated in the layout). Three modules have contact interfaces: the ball return module, return positioning module, and return control module. As a result the interactions between these modules will have to be analyzed and

1. Adapted from Josiah Davis, Jamil Decker, James Maresco, Seth McBee, Stephen Phillips, and Ryan Quinn, *"JSR Design Final Report: Shot-Buddy,"* unpublished, ENME 472, University of Maryland, May 2010.

planned. Vibration and electromagnetic interference will have to be carefully considered in order to prevent any harmful effect on the sensing or positioning components. Tolerances and geometries will also have to be considered to ensure all parts fit together. Interactions with the other three modules will still have to be considered in terms of energy flows and material flows, but there should be no direct interference issues.

An acceptable layout is one in which all modules (roughly sized) fit into the envelope of the final design. If there are objects in the use environment that will interact with the final design, it is good to include a representation of them in the layout. During a review of the layout, designers should indicate motion direction to assure there is no physical interference in the operation. Sometimes it is not possible to arrive at a geometrically feasible layout, even after trying several alternatives. This means it is necessary to go back to the previous step and change the assignment of elements to modules until an acceptable layout is achieved.

8.3.4 Define Interactions and Determine Performance Characteristics

The most critical task in determining a product's architecture is accurately modeling the interactions between the modules and setting the performance characteristics for the modules. Function happens primarily at the interfaces between modules, and unless modules are carefully thought out, complexity can build up at these interfaces. Therefore at the conclusion of the embodiment design phase of the product development process, each product module must be described in complete detail. The documentation on each module should include:

- Functional requirements
- Drawings or sketches of the module and its component parts
- Preliminary component selection for the module
- Detailed description of placement within the product
- Detailed descriptions of interfaces with neighboring modules
- Accurate models for expected interactions with neighboring modules

The most critical items in the module description are the descriptions of the interfaces and the modeling of interactions between neighboring modules. There are four types of interactions possible between component modules—spatial, energy, information, and material.

- Spatial interactions describe physical interfaces between modules. These exist between mating parts and moving parts. The engineering details necessary for describing spatial interactions include information on mating geometry, surface finish, and tolerancing. A good example of a spatial interface between two moving parts is the relationship between the padded headrest and the notched metal supports connecting it to the car seat.
- Energy flows between modules represent another important type of interaction. These flows may be intentional, like the need to route electrical current from a switch to a motor, or they may be unavoidable, like the generation of heat by a motor contacting the case of a drill. Both planned and secondary types of energy interactions must be anticipated and described.

- Information flow between modules often takes the form of signals to control the product's operation or feedback relative to that operation. Sometimes these signals must branch out to trigger multiple functions simultaneously.
- Material can flow between product modules if required by product's functionality. For example, the paper path for a laser printer involves moving the paper through many different modules of the printer.

The design of modules may often proceed independently after the product architecture is completed. This allows the module design tasks to be given to teams specializing in the design of one particular type of subsystem. For example, a major manufacturer of power hand tools has defined motor design as one of the company's core competencies and has an experienced design team proficient in small motor design. In this case, the motor module description becomes the design specification for the motor design team. The fact that product design is divided into a group of module design tasks reemphasizes the need for clear communication between design teams working on separate modules.

There are two important issues with respect to the arrangement of the modules. The first is to ensure that the interfaces between the modules are designed to enable proper functioning of the adjacent components. The second issue is that the components at the interfaces can be assembled properly as discussed in Section 8.5.2. Guidelines on the design for assembly can be found in Chap. 13.

8.4
CONFIGURATION DESIGN

In configuration design we establish the shape and general dimensions of components. Exact dimensions and tolerances are established in parametric design (Sec. 8.6). The term *component* is used in the generic sense to include special-purpose parts, standard parts, and standard assemblies.[1] A part is a designed object that has no assembly operations in its manufacture. A part is characterized by its geometric *features* such as holes, slots, walls, ribs, projections, fillets, and chamfers. The *arrangement of features* includes both the location and orientation of the geometric features. Figure 8.6 shows four possible physical configurations for a component whose purpose is to connect two plates at right angles to each other. Note the variety of geometric features, and their much different arrangement in each of the designs.

A *standard part* is one that has a generic function and is manufactured routinely without regard to a particular product. Examples are bolts, washers, rivets, and I-beams. A *special-purpose part* is designed and manufactured for a specific purpose in a specific product line, as in Fig. 8.6. An *assembly* is a collection of two or more parts. A *subassembly* is an assembly that is included within another assembly or subassembly. A *standard assembly* is an assembly or subassembly that has a generic function and is manufactured routinely. Examples are electric motors, pumps, and gearboxes.

---

1. J. R. Dixon and C. Poli, *Engineering Design and Design for Manufacturing*, pp. 1–8, Field Stone Publishers, Conway, MA, 1995.

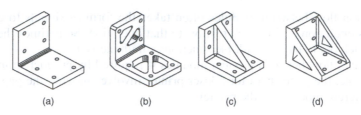

FIGURE 8.6
Four possible configurations of features for a right-angle bracket. (a) Bent from a flat plate.
(b) Machined from a solid block. (c) Bracket welded from three pieces. (d) Cast bracket.

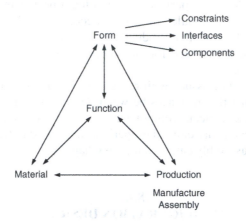

FIGURE 8.7
Schematic illustrating the close interrelationship between function and form and, in turn,
their dependence on the material and the method of production. (*After Ullman*)

As already stated several times in previous chapters, the form or configuration
of a part develops from its function. However, the possible forms depend strongly on
available materials and production methods used to generate the form from the material. Moreover, the possible configurations are dependent on the spatial constraints
that define the envelope in which the product operates and the product architecture.
This set of close relationships is depicted in Fig. 8.7.

Generally, detailed decisions about the design of a component cannot proceed
very far without making decisions about the material and the manufacturing process
from which it will be made. These vital topics are considered in detail in Chaps. 11,
12, and 13.

In starting configuration design we should follow these steps:[1]

• Review the product design specification and any specifications developed for the
particular subassembly to which the component belongs.

---

1. J. R. Dixon and C. Poli, op. cit., Chap. 10; D. G. Ullman, *The Mechanical Design Process*, 4th ed.,
McGraw-Hill, 2010.

- Establish the spatial constraints that pertain to the product or the subassembly being designed. Most of these will have been set by the product architecture (Sec. 8.3). In addition to physical spatial constraints, consider the constraints of a human working with the product (see Sec. 8.9) and constraints that pertain to the product's life cycle, such as the need to provide access for maintenance or repair or to dismantle it for recycling.

- Create and refine the interfaces or connections between components. Again, the product architecture should give much guidance in this respect. Much design effort occurs at the connections between components, because this is the location where failure often occurs. Identify and give special attention to the interfaces that transfer the most critical functions.

- Before spending much time on the design, answer the following questions: Can the part be eliminated or combined with another part? Studies of design for manufacture (DFM) show that it is almost always less costly to make and assemble fewer, more complex parts than it is to design with a higher part count.

- Can a standard part or subassembly be used? While a standard part is generally less costly than a special-purpose part, two standard parts may not be less costly than one special-purpose part that replaces them.

Generally, the best way to get started with configuration design is to just start sketching alternative configurations of a part. The importance of hand sketches should not be underestimated.[1] Sketches are an important aid in idea generation and a way for piecing together unconnected ideas into design concepts. Later as the sketches become scale drawings they provide a vehicle for providing missing data on dimensions and tolerances, and for simulating the operation of the product (3-D solid modeling, Fig. 8.8). Drawings are essential for communicating ideas between design engineers and between designers and manufacturing people, and as a legal document for archiving the geometry and design intent.

Consider the task of applying configuration design to create a special-purpose part to connect two plates with a bolted joint. Figure 8.9 portrays the images of possible solutions that would go through the mind of an experienced designer as he or she thinks about this design. Note that such issues as alternate bolt designs, the force distribution in the joint, the relationship of the design to surrounding components, and the ability to assemble and disassemble are considerations. Of special prominence in the designer's mind would be visualization of how the design would actually be manufactured.

8.4.1 Generating Alternative Configurations

As in conceptual design, generally the first attempt at a configuration design does not yield the best that you can do, so it is important to generate a number of alternatives for each component or subassembly. Ullman[2] characterizes configuration design as

1. J. M. Duff and W. A. Ross, *Freehand Sketching for Engineering Design,* PWS Publishing Co., Boston, 1995; G. R. Bertoline and E. N. Wiebe, *Technical Graphics Communication,* 5th ed., McGraw-Hill, New York, 2007.
2. D. G. Ullman, op.cit. pp. 260–264.

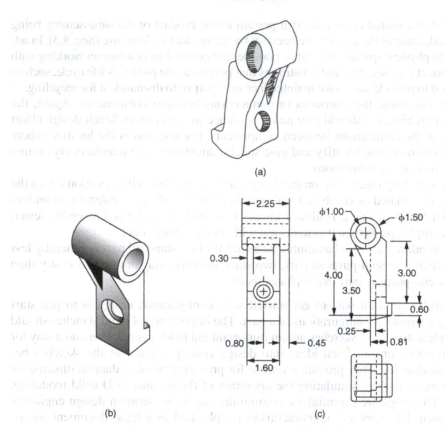

(a)

(b)   (c)

FIGURE 8.8
Showing the progression of a design configuration from a rough sketch (a) to a 3-D computer model (b) to a detailed three-view engineering drawing (c) Note the increase in detail from (a) to (b) to (c).

refining and patching. *Refining* is a natural activity as we move through the design process in which we develop more specificity about the object as we move from an abstract to a highly detailed description. Figure 8.8 illustrates the increase in detail as we refine the design. At the top is a rough sketch of a support bracket, while at the bottom is a detailed drawing showing the final dimensions after machining. *Patching* is the activity of changing a design without changing its level of abstraction. Refining and patching leads to a succession of configurational arrangements that hopefully improve upon the deficiencies of the previous designs.

Patching can be facilitated by applying the aids for brainstorming listed in Table 6.2.

- *Substituting* looks for other concepts, components, or features that will work in place of of the current idea.
- *Combining* aims to make one component replace multiple components or serve multiple functions. This is a move toward integral architecture, which we have seen

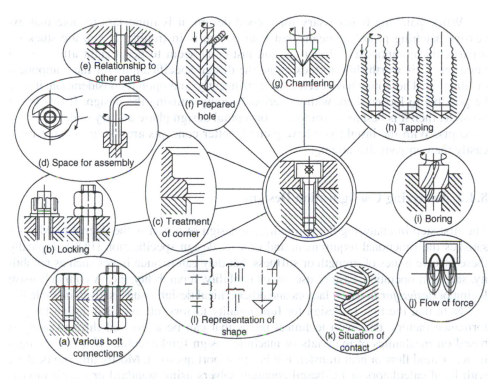

FIGURE 8.9
Images that come to a designer's mind when making a design of a bolted connection.

(From Y. Hatamura, *The Practice of Machine Design,* Oxford University Press, Oxford, UK, 1999, p. 78. Used with permission.)

is beneficial in reducing part count, and therefore lowering manufacturing and assembly costs.

- *Decomposing* is the opposite approach from combining. As new components and assemblies are developed through decomposing, it is important to consider whether the new configurations affect your understanding of the constraints on and connections between each component.
- *Magnifying* involves making some feature of a component larger relative to adjacent components.
- *Minifying* involves making some feature of a component smaller. In the limit, it means eliminating the component if its function can be provided for in some other way.
- *Rearranging* involves reconfiguring the components or their features. Changes in shape force rethinking of how the component carries out its functions. Another way to stimulate new ideas is to rearrange the order of the functions in the functional flow.

Another way to stimulate ideas for patching is to apply the 40 Inventive Principles of TRIZ presented in Sec. 6.7.

While patching is necessary for a good design, it is important to note that excessive patching probably means that your design is in trouble. If you are stuck on a particular component or function, and just can't seem to get it right after several iterations, it is worthwhile to reexamine the design specifications for the component or function. These may have been set too stringently, and upon reconsideration, it may be possible to loosen them without seriously compromising the design. If this is not possible, then it is best to return to the conceptual design phase and try to develop new concepts. With the insight you have gained, better concepts are likely to come more easily than on your first attempt.

8.4.2 Analyzing Configuration Designs

The first step in analyzing the configuration design of a part is the degree to which it satisfies the functional requirement and product design specification (PDS). Typically these involve issues of strength or stiffness, but they can include issues such as reliability, safety in operation, ease of use, maintainability, reparability, etc. A comprehensive listing of *design for function* factors and other critical design issues is given in Table 8.1.

Note that the first 14 design for functionality factors, often called design for performance factors, deal with technical issues that can be addressed through analysis based on mechanics of materials or machine design fundamentals, if it is a strength issue, or fluid flow or heat transfer, if it is a transport question. Mostly this can be done with hand calculators or PC-based equation solvers using standard or simple models of function and performance. More detailed analysis of critical components is carried out in the parametric design step. Typically this uses the field-mapping capabilities of finite-element methods and more advanced computational tools. The rest of the factors are all product or design characteristics that need special explanation as to their meaning and measurement. These factors are all discussed in detail elsewhere in this text.

8.4.3 Evaluating Configuration Designs

Alternative configuration designs of a part should be evaluated at the same level of abstraction. We have seen that design for function factors are important, because we need some assurance that the final design will work. The analysis used for this decision is fairly rudimentary, because the objective at this stage is to select the best of several possible configurations. More detailed analysis is postponed until the parametric design stage. The second most important criterion for evaluation is to answer the question, "Can a quality part or assembly be made at minimum cost?" The ideal is to be able to predict the cost of a component early in the design process. But because the cost depends on the material and processes that are used to make the part, and to a greater degree on the tolerances and surface finish required to achieve functionality, this is difficult to do until all of the specifications have been determined for the part. Accordingly, a body of guidelines that result in best practice for *design for manufacture* and *design for assembly* have been developed to assist designers in this area. Chapter 13 is devoted to this topic, while Chap. 17 covers cost evaluation in considerable detail.

TABLE 8.1
Typical *Design for Function* and Other Critical Design Issues

Factor	Issues
Strength	Can the part be dimensioned to keep stresses below yield levels?
Fatigue	If cyclic loads, can stresses be kept below the fatigue limit?
Stress concentrations	Can the part be configured to keep local stress concentration low?
Buckling	Can the part be configured to prevent buckling under compressive loads?
Shock loading	Will the material and structure have sufficient fracture toughness?
Strain and deformations	Does part have required stiffness or flexibility?
Creep	If creep is a possibility, will it result in loss of functionality?
Thermal deformation	Will thermal expansion compromise functionality? Can this be handled by design?
Vibration	Has design incorporated features to minimize vibration?
Noise	Has frequency spectrum been determined, and noise abatement considered in design?
Heat transfer	Will heat generation/transfer be an issue to degrade performance?
Fluids transport/storage	Has this been adequately considered in design? Does it meet all regulations?
Energy efficiency	Has the design specifically considered energy consumption and efficiency?
Durability	Estimated service life? How has degradation from corrosion and wear been handled?
Reliability	What is the predicted mean time to failure?
Maintainability	Is the prescribed maintenance typical for this type of design? Can it be done by the user?
Serviceability	Has a specific design study been done for this factor? Is cost for repair reasonable?
Life-cycle costs	Has a credible study been done on LCC?
Design for environment	Has reuse and disposal of product been explicitly considered in the design?
Human factors/ergonomics	Are all controls/adjustments logically labeled and located?
Ease of use	Are written installation and operating instructions clear?
Safety	Does design go beyond safety regulations in preventing accidents?
Styling/aesthetics	Have styling consultants adequately determined customer taste and wants?

The Pugh chart or weighted decision matrix, as discussed in Chap. 7, are useful tools for selecting the best of the alternative designs. The criteria are a selection of the design for function factors in Table 8.1 determined by management or the design team to be critical to quality plus the cost-related factors of design for manufacture (DFM) and design for assembly (DFA). Because these factors are not equally important, the weighted decision matrix is preferred for this task.

8.5
BEST PRACTICES FOR CONFIGURATION DESIGN

It is more difficult to give a prescribed set of methods for configuration design than for conceptual design because of the variety of issues that enter into the development of the product architecture and performance of components. In essence, the rest of this text is about these issues, like selection of materials, design for manufacture, and design for robustness. Nevertheless, many people have thought carefully about what constitutes the best practice of embodiment design. We record some of these insights here.

The general objectives of the embodiment phase of design are the fulfillment of the required technical functions, at a cost that is economically feasible, and in a way that ensures safety to the user and to the environment. Pahl and Beitz[1] give the basic guidelines for embodiment design as clarity, simplicity, and safety.

- *Clarity of function* pertains to an unambiguous relationship between the various functions and the appropriate inputs and outputs of energy, material, and signal flow. This means that various functional requirements remain uncoupled and do not interact in undesired ways, as if the braking and steering functions of an automobile would interact.
- *Simplicity* refers to a design that is not complex and is easily understood and readily produced. This goal is often expressed as a design with *minimum information content*. One way to minimize information content is to reduce the number and complexity of the components.
- *Safety* should be guaranteed by direct design, not by secondary methods such as guards or warning labels.
- *Minimal impact on the environment* is of growing importance, and should be listed as a fourth basic guideline.

8.5.1 Design Guidelines

In the extensive list of principles and guidelines for embodiment design, along with detailed examples, that are given by Pahl and Beitz,[2] four stand out for special mention.

- Force transmission
- Division of tasks
- Self-help
- Stability

Force Transmission

In mechanical systems the function of many components is to transmit forces and moments between two points. This is usually accomplished through a physical connection between components. In general, the force should be accommodated in

---

1. G. Pahl and W. Beitz, *Engineering Design: A Systematic Approach,* 2d ed. English translation by K. Wallace, Springer-Verlag, Berlin, 1996.
2. G. Pahl and W. Beitz, op. cit., 199–403.

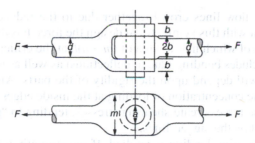

Side and top views of yoke connection, consisting of fork (left), pin (center), and blade (right).

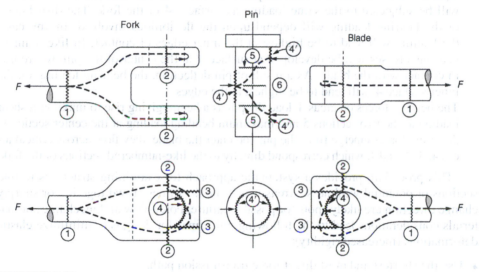

FIGURE 8.10
Force-flow lines and critical sections in a yoke connection. (After Juvinal.)

such a way as to produce a uniformly distributed stress on the cross section of the part. However, the design configuration often imposes nonuniform stress distributions because of geometric constraints. A method for visualizing how forces are transmitted through components and assemblies called *force-flow visualization* is to think of forces as flow lines, analogous to low-turbulence fluid flow streamlines or magnetic flux. In this model, the force will take the path of least resistance through the component.

Figure 8.10 shows the force flow through a yoke connection. Use sketches to trace out the path of the flow lines through the structure, and use your knowledge of mechanics of materials to determine whether the major type of stress at a location is tension (T), compression (C), shear (S), or bending (B). The flow of force through each member of the joint is indicated diagrammatically by the dashed lines in Fig. 8.10. Following along the path from left to right, the critical areas are indicated by jagged lines and numbered consecutively:

a. Tensile loading exists at section 1 of the fork. If there are ample material and generous radii at the transition sections, the next critical location is 2.

b. At 2 the force flow lines crowd together due to the reduced area caused by the holes. Note that with this symmetrical design the force F is divided into four identical paths, each of which has an area of $(m - a)b$ at the critical section. The loading at section 2 includes bending (due to deflections) as well as tension. The amount of bending load will depend upon the rigidity of the parts. Also, bending of the pin will cause some concentration of loading at the inside edges of the fork tines.

c. At section 3 the forces create shearing stresses, tending to "push out" the end segments bounded by the jagged lines.

d. At location 4 bearing loading is applied. If the strength at locations 1 to 4 is adequate, the force will flow into the pin. Surfaces 4' of the outer portions of the pin will be subjected to the same loading as surfaces 4 of the fork. The distribution of the bearing loading will depend upon the flexibilities involved. In any case, the loading will tend to be highest at the inner edges of contact. In like manner, bearing stresses will be developed at surface 4' at the center of the pin, where it is in contact with the blade. As a result of pin deflection, the bearing loading on the inner surface 4' will tend to be highest at the edges.

e. The bearing forces on areas 4' load the pin as a beam, giving rise to maximum shear loading at the two sections 5 and maximum bending loading at the center section 6. After the forces emerge from the pin and enter the blade, they flow across critical areas 4, 3, 2, and 1, which correspond directly to the like-numbered sections of the fork.

This procedure provides a systematic approach for examining structures to find sections of potential weakness. Areas where the flow lines crowd together or sharply change direction are likely spots for possible failure. Force-flow and mechanics of materials considerations lead to the following guidelines for designs to minimize elastic deformations (increased rigidity):

- Use the shortest and most direct force transmission path.
- Bodies that are shaped such that the material is uniformly stressed throughout will be the most rigid. The use of structures of tetrahedron or triangle shapes results in uniform stresses in tension and compression.
- The rigidity of a machine element can be increased by increasing its cross section or making the element shorter.
- To avoid sudden changes in the direction of force-flow lines, avoid sudden changes in cross section and use large radii at fillets, grooves, and holes.
- When there is a choice in the location of a discontinuity (stress raiser), such as a hole, it should be located in a region of low nominal stress.

Mismatched deformation between related components can lead to uneven stress distributions and unwanted stress concentrations. This usually occurs in redundant structures, such as in weldments. A redundant structure is one in which the removal of one of the load paths would still leave the structure in static equilibrium. When redundant load paths are present, the load will divide in proportion to the stiffness of the load path, with the stiffer path taking a proportionately greater fraction of the load. If problems are to be avoided with uneven load sharing, the design must be such that the strength of each member is approximately proportional to its stiffness. Note that stiffness mismatch can lead to high stress concentrations if mating parts are poorly matched in deformation.

Division of Tasks

The question of how rigorously to adhere to the principle of clarity of function is ever present in mechanical design. A component should be designed for a single function when the function is deemed critical and will be optimized for robustness. Assigning several functions to a single component (integral architecture) results in savings in weight, space, and cost but may compromise the performance of individual functions, and it may unnecessarily complicate the design.

Self-Help

The idea of self-help concerns the improvement of a function by the way in which the components interact with each other. A *self-reinforcing element* is one in which the required effect increases with increasing need for the effect. An example is an O-ring seal that provides better sealing as the pressure increases. A *self-damaging effect* is the opposite. A *self-protecting element* is designed to survive in the event of an overload. One way to do this is to provide an additional force-transmission path that takes over at high loads, or a mechanical stop that limits deflection.

Stability

The stability of a design determines whether the system will recover appropriately from a disturbance to the system. The ability of a ship to right itself in high seas is a classic example. Sometimes a design is purposely planned for instability. The toggle device on a light switch, where we want it to be either off or on and not at a neutral position, is an example. Issues of stability are among those that should be examined with the Failure Modes and Effects Analysis, Secs. 8.6.4 and 14.5.

Additional Design Suggestions

In this section additional design suggestions for good practice are presented.[1]

- *Tailor the shape to the stress or load distribution.* Loading in bending or torsion results in nonuniform distributions of stress. For example, a cantilever beam loaded at its free end has maximum stress at its clamped end and none at the point of load application. Thus, most of the material in the beam contributes very little to carrying the load. In situations such as this, think about changing the dimensions of the cross section to even out the stress distribution, thereby minimizing the material used, which will reduce the weight and the cost.
- *Avoid geometry that is prone to buckling.* The critical Euler load at which buckling occurs is proportional to the area moment of inertia (I), for a given length. But I is increased when the shape of the cross section is configured to place most of the material as far as possible from the axis of bending. For example, a tube with cross-sectional area equal to that of a solid of the same area has three times the resistance to buckling.

---

1. J. A. Collins, *Mechanical Design of Machine Elements and Machines*, John Wiley & Sons, 2003, Chap. 6.

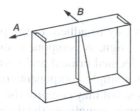

FIGURE 8.11
The use of a triangulated component to improve stiffness.

- *Use triangular shapes and structures.* When components need to be strengthened or stiffened, the most effective way is to use structures employing triangle shapes. In Fig. 8.11, the box frame would collapse without the *shear web* to transmit the force *A* from the top to the bottom surface. The *triangular rib* provides the same function for the force *B*.
- *Don't ignore strain considerations in design.* There is a tendency to give greater emphasis to stress considerations than strain in courses on mechanics of materials and machine design. Remember that otherwise good designs can become disasters by wobbly shafts or fluttering panels. At interfaces where load is transferred from one component to another, the goal should be to configure the components so that as load is applied and deformation occurs, the deformation of one component will be matched by the others in both magnitude and direction. Figure 8.12 shows a shaft surrounded by a journal bearing. In Fig. 8.12a, when the shaft bends under

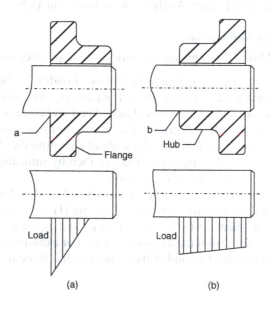

FIGURE 8.12
Journal bearings with mismatched and matched deformation.

(From J. G. Skakoon, "Detailed Mechanical Design," ASME Press, New York, 2000, p. 114. Used with permission.)

load it will be supported by the bearing chiefly at point (a) because the flange is thick at that end and allows minimal deflection of the bearing out along the axis from this point. However, in the design shown in Fig. 8.12b, the bending of the shaft is matched well by the deflection in the bearing because the bearing hub is less stiff at point (b). Therefore, the hub and shaft can deflect together as load is applied, and this results in more uniform load distribution.

One of the famous Augustine's Laws[1] is that "the last 10 percent of product performance generates one-third of the cost and two-thirds of the problems." Although developed from designs for military aircraft, the law carries a strong message for civilian products and systems.

8.5.2 Interfaces and Connections

We have mentioned several times in this section that special attention needs to be paid to the interfaces between components. Interfaces are the surfaces forming a common boundary between two adjacent objects. Often an interface arises because of the connection between two objects. Interfaces must always support force equilibrium and provide for a consistent flow of energy, material, and signal. Much design effort is devoted to the design of interfaces and connections between components.

Connections between components can be classified into the following types:[2]

- *Fixed, nonadjustable connection.* Generally one of the objects supports the other. These connections are usually fastened with rivets, bolts, screws, adhesives, welds, or by some other permanent method.
- *Adjustable connection.* This type must allow for at least one degree of freedom that can be locked. This connection may be field-adjustable or intended for factory adjustment only. If it is field-adjustable, the function of the adjustment must be clear and accessibility must be provided. Clearance for adjustability may add spatial constraints. Generally, adjustable connections are secured with bolts or screws.
- *Separable connection.* If the connection must be separated, the functions associated with it need to be carefully explored.
- *Locator connection.* In many connections the interface determines the location or orientation of one of the components relative to another. Care must be taken in these connections to account for errors that can accumulate in joints.
- *Hinged or pivoting connection.* Many connections have one or more degrees of freedom. The ability of these to transmit energy and information is usually key to the function of the device. As with the separable connections, the functionality of the joint itself must be carefully considered.

In designing connections at interfaces it is important to understand how geometry determines one or more constraints at the interface. A constrained connection is one

1. N. R. Augustine, *Augustine's Laws*, 6th ed., American Institute of Aeronautics and Astronautics, Reston, VA, 1997.
2. D. G. Ullman, op. cit., pp. 249–253.

that can move only in its intended direction. Every connection at an interface has potentially six degrees of freedom, translations along the x, y, and z-axes and rotation about these axes. If two components meet in a planar interface, six degrees of freedom are reduced to three—translation in the x and y directions (in both the positive and negative directions), and rotation about the z-axis (in either direction). If the plate is constrained in the positive x direction by a post, and the plate is kept in contact with the post by a nesting force, the plate has lost one degree of freedom (Fig. 8.13a). However, the plate is still free to translate along y and to rotate about the z-axis. Placing a second post, as in Fig. 8.13b, adds the additional constraint against rotation, but if the post is moved as in Fig. 8.13c the constraint is placed on translation along the y-axis, but rotation about the z-axis is allowed. It is only when three constraints (posts) are applied, and the nesting force is great enough to resist any applied forces, that the plate is perfectly fixed in a 2-D plane with zero degrees of freedom. The nesting force is a force vector that has components that are normal to the contacting surface at each contact point. It is usually provided by the weight of a part, locking screws, or a spring.

Figure 8.13 illustrates the important point that it takes three points of contact in a plane to provide exact constraint. Moreover, the nesting forces for any two constraints

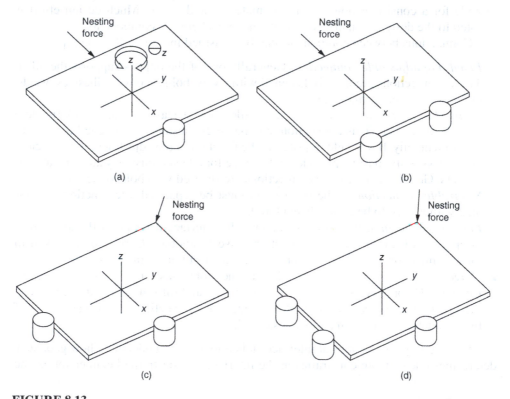

FIGURE 8.13
Illustration of the geometrical constraint in 2-D.

(From J. G. Skakoon, *Detailed Mechanical Design*, ASME Press, New York, 2000. Used with permission.)

must not act along the same line. In three dimensions it takes six constraints to fix the position of an object.[1]

Suppose in Fig. 8.13a we attempted to contain movement in the x-axis by placing a post opposite the existing post in the figure. The plate is now constrained from moving along the x-axis, but it actually is *overconstrained*. Because parts with perfect dimensions can be made only at great cost, the plate will be either too wide and not fit between the posts, or too small and therefore provide a loose fit. Overconstraint can cause a variety of design problems, such as loose parts that cause vibration, tight parts that cause surface fracture, inaccuracies in precision movements, and difficulties in part assembly. Usually it is difficult to recognize that theses types of problems have their root cause in an overconstrained design.[2]

Conventional mechanical systems consist of many overconstrained designs, such as bolted flange pipe connections and the bolts on a cylinder head. Multiple fasteners are used to distribute the load. These work because the interfaces are flat surfaces, and any flatness deviations are accommodated by plastic deformation when tightening down the mating parts. A more extreme example of the role of deformation in converting an overconstrained design into one with inconsequential overconstraint is the use of press fit pins in machine structures. These work well because they must be inserted with considerable force, causing deformation and a perfect fit between parts. Note however, with brittle materials such as some plastics and all ceramics, plastic deformation cannot be used to minimize the effects of an overconstrained design.

The subject of design constraint is surprisingly absent from most machine design texts. Two excellent references present the geometrical approach[3] and a matrix approach.[4]

8.5.3 Checklist for Configuration Design

This section, an expansion of Table 8.1, presents a checklist of design issues that should be considered during configuration design.[5] Most will be satisfied in configuration design, while others may not be completed until the parametric design or detail design phases.

Identify the likely ways the part might fail in service.
- Excessive plastic deformation. Size the part so that stresses are below the yield strength.

---

1. Of course, in dynamic mechanisms one does not want to reduce the design to zero degrees of freedom. Here one or more degrees of freedom must be left unconstrained to allow for the desired motion of the design.
2. J. G. Skakoon, *The Elements of Mechanical Design,* ASME Press, New York, 2008, pp. 8–20.
3. D. L. Blanding, *Exact Constraint: Machine Design Using Kinematic Principles,* ASME Press, New York, 1999.
4. D. E. Whitney, *Mechanical Assembly,* Chap. 4, Oxford University Press, New York, 2004. (Available at knovel.com.)
5. Adapted from J. R. Dixon, Conceptual and Configuration Design of Parts, *ASM Handbook Vol. 20,* Materials Selection and Design, pp. 33–38, ASM International, Materials Park, OH, 1997.

- Fatigue failure. If there are cyclic loads, size the part so that stresses are below the fatigue limit or fatigue strength for the expected number of cycles in service.
- Stress concentrations. Use generous fillets and radii so that stress raisers are kept low. This is especially important where service conditions are susceptible to fatigue or brittle failure.
- Buckling. If buckling is possible, configure the part geometry to prevent buckling.
- Shock or impact loads. Be alert to this possibility, and configure the part geometry and select the material to minimize shock loading.

Identify likely ways that part functionality might be compromised.

- Tolerances. Are too many tight tolerances required to make the part work well? Have you checked for tolerance stack-up in assemblies?
- Creep. Creep is change of dimensions over time at elevated temperature. Many polymers exhibit creep above 100°C. Is creep a possibility with this part, and if so, has it been considered in the design?
- Thermal deformation. Check to determine whether thermal expansion or contraction could interfere with the functioning of a part or assembly.

Materials and manufacturing issues.

- Is the material selected for the part the best one to prevent the likely failure modes in service?
- Is there a history of use for the material in this or similar applications?
- Can the form and features of the part be readily made on available production machines?
- Will material made to standard quality specifications be adequate for this part?
- Will the chosen material and manufacturing process meet the cost target for the part?

Design knowledge base.

- Are there aspects of the part design where the designer or design team is working without adequate knowledge? Is the team's knowledge of forces, flows, temperatures, environment, and materials adequate?
- Have you considered every possible unfortunate, unlikely, or unlucky event that could jeopardize the performance of the design? Have you used a formal method like FMEA to check for this?

8.5.4 Design Catalogs

Design catalogs are collections of known and proven solutions to design problems. They contain a variety of information useful to design, such as physical principles to achieve a function, solutions of particular machine design problems, standard components, and properties of materials. These are generally different in purpose and scope than the catalogs available from suppliers of components and materials. They provide quick, more problem-oriented solutions and data to design problems, and because they aim to be comprehensive, they are excellent places to find a broad range of design suggestions and solutions. Some catalogs, like the sample shown in Fig. 8.14 provide

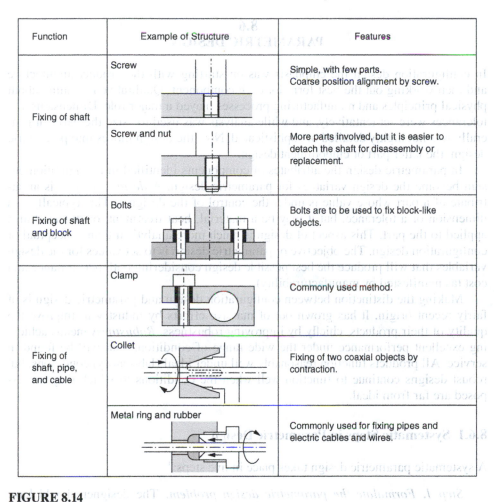

Function	Example of Structure		Features
Fixing of shaft	Screw		Simple, with few parts. Coarse position alignment by screw.
	Screw and nut		More parts involved, but it is easier to detach the shaft for disassembly or replacement.
Fixing of shaft and block	Bolts		Bolts are to be used to fix block-like objects.
Fixing of shaft, pipe, and cable	Clamp		Commonly used method.
	Collet		Fixing of two coaxial objects by contraction.
	Metal ring and rubber		Commonly used for fixing pipes and electric cables and wires.

FIGURE 8.14
Designs for fixing and connecting two components.

(From Y. Hatamura, *The Practice of Machine Design*, Oxford University Press, Oxford, 1999. Used with permission.)

specific design suggestions for a detailed task and are very useful in embodiment design. Most available design catalogs have been developed in Germany and have not been translated into English.[1] Pahl and Beitz list 51 references to the German literature for design catalogs.[2]

---

1. While they are not strictly design catalogs, two useful references are R. O. Parmley, *Illustrated Sourcebook of Mechanical Components*, 3d ed, McGraw-Hill, New York, 2005. Available online at knovel.com and N. Sclater and N. P. Chironis, *Mechanisms and Mechanical Devices Sourcebook*, 4th ed., McGraw-Hill, New York, 2007. (Available online at knovel.com)
2. G. Pahl and W. Beitz, op. cit.

8.6
PARAMETRIC DESIGN

In configuration design the emphasis was on starting with the product architecture and then working out the best form for each component. Qualitative reasoning about physical principles and manufacturing processes played a major role. Dimensions and tolerances were set tentatively, and while analysis was used to "size the parts" it generally was not highly detailed or sophisticated. Now the design moves into parametric design, the latter part of embodiment design.

In parametric design the attributes of components identified in configuration design become the design variables for parametric design. A *design variable* is an attribute of a part whose value is under the control of the designer. This typically is a dimension or a tolerance, but it may be a material, heat treatment, or surface finish applied to the part. This aspect of design is much more analytical than conceptual or configuration design. The objective of parametric design is to set values for the design variables that will produce the best possible design considering both performance and cost (as manifested by manufacturability).

Making the distinction between configuration design and parametric design is of fairly recent origin. It has grown out of massive efforts by industry to improve the quality of their products, chiefly by improving robustness. *Robustness* means achieving excellent performance under the wide range of conditions that will be found in service. All products function reasonably well under ideal (laboratory) conditions, but robust designs continue to function well when the conditions to which they are exposed are far from ideal.

8.6.1 Systematic Steps in Parametric Design

A systematic parametric design takes place in five steps:[1]

> *Step 1. Formulate the parametric design problem.* The designer should have a clear understanding of the function or functions that the component to be designed must deliver. This information should be traceable back to the PDS and the product architecture. Table 8.1 gives suggestions in this respect, but the product design specification (PDS) should be the guiding document. From this information we select the engineering characteristics that measure the predicted performance of the function. These *solution evaluation parameters* (SEPs) are often metrics like cost, weight, efficiency, safety, and reliability.
>
> Next we identify the design variables. The *design variables* (DVs) are the parameters under the control of the designer that determine the performance of the component. Design variables most influence the dimensions, tolerances, or choice of materials for the component. The design variables should be identified with variable name, symbol, units, and upper and lower limits for the variable.

---

1. J. R. Dixon and C. Poli, op. cit., Chap. 17; R. J. Eggert, *Engineering Design,* Pearson/Prentice Hall, Upper Saddle River, NJ, 2005, pp. 183–99.

Also, we make sure we understand and record the *problem definition parameters* (PDPs). These are the operational or environmental conditions under which the component or system must operate. Examples are loads, flow rate, and temperature increase.

Finally, we develop a *Plan for Solving the Problem*. This will involve some kind of analysis for stresses, or vibration, or heat transfer. Engineering analysis encompasses quite a spectrum of methods. These range from the educated guess by a very smart and experienced engineer to a very complex finite element analysis that couples stress analysis, fluid flow, and heat transfer. In conceptual design you used elementary physics and chemistry, and a "gut feel" for whether the concept would work. In configuration design you used simple models from engineering science courses, but in parametric design you will most likely use more detailed models, including finite-element analysis on critical components. The deciding factors for the level of detail in analysis will be the time, money, and available analysis tools, and whether, given these constraints, the expected results are likely to have sufficient credibility and usefulness. Often there are too many design variables to be comfortable with using an analytical model, and a full-scale proof test is called for. Testing of designs is discussed in Sec. 8.11.5.

Step 2. Generate alternative designs. Different values for the design variables are chosen to produce different candidate designs. Remember, the alternative configurations were narrowed down to a single selection in configuration design. Now, we are determining the best dimensions or tolerances for the critical-to-quality aspects of that configuration. The values of the DVs come from your or the company's experience, or from industry standards or practice.

Step 3. Analyze the alternative designs. Now we predict the performance of each of the alternative designs using either analytical or experimental methods. Each of the designs is checked to see that it satisfies every performance constraint and expectation. These designs are identified as *feasible designs*.

Step 4. Evaluate the results of the analyses. All the feasible designs are evaluated to determine which one is best using the solution evaluation parameters. Often, a key performance characteristic is chosen as an *objective function*, and optimization methods are used to either maximize or minimize this value. Alternatively, design variables are combined in some reasonable way to give a *figure of merit*, and this value is used for deciding on the best design. Note that often we must move back and forth between analysis and evaluation, as is seen in the example in Sec. 8.6.2.

Step 5. Refine/Optimize. If none of the candidate designs are feasible designs, then it is necessary to determine a new set of designs. If feasible designs exist, it may be possible to improve their rating by changing the values of the design variables in an organized way so as to maximize or minimize the objective function. This involves the important topic of design optimization discussed in Chap. 15.

It is worthwhile to note that the process followed in parametric design is the same as followed in the overall product design, but it is done with a narrower scope. This is evidence of the recursive nature of the design process.

8.6.2 A Parametric Design Example: Helical Coil Compression Spring

Design Problem Formulation

Figure 8.15 shows a brake for an electric hoist that is actuated by a helical coil compression spring.[1] The brake must provide 850 ft-lb stopping torque. Given the geometry of the brake drum and the frictional characteristics of the brake shoes, it was determined that the required compressive force applied by the spring should be $P = 716 \pm 34$ lb. The design of the spring should allow for 1/8 in. break pad wear, and the brake shoe must clear the drum by an additional 1/8 in.

The problem situation describes a service environment that is essentially static loading. However, as the brake pad wears it will cause a change in the length of the spring. Assuming a maximum wear of 1/8 in. before the pad is replaced, the change in spring deflection will be $\Delta\delta = [(9.5 + 8.5)/8.5] \times 0.125 = 0.265$ in. Also, during brake wear the allowable change in spring force is $\Delta P = (716 + 34) - (716 - 34) = 68$ lb. Therefore, the required spring constant (spring rate) is $k = \Delta P/\Delta\delta = 68/0.265 = 256$ lb/in. In addition, the brake shoe must clear the drum by 1/8 in. This causes the spring to be compressed by an additional 0.265 in. Therefore, the force the spring must deliver is $P = (716 + 34) + 256 \times 0.265 = 820$ lb.

The geometrical constraints on the spring are as follows: The ID of the spring must fit readily over the 2-inch-diameter tie rod. To allow for this we will use a 10% clearance. There is ample space for the free length of the spring, and the compressed length is not critical so long as the coils of the spring do not close on themselves to produce "solid stacking." The OD of the spring could be as large as 5 inches.

Solution Evaluation Parameters

The metrics that determine whether the design is performing its intended function are listed in Table 8.2.

Limits are placed on the spring geometry by the spring index $C = D/d$. Since this is a static loading situation with normal temperature and corrosive conditions, a low safety factor is called for. For more on safety factors, see Chap. 14. Also, since the load is essentially static, we are not concerned with designs for alternating fatigue stresses, or for resonance conditions due to vibrations. The expected failure mode is gross yielding at the inside surface of the spring (see the following discussion of stresses).

Design Variables

We define the following design variables:

Geometry of spring: d, wire diameter; *D*, mean coil diameter. *D* is measured from the center of one coil to the center of the opposite coil along a plane normal to the spring axis; see Fig. 8.16*a*.
C, spring index: C = D/d. The spring index typically ranges from 5 to 12. Below $C = 5$ the spring will be difficult to make because of the large diameter wire. C above 12 means the wire diameter is small, so the springs tend to buckle in service or tangle together when placed in a bin.

---

1. D. J. Myatt, *Machine Design*, McGraw-Hill, New York, 1962, pp. 181–185.

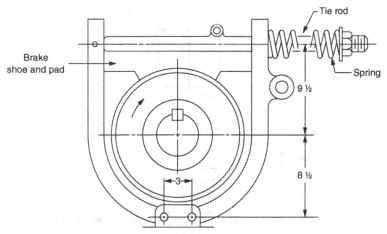

FIGURE 8.15
Drawing of hoist brake, showing brake block (at top) bearing on hoist drum.

TABLE 8.2
Solution Evaluation Parameters for the Helical Coil Spring

Parameter	Symbol	Units	Lower Limit	Upper Limit
Spring force	P	lb	820	
Spring deflection	δ	in.		0.265
Spring index	C		5	12
Spring inside diameter	ID	in.	2.20	2.50
Safety factor	FS		1.2	

Outside spring diameter: $OD = d\,(C + 1) = D + d$
Inside diameter of spring: $ID = d\,(C - 1) = D - d$

N, number of active turns: Active coils are those that are effective in giving "spring action." Depending on the modifications in the geometry at the ends of the spring made to ensure the spring seats squarely, there can be between 0.5 and 2 inactive turns that do not participate in the spring action.

N_t, total number of turns: It equals the sum of the active and inactive turns.

L_s, the solid length of the spring: It equals total turns times wire diameter. This is the length (height) when the coils are compressed tight.

K_w, the Wahl factor: It corrects the torsional shear stress in the wire for a transverse shear stress induced by the axial stress and the curvature of coils. See Eq. (8.5).

Spring wire material: Because the service conditions are mild we will limit consideration to ordinary hard-drawn steel spring wire. This is the least costly spring wire material.

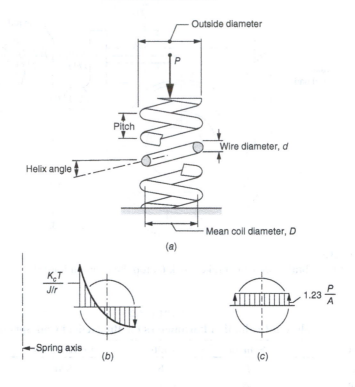

FIGURE 8.16
(a) Details of the spring. (b) Torsional stress distribution. (c) Transverse shear stress distribution.

Plan for Solving the Problem

The constraints imposed by the design are given in the design problem formulation. We have selected the least expensive material, and we will upgrade if the design requires it. We will start by making an initial selection of wire diameter based on $C = 7$, near the mid-range of allowable values of C. We will check that the constraints are not violated, particularly that the spring load falls within the required limits. The initial design criterion will be that the yield strength of the spring wire is not exceeded at the critical failure site. This will constitute a feasible design. Then we will check that the spring is not compressed to its solid height or in danger of buckling. The design goal will be to minimize the mass (cost) of the spring within all of these design constraints.

Generate Alternative Designs Through Analysis

This analysis follows that in standard machine design texts.[1] Figure 8.16 shows the stresses developed in a helical spring loaded axially in compression. They consist of both a torsional shear stress and a transverse shear stress.

1. J. E. Shigley and C. R. Mishke, *Mechanical Engineering Design,* 6th ed., McGraw-Hill, New York; J. A. Collins, *Mechanical Design of Machine Elements and Machines,* John Wiley & Sons, New York, 2003.

The primary stress is produced by a torsional moment $T = PD/2$, which produces a torsional shear stress on the outer fiber of the wire.

$$\tau_{torsion} = \frac{Tr}{J} = \frac{(PD/2)\dfrac{d}{2}}{\dfrac{\pi d^4}{32}} = \frac{8PD}{\pi d^3} \tag{8.1}$$

In addition, a transverse shear stress is induced in the wire by the axial stress. This shearing stress reaches a maximum value at the mid-height of the wire cross section, with a magnitude given by[1]

$$\tau_{transverse} = 1.23 \frac{P}{A_{wire}} \tag{8.2}$$

Also, because of the curvature of the coils in the spring, a slightly larger shearing strain is produced by the torsion at the inner fiber of the coil than at the outer fiber. This curvature factor, K_C, is given by

$$K_C = \frac{4C-1}{4C-4} \quad \text{where } C \text{ is the spring index} \tag{8.3}$$

Therefore, the critical failure site is the mid-height of the wire on the *inner* coil radius. Because the two shear stresses are in alignment at the inner surface, we can add them to find the maximum shearing stress.

$$\tau_{max} = \left(\frac{4C-1}{4C-4}\right)\left(\frac{8PD}{\pi d^3}\right) + 1.23\frac{P}{A} \quad \text{which can be rewritten as} \tag{8.4}$$

$$\tau_{max} = \left(\frac{4C-1}{4C-4}\right)\left(\frac{8PD}{\pi d^3}\right) + 1.23\left(\frac{P}{\pi d^2/4}\right)\left(\frac{D}{D}\right)\left(\frac{d}{d}\right) = \left[\frac{4C-1}{4C-4} + \frac{0.615}{C}\right]\left(\frac{8PD}{\pi d^3}\right) \tag{8.5}$$

The term in brackets is called the Wahl factor, K_w. Thus, the max shear stress can be written

$$\tau_{max} = \frac{8PD}{\pi d^3} \times K_w \tag{8.6}$$

Solution Plan

1. Initial Criterion: Provide for the maximum load without yielding.

   Finding the design parameters for a spring is an inherently iterative process. We start by selecting $C = 7$ and $D = 5$ in. Therefore, our first trial will be using a wire of diameter $d = D/C = 5/7 = 0.714$ in. We use Eq. (8.6) to determine if a

---

1. A. M. Wahl, "Mechanical Springs," McGraw-Hill, New York, 1963.

TABLE 8.3
Load at Yielding Calculated from Equation (8.6)

Iteration	C	D	d	K_w	P
1	10	5.0	0.5	1.145	572
2	7	3.5	0.5	1.213	771
3	6	3.0	0.5	1.253	870

hard-drawn steel spring wire is strong enough to prevent yielding at the failure site. This steel is covered by ASTM Standard A227. A227 steel is a high-carbon, plain carbon steel that is sold in the drawn condition. The ultimate tensile strength is the mechanical property most readily available for spring steel, but to use Eq. (8.6) we need to know a typical value for the *yield strength in shear*. Also, because of the process that is used to make wire, the value of the strength decreases with increasing size of the wire.

Fortunately, machine design texts give data on properties of spring wire as a function of wire diameter, but most data do not extend much beyond 0.6 in. diameter. Therefore, as a first compromise, we will try a wire with $d = 0.5$ in. giving $C = 10$. Also, 0.50 is the upper limit for commercially available hard-drawn wire. An empirical equation giving the tensile strength, S_u versus wire diameter is, $S_u = 140d^{-0.190}$, which gives a value of 160 ksi.[1] The same reference also tells us that for this steel, torsional yield stress is 50% of the ultimate tensile strength, which is 80,000 psi. But we have decided to use a factor of safety of 1.2, so the allowable stress that cannot be exceeded by Eq. (8.6) is 80,000/1.2 = 66,666 psi.

We can now use Eq. (8.6) to solve for the allowable compressive load on the spring. From Table 8.2 we see that the spring must be able to carry a load of 820 lb without yielding. Table 8.3 shows the results of the first three iterations. Note that in three iterations we have found a *feasible design* based on the load-carrying capacity of the spring, $C = 6, D = 3.0, d = 0.5$.

2. Second Criterion: Deformation of the spring.

The deformation of the spring from force P is given by:

$$\delta = \frac{PD^2L}{4JG} = \frac{PD^2(\pi DN)}{4\left(\dfrac{\pi}{32}d^4\right)G} = \frac{8D^3PN}{d^4G} \qquad (8.7)$$

where $L = \pi DN$ is the active spring length.

G is the elastic modulus in shear, equal to 11.5×10^6 lb/in.[2] for hard-drawn spring wire. Solving for the number of active coils, N:

$$N = \frac{d^4G\delta}{8D^3P} = \frac{d^4G}{8D^3k} \quad \text{where } k \text{ is the spring constant, } \frac{1}{k} = \frac{\delta}{P} \qquad (8.8)$$

---

1. J. E. Shigley and C. R. Mischke, op. cit., p. 600.

Substituting into Eq. (8.8) to find the number of free coils:

$$N = \frac{(0.5)^4 (11.5 \times 10^6)}{8(3.0)^3 \times 256} = 13 \text{ coils} \tag{8.9}$$

We decide that the spring requires squared ends to facilitate axial loading. This requires two inactive coils, so the total number of coils in the spring is $N_t = N + N_i = 13 + 2 = 15$

The *solid height,* when the coils are closed tight on each other, is given by

$$L_s = N_t d = 15(0.5) = 7.5 \text{ in.} \tag{8.10}$$

To ensure that the spring will operate in the linear portion of the $P - \delta$ curve, we add 10% to the solid height. This is often called a "clash allowance" and the length at this condition is called the load height, $L_P = 1.10 \, (7.5) = 8.25$ in. Next we determine the amount the spring deflects from its original length to reach the maximum load of 820 lb. $\delta_p = P/k = 820/256 = 3.20$. If we add this length to the load height we have the original length of the spring in the unloaded condition. This length is the *free length* of the spring.

$$L_f = L_P + \delta_p = 8.25 + 3.20 = 11.45 \text{ in.} \tag{8.11}$$

3. Third Criterion: Buckling under the compressive load.

We have provided for square ends to assist with maintaining an axial load. We could go to the more expensive ground ends if buckling is a problem. Collins[1] presents a plot of critical deflection ratio, δ/L_f versus slenderness ratio, L_f/D. For the spring designed in the third iteration, these values are:
Critical deflection ratio: $3.20/11.45 = 0.28$
Slenderness ratio: $11.45/3.00 = 3.82$
For a fixed-end spring, this is well within the stable region. If the ends were to slide for some reason, it would place the spring close to the region of buckling, but because the rod used to apply the force through the spring goes through the center of the spring, it will also serve as a guide rod to minimize buckling.

Specification of the Design

We have found a feasible design for a helical compression spring with the following specification:

Material: ASTM 227 hard-drawn spring wire
Wire diameter, d: 0.500 This is a standard wire size for ASTM 227 spring steel.
Outside diameter, OD: $OD = D + d = 3.00 + 0.50 = 3.50$ in.
Inside diameter, ID: $ID = D - d = 3.00 - 0.50 = 2.50$ in.

---

1. J. A. Collins, op. cit., p. 528.

Spring ratio, C: 6
Clearance between ID and tie rod: $(2.5 - 2.0)/2 = 0.25$ in.
Maximum load to produce yielding with SF = 1.2: 870 lb
Number of coils, N_t: 15(13 active coils and 2 inactive coils due to squared ends)
Free length, L_f: 11.45 in.
Solid height, L_s: 7.5 in.
Compressed length at maximum load, L_p: 8.25
Spring constant(spring rate), k: 256 lb/in.
Critical deflection ratio: 0.28
Slenderness ratio: 3.82

Refinement

Although we have found a feasible design, it may not be the best design that could be achieved for the problem conditions. We have kept the wire diameter constant in finding this design. By changing this and other design variables we might be able to create a better design. An obvious criterion for evaluating further designs is the cost of a spring. A good surrogate for the cost is the mass of the spring, since within a class of springs and spring materials, the cost will be directly proportional to the amount of material used in the spring. The mass of a spring is given by

$$m = (density)(volume) = \rho \frac{\pi d^2 L}{4} = \rho \frac{\pi d^2}{4} \frac{\pi D N_t}{} = \rho \frac{\pi^2}{4} C d^3 N_t \qquad (8.12)$$

Since the first two terms in Eq. (8.12) are common to all spring steels, we can define a *figure of merit*, f.o.m., for evaluating alternative spring designs as $Cd^3 N_t$. Note that in this situation, smaller values of f.o.m. are preferred. Eq. (8.12) suggests that lower mass (cost) springs will be found with smaller diameter wire.

Equation (8.6) can be written as $P = \dfrac{\tau_{max} \pi d^2}{8 C K_w}$ $\qquad (8.13)$

If we decide that we shall continue to use the least expensive spring wire, ASTM A227, then Eq. (8.13) becomes

$$P = \frac{26,189 d^2}{C K_w} \qquad (8.14)$$

Since K_w does not vary much, Eq. (8.14) indicates that the highest load-carrying capacity springs will be found with large-diameter wires and low values of spring index C. There is a trade-off between d and load capacity, Eq. (8.13), and cost, Eq. (8.12). However, reducing C is beneficial in both instances. As noted previously, there is a manufacturing limitation in drawing wire larger than 0.5 in. without incurring extra costs, and C can only vary from about 4 or 5 to 12 for reasons discussed earlier.

TABLE 8.4
Maximum Applied Load (Limited by Yielding) and Relative Cost

Iteration	C	D	d	$ID = D - d$	K_w	P	N_t	f.o.m.
1	10	5.00	0.5	4.5	1.145	572 lb	16	20
2	7	3.5	0.5	3.0	1.213	771	15	13.13
3	6	3.00	0.5	2.5	1.235	870	15	11.25
4	4	2.00	0.5	1.5	Not feasible based on ID constraint			
5	7	2.80	0.4	2.40	1.213	493	9	4.03
6	6.5	2.60	0.40	2.20	1.231	523	10	4.16
7	6.03	2.637	0.437	2.20	1.251	663	13	6.54

Also, we have a constraint on the inside diameter of the spring. Therefore, we reduce the clearance for the 2-inch diameter tie rod from 0.25 to 0.10 in. Clearance = $(ID - 2.0)/2 = 0.10$ in. The minimum ID can now be as small as 2.20 in. Alternatively, we could keep the original clearance and reduce the diameter of the tie rod. This shows how modest changes in specification can often lead to improved designs.

Table 8.4 shows the design variables and problem definition parameters for the spring design for the variations of C in the design to this point.

As previously stated, the first feasible design was found in iteration 3. It was the only one that could sustain the required 820 lb load without yielding. However, the choice of D and d resulted in a rather large spring with 15 coils at 3.5 OD. Because of this the relative cost is high. We then reduced d to 0.40 in., and as expected the relative cost decreased substantially, and although P was increasing nicely with decreasing C we soon ran into the constraint on the ID of the spring. In iteration 7 we selected a standard wire size between 0.4 and 0.5, to see whether this would be a good compromise.

It is clear that the constraint on the ID limits how far we can raise the load capacity. Iteration 7 is as far as we can go with a wire diameter less than 0.5 in. We are approaching the target of 820 lb, but we are still not there.

Realizing now the extent of the constraint imposed by the ID criterion, it is now worth removing the design restriction on using only hard-drawn spring steel. Let us now see whether the increased cost of wire, with higher yield strength, would be offset by the ability to reach the required load with a smaller diameter wire, resulting in a spring that is less costly than the one given by iteration 3 in Table 8.4.

The class of steel spring wire that is next stronger than hard-drawn wire is oil quenched and tempered wire, ASTM Standard A229. A standard machine design text[1] gives its tensile strength as a function of wire diameter as $S_u = 147d^{-0.187} = 174$ ksi. The yield strength in shear is 70% of the ultimate tensile strength, whereas the yield strength in shear was 0.5 ultimate tensile strength for the hard-drawn spring wire. Thus, $\tau_{max} = 121.8$ ksi, and applying the safety factor of 1.2, the working value of τ_{max} is 100 ksi. Using Eq. (8.13) with the new value of $\tau_{max} = 100/66.7 = 1.5$ times larger raises the calculated values of P in Table 8.4 by 50%. The cost of A229 is

---

1. J. E. Shigley and C. R. Mischke, op. cit., p. 600.

TABLE 8.5
Maximum Applied Load (Limited by Yielding) and
Relative Cost (f.o.m.) for Quenched and Tempered Steel Spring Wire

Iteration	C	D	d	ID = $D - d$	K_w	P	N_t	f.o.m.
8	8.03	2.512	0.312	2.20	1.183	401 lb		
9	6.42	2.606	0.406	2.20	1.236	815	20	8.59
10	6.03	2.637	0.437	2.20	1.251	994	13	6.55

given as 1.3 times the cost of A227 spring wire. This opens up new opportunities to find design parameters that satisfy the load conditions but have lower costs than iteration 3.

We first go for a large reduction in wire diameter, to a standard size of 0.312 in., iteration 8, Table 8.5. However, even with a 50% increase in wire strength, this size wire will support only 401 lbs before yielding. Therefore, we return to wire diameters greater than 0.40, and select the smallest standard wire diameter in this range, 0.406 in. (iteration 9). This results in a load-carrying capacity of 815 lb, only 0.6 % less than the 820 lb requirement. The next standard wire size, 0.437, gives a load-carrying capacity of 994 lb. This is well above the load-carrying requirement, and even including the 30% increase in cost in the figure of merit, the relative cost is less than the previous feasible design, iteration 3. Table 8.5 records these results.

Design 10 is an attractive alternative to Design 3 because it offers the possibility of significant reduction in cost. It will need to be explored in greater detail by first checking on the buckling of the spring and other spring parameters such as solid height and free length. Then the cost estimate needs to be verified by getting quotations from possible suppliers.

8.6.3 Design for Manufacture (DFM) and Design for Assembly (DFA)

It is imperative that during embodiment design decisions concerning shape, dimensions, and tolerances be closely integrated with manufacturing and assembly decisions. Often this is achieved by having a member of the manufacturing staff as part of the design team. Since this is not always possible, all design engineers need to be familiar with manufacturing and assembly methods. To assist in this, generalized DFM and DFA guidelines have been developed, and many companies have specific guidelines in their design manuals. Design software, to aid in this task, has been developed and is being used more widely. Chapter 13 deals with DFM and DFA in considerable detail, and should be consulted during your embodiment design activities.

The reason for the strong emphasis on DFM/DFA is the realization by U.S. manufacturers in the 1980s that manufacturing needs to be linked with design to produce quality and cost-effective designs. Prior to this time there was often a separation between the design and manufacturing functions in manufacturing companies. These disparate cultures can be seen by the statement, often made in jest by the design engineers, "we finished the design and threw it over the wall for the manufacturing engineers to do with it what they will." Today, there is recognition that integration of these functions is the only way to go.[1]

8.6.4 Failure Modes and Effects Analysis (FMEA)

A *failure* is any aspect of the design or manufacturing process that renders a component, assembly, or system incapable of performing its intended function. FMEA is a methodology for determining all possible ways that components can fail and establishing the effect of failure on the system. FMEA analysis is routinely performed during embodiment design. To learn more about FMEA, see Sec. 14.5.

8.6.5 Design for Reliability and Safety

Reliability is a measurement of the ability of a component or system to operate without interruption of service or failure in the service environment. It is expressed as the probability of the component functioning for a given time without failure. Chapter 14 gives considerable detail on methods for predicting and improving reliability. *Durability* is the amount of use that a person gets out of a product before it deteriorates—that is, it is a measure of the product lifetime. While durability, like reliability, is measured by failure, it is a much more general concept than reliability, which is a technical concept using probabilities and advanced statistical modeling. However, it is more likely to be able to estimate product lifetime than reliability.

Safety involves designing products that will not injure people or damage property. A safe design is one that instills confidence in the customer and does not incur product liability costs. To develop a safe design one must first identify the potential hazard, and then produce a design that keeps the user free from the hazards. Developing safe designs often requires trade-offs between safe design and wanted functions. Details of design for safety can be found in Sec. 14.8.

8.6.6 Design for Quality and Robustness

Achieving a quality design places great emphasis on understanding the needs and wants of the customer, but there is much more to it than that. In the 1980s there was the realization that the only way to ensure quality products is to *design quality into the product*, as opposed to the then-current thinking that quality products were produced by careful inspection of the output of the manufacturing process. Other contributions to design from the quality movement are the simple total quality

---

1. In fact, in Japan, which has been recognized as a leader in manufacturing and product design, it is common for all university engineering graduates taking employment with a manufacturing company to start their careers on the shop floor.

management tools, presented in Chap. 4, that can be quickly learned and used to simplify team understanding of various issues in the design process, and QFD, in Chap. 6, for aligning the needs of the customer with the design variables. Another important tie between quality and design is the use of statistics to set the limits on tolerances in design and the relationship to the capability of a manufacturing process to achieve a specified quality (defect) level. These topics are discussed in detail in Chap. 15.

A *robust design* is one whose performance is insensitive to variations in the manufacturing processes by which it has been made or in the environment in which it operates. It is a basic tenet of quality that variations of all kinds are the enemy of quality, and a guiding principle to achieving quality is to reduce variation. The methods used to achieve robustness are termed *robust design*. These are basically the work of a Japanese engineer, Genichi Taguchi, and his co-workers, and have been adopted by manufacturing companies worldwide. They employ a set of statistically designed experiments by which alternative designs are generated and analyzed for their sensitivity to variation. The parametric design step is the place where design for robustness methods are applied to critical-to-quality parameters. Methods for robust design, especially Taguchi's methods, are presented in Chap. 15.

8.7
DIMENSIONS AND TOLERANCES

Dimensions are used on engineering drawings to specify size, location, and orientation of features of components. Since the objective of product design is to market a profitable product, the design must be manufactured and to make that product the design must be described in detail with engineering drawings. Dimensions are as important as the geometric information that is conveyed by the drawing. Each drawing must contain the following information:

- The size of each feature
- The relative position between features
- The required precision (tolerance) of sizing and positioning features
- The type of material, and how it should be processed to obtain its expected mechanical properties

A *tolerance* is the acceptable variation in the dimension. Tolerances must be placed on a dimension or geometric feature of a part to limit the permissible variations in size because it is impossible to repeatedly manufacture a part exactly to a given dimension. A small (tight) tolerance results in greater ease of interchangeability of parts and improved functioning. Tighter tolerances result in less play or chance for vibration in moving parts. However, smaller (tighter) tolerances are achieved at an increased cost of manufacture. Larger (looser) tolerances reduce the cost of manufacture and make it easier to assemble components, but often at the expense of poorer system performance. An important responsibility of the designer is to make an intelligent choice of tolerances considering the trade-off between cost and performance.

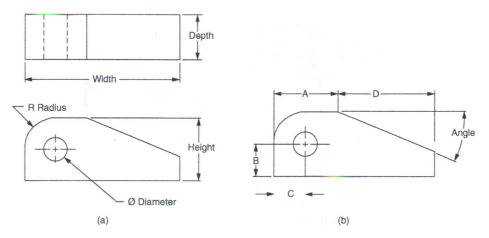

FIGURE 8.17
(a) Proper way to give dimensions for size and features; (b) proper way to give dimensions for location and orientation of features.

8.7.1 Dimensions

The dimensions on an engineering drawing must clearly indicate the size, location, and orientation of all features in each part. Standards for dimensioning have been published by the American Society of Mechanical Engineers (ASME).[1]

Figure 8.17a shows that the overall dimensions of the part are given. This information is important in deciding how to manufacture the part, since it gives the size and weight of the material needed for making the part. Next, the dimensions of the features are given: the radius of the corner indicated by R and the diameter of the hole by a circle with forward leaning slash through it, ∅. In Fig. 8.17b the centerline of the hole is given by dimensions B and C. A and D are the horizontal position dimensions that locate the beginning of the sloping angle. The orientation dimension of the sloping portion of the part is given by the angle dimension measured from the horizontal reference line extending out from the top of the part.

Section views, drawings made as if a portion of the part were cut away, are useful to display features that are hidden inside the part. A section view in Fig. 8.18 presents a clear understanding of the designer's intent so that an unequivocal message is sent to the machine operator who will make the part. Section views are also useful in specifying position dimensions.

Figure 8.19 illustrates the importance of removing redundant and unnecessary dimensions from chained dimensions on a drawing. Since the overall dimensions are given, it is not necessary to give the last position dimension. With all four position

1. ASME Standard Y14.5 2009; P. J. Drake Jr., *Dimensioning and Tolerancing Handbook,* McGraw-Hill, New York, 1999. (Available online at knovel.com.)

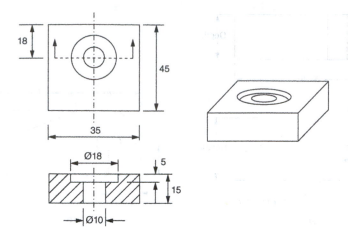

FIGURE 8.18

Use of section view to clarify dimensioning of internal features.

(Courtesy of Professor Guangming Zhang, University of Maryland.)

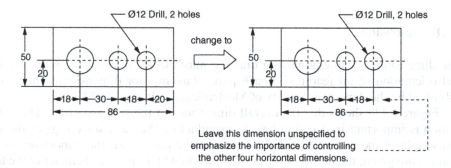

FIGURE 8.19

Elimination of redundant dimension.

(Courtesy of Professor Guangming Zhang, University of Maryland.)

dimensions given, the part is overconstrained because of overlap of tolerances. Fig. 8.19 also illustrates the good practice of laying out the overall part dimensions from a common datum reference, in this case datum planes in the x and y directions that intersect at the lower left corner of the part.

8.7.2 Tolerances

A *tolerance* is the permissible variation from the specified dimension. The designer must decide how much variation is allowable from the basic dimension of the component to accomplish the desired function. The design objective is to make the tolerance no tighter than necessary, since smaller tolerances increase manufacturing cost and make assembly more difficult.

The tolerance on a part is the difference between the upper and lower allowable limits of a *basic size* dimension. Note that so long as the dimension falls within the tolerance limits the part is acceptable and "in spec." The *basic size* is the theoretical dimension, often a calculated size, for a component. As a general rule, the basic size of a hole is its minimum diameter, while the basic size for its mating shaft is the maximum diameter. Basic size is not necessarily the same as *nominal size*. For example, a ½ in. bolt has a nominal diameter of ½ inch, but its basic size may be different, for example, 0.492 in. The American National Standards Institute (ANSI) gives tables of "preferred" basic sizes which can be found in all machine component design books and handbooks. The object of a preferred series of basic sizes is to make possible the use of standard components and tools.[1]

Tolerances may be expressed in several ways.

- *Bilateral tolerance.* The variation occurs in both directions from the basic dimension. That is, the upper limit exceeds the basic value and the lower limit falls below it.

  - Balanced bilateral tolerance: The variation is equally distributed around the basic dimension: 2.500 ± 0.005. This is the most common way of specifying tolerances. Alternatively, the limits of allowable variation may be given: $\dfrac{2.505}{2.495}$

  - Unbalanced bilateral tolerance: The variation is not equal around the basic dimension: $2.500^{+0.070}_{-0.030}$

- *Unilateral tolerance:* The basic dimension is taken as one of the limits, and variation is in only one direction: $2.500^{+0.000}_{-0.010}$

Each manufacturing process has an inherent ability to maintain a certain range of tolerances, and to produce a certain surface roughness (finish). To achieve tolerances outside of the normal range requires special processing that typically results in an exponential increase in the manufacturing cost. For further details refer to Sec. 13.4.5. Thus, the establishment of the needed tolerances in embodiment design has an important influence on the choice of manufacturing processes and the cost. Fortunately, not all dimensions of a part require tight tolerances. Typically those related to critical-to-quality functions require tight tolerances. The tolerances for the noncritical dimensions should be set at values typical for the process used to make the part.

An engineering drawing must indicate the required tolerance for all dimensions. Usually, only the critical dimensions have labeled tolerances. The other dimensions gain their tolerance from a general (default) tolerance statement like "All dimensions have a tolerance of ±0.010 unless otherwise specified." Often this information is given in the title block of the drawing.

---

1. It would be ridiculous if a machine shop had to keep in its tool room every decimal size drill in increments of 0.001 in. Using standard sizes keeps this to a manageable number.

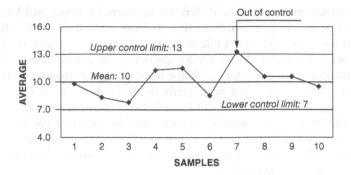

FIGURE 8.20
Quality control chart based on sampling the diameter of shafts.

A second use for tolerance information is to set the upper and lower limits for quality control of the manufacturing process. Figure 8.20 shows a *quality control chart* for the machining of shafts on a CNC lathe. Every hour the diameter of four samples made that hour is measured and their average is plotted on the chart. The upper and lower control limits are based on the tolerances adjusted by a statistically relevant multiplier. When a sample mean exceeds one of the control limits it tells the operator that something is out of control with the process, perhaps a worn cutting tool, and that an adjustment must be made. This procedure helps to produce a product having minimum variability, but it is never a substitute for designing quality into the product with robust design, see Chap. 15. For more information on quality control, see Sec. 15.3.

There are generally two classes of issues in parametric design associated with tolerances on parts when they must be assembled together. The first deals with *fit,* how closely the tolerances should be held when two components fit together in an assembly. The second is *tolerance stackup*, the situation where several parts must be assembled together and interference occurs because the tolerances of the individual parts overlap.

Fit

A typical mechanical assembly where fit is of concern is a shaft running in a bearing or a piston sliding in a cylinder. The fit between the shaft and the bearing, as expressed by the *clearance,* is important to the functioning of the machine. Figure 8.21 illustrates the situation.

The *clearance* for the fit is the distance between the shaft and the inner race of the bearing. Because of the tolerances on the components, this will have an upper limit (when the bearing ID is at a maximum and the shaft OD is at a minimum) and a lower limit (when the bearing ID is at a minimum and the shaft OD is at a maximum limit). From Fig. 8.21:

$$\text{Maximum clearance} = A_{max} - B_{min} = 30.40 - 29.70 = 0.70 \text{ mm}$$
$$\text{Minimum clearance} = A_{min} - B_{max} = 30.00 - 29.80 = 0.20 \text{ mm}$$

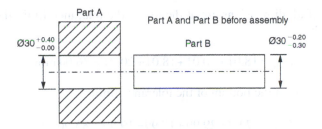

FIGURE 8.21
The bearing (Part A) and the shaft (Part B) before assembly.

Since tolerance is the permissible difference between maximum and minimum limits of size, the tolerance of the shaft/bearing assembly is 0.70 − 0.20 = 0.50 mm.

There are three zones of tolerance when dealing with fits.

- *Clearance fits.* As shown above, both the maximum and minimum clearances are positive. These fits always provide a positive clearance and allow for free rotation or sliding. ANSI has established nine classes of clearance fits, ranging from close sliding fits that assemble without perceptible play (RC 1) to loose running fits (RC 9).
- *Interference fits.* In this category of fits, the shaft diameter is always larger than the hole diameter, so that both the maximum and minimum clearance are negative. Such fits can be assembled by heating the outer body and/or cooling the shaft, or by press fitting. They provide a very rigid assembly. There are five ANSI classes of interference fits, ranging from FN 1, light drive fits, to FN 5, heavy shrink fits.
- *Transition fits.* In this category of fits the maximum clearance is positive and the minimum clearance is negative. Transition fits provide accurate location with either slight clearance or slight interference. ANSI class LC, LT, and LN fits apply in this case.

Another way of stating clearance fit is to give the *allowance*. Allowance is the tightest possible fit between two mating parts, that is, the minimum clearance or the maximum interference.

Stackup

Tolerance stackup occurs when two or more parts must be assembled in contact. Stackup occurs from the cumulative effects of multiple tolerances. This is called a stackup because as the dimensions and their tolerances are added together they "stack up" to add to the possible total variation. A stackup analysis typically is used to properly tolerance a dimension that has not been given a tolerance or to find the limits on a clearance (or interference) gap. Such an analysis allows us to determine the maximum possible variation between two features on a single component or between components in an assembly.

Refer to the drawing on the left side of Fig. 8.19. Assume that the tolerance on each dimension giving the location of the holes along the *x*-axis is ±0.01 mm. Then the dimensions from left to right would be A = 18 ± 0.01, B = 30 ± 0.01, C = 18 ± 0.01,

$D = 20 \pm 0.01$. If all dimensions are at the top of the tolerance limit, then the overall length is given by:

$$L_{max} = 18.01 + 30.01 + 18.01 + 20.01 = 86.04 \, mm$$

If all dimensions are at the bottom of the tolerance limit:

$$L_{min} = 17.99 + 29.99 + 17.99 + 19.99 = 85.96 \, mm$$

The tolerance on the overall length is $T_L = L_{max} - L_{min} = 86.04 - 85.96 = 0.08$ and $L = 86 \pm 0.04$ mm. We see that the tolerances "stack up", that is, they add together. The tolerance on the chain (assembly) of dimensions is

$$T_{assembly} = T_A + T_B + T_C + T_D = 0.02 + 0.02 + 0.02 + 0.02 = \sum_i T_i \qquad (8.15)$$

We can now see why it is good practice to not give all of the dimensions in a chain; see the right side of Fig. 8.19. Suppose we set the tolerance on the length dimension, $L = 86 \pm 0.01$. We keep L fixed at its tolerance limits and find the limits on the dimension at the right end, D, while keeping the other three dimensions at their limits.

$$D_{min} = 85.99 - 18.01 - 30.01 - 18.01 = 85.99 - 66.03 = 19.96$$
$$D_{max} = 86.01 - 17.99 - 29.99 - 17.99 = 86.01 - 65.97 = 20.04$$
$$T_D = 20.04 - 19.96 = 0.08 \quad \text{and} \quad D = 20.00 \pm 0.04$$

The tolerance on D is four times the tolerance on the other hole locations.

Note that if we laid out the centerlines of the three holes, starting with a datum plane at the left and moving successively to the right, the tolerance stackup would not have been an issue.

if we define $L3 = A + B + C$, then $T_{L3} = 0.02 + 0.02 + 0.02 = 0.06$
and $T_D = T_L - T_{L3} = 0.08 - 0.06 = 0.02$

But if we laid out the first hole at the left, and then moved to the hole on the far right, we would have encountered stack up problems that would have required a change in the tolerance to achieve the design intent. Therefore, using a dimensioning scheme of referring all dimensions to a datum reference eliminates tolerance stackup and preserves design intent.

Worst-Case Tolerance Design

In the worst-case tolerance design scenario the assumption is made that the dimension of each component is at either its maximum or minimum limit of the tolerance. This is a very conservative assumption, for in reality when a manufacturing process is running in control many more of the components will be closer to the basic dimension than will be close to the limits of the tolerance. Figure 8.22 shows one way of systematically determining the tolerance stackup.

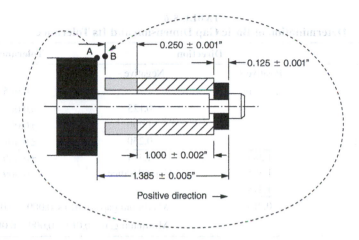

FIGURE 8.22
Finding tolerance stackup using a 2-D dimension chain.

EXAMPLE 8.1

Figure 8.22 shows an assembly consisting of a pin in a wall with a washer under its head and a sleeve and snap ring, going from right to left. Dimensions and tolerances are given on the sketch. Use worst-case tolerance design to find the mean gap **A-B** between the wall and the snap ring and the limits on the gap.

The steps for solving problems of this type are:[1]

1. Select the gap or dimension whose variation needs to be determined.
2. Label one end of the gap A and the other B.
3. Select a dimension that spans the gap to be analyzed. Establish the positive direction (usually to the right) and label it on the drawing.
4. Follow the chain of dimensions from point A to point B: see dashed line on Fig. 8.22. You should be able to follow a continuous path. For this example it is: wall to head of pin interface; right surface of washer to left surface of washer; right end of sleeve to left end of sleeve; right end of snap ring to point B; point B to point A.
5. Convert all dimensions and tolerances to equivalent balanced bilateral format, if they are not in this format already.
6. Set up Table 8.6, being careful to include all dimensions and their tolerances in the chain and paying attention to their direction.

Note that to use this method of tolerance analysis requires that the tolerance must be in balanced bilateral format. To make this conversion from unequal bilateral or unilateral, first find the limits of the tolerance range. For example, $8.500^{+0.030}_{-0.010} = 8.530 - 8.490 = 0.040$. Divide this tolerance range by 2 and add it to the lower limit to get the new basic dimension $8.490 + 0.020 = 8.510 \pm 0.020$.

---

1. B. R. Fischer, *Mechanical Tolerance Stackup and Analysis,* Chap. 7, Marcel Dekker, New York, 2004.

TABLE 8.6
Determination of Basic Gap Dimension and Its Tolerance

	Direction		Tolerance
	Positive +	Negative −	
Wall to washer	1.385 in.		±0.005
Across washer		0.125	±0.001
Across sleeve		1.000	±0.002
Across snap ring		0.250	±0.001
Totals	1.385	1.375	±0.009
Positive total	1.385	Gap tolerance	±0.009
Negative total	1.375		
Basic gap	0.010	Maximum gap = 0.010 + 0.009 = 0.019	
		Minimum gap = 0.010 − 0.009 = 0.001	

Statistical Tolerance Design

An important method used to determine assembly tolerances is based on statistical interchangeability. This approach assumes that a manufacturing process will more likely produce parts for which each dimension is a normal distribution with a mean μ and standard deviation σ. Thus, a very large percentage of the available parts are interchangeable. As a result, this approach results in larger allowable tolerances at the expense of having a small percentage of mating parts that cannot be assembled during the first attempt. The method is based on the following additional assumptions:

• The manufacturing process for making the components is in control, with no parts going outside of the statistical control limits. In effect, the basic manufacturing dimension is the same as the design basic dimension. This also requires that the center of the tolerance band coincides with the mean of the basic dimension produced by the production machine. For more on process capability, see Chap. 15.
• The dimensions of the components produced by the manufacturing process follow a normal or Gaussian frequency distribution.
• The components are randomly selected for the assembly process.
• The product manufacturing system must be able to accept that a small percentage of parts produced will not be able to be easily assembled into the product. This may require selective assembly, reworking, or scrapping these components.

The *process capability index*, C_p, is commonly used to express the relationship between the tolerance range specified for the component and the variability of the process that will make it. Variability is given by the standard deviation, σ, of a critical dimension that is produced by the process. It is also considered that the *natural tolerance limits* represent plus or minus three standard deviations from the mean of the distribution of the dimension. For a normal distribution, when design tolerance limits are set at the natural tolerance limits, 99.74% of all dimensions would fall within tolerance and 0.26% would be outside the limits; see Sec. 15.5 for more details. Thus,

$$C_p = \frac{\text{desired process spread}}{\text{actual process spread}} = \frac{\text{tolerance}}{3\sigma + 3\sigma} = \frac{USL - LSL}{6\sigma} \tag{8.16}$$

where USL and LSL are the upper and lower specification limits, respectively. A capable manufacturing process has a C_p at least equal to unity (1). Equation (8.16) provides a way to estimate what the tolerance should be based on the standard deviation of the parts coming off the production machine.

The relationship between the standard deviation of a dimension in an *assembly of components* and the standard deviation of the dimensions *in separate components* is

$$\sigma^2_{assembly} = \sum_{i=1}^{n} \sigma_i^2 \tag{8.17}$$

where n is the number of components in the assembly and σ_i is the standard deviation of each component. From Eq. (8.16), when $C_p = 1$, the tolerance is given by $T = 6\sigma$ and the tolerance on an assembly is

$$T_{assembly} = \sqrt{\sum_{i=1}^{n} T_i^2} \tag{8.18}$$

Because the tolerance of an assembly varies as the square root of the sum of the squares of the tolerance of the individual components, the statistical analysis of tolerances is often referred to as the root sum of the squares, RSS, method.

EXAMPLE 8.2

We can now apply these ideas to the tolerance design problem given in Fig. 8.22. We proceed in exactly the same way as in Example 8.1, determining a positive direction, and writing down the chain of dimensions and their tolerances. The only difference is that in the solution table, Table 8.7, we must add a column for the square of the tolerances.

TABLE 8.7
Determination of Gap and Its Tolerance Using Statistical Method

	Direction		Tolerance	(Tolerance)^2
	Positive +	Negative −		
Wall to washer	1.385 in.		±0.005	25×10^{-6}
Across washer		0.125	±0.001	1×10^{-6}
Across sleeve		1.000	±0.002	4×10^{-6}
Across snap ring		0.250	±0.001	1×10^{-6}
Totals	1.385 in.	1.375 in.	±0.009 in.	31×10^{-6}
Positive total	1.385			
Negative total	1.375			
Basic gap	0.010			

$$T_{assembly} = \sqrt{31 \times 10^{-6}} = 5.57 \times 10^{-3} = \pm 0.006 \text{ in.}$$

Maximum gap = $0.010 + 0.006 = 0.016$

Minimum gap = $0.010 - 0.006 = 0.004$

TABLE 8.8
Determination of Variation Contribution of Each Part in Assembly

Part	T	Tolerance range	σ	σ^2	% Contribution To Variation
Pin	±0.005	0.010	1.666×10^{-6}	2.777×10^{-6}	80.6
Washer	±0.001	0.002	0.333×10^{-6}	0.111×10^{-6}	3.2
Sleeve	±0.002	0.004	0.667×10^{-6}	0.445×10^{-6}	13.0
Snap ring	±0.001	0.002	0.333×10^{-6}	0.111×10^{-6}	3.2
				3.444×10^{-6}	

We see that by using statistical tolerance design the tolerance on the clearance gap has been significantly reduced compared with that found using worst-case tolerance design, 0.012, compared with 0.018 for the worst-case design. The risk one runs by using this scenario is the possibility that 0.24% of the parts would present a problem in assembly.

Suppose that the designer decides that the clearance gap is not all that critical to quality, but she would rather use statistical tolerance design to relieve some of the tolerance requirements for the components in the assembly while maintaining the gap tolerance at ± 0.009 in. So long as the gap width does not go negative, it will not affect the function. The question is, which part in the assembly should be considered for an increase in tolerance? A quick took at the tolerances shows that the tolerance on the length of the pin is the largest, but to be sure to determine which tolerance makes the greatest contribution to the clearance gap tolerance she needs to make a sensitivity analysis. Table 8.8 shows the method and results.

The standard deviation of a part was determined by dividing the tolerance range by 6, in agreement with Eq. (8.16). The percent variation attributed to each part was found by dividing the total square of the standard deviation into that for each part. The result shows overwhelmingly that the tolerance on the length of the pin contributes in the greatest degree to the tolerance in the gap.

Now the designer decides to find out how much the tolerance on the pin length could be loosened without putting the clearance into interference. As a safety factor, she decides to keep the clearance at 0.009 in., as found in Example 8.1. Then setting $T_{assembly} = 0.009$ in Table 8.7, and solving for the new tolerance on the pin, it turns out that the tolerance can be increased from ± 0.005 to ± 0.008. This is just enough increase in tolerance to allow a cheaper cold heading process to substitute for the screw machine manufacturing process that was necessary to achieve the original tolerance on the pin length. This is an example of a typical trade-off that is common in engineering design, substituting one model of reality for another (worst-case versus an allowable small level of defects) by deciding how much additional analysis is justified to achieve a modest cost savings.

There is one last step in the statistical tolerance design. Having established the mean and tolerance on the clearance gap, we need to determine how many parts would be expected to produce defects in manufacturing. Given a mean gap of $\bar{g} = 0.010$ in and a tolerance of ± 0.009, the standard deviation is obtained from Eq. (8.16) as $C_p = 1 = \frac{0.019 - 0.001}{6\sigma}$, and $\sigma = 0.003$ in. Since the dimensions are random variables that follow a normal frequency distribution, we can use the table for the area under the

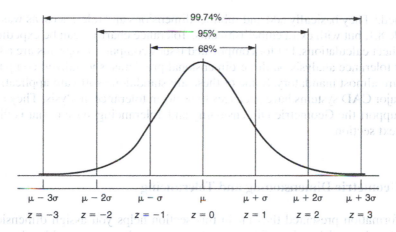

FIGURE 8.23
Normal distribution in terms of z.

normal distribution when problem variables are transformed into the *standard normal distribution*, z, according to

$$z = \frac{x - \mu}{\sigma} \tag{8.19}$$

where μ is the mean of the clearance, in this case $\bar{g} = 0.010$ in., $\sigma = 0.003$, and x is any cutoff point along the axis of z. There are two cutoff points that constitute failure of the design. The first is if $x = 0$, the clearance disappears. As Fig. 8.23 shows, this represents a point at $z = -3.33$.

When $x = \bar{g} = 0$, $z = \dfrac{0 - 0.010}{0.003} = -3.33$. The probability of $z \leq -3.33$ is very small. From tables of the area under the z-distribution (see Appendix B), we see the probability is 0.00043 or 0.043%.

When $x = \bar{g} = 0.019$, the value of z is $z = \dfrac{0.019 - 0.010}{0.003} = 3.0$. Once again, the probability of exceeding 0.0019 is small, 0.14%. We conclude that the probability of encountering these types of design failures with the mean and tolerance of the clearance gap as shown above is indeed very low.

Advanced Tolerance Analysis

The example given in Fig. 8.22 is a relatively simple problem involving only variation along one axis and only four dimensions in the stackup. If you ever looked in the gear case of your car, you can appreciate that many mechanical systems are much more complicated. When many dimensions are involved, and the mechanism is definitely three-dimensional, it is helpful to have a better way of keeping track of what you are doing. To accomplish this, a system of *tolerance charts* has been

---

1. D. H. Nelson and G. Schneider, Jr., *Applied Manufacturing Process Planning,* Chap. 7, Prentice Hall, Upper Saddle River, NJ, 2001; B. R. Fischer, op. cit., Chap. 14.

developed.[1] They basically add and subtract dimensions and tolerances, as was done in Example 8.1, but with extra embellishments. Tolerance charting can be expedited with spreadsheet calculations, but for complicated issues computer programs are advisable.

For tolerance analysis on three-dimensional problems, specialized computer programs are almost mandatory. Some of these are standalone software applications, but most major CAD systems have packages to perform tolerance analysis. They also typically support the Geometric Dimensioning and Tolerancing system that is discussed in the next section.

8.7.3 Geometric Dimensioning and Tolerancing

The information presented thus far in this section helps you assign dimensions that define the *size* and location of features. However, it does not consider the variation in the *form* of the component, which involves such geometric aspects as flatness or straightness. For example, the diameter of the pin in Fig. 8.22 could be completely in tolerance on its diameter, but not fit inside the sleeve because the diameter was slightly bowed so it was outside the tolerance band for straightness. In engineering practice this and many other tolerance issues are described and specified by a system of *Geometric Dimensioning and Tolerancing* (GD&T) based on ASME standard Y14.5–2009. GD&T is a universal design language to precisely convey design intent. It avoids ambiguous situations that arise when only size tolerances are used.

Geometric Dimensioning and Tolerancing introduces two important pieces of information to an engineering drawing: (1) it clearly defines the *datum surfaces* from which dimensions are measured, and (2) it specifies a *tolerance zone* that must contain all points of a geometric feature.

Datums

Datums are theoretically perfect points, lines, and planes that establish the origin from which the location of geometric features of a part is determined. In Fig. 8.19 the datums were *implied* as the x-z and y-z planes, where z is the direction normal to the plane of the page. However, most engineering drawings are not as simple as Fig. 8.19, so a system of clearly identifying the datum surfaces is necessary. Datums serve the purpose of explicitly telling the machinist or inspector the point from which to take measurements. In assigning datums the designer should consider how the part will be manufactured and inspected. For example, the datum surface should be one that can be defined by the machine table or vise used in making the part, or the precision surface plate used to inspect the part.

A part has six degrees of freedom in space. It may be moved up or down, left or right, and forward or backward. Depending on the complexity of the part shape there may be up to three datums. The primary datum, A, is usually a flat surface that predominates in the attachment of the part with other parts in the assembly. One of the other datums, B or C, must be perpendicular to the primary datum. The datum surfaces are shown on the engineering drawing by datum feature identifiers in which a triangle identifies the surface and a boxed letter identifies the order of the datums, Fig. 8.24.

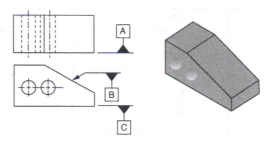

FIGURE 8.24
Datum feature identifiers.

Geometric Tolerances

Geometric tolerances can be defined for the following characteristics of geometric features:

- Form—flatness, straightness, circularity, cylindricity
- Profile—line or surface
- Orientation—parallelism, angularity
- Location—position, concentricity
- Runout—circular runout or total runout

Figure 8.25 shows the symbol for each geometric characteristic and how a geometric tolerance is shown on the engineering drawing. The sketches at the right side of the figure show how the tolerance zones are defined.

For example, if the tolerance for flatness is given as 0.005 in. it means that the surface being controlled by this tolerance must lie within a tolerance zone consisting of two parallel planes that are 0.005 inches apart. In addition to the geometric tolerance, the part must also conform to its size tolerance.

Circularity refers to degree of roundness, where the tolerance zone is represented by the annulus between two concentric circles. In the example shown in Fig. 8.25 the first circle is 0.002 outside of the basic dimension, and the second circle is 0.002 inside of the basic circle. Cylindricity is the three-dimensional version of circularity. The tolerance zone lies between two coaxial cylinders in which the radial distance between them is equal to the tolerance. Cylindricity is a composite form tolerance that simultaneously controls circularity, straightness, and taper of a cylinder. Another combined geometric tolerance is circular runout. To measure runout, a cylindrical part is rotated about its axis and the "wobble" is measured to see if it exceeds the tolerance. This measure controls both circularity and concentricity (coaxiality).

Material Condition Modifiers

Another aspect of GD&T is the ability to modify the size of the tolerance zone of a feature depending on the size of the feature. There are three possible material condition modifiers.

- *Maximum material condition* (MMC) is the condition in which an external feature like a shaft is at its largest size allowable by the size tolerance. MMC also means that an internal feature like a hole is at its smallest allowable size. The symbol for MMC is an M inside a circle.

Geometric Symbol	As Shown on Drawing	Depicted Tolerance Conditions
Flatness	.005	Two parallel planes .005 apart
Straightness (of an axis)	.500 ± .003 — .001	.001 diameter same length as shaft
Straightness (surface element)	— .002	Two parallel lines .002 apart
Circularity (roundness)	.500 ± .002 ○ .001	Two concentric circles .004 apart
Cylindricity	.500 ± .003 ⌭ .001	Two concentric cylinders .001 apart
Perpendicularity	.500 ± .003 ⊥ .002 -C- / -C-	90° .002
Angularity	∠ .004 A / 30° / -A-	Two parallel planes .004 apart / 30°
Parallelism	// .003 B / -B-	.003 - wide tolerance zone / -B-

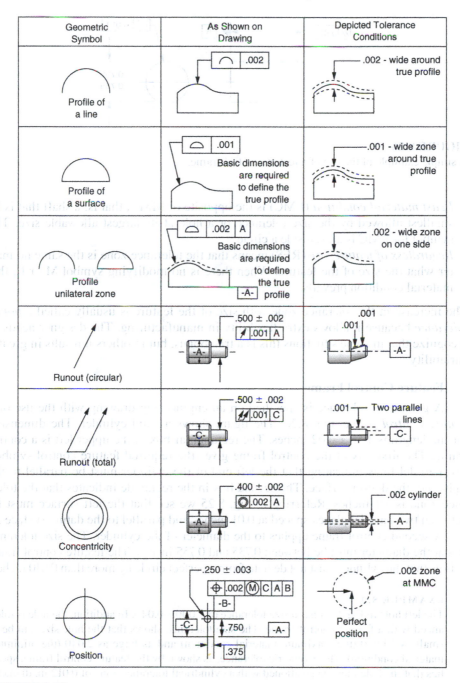

Geometric Symbol	As Shown on Drawing	Depicted Tolerance Conditions
Profile of a line	⌒ .002	.002 - wide around true profile
Profile of a surface	⌒ .001 Basic dimensions are required to define the true profile	.001 - wide zone around true profile
Profile unilateral zone	⌒ .002 A Basic dimensions are required to define the true profile -A-	.002 - wide zone on one side
Runout (circular)	.500 ± .002 ⟋ .001 A -A-	.001 .001 -A-
Runout (total)	.500 ± .002 ⟋⟋ .001 C -C-	.001 Two parallel lines -C-
Concentricity	.400 ± .002 ⌽ .002 A -A-	.002 cylinder -A-
Position	.250 ± .003 ⊕ .002 Ⓜ C A B -B- -C- .750 -A- .375	.002 zone at MMC Perfect position

FIGURE 8.25

Geometric Dimensioning and Tolerancing symbols and interpretation.

(From D. H. Nelson and G. Schneider, Jr., *Applied Manufacturing Process Planning,* Prentice-Hall, Inc., Upper Saddle River, NJ, 2001, p. 95. Used with permission.)

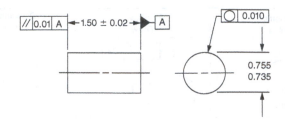

FIGURE 8.26
A simple example of the use of a feature control frame.

- *Least material condition* (LMC) is the opposite of MMC, that is, a shaft that is its smallest allowed by the size tolerance or a hole at its largest allowable size. The symbol for LMC is an L inside a circle.
- *Regardless of feature size* (RFS) means that the tolerance zone is the same no matter what the size of the feature. When there is no modifying symbol M or L, this material condition prevails.

The increase in the tolerance zone with size of the feature is usually called a *bonus tolerance* because it allows extra flexibility in manufacturing. The designer needs to recognize that in some situations this is a true bonus, but in others it results in greater variability.[1]

Feature Control Frame

A geometric tolerance is specified on an engineering drawing with the use of a *feature control frame*, Fig. 8.26. The figure shows a solid cylinder. The dimension for the length is 1.50 ± 0.02 inches. The rectangular box at the upper left is a control frame. The first box of the control frame gives the required feature control symbol, two parallel lines indicating that the left end of the cylinder must be parallel to the right end, the datum surface. The second box in the rectangle indicates that the tolerance zone is .01 inches. Referring to Fig. 8.25 we see that the left surface must lie between two parallel planes spaced at 0.01 inches and parallel to the datum surface A.

A second control frame applies to the diameter of the cylinder. The size tolerance is that the diameter must be between 0.735 and 0.755 inches. The feature control frame tells us that the cylinder must not deviate from a perfect circle by more than 0.010 inches.

EXAMPLE 8.3
The left hole in Fig. 8.24 has a size tolerance of 2.000 ± 0.040. In addition, the hole is toleranced with a feature control frame. The size tolerance shows that the hole size can be as small as $\varnothing 1.960$ (the maximum material condition) and as large as 2.040 (the minimum material condition). The geometric tolerance, as shown by the feature control frame, specifies that the hole must be positioned with a cylindrical tolerance zone of 0.012 in. diameter

---

1. B. R. Fischer, *Mechanical Tolerance Stackup and Analysis*, Chap. 12, Marcel Dekker, New York, 2004. G. Henzold, *Geometrical Dimensioning for Design, Manufacturing, and Inspection*, 2d ed., Butterworth-Heinemann, Boston, 2006.

(see last row in Fig. 8.25). The circle M symbol also specifies that that this tolerance holds when the hole is produced at its maximum material condition (MMC).

$$\boxed{\oplus \;|\; \varnothing\,.012 \;|\; \text{M} \;|\; \text{A} \;|\; \text{B}}$$

If the hole size falls below MMC, additional tolerance on hole location, called *bonus tolerance*, is allowed. If the hole is actually made with a diameter of 2.018, then the total tolerance on the hole position would be:

Actual hole size	2.018
Minus maximum material condition	−1.960
Bonus tolerance	0.058
Geometric tolerance on the feature (hole)	+0.012
Total tolerance	0.070

Note that the use of the maximum material modifier to the geometric tolerance allows the designer to take advantage of all available tolerance.

There are many other geometrical features that can be specified precisely with GD&T. Understanding GD&T is detailed but straightforward. Space considerations do not begin to allow a thorough discussion. Any engineer involved in detailed design or manufacturing will have to master this information. A quick search of the library or the World Wide Web will yield many training courses and self-study manuals on GD&T.[1]

8.7.4 Guidelines for Tolerance Design

The following guidelines summarize much of this section.

- Focus on the critical-to-quality dimensions that most affect fit and function. This is where you should spend most of your efforts on tolerance stackup analysis.
- For the noncritical dimensions, use a commercial tolerance recommended for the production process of the components.
- A possible alternative for handling a difficult tolerance problem might be to redesign a component to move it to the noncritical classification.
- A difficult problem with tolerance stackup often indicates that the design is overconstrained to cause undesirable interactions between the assembled components. Go back to the configuration design step and try to alleviate the situation with a new design.
- If tolerance stackup cannot be avoided, it often is possible to minimize its impact by careful design of assembly fixtures.

---

1. G. R. Cogorno, *Geometric Dimensioning and Tolerancing for Mechanical Design,* McGraw-Hill, New York, 2006. G. Henzold, Geometrical *Dimensioning and Tolerancing for Design, Manufacturing, and Inspection,* 2d ed., Butterworth-Heinemann, Boston, 2006. A short, well illustrated description of the GD&T control variables, including how they would be measured in inspection, is given in G. R. Bertoline, op. cit., pp. 731–44.

- Use selective assembly where critical components are sorted into narrow dimensional ranges before assembling mating components. Before doing this, give careful consideration to possible customer repercussions with future maintenance problems.
- Before using statistical tolerancing make sure that you have the agreement from manufacturing that the product is receiving components from a well-controlled process with the appropriate level of process capability.
- Consider carefully the establishment of the datum surfaces, since the same datums will be used in manufacture and inspection of the part.

8.8
INDUSTRIAL DESIGN

Industrial design, also often called just product design, is concerned with the visual appearance of the product and the way it interfaces with the customer. The terminology is not precise in this area. Up until now, what we have called product design has dealt chiefly with the function of the design. However, in today's highly competitive marketplace, performance alone may not be sufficient to sell a product. The need to tailor the design for aesthetics and human usability has been appreciated for many years for consumer products, but today it is being given greater emphasis and is being applied more often to technically oriented industrial products.

Industrial design[1] deals chiefly with the aspects of a product that relate to the user. First and foremost is its aesthetic appeal. Aesthetics deal with the interaction of the product with the human senses—how it looks, feels, smells, or sounds. For most products the visual appeal is most important. This has to do with whether the shape, proportion, balance, and color of the elements of the design create a pleasing whole. Often this goes under the rubric of *styling*. Proper attention to aesthetics in design can instill a pride of ownership and a feeling of quality and prestige in a product. Appropriate styling details can be used to achieve product differentiation in a line of similar products. Also, styling often is important in designing the packaging for a product. Finally, proper attention to industrial design is needed to develop and communicate to the public a corporate image about the products that it makes and sells. Many companies take this to the point where they have developed a corporate style that embodies their products, advertising, letterheads, and so on. Aspects of the style can include colors, color ratios, and shapes.[2]

The second major role of industrial design is in making sure that the product meets all requirements of the user human interface, a subject often called *ergonomics* or usability.[3] This activity deals with the user interactions with the product and making sure that it is easy to use and maintain. The human interface is discussed in Sec 8.9.

---

1. P. S. Jordan, *Design of Pleasurable Products,* Taylor & Francis, 2000. B. E. Bürdek, *Design: History, Theory and Practice of Product Design,* Birkauser Publishers, Basel, 2005.
2. To explore a world of industrial design, go to Google, select Images, and type in industrial design.
3. A. March, "Usability: The New Dimension of Product Design," *Harvard Business Review,* September–October 1994, pp. 144–49.

The industrial designer is usually educated as an applied artist or architect. This is a decidedly different culture than that of the education of the engineer. While engineers may see color, form, comfort, and convenience as minor issues in the product design, the industrial designer is more likely to see these features as intrinsic in satisfying the needs of the user. The two groups have roughly opposite styles. Engineers work from the inside out. They are trained to think in terms of technical details. Industrial designers, on the other hand, work from the outside in. They start with a concept of a complete product as it would be used by a customer and work back into the details needed to make the concept work. Industrial designers often work in independent consulting firms, although large companies may have their own in-house staff. Regardless, it is important to have the industrial designers involved at the beginning of a project, for if they are called in after the details are worked out, there may not be room to develop a proper concept.

Apple Computer has for a long time maintained a profitable niche market in personal computers and digital appliances because of its superior industrial design. Once a leader in software technology, in recent times it has been industrial design that has kept Apple profitable. In the late 1990s the translucent iMac in an array of eye-catching colors made it an instant success. Then the Power Mac G4 cube, looking more like a postmodern sculpture than a piece of office equipment, became a high-status must have product. But the iPod and iPhone are the biggest hits of all.

8.8.1 Visual Aesthetics

Aesthetics relate to our emotions. Since aesthetic emotions are spontaneous and develop beneath our level of consciousness, they satisfy one of our basic human needs. Visual aesthetic values can be considered as a hierarchy of human responses to visual stimuli.[1] At the bottom level of the hierarchy is order of visual forms, their simplicity, and clarity—our visual neatness. These values are derived from our need to recognize and understand objects. We relate better to symmetric shapes with closed boundaries. Visual perception is enhanced by the repetition of visual elements related by similarity of shape, position, or color (rhythm). Another visual characteristic to enhance perception is homogeneity, or the standardization of shapes. For example, we relate much more readily to a square shape with its equal angles than to a trapezoid. Designing products so that they consist of well-recognized geometric shapes (geometrizing) greatly facilitates visual perception. Also, reducing

---

1. Z. M. Lewalski, *Product Esthetics: An Interpretation for Designer,* Design & Development Engineering Press, Carson City, NV, 1988.

FIGURE 8.27
Note how the design of the four-wheel-drive agricultural tractor projects rugged power. The
clearly defined grid of straight lines conveys a sense of unity. The slight forward tilt of the
vertical lines adds a perception of forward motion.

(From Z. M. Lewalski, *Product Esthetics*, Design & Development Engineering Press, Carson City, NV. Used with
permission.)

the number of design elements and clumping them into more compact shapes aids
recognition.

The second level of visual aesthetics is concerned with recognition of the functionality or utility of the design. Our everyday knowledge of the world around us gives
us an understanding of the association between visual patterns and specific functions.
For example, symmetrical shapes with broad bases suggest inertness or stability.
Patterns showing a tendency toward visual separation from the base suggest a sense
of mobility or action (see Fig. 8.27). A streamlined shape suggests speed. Looking
around, you can observe many visual symbols of function.

The highest level of the visual aesthics hierarchy deals with the group of aesthetic
values derived from the prevailing fashion, taste, or culture. These are the class of
values usually associated with styling. There is a close link between these values and
the state of available technology. For example, the advent of steel beams and columns
made the high-rise building a possibility, and high-strength steel wire made possible
the graceful suspension bridge. A strong driver of prevailing visual taste traditionally
has been the influence of people in positions of power and wealth. In today's society
this is most likely to be the influence of media stars. Another strong influence is the
human need and search for newness.

8.9
HUMAN FACTORS DESIGN

Human factors is the study of the interaction between people, the products and systems they use, and the environments in which they work and live. This field also is described by the terms *human factors engineering* and *ergonomics*.[1] Human factors design applies information about human characteristics to the creation of objects, facilities, and environments that people use. It considers the product as part of a human and machine system in which the operator, the machine, and the environment in which it operates must all function effectively. Human factors goes beyond the issues of usability to consider design for ease of maintenance and for safety. Human factors expertise is found in industrial designers, who focus on ease of use of products, and in industrial engineers, who focus on design of production systems for productivity.

We can relate the human interaction with a product to the inputs used in Chap. 6 to describe the function structure of a design. A person provides energy to the system by applying forces and torques with human muscle power. People also provide signal information through their senses of sight, hearing, touch, and to a limited degree, with taste and smell. They provide material input when their body must be contained by the product (the door must be large enough for the shoulders of the body, or the light switch must be within reach). Thus, it is important to understand more about human factors design to achieve a harmonious interaction with human functions. Products that rate high in human factors engineering are generally regarded as high-quality products since they are perceived to work well by the user. Table 8.9 shows how various important product characteristics can be achieved by focusing on key human factors characteristics.

8.9.1 Human Physical Effort

Measurement of the physical effort that a man could perform in the manual handling of materials (shoveling coal) and supplies was one of the first studies made in human factors engineering. Such studies involve not only measurement of the force that can be applied by ligaments and muscles but also measurement of the cardiovascular and respiratory systems of the body to assess the physiological distress (energy expenditure) that occurs during sustained work. In today's mechanized workplace this information is less important than knowing the magnitude of forces and torques that can be applied by the human body, Fig. 8.28.

Figure 8.28 is just one example of information that is available.[2] Note that it is for males who are at the 5th percentile of the strength distribution, meaning that it

---

1. From the Greek words *ergon* (work) and *nomos* (study of).
2. *Human Engineering Design Criteria for Military Systems and Facilities,* MIL-STD 1472F, http://hfetag.dtic.mil/docs-hfs/mil-std-1472f.pdf; *Human Factors Design Guide,* DOT/FAA/CT-96/1, www.asi.org/adb/04/03/14/faa-hf-design-guide.pdf; N. Stanton et al., *Handbook of Human Factors and Ergonomic Methods,* CRC Press, Boca Raton, FL, 2004; M. S. Sanders and E. J. McCormick, *Human Factors in Engineering and Design,* 7th ed., McGraw-Hill, New York, 1993. J. H. Burgess, *Designing for Humans: Human Factors in Engineering,* Petrocelli Books, Princeton, NJ, 1986.

TABLE 8.9
Correspondance Between Human Factors Characteristics and Product Performance

Product Performance	Human Factors Characteristic
Comfortable to use	Good match between product and person in the workspace
Easy to use	Requires minimal human power; clarity of use
Operating condition easily sensed	Human sensing
Product is user-friendly	Control logic is natural to the human

represents only the weakest 5 percent of the male population. It is characteristic of data on human performance that there is a wide deviation from the mean. The data for females is different from that for men. In addition, the force or torque that can be applied depends on the range of motion and position of the various joints of the human body. For example, Fig. 8.28 shows that the force that can be applied depends on the angle that the elbow makes with the shoulder. This gets us into the topic of *biomechanics*. The force that can be exerted also depends on whether the person is seated, standing, or lying down. Thus, the references noted here need to be consulted for data referring to the specific type of action or motion.

Human muscle output is typically applied to a machine at a control interface, like a brake pedal or a selector switch. These control interfaces can take many forms: a handwheel, rotary knob, thumbwheel, rollerball, lever, joystick, toggle switch, rocker switch, pedal, handle, or slide. These devices have been studied[1] to determine the force or moment needed for their operation, and whether they are best suited for on-off control, or more precise control.

In designing control interfaces it is important to avoid awkward and extreme motions for the product user. Controls should not require a large actuation force unless they are used in emergencies. It is particularly important to design the location of controls so that bending and movements of the spine are not required, particularly if these motions will be repetitive. This can lead to cumulative trauma disorders, where stresses cause nerve and other damage. Such situations will lead to operator fatigue and errors.

8.9.2 Sensory Input

The human senses of sight, touch, hearing, taste, and smell are chiefly used for purposes of controlling devices or systems. They provide signals to the user of the design. Visual displays are commonly used, Fig. 8.29. In selecting visual displays remember that individuals differ in their ability to see, so provide sufficient illumination. As shown in Fig. 8.30, different types of visual displays differ in their ability to provide on-off information, or exact values and rate of change information.

---

1. G. Salvendy (ed.), *Handbook of Human Factors,* 3rd ed., John Wiley & Sons, New York, 2006.

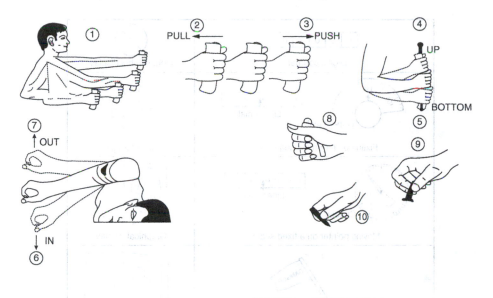

Arm Strength													
(1)	(2)		(3)		(4)		(5)		(6)		(7)	
Degree of Elbow Flexion (deg)	Pull		Push		Up		Down		In		Out	
	L	R*	L	R	L	R	L	R	L	R	L	R
180	50	52	42	50	9	14	13	17	13	20	8	14
150	42	56	30	42	15	18	18	20	15	20	8	15
120	34	42	26	36	17	24	21	26	20	22	10	15
90	32	37	22	36	17	20	21	26	16	18	10	16
60	26	24	22	34	15	20	18	20	17	20	12	17

Hand and Thumb-Fingers Strength (lb)				
	(8)		(9)	(10)
	Hand Grip		Thumb-Finger Grip (Palmar)	Thumb-Finger Grip (Tips)
	L	R		
Momentary Hold	56	59	13	13
Sustained Hold	33	35	8	8

*L = Left; R = Right

FIGURE 8.28
Muscle strength of the arm, hand, and thumb for males at 5th percentile. From MIL-STD-1472F, p. 95.

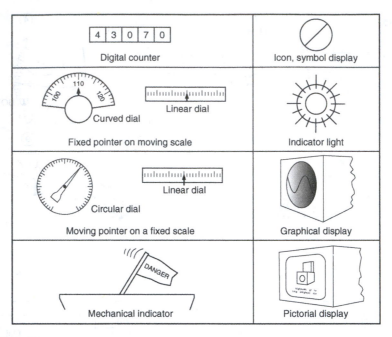

FIGURE 8.29
Types of visual displays.
(After Ullman.)

	Exact value	Rate of change	Trend, direction of change	Discrete information	Adjusted to desired value
Digital counter	●	○	○	●	◐
Moving pointer on fixed scale	●	●	●	●	◐
Fixed pointer on moving scale	●	●	○	○	○
Mechanical indicator	○	○	○	●	○
Icon, symbol display	○	○	○	●	○
Indicator light	○	○	○	●	○
Graphical display	◐	◐	●	●	●
Pictorial display	◐	●	●	●	●

○ Not suitable   ◐ Acceptable   ● Recommended

FIGURE 8.30
Characteristics of common visual displays.
(After Ullman.)

The human ear is effective over a frequency range from 20 to 20,000 Hz. Often hearing is the first sense that indicates there may be trouble, as in the repetitive thumping of a flat tire or the scraping sound of a worn brake. Typical auditory displays that are used in devices are bells, beeps (to acknowledge an action), buzzers, horns and sirens (to sound an alarm) and electronic devices to speak a few words.

The human body is especially sensitive to touch. With tactile stimulation we can feel whether a surface is rough or smooth, hot or cold, sharp or blunt. We also have a kinesthetic sense that uses receptors to feel joint and muscle motion. This is an ability that is highly developed in great athletes.

User-Friendly Design

Careful attention to the following design issues will create user-friendly designs:

- *Simplify tasks:* Control operations should have a minimum number of operations and should be straightforward. The learning effort for users must be minimal. Incorporating microcomputers into the product may be used to simplify operation. The product should look simple to operate, with a minimum number of controls and indicators.
- *Make the controls and their functions obvious:* Place the controls for a function adjacent to the device that is controlled. It may look nice to have all the buttons in a row, but it is not very user-friendly.
- *Make controls easy to use:* Shape knobs and handles of controls differently so they are distinguishable by look and by touch. Organize and group them to minimize complexity. There are several strategies for the placement of controls: (1) left to right in the sequence they are used, (2) key controls located near the operator's right hand, (3) most commonly used controls near the operator's hand.
- *Match the intentions of the human with the actions required by the system:* There should be a clear relationship between the human intent and the action that takes place on the system. The design should be such that when a person interacts with it there is only one obviously correct thing to do.
- *Use mapping:* Make the control reflect, or map, the operation of the mechanism. For example, the seat position control in an automobile could have the shape of a car seat, and moving it up should move the seat up. The goal should be to make the operation clear enough that it is not necessary to refer to nameplates, stickers, or the operator's manual.
- *Displays should be clear, visible, large enough to read easily, and consistent in direction:* Analog displays are preferred for quick reading and to show changing conditions. Digital displays provide more precise information. Locate the displays where viewing would be expected.
- *Provide feedback:* The product must provide the user with a clear, immediate response to any actions taken. This feedback can be provided by a light, a sound, or displayed information. The clicking sound and flashing dashboard light, in response to actuating an automobile turn signal, is a good example.
- *Utilize constraints to prevent incorrect action:* Do not depend on the user always doing the correct thing. Controls should be designed so that an incorrect movement or sequence is not possible. An example is the automatic transmission that will not go into reverse when the car is moving forward.

- *Standardize:* It pays to standardize on the arrangement and operation of controls because it increases the users knowledge. For example, in early days the placement of the brake, clutch, and accelerator pedals in an automobile was arbitrary, but once standardized they become part of the user knowledge base and should not be changed.

Norman contends that in order for a design to be truly user-friendly it must employ the general knowledge that many people in the population possess.[1] For example, a red light means stop, and the higher values on a dial should be in the clockwise direction. Be sure that you do not presume too much knowledge and skill on the part of the user.

Reaction Time

The *reaction time* is the time to initiate a response when a sensory signal has been received. The reaction time is made up of several actions. We receive information in the form of a sensory signal, interpret it in the form of a set of choices, predict the outcomes of each choice, evaluate the consequence of each choice, and then select the best choice—all in about 200 ms. To achieve this the product should very quickly provide clear visual and auditory signals. To achieve this in simple products, the controls must be intuitive. In complex systems, like a nuclear power plant, the human control interface must be very carefully designed in terms of the concepts mentioned in this section, but in addition, the operators must be disciplined and well trained.

8.9.3 Anthropometric Data

Anthropometrics is the field of human factors that deals with the measurements of the human body. Humans vary in size. On average, children are smaller than adults and men are taller than women. Variations in such factors as height when standing, shoulder width, length and width of fingers, arm reach (Fig. 8.31), and eye height on sitting need to be considered when designing products. This information is available online in MIL-STD-1472F and the FAA Human Factors Design Guide.

In design there is no such thing as an "average person." The choice of which percentile of the distribution of human dimensions to use depends upon the design task at hand. If the task is to make a decision on the placement of a critical emergency lever in a crowded aircraft cockpit, use the smallest expected reach, that for a woman in the 1st percentile. If you were designing the escape hatch in a submarine, use the 99th percentile of the shoulder width of men. Clothing manufacturers use a *close fit design* approach rather than the *extreme case* approach. They select their "off the rack" sizes to provide an acceptable fit for their customers in each size range. In other products it often is possible to design for an *adjustable fit*. Adjustable car seats, desk chairs, and stereo headphones are common examples.

---

1. D. A. Norman, *The Design of Everyday Things,* Doubleday, New York, 1988. This book is full of good and poor ways to practice human factors design.

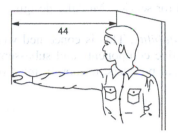

44 Functional (thumb-tip) reach, extended. Measured similarly to functional (thumb-tip) reach, except that the right shoulder is extended forward as far as possible, while the left shoulder is kept pressed firmly against the wall.

				Percentiles		
Sample		1st	5th	50th	95th	99th
A Men	cm	77.9	80.5	87.3	94.2	97.7
	(in.)	(30.0)	(31.7)	(34.4)	(37.1)	(38.5)
B Women	cm	71.2	73.5	79.6	86.2	89.0
	(in.)	(28.0)	(28.9)	(31.3)	(33.9)	(35.0)

FIGURE 8.31
Anthropometric data on the extended reach of men and women.
(From FAA Human Factors Design Guide.)

8.9.4 Design for Serviceability

Human factors issues are related to many of the design for X strategies mentioned in this chapter (see Sec. 8.12). *Serviceability* is concerned with the ease with which maintenance can be performed on a product.[1] Many products require some form of maintenance or service to keep them functioning properly. Products often have parts that are subject to wear and that are expected to be replaced at periodic intervals. There are two general classes of maintenance. *Preventive maintenance* is routine service required to prevent operating failures, such as changing the oil in your car. *Breakdown maintenance* is the service that must take place after some failure or decline in function has occurred.

It is important to anticipate the required service operations during the design of the product. Repair may only require replacing a gasket or filter, but if the part is not accessible without dismantling most of the machine, then maintenance costs will be excessive. Don't make a design like the automobile that requires the removal of a wheel to replace the battery. Also, remember that service often will be carried out in "the field" where special tools and fixtures used in factory assembly will not be available. Design for field service is not complete until a successful simulation of how the failed component will be repaired or replaced in the field has been carried out.

The best way to improve serviceability is to reduce the need for service by improving reliability. Reliability is the probability that a system or component will perform without failure for a specified period of time (see Chap. 14). Failing this, the product must be designed so that components that are prone to wear or failure, or require periodic maintenance, are easily visible and accessible. It means making covers, panels, and housings easy to remove and replace. It means locating components that must be serviced in accessible locations. Avoid press fits, adhesive bonding, riveting,

---

1. J. C. Bralla, *Design for Excellence*, Chap. 16., McGraw-Hill, New York, 1996; M. A. Moss, *Designing for Minimum Maintenance Expense*, Marcel Dekker, New York, 1985.

welding, or soldering for parts that must be removed for service. Modular design is a great boon to serviceability.

A concept closely related to serviceability is *testability*. This is concerned with the ease with which faults can be isolated in defective components and subassemblies. In complicated electronic and electromechanical products, testability must be designed into the product.

8.9.5 Design for Packaging

Packaging is related to visual aesthetics because attractive, distinctive product packaging is typically used to attract customers and to identify product brands. But there is a broader importance for careful design of packaging. Packaging provides physical protection against mechanical shock, vibration, and extreme temperatures in shipping and storage. Different packaging is required for liquids, gases and powders than for solid objects. Large mechanical equipment, such as jet engines, requires special packaging which is often reusable.

A shipping package provides information about the recipient, tracking information, instructions regarding hazardous materials, and disposal. Many types of packaging provide security against tampering, pilfering, and theft. Transport packaging can vary in size from a steel shipping container to a package directed to an individual consumer.

With the increasing use of plastics in packaging, for example, plastic shrink-wrapped pallets, environmentally safe disposal can be a problem since plastics do not degrade in a landfill. More traditional packaging materials like cardboard and wood crates and barrels are better environmentally, and they can be recycled or used as fuel. A general rule regarding package design is that packages should be made as inexpensively as possible consistent with providing the needed level of protection and security. With certain types of package contents, for example, hazardous materials and medicine, the packaging standards are proscribed by law. For more information on packaging and packaging design, see K. L. Yam, *The Wiley Encyclopedia of Packaging*, 3rd ed., 2009.

8.10
LIFE-CYCLE DESIGN

The worldwide concern over global warming coupled with concerns over energy supply and stability have moved design for the environment (DFE) to a top consideration in design for all types of engineering systems and consumer products. Greater concern for the environment places emphasis on *life-cycle design* in the PDP. Life-cycle design emphasizes giving attention in embodiment design to those issues that impact a long, useful service life. Life-cycle design is not the same thing as *product life cycle* which refers to the length of time a product remains in production before it is replaced by a better or competing design. Life-cycle design also refers to those aspects of design that are needed to get the product in the hands of its user, to keep it functioning while in service, and to dispose of it in an environmentally friendly way. The major issues of life-cycle design are:

- Design for packaging and shipping (Sec. 8.9.5)
- Design for serviceability and maintenance (Sec. 8.9.4)
- Design for testability
- Design for disposal

Design for disposal is an important issue in *design for the environment* (see Chapter 10). However, in a world of finite natural resources, any design modifications that can keep a product in service will benefit the environment in the long run because the product will not have to be disposed of, and therefore will not consume additional natural resources for its replacement. The following design strategies can be used to extend a product's useful life.

- *Design for durability*: Durability is the amount of use one gets from a product before it breaks down and replacement is preferred to repair. Durability depends on the skill of the designer in understanding service conditions, analyzing stresses and strains, and in selecting materials that minimize degradation over time due to corrosion or wear.
- *Design for reliability*: Reliability refers to interruptions in usage during service. It is a more technical performance characteristic than durability and is measured by the probability that a product will neither malfunction nor fail within a specified time period. See Chap.14 for details.
- *Create an adaptable design*: A modular design allows for continual replacement or improvement of its various functions.
- *Repair:* Concern for future repair in design can greatly facilitate the replacing of nonfunctioning components. While not always economical, there are instances where it pays to design-in sensors to tell the operator when it is time to replace parts before they fail.
- *Remanufacture*: Worn parts are restored to like-new condition.
- *Reuse:* Find another use for the product after it has been retired from its original service. The reuse of inkjet cartridges is a common example.

8.11
PROTOTYPING AND TESTING

We are nearing the end of the embodiment design phase. The product architecture has been decided, we have configured the components, determined the dimensions and tolerances on the features, and carried out parametric design on several critical-to-quality parts and assemblies. Careful decisions have been made on the selection of materials and manufacturing processes using DFM, DFA, and DFE. The design has been checked for possible failure modes using FMEA, the reliability of several critical subsystems has been discussed with suppliers, and the experts in human factors design have given their approval. Design for quality and robustness concepts have been employed in decisions on several critical parameters. Preliminary cost estimates look as if we will come under the target cost.

So, what is left yet undone? We need to assure ourselves that the product will really function the way it is expected to work. This is the role of the prototype.

Prototypes are physical models of the product that are tested in some way to validate the design decisions that have been made up to that point in the design process. As will be discussed in the next section, prototypes come in various forms and are used in different ways throughout the design process. A prototype is a physical model of the product, as opposed to a computer model (CAD model) of the product or other simulation of the design. Much attention has been given to computer modeling because it often provides insights faster and with less cost than building and testing a physical model or prototype. Also, using finite element analysis or some other CAE tool can provide technical answers that may not be available any other way. Both prototypes and computer models are valuable tools in carrying out the design process.

8.11.1 Prototype and Model Testing Throughout the Design Process

Up to this point we have not given much attention to how models and prototypes are used throughout the design process. We will start the discussion at the very beginning of the product development process, Phase Zero, where marketing and technical people are working to understand customer interest and need for a new product, and move all the way down to the point where the product is about to be introduced to the marketplace.

- **Phase Zero:** *Product Concept Model.* A full-scale or reduced-scale model of a new product is made to look like the final product. This often is prepared by technical designers and industrial designers working collaboratively. Emphasis is on appearance to gage customer reaction to a possible new product. For example, a defense contractor trying to stir up interest in a new fighter plane would make up glitzy models and pass them around to the generals and politicians.
- **Conceptual Design:** *Proof-of-Concept Prototype.* This is a physical model to show whether the concept performs the functions that satisfy the customer's needs and corresponding engineering specifications. There may have been a succession of proof-of-concept models, some physical and others rough sketches, that serve as learning tools until reaching the final proof-of-concept prototype. No attempt is made to make the proof-of-concept model look like the product as far as size, materials, or manufacturing methods are concerned. The emphasis is on showing that the concept will deliver the needed functions. It is sometimes known as a "string and chewing gum" model.
- **Embodiment Design:** *Alpha-Prototype Testing.* The end of the embodiment design phase is usually capped off by testing product prototypes. These are called alpha-prototypes because while the parts are made to the final design drawings with the same materials as the product, they are not made using the same manufacturing processes as the production-run parts. For example, parts that might be made as

castings or forgings in the production run will be machined from plates or bar stock because the tooling for the production parts is still being designed.

Embodiment design makes frequent use of computer-aided engineering (CAE) tools for various design tasks. Sizing of parts might require finite element analysis to find the stresses in a complex part, or the designer might use a fatigue design package to size a shaft, or use tolerance stackup design software.

- **Detail Design:** *Beta-Prototype Testing.* This involves full-size functional part or product testing using the materials and processes that will be used in production. This is a *proof-of-process prototype.* Often customers are enlisted to help run these tests. The results of the beta-prototype tests are used to make any remaining changes in the product, complete the production planning, and try out the production tooling.

- **Manufacturing:** *Preproduction Prototype Testing.* This represents the first several thousand of units of production from the actual production line using the assigned production workers. Therefore, the output from the line represents the product that will shortly be shipped and sold to the customer. The tests on these products are made to verify and document the quality of the design and production and assembly processes.

There is a trade-off between the number of prototypes that will be built for a product design and tested and the cost and length of the product development cycle. Prototypes help to verify the product but they have a high cost in money and time. As a result, there is a strong trend, particularly in large companies, to replace physical prototypes with computer models (virtual prototypes) because simulation is cheaper and faster. The opposing position, taken by many experienced engineers, is that computer modeling has been taken too far too fast, and that carefully planned and executed simulated service tests, and full-sized tests under extreme conditions should not be abandoned.

One place where physical models should not be completely replaced by computer modeling is in the early stages of conceptual design.[1] Here the goal is to gain insight about a design decision by physically building a quick-and-dirty physical model from common construction materials without waiting for a model shop to do the work for you. A hands-on approach where the designers actively build many simple prototypes is highly recommended as the best way to understand and advance the concept development activity. The approach has been called "just build it" by the highly successful product design firm IDEO. Others call this the *design-build-test-cycle.*[2]

8.11.2 Building Prototypes

It is highly recommended that the design team build its own physical models leading up to the proof-of-concept prototype. Product concept models, on the other hand, are often carefully crafted to have great visual appeal. These are traditionally made by

---

1. H. W. Stoll, *Product Design Methods and Practices,* Marcel Dekker, New York, 1999, pp. 134–35.
2. D. G. Ullman, 4th ed., p. 217.

firms specializing in this market or by industrial designers who are part of the design team. Computer modeling is rapidly overtaking the physical model, which by its nature is static, for this application. A 3-D computer model can show cutaway views of the product as well as dynamic animations, all on a DVD that can be easily produced in quantity. Nevertheless, an attractive physical model still has status appeal with important customers.

Models for alpha-prototype testing are typically made in the model shop, a small machine shop staffed with expert craftsmen and equipped with computer-controlled machine tools and other precision machine tools. To be effective it is important to use CAD software that interfaces well with the numerically controlled (NC) machine tools, and it is important that the shop personnel be well trained in its use. Most of the time required to make a prototype by NC machining is consumed not by metal cutting but in process planning and NC programming. Recent developments have reduced the time needed for these operations so that NC machining is becoming competitive with rapid prototyping methods for the simpler geometries. Beta-prototype models and preproduction test prototypes are made by the manufacturing department using the actual materials and processes in which the product will be produced.

8.11.3 Rapid Prototyping

Rapid prototyping (RP) is a technology that produces prototypes directly from computer-aided design (CAD) models in a fraction of the time required to make them by machining or molding methods.[1] Another name for RP is solid freeform fabrication. RP is used for producing the final proof-of-concept model and is used extensively in embodiment design to check form, fit, and function. The earliest applications of RP were as appearance models, but as dimensional control approached ± 0.005 inches in RP objects they began to be used for issues of fit and assembly. RP objects are often used to check the function of kinematic motion, but they are not generally strong enough to be used as prototypes where strength issues are important.

The steps in rapid prototyping are shown in Fig. 8.32.

- *Create a CAD model*: Any RP process starts with a three-dimensional CAD model, which can be considered a virtual prototype of the part. The only requirement on the model for using a RP process is that the model must be a fully closed volume. Thus, if we were to pour water into the model it would not leak.
- *Convert the CAD model to the STL file format.* In this format the surfaces of the component are converted to very small, triangular facets by a process called tessellation. When taken together, this network of triangles represents a polyhedral approximation of the surfaces of the component. CAD software has the capability to convert a CAD file to STL.
- *Slice the STL file into thin layers.* The tessellated STL file is moved to the RP machine, and its controlling software slices the model into many thin layers. This is required because most RP processes build up the solid body layer by layer. For

1. R. Noorani, *Rapid Prototyping: Principles and Applications,* John Wiley & Sons, New York, 2006.

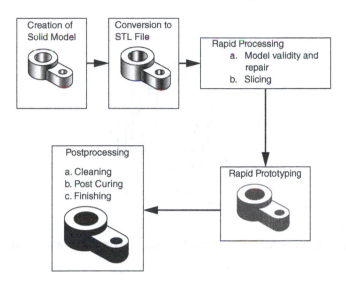

FIGURE 8.32
Steps in the rapid prototyping process.

(From R. Noorani, *Rapid Prototyping,* John Wiley & Sons, New York, 2006, p. 37. Used with permission.)

example, if a part is to be 2 inches high, and each layer is 0.005 inches thick, it requires the addition of material by a buildup of 400 layers. Thus, most RP processes are slow, taking hours to build out a part. They gain speed over numerically controlled machining by virtue of the fact that NC machining often takes many more hours of process planning and computer programming before metal cutting can start.

- *Make the prototype*: Once the sliced computer model is in the computer of the RP machine it runs without much attention until the part is completely built up.
- *Postprocessing*: All objects removed from RP machines need processing. This consists of cleaning, removal of any support structures, and light sanding of the surfaces to remove the edges from the layering process. Depending on the material used in the RP process, the object may need curing, sintering, or infiltration of a polymer to give it strength.

Note that the time to make a RP model may take from 8 to 24 hours, so the term *rapid* may be something of a misnomer. However, the time from detail drawing to prototype is typically shorter than if the part was made in a model shop due to issues of scheduling and programming the machine tools. Also, RP processes are able to produce very complex shapes in one step, although typically they are made from a plastic, not a metal.

8.11.4 RP Processes

There are a number of RP processes currently in use. They differ chiefly in whether they are liquid, solid, or powder-based systems, and whether they are capable of working with polymers, metals, or ceramics.

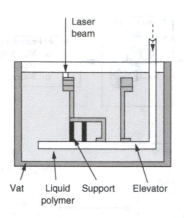

FIGURE 8.33
Rapid prototyping by stereolithography (SL). Note the supports needed for overhanging parts.
(After Schey.)

The first commercial RP process was *stereolithography* (SL). This process uses a UV laser beam to build up layers of solid polymer by scanning on the surface of a bath of photosensitive polymer. Where the laser strikes the liquid polymer it rapidly polymerizes and forms a solid networked polymer. After the layer of liquid is scanned, the platform holding the bath is lowered one layer of thickness, and the process is repeated, layer after layer, until the prototype object has been built up, Fig. 8.33. The laser beam is controlled by the sliced STL file in the memory of the RP machine's computer. The resulting prototype, Fig. 8.34, is much weaker than a metal prototype made by NC machining, but it has excellent dimensional control and a smooth surface finish. The promise shown by these early prototypes set off a search for other RP systems.

Selective laser sintering (SLS) was developed to use stronger, higher-melting-temperature materials than polymers in the RP process. The general layout of the process is shown in Fig. 8.33. In principle, any powder that can be fused together by sintering can be used. A thin layer of powder is spread and sintered by the passage of a high-energy laser beam. Then the platform is lowered, a new layer of powder is spread and sintered, and the process proceeds, layer by layer. The SLS process is mostly used with thermoset polymer particles, or metal particles coated with plastic to facilitate bonding.

Laminated object modeling (LOM) is an older method that continues to have useful applications because of the simplicity of the equipment that is needed. Thin layers of paper, polymer, or thin sheet steel that are cut with a NC knife or laser are glued together to form a laminate. After fabrication, preparation of the edges is required. All RP machines have definite size limitations, but LOM is most suitable for making prototypes of large components.

Fused-deposition modeling (FDM) is an example of several liquid-state deposition processes used to make prototypes. A continuous filament of thermoplastic polymer is heated and extruded through a nozzle, the movement of which is computer

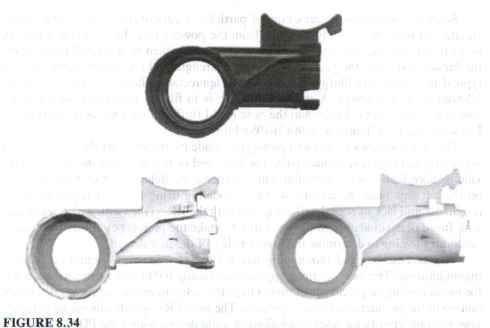

FIGURE 8.34
Examples of student-made prototypes. Top: Injection-molded part. Left: Paper prototype made
by laminated object modeling. Right: Plastic prototype made by fused deposition modeling.

(Courtesy of David Morgan.)

controlled along three axes. The polymer leaves the tip of the nozzle at just above the
melting temperature of the polymer and rapidly solidifies when it strikes the previous
layer. With proper control the extruded bead bonds to the previous layer. Strong and
tough engineered polymers like ABS and polycarbonate can be used in the FDM process, and this produces prototypes with better mechanical properties than those made
by stereolithography.

Three-dimensional printing (3DP) is a RP process that is based on the principle
of the inkjet printer.[1] A thin layer of metal, ceramic, or polymer powder is spread over
a part-build bed. Using inkjet printing technology, fine droplets of a binder material
are deposited on the powder in the two-dimensional geometry defined by the digital
slice of the three-dimensional part. The inkjet is under computer control as in the
other RP processes described previously. The droplets agglomerate powder particles,
bonding them together into a primitive volume element, or voxel. The binder droplets
also bond voxels together within the plane and to the plane below it. Once a layer is
deposited, the powder bed and part are lowered and a new layer of powder is spread
out and the binder is applied by the jet. This layer-by-layer process is repeated until
the part is completed and removed from the powder bed.

---

1. E. Sachs, et al., "Three Dimensional Printing," *ASME Jnl. of Engineering for Industry,* vol. 114, 1992,
pp. 481–88.

Since no mechanical interlocking of particles occurs in this process, the part is fragile and must be carefully removed from the powder bed. The as-printed density is 40 to 60 percent of a fully dense part. The part is then heat treated to drive off the binder and sinter the particles to improve strength and reduce the porosity, as is typical in powder metallurgy. This results in appreciable shrinkage, which must be allowed for in the design. A common practice is to fill the void space with a metal whose melting point is lower than the base metal (infiltration). One example is stainless steel that is infiltrated with 90Cu-10Sn bronze.

Figure 8.34 shows some rapid prototypes made by students. At the top (in black) is the original injection-molded part. The right end of this part was modified to provide a more leak-proof connection with another part. Below are two RPs. The one on the left was made by laminated object modeling using layers of paper. Note the roughness and lack of edge detail compared with the RP on the right, which was made with fused deposition of plastic. The ability to make the prototypes allowed for actual laboratory testing to determine the functionality of the new design.

The success of rapid prototyping has led to extensions into actual component manufacturing. The most common application is using RP to make patterns for molds for metal casting or polymer molding.[1] Often the delay in procuring molds is a major holdup in the product development process. The use of RP speeds this up, and also allows the opportunity for several iterations of mold design within the PDP schedule. A casting process starts with a pattern of the part to be made to make the shaped cavity that the fluid cast material will fill. Usually this is made from wood, plastic, or metal. Making the pattern can be a time-consuming process. With RP a pattern can be made in a day or two. Also, using the additive layered methods of RP allows the formation of patterns with undercuts, overhangs, or internal channels, features that would be impossible to make or would be excessively costly by conventional machining.

8.11.5 Testing

In Sec. 8.11.1 we discussed the sequence of prototypes that are typically used in the product development process. These prototype tests are used to verify the design decisions that are made along the way to launching a product or installing an engineered system. The marketplace validates the acceptability for a consumer product, while for many other types of engineered products there is a set of prescribed acceptance tests. For example, most military equipment and systems are governed by contracts that stipulate specific test requirements.

One of the important documents that is developed at the start of a major design program is the *test plan*. The test plan gives a description of the types of tests to be performed, when the test will be made in the design process, and the cost of the tests. It should be part of the PDS. All managers and engineers should be informed of the test plan because this is an important pacing activity for the design project.

---

1. R. Noorani, op. cit., Chap. 8.

There are many kinds of tests that may be needed in a design project. Some examples are:

- Testing of design prototypes, as discussed in Sec. 8.11.1.
- Modeling and simulations. See Sec. 7.4.
- Testing for all mechanical and electrical modes of failure. See Chap. 14.
- Specialized tests on seals, or for thermal shock, vibration, acceleration, or moisture resistance, as design dictates.
- Accelerated life testing. Evaluating the useful life of the critical-to-quality components.
- Testing at the environmental limits. Testing at specification extremes of temperature, pressure, humidity, etc.
- Human engineering and repair test. Evaluate all human interfaces with actual users. Check maintenance procedures and support equipment in a user environment.
- Safety and risk test. Determine likelihood of injury to users and prospect of product liability litigation. Check for compliance with safety codes and standards in all countries where product will be sold.
- Built-in test and diagnostics. Evaluate the capability and quality of built-in test, self-diagnosis, and self-maintenance systems.
- Manufacturing supplier qualification. Determine the capability of suppliers with regard to quality, on-time delivery, and cost.
- Packaging. Evaluate the ability of the packaging to protect the product.

There are two general reasons for conducting a test.[1] The first is to establish that the design meets some specification or contractual requirement (verification). For example, the motor must deliver a torque of 50 ft-lb at a speed of 1000 rpm with a temperature rise not to exceed 70 °F above room temperature. This is a test that is conducted with the expectation of a success. If the motor does not meet the requirement, then you must redesign the motor. Most of the kinds of tests listed above are of this type.

The other broad category of tests are planned to generate failures. Most tests of materials carry out the test to a point of failure. Likewise, tests of subsystems and products should be designed to overstress the product until it fails. In this way, we learn about the actual failure modes and gain insight into the weaknesses of the design. The most economical way to do this is with *accelerated testing*. This type of testing uses test conditions that are more severe than those expected to be encountered in service. A common way to do this is with *step testing*, in which the level of the test is progressively increased by increments until failure occurs. Accelerated testing is the most economical form of testing. The times to failure will be orders of magnitude shorter than tests at the worst expected service conditions.

Accelerated testing is used in the following way to improve a design. At the outset, determine what types of failure would be expected from the service conditions. The QFD and FMEA analyses will be helpful. Start testing at the design maximum, ramping up in steps until failure occurs. Using failure analysis methods, Chap. 14,

---

1. P. O'Connor, *Test Engineering,* John Wiley & Sons, New York, 2001.

determine the cause of failure and take action to strengthen the design so it can withstand more severe test conditions. Continue the step testing process until another failure occurs. Repeat the process until all transient and permanent failure modes have been eliminated, within limits of cost and practicability.

8.11.6 Statistical Design of Testing

In the discussion to this point it has been implied that the testing is carried out in such a way that only one design parameter is varied. However, we may have two or more parameters, such as stress, temperature, and rate of loading, which are critical and for which we would like to devise a test plan that considers their joint testing in the most economical way. The discipline of statistics has provided us with the tools to do just that in the subject called *Design of Experiments* (DoE). The most important benefit from statistically designed experiments is that more information per experiment will be obtained than with unplanned experimentation. A second benefit is that statistical design results in an organized approach to the collection and analysis of information. Conclusions from statistically designed experiments very often are evident without extensive statistical analysis, whereas with a haphazard approach the results often are difficult to extract from the experiment even after detailed statistical analysis. Still another advantage of statistically planned testing is the credibility that is given to the conclusions of an experimental program when the variability and sources of experimental error are made clear by statistical analysis. Finally, an important benefit of statistical design is the ability to confirm and quantify interactions between experimental variables.

Figure 8.35 shows the various ways that two parameters (factors) x_1 and x_2 can vary to give a joint response y. In this case the response y is the yield strength of an alloy as it is influenced by two factors, temperature x_1 and aging time x_2. In Fig. 8.35a the two factors have no effect on the response. In Fig. 8.35b only temperature x_1 has an effect on y. In Fig. 8.35c both temperature and time influence yield strength, but they vary in the same way, indicating no interaction between the two factors. However, in Fig. 8.35d at different values of temperature x_1 the effect of aging on the yield strength y with time x_2 is different, indicating an interaction between the two factors x_1 and x_2. Interactions between factors are determined by varying factors simultaneously under statistical control rather than one at a time.

There are three classes of statistically designed experiments.[1]

- Factorial designs are experiments in which all levels of each factor in an experiment are combined with all levels of all other factors. This results in a drastic reduction in the number of tests that need to be run at the expense of loss of some information about interaction between factors.
- Blocking designs use techniques to remove the effect of background variables from the experimental error. The most common designs are the randomized block plan and the balanced incomplete block.

---

1. G. E. P. Box, W. G. Hunter, and J. S. Hunter, *Statistics for Experimenters,* John Wiley & Sons, New York, 1978; D. C. Montgomery, *Design and Analysis of Experiments,* 7th ed., John Wiley & Sons, Hoboken, NJ, 2009. (Available online at knovel.com) 1996.

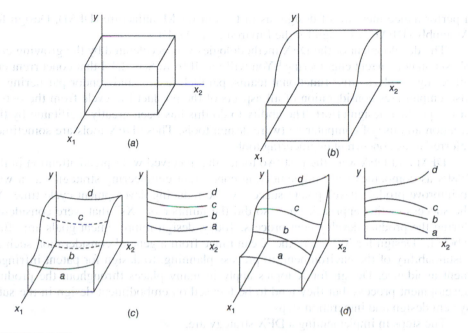

FIGURE 8.35
Different behavior of response y as a function of the parameters x_1 and x_2. (a) No effect of x_1 and x_2 on y. (b) Main effect of x_1 on y. No effect of x_2 on y. (c) Effect of x_1 and x_2 on y but no $x_1 - x_2$ interaction. (d) Main effects of x_1 and x_2. Interaction between x_1 and x_2.

- Response surface designs are used to determine the empirical relation between the factors (independent variables) and the response (performance variable). The composite design and rotatable designs are frequently used for this purpose.

Design of Experiments is facilitated by the use of many statistical design computer programs currently on the market. However, unless one is skilled in DoE it is advisable that a statistician be consulted during the development of the testing plan to be sure that you are getting the most unbiased information possible for the money that you can spend in testing. Today's engineers need a rudimentary understanding of DoE principles to make effective use of this software.

8.12
DESIGN FOR X (DFX)

A successful design must satisfy many requirements other than functionality, appearance, and cost. Reliability has been recognized as a needed attribute for many years. As more attention was focused on improving the design process, effort has been given to improving many other "ilities" such as manufacturability, maintainability, testability, and serviceability. As more life-cycle issues came under study, the terminology to describe a design methodology became known as Design for X, where X represents a performance measure of design, as in Design for Manufacture (DFM), Design for Assembly (DFA), or Design for the Environment (DFE).

The development of the DFX methodologies was accelerated by the growing emphasis on concurrent engineering.[1] You will recall from Sec. 2.4.4 that concurrent engineering involves cross-functional teams, parallel design, and vendor partnering. It also emphasizes consideration of all aspects of the product life cycle from the outset of the product design effort. The ability to do this has been greatly facilitated by the creation and use of computer software design tools. These DFX tools are sometimes referred to as concurrent engineering tools.

DFM and DFA were the first two topics that received widespread attention in the 1980s as companies were implementing concurrent engineering strategies as a way to improve product development success while reducing development cycle time. As the success of this approach grew, so did the number of "Xs" that were considered during the product development process. Today, design improvement goals are often labeled, "Design for X," where the X can range from a general consideration such as sustainability of the environment, to process planning, to design for patent infringement avoidance. Design for X topics apply in many places throughout the product development process, but they tend to be focused on embodiment design in the subsystem design and integration steps.

The steps in implementing a DFX strategy are:

- Determine the issue (X) targeted for consideration.
- Determine where to place your focus: the product as a whole, an individual component, a subassembly, or a process plan.
- Identify methods for measuring the X characteristics, and techniques to improve them. These techniques may include mathematical or experimental methods, computer modeling, or a set of heuristics.
- The DFX strategy is implemented by insisting the product development team focus on the X and by using parametric measurements and improvement techniques as early in the design process as possible.

Some of the DFX topics have been included in this chapter. Much of the rest of this text explains the DFX issues in greater detail. Also included are many other design issues not usually encompassed under the DFX rubric. The following list directs the reader to information on a variety of design issues throughout the text.

Issue	Location
Cost estimation of the product	Chaps. 16, 17
Design for X	
Assembly	Chap. 13
Environment	Chap. 10
Manufacture	Chap. 13

---

1. G. Q. Huang (ed.), *Design for X: Concurrent Engineering Imperatives,* Chapman & Hall, New York, 1996.

Quality	Chap. 15
Reliability	Chap. 14
Safety	Chap. 14
Serviceability	Sec. 8.9.4
Tolerances	Sec. 8.7
Failure modes and effect analysis	Sec. 14.5
Human factors design	Sec. 8.9
Industrial design	Sec. 8.8
Legal and regulatory issues	Chap. 18
Life-cycle cost	Secs. 8.10, 17.15
Materials selection	Chap. 11
Mistake-proofing	Sec. 13.8
Product liability	Chap. 18
Robust design	Chap. 15
Standardization in design and manufacturing	Sec. 13.7
Testing	Sec. 8.11.1
User-friendly design	Sec. 8.9

8.13

SUMMARY

Embodiment design is the phase in the design process where the design concept is invested with physical form. It is the stage where the 4Fs of design (function, form, fit, and finish) gain major attention. It is the stage where most analysis takes place to determine the physical shape and configuration of the components that make up the system. In accordance with a growing trend in the design community, we have divided embodiment design into three parts:

- Establishment of the product architecture: Involves arranging the functional elements of the product into physical units. A basic consideration is how much modularity or integration should be provided to the design.
- Configuration design: Involves establishing the shape and general dimensions of the components. Preliminary selection of materials and manufacturing processes. Design for manufacturability principles are applied to minimize manufacturing cost.
- Parametric design: Greater refinement takes place to set critical design variables to enhance the robustness of the design. This involves optimizing critical dimensions and the setting of tolerances.

By the conclusion of embodiment design a full-scale working prototype of the product will be constructed and tested. This is a working model, technically and visually complete, that is used to confirm that the design meets all customer requirements and performance criteria.

A successful design requires considering a large number of factors. It is in the embodiment phase of design that studies are made to satisfy these requirements. The physical appearance of the design, often called industrial design, affects the sales of consumer products. Human factors design determines the way that a human interfaces with and uses the design. This, too, often affects sales. Sometimes, it affects safety. Increasingly the acceptance of a product by the public is determined by whether the product is designed to be environmentally friendly. Governments, through regulation, also promote environmental design.

More issues remain to be considered in the rest of this text. A number of these are contained within the rubric DFX, such as design for assembly and design for manufacturability.

NEW TERMS AND CONCEPTS

Accelerated testing	Industrial design	Refining (in configuration design)
Assembly	Interference fit	Self-help
Clearance fit	Life-cycle design	Special-purpose component
Configuration design	Module	Stackup
Design for X	Overconstrained part	Standard assembly
Design of Experiments (DOE)	Parametric design	Standard part
Feature control frame	Patching	Subassembly
Force transmission	Preliminary design	Tolerance

BIBLIOGRAPHY

Embodiment Design

Avallone, E. A. and T. Baumeister, eds., *Marks' Standard Handbook for Mechanical Engineers,* 11th ed., McGraw-Hill, New York, 2007. (Available online at knovel.com.)

Dixon, J. R., and C. Poli: *Engineering Design and Design for Manufacturing,* Field Stone Publishers, Conway, MA, 1995, Part III.

Hatamura, Y.: *The Practice of Machine Design*, Oxford University Press, New York, 1999.

Pahl, G., W. Beitz, J. Feldhausen, and K. H. Grote: *Engineering Design: A Systematic Approach,* 3d ed., Springer, New York, 2007.

Pope, J. E. ed., *Rules of Thumb for Mechanical Engineers, Elsevier,* 1997. (A place to find practical design methods and calculations, written by experienced practicing engineers on such topics as drive motors, pumps and compressors, seals, bearings, gears, piping, stress analysis, finite element analysis, and engineering materials. Available online at knovel.com.)

Skakoon, J. G.: *The Elements of Mechanical Design,* ASME Press, New York, 2008.

Stoll, H. W.: *Product Design Methods and Practices,* Marcel Dekker, New York, 1999.

Young, W. C. and R. G. Budynas, *Roark's Formulas for Stress and Strain*, 7th ed., McGraw-Hill, New York, 2001. (Available online at knovel.com.)

PROBLEMS AND EXERCISES

8.1 Look around your environment to find some common consumer products. Identify which are primarily modular, integral, or mixed product architecture.

8.2 The standard fingernail clipper is an excellent illustration of the integral style of product architecture. The clipper system consists of four individual components: lever (A), pin (B), upper clipper arm (C), and lower clipper arm (D). Sketch a fingernail clipper, label its four components, and describe the functionality provided by each component.

8.3 Design a new fingernail clipper with totally modular product architecture. Make a sketch and label the function provided by each part. Compare the number of parts in this design with the original standard nail clipper.

8.4 Examine the various configuration designs for the right-angle bracket shown in Fig. 8.6. Make a sketch and label it to show the following forms or features: (a) solid form, (b) a rib feature, (c) a weld, (d) a cut-out feature, (e) webs.

8.5 A structure with redundant load paths is shown below. The force F causes the structure to elongate by an amount δL. Because the cross sections of the tie rods are not the same, their stiffness $k = \dfrac{\delta P}{\delta L}$ will be different. Show that the load will divide itself in proportion to the stiffness of the load path.

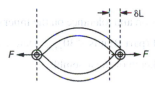

8.6 Design the ladle hooks to be used with the transfer ladle for a steel-melting furnace. The hook should be able to lift a maximum weight of 150 tons. The hook should be compatible with the interfaces shown for the ladle in the following sketch. The hook eye should receive an 8-inch-diameter pin for attaching to the crane.

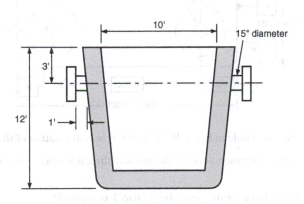

8.7 Make a three-dimensional freehand sketch of the part shown in Fig. 8.19.

8.8 Find the missing dimension AB and its tolerance.

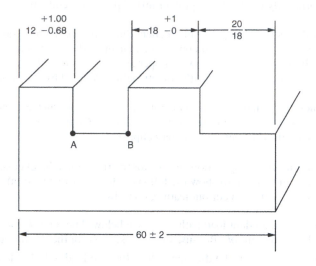

8.9 In Example 8.1, start with Point B and go clockwise around the circuit to find the gap at the wall and its tolerance.

8.10 Using Fig. 8.21, the dimension and tolerance on the inner diameter of the bearing (part A) is $\varnothing 30^{+0.20}_{-0.00}$ and for the shaft (part B) it is $\varnothing 30^{+0.35}_{+0.25}$. Determine the clearance and tolerance of the assembly. Make a sketch of the assembly.

8.11 What is the minimum distance from the holes at each end of the following part?

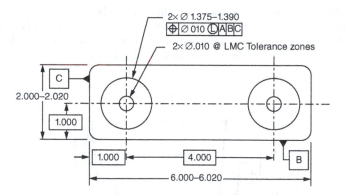

8.12 Consider the leftmost hole in Fig. 8.19. If the tolerance on location of the hole is ± 2 mm,

(a) What is the tolerance zone if the normal dimensioning system (non-GD&T) is applied?

(b) What would the tolerance zone be if GD&T is applied?

(c)   Sketch the tolerance zone for (a) and (b).

(d)   Write the feature control frame for (b) and discuss its advantages over the normal dimensional system.

8.13   Starting with Example 8.3, construct a table that shows how the tolerance zone on the position of the hole changes with the diameter of the hole if the hole is specified at the maximum material condition (MMC). Start at the MMC for the hole and change the hole size in units of 0.020 inches until it reaches the LMC. Hint: Determine the virtual condition of the hole, which is the MMC hole diameter minus the MMC positional tolerance.

8.14   Take photographs of consumer products, or tear pictures out of old magazines, to build a display of industrial designs that appeal to you, and designs that you feel need improvement. Be able to defend your decisions on the basis of aesthetic values.

8.15   Consider the design of a power belt sander for woodworking. (a) What functions of the tool depend on human use? (b) One of the features a user of this tool wants is light weight to reduce arm fatigue during prolonged use. Other than reducing the actual weight, how can the designer of this tool reduce arm fatigue for the user?

8.16   Look at the website http://www.baddesigns.com/examples.html for examples of poor user-friendly designs. Then, from your everyday environment, identify five other examples. How would you change these designs to be more user-friendly?

8.17   Diesel-powered trucks are a target for conversion to natural gas. Dig deeper into this subject to find out what has happened to bring this about.

8

9

DETAIL DESIGN

9.1
INTRODUCTION

We have come to *detail design,* the last of the three phases into which we have divided the design process. The boundary between embodiment design and detail design has become blurred and shifted forward in time by the emphasis on reducing the product development cycle time by the use of concurrent engineering methods (Design for X), enabled by computer-aided engineering (CAE). In many engineering organizations it is no longer correct to say that detail design is the phase of design where all of the dimensions, tolerances, and details are finalized. Nonetheless, detail design is the phase where *all of the details are brought together, all decisions are finalized,* and a decision is made by management to release the design for production.

Figure 9.1 shows the stages of design by which we have organized this book. The numbers of the Chaps. 8 through 17 have been superimposed in order to show you where in the process this knowledge is generally applied. Detail design is the lowest level in the hierarchy of design abstraction. It is a very specific and concrete activity. Many decisions have been made to get to this point. Most of these decisions are fundamental to the designed product and to change them now would be costly in time and effort. Poor detail design can ruin a brilliant design concept and lead to manufacturing defects, high costs, and poor reliability in service. The reverse is not true. A brilliant detail design will not rescue a poor conceptual design. Thus, as the name implies, detail design[1] is mainly concerned with confirming details and supplying missing ones to ensure that a proven and tested design can be manufactured into a quality and cost-effective product. An equally

---

1. Here *detail* is used as a noun. The team pulls together and confirms all details. Some writers have mistakenly called this stage *detailed design,* implying it is more like the embodiment design stage. This is not true, especially in the current business environment.

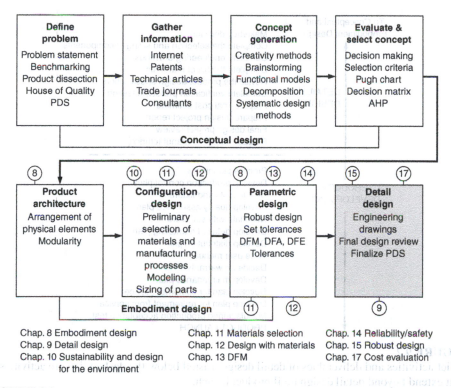

FIGURE 9.1

Steps in the design process, showing where Chaps. 8 through 17 are chiefly applied.

important task of detail design is communicating these decisions and data to the parts of the business organization that will carry on the product development process.

9.2
ACTIVITIES AND DECISIONS IN DETAIL DESIGN

Figure 9.2 shows the tasks to be completed as a result of activities in the detail design phase. These steps are the culmination of the decision made at the end of Phase 0, product planning (see Fig. 2.1), to allocate capital funding to proceed with the product development program. Below the dashed line in Fig. 9.2 are the main activities involved in the product development process that must be completed by other departments in the company once the design information is transmitted to them; see Sec. 9.5. The activities in the detail design phase are as follows:

Make/Buy Decision

Even before the design of all components is completed and the drawings finalized, meetings are held on deciding whether to make a component in-house or to buy it

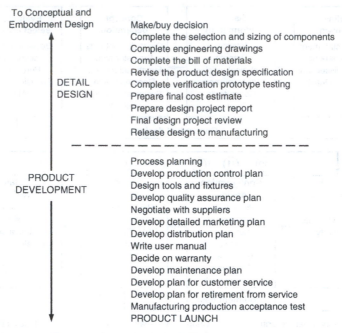

To Conceptual and
Embodiment Design

DETAIL
DESIGN

Make/buy decision
Complete the selection and sizing of components
Complete engineering drawings
Complete the bill of materials
Revise the product design specification
Complete verification prototype testing
Prepare final cost estimate
Prepare design project report
Final design project review
Release design to manufacturing

PRODUCT
DEVELOPMENT

Process planning
Develop production control plan
Design tools and fixtures
Develop quality assurance plan
Negotiate with suppliers
Develop detailed marketing plan
Develop distribution plan
Write user manual
Decide on warranty
Develop maintenance plan
Develop plan for customer service
Develop plan for retirement from service
Manufacturing production acceptance test
PRODUCT LAUNCH

FIGURE 9.2
Chief activities and deliverables of detail design. Listed below the dashed line are activities
that extend beyond detail design until product launch.

from an external supplier. This decision will be made chiefly on the basis of cost and
manufacturing capacity, with due consideration given to issues of quality and reliability of delivery of components. Sometimes the decision to manufacture a critical component in-house is based solely on the need to protect trade secrets concerned with a
critical manufacturing process. An important reason for making this decision early is
so you can bring the supplier into the design effort as an extended team member.

Complete the Selection and Sizing of Components

While most of the selection and sizing of components occurs in embodiment design,
especially for those components with parameters deemed to be critical-to-quality,
some components may not yet have been selected or designed. These may be standard
components that will be purchased from external suppliers or routine standard parts
like fasteners. Or, there may be a critical component for which you have been waiting
for test data or FEA analysis results. Regardless of the reason, it is necessary to complete these activities before the design can be complete.

If the product design is at all complex, it most likely will be necessary to impose
a *design freeze* at some point prior to completion. This means that beyond a certain
point in time no changes to the design will be permitted unless they go through a formal review by a design control board. This is necessary to prevent the human tendency
to continually make slight improvements, which unless controlled by some external

means results in the job never actually being completed. With a design freeze, only those last-minute changes that truly affect performance, safety, or cost are approved.

Complete Engineering Drawings

A major task in the detail design phase is to complete the engineering drawings. As each component, subassembly, and assembly is designed, it is documented completely with drawings (see Sec. 9.3.1). Drawings of individual parts are usually called *detail drawings*. These show the geometric features, dimensions, and tolerances of the parts. Sometimes special instructions for processing the part in manufacture, like heat treating or finishing steps, are included on the drawing. Assembly drawings show how the parts are put together to create the product or system.

Complete the Bill of Materials

The bill of materials (BOM) or parts list is a list of each individual component in the product, Sec. 9.3.2. It is used in planning for manufacture and in determining the best estimate of product cost.

Revise the Product Design Specification

When the Product Design Specification was introduced in Sec. 3.7 it was emphasized that the PDS is a "living document" that changes as the design team gains more knowledge about the design of the product. In detail design the PDS should be updated to include all current requirements that the design must meet.

We need to distinguish between the part specification and the product design specification. For individual parts the drawing and the specification are often the same document. When a part specification is issued it contains information on the technical performance of the part, its dimensions, test requirements, materials requirements, reliability requirement, design life, packaging requirement, and marking for shipment. The part specification should be sufficiently detailed to avoid confusion as to what is expected from the supplier.

Complete Verification Prototype Testing

Once the design is finalized, a beta-prototype is built and verification tested to ensure that the design meets the PDS and that it is safe and reliable. Recall from Sec. 8.11.1 that beta-prototypes are made with the same materials and manufacturing processes as the product but not necessarily from the actual production line. Later, before product launch, actual products from the production line will be tested. Depending on the complexity of the product, the verification testing may simply be to run the product during an expected duty cycle and under overload conditions, or it may be a series of statistically planned tests.

Final Cost Estimate

The detail drawings allow the determination of final cost estimates, since knowledge of the material, the dimensions, tolerances, and finish of each part are needed to determine manufacturing cost. To make these calculations a bill of materials (see Sec. 9.3.2) is utilized. Cost analysis also needs specific information about the particular machines and process steps that will be used to make each part. Note that cost estimates will have been made at each step of the product design process with successively smaller margins for error.

Prepare Design Project Report

A design project report usually is written at the conclusion of a project to describe the tasks undertaken and to discuss the design in detail. This is a vital document for passing on design know-how to a subsequent design team engaged in a product redesign project. Also, a design project report may be an important document if the product becomes involved in either product liability or patent litigation. Suggestions for preparing a design project report are given in Sec. 9.3.3.

Final Design Review

Many formal meetings or reviews will have preceded the final design review. These include an initial product concept meeting to begin the establishment of the PDS, a review at the end of conceptual design to decide whether to proceed with full-scale product development, and a review after embodiment design to decide whether to move into detail design. The latter may take the form of detailed partial reviews (meetings) to decide important issues like design for manufacturing, quality issues, reliability, safety, or preliminary cost estimates. However, the final design review is the most structured and comprehensive of the reviews.

The final design review results in a decision by management on whether the product design is ready for production, and the major financial commitment that this entails. Section 9.4 discusses the final design review.

Release Design to Manufacturing

The release of the product design to manufacturing ends the main activity of the design personnel on that product. The release may be done unconditionally, or under pressure to introduce a new product it may be done conditionally. In the latter case, manufacturing moves ahead to develop tooling while design works on an accelerated schedule to fix some design deficiencies. The increasing use of the concurrent engineering approach to minimize the product development time blurs the boundary between detail design and manufacturing. It is common to release the design to manufacturing in two or three "waves," with those designs that have the longest lead time for designing and making tooling being released first.

If the product is being managed by a project manager in a heavyweight matrix organization, such as that discussed in Sec. 2.4.3, this manager continues with the project as it passes from design to manufacturing and on to product launch. Also, design input does not necessarily stop once manufacturing takes over, for their technical expertise is needed in such areas as quality assurance, warranty issues, and deciding on maintenance requirements.

The activities shown in Fig. 9.2 below the dotted line occur after detail design. They are necessary to complete the product development process and are discussed in Sec. 9.5.

9.3
COMMUNICATING DESIGN AND MANUFACTURING INFORMATION

A design project generates a very large amount of data. A typical automobile has about 10,000 parts, each containing as many as 10 geometric features. Also, for every geometric feature on a mechanical part that must be manufactured, there are about 1000 geometric features related to the manufacturing equipment and support apparatus, such as fixtures. CAD representation of parts has become commonplace, and this permits the transfer of design drawings via the Internet from design centers to tool makers or manufacturing plants anywhere in the world. Design data consists of engineering drawings made for various purposes, design specifications, bills of material, final design reports, progress reports, engineering analyses, engineering change notices, results from prototype tests, minutes of design reviews, and patent applications. However, the interoperability and exchange among this broad range of design databases is far from optimum. "There is no common data architecture that can hold and exchange technical information such as part shapes, bills of materials, product configurations, functional requirements, physical behavior, and much else that is required for deep exploitation of virtual manufacturing."[1] Regardless of the sophistication of computer-based design, drawing, and planning tools, engineers will always need to communicate in the written and spoken word.

9.3.1 Engineering Drawings

The historical goal of detail design has been to produce drawings that contain the information needed to manufacture the product. These drawing should be so complete that they leave no room for misinterpretation. The information on a detail drawing includes:

- Standard views of orthogonal projection—top, front, side views
- Auxiliary views such as sections, enlarged views, or isometric views that aid in visualizing the component and clarifying the details
- Dimensions—presented according to the GD&T standard ANSI Y14.5M
- Tolerances
- Material specification, and any special processing instructions
- Manufacturing details, such as parting line location, draft angle, surface finish

---

1. *Retooling Manufacturing: Bridging Design, Materials, and Production,* National Academies Press, Washington, DC, 2004, p. 10.

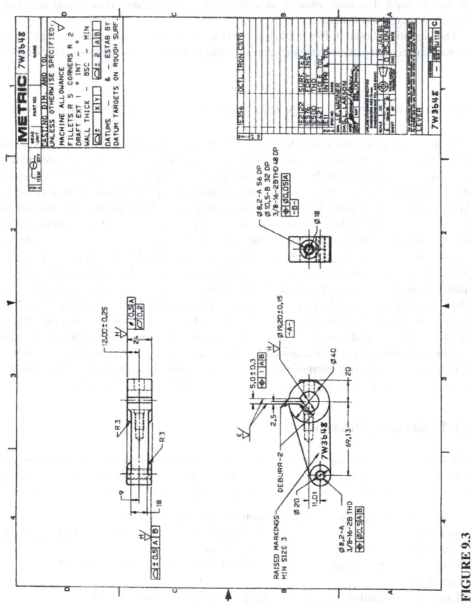

FIGURE 9.3
A detail design drawing of a lever.

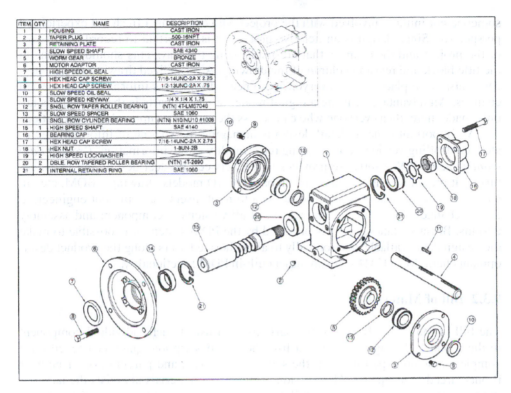

ITEM	QTY	NAME	DESCRIPTION
1	1	HOUSING	CAST IRON
2	2	TAPER PLUG	.500-16NPT
3	2	RETAINING PLATE	CAST IRON
4	1	SLOW SPEED SHAFT	SAE 4340
5	1	WORM GEAR	BRONZE
6	1	MOTOR ADAPTOR	CAST IRON
7	1	HIGH SPEED OIL SEAL	
8	4	HEX HEAD CAP SCREW	7/16-14UNC-2A X 2.25
9	8	HEX HEAD CAP SCREW	1/2-13UNC-2A X .75
10	2	SLOW SPEED OIL SEAL	
11	1	SLOW SPEED KEYWAY	1/4 X 1/4 X 1.75
12	2	SNGL. ROW TAPER ROLLER BEARING	(NTN) 4T-LM67048
13	2	SLOW SPEED SPACER	SAE 1060
14	1	SNGL. ROW CYLINDER BEARING	(NTN) N10/NU10 #1008
15	1	HIGH SPEED SHAFT	SAE 4140
16	1	BEARING CAP	
17	4	HEX HEAD CAP SCREW	7/16-14UNC-2A X 2.75
18	1	HEX NUT	1-8UN-2B
19	2	HIGH SPEED LOCKWASHER	
20	2	DBLE. ROW TAPERED ROLLER BEARING	(NTN) 4T-2690
21	2	INTERNAL RETAINING RING	SAE 1060

FIGURE 9.4
Exploded assembly drawing for a speed reducer.

Sometimes a specification sheet replaces the notes on the drawing and accompanies it. Figure 9.3 is an example of a detail drawing for a lever. Note the use of GD&T dimensions and tolerances. If the design is developed digitally in a CAD system, then the digital model becomes the governing authority for the component definition.

Two other common types of engineering drawings are the layout drawing and the assembly drawing. *Design layouts* show the spatial relationships of all components in the assembled product (the system). The design layout is developed fully in the product architecture step of embodiment design. It serves to visualize the functioning of the product and to ensure that there is physical space for all of the components.

Assembly drawings are created in detail design as tools for passing design intent to the production department, as well as the user. They show how the part is related in space and connected to other parts of the assembly. Dimensional information in assembly drawings is limited to that necessary for the assembly. Reference is made to the detail drawing number of each part for full information on dimensions and tolerances. Figure 9.4 is an exploded assembly drawing of a speed reducer.

When a detail drawing is finished, it must be checked to ensure that the drawing correctly portrays the function and fit of the design.[1] Checking should be performed by

1. G. Vrsek, "Documenting and Communicating the Design," *ASM Handbook,* vol. 20, ASM International, Materials Park, OH, 1998, pp. 222–30.

someone not initially involved with the project who can bring a fresh but experienced perspective. Since design is an iterative process, it is important to record the history of the project and the changes that are made along the way. This should be done in the title block and revision column of the drawing. A formalized drawing release process must be in place so that everyone who needs to know is informed about design changes. An advantage of using a digital model of design parts is that if changes are only made there, then everyone who can access the model has up-to-date information.

An important issue in detail design is managing the volume of information created, controlling versions, and assuring retrievability of the information. *Product data management* (PDM) software provides a link between product design and manufacturing. It provides control of design databases (CAD models, drawings, BOM, etc.) in terms of check-in and check-out of the data to multi-users, carrying out engineering design changes, and control of the release of all versions of component and assembly designs. Because data security is provided by the PDM system, it is possible to make the design data available electronically to all authorized users along the product development chain. Most CAD software has a built-in PDM functionality.

9.3.2 Bill of Materials

The bill of materials (BOM) or the parts list is a list of each individual component in the product. As Fig. 9.5 shows, it lists the part description, quantity needed for a complete assembly, part number, the source of the part, and purchase order number if outsourced to a supplier. This version of the bill of materials also lists the name of the engineer responsible for the detail design of each part, and the name of the project engineer who is responsible for tracking the parts through manufacture and assembly.

ENGINE PROGRAM PARTS LIST									
DOCUMENTING THE DESIGN									
Qty /		PART NUMBER					Delivery	RESPONSIBILITY	
Engine	PART DESCRIPTION	Prefix	Base	End	P.O. #	Source	Date	Design	Engineer
	PISTON								
6	PISTON (CAST/MACH)	SRLE	6110	24093	RN0694	Ace	11/17/95	S. LOPEZ	M. Mahoney
6	PISTON RING - UP COMPRESSION	SRLE	6150	AC	RN0694	Ace	rec'd FRL	S. LOPEZ	M. Mahoney
6	PISTON RING - LOWER COMPRESSION	SRLE	6152	AC	RN0694	Ace	rec'd FRL	S. LOPEZ	M. Mahoney
12	PISTON RING - SEGMENT OIL CONTROL	SRLE	6159	AC	RN0694	Ace	rec'd FRL	S. LOPEZ	M. Mahoney
6	PISTON RING - SPACER OIL CONTROL	SRLE	6161	AB	RN0694	Ace	rec'd FRL	S. LOPEZ	M. Mahoney
6	PIN - PISTON	SRLE	6135	AA		BN Inc.		S. LOPEZ	M. Mahoney
6	PISTON & CONNECTING ROD ASSY	SRLE	6100	AG				S. LOPEZ	M. Mahoney
6	CONNECTING ROD - FORGING	SRLE	6205	AA		Formall		S. LOPEZ	M. Mahoney
6	CONNECTING ROD ASSY	SRLE	6200	CI		MMR Inc.		S. LOPEZ	M. Mahoney
12	BUSHINGS - CONNECTING ROD	SRLE	6207	AE		Bear Inc.		S. LOPEZ	M. Mahoney
12	RETAINER - PISTON PIN	SRLE	6140	AC		Spring Co.		S. LOPEZ	M. Mahoney

FIGURE 9.5
An example of a bill of materials. (*ASM Handbook,* vol. 20, p. 228, ASM International. Used with permission.)

The bill of materials has many uses. It is essential for determining the cost of the product. A bill of materials will be started early in the embodiment design phase, when the product architecture has been established, as a way of checking whether the product costs are in line with that called for in the PDS. The bill of materials will be finalized in the detail design phase and will be used in the detailed cost analysis. The bill of materials is vital for tracking the parts during manufacture and assembly. It is an important archival document for the design that needs to be preserved and be available for retrieval.

9.3.3 Written Documents

Novice design engineers often are surprised at how much time is spent in writing tasks connected with a design project. Design is a complicated process with many stakeholders. There are many groups who provide input to the design process and many groups who participate in decision-making during the process. Often a current decision can only be made after reviewing work done earlier in the design process. Members of a design team on a complicated project may need to refresh their memories about work done at a prior stage in the process just to move into new stages. The importance of creating an accessible and correct collection of information on all aspects of the design process cannot be over emphasized.

The critical need for precise and formal documentation drives all design engineers to become effective at writing technical documents. Written documents create a lasting record of the author's work. Rightly or wrongly the quality of the documentation gives a lasting impression of the quality of the work and of the skill of the writer.

Design engineers prepare both informal and formal documents as part of their daily routines. Informal documentation includes e-mail messages, and brief memoranda and daily entries in a design journal. Formal written documentation usually takes the form of letters, formal technical reports (e.g., progress reports, laboratory reports, process descriptions), technical papers, and proposals.

Electronic Mail

No form of communication has grown so rapidly as electronic mail (e-mail). Well over eight trillion e-mail messages are sent each year. Electronic mail is invaluable for scheduling meetings, communicating between engineers who are continents apart, communicating with the office while on a trip, confirming decisions made and action items, keeping up with the activities of professional societies, to name a few common uses.

It is important to use e-mail appropriately. E-mail cannot take the place of a face-to-face meeting or a telephone call. You cannot assume that the recipient has read an e-mail, so it is not appropriate to use e-mail when you need assurance that a message is received by a particular time.

The following are guidelines to professional e-mail writing.

- For formal business correspondence, write as you would in a business letter. Use proper capitalization, spelling and sentence structure.
- Use informative and brief Subject lines in all your messages.
- Keep your messages short.

- Compress any attachments that are large and refrain from sending them to a colleague without advance warning.
- Do not use emoticons or other informal visuals better suited for instant messaging or text messaging in personal messages.
- In addition to an informal signature use a formal signature block that includes the same contact information one would have on a business card.
- Delete unnecessary or repetitive information in a response to a sender when you are including their original message in your reply.
- Include relevant detail when you are responding to a sender without including their original message in your reply.

E-mail is instant and personal so there is a tendency to treat it differently from other written communication. Users often treat e-mail with the informality of a telephone call. People feel free to write and send things they would never put in a business letter. E-mail seems to free people from their normal inhibitions. It is easy to "reply" to a message without thinking about the consequences. There are many documented instances of two business friends "having fun" in their e-mail exchange, only to discover to their embarrassment that the message inadvertently was given mass circulation. It is important to remember that e-mails can be saved and retrieved just like newspapers.

Naturally, there are many online sources for etiquette in using online communication technology. Most technical writing manuals include sections on e-mail composition. A good mindset for e-mail writing is to *expect to lose control over the dissemination or reproduction of any information you include in an e-mail message,* so compose e-mails thoughtfully.

Memorandum Reports

The memorandum report (a.k.a. a *memo*) usually is written to a specific person or group of persons concerning a specific topic with which both the writer and recipient(s) are familiar. A memorandum is a letter written to a colleague or group of colleagues within the same organization. A memo is transmitted to the recipient by internal means (not the U.S. Postal Service). A memorandum report is an effective way to communicate the same information to an entire business unit or all members of the same group. Since a memo is distributed within an organization, it does not require an individual address for each recipient. Memos are appropriate for reporting on a trip you took to observe a competitor's new product; the disclosure of a new idea you have for an improved product; or disseminating minutes of a meeting of the design team.

Memorandum reports are short (one to three pages). The purpose in writing a memorandum report is to get a concise report to interested parties as quickly as possible. The main emphasis is on results, discussion, and conclusions with a minimum of details unless, of course, those details are critical to the analysis of the data.

The Design Notebook

Unfortunately, there is not a strong tradition of recording the decisions made during design and capturing the broad picture of *design intent* so that the knowledge is not lost with the designer, and so that novices can learn from it. The place where this information most often is found is the *design notebook.* It should be an 8 by 11-inch bound notebook (not spiral bound), preferably with a hard cover. It should be the repository for all of your planning (including plans that were not carried out), all analytical calculations, all records of experimental data, all references to sources of information, and all significant thinking about your project.

Information should be entered directly into the notebook, not recopied from rough drafts. However, you should organize the information you enter in the notebook. Use main headings and subheadings; label key facts and ideas; liberally cross-reference your material; and keep an index at the front of the book to aid in your organization. Pages should be numbered consecutively. About once a week, review what you have done and write a summary of your progress that emphasizes the high points. Whenever you do anything that may seem the least bit patentable, have your notebook read and witnessed by a knowledgeable colleague.

The following are good rules[1] for keeping a design notebook.

- Keep an index at the front of the book.
- Entries should be made in ink and, of course, must be legible.
- Make your entries at the time you do the work. Include favorable and unfavorable results and things not fully understood at the time. If you make errors, just cross them out. Do not erase, and never tear a page out of the notebook.
- All data must be in their original primary form (strip charts, oscilloscope pictures, photomicrographs, etc), not after recalculation or transformation.
- Rough graphs should be drawn directly in the notebook, but more carefully prepared plots on graph paper also should be made and entered in the book.
- Give complete references to books, journals, reports, patents and any other sources of information.

A good engineering design notebook is one from which, several years after the project is completed, critical decisions will be apparent, and the reasons for the actions taken will be backed up by facts. It should be possible to show where every figure, statement and conclusion of the published report of the project can be substantiated by original entries in the design notebook.

Formal written documentation done by design engineers is of two basic types. Engineers write letters to persons outside their organization using the standard business guidelines. These are available in any text on writing. Engineers also must write technical reports. The definition of technical writing is that it is concise and precise, written for a specialty audience, and relies on visual and written data and analysis of results. The purpose of engineering writing is to present information, not to entertain or dazzle them with flowery language. Therefore, the information should be easy to find within the written document. Always when writing your report, keep in mind the busy reader who has limited time. A *design project report* is a standard technical report that is written at the conclusion of a project. It describes the tasks undertaken and discusses the design in detail. It is an important category of the formal technical report.

Formal Technical Reports

A formal technical report usually is written at the end of a project. Generally, it is a complete, stand-alone document aimed at persons having widely diverse backgrounds. Therefore, much more detail is required than for the memorandum report.

---

1. Adapted from T. T. Woodson, "Engineering Design," Appendix F, McGraw-Hill, New York, 1966.

The outline of a typical professional report[1] might be:

- Cover letter (letter of transmittal): The cover letter is provided so that persons who might receive the report without prior notification will have some introduction to it.
- Title page: The title page includes names, affiliations, and addresses of the authors.
- Executive summary (containing conclusions): The summary is generally less than a page in length and contains three paragraphs. The first briefly describes the objective of the study and the problems studied. Paragraph two describes your solution to the problem. The last paragraph addresses its importance to the business in terms of cost savings, improved quality, or new business opportunities.
- Table of contents, including list of figures and tables.
- Introduction: The introduction contains the pertinent technical facts that might be unknown to the reader but will be used in the report.
- Technical issue sections (analysis or experimental procedures, pertinent results, discussion of results):

  - The experimental procedure section is usually included to indicate how the data were obtained and to describe any nonstandard methods or techniques that were employed.
  - The results section describes the results of the study and includes relevant data analysis. Any experimental error allowances are included here.
  - The discussion section presents data analysis analyzing the data to make a specific point, develops the data into some more meaningful form, or relates the data to theory described in the introduction.

- Conclusions: The conclusion section states in as concise a form as possible the conclusions that can be drawn from the study. In general, this section is the culmination of the work and the report.
- References: References support statements in the report and lead the reader to more in-depth information about a topic.
- Appendixes: Appendixes are used for mathematical developments, sample calculations, etc., that are not directly associated with the subject of the report and that, if placed in the main body of the report, would seriously impede the logical flow of thought. Final equations developed in the appendixes are then placed in the body of the report with reference to the appendix in which they were developed.

9.3.4 Common Challenges in Technical Writing

The following suggestions are presented as a guide to writing and an aid in avoiding some of the most common mistakes. You also should avail yourself of one of the popular guides to English grammar and style.[2]

---

1. The contribution of Professor Richard W. Heckel for much of the material in this section is acknowledged.
2. W. Strunk and E. B. White, "The Elements of Style," 4th ed., Allyn & Bacon, Needham Heights, MA, 2000; S. W. Baker, "The Practical Stylist," 8th ed., Addison-Wesley, Reading, MA, 1997.

Tense

The choice of the tense of verbs is often confusing. The following simple rules are usually employed by experienced writers:

- Past tense: Use to describe work completed or in general to past events. "Hardness readings were taken on all specimens."
- Present tense: Use in reference to items and ideas in the report itself. "It is clear from the data in Figure 4 that the motor speed is not easily controlled" or "The group recommends that the experiment be repeated" (present opinion).
- Future tense: Use in making prediction from the data that will be applicable in the future. "The market data given in Table II indicate that the sales for the new product line will continue to increase in the next ten years."

References

References are usually placed at the end of the written text. Those to the technical literature (described as readily available on subscription and included in most library collections) are made by author and journal reference (often with the title of article omitted) as shown by the following example. There is no single universally accepted format for references. Each publishing organization has a preferred style for referencing material. Examples are given below.

- Technical Journal Article: C. O. Smith, Transactions of the ASME, Journal of Mechanical Design, 1980, vol. 102, pp. 787–792.
- Book: Thomas T. Woodson: "Introduction to Engineering Design," pp. 321–346. McGraw-Hill, New York, 1966.
- A private communication: J. J. Doe, XYZ Company, Altoona, PA, unpublished research, 2004.
- Internal reports: J. J. Doe, Report No. 642, XYZ Company, Altoona, PA, February 2001.

Many engineering journals use the style guidelines for referencing developed by the IEEE.[1]

9.3.5 Meetings

The business world is full of meetings that are held to exchange information and plan on a variety of levels and subjects. Most of these involve some kind of prepared oral presentation; see Sec. 9.3.6. At the lowest level of formality in this hierarchy is the *design team meeting.* Those present are focused on a common goal and have a generally common background. The purpose of the meeting is to share the progress that has been made, identify problems, and hopefully, find help and support in solving the problems. This is a group discussion, with an agenda and probably some visual aids, but the presentation is informal and not rehearsed. Detailed tips for effectively holding this type of meeting were given in Sec. 4.5.

---

1. *IEEE Editorial Style Manual* [Online]. Available: http://www.ieee.org/documents/stylemanual.pdf

Next up in the meeting hierarchy would be a *design briefing* or design review. The size and diversity of the audience would depend on the importance of the project. It could vary from 10 to 50 people and include company managers and executives. A design briefing for high-level management must be short and to the point. Such people are very busy and not at all interested in the technical details that engineers love to talk about. A presentation of this type requires extensive preparation and practice. Usually you will have only 5 to 10 minutes to get your point across to the top executive. If you are speaking to an audience of technical managers, they will be more interested in the important technical details, but don't forget also to cover information on schedule and costs. Generally, they will give you 15 to 30 minutes to get your points across.

A presentation similar to the design briefing on technical details is a talk before a professional or technical society. Here you will generally have 15 to 20 minutes to make your presentation before an audience of 30 to 100 people. Speaking at this kind of venue, whether at a national or local meeting, can be an important step in developing your career and in gaining professional reputation.

9.3.6 Oral Presentations

Impressions and reputations (favorable or unfavorable) are made most quickly by audience reaction to an oral presentation. There are a number of situations in which you will be called upon to give a talk. Oral communication has several special characteristics: quick feedback by questions and dialogue; impact of personal enthusiasm; impact of visual aids; and the important influence of tone, emphasis, and gesture. A skilled speaker in close contact with an audience can communicate far more effectively than the cold, distant, easily evaded written word. On the other hand, the organization and logic of presentation must be of a higher order for oral than for written communication. The listener to an oral communication has no opportunity to reread a page to clarify a point. Many opportunities for "noise" exist in oral communication. The preparation and delivery of the speaker, the environment of the meeting room, and the quality of the visual aids all contribute to the efficiency of the oral communication process.

The Design Briefing

The purpose of your talk may be to present the results of the past three months of work by a 10-person design team, or it may be to present some new ideas on computer-aided design to an audience of upper management who are questioning if their large investment in CAD has paid off. Whatever the reason, you should know the purpose of your talk and have a good idea of who will be attending your presentation. This information is vital if you are to prepare an effective talk.

The most appropriate type of delivery for most business-oriented talks is an *extemporaneous-prepared talk*. All the points in the talk are thought out and planned in detail. However, the delivery is based on a written outline, or alternatively, the text of the talk is completely written but the talk is delivered from an outline prepared from the text. This type of presentation establishes a closer, more natural contact with the audience that is much more believable than if the talk is read by the speaker.

Develop the material in your talk in terms of the interest of the audience. Organize it on a thought-by-thought rather than a word-by-word basis. Write your conclusions first. That will make it easier to sort through all the material you have and to select only the pieces of information that support the conclusions. If your talk is aimed at selling an idea, list all of your idea's strengths and weaknesses. That will help you counter arguments against adopting your idea.

The opening few minutes of your talk are vital in establishing whether you will get the audience's attention. You need to "bring them up to speed" by explaining the reason for your presentation. Include enough background that they can follow the main body of your presentation, which should be carefully planned. Stay well within the time allotted for the talk so there is an opportunity for questions. Include humorous stories and jokes in your talk only if you are very good at telling them. Also, avoid specialized technical jargon in your talk. Before ending your presentation, summarize your main points and conclusions. The audience should have no confusion as to the message you wanted to deliver.

Visual aids are an important part of any technical presentation; good ones can increase the audience retention of your ideas by 50 percent. The type of visual aid to use depends upon the nature of the talk and the audience. For a small informal meeting of up to 10 or 12 people, handouts of an outline, data, and charts usually are effective. Transparencies used with an overhead projector or PowerPoint slides with digital projection are good for groups from 10 to 200 people. Slides are the preferred visual aids for large audiences.

The usual reason a technical talk is poor is lack of preparation. It is a rare person who can give an outstanding talk without practicing it. Once you have prepared the talk, the first stage is individual practice. Give the talk out loud in an empty room to fix the thoughts in your mind and check the timing. You may want to memorize the introductory and concluding remarks. If at all possible, videotape your individual practice. The dry run is a dress rehearsal before a small audience. If possible, hold the dry run in the same room where you will give the talk. Use the same visual aids that you will use in your talk. The purpose of the dry run is to help you work out any problems in delivery, organization, or timing. There should be a critique following the dry run, and the talk should be reworked and repeated as many times as are necessary to do it right.

When delivering the talk, if you are not formally introduced, you should give your name and the names of any other team members. This information also should be on your first slide. You should speak loudly enough to be easily heard. For a large group, that will require the use of a microphone. Work hard to project a calm, confident delivery, but don't come on in an overly aggressive style that will arouse adversarial tendencies in your audience. Avoid annoying mannerisms like rattling the change in your pocket and pacing up and down the platform. Whenever possible, avoid talking in the dark. The audience might well go to sleep or, at worst, sneak out. Maintaining eye contact with the audience is an important part of the feedback in the communication loop.

The questions that follow a talk are an important part of the oral communication process; they show that the audience is interested and has been listening. If at all possible, do not allow interruptions to your talk for questions. If the "big boss" interrupts with a question, compliment him for his perceptiveness and explain that the point will be covered in a few moments. Never apologize for the inadequacy of your results. Let a questioner complete the questions before breaking in with an answer. Avoid being argumentative or letting the questioner see that you think the question is stupid. Do not prolong the question period unnecessarily. When the questions slack off, adjourn the meeting.

9.4
FINAL DESIGN REVIEW

The final design review should be conducted when the detail drawings are complete and ready for release to manufacturing. In most cases beta-prototype testing will have been completed. The purpose of the final design review is to compare the design against the most updated version of the product design specification (PDS) and a design review checklist, and to decide whether the design is ready for production.

The general conditions under which design reviews are held were discussed in Sec. 1.8. Since this is the last review before design release, a complete complement of personnel should be in attendance. This would include design specialists not associated with the project to constructively review that the design meets all requirements of the PDS. Other experts review the design for reliability and safety, quality assurance, field service engineering, and purchasing. Marketing people or the customer's representatives will be present. Manufacturing personnel will be in strong attendance, especially plant operating management responsible for producing the design, and DFM experts. Other experts, who might be called in, depending upon circumstances, are representatives from legal, patents, human factors, or R & D. Supplier representation is often desirable. The intent is to have a group comprised of people with different expertise, interests, and agendas. The chairperson of the final design review will be an important corporate official like the VP of engineering, the director of product development, or an experienced engineering manager, depending on the importance of the product.

An effective design review consists of three elements:[1] (1) input documents, (2) an effective meeting process, and (3) an appropriate output.

9.4.1 Input Documents

The input for the review consists of documents such as the PDS, the QFD analysis, key technical analyses like FEA and CFD, FMEAs, the quality plan, including robustness analysis, the testing plan and results of the verification tests, the detail and assembly drawings, the product specifications, and cost projections. This documentation can be voluminous, and it is not all covered in the final review. Important elements will have been reviewed previously, and they will be certified as satisfactory at the final review. Another important input to the meeting is the selection of the people who will attend the review. They must be authorized to make decisions about the design and have the ability and responsibility to take corrective action.

Everyone attending the design review must receive a package of information well before the meeting. An ideal way to conduct a review is to hold a briefing session at least 10 days before the formal review. Members of the design team will make presentations to review the PDS and design review checklist to ensure that the review

---

1. K. Sater-Black and N. Iverson, *Mechanical Engineering*, March 1994, pp. 89–92.

team has a common understanding of the design requirements. Then an overview of the design is given, describing how the contents of the design review information package relate to the design. Finally, members of the design review team will be assigned questions from the design checklist for special concentration. This is an informational meeting. Criticism of the design is reserved for the formal design review meeting.

9.4.2 Review Meeting Process

The design review meeting should be formally structured with a well-planned agenda. The final design review is more of an audit in contrast to the earlier reviews, which are more multifunctional problem-solving sessions. The meeting is structured so that it results in a documented assessment of the design. The review uses a checklist of items that need to be considered. Each item is discussed and it is decided whether it passes the review. The drawings, simulations, test results, FMEAs, and other elements are used to support the evaluation. Sometimes a 1–5 scale is used to rate each requirement, but in a final review an "up or down" decision needs to be made. Any items that do not pass the review are tagged as action items with the name of the individual responsible for corrective action. Figure 9.6 shows an abbreviated checklist for a final design review. A new checklist should be developed for each new product. While the checklist in Fig. 9.6 is not exhaustive, it is illustrative of the many details that need to be considered in the final design review.

   The design review builds a paper trail of meeting minutes, the decisions or ratings for each design requirement, and a clear action plan of what will be done by whom and by when to fix any deficiencies in the design. This is important documentation to be used in any future product liability or patent litigation, and for guidance when the time comes for a product redesign.

9.4.3 Output from Review

The output from the design review is a decision as to whether the product is ready to release to the manufacturing department. Sometimes the decision to proceed is tentative, with several open issues that need to be resolved, but in the judgment of management the fixes can be made before product launch.

9.5
DESIGN AND BUSINESS ACTIVITIES BEYOND DETAIL DESIGN

Figure 9.2 shows a number of activities that must be carried out after the end of the detail design phase in order to launch a product. In this section we briefly discuss each activity from the viewpoint of the engineering information that must be supplied to each of these business functions. These activities are divided into two groups: technical (manufacturing or design) and business (marketing or purchasing).

1. Overall requirements—does it meet:
   Customer requirements
   Product design specification
   Applicable industry and governmental standards
2. Functional requirements—does it meet:
   Mechanical, electrical, thermal loads
   Size and weight
   Mechanical strength
   Projected life
3. Environmental requirements—does it meet:
   Temperature extremes, in operation and storage
   Extremes of humidity
   Extremes of vibration
   Shock
   Foreign material contamination
   Corrosion
   Outdoor exposure extremes (ultraviolet radiation, rain, hail, wind, sand)
4. Manufacturing requirements—does it meet:
   Use of standard components and subassemblies
   Tolerances consistent with processes and equipment
   Materials well defined and consistent with performance requirements
   Materials minimize material inventory
   Have critical control parameters been identified?
   Manufacturing processes use existing equipment
5. Operational requirements
   Is it easy to install in the field?
   Are items requiring frequent maintenance easily accessible?
   Has serviceperson safety been considered?
   Have human factors been adequately considered in design?
   Are servicing instructions clear? Are they derived from FMEA or FTA?
6. Reliability requirements
   Have hazards been adequately investigated?
   Have failure modes been investigated and documented?
   Has a thorough safety analysis been conducted?
   Have life integrity tests been completed successfully?
   Has derating been employed in critical components?
7. Cost requirements
   Does the product meet the cost target?
   Have cost comparisons been made with competitive products?
   Have service warranty costs been quantified and minimized?
   Has value engineering analysis been made for possible cost reduction?
8. Other requirements
   Have critical components been optimized for robustness?
   Has a search been conducted to avoid patent infringement?
   Has prompt action been taken to apply for possible patent protection?
   Does the product appearance represent the technical quality and cost of the product?
   Has the product development process been adequately documented for defense in possible product
      liability action?
   Does the product comply with applicable laws and agency requirements?

FIGURE 9.6
Typical items on a final design review checklist.

Technical activities

- Process planning: Decisions on what manufacturing processes and process steps
  are needed to make parts that will be made in-house must be taken before a cost
  estimate can be made for the final design review. Additional process planning

finalizes these decisions and makes adjustments between make-buy determinations. This requires detail drawings with final dimensions and tolerances.

- Develop production control plan: Production control is concerned with routing, scheduling, dispatching, and expediting the flow of components, subassemblies, and assemblies for a product within a manufacturing plant in an orderly and efficient manner. This requires information on the BOM and the process plan for each part. One popular way of doing this today is just-in-time (JIT) manufacturing. With JIT a company minimizes inventory by receiving parts and subassemblies in small lots just as they are needed on the production floor. With this method of manufacturing the supplier is an extension of the production line. JIT manufacturing obviously requires close and harmonious relations with the supplier companies. The supplier must be reliable, ethical, and capable of delivering quality parts.

- Designing of tooling and fixtures: Tooling applies the forces to shape or cut the parts, and fixtures hold the parts for ease of assembly. In a concurrent engineering strategy of design, both of these first two activities would start in detail design before the final design review.

- Develop quality assurance plan: This plan describes how statistical process control will be used to ensure the quality of the product. This requires information on CTQs, FMEAs, and results of prototype testing that has been carried out to that point.

- Develop maintenance plan: Any specific maintenance will be prescribed by the design team. The extent of this varies greatly depending on the product. For large, expensive products like aircraft engines and land-based gas turbines the manufacturers usually perform the maintenance and overhaul functions. This can prove to be a very profitable business over the long expected life of such equipment.

- Develop plan for retirement from service: As discussed in Chap. 10, it is the responsibility of the design team to develop a safe and environmentally friendly way to retire the product after it has completed its useful life.

- Manufacturing production acceptance test: This testing of products produced from the actual production line is carried out in conjunction with members of the design team.

Business activities

- Negotiate with suppliers: Manufacturing in conjunction with purchasing decides which components or assemblies should be outsourced from the plant, the make/buy decision. Purchasing then negotiates with suppliers using complete specifications and drawings for the components.

- Develop distribution plan: A general idea about the distribution system for the product will be part of the original marketing plan that started the product development process. Now marketing/sales will develop a detailed plan for warehouses, supply points, and ways of shipping the product. The design team will provide any needed information about possible damage to the product in shipping or with regard to product shelf life.

- Write the user manual: Generally, this is the responsibility of marketing, with needed technical input from the design team.

- Decide on warranty: Marketing makes decisions about the warranty on a product because this is a customer-related issue. Input is obtained from the design team about expected durability and reliability of the product.

- Develop a plan for customer service: Again, marketing is responsible for this activity because it is customer related. They either develop a network of dealers who do maintenance, as with automobiles, or develop one or more repair depots to which the customer sends the product for repair. Customer service supplies design with information on the nature of product failures or weaknesses for consideration in product redesign. If a serious weakness is uncovered, then a design fix will be called for.

Just as successful testing of a qualification prototype ends the design phase of product development, the successful testing of the pilot runs from manufacturing ends the product development process. The proven ability to manufacture the product to specification and within cost budget makes possible the product launch in which the product is released to the general public or shipped to the customer. Often the product development team is kept in place for about six months after launch to take care of the inevitable "bugs" that will appear in a new product.

9.6
FACILITATING DESIGN AND MANUFACTURING
WITH COMPUTER-BASED METHODS

Engineering design is a complex process that produces large quantities of data and information. Moreover, we have seen that there is a strong imperative to reduce the product design cycle time, improve the quality of the product, and decrease manufacturing cost. Computer-aided engineering (CAE) has had an important and growing influence on these goals. Clearly the ability to make computer models and carry out computer-based simulation has greatly increased our ability to efficiently size parts and improve their durability. The ability to design for robustness (Chap. 15) has increased the quality of what we design. But it is in detail design, and beyond, where everything comes together, that CAE has the greatest economic impact. Detail design traditionally has involved the greatest commitment of personnel of the three phases of design because there is such a great volume of work to do. CAE has significantly reduced the drafting task of preparing engineering drawings. The ability to make changes quickly in a CAD system has saved countless hours of redrawing details. Similarly, the ability to store standard details in a CAD system for retrieval when needed saves much drafting labor.

Many companies have a product line that is generic but requires engineering decisions to tailor the product to the customer's needs. For example, a manufacturer of industrial fans will change the motor speed, propeller pitch, and structural supports depending on the required flow rate, static pressure, and duct size. Typically this requires standard engineering calculations and drawings and a bill of materials (BOM) in order to produce a quote to the customer. Using conventional methods this might require a two-week turnaround, but using modern integrated CAD software that automates the computation, drawing, and BOM generation, the quote can be developed in one day.[1]

---

1. T. Dring, *Machine Design*, Sept. 26, 1994, pp. 59–64.

CAD has evolved rapidly to include capabilities such as 3-D solid modeling performed on desktop workstations and integrated with such powerful CAE tools as finite element analysis (FEA) and computational fluid dynamics (CFD). Collaborative design, where different engineers located anywhere in the world can contribute to a common CAD design, is being practiced routinely, and virtual reality, where the viewer becomes an active part of the design model, is within the capability of any design office. In addition, CAD modeling places rapid prototyping within the reach of any design office.

9.6.1 Product Lifecycle Management (PLM)

Product lifecycle management (PLM) refers to a set of computer-based tools that has been developed to assist a company to more effectively perform the product design and manufacturing functions from conceptual design to product retirement; see Figs. 9.1 and 9.2. The software provides complete integration of the engineering workflow from start to finish of product design.

There are three major subsystems to PLM.

- *Product data management* (PDM) software provides a link between product design and manufacturing. It provides control of design databases (CAD models, drawings, BOM, etc.) in terms of check-in and check-out of the data to multiple users, carrying out engineering design changes, and control of the release of all versions of component and assembly designs. Because data security is provided by the PDM system, it is possible to make the design data available electronically to all authorized users along the product development chain. Most CAD software has a built-in PDM functionality.
- *Manufacturing process management* (MPM) bridges the gap between product design and production control. It includes such technologies as computer-aided process planning (CAPP), computer-aided manufacturing (NC machining and direct numerical control), and computer-aided quality assurance (FMEA, SPC, and tolerance stackup analysis). It also includes production planning and inventory control using *materials requirements planning* software (MRP and MRP II).
- *Customer relationship management* (CRM) software provides integrated support to marketing, sales, and the customer service functions. It provides automation of the basic customer contact needs in these functional areas, but it also provides analytical capabilities for the data collected from customers to provide information on such issues as market segmentation, measures of customer satisfaction, and degree of customer retention.

While PLM systems are specifically designed to increase the effectiveness of the product design process, *enterprise resource planning* (ERP) systems are aimed at integrating the basic business processes of an organization. Originally ERP dealt with manufacturing issues like order entry, purchasing execution, inventory management, and MRP. Today the scope of ERP is very broad and includes every aspect of the business enterprise. This includes human resources, payroll, accounting, financial management, and supply chain management.

9.7
SUMMARY

Detail design is the phase of the design process where all of the details are brought together, decisions finalized, and a decision is made by management whether to release the design for production. The first task of detail design is to complete configuration and parametric design and to develop the engineering drawings. These documents, together with the design specifications, should contain the information to unambiguously manufacture the product. Any drawings, calculations, and decisions not completed in the embodiment design phase need to be made. Often in order to complete all these myriad details it is necessary to impose a design freeze. Once a freeze has been imposed, no changes can be made to the design unless they have been approved by a formal design control authority.

The detail design phase also involves verification testing of a prototype, the generation of a bill of materials (BOM) from the assembly drawings, a final cost estimate, and decisions on whether to make each part in-house or to obtain it from an outside supplier. These activities are greatly facilitated by the use of CAD tools.

Detail design ends when the design is reviewed and accepted by a formal design review process. The review consists of comparing the design documentation (drawings, analyses, simulations, test results, HOQ, FMEAs, etc.) against a checklist of design requirements.

While detail design is the end of the design process, it is not the end of the product development process. Some of the tasks that must be completed before product launch are process planning, design of tooling, negotiating with suppliers, developing a quality assurance plan, marketing plan, distribution plan, customer service plan, maintenance plan, and a plan for retirement of the product from service. Product launch depends on the first batch of product from the production line passing a manufacturing prototype acceptance test. Product lifecycle management (PLM) software increasingly is being used in carrying out the many tasks needed to achieve a timely product launch.

The engineering design process, and in particular the detail design phase, requires considerable skill and effort in communication on the part of design team members. For both written and oral communication the most important rules for success are (1) understand your audience, and (2) practice, practice, practice. In writing a technical report this means understanding the various audiences that will read the report, and organizing it accordingly. It also means working the original draft into a polished communication by the hard work of several rewrites. In making an oral presentation it means understanding your audience and organizing the talk accordingly. It also requires the hard work of practice until you have mastered the talk.

NEW TERMS AND CONCEPTS

Assembly drawing	MPM software	ERP software
Design freeze	Collaborative design	PDM software
Layout drawing	Detail drawing	Design briefing
Bill of materials	Memorandum report	Exploded assembly
Design review	CRM software	PLM software

BIBLIOGRAPHY

Detail Design

AT&T: *Moving a Design into Production,* McGraw-Hill, New York, 1993.
Detail Design, The Institution of Mechanical Engineers, London, 1975.
Hales, C. and S. Gooch: *Managing Engineering Design,* Springer, New York, 2004.
Vrsek, G.: "Documenting and Communicating the Design" *ASM Handbook,* vol. 20: *Materials Selection and Design,* pp. 222–30, ASM International, Materials Park, OH, 1997.

Written Communication

Brusaw, C. T. (ed.): *Handbook of Technical Writing,* 5th ed., St. Martin's Press, New York, 1997.
Eisenberg. A.: *A Beginner's Guide to Technical Communication,* McGraw-Hill, New York, 1997.
Ellis, R.: *Communication for Engineers: Bridge the Gap,* John Wiley & Sons, New York, 1997.
Finkelstein, L.: *Pocket Book of Technical Writing for Engineers and Scientists,* 3d ed., McGraw-Hill, New York, 2006.
McMurrey, D., and D. F. Beer: *A Guide to Writing as an Engineer,* 2d ed., John Wiley & Sons, New York, 2004.

Oral Communication

Goldberg, D. E.: *Life Skills and Leadership for Engineers,* Chap. 3, McGraw-Hill, New York, 1995.
Hoff, R. : *I Can See You Naked: A Fearless Guide to Making Great Presentations,* Simon & Schuster, New York, 1992.
Wilder, L.: *Talk Your Way to Success,* Eastside Publishers, New York, 1991.

PROBLEMS AND EXERCISES

9.1 Examine the detail drawings for a product designed by a nearby manufacturing company. Be sure you can identify the actual shape, dimensions, and tolerances. What other information is contained in the drawing?

9.2 Look at an automotive mechanics manual. Identify a subassembly like a fuel-injection system or a front suspension. From the assembly drawings, write up a bill of materials.

9.3 It is important for an OEM to maintain a strong positive relationship with its suppliers. A key to achieving this is in understanding the goals that the supplier has for its business and aligning your organization with them. Make a list of four goals that would be typical for a supplier in a manufacturing industry.

9.4 The past 10 years have seen a growing trend for manufacturing operations to be moved off shore from the United States to Asian countries. Prepare a list of pros and cons concerning the off-shoring issue.

9.5 Visualize the impact of CAE in a world that is even more electronically connected than it is today. How might the practice of detail design change?

9.6 Prepare a final design review checklist for your design project.

9.7 Carefully read a technical paper from a journal in your field of interest and comment on whether it conforms with the outline for technical reports discussed in Sec. 9.3.3. If there are major differences, explain the reasons for these.

9.8 Write a memorandum to your supervisor justifying your project being three weeks late and asking for an extension.

9.9 Prepare a PowerPoint presentation for the first design review of your team project.

9.10 Prepare a poster for the final presentation for your design project. A poster is a large visual display, with a series of graphics, containing text, mounted on a large sheet of poster board. The display should be self-contained, such that a technical person will be able to understand what you did.

10

DESIGN FOR SUSTAINABILITY
AND THE ENVIRONMENT

The ecosystem of planet Earth is human-dominated. The standard of living of humans is due, in large part, to their ability to develop technology to dominate the ecosystem. Rather than adapting to the earth's environment, humans have found ways to alter their immediate environment for their own needs. So significant is the human alteration of the planet that some scientists declare present times to represent a new geological era: the *Anthropocene* era, also known as the age of the human.[1]

Today's designers and engineers are expected to move beyond design for recyclability (an end-of-life strategy), beyond design for the environment (a life-cycle objective), to design for sustainability—an all encompassing paradigm that reaches outside of physical behavior and beyond the physical life cycle. Sustainability impacts a product's complete supply chain and radiates from supply chain links into the social fabric of everyday life. Sustainable development can be generally defined as economic growth done in a way that is compatible with the environment.[2]

This chapter introduces the topic of sustainability after providing comments on its precursor, the environmental movement. Sustainability is defined by the report of a United Nations Commission in Sec. 10.2. How companies respond to being viewed through the lens of sustainability is introduced in Sec. 10.3. The remainder of the chapter presents the change that has occurred in engineering design methods moving from an emphasis on environmental concerns to an emphasis on sustainability concerns.

10.1
THE ENVIRONMENTAL MOVEMENT

This section will provide the briefest of introductions to the language of the environmental movement in order to contrast the scope of sustainability and follow some of its core principles. Understanding relies on precise definitions of terms, particularly

---

1. B. R. Allenby, *The Theory and Practice of Sustainable Engineering,* Prentice Hall, Upper Saddle River, NJ, 2011.
2. M. R. Chertow, The IPAT Equation and Its Variants," *Journal of Industrial Ecology,* October 2000, vol. 4, 4:13–29.

when branching into new fields of study; the definitions given in this section are from the Biology-Online Dictionary.[1] An *ecosystem* is "a system that includes all living organisms in an area as well as its physical environment (including the local atmosphere) functioning together as a unit." All ecosystems on the earth combine into what is known as the *biosphere*. Earth's biosphere is the part of the planet that can support life and includes the earth's crust, atmosphere, and water layers as well.

10.1.1 Ecosystems and Balance

Ecosystems can be modeled in engineering terms as control volumes with defined physical boundaries enclosing living organisms, natural elements (i.e., portion of the earth's surface and crust, air, water), people, and man-made structures. Ongoing ecosystems must achieve a balance among all the exchanges of energy and material among its inhabitants; these exchanges are made possible by signals passed between them. Members of an ecosystem can be modeled as a set of function structures interacting with each other and their environment by exchanging flows of energy, material and signals (see Sec. 6.5). *Ecosystem ecology* is the study of these flows.

An ecosystem survives as long as the resources within it are adequate to support the living systems it holds. The resources may be supplied from other members inside the system, from the outside, or restored through some cyclic action of the system itself. The earth supplies essential resources for plants and animals through sunlight and three natural cycles: carbon, nitrogen, and water. Disruption of these inputs causes imbalance in the ecosystems, which will ultimately impact the survivability of its inhabitants.

Basic biological ecosystems maintain balance by adhering to canons developed through evolution.[2]

1. Use waste as a resource
2. Diversify and cooperate to fully use the habitat
3. Gather and use energy efficiently
4. Optimize rather than maximize
5. Use material sparingly
6. Don't foul nests
7. Don't draw down resources
8. Remain in balance with the biosphere
9. Run on information
10. Shop locally

Canon 4 may seem odd to designers. In this case, optimize implies making decisions based on optimizing the outcome of the ecosystem as a whole, rather than maximizing the outcome of one particular member. Absent from the list of principles for balancing biological ecosystems is the action of inhabitants to restore resources. That role can only be filled by human inhabitants of an ecosystem.

___

1. Biology-Online.org, http://www.biology-online.org/dictionary.
2. J. M. Benyus, *Biomimicry*, Morrow, New York, 1997.

10.1.2 United States Environmental Movement

Throughout history there have been cases of naturalists, scientists, social workers, and politicians who have publicized issues of sanitation and improper waste handling, conservation of natural resources (e.g., President Theodore Roosevelt), and pollution. However, efforts to alleviate problems associated with these issues usually came only after their impact severely affected the populous. Rachel Carson's book *Silent Spring,*[1] published in 1962, is credited with raising awareness about beginning the environmental movement in the United States. *Silent Spring* documented the negative impact of pesticides, particularly DDT, on the environment. The title of the book projects the vision of a spring without the sound of birds.

The Environmental Protection Agency (www.epa.gov) was consolidated from smaller governmental units into one agency by President Richard Nixon in 1970. The EPA is endowed with the authority to write and enforce regulations based on legislation. Several landmark pieces of environmental legislation were passed in the late 1960s and 1970s. The EPA website includes descriptions of all of the laws and regulations created to protect human health and the environment. Most readers will be familiar with a popular and voluntary program for encouraging energy efficiency in consumer goods—Energy Star (www.energystar.gov). There is a rich literature base on the environmental movement in the United States[2] and around the world.[3]

10.1.3 Measures of Environmental Impact

Today there are both public and private organizations with goals of preserving and improving the environment for human habitation. These groups range in scale from the local level to the national level and on to the international level. Environmental Science and Environmental Engineering degree programs exist to train students in history, science, policy, and design to mitigate environmental problems.

Interdisciplinary sciences for the study of environmental issues are flourishing. A key concept in environmental science is the quantification of the impact of humans and technology on the environment. A simple equation was developed in the 1970s to model the interaction of population with the environment.[4] The factors in the equation are environmental impact (I), population (P), affluence[5] (A), and technology (T). The resulting *IPAT equation* is shown in Eq. (10.1).

$$I = P \times A \times T \tag{10.1}$$

1. R. Carson, *Silent Spring,* Houghton Mifflin, Boston, 1962.
2. See for example: P. Shabecoff, *A Fierce Green Fire: The American Environmental Movement,* Hill and Wang, New York, 1993.
3. See for example: R. Guha, *Environmentalism: A Global History,* Oxford University Press, New Delhi, 2000.
4. M. R. Chertow, "The IPAT Equation and Its Variants," *Journal of Industrial Ecology,* 2000, 4:9, pp. 13–29.
5. Affluence means a flow or supply of something. In the IPAT equation it usually means a measure of value on a per capita basis.

This equation is a model to indicate the influence of each of the terms on the others. Allenby's interpretation of the IPAT equation for the environment is given by Eq. (10.2).

$$Overall\ Environmental\ Impact = population \times \frac{resource\ use}{person} \times \frac{environmental\ impact}{unit\ of\ resource\ use}$$

(10.2)

Technology (T) is used to represent the influence that man-made systems can have on environmental impact. Strategies for reducing the overall environmental impact can be interpreted directly from Eq. (10.2). They are: (1) reduce population or slow its growth; (2) reduce the resource use per person; and (3) reduce the impact per unit of resource. Designers and engineers will naturally focus on developing technology that will reduce the T in the equation.

The *carbon footprint* is another popular metric for measuring environmental impact. Merriam-Webster gives the following definition:

> Carbon footprint—the amount of greenhouse gases and specifically carbon dioxide emitted by something (as a person's activities or a product's manufacture and transport) during a given period.[1]

The term "carbon footprint" was introduced in 1999 but is now in common usage.

In 2007 carbon footprint labels began appearing on some goods in Great Britain.[2] The process to measure the carbon footprint is complex, and boundaries must be set on how far back along the supply chain the calculation should be taken. The International Standards Organization is preparing a standard for carbon footprinting that is expected in 2012 (ISO 14067). One benefit of undergoing the carbon footprint determination process is the focus it brings to the environmental impact of each step of production.

It is no accident that the most popular metric for environmental impact is based on CO_2 emissions. Carbon dioxide is the most prevalent greenhouse gas after water vapor and has the most harmful impact of the other greenhouse gases (GHG). Carbon dioxide is generated by the combustion of fossil fuels—linking it directly to the behavior of the human inhabitants of the biosphere. The fact that climate change (formerly known as global warming) is a result of human habitation has been accepted into our culture.

10.1.4 Interaction of Energy Use with the Environment

Modern society has been built with a dependence on fossil fuels: coal, oil, and natural gas. The combustion of fossil fuels produces CO_2, the chief constituent of greenhouse gas. Not only is there a concern over the long-term supply of fossil fuels, and their

---

1. www.merriam-webster.com/dictionary, accessed July 27, 2011.
2. "Following the Footprints," *The Economist Technical Quarterly,* June 4, 2011, pp. 14–18.

spiraling cost, but there is a growing recognition of the need to reduce dependence on fossil fuels because of their impact on the environment.

There are also economic concerns about the stability of our energy supply (specifically oil). The increase in environmental awareness in the 1970s coincided with the first serious shock to the U.S. oil supply: the 1973 oil crisis. In response to U.S. support for Israel, Arab members of OPEC (Organization of Petroleum Exporting Countries) and Tunisia, Syria, and Egypt declared an embargo on oil exports to the United States, Japan, Great Britain, Canada and the Netherlands and reduced their production. The embargo lasted from October of 1973 to mid-March of 1974. At the end of the period the price of oil in the United States was $12 per barrel, about four times the price less than an year earlier and the risk of dependence on imported oil became clear.[1] The 1973 oil crisis was the first in a series of politically-motivated disruptions of the oil supply. Disruptions and price volatility can also be caused by natural disasters like Hurricane Katrina (2005) and technical disasters like the Deepwater Horizon oil spill (2010).

At the time of the writing of this book, oil prices are hovering around $100 per barrel, and the United States is still heavily dependent on oil imports. Both environmental and economic concerns have placed a great emphasis on the development of renewable energy sources such as agriculture-based fuel sources (biofuels), solar power, wind power, and tidal power. Many believed nuclear power would undergo a major revival once the waste fuel storage problem was clarified. Nuclear power does not produce greenhouse gases. However, the Fukushima Daiichi nuclear disaster (March 11, 2011) brought the potential risks inherent with nuclear reactors to the forefront of the public psyche. The nuclear plant failures were the result of the 2011 Tōhoku 9.0 earthquake and subsequent tsunami. The future of nuclear power is therefore clouded.

10.1.5 Behavior Changes Started by the U.S. Environmental Movement

The environmental movement has already produced great impact on changing attitudes and behaviors of citizens, governments, and businesses. Environmental science principles are now engrained in the education system and taught to children in K–12 classrooms. Elementary school children learn about GHG, threats to the ozone layer, and that some aerosol sprays are not good for the planet. The habit of recycling household trash was born out of the environmental movement.

Most U.S. citizens view *recycling* as separating their waste into different streams as it comes out of their homes and businesses.[2] The EPA tracks the generation of solid waste and how it is handled using a mass balance methodology based on data from a variety of sources including suppliers, industry associations, municipal governments, and other agencies (e.g., U.S. Census Bureau and Department of Commerce). Figure 10.1 displays a breakdown of the handling of material solid waste throughout the United States in 2009. The actions of local governments and adaptation by waste collection businesses have resulted in the creation of 9000 recycling programs (reported in 2009). According

---

1. R. Mabro, "On the Security of Oil Supplies, Oil Weapons, Oil Nationalism and All That," *OPEC Energy Review,* March 2008, vol. 32:1, pp. 1–12.
2. The United States has a history of recycling metals from their experiences in WWII.

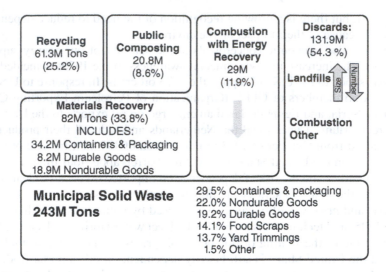

FIGURE 10.1
2009 Data on disposition of municipal solid waste (MSW) in the United States. (The sizes of the boxes in the figure do not represent the values of the categories.)[1]

to the 2009 figures from the EPA, about 12 percent of the municipal solid waste (MSW) consists of containers and packaging (i.e., 29.5M out of 243M tons). Nearly one-half of the packaging MSW (or 47.8 percent) is recovered through the recycling process. The 2009 percentage recycled packaging by container material type were: 66.2 percent steel, 62.4 percent paperboard, 51 percent aluminum cans, 37.5 percent aluminum, 31 percent glass, 22 percent wood, and 14 percent plastic. Section 10.4 will introduce how this stream of recycled waste becomes resources available to industries for their needs.

U.S. government regulations continue to motivate businesses to modify production methods to reduce the creation of harmful emissions into the air, the wastewater, and the solid waste. In many areas of environmental regulation the European Union (EU) has moved more aggressively than the United States. Because of world trade this has required many U.S. companies to adopt EU regulations. Engineers and designers respond to these incentives by creating technology to reduce air, water, and solid waste pollution. Section 10.5 discusses Design for the Environment (DfE) strategies for creating products that align with the established goals of the environmental movement.

10.2
SUSTAINABILTY

To sustain means to continue on into the future, usually without significant change. *Sustainability* is a term that has come into popular culture as a characteristic of national fiscal policy, personal budgetary policy, and environmental policy. A formal

---

1 "Municipal Solid Waste Generation, Recycling, and Disposal in the United States: Facts and Figures for 2009," EPA-530-F-010-012, U.S. EPA, December 2010, www.epa.gov/wastes.

definition of *sustainable development* (used interchangeably with *sustainability* in much of the literature) was established by a report titled, "Our Common Future," from the United Nations World Commission on Environment and Economic Development (WCED).[1] The report states: *"Sustainable development is development that meets the needs of the present without compromising the ability of future generations to meet their own needs."*[2] This is a statement of *social equity* for future generations. The statement's language is clear, but the true meaning and interpretation requires elaboration in order to extract operational principles.

10.2.1 WCED Report on Sustainability

In 1983 the UN General Assembly created the WCED to examine economic and environmental conditions in order to recommend strategies for managing global resources and preserving the environment to meet the needs of the rapidly expanding population. This commission was chaired by Gro Harlem Brundtland; its report, known also as the "Brundtland Report," was issued in 1987.

The WCED Report described characteristics of activities that would meet the definition of sustainable and, by extension, contribute to sustainable development. Critical objectives for environment and development policies that follow from the concept of sustainable development include:

- *Revive growth in an economic sense*—particularly in developing countries where increases in population represent unused human capacity.
- *Change the quality of growth*—Sustainable development ". . . requires a change in the content of growth, to make it less material- and energy-intensive and more equitable in its impact."[3] Any activities undertaken for economic growth must weigh expected financial gain against the impact on the environment and human population.
- *Meet essential needs for jobs, food, energy, water, and sanitation*—This objective is a restatement of Maslow's most basic human needs (see Chap. 6) with the addition of jobs as an enabling condition. The WCED report further states that "overriding priority should be given" to meeting needs of the world's poor.
- *Ensure a sustainable level of population*—Every ecosystem has an intrinsic limit on the size of the population it can support.
- *Conserve and enhance the resource base*—The earth has a set of finite resources (e.g., potable water, fossil fuels, and minerals). The earth also has renewable resources (e.g., forests) and, to some extent, unlimited resources (e.g., sunlight, air, and water). Some resources can be reused, but there may be temporary degradation of the resource until the ecosystem can restore it. The best way to conserve the resource base is to reduce per-capita consumption.

---

1. The World Commission on Environment and Development, *Our Common Future,* Oxford University Press, 1987.
2. The World Commission, op. cit., paragraph 1.
3. The World Commission, op. cit., paragraph 35.

- *Reorient technology and manage risk*—The technology that defines the Anthropocene age must be focused on solving problems of sustainability. It has been acknowledged that technology in the developed world has caused many of the sustainability problems. New attitudes, technology, and infrastructure are being used in the developed world to promote sustainability. The currently developing countries are not always economically able or willing to adopt the newer, sustainable technologies. Finally there are a large number of underdeveloped countries who lack the resources (social and environmental) to move in the direction of sustainability. Thus, the most economically vulnerable peoples of the world are most at risk from environmental hazards.
- *Merge environment and economics in decision making*—This objective is a direct instruction to decision-makers. The environmental objectives included in decision-making are typically assessed over a local or regional impact area rather than the global scale required for sustainability. The WCED report declares that economic and ecological goals are not naturally in opposition. However, the decision-making process must be able to articulate and quantify the impacts of alternatives on a vastly broader spectrum of objectives than required at the product level.

The themes of sustainability emerging from the WCED report can seem intimidating and unrealistic to countries established on market-based economies and equality of opportunity as opposed to equality of outcome. The guidelines appear to support the belief that sustainability will only be achieved by equalizing resource consumption on a global basis and that the level of equilibrium will be below the resource consumption currently enjoyed by several industrialized nations, a group the largest of which is the United States. Another potentially unsettling aspect of the sustainability movement to some in industrialized countries is that it is driven by policy makers outside of their own sphere of political influence.

10.2.2 Twenty Years After the WCED Report on Sustainability

The 2011 public debate on legislation to raise the U.S. national debt ceiling limit offers an example of the concept of sustainability and potential resistance to it. The U.S. national debt accumulated due to annual budget shortfalls increasingly due to the expanding costs of benefits for the elderly, unemployed, and disabled. Many fiscal conservatives labeled the level of government spending as *unsustainable* and sought to reduce it rather than raise the debt ceiling. Many liberals held that revenues to the U.S. government should be increased to support continued government spending on benefits programs by raising taxes on individuals who were "rich," thereby *redistributing income*. Proponents of tax increases argued that fairness directed the government to require more money from citizens (who had more than they needed) to raise the quality of life for others. The conservative position of self-reliance on individual economic responsibility stood in stark contrast to the liberal position of social equity enforced by a governing authority.

Writers of the 1987 WCED Report gave a sense of urgency to the need for adoption of recommended sustainable development guidelines. More than 20 years have passed since that time, and indicators mentioned later have not changed significantly.

However, globalization and global scale thinking are much more common. In addition, changes in means and modes of communication (i.e., the Internet, social networking, mobile devices) have served to raise awareness of disparities in socioeconomic status for people at both ends of the spectrum. Changes in indicators of global sustainability are as follows:[1]

- Population growth rate is now 1.2 percent rather than 1.7 percent (yet will still exceed estimates of the maximum population capacity)
- Malnutrition has increased
- HIV/AIDS cases have increased from 10M to 40M
- CO_2 atmospheric concentration has increased from 325 ppm to 385 ppm at an accelerating rate

On balance, the changes cited indicate that the importance of personal health to sustainability should be emphasized more than in the previous report.

10.2.3 Measures of Sustainability

Sustainability requires metrics that incorporate a more holistic assessment of the state of the population than the environmental metrics of Sec. 10.1.3. That equation serves more as a model for understanding than a formula for calculation. Allenby revises the former metric, Eq. (10.2), into a sustainability impact equation[2] as shown below.

$$Sustainability\ Impact = population \times \frac{quality\ of\ life}{person} \times \frac{sustainability\ load}{quality\ of\ life\ unit} \quad (10.3)$$

There are two changes. The affluence term has been broadened from a consumption measure (e.g., resources per capita) to a more expansive term "quality of life." This revision further emphasizes the social nature of sustainability. Allenby makes clear that high quality of life requires reasonable levels of living conditions (formerly called out as environment), health, and personal economy. Since the affluence term has expanded it puts additional importance on using technology to make the sustainability load (the second change) as small as possible in order to accommodate the difference.

The WCED Report has stimulated a great deal of discourse on development across the globe. The rhetoric and reality of the discourse, the increasing awareness of resource limitations, and anxiety over climate change are all acting together to shape a new world view. Globalization is the reality. Socioeconomic imbalance is evident. Quality of life disparities are apparent. In this context sustainability represents a paradigm shift with an impact that will be more far-reaching than even the environmental movement.

---

1. A. J. McMichael, "Population, Human Resources, Health, and the Environment: Getting the Balance Right," *Environment*, 2008, vol. 50, no. 1, pp. 48–59.
2. B. R. Allenby, *The Theory and Practice of Sustainable Engineering*, Prentice Hall, Upper Saddle River, NJ, 2011, p. 55.

10

10.3
CHALLENGES OF SUSTAINABILITY FOR BUSINESS

Protection of the earth's environment is high on the value scale of most citizens of the world's developed countries. Investment firms now offer options of stocks and mutual funds of companies that meet thresholds of corporate social responsibility (CSR). Measures used to categorize firms based on corporate social responsibility include (but are not limited to): environmental performance based on impact reports of toxins and emission releases, regulatory compliance violations, and organization processes in place (e.g., environmental managements systems).[1]

Most corporations realize that it is in their best interest to take a strong pro-environment approach to their business. Publically traded corporations include corporate environmental goals and sustainability statements on their websites and in their annual reports.[2] Some provide additional detail on sustainability efforts. For example, General Electric provides an annual citizenship report.[3]

Being pro-environment can have repercussions for the bottom line. Consider a simple example of the creation of a new product with recycled content.

EXAMPLE 10.1
Typical letter-sized paper purchased for use in copiers is of 20 lb. weight and has a brightness rating of 92, according to U.S. standards (100–104 Euro Bright scale). The online retail price for one case (5000 sheets) of Staples® brand paper meeting these specifications in July 2011 was $39.99 for nonrecycled Staples Copy Paper.[4] Costs for cases of recycled paper meeting the same specifications were $45.99 for Staples 30% Recycled Copy Paper and $55.99 for Staples 100% Recycled Copy Paper (this paper had a brightness rating of 90 U.S.). The 100% recycled paper product displays the FSC logo of the Forest Stewardship Council, an international nonprofit organization championing forest management (including consumption) aligned with sustainability objectives.[5] This example illustrates the impact that design for sustainability policies can have on a product's quality and cost. Describing the reasons why adding recycled content to paper increases its cost is left for a homework assignment.

Example 10.1 illustrates the counter-intuitive pricing reality for recycled paper. It costs more than regular paper. The economic success of recycled paper relies on customers choosing to pay a premium to support use of recycled content. However, not all recycled products will be more expensive than their counterparts. Companies must develop sophisticated decision-making strategies to predict accordingly. Companies must also continue to develop improved processes to reduce the cost of including recycled content in goods.

Businesses operating in the United States necessarily adhere to all legislation, including the growing body of environmental regulations. Companies with global markets must also meet regulations in the countries of their markets. This may mean creating product variants for different markets. Many corporations have an Environment Health

---

1. M. Delmas and V. D. Blass, "Measuring Environmental Performance: The Trade-Offs of Sustainability Ratings," *Business Strategy and the Environment,* 2010, vol. 19, 245–260.
2. "Sustainability," Stanley Black & Decker, 2011, www.stanleyblackanddecker.com.
3. "Sustainable Growth: GE 2010 Citizenship Report," July 24, 2011, www.gecitizenship.com.
4. Prices found online at www.staples.com on July 23, 2011.
5. "Forest Stewardship Council," www.fsc.org.

and Safety (EHS) unit that monitors environmental regulations in relevant locations and develops strategies for dealing with differences in laws between countries. These groups must also review legislation developing in other countries that is likely to influence the United States in the future.

The sustainable development paradigm provides a new set of challenges to businesses by asserting that they are now accountable for three impacts of their actions—environmental, social, and economic—on the population. If a corporation's actions would result in improvement on all three fronts at the same time, sustainable development would be easy. However, corporate actions routinely are taken for the overall economic benefit of their owners after taking into account primarily economic and performance trade-offs.[1]

The majority of sustainable development decision-making scenarios will include some negative movement in one of the environment, economic, or social objectives. The state of the Cape Wind, wind-power farm provides an example of the difficult trade-offs to businesses (and governments) when implementing sustainable development projects.

EXAMPLE 10.2 The Cape Wind™ Project

Cape Wind is the name for the first off-coast *wind farm* proposed in the United States. Cape Wind is planned to comprise 130 wind turbines, covering about 24 acres on Horseshoe Shoal, a shallow portion of the Nantucket Sound. The wind turbine towers will be about 258 feet high with base diameters of 16 feet. The propeller tips will reach 440 feet from the water level. The farm will be built for maximum energy production of about 450 MW with an average of nearly 170 MW. This would offset the use of 113 million gallons of oil each year. The overall cost: $2.5 billion.

Massachusetts and federal government agencies have given approval and permits for the project and the infrastructure required for Cape Wind's development and use. Readers are referred to Wikipedia[2] and eCape[3] websites for information on the battles waged during the development process. Some environmentalists have opposed Cape Wind because of the projected negative impact on fish, fishing, real estate, and tourism.

Robert F. Kennedy, Jr., a well-known environmental attorney, wrote a *Wall Street Journal* opinion page[4] calling the project a "rip-off," as it will result in an estimated cost to consumers of $ 0.25 per kilowatt hour (kwh) when hydropower is available from Quebec producers at $0.06 kwh. Kennedy claims that a Massachusetts energy company, NSTAR, is being pressured to agree to buy power from Cape Wind in order to meet state regulations that utilities obtain 3.5 percent of their power from green sources. The eCape website provides many rebuttals to the Kennedy editorial. One reports that the Massachusetts Department of Public Utilities estimates the impact of Cape Wind product to average consumers would be $1.25 per month. Another is from the American Wind Energy Association (AWEA) Blog[5] in which the writer implies that the need to develop wind power sources outweighs other economic considerations.

1. R. B. Pojasek, "Sustainability: The Three Responsibilities," *Environmental Quality Management,* Spring 2010, pp. 87–94.
2. Cape Wind, http://en.wikipedia.org/wiki/Cape_Wind.
3. "Cape Wind: Energy for Life," eCape, Inc., http://www.capewind.org, accessed July 30, 2011. (Note: It seems that the website company eCape, Inc. is affiliated with Cape Wind Associates, LLC, the company that set up the joint business venture.)
4. Robert F. Kennedy, Jr., "Nantucket's Wind Power Rip-Off", *Wall Street Journal, Opinion Section,* July 18, 2011. Retrieved from http://www.wsj.com.
5. T. Gray, "Why Cape Wind? Investing in America's Energy Future Not Just Canada's," *Into the Wind,* posted July 20, 2011, http://www.awea.org/blog.

Example 10.2 illustrates the complexities of corporations moving forward on even the most seemingly advantageous sustainable development actions.

The intent of the WCED report would mean that the benefits of a project like Cape Wind would include more quantifiable outcomes. Outcomes would include the economic impacts to those receiving power from the facility, the result of jobs it would create, and estimates of any losses in tourism or property values by the change in the view. Ideally, the changes in quality of life for all those affected by the project would be included. However, the concepts are too new and too ambiguous for a consensus definition to exist. The search for a way to quantify the impact a company has on all sustainability outcomes has made some progress.

The term *triple bottom line* (TBL) was first used by John Elkington in 1994.[1] It is an accounting-like term to model the impacts of the three objectives of sustainability: economics and environmental and social improvement.[2] It is common to regard the TBL as a tool to measure impact on profits, planet, and people. Elkington's contention was that businesses should keep three separate balance sheets, one for each set of stakeholders impacted by its actions. At the time it was introduced, the TBL is not intended to be used to literally combine and sum impacts on profits, planet, and people. It is likely that implementations of the TBL will begin with a tool called a *balanced scorecard,* a way of assessing performance toward targets on objectives on non-commensurate scales. General Electric is one of the early adopters of the balanced scorecard approach to monitoring progress on non-financial targets. GE's 2010 Citizenship Report includes an EHS Performance Against Commitments Table (p. 37), in which one row reads as follows:

> 2010 Commitment: *"Continue long-term GHG and energy use reduction trend and drive to the following goals: 50% improvement in energy intensity by 2015."*
> Progress: *"GE continued to make progress on these goals; GHGs were reduced by 24% and energy intensity improved by 33% from the 2004 baselines."*
> 2011 Commitment: *"Implement an ecomagination scorecard for GE's internal environmental footprint against which activities that drive the goals will be measured."*

This example provides an indication of how General Electric articulates their corporate goals on environmental issues in a format that makes a statement about sustainability.

10.4
END-OF-LIFE PRODUCT TRANSFORMATIONS

Essentially, all products, devices and systems that will degrade throughout their lifetime until they are no longer useful should be rendered into a form that supports sustainability. Putting nonworking items into a landfill is the least desirable outcome. Ashby puts it succinctly: "When stuff is useful, we show it respect and call it material.

1. J. Elkington, *Cannibals with Forks: The Triple Bottom Line of 21st Century Business,* Capstone, 1997.
2. "Triple Bottom Line," *The Economist,* November 17, 2009.

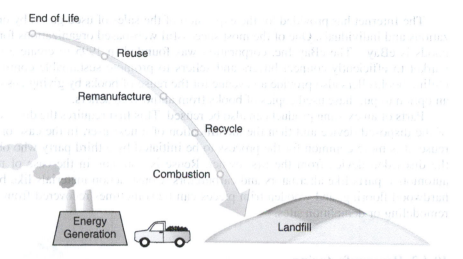

FIGURE 10.2
Options for transformation at product end of life.

When the same stuff ceases to be useful, we lose respect for it and call it waste."[1]
There are limited options for handling the waste of products that an end-user no
longer wants. The options are reuse, remanufacture, recycle, combustion, or landfill
(Figure 10.2).

10.4.1 Reuse

Reuse means identifying a new end-user who sees value in the product as it exists at
the time of the original user's plan of disposal. Allenby calls this point in a product
life cycle the "end of first life." There is a challenge to getting the two parties together.
Proponents of capitalism point out that markets emerge when there is a clear demand
for the used product and a means for the buyer (first user) and seller (next user) to
communicate and exchange the product for cash or other goods.

Planned meetings for people to exchange goods used to be called, in some areas,
swap meets. These events were scheduled in advance to allow exchangers to meet
face-to-face to negotiate the transaction. Today there are more venues for the sale of
used goods, either by the first-owner or by an intermediary. They include garage sales,
rummage sales (usually organized by volunteer members of a nonprofit organization),
and retail stores supported by donations and run by charitable or philanthropic organizations (e.g., Goodwill Industries International and Habitat for Humanity operate
stores for the sales of donated used goods).

---

1. M. F. Ashby, *Materials and the Environment: Eco-Informed Material Choice,* Elsevier, Boston, 2009,
p. 65.

The Internet has provided for the expansion of the sales of used goods by organizations and individuals. One of the most successful web-based organizations for used goods is eBay®. The eBay Inc. corporation was founded in 1995 to create a global market to efficiently connect buyers and sellers to promote sustainable commerce.[1] Online booksellers also provide an avenue for the reuse of books by giving customers an option to purchase used copies of books from authorized sellers.

Parts of an existing product can also be reused. This first requires the disassembly of the disposed device and then the identification of a next user. In the case of parts reuse, it is more common for the process to be initiated by a third party who obtains the discarded device from the first owner. Reuse is common in the case of rebuilt automotive parts like alternators and carburetors. Construction materials like bricks, hardwood flooring, and wooden trim pieces can be sometimes recovered from home remodeling or demolition sites.

10.4.2 Remanufacturing

An alternative to recycling is remanufacturing. *Remanufacturing* is the refurbishing of an existing system by restoring it to near new condition. This process can be limited to the cleaning and replacement of worn parts (*refurbishing*) or the harvesting of key subassemblies for placement into new parts. Remanufacturing saves energy by reducing the need for the processing of raw materials into new products.

The refilling of inkjet cartridges is a common example of remanufacturing. Office supply companies call this practice of returning the empty cartridges "recycling," but the process is a reconditioning of the product so that it can be restored to its original use. In this instance, end-users are surrendering their property (empty or near-empty printer ink cartridge). There are drawbacks to the remanufacturing process from the producer's point of view: (1) it can cut into the new product market share and (2) the quality of the remanufactured product may be lower than that of the original product. The latter concern may deter the OEM from engaging in remanufacturing, leaving the market open for third-parties. In other cases, OEM firms do remanufacture their own used goods such as copiers and printers, and some larger engineered products like diesel engines and construction equipment.

10.4.3 Recycling

Recycling is the recovery from the waste stream, products or goods that can be used in the raw materials stream to make the same or similar material. Recycling is the end-of-life strategy that is best suited for deriving profit from the waste stream. The benefits of materials recycling are the contribution to the supply of materials, with corresponding reduction in the consumption of natural resources, and the reduction in the volume of solid waste. Moreover, recycling contributes to environmental improvement through

---

1. "Who We Are," eBay Inc., http://www.ebayinc.com.

the amount of energy saved by producing the material from recycled (secondary) material rather than primary sources (ore or chemical feedstock). Recycling requires energy, with its accompanying gas emissions, but recycle energy is small compared with the original energy required to make the material. For example, recycled aluminum requires only 5 percent of the energy required to produce it from ore. Between 10 to 15 percent of the total energy used in the United States is devoted to the production of steel, aluminum, plastics, or paper. Since most of this energy is generated from fossil fuels, the reduction of carbon dioxide and particulate emissions due to recycling is appreciable. Recycling of materials also directly reduces pollution. For example, the use of steel scrap in making steel bypasses the blast furnace at a considerable economic benefit. Bypassing the blast furnace in processing also eliminates the heavy pollution associated with coke making.

The steps in recycling a material are (1) collection and transport, (2) separation, and (3) identification and sorting.[1]

Collection and Transport

Collection for recycling is determined by the location in the material cycle where the discarded material is found. *Home scrap* is residual material from primary material production, such as cropped material from ingots or sheared edges from plates, which can be returned directly to the production process. Essentially all home scrap is recycled. *Prompt industrial scrap* or new scrap is that generated during the manufacture of products, for example, compressed bundles of lathe turnings or stamping discard from sheets. This type of scrap is sold directly in large quantities by the manufacturing plant to the material producer. *Old scrap* is scrap generated from a product which has completed its useful life, such as a scrapped automobile or refrigerator. These products are collected and processed in a scrap yard and sold to material producers. The collection of recycled material from consumers can be a more difficult proposition because the material is widely distributed. Materials can be economically recycled only if an effective collection system can be established, as with aluminum cans. Collection methods include curbside pickup, buy-back centers (for some containers), and resource recovery centers where solid waste is sorted for recyclables and the waste is burned for energy.

Separation

Separation of economically profitable recyclable material from scrap typically follows one of two paths. In the first path, selective dismantling takes place. Toxic materials like engine oil are removed, and high-value materials like gold and copper are removed and segregated. Dismantling leads naturally to sorting of materials into like categories. In the second path, the product is subjected to multiple high-energy impacts to shred it into small, irregular pieces. For example, automobile hulks are routinely processed by shredding. Shredding creates a material form that assists in separation. Ferrous material can be removed with large magnets, leaving behind debris that must be disposed of, sometimes by incineration.

---

1. "Design for Recycling and Life Cycle Analysis," *Metals Handbook, Desk Edition,* 2d ed., ASM International, Materials Park, OH, 1998, pp. 1196–99.

Identification and Sorting

The economic value of recycling is largely dependent on the degree to which materials can be identified and sorted into categories. Material that has been produced by recycling is generally called *secondary material*. The addition of secondary material to *virgin material* in melting or molding can degrade the properties of the resultant material if the chemical composition of the secondary material is not carefully controlled. For example, in steel more than 0.20 percent copper or 0.06 percent tin cause cracking in hot working. There is concern about buildup of these *tramp elements* as steel is recycled, sometimes multiple times. Thus, the price of metal scrap depends on its freedom from tramp elements, which is determined by the effectiveness of the identification and separation process.

Degradation of plastics from secondary materials is more critical than in metals, since it is often difficult to ensure that different types of polymers are not mixed together. Only thermoplastic polymers can be recycled. Often, recycled material is used for a less critical application than its original use. There is an intensive effort to improve the recycling of plastics, and it is claimed that under the best of conditions engineered plastics can be recycled three or four times without losing more than 5 to 10 percent of their original strength.

Other materials that may not be recycled economically are zinc-coated steel (galvanized), ceramic materials (except glass), and parts with glued identification labels made from a different material than the part. Composite materials consisting of mixtures of glass and polymer represent an extreme problem in recycling.

Metals are identified by chemical spot testing or by magnetic or fluorescence analysis. Different grades of steel can be identified by looking at the sparks produced by a grinding wheel. Identification of plastics is more difficult. Fortunately, most manufacturers have adopted the practice of molding or casting a standard Society for the Plastic Industry identification symbol into the surface of plastic parts. This consists of a triangle having a number in the middle to identify the type of plastic.[1] Much effort is being given to developing devices that can identify the chemical composition of plastic parts at rates of more than 100 pieces per second, much like a bar-code scanner.

Ferrous metals are separated from other materials by magnetic separation. For nonferrous metals, plastics, and glass, separation is achieved by using such methods as vibratory sieving, air classification, and wet flotation. However, much hand sorting also is used. After sorting, the recycled material is sold to a secondary materials producer. Metals are remelted into ingots; plastics are ground and processed into pellets. These are then introduced into the materials stream by selling the recycled material to parts manufacturers.

10.4.4 Energy Recovery Through Combustion

Incineration reduces the volume of solid waste that needs to go to the landfill by burning the combustible organic material. The combustion converts the retained energy of the solid waste into heat, which can be used to generate steam and thus electricity. The volume of solid waste is reduced by 50 to 95 percent depending on the effectiveness of

---

1. The following is the identification scheme: 1, polyethylene; 2, high-density polyethylene; 3, vinyl; 4, low-density polyethylene; 5, polypropylene; 6, polystyrene; 7, other. This coding scheme has been extended to show a wider range of polymers and polymer blends. See Ashby, pp. 79–81.

the sorting process. The products of the incineration process are ash (the noncombustible inorganic waste), flue gases, and heat.

A modern municipal incineration plant employs considerable technology in order to perform its function without causing objectionable pollution. The combustion process must be capable of handling a large flow of solid waste and produce controlled high temperatures. The flue gas contains significant amounts of fine particulates in the form of heavy metals, dioxins, furans, SO_2, CO_2, and HCl. Thus, it is vital to control the flue gases by means of temperature control, particle filtration, and gas scrubbers.

10.4.5 Landfill

A landfill[1] is not a trash dump. A landfill is an airtight, lined (usually with polymer sheeting), structured, containment making for a permanent underground burial site for compacted solid waste. The waste is set into the ground and isolated from the elements (e.g., air, rain, groundwater, and animals) so that it can remain in its original form and decompose very slowly.[2] Landfills literally fill in space to create more land surface. Some landfills are enclosed in a clay liner that serves to separate the waste more thoroughly from its physical surroundings.

The two challenges to maintaining a landfill are controlling leaching and methane gas. The penetration of some water into a landfill is inevitable. As water moves through the waste, it is contaminated by organic material, metals, and products of the slow decomposition to create a mixture of water and *leachate*. Landfills are built with systems to direct leachate drainage into a collection pond where the contaminated water can be treated. If leachate breaks through a spot in a landfill, that area is patched.

The trash in a landfill will decompose in an anaerobic process. This creates a mixture of gases that is composed of about 50 percent methane and 50 percent carbon dioxide (both greenhouse gases). These gases must be released through a piping system. In the United States, solid waste landfills rank third in the source of human-related methane gas emissions.

The gases produced by a landfill (LFGs) can be used as an energy source.[3] There are methods to use LFG in combustion as a fuel for producing electricity and running some heavy equipment. Methane is the major component of liquid natural gas and can be recovered from the LFG for separate use. Naturally, LFGs require processing in preparation for their use, and the combustion will produce some pollution. Converting LFGs to fuel or heat requires conditions that make it economically feasible, but when those conditions are met landfills become renewable energy sources.

10.5
ROLE OF MATERIAL SELECTION IN DESIGN FOR ENVIRONMENT

Material selection has a unique role in Design for Environment tools, practices, and methods. Design for Environment methods are those that bring the consideration of the entire life cycle of the product into the earliest stages of the design process.

---

1. C. Freudenrich, "How Landfills Work," 2008, http://science.howstuffworks.com.
2. A composting site is designed to decompose its waste quickly. This is the opposite of a landfill.
3. "Landfill Methane Outreach Program," U.S. EPA, July 25, 2011, http://www.epa.gov/lmop/basic-info.

The material of which a product is made greatly influences the impact of a product on the environment in terms of natural resource use and impact and end-of-life options.

10.5.1 Material Life Cycle

Materials have a life cycle that begins before the products in which they are used (see Fig. 1.7). A material's life cycle begins with its removal from the earth and subsequent refining and shaping into stocks of engineering materials. These materials are further processed into final products and put into service. Once the products are retired from service the materials must be disposed of in an environmentally safe manner.

Designers can use their knowledge of materials and selection methods (see Chaps. 11 and 12) to make decisions as to which materials to use for a design once the major details of the design are known. Example 10.3 provides an example of this process.

> **EXAMPLE 10.3**
> A number of federal laws have mandated radical changes in automotive design including the material used for gas tanks. First is the act mandating Corporate Average Fuel Economy, which creates an incentive for weight reduction to increase gas mileage. Next are the Alternative Motor Fuels Act of 1988 and the Clear Air Act Amendments of 1990 requiring that auto makers prepare gas tanks for the wider use of alternative-fuels to reduce oil imports and to increase the use of U.S. sourced renewable fuels such as ethanol. Finally, the EPA has introduced fuel-permeation standards that challenge the designs and materials traditionally used in automotive fuel tanks. Terne-coated steel (8 percent tin-lead coating) has been the traditional material selection for automotive gas tanks. Tests have shown that neither painted nor bare terne-coated steel will resist the corrosive effects of alcohol for the 10-year expected life of the fuel tank.[1] In the selection of a material for an automotive fuel tank, the following factors are most important: manufacturability, cost, weight, corrosion resistance, permeability resistance, and recyclability. Another critical factor is safety and the ability to meet crash requirements.
>
> Two new competing materials have emerged to replace terne steel: electro-coated zinc-nickel steel sheet and high-density polyethylene (HDPE). The steel sheet is painted on both sides with an aluminum-rich epoxy. The epoxy is needed to provide exterior protection from road-induced corrosion. Stainless steel performs admirably for this application but at nearly five times the cost of terne steel. HDPE is readily formed by blow-molding it into the necessary shape, and it has long-term structural stability, but it will not meet the permeability requirement. Two approaches have been used to overcome this problem. The first is multilayer technology, in which an inner layer of HDPE is adhesively joined to a barrier layer of polyamide. The second approach is a barrier to permeability that involves treating the HDPE with fluorine.
>
> The first step in arriving at a material selection decision is to make a competitive analysis, laying out the advantages and disadvantages of each candidate.

---

1. P. J. Alvarado, "Steel vs. Plastics: The Competition for Light-Vehicle Fuel Tanks," *JOM,* July 1996, pp. 22–25.

Material Option	Advantages	Disadvantages
Terne-coated steel	Low cost at high volume Modest material cost Meets permeability requirement	Poor shape flexibility Poor corrosion protection from alcohol fuels Lead-containing coating gives problems with recycling or disposal
Electrocoated Zn-Ni steel	Low cost at high volumes Effective corrosion protection Meets material cost Meets permeability requirement	Poor weldability Low shape flexibility
Stainless steel	High corrosion resistance Recyclable Meets permeability requirements	High cost at all volumes, Low formability Poor weldability
HDPE (high-density polyethylene)	Good shape flexibility Low tooling costs at low volumes Low weight Good corrosion resistance	High tooling costs at high volumes High material cost Fails permeability and recyclability requirements.
Multilayer and barrier HDPE	Same as HDPE Meets permeability requirement	Compared to HDPE Higher tooling costs at high volume Higher material cost Hard to recycle

The next step would be to use one of the matrix methods discussed in Chapter 7 to arrive at the decision. Two of the Big Three automobile producers are changing to some variant of the Zn-Ni coated steel. The other is going with HDPE fuel tanks. The fact that there is not a clearly superior choice demonstrates the complexity of selection decisions in the modern day when there often are so many competing issues.

10.5.2 Selection of Eco-Efficient Materials

Metallic and ceramic materials originate from minerals (ores) obtained from the earth, which are refined using energy. Polymers are made from fossil fuel feedstock, chiefly petroleum and natural gas, and require energy to turn them into engineering plastics.

We can divide the product life-cycle into five phases: (1) material production, (2) part manufacture, (3) transport between phases, (4) service in use, and (5) disposal. Energy is consumed at each phase in this cycle, and some emissions are produced (heat, liquids, solid wastes, and gases, chiefly CO_2). Generally one of these phases consumes the preponderance of the lifetime energy use. With the aluminum beverage can it is the refining of aluminum from the bauxite ore (phase 1). With a transport plane it is the jet fuel used in service (phase 4).

Materials selection plays a major role in each of the phases of product life cycle.

1. Material production: reduce mass of material and choose material with low eco-indicators.
2. Part manufacture: select a process with low energy requirement and CO_2 footprint.
3. Transportation: low mass reduces energy consumption.

4. Service in use: thermal and electrical losses are often important and are material dependent. Low mass is important in dynamic products, for example, automobiles.
5. Disposal: High recyclability is a strong benefit and toxins must not be produced during combustion or recycling processes.

Life-Cycle Assessment (LCA). A *life-cycle assessment* (LCA) is the ideal way to assess alternatives in design for environment issues and the broader set of sustainability issues in design. An LCA determines all of the resource consumption and emissions involved with a product, and then assesses the impacts in terms of such categories as potential for global warming, ozone depletion, acidification of streams and rivers, human toxicity, and so on. Creating an LCA is a resource-intensive activity in terms of data collection and interpretation, requiring real expertise in the field for meaningful results.[1]

Life-cycle assessment proceeds in three stages:

1. Inventory analysis: The flows of energy and materials to and from the product during its life are determined quantitatively.
2. Impact analysis: All potential environmental consequences of the flows catalogued above are considered.
3. Improvement analysis: Results of steps 1 and 2 are translated into specific actions that reduce the impact of the product or the process on the environment.

LCA is always preferred when making engineering decisions on existing products or on a larger scale (e.g., factory remodeling). However, LCA is not well suited for making decisions in the time frame needed in product-focused engineering design. As a result, efforts have been made to find simple *eco-indicators* that could serve as useful surrogates for a LCA.

A widely used factor in material selection in design is the *embodied energy,* sometimes called the production energy. This is the energy per unit mass consumed in making the material from its ores or feedstock. Values of embodied energy and CO_2 footprint, along with energies and footprints involved in manufacturing and recycling are given for 47 commonly used materials by Ashby.[2]

A second key eco-indicator is the *carbon footprint.* This is actually a measure of the carbon dioxide, CO_2, in kg of gas produced per kg of material. Carbon dioxide is chosen because of its great importance in global warming and climate change. By multiplying the footprint by the annual world production we can rank the four worst contributors of CO_2 from material production.[3] They are: (1) steel, (2) aluminum, (3) cement, and (4) paper and cardboard.

A number of methods and tools have been proposed for simplifying the LCA.[4] One of the more useful and less complicated of these is the *eco-audit* method by Ashby.[5] An eco-audit is a scaled-down, simplified calculation to estimate environmental impact.

1. T. E. Graedel and B. R. Allenby, *Design for Environment,* Prentice Hall, Upper Saddle River, NJ, 1996.
2. M. F. Ashby, op. cit., Chapter 12.
3. M. F. Ashby, *Materials and the Environment,* Butterworth-Heinemann, Boston, 2009, p. 118.
4. K. Ramani, D. Ramanujan, W. Z. Bernstein, F. Zhao, J. Sutherland, C. Handwerker, J-K. Choi, H. Kim, and D. Thurston, "Integrated Sustainable Life Cycle Design: A Review," *Journal of Mechanical Design,* September 2010, vol. 132, 091004.
5. M. F. Ashby, op. cit., Chapter 7.

Each of the five phases of product life-cycle is evaluated for energy requirements and CO_2 production.

In the stage of manufacturing the parts, eco-indicators are those specific to the manufacturing process under study. In the transport phase, energy consumed in transportation is multiplied by the mass of the product. In the service-in-use phase, a simple engineering model is constructed for the major energy consuming activities and their corresponding emissions. In the disposal phase, if the product is remanufactured or recycled there will be a negative value for the eco-indicator. A more extensive collection of eco-properties can be found in the software CES EduPack™ from Granta Design. The use of this software facilitates the preparation of eco-audits.

10.6
TOOLS TO AID DESIGN FOR THE
ENVIRONMENT AND SUSTAINABILITY

State-of-the art engineering design practice has been broadened to consider the product as it will exist throughout its entire life-cycle. Sustainability forces designers to consider a product as having a first life-cycle and then being retired in such as way so that the used components or material can be transformed into new products. The proliferation of Design for X guidelines (see Sec. 8.12) is a well-known phenomenon. No one resource can provide a definitive list of the guidelines, especially in a field as broad as Design for Sustainability and its subset Design for Environment.[1] Figure 10.3 provides a schematic diagram showing the relationship between sustainability and the major Design for X topics of this section. Several strategies are included in this section and references are given for many others.

10.6.1 Design for Life Cycle

Design for the Life Cycle emphasizes in embodiment design those issues that impact a long, useful service life of the product. It also means designing for eventual replacement or disposal. Design modifications that can keep a product in service will benefit the environment in the long run because the product will not have to be disposed of, and will not consume additional natural resources to be replaced. The following life-cycle design strategies can be used to protect the environment and increase a product's sustainability.

- Minimize emissions and waste in the manufacturing process. Examine all of the ways that the product negatively impacts the environment and eliminate or minimize them using design. A polluting product is a defective product.
- Substitute recyclable materials where possible, and use design for disassembly guidelines to improve chances for recycling.
- Increase the useful life of the product, thereby prolonging the time when new material and energy resources need to be committed to a replacement of the product. The useful life may be limited by degraded performance due to wear and corrosion,

---

1. M. Kutz, ed., *Environmentally Conscious Mechanical Design,* John Wiley & Sons, Hoboken, NJ, 2007.

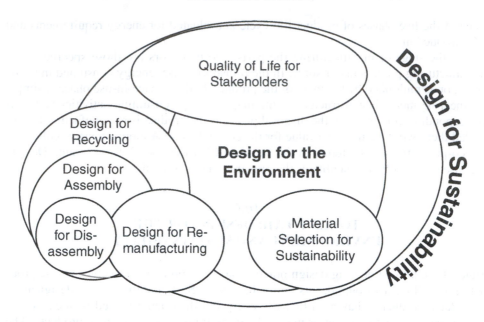

FIGURE 10.3
Conceptual diagram of relationship of Design for Sustainability and other Design for X topics.

damage (either accidental or because of improper use), or environmental degradation. Other reasons to terminate the useful life not related to life-cycle issues are technological obsolescence (something better has come along) or styling obsolescence.

There are a variety of design strategies to extend a product's useful life appearing in later sections of this text. A list locating the material can be found in Sec. 8.12. Design for reliability (Chap. 14), durability (Chap. 14), and serviceability (Chap. 8) are among them.

10.6.2 Design for Sustainability During Conceptual Design

New products need to meet the sustainability criteria as much as possible given the state of technology, regulation, and business policy of the company. Design tools and methods are being proposed to meet these criteria. It will be some time before a definitive set of principles can be consolidated from the influx of new design aids. Table 10.1 is a set of proposed guidelines for design that would meet the environmental objectives of sustainability as described by the Brundtland Report in Sec. 10.2. Many new methods for sustainable design are adaptations of existing methods (e.g., design for analogy[1]).

---

1. D. P. Fitzgerald, J. W. Herrmann and L. C. Schmidt, "A Conceptual Design Tool for Resolving Conflicts Between Product Functionality and Environmental Impact," *Journal of Mechanical Design,* 2010, vol. 132, 091006.

TABLE 10.1
Suggested Sustainability Design Guidelines for Conceptual Design

Proposed Design Guideline	Sample Products
1. Minimize quantity of resource use by optimizing its rate and duration	• Tankless water heaters (heat water as it is demanded) • Low-flow (or time-limited) shower heads
2. Incorporate automatic or manual tuning capabilities	• Clothes dryers with moisture sensors • Multi-flush toilets • Programmable thermostats
3. Use feedback mechanism to inform end-user of current status of process	• Temperature probes with readouts in ovens
4. Create separate modules for behaviors with conflicting requirements	• Hybrid electric vehicles

Adapted from C. Telenko and C. C. Seepersad, "A Methodology for Identifying Environmentally Conscious Guidelines for Product Design," *Journal of Mechanical Design,* vol. 132, 2010, 091009.

10.6.3 Design Guidelines Applying to Embodiment Design

1. ***Design for the minimal use of materials and energy.*** First, achieve minimum weight without affecting quality, performance, and cost. Automotive manufacturers have made it a long-term goal to achieve greater fuel economy and have done so with weight reduction (often by substitution of materials). Second, reduce waste of all types: scrap in manufacture, defective components in assembly, damaged goods in shipping. Thus, good design for quality practice will invariably result in a reduction in material consumption. Third, look hard at the design and use of packaging. Recognize changes in polymer packaging materials that allow for the recovery, recycling, and reuse of packaging materials. The substitution of cardboard for Styrofoam in fast food packaging is a common example. Look for ways to design shipping containers so they can be reused.

2. ***Design for Disassembly.*** Remanufacture, reuse, and recycling require the ability to economically remove the most valuable components when the product reaches the end of its useful life.

   • Minimize the number of adhesive and welded joints when it makes sense
   • Use removable fasteners and those that are not prone to breakage (i.e. avoid snap fits)
   • Increase the corrosion resistance of fasteners

3. ***Design for Maintainability.*** Most products need to be opened to be maintained. Thus, maintainability guidelines include the previous category of Design for Disassembly and Design for Serviceability (Sec. 8.9.4).

10.6.4 Design Guidelines Applying to End-of-Life Transformations

Engineers design products and systems for a certain behavior during their useful life. Design for the Environment or Green Design, or eco-design, or design for sustainability requires designing a product or system for a particular behavior at the end of its useful life. These design approaches are not necessarily aligned with performance, which requires the design team to have expertise in the planned transformation of the product at the end of its first useful life (Sec. 10.4).

1. *Design for Remanufacturing.* One challenge to remanufacturing is that the original product was not designed with this goal in mind. Not every product is a good candidate for remanufacturing. An MIT group studied 25 products and found that the energy used to remanufacture them was typically less than to create new products, as expected. Yet the remanufactured products were likely to be less efficient in energy use than new products or their replacement versions. In about half of the cases studied, there were no net energy savings; in the other half the savings were minimal.[1]

Characteristics of products that are good candidates for remanufacturing include the following:[2]

- Technology that will be relevant (stable) for 7 to 10 years
- Product redesign cycle of 1 to 4 years
- Rate of return for remanufacture at 15 percent or more (an option some companies employ)
- 50 to 75 percent of parts to be remanufactured
- Modular architecture with good separation of materials into different modules

Guidelines to employ while originally designing a product that may be remanufactured are as follows:

- Increase damage resistance throughout life cycle
- Make location of wear detectable
- Use components that can be assembled by commonly available tools
- Eliminate part features that can collect dirt and debris

Unfortunately, some of these guidelines are contrary to guidelines for good Design for Assembly (see Sec. 13.6), and a trade-off decision is necessary.

2. *Design for Recycling.* There are several steps that the designer can follow to enhance the recyclability of a product.

- Make it easier to disassemble the product and thus enhance the yield of the separation step
- Minimize the number of different materials in the product to simplify the identification and sorting issue
- Choose materials that are compatible and do not require separation before recycling (e.g., a bronze bushing embedded in a steel part will cause severe processing difficulty during hot working recycled steel)

---

1. "When is it worth remanufacturing?" *Advanced Materials and Processes,* July 2011, p. 14.
2. P. Zwokinski and D. Brissaud, "Remanufacturing Strategies to Support Product Design and Redesign," *Journal of Engineering Design,* vol. 19, no. 4, August 2008, pp. 321–355; B. Bras, "Design for Remanufacturing Processes," in M. Kutz, ed., *Environmentally Conscious Mechanical Design,* John Wiley & Sons, Hoboken, NJ, 2007.

- Identify the material that the part is made from right on the part. Use the identification symbols for plastics.

Applying these guidelines can require serious trade-offs. Minimizing the number of materials in the original product may require a compromise in performance from the use of a material with less-than-optimum properties. A clad metal sheet or chromium-plated metal provides the desired attractive surface at a reasonable cost, yet it cannot be readily recycled. In the past, decisions of this type would be made exclusively on the basis of cost.

3. ***Design for waste recovery and reuse in processing.*** The waste associated with a product can be a small fraction of the waste generated by the processes that produced the product. Be alert to ways of reducing process waste. Avoid the use of hazardous or undesirable materials. Keep current on changes in government regulations and lists of hazardous materials. For example, avoid the use of CFC refrigerants, use aqueous solvents for cleaning instead of chlorinated solvents, and use biodegradable materials whenever possible.

10.7
SUMMARY

The question arises, What is the impact of sustainability on the average design engineer? The answer depends on the type of business in which the engineer is employed. If it is an industry that is heavily impacted by governmental regulations like oil production, chemicals, or automobiles, then regulations that are intended to protect the environment or safety of the public already strongly influence products design or manufacturing operations. If it is in the business of energy production, there is no question that major changes will take place as fossil fuels become depleted and global warming becomes more widely recognized as a threat to life on our planet. These will present major challenges and opportunities for design engineers. Many other businesses recognize the importance of sustainability and view it as an opportunity to differentiate their products as green products. Thus, the design engineer can be assured that sustainability, and its engineering embodiment, design for the environment, will be of increasing importance into the future.

It is generally believed that improvement of the environment is the joint responsibility of all citizens in partnership with business and government. Government plays a crucial role, usually through regulation, to ensure that all businesses share equitably in the cost of an improved environment. Since these increased product costs often are passed on to the customer, it is the responsibility of government to use the tool of regulation prudently and wisely. Here the technical community can play an important role by providing fair and timely technical input to government. Finally, many visionaries see a future world based on sustainable development in which the world's resources will no longer be depleted because the rate of resource consumption will be balanced by the rate of resource regeneration.

10

NEW TERMS AND CONCEPTS

Carbon footprint	IPAT	Remanufacturing
Ecosystem	Life-cycle assessment	Reuse
Embodied energy	Municipal solid waste	Sustainability
Environmental impact	Recycling	Triple bottom line

BIBLIOGRAPHY

"20 Years: Into Our Common Future," *Environment,* 50(1), 46–59, 2008.

Allenby, B. R: *The Theory and Practice of Sustainable Engineering,* Prentice Hall, Upper Saddle River, NJ, 2011.

Ashby, M. F.: *Materials and the Environment: Eco-Informed Material Choice,* Elsevier, Boston, 2009.

Azapagic, A. and P. Slobodan: *Sustainable Development in Practice: Case Studies for Engineers and Scientists,* 2d ed, John Wiley and Sons, New York, 2011.

de Steiguer, J. E.: *The Origins of Modern Environmental Thought,* University of Arizona Press, Tucson, 2006.

Graedel, T. E. and B. R. Allenby: *Design for Environment,* Prentice Hall, Upper Saddle River, NJ, 1996.

Kates, R. W., T. M. Parris, and A. A. Leiserowitz: "What Is Sustainable Development?," *Environment,* 47(3), 8–21, 2005.

Kutz, M. ed.: *Environmentally Conscious Mechanical Design,* John Wiley and Sons, Hoboken, NJ, 2007.

PROBLEMS AND EXERCISES

10.1 Count the number of light bulbs in your residence. Calculate the power use for one day. What is the power difference between using incandescent and fluorescent bulbs in your residence for one year? Where can you safely dispose of the fluorescent bulbs?

10.2 Example 10.1 gives the current cost for standard size and quality copy paper with varying levels of recycled material content. Do further research to see what differences in the production of the paper lead to the cost differences.

10.3 Write an essay answering the question, Is Walmart's business practice of locating in small communities a good example of sustainable development?

10.4 There has been a major shift toward outsourcing manufacturing from the United States to foreign countries. One unintended consequence of this situation is the appearance of counterfeit parts. This has occurred in both electrical components and mechanical parts. Find an example of this phenomenon and write a brief essay.

10.5 Use the concept of a force field diagram (Chap. 4) to show the main factors that help and hinder the reduction of the use of energy in product development. (*Hint:* A "help" would be miniaturization; a hindrance would be increasing world wealth.)

10.6 Enthusiastic environmentalists often take the position that in the interest of saving the world, products should be designed to be as durable as possible, with major emphasis on modularity so that worn parts can be easily replaced. Discuss the advantages and disadvantages of this approach to design for the environment.

11

MATERIALS SELECTION

11.1
INTRODUCTION

This chapter provides a comprehensive treatment of the selection of materials for manufacturing the design. Chapter 12 presents some advanced topics in the mechanical behavior of materials that are relevant to design but not generally taught in mechanics of materials courses. The discussion assumes the reader has a working knowledge of the mechanical behavior of materials. Additional topics dealing with making products and parts from materials are considered in Chap. 13, Design for Manufacturing.

Materials and the manufacturing processes that convert them into useful parts underlie all of engineering design. There are over 100,000 engineering materials to choose from. The typical design engineer should have ready access to information on 30 to 60 materials, depending on the range of applications he or she deals with.

The recognition of the importance of materials selection in design has increased in recent years. Concurrent engineering practices have brought materials specialists into the design process at an earlier stage. The importance given to quality and cost aspects of manufacturing in present-day product design has emphasized the fact that materials and manufacturing are closely linked in determining final product performance. Moreover, the pressures of global competition have increased the level of automation in manufacturing to the point where material costs often comprise 60 percent or more of the cost of a product. Finally, the extensive activity in materials science worldwide has created a variety of new materials and focused our attention on the competition between six broad classes of materials: metals, polymers, elastomers, ceramics, composites, and electronic materials. Thus, the range of materials available to the engineer is much broader than ever before. This presents the opportunity for innovation in design by utilizing these materials to provide greater performance at lower cost. Achieving these benefits requires a rational process for materials selection.

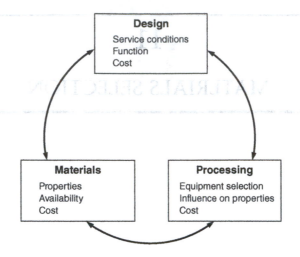

FIGURE 11.1
Interrelations of design, materials, and processing to produce a product.

11.1.1 Relation of Materials Selection to Design

An incorrectly chosen material can lead not only to part failure but also to excessive life-cycle cost. Selecting the best material for a part involves more than choosing both a material that has the properties to provide the necessary performance in service and the processing methods used to create the finished part (Fig. 11.1). A poorly chosen material can add to manufacturing cost. Properties of the material can be enhanced or diminished by processing, and that may affect the service performance of the part. Chapter 13 focuses on the relationship between materials processing and manufacturing and design.

Faced with the large number of combinations of materials and processes from which to choose, the materials selection task can only be done effectively by applying simplification and systemization. As design proceeds from concept design, to configuration and parametric design (embodiment design), and to detail design, the material and process selection becomes more detailed.[1] At the concept level of design, essentially all materials and processes are considered in broad detail. The materials selection charts and methodology developed by Ashby[2] are highly appropriate at this stage (see Sec. 11.3). The task is to determine whether each design concept will be made from metal, plastics, ceramic, composite, or wood, and to narrow it to a group of materials within that material family. The required precision of property data is rather low. Note that if an innovative choice of material is to be made it must be done

1. M. F. Ashby, *Met. Mat. Trans.*, 1995, vol. 26A, pp. 3057–3064.
2. M. F. Ashby, *Materials Selection in Mechanical Design,* 4th ed., Elsevier Butterworth-Heinemann, Oxford, UK, 2010.

at the conceptual design phase because later in the design process too many decisions have been made to allow for a radical change. The emphasis at the embodiment phase of design is on determining the shape and size of a part using engineering analysis. The designer will have decided on a class of materials and processes, such as a range of aluminum alloys, wrought and cast. The material properties must be known to a greater level of precision. At the parametric design step the alternatives will have narrowed to a single material and only a few manufacturing processes. Here the emphasis will be on deciding on critical tolerances, optimizing for robust design (see Chap. 15), and selecting the best manufacturing process using quality engineering and cost modeling methodologies. Depending on the importance of the part, materials properties may need to be known to a high level of precision. This may require the development of a detailed database based on an extensive materials testing program. Thus, material and process selection is a progressive process of narrowing from a large universe of possibilities to a specific material and process.

11.1.2 General Criteria for Selection

Materials are selected on the basis of four general criteria:

- Performance characteristics (properties)
- Processing (manufacturing) characteristics
- Environmental profile
- Business considerations

Selection on the basis of performance characteristics is the process of matching values of the properties of the material with the requirements and constraints imposed by the design. Most of this chapter and Chap. 12 deal with this issue.

Selection on the basis of processing characteristics means finding the process that will form the material into the required shape with a minimum of defects at the least cost. Chapter 13 is devoted exclusively to this topic.

Selection on the basis of an environmental profile is focused on predicting the impact of the material throughout its life cycle on the environment. Environmental considerations are growing in importance because of greater societal awareness and governmental regulation caused by concerns with global warming and the role that energy production and use play in it. These issues have been raised in Chap. 10.

The chief business consideration that affects materials selection is the cost of the part that is made from the material. This includes both the purchase cost of the material and the cost to process it into a part. A more exact basis for selection is life-cycle cost, which includes the cost of replacing failed parts and the cost of disposing of the material at the end of its useful life. Issues concerning cost of materials are considered in Sec. 11.5. Chapter 13 presents information on estimating costs as an aid in selecting the best manufacturing process. Chapter 17 deals with cost evaluation in further detail.

In Sec. 11.2 we will consider the important issue in materials selection of identifying the appropriate material properties that allow the prediction of failure-free functioning of the component. The equally important task of identifying a process

11

to manufacture the part with the material is discussed in Chap. 13. While these are important considerations, they are not the only issues in materials selection. The following business issues must also be considered. Failure to get a positive response in any of these areas can disqualify a material from selection.

1. Availability
    a. Are there multiple sources of supply?
    b. What is the likelihood of availability in the future?
    c. Is the material available in the forms needed (tubes, wide sheet, etc.)?

2. Size limitations and tolerances on available material shapes and forms, e.g., sheet thickness or tube wall concentricity
3. Excessive variability in properties
4. Low environmental impact, including ability to recycle the material
5. Cost. Materials selection comes down to buying properties at the best available price.

11.2
PERFORMANCE REQUIREMENTS OF MATERIALS

The performance requirements of a material usually are expressed in terms of physical, mechanical, thermal, electrical, or chemical properties. Material properties are the link between the basic structure and composition of the material and the service performance of the part (Fig. 11.2). The performance requirements follow from the function of a part. For example, the function of a connecting rod in an IC engine is to connect the piston to the crank shaft. The performance requirement is that it should deliver the required power without failing during the useful life of the engine. The essential material properties are tensile yield strength and fatigue strength along with sufficient resistance to the operating environment that these properties do not degrade during service.

Materials science predicts how to improve the properties of materials by understanding how to control their structure. Structure can vary from atomic dimensions to dimensions of several millimeters. The chief methods of altering structure are through composition control (alloying), heat treatment, and controlling the processing of the

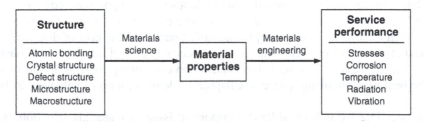

FIGURE 11.2
Material properties, the link between structure and performance.

material. A general background in the way structure determines the properties of solid materials usually is obtained from a course in materials science or fundamentals of engineering materials.[1] The materials engineer specializes in linking properties to design through a deep understanding of material properties and the processing of materials.

Since structure determines properties, everything about materials is *structure*. The term *structure* has different meanings as we change the scale of observation. To materials scientists, structure describes the way atoms and larger configurations of atoms arrange themselves, but to the design engineer structure refers to the form of a component and how the forces are applied to it. At the atomic level, materials scientists are concerned with basic forces between atoms, which determine the density, inherent strength, and Young's modulus. Moving upward in scale, they deal with the way the atoms arrange themselves in space, that is, the *crystal structure*. Crystal type and lattice structure determine the slip plane geometry and ease of plastic deformation. Superimposed on the crystal structure is the *defect structure* or the imperfections in the perfect three-dimensional atomic pattern. For example, are there lattice points where atoms are missing (vacancies), or are there missing or extra planes of atoms (dislocations)? All of these deviations from perfect atomic periodicity can be studied with sophisticated tools like an electron microscope. The defect structure greatly influences the properties of materials. At a higher scale of observation, such as that seen through an optical microscope, we observe the *microstructure* features such as grain size and the number and distribution of individual crystal phases. Finally, with a low-power microscope, we may observe porosity, cracks, seams, inclusions, and other gross features of the *macrostructure*.

11.2.1 Classification of Materials

We can divide materials into metals, ceramics, and polymers. Further division leads to the categories of elastomers, glasses, and composites. Finally, there are the technologically important classes of optical, magnetic, and semiconductor materials. An *engineering material* is a material that is used to fulfill some technical functional requirement, as opposed to being used just for decoration. Those materials that are typically used to resist forces or deformations in engineering structures are called *structural materials*. Other materials are used primarily for their electrical, semiconductor, or magnetic properties.

Engineering materials usually are not made up of a single element or one type of molecule. Many elements are added together in a metal to form an alloy with specially tailored properties. For example, pure iron (Fe) is rarely used in the elemental

1. W. D. Callister, *Materials Science and Engineering*, 8th ed., John Wiley & Sons, New York, 2010; J. F. Shackelford, *Introduction to Materials Science for Engineers*, 7th ed., Prentice Hall, Upper Saddle River, NJ; 2009; W.E. Smith and J. Hashemi, *Foundation of Materials Science and Engineering*, 5th ed, McGraw-Hill, New York, 2010; M. Ashby, H. Shercliff, and D. Cebon, *Materials: Engineering Science, Processing, and Design*, 2nd ed., Butterworth-Heinemann, Oxford, UK, 2009; T. H. Courtney, "Fundamental Structure-Property Relationships in Engineering Materials," *ASM Handbook*, vol. 20, pp. 336–356.

state, but when it is alloyed with small amounts of carbon to form steel its strength
is improved markedly. This is brought about by the formation throughout the solid
of strong intermetallic iron carbide Fe_3C particles. The degree of strengthening increases with the amount of iron carbide, which increases with the carbon content.
However, an overriding influence is the distribution and size of the carbide particles
in the iron matrix. The distribution is controlled by such processing operations as
the hot rolling or forging of the steel, or by its thermal treatment such as quenching
or annealing. Thus, there are a great variety of properties that can be obtained in a
given class of alloys. The same applies to polymers, where the mechanical properties depend upon the types of chemical groups that make up the polymer chain, how
they are arranged along the chain, and the average length of the chain (molecular
weight).

Thus, there is a material classification hierarchy,[1] starting with the **Materials
Kingdom** (all materials) → **Family** (metals, polymers, etc.) → **Class** (for metals:
steels, aluminum alloys, copper alloys, etc.) → **Subclass** (for steels: plain carbon, low-
alloy, heat treatable, etc.) → **Member** (a particular alloy or polymer grade). A member
of a particular family, class, and subclass of materials has a particular set of attributes
that we call its material properties. The classification does not stop here, because for
most materials the mechanical properties depend upon the mechanical (plastic deformation) or thermal treatment it has last been given. For example, the yield strength
and toughness of AISI 4340 steel will depend strongly on the tempering temperature
to which it has been subjected after oil quenching from an elevated temperature.

Figure 11.3 shows a selection of engineering materials commonly used in structural applications. Information on the general properties and applications for these
materials can be found in your materials science text and any one of a number of
specialized sources.[2]

11.2.2 Properties of Materials

The performance or functional capabilities of a material are usually given by a definable and measurable set of material properties. The first task in materials selection is to determine which material properties are relevant to the application. We
look for material properties that are easy and inexpensive to measure, are reproducible, and are associated with a material behavior that is well defined and related to
the way the material performs in service. For reasons of technological convenience
we often measure something other than the most fundamental material property.
For example, the elastic limit measures the first significant deviation from elastic

1. M. F. Ashby, *Materials Selection in Mechanical Design,* 4th ed., Elsevier Butterworth-Heinemann,
Oxford, UK, 2011.
2. K. G. Budinski and M. K. Budinski, *Engineering Materials,* 9th ed., Prentice Hall, Upper Saddle
River, NJ, 2010; P. L. Mangonon, *The Principles of Materials Selection in Design,* Prentice Hall, Upper Saddle River, NJ, 1999; *Metals Handbook, Desk Edition,* 2nd ed., ASM International, Materials
Park, OH, 1998; *Engineered Materials Handbook, Desk Edition,* ASM International, Materials Park,
OH, 1995.

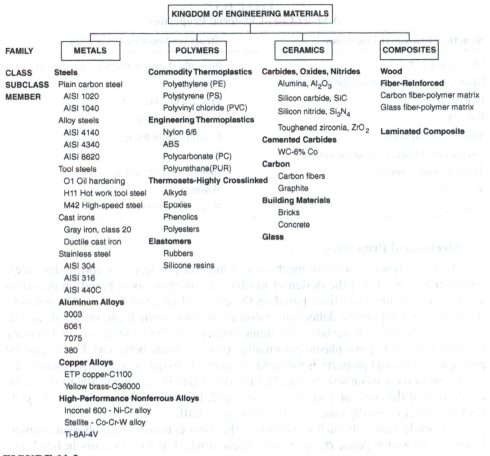

	KINGDOM OF ENGINEERING MATERIALS			
FAMILY	**METALS**	**POLYMERS**	**CERAMICS**	**COMPOSITES**
CLASS	Steels	Commodity Thermoplastics	Carbides, Oxides, Nitrides	Wood
SUBCLASS	Plain carbon steel	Polyethylene (PE)	Alumina, Al_2O_3	Fiber-Reinforced
MEMBER	AISI 1020	Polystyrene (PS)	Silicon carbide, SiC	Carbon fiber-polymer matrix
	AISI 1040	Polyvinyl chloride (PVC)	Silicon nitride, Si_3N_4	Glass fiber-polymer matrix
	Alloy steels	Engineering Thermoplastics	Toughened zirconia, ZrO_2	
	AISI 4140	Nylon 6/6	Cemented Carbides	Laminated Composite
	AISI 4340	ABS	WC-6% Co	
	AISI 8620	Polycarbonate (PC)	Carbon	
	Tool steels	Polyurethane(PUR)	Carbon fibers	
	O1 Oil hardening	Thermosets-Highly Crosslinked	Graphite	
	H11 Hot work tool steel	Alkyds	Building Materials	
	M42 High-speed steel	Epoxies	Bricks	
	Cast irons	Phenolics	Concrete	
	Gray iron, class 20	Polyesters	Glass	
	Ductile cast iron	Elastomers		
	Stainless steel	Rubbers		
	AISI 304	Silicone resins		
	AISI 316			
	AISI 440C			
	Aluminum Alloys			
	3003			
	6061			
	7075			
	380			
	Copper Alloys			
	ETP copper-C1100			
	Yellow brass-C36000			
	High-Performance Nonferrous Alloys			
	Inconel 600 - Ni-Cr alloy			
	Stellite - Co-Cr-W alloy			
	Ti-6Al-4V			

FIGURE 11.3
Commonly used engineering materials for structural applications.

behavior, but it is tedious to measure, so we substitute the easier and more reproducible 0.2% offset yield strength. That, however, requires a carefully machined test specimen, so the yield stress may be approximated by the exceedingly inexpensive hardness test.

A first step in classifying material properties is to divide them into *structure-insensitive properties* and *structure-sensitive properties,* Table 11.1. Both types of properties depend on the atomic binding energy and arrangement and packing of the atoms in the solid, but the structure-sensitive properties also depend strongly on the number, size, and distribution of the imperfections (dislocations, solute atoms, grain boundaries, inclusions, etc.) in the solid. Except for modulus of elasticity and corrosion rate in this table, all of the structure-insensitive properties are classified as *physical properties*. All of the properties listed as structure sensitive are *mechanical properties;* that is, they measure the response of the material to some kind of force.

11

TABLE 11.1
A Short List of Material Properties

Structure-Insensitive Properties	Structure-Sensitive Properties
Melting point, T_m	Strength, σ_f, where f denotes a failure mode
Glass transition temperature, for polymers, T_g	Ductility
Density, ρ	Fracture toughness, K_{Ic}
Porosity	Fatigue properties
Modulus of elasticity, E	Damping capacity, η
Coefficient of linear thermal expansion, α	Creep
Thermal conductivity, k	Impact or shock loading resistance
Specific heat, c_p	Hardness
Corrosion rate	Wear rate or corrosion rate

Mechanical Properties

We know from a course in mechanics of materials or from the spring design example in Sec. 8.6.1 that the design of mechanical components is based on the stress level not exceeding some limit based on the expected mode of failure. Alternatively, we design for keeping the deflection or distortion below some limit. In ductile metals and polymers (those materials with about greater than 10% elongation at fracture), the failure mode is gross plastic deformation (loss of elastic behavior). For metals the appropriate material property is the *yield strength*, σ_0, based on a 0.2% permanent deformation in the tension test. In Fig. 11.4 the offset line is drawn parallel to the linear elastic part of the curve at a strain offset of 0.002. For ductile thermoplastics the yield strength offset is usually taken at a larger strain of 0.01.

For brittle materials such as ceramics, the most common strength measurement is the modulus of rupture, σ_r, the tensile stress at which fracture occurs in bending a flat beam. Strength values obtained this way are about 30 percent higher than those

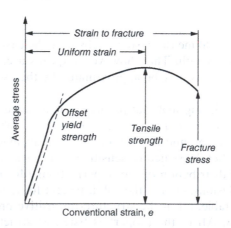

FIGURE 11.4
A typical stress-strain curve for a ductile metal.

measured in direct tension, but they are more consistent values. In fiber-reinforced composite materials, yielding is typically taken at a 0.5% deviation from linear elastic behavior. Composites with fibers are weaker in compression than in tension because the fibers buckle. Fiber-reinforced composites are also highly *anisotropic; that is, the* properties vary considerably with orientation of the loading direction to the fibers.

- *Ultimate tensile strength,* σ_u, is the maximum tensile stress that a material can withstand in the tension test, measured by load divided by the original area of the specimen. While it has little fundamental relevance to design, it is a simple property to measure in a tension test since it requires no extensometer to measure strain. Therefore, it is often reported and correlated with other properties as a surrogate for the overall strength of a material. For brittle materials it is the same as their fracture strength, but for ductile materials it is larger by a factor of 1.3 to 3 because of strain-hardening.
- *Modulus of elasticity* (Young's modulus), *E,* is the slope of the stress-versus-strain curve where it initially shows linear behavior, Fig. 11.4. A material with a high E is *stiffer* than a material with a lower E and resists deformation by bending or twisting to a greater extent.
- *Ductility* is the opposite of strength. It is the ability of a material to plastically deform before it fractures. It is usually measured by the percent elongation of a gage length inscribed on the test section of a tension test specimen or by the reduction in area of the tensile specimen at fracture.
- *Fracture toughness,* K_{Ic}, is a measure of the resistance of a material to the propagation of a crack within it. The use of this important engineering property in design is presented in Sec. 12.2. Other less sophisticated ways of measuring the tendency for brittle fracture are the Charpy V-notch impact test and using other notched specimens loaded in tension.
- *Fatigue properties* measure the ability of a material to resist many cycles of alternating stress. Fatigue failure, in all of its variations (high-cycle, low-cycle, and corrosion fatigue) is the number one cause of mechanical failure. See Sec. 12.3 for more information.
- *Damping capacity* is the ability of a material to dissipate vibrational energy by internal friction, converting the mechanical energy into heat. It is measured by the loss coefficient, η, which expresses the fractional energy dissipated in a stress-strain cycle.
- *Creep* is the time-dependent strain that occurs under constant stress or load in materials at temperatures greater than half of their melting point.
- *Impact resistance* is the ability of a material to withstand sudden shock or impact forces without fracturing. It is measured by the Charpy impact test or various kinds of drop tests. A material with high impact resistance is said to have high *toughness.*
- *Hardness* is a measure of the resistance of the material to surface indentation. It is determined by pressing a pointed diamond or steel ball into the surface under a known load.[1] Hardness is usually measured on arbitrary scales using the Rockwell, Brinell, or Vickers hardness tests. Hardness is a surrogate for yield stress. As a rough approximation, the higher the hardness number, the greater the yield stress.

---

1. *ASM Handbook,* vol. 8, *Mechanical Testing and Evaluation,* ASM International, Materials Park, OH, 2000, pp. 198–287.

Hardness measurements are used extensively as a quality control test because they are quick and easy to make and the test can be made directly on the finished component.

- *Wear rate* is the rate of material removal from two sliding surfaces in contact. Wear, an important failure mode in mechanical systems, is considered in Sec. 12.5.

Table 11.2 gives an overview of the most common types of failure modes that are likely to be encountered in various service environments. To identify the appropriate mode for designing a part, first decide whether the loading is static, repeated (cyclic), or dynamic (impact). Then decide whether the stress state is primarily tension, compression, or shear, and whether the operating temperature is well above or below room temperature. This will narrow down the types of failure mechanisms or modes, but in general it will not lead to a single type of failure mode. This will require consultation with a materials expert, or some further study by the design team.[1]

The mechanical property that is most associated with each mode of failure is given in the rightmost column of Table 11.2. However, the service conditions met by materials in general are often more complex than the test conditions used to measure material properties. The stress level is not likely to be a constant value; instead, it is apt to fluctuate with time in a random way. Or the service condition consists of a complex superposition of environments, such as a fluctuating stress (fatigue) at high temperature (creep) in a highly oxidizing atmosphere (corrosion). For these extreme service conditions, specialized simulation tests are developed to "screen materials." Finally, the best candidate materials must be evaluated in prototype tests or field trials to evaluate their performance under actual service conditions.

Table 11.3 gives typical room temperature mechanical properties for several engineering materials selected from Fig. 11.3. Examination of the properties allows us to learn something about how the processing, and thus the structure of the material, affects the mechanical properties.

First look at the values for elastic modulus, E, over the range of materials shown in Table 11.3. E varies greatly from 89×10^6 psi for tungsten carbide particles held together with a cobalt binder, a cemented carbide composite, to 1.4×10^2 psi for a silicone elastomer. Elastic modulus depends on the forces between atoms, and this very large range in E reflects the strong covalent bonding in the ceramic carbide and the very weak bonding of van der Waals forces in the polymeric elastomer.

Next, turn your attention to the values of yield strength, hardness, and elongation. The properties of the plain carbon steels, 1020 and 1040, well illustrate the influence of microstructure. As the carbon content is increased from 0.2% carbon to 0.4%, the amount of hard carbide particles in the soft iron (ferrite) matrix of the steel increases. The yield strength increases and the elongation decreases as dislocations find it more difficult to move through the ferrite grains. The same effect is observed in the alloy steel 4340, which is heated to the austenite region of the Fe-C phase diagram and then quenched rapidly to form the strong but brittle martensite phase. Tempering the

1. G. E. Dieter, *Mechanical Metallurgy,* 3rd ed., McGraw-Hill, New York, 1986; N. E. Dowling, *Mechanical Behavior of Materials,* 3rd ed., Pearson Prentice Hall, Upper Saddle River, NJ, 2007; *ASM Handbook,* vol. 8, *Mechanical Testing and Evaluation,* ASM International, Materials Park, OH, 2000; *ASM Handbook,* vol. 11, *Failure Analysis and Prevention,* ASM International, Materials Park, OH, 2002.

TABLE 11.2
Guide for Selection of Material Based on Possible Failure Modes, Types of Loads, Stresses, and Operating Temperature

Failure Mechanisms	Types of Loading			Types of Stress			Operating Temperatures			Criteria Generally Useful for Selection of Material
	Static	Repeated	Impact	Tension	Compression	Shear	Low	Room	High	
Brittle fracture	X	X	X	X	...	...	X	X	...	Charpy V-notch transition temperature. Notch toughness K_{Ic} toughness measurements
Ductile fracture(a)	X	...	...	X	...	X	...	X	X	Tensile strength. Shearing yield strength
High-cycle fatigue(b)	...	X	...	X	...	X	X	X	X	Fatigue strength for expected life, with typical stress raisers present
Low-cycle fatigue	...	X	...	X	...	X	X	X	X	Static ductility available and the peak cyclic plastic strain expected at stress raisers during prescribed life
Corrosion fatigue	...	X	...	X	...	X	...	X	X	Corrosion-fatigue strength for the metal and contaminant and for similar time(c)
Buckling	X	...	X	...	X	...	X	X	X	Modulus of elasticity and compressive yield strength
Gross yielding(a)	X	...	...	X	X	X	X	X	X	Yield strength
Creep	X	...	...	X	X	X	...	...	X	Creep rate of sustained stress-rupture strength for the temperature and expected life(c)
Caustic or hydrogen embrittlement	X	...	...	X	...	...	...	X	X	Stability under simultaneous stress and hydrogen or other chemical environment(c)
Stress corrosion cracking	X	...	...	X	...	X	...	X	X	Residual or imposed stress and corrosion resistance to the environment. K_{ISCC} measurements(c)

K_{Ic}, plane-strain fracture toughness; K_{ISCC}, threshold stress intensity to produce stress-corrosion cracking. (a) Applies to ductile metals only. (b) Millions of cycles. (c) Items strongly dependent on elapsed time.

Source: B. A. Miller, "*Materials Selection for Failure Prevention,*" *Failure Analysis and Prevention, ASM Handbook,* vol. 11, ASM International, Materials Park, OH, 2002, p. 35.

447

11

TABLE 11.3
Typical Room Temperature Mechanical Properties of Selected Materials

Material Class	Class Member	Heat Treatment or Condition	Elastic Modulus 10^6 (psi)	Yield Strength 10^3 (psi)	Elonga-tion %	Hardness
Steels	1020	Annealed	30.0	42.8	36	HB111
	1040	Annealed	30.0	51.3	30	HB149
	4340	Annealed	30.0	68.5	22	HB217
	4340	Q&temper 1200 F	30.0	124.0	19	HB280
	4340	Q&temper 800 F	30.0	135.0	13	HB336
	4340	Q&temper 400 F	30.0	204.0	9	HB482
Cast iron	Gray iron, class 20	As cast	10.0	14.0	0	HB156
	Ductile cast iron	ASTM A395	24.4	40.0	18	HB160
Aluminum	6061	Annealed	10.0	8.0	30	HB30
	6061	T4	10.0	21.0	25	HB65
	6061	T6	10.0	40.0	17	HB95
	7075	T6	10.4	73.0	11	HB150
	A380	As die cast	10.3	23	3	HB80
Thermoplastic polymers	Polyethylene (LDPE)	Low density	0.025	1.3	100	HRR10
	Polyethylene (HDPE)	High density	0.133	2.6	170	HRR40
	Polyvinyl chloride (PVC)	Rigid	0.350	5.9	40	
	ABS	Medium impact	0.302	5.0	5	HRR110
	Nylon 6/6	unfilled	0.251	8.0	15	HRR120
	Nylon 6/6	30% glass fiber	1.35	23.8	2	
	Polycarbonate (PC)	Low viscosity	0.336	8.5	110	HRM65
Thermosets	Epoxy resin	Unfilled	0.400	5.2	3	
	Polyester	Cast	0.359	4.8	2	
Elastomers	Butadiene	Unfilled	0.400	4.0	1.5	
	Silicone		1.4×10^{-4}	0.35	450	
Ceramics	Alumina	Pressed & sintered	55.0	71.2	0	
	Silicon nitride	Hot pressed	50.7	55.0	0	
	WC + 6% Co	Hot pressed	89.0	260	0	
	Concrete	Portland cement	2.17	0.14	0	
Composites	Wood	Pine—with the grain	1.22	5.38	2	
	Wood	Pine—across grain	0.11	0.28	1.3	
	Epoxy matrix-glass fiber	Longitudinal— parallel to fiber	6.90	246	3.5	
	Epoxy-glass fiber	Transverse to fiber	1.84	9.0	0.5	

Hardness: HB Brinell test; HR Rockwell hardness test; HRR Rockwell test using R scale; HRM Rockwell test using M scale. Metals data taken from *Metals Handbook, Desk Edition,* 2nd ed., ASM International, Materials Park, OH, 1998. Other data were taken from the Cambridge Engineering Selector software, Granta Design, Cambridge, UK. Where a range of values is given, only the lowest value was used.

psi = lb/in.^{2} = 6895 Pa = 6895 N/m^2

10^3(psi) = ksi = kip/in.^{2} = 6.895 MPa = 6.895 MN/m^2 = 6.895 N/mm^2

quenched steel causes the martensite to break down into a dispersion of fine carbide particles. The higher the tempering temperature, the larger is the particle size and the greater the average distance between them, which means that dislocations can move more easily. Thus, yield strength and hardness decrease with increasing tempering temperature, and elongation (ductility) varies inversely with yield strength. Note that elastic modulus does not vary with these changes in carbon content and heat treatment, because it is a structure-insensitive property that depends only on atomic bonding forces. This discussion illustrates the way that materials engineers can significantly alter the structure of materials to change their properties.

While viewing Table 11.3 it is instructive to examine how yield strength and ductility vary between families of materials. Ceramics are very strong because their complex crystal structures make it difficult for plastic deformation by dislocation motion (slip) to occur. Unfortunately, this also means that they are very brittle, and they cannot practically be used as monolithic structural materials in machine components. Polymers are very weak compared with metals, and they are subject to creep at or near room temperature. Nevertheless, because of many attractive attributes polymers are increasingly finding applications in consumer and engineered products. The special precautions that must be taken in designing with plastics (polymers) are discussed in Sec. 12.6.

Composite materials are hybrids that combine the best properties from two families of materials. The most common composites combine high modulus glass or carbon (graphite) fibers with a polymer matrix to improve both its modulus and its strength. Composite materials have reached such a high state of development that a large portion of Boeing's latest airliner is being made from polymer-based composites. However, as shown in Table 11.3, fiber-reinforced composite (FRP) materials exhibit much different properties when tested parallel (longitudinal direction) to the fiber, or at 90° (transverse) to the fiber. This type of *anisotropy* in mechanical properties is present in all materials, but it is extreme with FRP composites. To compensate for this, sheets of composite material are stacked up in different orientations of fiber to create laminates, much as with plywood. Because of the anisotropy of properties, design with composite materials requires special methods not generally covered in design courses.[1]

11.2.3 Specification of Materials

The material properties required in a part usually are formalized through specifications. Sometimes this is done by listing the material designation, AISI 4140 steel—for example, on the detail drawing of the part, along with processing instructions, such as the heat treatment temperatures and times. In this case the designer depends on generally accepted specifications established through organizations such as the Society of Automotive Engineers (SAE), ASTM, or ISO to give the requirements on chemical composition, grain size, surface finish, and other material descriptors.

Often companies find that using common standards, which are "consensus standards" agreeable to a wide sector of a material producing industry, do not provide the material quality they need for particularly sensitive manufacturing operations. For

---

1. *ASM Handbook,* vol. 21, *Composites,* ASM International, Materials Park, OH, 2001.

example, they may learn through a painful series of failures in production that the chemical limits on a minor element in a material must be held to a tighter tolerance range on chemical composition if they are to get an acceptable yield for a critical spot-welded part. The company will then issue their own specification for the material, which legally requires the supplier to supply material within a narrower range of chemistry. If the company is a large purchaser of the material, its supplier will generally accept the business and deliver material to the company specification, but if it is only a small customer the company will have to pay a "quality premium" for material made to its tighter specifications. The designer must make a trade-off between the cost of scrapped parts in production and the cost for the premium grade material.

11.2.4 Ashby Charts

Ashby[1] has created materials selection charts that are very useful in comparing a large number of materials during conceptual design. These charts are based on a large computerized material property database.[2] A typical chart is shown in Figure 11.5. It displays the elastic modulus of polymers, metals, ceramics, and composites plotted against density. Note that the elastic modulus of solid materials spans seven decades, from foam polymers to hard ceramics. Note how the classes of materials group into common regions with ceramics and metals in the upper right, polymers in the middle, and cellular materials such as polymer foams and cork in the lower left. While it is difficult to discern small differences in the location of materials in crowded sections of the chart, when we use the computer program this is no longer a problem.

In the lower right corner of the Ashby chart in Fig. 11.5 are dotted lines of various slopes. Depending on the type of loading, different slopes are appropriate to use. This will become clearer after reading Sec. 11.7. If we need to find the lightest tie rod loaded in axial tension to resist elongation, the line E/ρ = Constant would be chosen. Starting at the lower right corner of the chart, move a straightedge up toward the opposite corner parallel to this slope. At any instant, all materials lying on the straightedge would be equal candidates for selection, while all those lying below the straightedge would have been discarded. All those above the straightedge would be superior candidates.

EXAMPLE 11.1
Move up four of the dotted lines in Fig. 11.5 to $E = 10^{-1}$ GPa. We have exceeded the properties of most of the polymers and lead alloys, but zinc-based alloys and graphite fiber–reinforced polymers (GFRP) are on the line. Steels, titanium, and aluminum alloys lie above the line, and close examination of the chart shows that titanium alloys are the best selection. However, using actual numbers, the ratio E/ρ for plain carbon steel/aluminum alloy/titanium alloy is 104.9/105.5/105.9. This shows that to withstand a given elastic deformation the titanium alloy would be the lightest tie rod. However, the difference is so small that the much less expensive plain carbon steel would be selected. Note that E/ρ for Al_2O_3 is 353. Why wouldn't this be the chosen material?

1. M. F. Ashby, *Materials Selection in Mechanical Design*, 4th ed., Butterworth-Heineman, Oxford, UK, 2011.
2. Cambridge Engineering Selector, Granta Design, www.grantadesign.com.

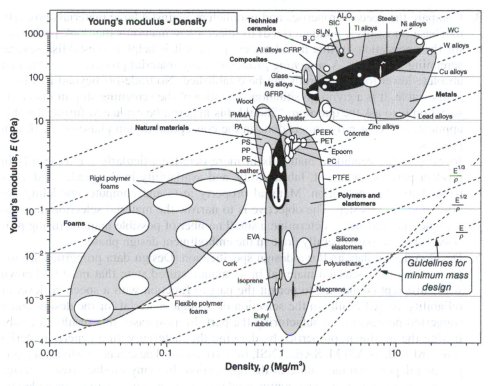

FIGURE 11.5
Ashby materials selection chart: Elastic modulus versus density. (From M. F. Ashby, *Materials Selection in Mechanical Design,* 3rd ed., p. 51, Copyright Elsevier, 2005. Used with permission.)

11.3
THE MATERIALS SELECTION PROCESS

Material choices will always be governed by material properties and manufacturing issues. However, the material selection process for a new product development differs slightly from the process for material substitution in an existing design. Each process is outlined in this section.

Materials Selection for a New Product or New Design

In this situation the materials selection steps are:

1. Define the functions that the design must perform and translate these into required materials properties such as stiffness, strength, and corrosion resistance, and such business factors as the cost and availability of the material.
2. Define the manufacturing parameters, such as the number of parts to be produced, the size and complexity of the part, its required tolerance and surface finish, general quality level, and overall manufacturability of the material.

3. Compare the needed properties and parameters against a large materials property database (most likely computerized) to select a few materials that look promising for the application. In this initial review process it is helpful to establish several screening properties. A *screening property* is any material property for which an absolute lower (or upper) limit can be established. No trade-off beyond this limit is allowable. It is a go/no-go situation. The idea of the screening step in materials selection is to ask the question: "Should this material be evaluated further for this application?" Generally, this is done in the conceptual design phase of the design process.

4. Investigate the candidate materials in more detail, particularly for trade-offs in product performance, cost, fabricability, and availability in the grades and sizes needed for the application. Material property tests and computer modeling are often done in this step. The objective is to narrow the material selection down to a single material and to determine a small number of possible manufacturing processes. This step is generally done in the embodiment design phase.

5. Develop design data and/or a design specification. Design data properties are the properties of the selected material in its manufactured state that must be known with sufficient confidence to permit the part to function with a specified level of reliability. Step 4 results in the selection of a single material for the design and a suggested process for manufacturing the part. In most cases this results in establishing the minimum properties by defining the material with a generic material standard such as ASTM, SAE, ANSI, or a MIL spec. The extent to which step 5 is pursued depends on the nature of the application. In many product areas, service conditions are not severe, and commercial specifications such as those provided by ASTM may be used without adopting an extensive testing program. In other applications, such as the aerospace and nuclear areas, it may be necessary to conduct an extensive testing program to develop design data that are statistically reliable.

Materials Substitution in an Existing Design

In this situation the following steps pertain:

1. Characterize the currently used material in terms of performance, manufacturing requirements, and cost.
2. Determine which properties must be improved for enhanced product function. Often failure analysis reports play a critical role in this step (see Sec. 14.6).
3. Search for alternative materials and/or manufacturing routes. Use the idea of screening properties to good advantage.
4. Compile a short list of materials and processing routes, and use these to estimate the costs of manufactured parts. Use the methods discussed in Sec. 13.9 or the method of value analysis in Sec. 17.13.
5. Evaluate the results of step 4 and make a recommendation for a replacement material. Define the critical properties with specifications or testing, as in step 5 of the previous section.

It generally is not possible to realize the full potential of a new material unless the product is redesigned to exploit both the properties and the manufacturing characteristics of the material. In other words, a simple substitution of a new material without changing the design rarely provides optimum utilization of the material. Most often the crux of materials selection is not that one material competes against another; rather, it is that the processes associated with the production or fabrication of one material compete with the processes associated with the other. For example, the pressure die casting of a zinc-based alloy may compete with the injection molding of a polymer. Or a steel forging may be replaced by sheet metal because of improvements in laser welding sheet-metal components into an engineering part. Thus materials selection is not complete until the issues discussed in Chap. 13 are fully considered.

11.3.1 Two Different Approaches to Materials Selection

There are two approaches[1] to settling on the material-process combination for a part. In the *material-first approach,* the designer begins by selecting a material class and narrowing it down as described previously. Then manufacturing processes consistent with the selected material are considered and evaluated. Chief among the factors to consider are production volume and information about the size, shape, and complexity of the part. With the *process-first approach,* the designer begins by selecting the manufacturing process, guided by the same factors. Then materials consistent with the selected process are considered and evaluated, guided by the performance requirements of the part. Both approaches end up at the same decision point. Most design engineers and materials engineers instinctively use the materials-first approach, since it is the method taught in strength of materials and machine design courses. Manufacturing engineers and those heavily involved with process engineering naturally gravitate toward the other approach.

11.3.2 Materials Selection in Embodiment Design

A more comprehensive materials selection process than is done in conceptual design is carried out in the embodiment design phase using the process shown in Fig. 11.6. At the beginning there are parallel materials selection and component design paths to follow. The input to the material selection process is a small set of tentative materials chosen in conceptual design based on the Ashby charts and sources of data described in Sec. 11.4. At the same time in the configuration design step of embodiment design, a tentative component design is developed that satisfies the functional requirements, and, using the material properties, an approximate stress analysis is carried out to calculate stresses and stress concentrations. The two paths merge in an examination of whether the best material, fabricated into the component by its expected manufacturing process, can bear the loads, moments, and torques that the component is expected

1. J. R. Dixon and C. Poli, *Engineering Design and Design for Manufacturing,* Field Stone Publishers, Conway, MA, 1995.

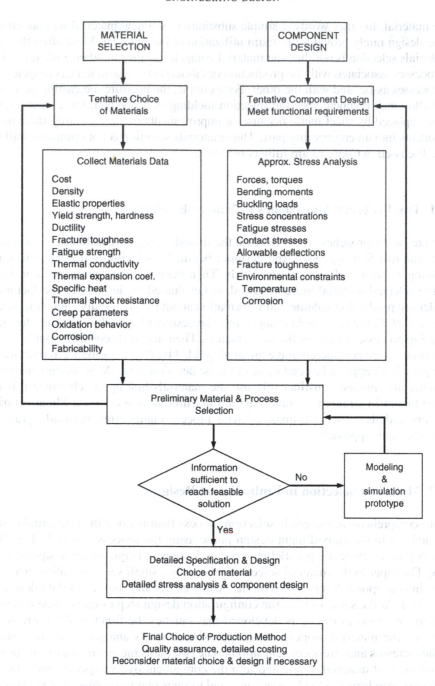

FIGURE 11.6
Steps in materials selection at the embodiment and detail design phases.

to withstand. Often the information is inadequate to make this decision with confidence and finite element modeling or some other computer-aided predictive tool is used to gain the needed knowledge. Alternatively, a prototype component is made and subjected to testing. Sometimes it becomes clear that the initial selections of materials are just inadequate, and the process iterates back to the top and the selection process starts over.

Before making a final material selection based on screening and ranking, as discussed in Secs. 11.6 through 11.9, it is important to determine that your selection does not result in any unpleasant surprises. This requires getting documentation about the material concerning such issues as failure analysis, case studies about its use, possible corrosion issues, prices, availability in needed sizes, and so on. This information generally is not available in databases. The information sources discussed in Sec. 11.4 , as well as contacts with suppliers, should prove helpful.

When the material-process selection is deemed adequate for the design, the choice passes to a detailed specification of the material and the design. This is the parametric design step discussed in Chap. 8. In this design step, an attempt should be made to optimize the critical dimensions and tolerances to achieve a component that is robust to its service environment, using an approach such as the Taguchi robust design methodology (see Chap. 15). The next step is to finalize the choice of the production method. This is based chiefly on a detailed calculation of the cost to manufacture the component (Chaps. 13 and 17). The material cost and the inherent workability and formability of the material, to reduce scrapped parts, are a major part of this determination. Another important consideration is the quality of the manufactured component, again strongly influenced by the choice of material. Still other considerations are the heat treatment, surface finishing, and joining operations that will be required.

A shortcut approach to materials selection that is often used is to select a material based on a component that has been used before in a similar application. This imitative approach results in a quick decision but it may not lead to a superior design if the service conditions are slightly different from those of the previous application, or if improvements in materials or the cost of manufacturing with the materials have changed from the date of the previous application. As an aid in starting the materials selection process, a listing of materials commonly used in various components is given in Appendix D.

11.4
SOURCES OF INFORMATION ON MATERIAL PROPERTIES

Most practicing engineers develop a file (paper or electronic) of trade literature, technical articles, and company reports. Material property data comprise an important part of this personal data system. In addition, many large corporations and government agencies develop their own compendiums of data on materials properties.

The purpose of this section is to provide a guide to material property data that are readily available in the published technical literature. There are several factors to have clearly in mind when using property data in handbooks and other open-literature sources. Usually a single value is given for a property, and it must be assumed that the

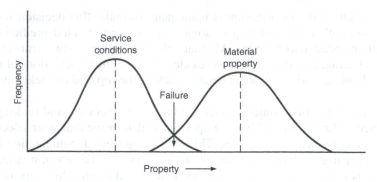

FIGURE 11.7
Overlapping distributions of material property and service requirement.

value is "typical." When scatter or variability of results is considerable, the fact may be indicated in a table of property values by a range of values (i.e., the largest and smallest values) or be shown graphically by scatter bands. Unfortunately, it is rare to find property data presented in a proper statistical manner by a mean value and the standard deviation. Obviously, for critical applications in which reliability is of great importance, it is necessary to determine the frequency distribution of both the material property and the parameter that describes the service behavior. Figure 11.7 shows that when the two frequency distributions overlap, there will be a statistically predictable number of failures. For more on variability of material properties see Sec. 14.2.3.

It is important to realize that a new material cannot be used in a design unless the engineer has access to reliable material properties and cost data. This is a major reason why the tried and true materials are used repeatedly for designs even though better performance could be achieved with advanced materials. At the start of the design process, low-precision but all-inclusive data is needed.

At the end of the design process, data is needed for only a single material, but it must be accurate and very detailed. Following is information on some widely available sources of information on materials properties. A significant start has been made in developing computerized materials databases, and in converting handbook data to CD-ROM for easier searching and retrieval.

11.4.1 Conceptual Design

Metals Handbook Desk Edition, 2nd ed., ASM International, Materials Park, OH, 1998. A compact compilation of metals, alloys, and processes.

Engineered Materials Handbook Desk Edition, ASM International, Materials Park, OH, 1995. A compact compilation of data for ceramics, polymers, and composite materials.

M.F. Ashby, *Materials Selection in Mechanical Design,* 4th ed., Butterworth-Heinemann, Oxford, UK, 2011. Extensive discussion of Ashby charts and materials selection, along with tables of property data suitable for screening at conceptual design level. Appendix D in this text provides 25 pages on sources of material property data.

Cambridge Materials Selector, CES 06, Granta Design Ltd., Cambridge, UK. This software implements the Ashby materials selection scheme and provides data on 3000 materials. http://www.granta.com.uk.

K. G. Budinski and M. K. Budinski, *Engineering Materials: Properties and Selection,* 9th ed., Pearson Prentice Hall, Upper Saddle River, NJ, 2010. Broad-based, practically oriented.

11.4.2 Embodiment Design

At this phase of design, decisions are being made on the layout and sizes of parts and components. The design calculations require materials properties for a member of a subclass of materials but specific to a particular heat treatment or manufacturing process. These data are typically found in handbooks and computer databases, and in data sheets published by trade associations of materials producers. The following is a list of handbooks commonly found in engineering libraries. The series of handbooks published by ASM International, Materials Park, OH, are by far the most complete and authoritative for metals and alloys. They also are available online and from knovel.com.

Metals

ASM Handbook, Vol. 1, *Properties and Selection: Irons, Steels, and High-Performance Alloys,* ASM International, 1990.

ASM Handbook, Vol. 2, *Properties and Selection: Nonferrous Alloys and Special-Purpose Alloys,* ASM International, 1991.

SAE Handbook, Part 1, "Materials, Parts, and Components," Society of Automotive Engineers, Warrendale, PA, published annually. Similar but different European design allowables are available from ESDU as ESDU 00932.

MMPDS-04: *Metallic Materials Properties Development and Standardization.* This is the successor to MIL-HDBK-5, the preeminent source of design allowables for aerospace materials. Published by Battelle Memorial Institute under license from U.S. Federal Aviation Administration. Internet access to MIL-HDBK-5 is provided at http://www.barringer1.com/mil.html.

Woldman's Engineering Alloys, 9th ed., L. Frick (ed.), ASM International, 2000. References on approximately 56,000 alloys. Use this to track down information on an alloy if you know only the trade name. Available in electronic form.

Ceramics

ASM Engineered Materials Handbook, Vol. 4, *Ceramics and Glasses,* ASM International, 1991.

R. Morrell, *Handbook of Properties of Technical and Engineering Ceramics,* HMSO, London, Part 1, 1985, Part 2, 1987.

C. A. Harper, ed., *Handbook of Ceramics, Glasses, and Diamonds,* McGraw-Hill, New York, 2001.

R. W. Cahn, P. Hassen, and E. J. Kramer, eds., *Materials Science and Technology,* Vol. 11, *Structure and Properties of Ceramics,* Weinheim, New York, 1994.

Polymers

ASM Engineered Materials Handbook, Vol. 2, *Engineered Plastics,* ASM International, 1988.

ASM Engineered Materials Handbook, Vol. 3, *Adhesives and Sealants,* ASM International, 1990.

A. B. Strong, *Plastics: Materials and Processing,* 3rd ed., Pearson Prentice Hall, Upper Saddle River, NJ, 2006.

J. M. Margolis, ed., *Engineering Plastics Handbook,* McGraw-Hill, New York, 2006.

Dominic V. Rosato, Donald V. Rosato, and Marlene G. Rosato, *Plastics Design Handbook,* Kluwer Academic Publishers, Boston, 2001.

Composites

ASM Handbook, Vol. 21, *Composites,* ASM International, 2001.

"Polymers and Composite Materials for Aerospace Vehicle Structures," MILHDBK-17, U.S. Department of Defense.

P. K. Mallick, ed., *Composites Engineering Handbook,* Marcel Dekker, Inc., 1997.

S. T. Peters, ed., *Handbook of Composites,* 2nd ed., Chapman & Hall, New York, 1995.

Electronic Materials

C. A. Harper, ed., *Handbook of Materials and Processes for Electronics,* McGraw-Hill, New York, 1970.

Electronic Materials Handbook, Vol. 1, *Packaging,* ASM International, 1989.

Springer Handbook of Electronic and Photonic Materials, Springer-Verlag, Berlin, 2006.

Thermal Properties

Thermophysical Properties of High Temperature Solid Materials, Vols. 1 to 9, Y. S. Touloukian (ed.), Macmillan, New York, 1967.

Chemical Properties

ASM Handbook, Vol. 13A, *Corrosion: Fundamentals, Testing, and Protection,* ASM International, 2003.

ASM Handbook, Vol. 13B, *Corrosion: Materials,* ASM International, 2005.

ASM Handbook, Vol. 13C, *Corrosion: Environment and Industries,* ASM International, 2006.

R. Winston Revie, ed., *Uhlig's Corrosion Handbook,* 2nd ed., John Wiley & Sons, New York, 2000.

11

Internet

Many sites provide Internet information on materials and materials properties. Most of those with useful data are subscription-only sites. Sites that provide some free information are:

- www.matdata.net: Provides direct link to ASM International handbooks and metals databases, as well as Granta Design databases. Most of these are subscription services, but this site will provide locations in the 22 volumes of ASM handbooks for data on specific materials. Most engineering libraries will have these handbooks.
- www.matweb.com: Provides 80,000 material data sheets for free. Registered viewers can make searches for materials for free. For more advanced searches a subscription is required.
- www.campusplastics.com: The "Computer Aided Materials Preselection by Uniform Standards" is a database of polymers properties sponsored by a network of worldwide plastic resin producers. In order to provide comparability between the data of different suppliers, each participant is required to use a uniform standard for the generation of the data. Use of the database is free.
- www.custompartnet.com: Provides a diverse property database for a wide spectrum of metals and plastics.

11.4.3 Detail Design

At the detail design phase, very precise data are required. These are best found in data sheets issued by materials suppliers or by conducting materials testing within the organization. This is particularly true for polymers, whose properties vary considerably depending upon how they are manufactured. For all materials for critical parts, tests on the actual material from which they will be made are a requirement.

There is a wide range of material information that may be needed in detail design. This goes beyond just material properties to include information on manufacturability, including final surface finish and tolerances, cost, the experience in using the material in other applications (failure reports), availability of the sizes and forms needed (sheet, plate, wire, etc.), and issues of repeatability of properties and quality assurance. Two often-overlooked factors are whether the manufacturing process will produce different properties in different directions in the part, and whether the part will contain a detrimental state of residual stress after manufacture. These and other issues that influence the cost of the manufactured part are considered in detail in Chap. 13.

Many databases include disclaimer statements about not using the data for design. This is just another manifestation of our litigious society. Actually, most data is presented in good faith and can be used for screening purposes. However, if there are questions about the type of test used to measure the property, or if the range of given values is large (particularly important with polymers), the material supplier should be contacted before making a final decision. The data presented in this text is believed to be realistic, but it is for educational purposes and should not be used for final designs.

11

11.5
COST OF MATERIALS

Ultimately the material-process decision on a particular design will come down to a trade-off between performance and cost. There is a continuous spectrum of applications, varying from those where performance is paramount (aerospace and defense are good examples) to those where cost clearly predominates (household appliances and low-end consumer electronics are typical examples). In the latter type of application the manufacturer does not have to provide the highest level of performance that is technically feasible. Rather, the manufacturer must provide a value-to-cost ratio that is no worse, and preferably better, than the competition. By value we mean the extent to which the performance criteria appropriate to the application are satisfied. Cost is what must be paid to achieve that level of value.

11.5.1 Cost of Materials

Cost is such an overpowering consideration in many materials selection situations that we need to give this factor additional attention. The basic cost of a material depends upon (1) scarcity, as determined by either the concentration of the metal in the ore or the cost of the feedstock for making a polymer, (2) the cost and amount of energy required to process the material, and (3) the basic supply and demand for the material. In general, large-volume-usage materials like stone and cement have very low prices, while rare materials, like industrial diamonds, have very high prices. Figure 11.8 shows the range of price for some common engineering materials.

As is true of any commodity, as more work is invested in the processing of a material, the cost increases. Table 11.4 shows how the relative price of various steel products increases with further processing steps. Improvement in properties, like yield strength, beyond those of the basic material are produced by changes in structure brought about by compositional changes and additional processing steps. For example, increases in the strength of steel are achieved by expensive alloy additions such as nickel, by heat treatment such as quenching and tempering, or by vacuum treatment of the liquid steel to remove gaseous impurities. However, the cost of an alloy may not simply be the weighted average of the cost of the constituent elements that make up the alloy. Often, a high percentage of the cost of an alloy is due to the need to control one or more impurities to very low levels. That could mean extra refining steps or the use of expensive high-purity raw materials.

Because most engineering materials are produced from nonrenewable resources, mineral ores or oil and natural gas, there is a continuous upward trend of cost over time. As commodities, materials fluctuate in price due to temporary over- or under-supply. Over the long term the cost of materials has risen at a rate about 10 percent greater than the costs of goods and services in general. Therefore, conservation in the use of materials is increasingly important.

It is difficult to get current prices for materials from published sources. Several sites are available on the Internet, but only on a subscription basis. Two sources

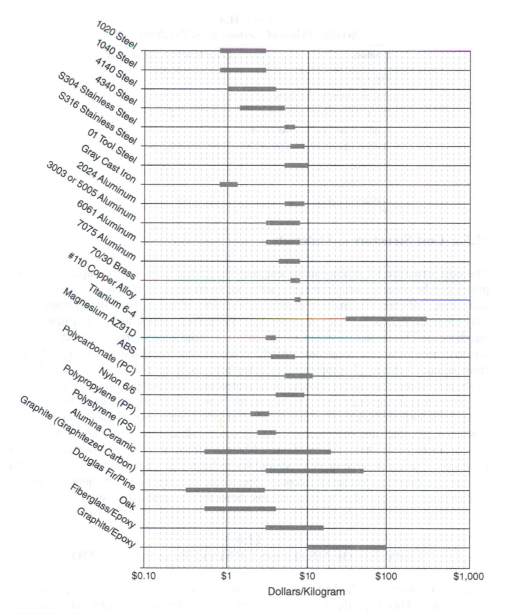

FIGURE 11.8
Price ranges for different materials purchased in bulk at 2007 prices. (From K.T. Ulrich and
S. D. Eppinger, *Product Design and Development,* 4th ed. 2008.)

useful for student design projects are the Cambridge Engineering Selector software
and www.custompartnet.com. To compensate for the change in the prices of materials
over time, costs are often normalized relative to a common inexpensive material such
as a steel reinforcing bar or a plain carbon steel plate.

TABLE 11.4
Relative Prices of Various Steel Products

Product	Price Relative to Pig Iron
Pig iron	1.0
Billets, blooms, and slabs	1.4
Hot-rolled carbon steel bars	2.3
Cold-finished carbon steel bars	4.0
Hot-rolled carbon steel plate	3.2
Hot-rolled sheet	2.6
Cold-rolled sheet	3.3
Galvanized sheet	3.7

11.5.2 Cost Structure of Materials

The cost structure for pricing many engineering materials is quite complex, and true prices can be obtained only through quotations from vendors. Reference sources typically give only the nominal or baseline price. The actual price depends upon a variety of price extras in addition to the base price (very much as when a new car is purchased). The actual situation varies from material to material, but the situation for steel products is a good illustration.[1] Price extras are assessed for any changes from standard chemical composition, for vacuum melting or degassing, special sizes or shapes, tighter tolerances on size, heat treatment or surface preparation, and so on.

From this listing of price extras we can see how inadvertent choices by the designer can significantly influence material cost. Standard chemical compositions should be used whenever possible, and the number of alloy grades should be standardized to reduce the cost of stocking many grades of steel. Manufacturers whose production rates do not justify purchasing in large quantity should try to limit their material use to grades that are stocked by local steel service centers. Special section sizes and tolerances should be avoided unless a detailed economic analysis shows that the cost extras are really justified.

11.6
OVERVIEW OF METHODS OF MATERIALS SELECTION

There is no single method of materials selection that has evolved to a position of prominence. This is partly due to the complexity of the comparisons and trade-offs that must be made. Often the properties we are comparing cannot be placed on comparable terms so that a clear decision can be made.

A variety of approaches to materials selection are followed by designers and materials engineers. A common path is to critically examine the service of existing designs in environments similar to the one of the new design. Information on service failures can be very helpful. The results of accelerated laboratory screening tests or short-time experience with a pilot plant can also provide valuable input. Often a minimum

1. R. F. Kern and M. E. Suess, *Steel Selection,* John Wiley & Sons, New York, 1979.

innovation path is followed and the material is selected on the basis of what worked before or what is used in the competitor's product. Appendix D gives suggestions.

Some of the more common and more analytical methods of materials selection are:

1. Performance indices (11.7)
2. Decision matrices (11.8)
   Pugh selection method (11.8.1)
   Weighted property index (11.8.2)
3. Computer-aided databases (11.9)

These materials selection methods are especially useful for making the final selection of a material in the embodiment design phase.

A rational way to select materials is by using a material performance index (Sec. 11.7). This is an important adjunct to the use of the Ashby selection charts during the initial screening in the conceptual design phase and as a design framework for comparing the behavior of materials in different applications.

Various types of decision matrices were introduced in Chap. 7 to evaluate design concepts. These can be used to good advantage to select materials when it is necessary to satisfy more than one performance requirement. The weighted property index, considered in Sec. 7.6, has been the most commonly used method.

With the growing access to computer-aided materials databases, more engineers are finding the materials they need with a computerized search. Section 11.9 discusses this popular method of materials selection and suggests some cautions that should be observed.

Another rational way to select materials is to determine the way in which actual parts, or parts similar to a new design, fail in service. Then, on the basis of that knowledge, materials that are unlikely to fail are selected. The general methodology of failure analysis is considered in Chap. 14.

Regardless of how well a material has been characterized and how definitive the performance requirements and the PDP schedule are, there will always be a degree of uncertainty about the ability of the material to perform. For high-performance systems where the consequences of failure can be very severe, material selection based on risk analysis can be very important. Some of the ideas of risk analysis are discussed in Chap. 14.

11.7
MATERIAL PERFORMANCE INDICES

A material performance index is a group of material properties that governs some aspect of the performance of a component.[1] If the performance index is maximized, it gives the best solution to the design requirement. Consider the tubular frame of a bicycle.[2] The design requirement calls for a light, strong tubular beam of fixed outer diameter. Its function is to carry bending moments. The objective is to minimize the mass m of the frame. The mass per unit length m/L can expressed by

$$\frac{m}{L} = 2\pi r t \rho \tag{11.1}$$

---

1. M. F. Ashby, *Acta Met.*, 1989, vol. 37, p. 1273.
2. M. F. Ashby, *Met. Mat. Trans.*, 1995, vol. 26A, pp. 3057–64.

where r is the outer tube radius, t is the wall thickness, and ρ is the density of the material from which it is made. Equation (11.1) is the *objective function,* the quantity to be minimized. This optimization must be done subject to several constraints. The first constraint is that the tube strength must be sufficient so it will not fail. Failure could occur by buckling, brittle fracture, plastic collapse, or fatigue caused by repeated cyclic loads. If fatigue is the likely cause, then the cyclic bending moment M_b the tube can withstand with infinite life is

$$M_b = \frac{I\sigma_e}{r} \tag{11.2}$$

where σ_e the endurance limit in fatigue loading and $I = \pi r^3 t$ is the second moment of inertia for a thin-walled tube. The second constraint is that r is fixed. However, the wall thickness of the tube is free, and this should be chosen so that it will just support M_b. Substituting Eq. (11.2) into Eq. (11.1) gives the mass per unit length in terms of the design parameters and material properties.

$$m = \frac{2M_b L}{r}\left(\frac{\rho}{\sigma_e}\right) = (2M_b)\left(\frac{L}{r}\right)\left(\frac{\rho}{\sigma_e}\right) \tag{11.3}$$

In Eq. (11.3) m is a *performance metric* for the design element, the bicycle tubular beam. The smaller the mass of a part, the less its cost and the lower the energy expended in pedaling the bike. Equation (11.3) has been written in the second form to illustrate a general feature of *performance metrics, P.*

$$P=[(functional\ requirements),(geometric\ parameters),(material\ properties)] \tag{11.4}$$

In this example, the functional requirement is to resist a certain bending moment, but in other problems it could be to resist a compressive buckling force, or to transmit a certain heat flux. The geometric parameters in this example are L and r. The third component of Eq. (11.3) is a ratio of material parameters, density, and fatigue endurance limit. We see that to reduce m this ratio should be as small as possible. This is the *material index, M.*

Generally, the three components of the performance metric are separable functions, so Eq. (11.4) can be written as

$$P = f_1(F) \times f_2(G) \times f_3(M) \tag{11.5}$$

Thus, the choice of material to optimize P is not dependent on the values for function F or geometry G, and a search for the best material can be carried out without the need for the details of F or G, provided that the material index has the proper form for the function and geometry.

11.7.1 Material Performance Index

Equation (11.3) indicates that best performance is achieved when mass is low. This requires in the search for best materials that those with low values of the index M be

TABLE 11.5
Material Performance Indices

Design Objective: Minimum Weight for Different Shapes and Loadings	To Maximize Strength	To Maximize Stiffness
Bar in tension: load, stiffness, length are fixed; section area is variable	σ_f/ρ	E/ρ
Torsion bar: torque, stiffness, length are fixed; section area is variable	$\sigma_f^{2/3}/\rho$	$G^{1/2}/\rho$
Beam in bending: loaded with external forces or self-weight; stiffness, length fixed; section area free	$\sigma_f^{2/3}/\rho$	$E^{1/2}/\rho$
Plate in bending: loaded by external forces or self-weight; stiffness, length, width fixed; thickness free	$\sigma_f^{1/2}/\rho$	$E^{1/3}/\rho$
Cylindrical vessel with internal pressure: elastic distortion, pressure, and radius fixed; wall thickness free	σ_f/ρ	E/ρ
Other design objectives, as stated below		
Thermal insulation: minimize heat flux at steady state; thickness given	$1/k$	
Thermal insulation: minimum temperature after specified time; thickness given	$C_p\rho/k$	
Minimize thermal distortion	k/α	
Maximize thermal shock resistance	$\sigma_f/E\alpha$	

α_f = failure strength (yield or fracture stress as appropriate to problem); E = Young's modulus; G = shear modulus; ρ = density; C_p = specific heat capacity; α = thermal expansion coefficient; k = thermal conductivity.

selected. However, it is usual practice to select materials with the largest values of the index, which is often called the *materials performance index*,[1] M_1, where $M_1 = 1/M$.

However, the form of the material performance index depends on the functional requirements and the geometry. Table 11.5 gives a short list of material performance indices for different types of loading and for several thermally related design objectives. Ashby gives a much more detailed listing.[2]

EXAMPLE 11.2 Selection of Materials for Automobile Cooling Fans[3]

Problem Statement/ Selection of Design Space

The radiator cooling fan in automobiles has typically been driven by a belt from the main drive shaft of the engine. Sudden acceleration of the engine causes high bending moments and centrifugal forces on the fan blades. On several occasions blades have broken, causing serious injury to mechanics working on the engine. Find a better material than the sheet steel used in the blades.

Boundaries of the Problem

The redesign will be limited to the selection of a cost-effective material that has more resistance to the propagation of small cracks than the current material.

---

1. The materials performance index is always such that the ratio is greater than unity.
2. M. F. Ashby, *Materials Selection in Mechanical Design,* 4th ed., Butterworth-Heinemann, Oxford, UK, 2011, Appendix C, pp. 559–564.
3. M. F. Ashby and D. Cebon, *Case Studies in Materials Selection,* Granta Design Ltd, Cambridge, UK, 1996.

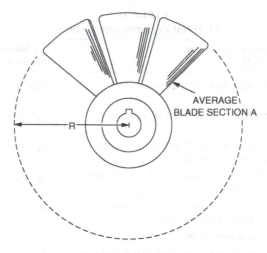

FIGURE 11.9
Sketch of the fan blades and hub.

Available Information

Published Ashby charts and the database of material properties available in the CES software will be used.

Physical Laws/Assumptions

Basic mechanics of materials relationships will be used. It is assumed that the radius of the fan is determined by the needed flow rate of air, so the size of the fan hub and blade remain the same for all design options. It also is assumed that all fan blades will be damaged by impact of road debris, so that some blades will contain small cracks or other defects. Therefore, the basic material property controlling service performance is fracture toughness, K_{Ic}; see Sec. 12.2.

Construct the Model for the Material Performance Index

Figure 11.9 shows a sketch of the fan hub with blades attached. The centrifugal force is

$$F = ma = [\rho(AcR)](\omega^2 R) \tag{11.6}$$

where ρ is the density, A is the cross-sectional area of a blade, c is the fraction of the radius that is blade, not hub, and R is the total radius to the centerline of the fan shaft. $\omega^2 R$ is the angular acceleration. The likely place for the blade to fail is at the root location. The stress at this location is

$$\sigma = \frac{F}{A} = c\rho\omega^2 R^2 \tag{11.7}$$

We have assumed that the most likely cause of blade failure is the initiation of a crack where the blade meets the hub, either by road debris damage or from a manufacturing defect, which propagates at some point into a fast-moving, brittle fracture type of crack. Therefore, the critical value of stress is controlled by the fracture toughness of the blade material, Sec. 12.2. Fracture toughness is given by $K_{Ic} = \sigma\sqrt{\pi a_c}$, where a_c is the critical crack length that causes fracture and K_{Ic} is the material property plane strain

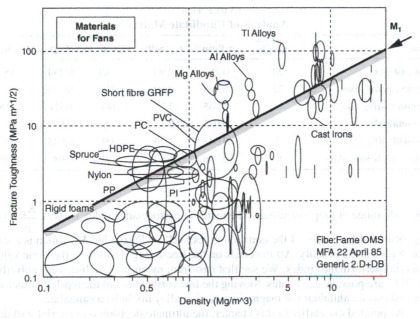

FIGURE 11.10
Chart of fracture toughness versus density. (From Cebon and Ashby. Used with permission of Granta Design, Inc.)

fracture toughness. Thus, a safe condition exists when stress due to centrifugal force is less than that required to propagate a crack to failure.

$$c\rho\omega^2 R^2 \le \frac{K_{Ic}}{\sqrt{\pi a_c}} \qquad (11.8)$$

We are trying to prevent the blade from failing when the fan overspeeds. Equation (11.7) shows that the centrifugal stress is proportional to the square of the angular velocity, so an appropriate performance metric is ω. Therefore,

$$\omega \le \left(\frac{1}{\sqrt{\pi a_c}}\right)^{1/2} \left(\frac{1}{\sqrt{c}\,R}\right) \left(\frac{K_{Ic}}{\rho}\right)^{1/2} \qquad (11.9)$$

R and c are fixed parameters. Critical crack length, a_c varies somewhat with material, but can be considered a fixed parameter if we define it as the smallest crack that can be detected by a nondestructive inspection technique such as eddy current testing. Thus, the materials performance index is $(K_{Ic}/\rho)^{1/2}$. But when comparing a group of materials we can simply use K_{Ic}/ρ, since the ranking will be the same. In this case we did not need to take the reciprocal of M because the ratio is greater than 1.

Analysis

In this situation the first step in analysis consists of searching material property databases. For initial screening, the Ashby chart shown in Fig. 11.10 provides useful information. We note that the chart is plotted to a log-log scale in order to accommodate

TABLE 11.6
Analysis of Candidate Materials

Material	K_{Ic} ksi$\sqrt{\text{in.}}$	ρ, lb/in.^{3}	C_m, \$/lb	$M_1 = K_{Ic}/\rho$	$C_m\rho$	$M_2 = K_{Ic}/C_m\rho$
Nodular cast iron	20	0.260	0.90	76.9	0.234	85.5
Aluminum casting alloy	21	0.098	0.60	214	0.059	355
Magnesium casting alloy	12	0.065	1.70	184	0.111	108
HDPE—unfilled	1.7	0.035	0.55	48	0.019	89.5
HDPE—with 30% glass fiber	3	0.043	1.00	69	0.043	69.7
Nylon 6/6—with 30% glass	9	0.046	1.80	195	0.083	108

the wide range of property values. The material performance index is $M_1 = \dfrac{K_{Ic}}{\rho}$. Taking logarithms of both sides of the equation gives $\log K_{Ic} = \log \rho + \log M_1$, which is a straight line with a slope of unity. All materials on the line in Fig. 11.10 have the same values of material performance index. We see that cast iron, nylon, and high-density polyethylene (HDPE) are possible candidates. Moving the line further toward the top-left corner would suggest that an aluminum or magnesium casting alloy might be a candidate.

As pointed out earlier in this chapter, the ultimate decision on material will depend on a trade-off between performance and cost. Most likely the blades will be made by a casting process if a metal and a molding process if a polymer.

The cost of a blade is given by $C_b = C_m\rho V$, where C_m is the material cost in \$/lb, density is lb/in.^{3}, and volume is in cu. in. However, the volume of material is essentially determined by R, which is set by the required flow rate of air, so V is not a variable in this cost determination. From a cost viewpoint, the best material has the lowest value of $C_m\rho$. Note that this cost discussion has considered only the cost of material. Since all materials are suitable for use in either casting or injection molding processes, it is assumed that the manufacturing costs would be equivalent across all candidate materials. More detailed analysis would require the methods discussed in Chap. 13.

To introduce cost into the material performance index, we divide M_1 by C_m to give $M_2 = K_{Ic}/C_m\rho$.

Typical values of material properties and material costs were obtained from the CES database. The results are shown in Table 11.6. Based on the main performance criterion, an aluminum casting alloy is the best material for the fan blade. A possible concern is whether it can be cast in the thin sections required for a blade with suitable control of dimensions, warping, and surface finish. Injection molded nylon with 30% chopped glass fiber is tied for second place on a cost-property basis with a magnesium casting alloy.

Validation

Clearly, extensive prototype testing will be required whatever the final decision on material may be.

In this section we have shown how the material index, M, in Eq. (11.5) can be used to improve the performance metric, P, through the optimal selection of materials using the materials performance index. Since the three terms in Eq. (11.5) are multiplied to determine P, changes in geometry as well as material properties can be used to enhance performance. We know from mechanics of materials that better stiffness can be achieved in a beam if it is in the shape of an I-section compared with a square cross-section. This leads to the concept of shape factor as another way of improving the load, torque, or buckling capacity of structural members.[1] For further details see Shape Factor in the website for this text, www.mhhe.com/dieter.

11.7.2 Material Selection for Eco-Attributes

The material performance index can be extended to consider the environmental issues covered in Chap. 10.[2] The following stages of the materials cycle must be considered: (1) material production, (2) product manufacture, (3) product use, and (4) product disposal.

Energy and Emissions Associated with Material Production

The greatest amount of environmental damage is done in producing a material. Most of the energy used in this process is obtained from fossil fuels: oil, gas, and coal. In many cases the fuel is part of the production process, in other cases it is first converted to electricity that is used in production. The pollution produced during material production takes the form of undesirable gas emissions, chiefly CO_2, NO_x, SO_x, and CH_4. For each kilogram of aluminum produced from fossil fuels there is created 12 kg of CO_2, 40 g of NO_x, and 90 g of SO_x. Also, individual processes may create toxic wastes and particulates that should be treated at the production site.

The fossil-fuel energy required to make one kilogram of material is called its *production energy*. Table 11.7 lists some typical values of production energy, H_p, as well as data on the production of CO_2 per kg of material produced.

Energy and Emissions Associated with Product Manufacture

The energy involved in manufacturing the product is at least an order of magnitude smaller than that required for producing the material. The energy for metal deformation processes like rolling or forging are typically in the range 0.01 to 1 MJ/kg. Polymer molding processes are 1 to 4 MJ/kg, while metal casting processes are 0.4 to 4 MJ/kg. While saving energy in manufacturing is important, of greater concern ecologically is eliminating any toxic wastes and polluted discharges created in manufacturing.

Energy Associated with Product Use

The energy consumed in product use is determined by the mechanical, thermal, and electrical efficiencies achieved by the design of the product. Fuel efficiency in a motor vehicle is achieved chiefly by reducing the mass of the vehicle, along with improving mechanical efficiency of the transmission and minimizing aerodynamic losses.

Energy and Environmental Issues in Product Disposal

It is important to realize that some of the energy used in producing a material is stored in the material and can be reused in the recycling or disposal process. Wood and paper products can be burned in an incinerator and energy recovered. While some energy is required for recycling, it is much less than H_p. Recycled aluminum requires much less energy to melt it than is required to extract it from its ore.

11

---

1. M.F. Ashby, op. cit., Chap. 11.
2. M.F. Ashby, op. cit., Chap. 16.

TABLE 11.7
Values for Production Energy and Amount of CO$_2$ Produced

Material	Production Energy (H_p) MJ/kg	CO$_2$ Burden, [CO$_2$] kg/kg
Low-carbon steels	22.4–24.8	1.9–2.1
Stainless steels	77.2–80.3	4.8–5.4
Aluminum alloys	184–203	11.6–12.8
Copper alloys	63.0–69.7	3.9–4.4
Titanium alloys	885–945	41.7–59.5
Borosilicate glass	23.8–26.3	1.3–1.4
Porous brick	1.9–2.1	0.14–0.16
CFRP composites	259–286	21–23
PVC	63.5–70.2	1.85–2.04
Polyethylene (PE)	76.9–85	1.95–2.16
Nylons (PA)	102–113	4.0–4.41

From M.F. Ashby, *Materials Selection in Mechanical Design,* 3rd ed.
(Used with permission of Butterworth-Heinemann.)

Important information to know when selecting an environmentally responsible material is:

- Is there an economically viable recycling market for the material? This can be determined by checking websites for recycled materials, and by determining the fraction of the market for the material that is made from recycled materials.
- It is important to know how readily the material can be added to virgin material without deleterious effects on properties. This is true recycling. Some materials can only be recycled into lower-grade materials.
- Some materials can be disposed of by biodegradation in landfills, while the bulk of materials that are not recycled go into landfills. However, some materials like lead, cadmium, and some of the heavy metals are toxic, especially in finely divided form, and must be disposed of by methods used for hazardous materials.

Material Performance Indices

Some of these environmental issues can be readily incorporated into the material performance index. Suppose we wanted to select a material with minimum production energy to provide a given stiffness to a beam. We know from Sec. 11.7.1 that the minimum mass beam would be given by the largest value of $M = E^{1/2}/\rho$ from among the candidate materials. To also accommodate production energy in this decision the material performance index would be written as

$$M = \frac{E^{1/2}}{H_p \rho}$$

(11.10)

and to select a material with minimum CO_2 production for a beam of given bending strength

$$M = \frac{\sigma_f^{2/3}}{[CO_2]\rho}$$

(11.11)

Production energy, H_p, or CO_2 produced in smelting the material could be included in any of the material performance indices listed in Table 11.6. These terms are placed in the denominator because convention requires that high values of M determine the material selection.

11.8
MATERIALS SELECTION WITH DECISION MATRICES

In most applications it is necessary that a selected material satisfy more than one performance requirement. In other words, compromise is needed in materials selection. We can separate the requirements into three groups: (1) go/no-go parameters, (2) nondiscriminating parameters, and (3) discriminating parameters. *Go/no-go parameters* are those requirements that must meet a certain fixed minimum value. Any merit in exceeding the fixed value will not make up for a deficiency in another parameter. Examples of go/no-go parameters are corrosion resistance or machinability. *Nondiscriminating parameters* are requirements that must be met if the material is to be used at all. Examples are availability or general level of ductility. Like the previous category, these parameters do not permit comparison or quantitative discrimination. *Discriminating parameters* are those requirements to which quantitative values can be assigned.

The decision matrix methods that were introduced in Chap. 7 are very useful in materials selection. They organize and clarify the selection task, provide a written record of the selection process (which can be useful in redesign), and improve the understanding of the relative merit among alternative solutions.

Three important factors in any formalized decision-making process are the alternatives, the criteria, and the relative weight of the criteria. In materials selection, each candidate material, or material-process pair, is an alternative. The selection criteria are the material properties or factors that are deemed essential to satisfy the functional requirements. The weighting factors are the numerical representations of the relative importance of each criterion. As we saw in Chap. 7, it is usual practice to select the weighting factors so that their sum equals unity.

11.8.1 Pugh Selection Method

The Pugh concept selection method is the simplest decision method discussed in Chap. 7. This method involves qualitative comparison of each alternative to a reference or datum alternative, criterion by criterion. No *go/no-go parameters* should be used as decision criteria. They have already been applied to screen out infeasible alternatives. The Pugh concept selection method is useful in conceptual design because it requires the least amount of detailed information. It is also useful in redesign, where the current material serves automatically as the datum.

EXAMPLE 11.3
The Pugh decision method is used to select a replacement material for a helical steel spring in a wind-up toy train.[1] The alternatives to the currently used ASTM A227 class I hard-drawn steel wire are the same material in a different design geometry, ASTM A228 music spring-quality steel wire, and ASTM A229 class I steel wire, quenched and oil tempered. In the decision matrix that follows, if an alternative is judged better than the datum, it is given a+, if it is poorer it gets a–, and if it is about the same it is awarded an S, for "same."[2] The +, –, and S responses are then totaled and discussed.

USE OF PUGH DECISION MATRIX FOR REDESIGN OF HELICAL SPRING

	Alternative 1 Present Material Hard-Drawn Steel ASTM A227	Alternative 2 Hard-Drawn Steel Class I ASTM A227	Alternative 3 Music Wire Quality Steel ASTM A228	Alternative 4 Oil-Tempered Steel Class I ASTM A229
Wire diameter, mm	1.4	1.2	1.12	1.18
Coil diameter, mm	19	18	18	18
Number of coils	16	12	12	12
Relative material cost	1	1	2.0	1.3
Tensile strength, MPa	1750	1750	2200	1850
Spring constant	D	–	–	–
Durability	A	S	+	+
Weight	T	+	+	+
Size	U	+	+	+
Fatigue resistance	M	–	+	S
Stored energy		–	+	+
Material cost (for one spring)		+	S	S
Manufacturing cost		S	+	–
Σ+		3	6	4
ΣS		2	1	2
Σ–		3	1	2

Both the music spring-quality steel wire and the oil-tempered steel wire are superior to the original material selection. The music wire is selected because it ranks highest in advantages over the current material, especially with regard to manufacturing cost.

---

1. D. L. Bourell, "Decision Matrices in Materials Selection," in *ASM Handbook,* vol. 20, *Materials Selection and Design,* ASM International, Materials Park, OH, 1997.
2. Note: Do not sum the + and – ratings as though they were +1 and –1 scores. This invalidates the selection method because it presumes all criteria have equal weight. They do not.

11.8.2 Weighted Property Index

The weighted decision matrix that was introduced in Chap. 7 is well suited to materials selection with discriminating parameters.[1] In this method each material property is assigned a certain weight depending on its importance to the required service performance. Techniques for assigning weighting factors are considered in Sec. 7.6. Since different properties are expressed in different ranges of values or units, the best procedure is to normalize these differences by using a scaling factor. Otherwise the property with the highest numerical value would have undue influence in the selection. Since different properties have widely different numerical values, each property must be so scaled that the largest value does not exceed 100.

$$\beta_i = \text{scaled property } i = \frac{\text{numerical value of property } i}{\text{largest value of } i \text{ under consideration}} 100 \qquad (11.12)$$

With properties for which it is more desirable to have low values, such as density, corrosion loss, cost, and electrical resistance, the scaled property is formulated as follows:

$$\beta_i = \text{scaled property } i = \frac{\text{lowest value of } i \text{ under consideration}}{\text{numerical value of property } i} 100 \qquad (11.13)$$

For properties that are not readily expressed in numerical values, like weldability and wear resistance, some kind of subjective rating is required. A common approach is to use a 5–point scale in which the property is rated excellent (5), very good (4), good (3), fair (2), or poor (1). Then the scaled property would be excellent (100), very good (80), good (60), fair (40), or poor (20).

The weighted property index γ is given by

$$\gamma = \Sigma \beta_i w_i \qquad (11.14)$$

where β_i is summed over all the properties (criteria) and w_i is the weighting factor for the i th property.

There are two ways to treat cost in this analysis. First, cost can be considered to be one of the properties, usually with a high weighting factor. Alternatively, the weighted property index can be divided by the cost of a unit mass or volume of material. This approach places major emphasis on cost as a material selection criterion.

EXAMPLE 11.4
The material selection for a cryogenic storage vessel for liquefied natural gas is being evaluated on the basis of the following properties: (1) low-temperature fracture toughness, (2) low-cycle fatigue strength, (3) stiffness, (4) coefficient of thermal expansion (CTE),

---

1. M. M. Farag, *Materials Selection for Engineering Design*, Prentice Hall Europe, London, 1997.

and (5) cost. Since the tank will be insulated, thermal properties can be neglected in the selection process.

First determine the weighting factors for these properties using pairwise comparison. There are $N = 5(5 - 1)/2 = 10$ possible comparisons of pairs. For each comparison, decide which is the more important property (decision criterion). Assign a 1 to the more important property and a 0 to the other. In this example we decided that fracture toughness is more important than each of the other properties, even cost, because a brittle fracture of a cryogenic tank would be disastrous. If a 1 goes in the (1)(2) position, then a 0 goes in the (2)(1) location, etc. In deciding between fatigue strength and stiffness, we decided that stiffness is more important, so a 0 goes in the (2)(3) and a 1 in the (3)(2) box.

Pairwise Comparison of Properties

Property	1	2	3	4	5	Row Total	Weighting Factor, w_i
1. Fracture toughness	–	1	1	1	1	4	0.4
2. Fatigue strength	0	–	0	1	0	1	0.1
3. Stiffness	0	1	–	0	0	1	0.1
4. Thermal expansion	0	0	1	–	0	1	0.1
5. Cost	0	1	1	1	–	3	0.3
					Totals	10	1.0

The pairwise comparison shows that out of the 10 choices made, the property fracture toughness received four positive (1) decisions, so that its weighting factor $w_1 = 4/10 = 0.4$. In the same way, the values of w for the other four properties are $w_2 = 0.1$; $w_3 = 0.1$; $w_4 = 0.1$; $w_5 = 0.3$.

Using pairwise comparison to establish the weighting factors is quick, but it has two deficiencies: (1) it is difficult to make a series of comparisons in a completely consistent way, and (2) each comparison is a binary decision (there are no degrees of difference). We have seen in Sec. 7.7 that AHP is a superior method for making this type of decision. When AHP was used in Example 11.4 to determine the weighting factors from fracture toughness to cost, the values were 0.45, 0.14, 0.07, 0.04, and 0.30.

The chart for selecting a material based on the weighted property index is shown in Table 11.8. Four candidate materials were identified from the preliminary screening. Several go/no-go screening parameters are included. On further investigation it was found the aluminum alloy is not available in the required plate thickness, so that material was dropped from further consideration. The body of the table shows both the raw data and the data in scaled form. The β values for toughness, fatigue strength, and stiffness were determined from Eq. (11.12). The β values for thermal expansion and cost were determined from Eq. (11.13) because for these properties a smaller value ranks higher. Since no comparable fracture toughness data was available for the candidate materials, a relative scale 1 to 5 was used. The weighting factors developed in the previous table are given beside the listing for each of the properties.

The best material among these choices for the application is the 9 percent nickel steel, which has the largest value of weighted property index.

11

TABLE 11.8
Weighted Property Index Chart for Selection of Material for Cryogenic Storage

Material	Go/No-Go Screening			Toughness (0.4)		Fatigue Strength (0.1)		Stiffness (0.1)		Thermal Expansion (0.1)		Cost (0.3)		Weighted Property Index
	Corrosion	Weldability	Available in Thick Plate	Rel. Scale	β	ksi	β	10⁶ psi	β	μin/in °F	β	$/lb	β	γ
304 stainless	S	S	S	5	100	30	60	28.0	93	9.6	80	3.00	50	78.3
9% Ni steel	S	S	S	5	100	50	100	29.1	97	7.7	100	1.80	83	94.6
3% Ni steel	S	S	S	4	80	35	70	30.0	100	8.2	94	1.50	100	88.4
Aluminum alloy	S	S	U											

S = satisfactory
U = unsatisfactory
Reixive Scale 5 = excellent, 4 = very good
Sample calculation: 304 stainless: $\gamma = 0.4(100) + 0.1(60) + 0.1(93) + 0.1(80) + 0.3(50) = 78.3$

11

11.9
SELECTION WITH COMPUTER-AIDED DATABASES

The use of computer-aided tools allows the engineer to minimize the materials selection information overload. A computerized materials search can accomplish in minutes what may take hours or days by a manual search. Over 100 materials databases are available worldwide. All materials property databases allow the user to search for a material match by comparing a number of property parameters, each of which can be specified as below, above, or within a stated range of values. Some databases have the ability to weight the importance of the various properties. The most advanced databases allow the materials property data to be transmitted directly to a design software package, such as finite element analysis, so that the effect of changing material properties on the geometry and dimensions of a part can be directly observed on the computer monitor.

Most existing databases provide numerical material properties as opposed to qualitative rankings. Usually mechanical and corrosion properties are well covered, with less extensive coverage of magnetic, electrical, and thermal properties. Since it is unlikely that any database will be sufficiently comprehensive for a specific user, it is vital that the search system be designed so that users may easily add their own data, and subsequently search, manipulate, and compare these values along with the entire collection of data.

To compare different materials using a computerized database, it is useful to employ limits on properties. For example, if it is necessary to have a stiff, light material, we would put a lower limit on Young's modulus and an upper limit on density. After screening, the remaining materials are those whose properties are above the lower limits and below the upper limits

EXAMPLE 11.5

In selecting a material for a design at the conceptual design phase, we know that we need a material with a yield strength of at least 60,000 psi and with both good fatigue strength and fracture toughness. The Cambridge Engineering Selector (CES), an extensive database for about 3000 engineering materials, is a very useful source of information.[1] Entering the software in Select Mode, we click on "All bulk materials" and go to "Limit stage" so we can set upper and lower limits, as desired. In the selection boxes we enter the following values:

General	Minimum	Maximum
Density, lb/in.^{3}	0.1	0.3
Mechanical		
Elastic limit, ksi	60	
Endurance limit, ksi	40	
Fracture toughness, ksi$\sqrt{\text{in.}}$	40	
Young's modulus, 10^6 psi	10	30

---

1. Cambridge Engineering Selector, v. 4 from Granta Design Ltd., Cambridge, UK, 2006 www.grantadesign .com.uk

These decisions reduced the possible selections from 2940 to 422, mostly steels and titanium alloys. Next, setting a maximum value on price at 1.00 $/lb reduced the options to 246 by eliminating all but the steels.

Introducing a maximum carbon content of 0.3% to minimize problems with cracking in either welding or heat treatment reduced the selection to 78 steels—plain carbon, low-alloy steels, and stainless steels. Since the application did not require resistance to other than a normal room temperature oil mist environment, the stainless steels were eliminated by specifying a chromium content not to exceed 0.5%. Now we are down to 18 plain carbon and low-alloy steels. The normalized AISI 4320 steel was selected because we wanted a material with better fatigue and fracture toughness properties than plain carbon steel, and being able to get these properties in the normalized condition, which means that no further heat treatment other than that given at the steel mill is necessary, was worth the small price differential. Moreover, we found that our local steel supply warehouse stocked this alloy grade in a convenient bar diameter.

By its very nature, a computerized database search uses screening properties in which upper or lower values of properties are employed. Once the search has narrowed the candidate materials it is important to use the decision methods discussed in Sec. 11.8 to arrive at a final material selection. In arriving at the list of desired properties used to make the decision it is very useful to frame them as material performance indices, Sec. 11.7.

11.10
DESIGN EXAMPLES

Engineered systems contain many components, and for each a material must be selected. The automobile is our most familiar engineering system and one that exhibits major changes in the materials used for its construction. These trends in materials selection reflect the great effort that is being made to decrease the fuel consumption of cars by downsizing the designs and adopting weight-saving materials. Prior to 1975, steel and cast iron comprised about 78 percent of the weight of a car, with aluminum and plastics each at slightly less than 5 percent. Today, ferrous materials comprise about 57 percent of the total weight, with plastics at about 20 percent and aluminum at about 8 percent. Aluminum is in an ongoing battle with steel to take over the structural frame and part of the sheet panels.

Complex and severe service conditions can be economically withstood only by combining several materials in a single component. The surface hardening of gears and other automotive components by carburizing or nitriding[1] is a good example. Here the high hardness, strength, and wear resistance of a high-carbon steel is produced in the surface layers of a ductile and tougher low-carbon steel.

EXAMPLE 11.6 Complex Materials System
Automobile manufacturers often use their high-end, high-performance cars as a test bed for the application of new materials and manufacturing processes. The Chevrolet Z06 Corvette is a good example where increased performance in speed, acceleration, and fuel economy were achieved by major changes in materials.[2] This was accomplished by

---

1. *Metals Handbook: Desk Edition,* 2nd ed., "Case Hardening of Steel," ASM International, 1998. pp. 982–1014.
2 D. A. Gerard, *Advanced Materials and Processes,* January 2008, pp. 30–33.

significant modifications to the body and powertrain architectures. The modifications included substantial reduction in vehicle mass, improvement of the mass distribution between front and rear of the vehicle, and incorporation of a newly designed high-performance 7-liter small-block engine.

Structural Modifications

The standard Corvette had a steel space frame made mainly from stamped parts joined by welding. This frame was replaced by a structure of twenty-one 6063 aluminum alloy extrusions that were formed into special shapes by the hydroforming process.[1] A key part of the frame is the 4.8 m long rail, weighing 24 kg, the largest hydroformed aluminum component in the world. Other components of the space frame include eight A356 aluminum castings, a 6061 T6 extruded beam, and several 5754 aluminum stampings. The completed space frame is a 33 percent mass reduction over the steel frame.[2]

Since aluminum has an elastic modulus (E) only one-third that of steel, major redesign was needed to achieve required vehicle stiffness. In addition, the cost of aluminum is about three times that of steel. Finite element analysis (FEA) was critical in making it possible to use aluminum alloys at an acceptable cost. A key design breakthrough made possible by using FEA was a reduction of the forces on the aluminum frame by transferring part of the load to a lightweight magnesium roof frame. Also, in designing the new aluminum frame, FEA facilitated the redistribution of weight from front to rear of the vehicle.

The Z06 is the first vehicle in the industry to use a large magnesium diecast engine cradle, a 35 percent mass reduction over the previous aluminum cradle. This is a major structural member (10.5 kg) that not only supports the engine and front bumper beam, but also ties the ram rails together and acts as the mounting point for certain front suspension systems. Since the cradle interfaces with several dissimilar metals, it was important to solve potential issues with corrosion of dissimilar metals in contact, as well as joining of dissimilar metals. Since Mg has a lower density than Al, its use as an engine cradle was motivated by the design objective of moving mass toward the rear of the vehicle. Several other material changes were made to achieve the same goal. Polymer-carbon fiber front fenders and wheel houses replaced metal components, and a floor pan consisting of a balsawood core with a carbon fiber skin replaced a metal pan.

LS 7 Engine

The LS 7 engine is a new high-performance internal combustion (IC) engine that delivers 505 hp and 7100 rpm while achieving a 24 mpg EPA highway rating. It is the first 500+ hp engine exempt from the gas guzzler tax in the United States. The new material and process innovations introduced in the engine are largely responsible for this result.

- A three-piece polymer composite manifold assembled by friction welding resulted in 20 percent reduction in air flow restriction to deliver the higher airflow needed for the larger horsepower engine.
- The engine has CNC-ported cylinder heads that deliver the required high air flow. Cylinder head porting refers to the process of modifying the profile of the intake and exhaust ports of an engine to improve the quality and quantity of air flow. This is usually done with 5-axis CNC machining.[3]

1. http://en.wikipedia.org/wiki/Hydroforming.
2. B. Deep, L. Decker, E. Moss, M. P. Kiley, R. Thomure, and J. Turczynski, SAE Technical Paper 2005-01-0465, Society of Automotive Engineers, 2005.
3. http://en.wikipedia/wiki/Cylinder_head_porting.

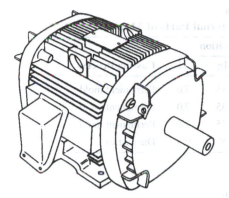

FIGURE 11.11
Horizontal aluminum alloy motor.
(Courtesy of General Electric Company.)

- The intake valves are Ti-6Al-2Sn-4Zr-2Mo, a high-strength, high-modulus, low-density material. The lower valve mass allows a larger valve head, needed for a larger inlet area required for the airflow to achieve 505 hp. The lighter valve permits achieving 7100 rpm without overstressing.
- The exhaust valves are made from two stainless steel parts, friction-welded together. The upper stem is 422 stainless (12 Cr, 1Ni, 1 Mo, 1.2 W), while the lower hotter part of the valve, which includes the valve head, is made from a high temperature valve steel SAE J775. The upper valve stem is hollow and contains sodium (mp. 140°C). The sodium serves as a heat transfer medium to carry heat from the hotter valve head to the stem, where it is dissipated by passing through the valve guides into the cylinder head.
- Other material technologies have further improved the powertrain. The aluminum piston is coated with an anti-seizure polymer to reduce friction and noise. A forged 4140 steel crankshaft has replaced the cast crankshaft. This provides improved stiffness and is better able to handle the increased loads resulting from the higher engine speed. A forged Ti-6Al-4V alloy connecting rod replaced one made from steel. The combination of tensile strength, fatigue strength and stiffness results in a 30 percent reduction in weight. As a consequence, the lighter titanium connecting rods produce lower loads on the rod ends and main bearings, thus allowing the bearings to be designed for minimal friction. A significant increase in bearing life is expected.
- Finally, a major redesign of the exhaust manifold, employing CFD modeling, resulted in better airflow into the catalytic converter. Hydroforming permitted the stainless steel exhaust tubes to be made with a complex pattern of inside diameters, based on CFD, that controlled pumping losses and kept airflow restriction to a minimum.

EXAMPLE 11.7 Material Substitution
This design example illustrates the common problem of substituting a new material for one that has been used for some time. It illustrates that material substitution should not be undertaken unless appropriate design changes are made. Also, it illustrates some of the practical steps that must be taken to ensure that the new material and design will perform adequately in service.

Aluminum alloys have been substituted for gray cast iron[1] in the external supporting parts of integral-horsepower induction motors (Fig. 11.11). The change in materials was

1. T. C. Johnson and W. R. Morton, IEEE Conference Record 76CH1 109-8-IA, Paper PCI-76-14, General Electric Company Report GER-3007.

TABLE 11.9
Aluminum Alloys Used in External Parts of Motors

| Part | Alloy | Composition | | | Casting Process |
		Cu	Mg	Si	
Motor frame	356	0.2 max	0.35	7.0	Permanent mold
End shields	356	0.2 max	0.35	7.0	Permanent mold
Fan casing	356	0.2 max	0.35	7.0	Permanent mold
Conduit box	360	0.6 max	0.50	9.5	Die casting

TABLE 11.10
Comparison of Typical Mechanical Properties

Material	Yield Strength, ksi	Ultimate Tensile Strength, ksi	Shear Strength, ksi	Elongation in 2 in., percent
Gray cast iron	18	22	20	0.5
Alloy 356 (as cast)	15	26	18	3.5
Alloy 360 (as cast)	25	26	45	3.5
Alloy 356-T61 (solution heat-treated and artificially aged)	28	38		5

brought about by increasing cost and decreasing availability of gray-iron castings. There has been a substantial reduction in gray-iron foundries, partly because of increased costs resulting from the more stringent environmental pollution and safety regulations imposed in recent years by governmental agencies. The availability of aluminum castings has increased owing to new technology to increase the quality of aluminum castings. Also, with aluminum castings there are fewer problems in operating an aluminum foundry, which operates at a much lower temperature than those required for cast iron.

There are a variety of aluminum casting alloys.[1] Among the service requirements for this application, strength and corrosion resistance were paramount. The need to provide good corrosion resistance to water vapor introduced the requirement to limit the copper content to an amount just sufficient to achieve the necessary strength. Actual alloy selection was dependent on the manufacturing processes used to make the part. That in turn depended chiefly (see Chap. 13) on the shape and the required quantity of parts. Table 11.9 gives details on the alloys selected for this application.

Since the motor frame and end-shield assemblies have been made successfully from gray cast iron for many years, a comparison of the mechanical properties of the aluminum alloys with cast iron is important (Table 11.10).

The strength properties for the aluminum alloys are approximately equal to or exceed those of gray cast iron. If the slightly lower yield strength for alloy 356 cannot be tolerated, it can be increased appreciably by a solution heat treatment and aging (T6 condition) at a slight penalty in cost and corrosion resistance. Since the yield and shear strength of the

1. *Metals Handbook: Desk Edition,* 2nd ed., "Aluminum Foundry Products," ASM International, 1998. pp. 484–96.

aluminum alloys and gray cast iron are about equal, the section thickness of aluminum to withstand the loads would be the same. However, since the density of aluminum is about one-third that of cast iron, there will be appreciable weight saving. The complete aluminum motor frame is 40 percent lighter than the equivalent cast iron design. Moreover, gray cast iron is essentially a brittle material, whereas the cast-aluminum alloys have enough malleability that bent cooling fins can be straightened without breaking them.

The aluminum alloys are inferior to cast iron in compressive strength. In aluminum, as with most alloys, the compressive strength is about equal to the tensile strength, but in cast iron the compressive strength is several times the tensile strength. That becomes important at bearing supports, where, if the load is unbalanced, the bearing can put an appreciable compressive load on the material surrounding and supporting it. That leads to excessive wear with an aluminum alloy end shield.

To minimize the problem, a steel insert ring is set into the aluminum alloy end shield when it is cast. The design eliminates any clearance fit between the steel and aluminum, and the steel insert resists wear from the motion of the bearing just as the cast iron always did. The greater ease of casting aluminum alloys permits the use of cooling fins thinner and in greater number than in cast iron. Also, the thermal conductivity of aluminum is about three times greater than that of cast iron. Those factors result in more uniform temperature throughout the motor, and this results in longer life and higher reliability. Because of the higher thermal conductivity and larger surface area of cooling fins, less cooling air is needed. With the air requirements thus reduced, a smaller fan can be used, also resulting in a small reduction of noise.

The coefficient of expansion of aluminum is greater than that of cast iron, and that makes it easier to ensure a tight fit of motor frame to the core. Only a moderate temperature rise is needed to expand the aluminum frame sufficiently to insert the core, and on cooling the frame contracts to make a tight bond with the core. That results in a tighter fit between the aluminum frame and the core and better heat transfer to the cooling fins. Complete design calculations need to be made when aluminum is substituted for cast iron to be sure that clearances and interferences from thermal expansion are proper.

Since a motor design that had many years of successful service was being changed in a major way, it was important to subject the redesigned motor to a variety of simulated service tests. The following were used:

- Vibration test
- Navy shock test (MIL-Std-901)
- Salt fog test (ASTM B 1 17-57T)
- Axial and transverse strength of end shield
- Strength of integral cast lifting lugs
- Tests for galvanic corrosion between aluminum alloy parts and steel bolts

This example illustrates the importance of considering design and manufacturing together in a material substitution situation.

11.11
SUMMARY

This chapter has shown that there are no magic formulas for materials selection. Rather, the solution of a materials selection problem is every bit as challenging as any other aspect of the design process and follows the same general approach of problem solving and decision making. Successful materials selection depends on the answers to the following questions.

1. Have performance requirements and service environments been properly and completely defined?
2. Is there a good match between the performance requirements and the material properties used in evaluating the candidate materials?
3. Has the material's properties and their modification by subsequent manufacturing processes been fully considered?
4. Is the material available in the shapes and configurations required and at an acceptable price?

The steps in materials selection are:

1. Define the functions that the design must perform and translate these into required materials properties, and to business factors such as cost and availability.
2. Define the manufacturing parameters such as number of parts required, size and complexity of the part, tolerances, quality level, and fabricability of the material.
3. Compare the needed properties and process parameters with a large materials database to select a few materials that look promising for the application. Use several screening properties to identify the candidate materials.
4. Investigate the candidate materials in greater detail, particularly in terms of trade-offs in performance, cost, and manufacturability. Make a final selection of material.
5. Develop design data and a design specification.

Materials selection can never be separated from the consideration of how the part will be manufactured. This large topic is covered in Chap. 13. The Ashby charts are very useful for screening a wide number of materials at the conceptual design stage, and should be employed with materials performance indices. Computer screening of materials databases is widely employed in embodiment design. Many of the evaluation methods that were introduced in Chap. 7 are readily applied to narrowing down the materials selection. The Pugh selection method and weighted decision matrix are most applicable. Failure analysis data (see Chap. 14) are an important input to materials selection when a design is modified. Life-cycle issues should always be considered, especially those having to do with recycling and disposal of materials.

NEW TERMS AND CONCEPTS

Anisotropic property	Defect structure	Scaled property
ASTM	Go-no go material property	Secondary material
Composite material	Material performance index	Structure-sensitive property
Crystal structure	Polymer	Thermoplastic material
Damping capacity	Recycling	Weighted property index

BIBLIOGRAPHY

Ashby, M. F.: *Materials Selection in Mechanical Design,* 4th ed., Elsevier, Butterworth-Heinemann, Oxford, UK, 2011.

"ASM Handbook," vol. 20, *Materials Selection and Design,* ASM International, Materials Park, OH, 1997.

Budinski, K.G.: *Engineering Materials: Properties and Selection,* 8th ed., Prentice Hall, Upper Saddle River, NJ, 2010.

Charles, J. A., F. A. A. Crane, and J. A. G. Furness: *Selection and Use of Engineering Materials*, 3rd ed., Butterworth-Heinemann, Boston, 1997.

Farag, M. M.: *Materials Selection for Engineering Design,* Prentice-Hall, London, 1997.

Kern, R. F., and M. E. Suess: *Steel Selection,* John Wiley, New York, 1979.

Kurtz, M. ed.: *Handbook of Materials Selection,* John Wiley & Sons, 2002

Mangonon, P. L.: *The Principles of Materials Selection for Engineering Design,* Prentice Hall, Upper Saddle River, NJ, 1999.

PROBLEMS AND EXERCISES

11.1 Think about why books are printed on paper. Suggest a number of alternative materials that could be used. Under what conditions (costs, availability, etc.) would the alternative materials be most attractive?

11.2 Consider a soft drink can as a materials system. List all the components in the system and consider alternative materials for each component.

11.3 Which material property would you select as a guide in material selection if the chief performance characteristic of the component was: (a) strength in bending; (b) resistance to twisting; (c) the ability of a sheet material to be stretched into a complex curvature; (d) ability to resist fracture from cracks at low temperatures; (e) ability to resist shattering if dropped on the floor; (f) ability to resist alternating cycles of rapid heating and cooling?

11.4 Rank-order the following materials for use as an automobile radiator: copper, stainless steel, brass, aluminum, ABS, galvanized steel.

11.5 Select a tool material for thread-rolling mild-steel bolts. In your analysis of the problem you should consider the following points: (1) functional requirements of a good tool material, (2) critical properties of a good tool material, (3) screening process for candidate materials, and (4) selection process.

11.6 Table 11.3 gives a range of tensile properties for aluminum alloy 6061. Look up information about this alloy and write a brief report about what processing steps are used to achieve these properties. Include a brief discussion of the structural changes in the material that are responsible for the change in tensile properties.

11.7 Determine the material performance index for a light, stiff beam. The beam is simply supported with a concentrated load at midlength.

11.8 Determine the material performance indices for a connecting rod in a high-performance engine for a racing car. The most likely failure modes are fatigue failure and buckling at the critical section, where the thickness is b and the width is w. Use the CES software to identify the most likely candidates in a material selection at the conceptual design stage.

11.9 Develop the materials performance index for an energy-storing flywheel. Consider the flywheel as a solid disk of radius r and thickness t rotating at an angular velocity ω. The kinetic energy stored in the flywheel is:

$$U = \frac{1}{2} J\omega^2 = \frac{1}{2}\left(\frac{\pi}{2}\rho r^2 t\right)\omega^2 \text{ where } J \text{ is the polar moment of inertia}$$

The quantity to be maximized is the kinetic energy per unit mass. The maximum centrifugal stress in the spinning disk is:

$$\sigma_{max} = \left(\frac{3+\nu}{8}\right)\rho r^2 \omega^2$$

Compare a high-strength aluminum alloy and high-strength steel, along with composite materials, as candidate materials. Discuss your results. Flywheels have been considered as a range extender in hybrid electric automobiles. Compare their capability against the energy density of gasoline (about 20,000 kJ/kg).

11.10 Two materials are being considered for an application in which electrical conductivity is important.

Material	Working Strength MN/m²	Electrical Conductance %
A	500	50
B	1000	40

The weighting factor on strength is 3 and 10 for conductance. Which material is preferred based on the weighted property index?

11.11 An aircraft windshield is rated according to the following material characteristics. The weighting factors are shown in parentheses.

Resistance to shattering (10)　　　The candidate materials are:
Fabricability (2)　　　　　　　　　*A*　plate glass
Weight (8)　　　　　　　　　　　　*B*　PMMA
Scratch resistance (9)　　　　　　　*C*　tempered glass
Thermal expansion (5)　　　　　　　*D*　a special polymer laminate

The properties are evaluated by a panel of technical experts, and they are expressed as percentages of maximum achievable values.

Property	Candidate Material			
	A	B	C	D
Resistance to shattering	0	100	90	90
Fabricability	50	100	10	30
Weight	45	100	45	90
Scratch resistance	100	5	100	90
Thermal expansion	100	10	100	30

Use the weighted property index to select the best material.

11.12 The materials used in a product can importantly influence the aesthetic responses produced by the product. For example, metals give a cold feel because of their high thermal conductivity, while polymers feel warmer because of their much lower conductivity.

Complete the matrix (by adding more columns) for sight, touch, and hearing by filling in with descriptive attributes, and give example materials. Try to find three or four additional attributes for each matrix.

Sight		Touch		Hearing	
Optically clear	optical glass	Warm	copper	Muffled	plastic foam
Textured	plywood	Stiff	steel plate	Low-pitched	cinder block

11.13 A cantilever beam is loaded with force P at its free end to produce a deflection $\delta = PL^3/3EI$. The beam has a circular cross section, $I = \pi r^4/4$. Develop a figure of merit for selecting a material that minimizes the weight of a beam for a given stiffness (P/δ). By using the following material properties, select the best material (a) on the basis of performance and (b) on the basis of cost and performance.

Material	E		P_1 Mgm^{-3}	Approx. Cost, $/ton (1980)
	GNm^{-2}	ksi		
Steel	200	29×10^3	7.8	450
Wood	9–16	1.7×10^3	0.4–0.8	450
Concrete	50	7.3×10^3	2.4–2.8	300
Aluminum	69	10×10^3	2.7	2,000
Carbon-fiber-reinforced plastic (CFRP)	70–200	15×10^3	1.5–1.6	200,000

11.14 Select the most economical steel plate to construct a spherical pressure vessel in which to store gaseous nitrogen at a design pressure of 100 psi at ambient weather conditions down to a minimum of −20°F. The pressure vessel has a radius of 138 in. Your selection should be based on the steels listed in the following table and expressed in terms of cost per square foot of material. Use a value of 489 lb/ft^3 for the density of steel (Add Table below Problem 11.13).

ASTM spec.	Grade	Allowable stress, psi	Pricing, ¢/lb (estimated 1997 prices)						
			Base	Special grade	Quality extra	Width extra	Testing	Heat-treat	Total
A-36		12,650	29.1	0.40	—	3.0	—	—	32.5
A-285	C	13,750	29.1	4.00	—	3.0	—	—	36.1
A-442	60	15,000	29.1	—	4.0	4.0	0.70	—	37.8
A-533	B	20,000	40.0	15.60	3.20	6.2	0.70	18.2	83.9
A-157	B	28,750	40.0	11.70	3.20	8.2	3.00	18.2	84.3

12

DESIGN WITH MATERIALS

12.1
INTRODUCTION

This chapter deals with topics of material performance that are not usually covered in courses in strength of materials but with which the mechanical designer needs to be familiar. Specifically we consider the following topics:

- Design for Brittle Fracture (Sec. 12.2)
- Design for Fatigue Failure (Sec. 12.3)
- Design for Corrosion Resistance (Sec. 12.4)
- Design for Wear Resistance (Sec. 12.5)
- Designing with Plastics (Sec. 12.6)

While this chapter on materials in design is quite comprehensive, it does not consider all of the failure mechanisms that can occur. The most prominent omissions when predicting design performance are high temperature creep and rupture,[1] oxidation,[2] and a variety of embrittling mechanisms.[3] The environmental conditions that cause these failure mechanisms occur less frequently in general engineering practice then those considered in the rest of this chapter, but if one is designing for high-temperature applications then they definitely need first-order consideration. It is expected that the references given to broad review articles will provide an entrance to the needed information.

---

1. D. A. Woodford, "Design for High-Temperature Applications," *ASM Handbook,* Vol. 20, pp. 573–88, ASM International, Materials Park, OH, 1997.
2. J. L. Smialek, C. A. Barrett, and J. C. Schaeffer, "Design for Oxidation Resistance," *ASM Handbook,* vol. 20, pp. 589–602, 1997.
3. G. H. Koch, "Stress-Corrosion Cracking and Hydrogen Embrittlement," *ASM Handbook,* Vol. 19, pp. 483–506, 1997.

12.2
DESIGN FOR BRITTLE FRACTURE

Brittle fracture is fracture that occurs with little accompanying plastic deformation and energy absorption. It generally starts at a small flaw that occurs during manufacture or develops from fatigue or corrosion. The flaw propagates slowly as a crack, often undetected, until it reaches a critical size depending on the loading conditions, at which it propagates rapidly as a catastrophic failure.

An important property of a material is *toughness*, the ability to absorb energy without failure. In a simple tension test loaded slowly, toughness is the area under the stress-strain curve. Notch toughness is the ability of a material to absorb energy in the presence of the complex stress state created by a notch. Conventionally, notch toughness is measured with the Charpy V-notch impact test, which has been very useful in delineating the transition from ductile-to-brittle behavior in steels and other materials as the test temperature decreases. However, this test does not readily lend itself to quantitative analysis. An important advance in engineering knowledge has been the ability to predict the influence of cracks and cracklike defects on the brittle fracture of materials through the science of *fracture mechanics*.[1] Fracture mechanics originated in the ideas of A. A. Griffiths, who showed that the fracture strength of a brittle material, like glass, is inversely proportional to the square root of the crack length. G. R. Irwin proposed that fracture occurs at a fracture stress, σ_f, corresponding to a critical value of crack-extension force, G_c, according to

$$\sigma_f = \left(\frac{EG_c}{\pi a} \right)^{1/2} \tag{12.1}$$

where G_c is the crack extension force, in-lb/in.^{2}
E is the modulus of elasticity of the material, lb/in.^{2}
a is the length of the crack, inches

An important conceptualization was that the elastic stresses in the vicinity of a crack tip (Fig. 12.1*a*) could be expressed entirely by a stress field parameter K called the *stress intensity factor*.

The equations for the stress field at the end of the crack can be written

$$\sigma_x = \frac{K}{\sqrt{2\pi r}} \left[\cos\frac{\theta}{2} \left(1 - \sin\frac{\theta}{2}\sin\frac{3\theta}{5} \right) \right]$$

$$\sigma_y = \frac{K}{\sqrt{2\pi r}} \left[\cos\frac{\theta}{2} \left(1 + \sin\frac{\theta}{2}\sin\frac{3\theta}{2} \right) \right] \tag{12.2}$$

$$\tau_{xy} = \frac{K}{\sqrt{2\pi r}} \left(\sin\frac{\theta}{2}\cos\frac{\theta}{2}\cos\frac{3\theta}{2} \right)$$

1. S. T. Rolfe and J. M. Barsom, *Fracture and Fatigue Control in Structures,* 2nd ed., Prentice Hall, Englewood Cliffs, NJ, 1987; T. L. Anderson, *Fracture Mechanics Fundamentals and Applications,* 3rd ed., Taylor & Francis, Boca Raton, FL, 2005; R.J. Sanford, *Principles of Fracture Mechanics,* Prentice Hall, Upper Saddle River, NJ, 2003; A. Shukla, *Practical Fracture Mechanics in Design,* 2nd ed., Marcel Dekker, New York, 2002.

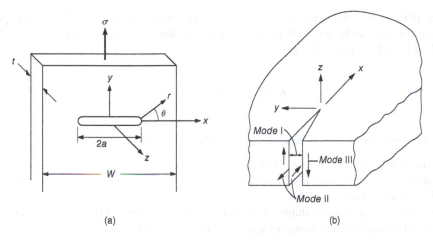

(a) (b)

FIGURE 12.1
(a) Model for equations for stress at a point near a crack. (b) The basic modes of crack surface displacement.

Equations (12.2) show that the elastic normal and elastic shear stresses in the vicinity of the crack tip depend only on the radial distance from the tip r, the orientation θ, and K. Thus, the magnitudes of these stresses at a given point are dependent completely on the *stress intensity factor K*. However, the value of K depends on the type of loading (tension, bending, torsion, etc.), the configuration of the stressed body, and the mode of crack displacement. Figure 12.1*b* shows the three modes of fracture that have been identified: Mode I (opening mode where the crack opens in the y direction and propagates in the x-z plane), mode II (shearing in the x direction), or type III (tearing in the x-z plane). Mode I is caused by tension loading in the y direction while the other two modes are caused by shearing in different directions. Most brittle fracture problems in engineering are caused by tension stresses in Mode I crack propagation. At some critical stress state given by K_{Ic} a flaw or crack in the material will suddenly propagate as a fast-moving brittle crack, according to Eq. (12.3).

For a crack of length $2a$ centered in an infinitely wide thin plate subjected to a uniform tensile stress σ, the stress intensity factor K is given by

$$K = \sigma \sqrt{\pi a} = GE \tag{12.3}$$

where K is in units of $ksi\sqrt{in}.\ or\ \mathrm{MPa}\sqrt{m}$ and σ is the nominal stress based on the gross cross section. Values of K have been determined for a variety of situations by using the theory of elasticity, often combined with numerical methods and experimental techniques.[1] For a given type of loading, Eq. (12.3) usually is written as

$$K = \alpha \sigma \sqrt{\pi a} \tag{12.4}$$

1. G. G. Sih, *Handbook of Stress Intensity Factors,* Institute of Fracture and Solid Mechanics, Lehigh University, Bethlehem, PA, 1973; Y. Murakami et al., (eds.), *Stress Intensity Factors Handbook,* (2 vols.), Pergamon Press, New York, 1987; H. Tada, P.C. Paris, and G.R. Irwin, *The Stress Analysis of Cracks Handbook,* 3rd ed., ASME Press, New York, 2000; A. Liu, "Summary of Stress-Intensity Factors," *ASM Handbook,* Vol. 19, *Fatigue and Fracture,* pp. 980–1000, ASM International, Materials Park, OH, 1996.

where α is a parameter that depends on the specimen, crack geometry, and type of loading. For example, for a plate of width w containing a *central through-thickness crack* of length $2a$ (Fig. 12.1a),

$$K = \left(\frac{w}{\pi a} \tan \frac{\pi a}{w} \right)^{1/2} \sigma \sqrt{\pi a} \tag{12.5}$$

Common geometries where fracture mechanics is often applied are:

- A crack in a sheet or a plate. This can be a through-thickness crack, a surface crack, or a crack embedded in the plate thickness. Each situation will result in a different relationship for α in Eq. (12.4).
- Similarly, changing the shape from plates to beams, solid or hollow cylinders, or a tapered lifting lug will result in a new relationship for α.
- Also, loading in bending, torsion, or point tension loading at the site of the crack will all result in very different relationships compared to the uniaxial uniform loading applied at a far distance from the crack, shown in Fig. 12.1a.

A readily available source of stress intensity factors is the chapter by A. Liu referenced in the footnote on the previous page.

12.2.1 Plane Strain Fracture Toughness

Since the crack tip stresses can be described by the stress intensity factor K, a critical value of K can be used to define the conditions that produce brittle fracture. The tests usually used subject the specimen to the crack opening mode of loading (Mode I) under a condition of plane strain at the crack front. The critical value of K that produces fracture is K_{Ic}, the *plane-strain fracture toughness*. The "*I*" in the subscript denotes a Mode I fracture. A great deal of engineering research has gone into standardizing tests for measuring fracture toughness.[1] If a_c is the critical crack length at which failure occurs, then

$$K_{Ic} = \alpha \sigma \sqrt{\pi a_c} \tag{12.6}$$

This is the same as Eq. (12.4), only now $a = a_c$ and $K = K_{Ic}$, the stress intensity factor required to trigger the fast-moving brittle fracture.

K_{Ic} is a basic material property called *plane-strain fracture toughness* or often called just *fracture toughness*. Some typical values are given in Table 12.1. Note the large difference in K_{Ic} values between the metallic alloys and the polymers and the ceramic material silicon nitride. Also, note that the fracture toughness and yield strength

---

1. J.D. Landes, "Fracture Toughness Testing," *ASM Handbook,* Vol. 8, *Mechanical Testing and Evaluation,* ASM International, 2000, pp. 576–85; The basic procedure for K_{Ic} testing is ASTM Standard E 399, "Standard Test Method for Plane Strain Fracture Toughness of Metallic Materials."

TABLE 12.1
**Some Typical Values of Plane-Strain Fracture Toughness
at Room Temperature**

	K_{Ic}		Yield Strength	
	MPa$\sqrt{m}$	ksi$\sqrt{in.}$	MPa	ksi
Plain carbon steel AISI 1040	54.0	49.0	260	37.7
Alloy steel AISI 4340				
Tempered @ 500°F	50.0	45.5	1500	217
Tempered @ 800°F	87.4	80.0	1420	206
Aluminum alloy 2024-T3	44.0	40.0	345	50
Aluminum alloy 7075-T651	24.0	22.0	495	71
Nylon 6/6	3.0	2.7	50	7.3
Polycarbonate (PC)	2.2	2.0	62	9.0
Polyvinyl chloride (PVC)	3.0	2.2	42	6.0
Silicon nitride—hot pressed	5.0	4.5	800	116

vary inversely for the alloy steel and the aluminum alloys. This is a general relationship: as yield strength increases, fracture toughness decreases. It is one of the major constraints in selecting materials for high-performance mechanical applications.

Although K_{Ic} is a basic material property, in the same sense as yield strength, it changes with important variables such as temperature and strain rate. The K_{Ic} of materials with a strong temperature and strain-rate dependence usually decreases with decreased temperature and increased strain rate. The K_{Ic} of a given alloy is strongly dependent on such variables as heat treatment, texture, melting practice, impurity level, and inclusion content.

Equation 12.6 contains the three design variables that must be considered in designing against fracture of a structural component—the fracture toughness, the imposed stress, and the crack or flaw size. Figure 12.2 shows the relationship between them. If K_{Ic} is known because the material has been selected, then it is possible to compute the maximum allowable stress to prevent brittle fracture for a given flaw size. Generally the flaw size is determined by actual measurement or by the smallest detectable flaw for the nondestructive inspection method that is available for use.[1] As Fig. 12.2 shows, the allowable stress in the presence of a crack of a given size is directly proportional to K_{Ic}, and the allowable crack size for a given stress is proportional to the square of the fracture toughness. Therefore, increasing K_{Ic} has a much larger influence on allowable crack size than on allowable stress.

To obtain a proper value of K_{Ic} it must be measured under plane-strain conditions to obtain a maximum constraint or material brittleness. Figure 12.3 shows

12

---

1. Crack detection sensitivity can vary from 0.5 mm for magnetic particle methods to 0.1 mm for eddy current, acoustic emission, and liquid penetrant methods. *ASM Handbook,* Vol. 17, p. 211.

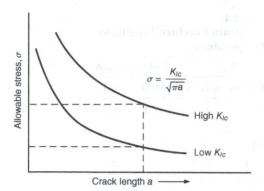

FIGURE 12.2
Relation between fracture toughness and allowable stress and crack size.

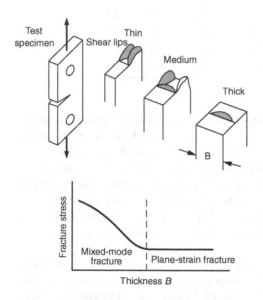

FIGURE 12.3
Effect of specimen thickness on fracture stress and mode of failure. Sketches depict appearance of fracture surface. Note the presence of shear lips.

how the measured fracture stress varies with specimen thickness B. A mixed-mode, ductile brittle fracture with 45° shear lips is obtained for thin specimens. Once the specimen has the critical thickness for the toughness of the material, the fracture surface is flat and the fracture stress is constant with increasing specimen thickness. The minimum thickness to achieve plane-strain conditions and valid K_{Ic} measurement is

$$B \geq 2.5 \left(\frac{K_{Ic}}{\sigma_y} \right)^2 \tag{12.7}$$

where B is the section thickness and σ_y is the allowable design stress, typically the yield stress decreased by a factor of safety.

EXAMPLE 12.1

An aircraft skin panel that is part of a commercial airliner is to be made from aluminum alloy 7075-T651. During construction of the plane an inspector noted a 10 mm deep surface crack along one of the long sheared edges of the panel. The panel is 20 cm wide, 100 cm tall, and 50 mm thick. It is subjected to a 200 MPa tensile stress. From Table 12.1, the mechanical properties of the material are

$$K_{Ic} = 24\,\text{MPa}\sqrt{\text{m}} \text{ and the yield strength } \sigma_y = 495 \text{ MPa}$$

In analyzing the design for fracture, first check to see whether the average stress is below the yield strength of the material.

Is $\sigma_{applied} \le \sigma_y$? In this case, 200 MPa ≤ 495 MPa, but the crack could cause a stress concentration of approximately $K_t = 2$, so there may be some local yielding at the crack. However, the panel acts elastically overall.

The equation for single-edge crack in a plate loaded in tension is found on page 983 of the chapter by Liu in *ASM Handbook* 17. The equation uses a different format from Eq. (12.4) by replacing α by φ as the geometric factor.

$$\varphi = \sec \beta \left[\frac{\tan \beta}{\beta} \right]^{1/2} \left[0.752 + 2.02 \left(\frac{a}{w} \right) + 0.37(1 - \sin \beta)^3 \right]$$

where $\beta = (\pi a / 2w)$

Substituting $a = 10$ mm; $w = 20 \times 10^2 = 2000$ mm; $a/w = 0.005$ we find $\varphi = \alpha = 1.122$. Now using Eq. (12.4) to calculate the stress intensity factor,

$$K = 1.122\sigma\sqrt{\pi a} = 1.122 \left[200\sqrt{\pi(0.010)} \right] = 39.8 \text{ MPa}\sqrt{\text{m}}$$

Since $K \ge K_{Ic}$, that is, 39.8 ≥ 24, the panel is expected to fail by a rapid brittle fracture, especially since at −50°F at high altitude the value of K_{Ic} is likely to be much lower than the room temperature value used here.

Solution

Switch to the lower-strength but tougher aluminum alloy 2024 with $K_{Ic} = 44\,\text{MPa}\sqrt{\text{m}}$. For 2024 alloy the yield strength is 345 MPa. This is above the 200 MPa general applied stress, so gross section yielding will not occur. The greater ductility of this alloy will allow more local plastic deformation at the crack and blunt the crack so the stress concentration will not be severe.

As a final check we use Eq. (12.7) to see if the plane-strain condition holds for the panel.

$$B \ge 2.5 \left(\frac{K_{Ic}}{\sigma_y} \right)^2 = 2.5 \left(\frac{44}{345} \right)^2 = 0.041 \text{ m} = 41 \text{ mm}$$

This says that the thickness of the panel must be greater than 41 mm to maintain the maximum constraint of the plane-strain condition. A panel with a thickness of 50 mm exceeds this condition.

Another way to look at the problem is to determine the critical crack size at which the crack would propagate to fracture at an average stress of 200 MPa. From Eq. (12.6), and using a value of $\alpha = 1.122$, the critical crack length a_c is 3.6 mm for the 7075 aluminum

12

alloy and 12.2 mm for the 2024 alloy. This means that a 7075 alloy sheet with a 10 mm long surface crack would fail in brittle fracture when the 200 MPa stress was applied because its critical crack size is less than 10 mm. However, a 10 mm crack will not cause fracture in the tougher 2024 alloy with a critical crack size of 12.2 mm.

12.2.2 Limitations on Fracture Mechanics

The fracture mechanics concept is strictly correct only for linear elastic materials (those materials in which no yielding occurs before fracture). Reviewing Eqs. (12.2) shows that as r approaches zero, the stress at the crack tip approaches infinity. Thus, in all but the most brittle material, local yielding occurs at the crack tip and the elastic solution should be modified to account for crack tip plasticity. However, if the plastic zone size, r_y, at the crack tip is small relative to the local geometry, for example, if r_y/t or $r_y/a \leq 0.1$, crack tip plasticity has little effect on the stress intensity factor. That limits the strict use of fracture mechanics to high-strength materials. Moreover, the width restriction to obtaining valid measurements of K_{Ic}, as described by Eq. (12.7), makes the use of linear elastic fracture mechanics (LEFM) impractical for low-strength materials. The criteria for LEFM behavior is given by[1]

$$a, (w-a), h = \geq \frac{4}{\pi}\left(\frac{K}{\sigma_y}\right)^2 \tag{12.8}$$

where a is the crack length, w is the specimen width, so $(w - a)$ is the uncracked width and h is the distance from the top of the specimen to the crack. Each of these parameters must satisfy Eq. (12.8). Otherwise the situation too closely approaches gross yielding with the plastic zone extending to one of the boundaries of the specimen. A value of K determined beyond the applicability of LEFM underestimates the severity of the crack.

Considerable activity has gone into developing tests for measuring fracture toughness in materials that have too much ductility to permit the use of LEFM testing methods.[2] The best approach uses the J-integral, which is obtained by measurements of the load versus the displacement of the crack.[3] These tests result in a valid measure of fracture toughness J_{Ic} which serves in the same way as K_{Ic} does for linear elastic materials.

12.3
DESIGN FOR FATIGUE FAILURE

Materials subjected to repetitive or fluctuating stress cycles will fail at a stress much lower than that required to cause fracture on a single application of load. Failures occurring under conditions of fluctuating stresses or strains are called

---

1. N.E. Dowling, *Mechanical Behavior of Materials,* 2nd ed., Prentice Hall, Upper Saddle River, NJ, 1999, p. 333.
2. A. Saxena, *Nonlinear Fracture Mechanics for Engineers,* CRC Press, Boca Raton, FL, 1998.
3. ASTM Standard E 1820.

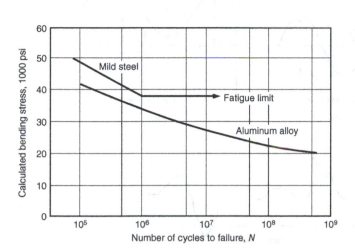

FIGURE 12.4
Typical fatigue S-N curves for ferrous and nonferrous metals.

fatigue failures.[1] Fatigue accounts for the majority of mechanical failures in machinery.

A fatigue failure is a localized failure that starts in a limited region and propagates with increasing cycles of stress or strain until the crack is so large that the part cannot withstand the applied load, and it fractures. Plastic deformation processes are involved in fatigue, but they are highly localized.[2] Therefore, fatigue failure occurs without the warning of gross plastic deformation. Failure usually initiates at regions of local high stress or strain caused by abrupt changes in geometry (stress concentration), temperature differentials, tensile residual stresses, or material imperfections. Much basic information has been obtained about the mechanism of fatigue failure, but at present the chief opportunities for preventing fatigue exist at the engineering design level. Fatigue prevention is achieved by proper choice of material, control of residual stress, and minimization of stress concentrations through careful design.

Basic fatigue data are presented in the S-N curve, a plot of stress,[3] S, versus the number of cycles to failure, N. Figure 12.4 shows the two typical types of behavior. The curve for an aluminum alloy is characteristic of all materials except ferrous metals (steels). The S-N curve is chiefly concerned with fatigue failure at high numbers of cycles ($N > 10^5$ cycles). Under these conditions the gross stress is elastic, although fatigue failure results from highly localized plastic deformation. Figure 12.4 shows the number of cycles of stress that a material can withstand before failure increases

---

1. L. Pook, *Metal Fatigue: what it is, why it matters,* Springer, 2007; S. Suresh, *Fatigue of Materials,* 2nd ed., Cambridge University Press, Cambridge, 1998; N. E. Dowling, *Mechanical Behavior of Materials,* 2nd ed., Chaps, 9, 10, 11, Prentice Hall, Englewood Cliffs, NJ, 1999; *ASM Handbook,* Vol. 19, *Fatigue and Fracture,* ASM International, Materials Park, OH, 1996.
2. ASM Handbook, Vol. 19, *Fatigue and Fracture,* pp. 63–109, ASM International, Materials Park, OH, 1996.
3. It is conventional to denote nominal stresses in fatigue by S rather than σ.

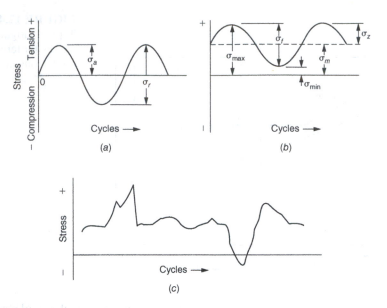

FIGURE 12.5
Typical fatigue stress cycles: (a) reversed stress; (b) repeated stress; (c) irregular or random stress cycle.

with decreasing stress. For most materials, the S-N curve slopes continuously downward toward increasing cycles of stress to failure with decreasing stress. At any stress level there is some large number of cycles that ultimately causes failure. This is called the *fatigue strength*. For steels in the absence of a corrosive environment, however, the S-N curve becomes horizontal at a certain limiting stress. Below that stress, called the *fatigue limit* or *endurance limit*, the steel can withstand an infinite number of cycles.

12.3.1 Fatigue Parameters

Typical cycles of stress that produce fatigue failure are shown in Fig. 12.5. Figure 12.5*a* illustrates a completely reversed cycle of stress of sinusoidal form. This is the type of fatigue cycle for which most fatigue property data is obtained in laboratory testing.[1] It is found in service by a rotating shaft operating at constant speed without overloads. For this type of stress cycle the maximum and minimum stresses are equal. In keeping with convention, the minimum stress is the lowest algebraic stress in the cycle. Tensile stress is considered positive, and compressive stress is negative. Figure 12.5*b* illustrates a repeated stress cycle in which the maximum stress σ_{max} and minimum stress σ_{min} are not equal. In this illustration they are both tension, but a repeated stress cycle could just as well contain maximum and minimum stresses of opposite signs or

---

1. Common types of fatigue testing machines are described in many references in this section and in *ASM Handbook,* vol. 8, pp. 666–716, ASM International, 2000.

both in compression. Figure 12.5c illustrates a complicated stress cycle that might be encountered in a part such as an aircraft wing, which is subjected to periodic unpredictable overloads due to wind gusts.

A fluctuating stress cycle can be considered to be made up of two components, a *mean*, or steady, stress, σ_m, and an *alternating*, or variable, stress, σ_a. We must also consider the range of stress, σ_r. As can be seen from Fig. 12.5b, the range of stress is the algebraic difference between the maximum and minimum stress in a cycle.

$$\sigma_r = \sigma_{max} - \sigma_{min} \tag{12.9}$$

The alternating stress, then, is one-half of the range of stress.

$$\sigma_a = \frac{\sigma_r}{2} = \frac{\sigma_{max} - \sigma_{min}}{2} \tag{12.10}$$

The mean stress is the algebraic mean of the maximum and minimum stress in the cycle.

$$\sigma_m = \frac{\sigma_{max} + \sigma_{min}}{2} \tag{12.11}$$

A convenient way to denote the fatigue cycle is with the *stress ratio, R*.

$$R = \sigma_{min} / \sigma_{max} \tag{12.12}$$

For a completely reversed stress cycle, $\sigma_{max} = -\sigma_{min}$, $R = -1$, and $\sigma_m = 0$. For a fully tensile repeated stress cycle with $\sigma_{min} = 0$, $R = 0$. The influence of mean stress on the S-N diagram can be expressed by the Goodman equation.

$$\sigma_a = \sigma_e \left[1 - \left(\frac{\sigma_m}{\sigma_{uts}} \right) \right] \tag{12.13}$$

12

where σ_e is the endurance limit in a fatigue test with a completely reversed stress cycle, and σ_{uts} is the ultimate tensile strength.

Fatigue is a complex material failure process. The fatigue performance of a component depends on the following important engineering factors:

Stress Cycle

- Repeated or random applied stress
- Mean stress. Most fatigue test data has been obtained under completely repeated stress with a mean stress of zero. The safe alternating stress decreases with increasing mean stress, as shown by Eq. (12.13).
- Combined stress state. See *ASM Handbook*, vol. 19, pp. 263–273.
- Stress concentration. Most fatigue cracks start at points of elevated stress. This is expressed by a stress *concentration factor* (for the geometry) and a *notch sensitivity factor* for the material's sensitivity for stress concentrations.

- Statistical variation in fatigue life and fatigue limit. There is more scatter in fatigue life than in any other mechanical property of materials. Often a probabilistic approach is needed. See *ASM Handbook,* vol. 19, pp. 295–313.
- Cumulative fatigue damage. The conventional fatigue test subjects a specimen to a fixed amplitude of stress until the specimen fails. However, in practice there are many situations where the cyclic stress does not remain constant, but instead there are periods when the stress is either above or below some average design level. Taking into consideration irregular stress versus cycle issues is an important area of fatigue design for which additional concepts are needed.
- Fatigue failure can occur without externally applied forces from thermal gradients that vary with time so as to produce cyclic thermal stresses, that is, *thermal fatigue.* Thermal stresses arise from the strain produced by thermal expansion and contraction.

Component or Specimen-Related Factors

- Size effect. The larger the section size of the part, the lower its fatigue properties. This is related to the higher probability of finding a critical crack-initiating flaw in a larger volume of material.
- Surface finish. Most fatigue cracks start at the surface of a component. The smoother the surface roughness, the higher is the fatigue property.
- Residual stress. Residual stresses are locked-in stresses that are present in a part when it is not subjected to loads, and which add to the applied stresses when under load. The formation of a compressive residual distribution on the surface of a part is the most effective method of increasing fatigue performance. The best way of doing this is shot blasting or rolling of the surface to cause localized plastic deformation.
- Surface treatment. As noted above, surface treatments that increase the compressive residual stresses at the surface are beneficial. Increasing the surface hardness by carburizing or nitriding often improves the fatigue performance of steel parts. However, losing carbon from the surface of steel by poor heat treatment practice is detrimental to fatigue properties. See *ASM Handbook,* vol. 19, pp. 314–320.

Environmental Effects

- Corrosion fatigue. The simultaneous action of cyclic stress and chemical attack is known as *corrosion fatigue.* Corrosive attack without superimposed stress often produces pitting of metal surfaces. The pits act as notches and produce a reduction in fatigue strength. However, when corrosive attack occurs simultaneously with fatigue loading, a very pronounced reduction in fatigue properties results that is greater than that produced by prior corrosion of the surface. When corrosion and fatigue occur simultaneously, the chemical attack greatly accelerates the rate at which fatigue cracks propagate. Steels that have a fatigue limit when tested in air at room temperature show no indication of a fatigue limit in corrosion fatigue. See *ASM Handbook,* vol. 19, pp. 193–209.
- Fretting fatigue. *Fretting* is the surface damage that results when two surfaces in contact experience slight periodic relative motion. The phenomenon is more related to wear than to corrosion fatigue. However, it differs from wear by the facts that the

relative velocity of the two surfaces is much lower than is usually encountered in wear and that since the two surfaces are never brought out of contact, there is no chance for the corrosion products to be removed. Fretting is frequently found on the surface of a shaft with a press-fitted hub or bearing. Surface pitting and deterioration occur, usually accompanied by an oxide debris (reddish for steel and black for aluminum). Fatigue cracks often start in the damaged area. See *ASM Handbook,* vol. 19, pp. 321–330.

12.3.2 Information Sources on Design for Fatigue

There is a considerable literature on design methods to prevent fatigue failure. Most machine design texts devote a chapter to the subject. In addition, the following specialized texts add a great deal of detail.

Fatigue design for infinite life is considered in:

> R. C. Juvinall, *Engineering Consideration of Stress, Strain, and Strength,* McGraw-Hill, New York, 1967. Chapters 11 to 16 cover fatigue design in considerable detail.
>
> L. Sors, *Fatigue Design of Machine Components,* Pergamon Press, New York, 1971. Translated from the German, this presents a good summary of European fatigue design practice.
>
> C. Ruiz and F. Koenigsberger, *Design for Strength and Production,* Gordon & Breach Science Publishers, New York, 1970. Pages 106 to 120 give a concise discussion of fatigue design procedures.

Detailed information on stress concentration factors and the design of machine details to minimize stress can be found in:

> W. D. Pilkey and D. F. Pilkey, *Peterson's Stress Concentration Factors,* 3rd ed., John Wiley & Sons, New York, 2008.
>
> R. B. Heywood, *Designing Against Fatigue of Metals,* Reinhold, New York, 1967.

The most complete books on fatigue design, including the more modern work on strain-life design and damage-tolerant design, are:

> R. I. Stephens, A. Fatem, R. R. Stephens, and H. O. Fuchs, *Metal Fatigue in Engineering,* 2nd ed., John Wiley & Sons, Hoboken, NJ, 2000.
>
> *Fatigue Design Handbook,* 3rd ed., Society of Automotive Engineers, Warrendale, PA, 1997.
>
> E. Zahavi, *Fatigue Design,* CRC Press, Boca Raton, FL, 1996.

A useful website for fatigue data and helpful calculators is www.efatigue.com. This website also provides calculators for determining stress intensity factors (see Sec. 12.2).

There are several different fatigue design strategies that must be understood to put the vast collection of fatigue design literature into perspective. Discussed in order in the next three sections are Stress-life design, Strain-life design, and Damage-tolerance design.

12

12.3.3 Stress-Life Design

Stress-life fatigue design is based on keeping the nominal stress below some value of fatigue strength, or more commonly below the endurance limit for steels. This is the oldest design methodology for fatigue and is found in most machine design texts.[1] It starts with a value for the fatigue life or fatigue limit based on laboratory fatigue test results from smooth (unnotched) specimens and reduces (derates) the value of the fatigue limit for all of the parameters described in Sec.12.3.1. Particularly important are the *fatigue notch factor*,[2] size effect, mean stress, surface roughness, and correction for variable amplitude loading (see *ASM Handbook* Vol. 19, p. 110). For discussion of fatigue notch factor and an example of design using the stress-life strategy, see Example for Infinite-Life Design under Chap. 12 at www.mhhe.com/dieter.

12.3.4 Strain-Life Design Strategy

Strain-life design based on failure at a finite number of cycles is based on strain-life curves. These are often called *low-cycle fatigue curves* because much of the data is obtained in less than 10^5 cycles. When fatigue occurs at a relatively low number of cycles, the stresses that produce failure often exceed the yield strength. Even when the gross stress remains elastic, the localized stress at a notch is inelastic. Under these conditions it is better to carry out fatigue tests under fixed amplitude of strain (strain control) rather than fixed amplitude of stress. Typically these tests are conducted in push-pull tension. Figure 12.6 shows the *stress-strain loop* that is produced by strain control cycling. Note that for the first hundred cycles or so the enclosed area of the cyclic stress-strain curve either increases (strain hardening) or decreases (strain softening) until it reaches a stable area as shown in Fig. 12.6. The strain amplitude is given by

$$\varepsilon_a = \frac{\Delta\varepsilon}{2} = \frac{\Delta\varepsilon_e}{2} + \frac{\Delta\varepsilon_p}{2} \tag{12.14}$$

where the total strain $\Delta\varepsilon$ is the sum of the elastic and plastic strains and the strain amplitude is one-half of the total strain.

A plot of the stress at the tip of the stress-strain loop (point B) for various strain amplitudes yields a cyclic true stress-strain curve. When stress vs. plastic strain are plotted on log-log coordinates this yields a straight line. The stress amplitude is given by

$$\frac{\Delta\sigma}{2} = k'\left(\frac{\Delta\varepsilon_p}{2}\right)^{n'} \tag{12.15}$$

1. For example, R.G. Budynas and J.K. Nisbett, *Shigley's Mechanical Engineering Design,* 9th ed., pp. 286–348, McGraw-Hill, 2011.
2. N. E. Dowling, *Mechanical Behavior of Materials,* 2nd ed., Chap. 10, Prentice-Hall, 1999; Y-L Lee, J. Pan, R. B. Hathaway, and M. E. Barkey, *Fatigue Testing and Analysis,* Chap. 4, Butterworth-Heinemann, 2005.

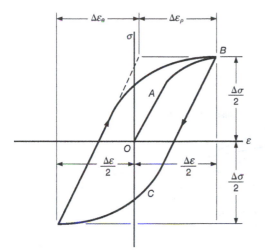

FIGURE 12.6
Typical cyclic stress-strain loop.

where k' is the *cyclic strength coefficient* (the intercept of the line at unit plastic strain) and n', the slope of the line, is the *cyclic strain hardening exponent*. Substituting into Eq. (12.14) from Eq. (12.15) gives

$$\frac{\Delta\varepsilon}{2} = \frac{\Delta\sigma}{2E} + \left(\frac{\Delta\sigma}{2k'}\right)^{1/n'} \tag{12.16}$$

Equation (12.16) allows the calculation of the cyclic stress-strain curve. Values of the material properties E, k' and n' are available in the literature.[1]

The strain-life curve plots the total strain amplitude versus the number of *strain reversals* to failure (one cycle equals 2 reversals), Fig 12.7. Note that the number of strain reversals equals twice the number of strain cycles. Both the elastic and plastic curves are approximated as straight lines. At small strains or long lives, the elastic strain predominates, and at large strains or short lives the plastic strain is predominant. The plastic curve has a negative slope of c and an intercept at $2N = 1$. The elastic curve has a negative slope of b and an intercept of σ'_f / E. The equation for the elastic line is

$$\log\left(\frac{\Delta\varepsilon_e}{2}\right) = \log\left(\frac{\sigma'_f}{E}\right) + b\log\left(2N_f\right) \text{ or } \frac{\Delta\varepsilon_e}{2} = \frac{\sigma'_f}{E}\left(2N_f\right)^b$$

Similarly, the plastic line is given by $\frac{\Delta\varepsilon_p}{2} = \varepsilon'_f\left(2N_f\right)^c$. But, from Eq. (12.14),

$$\varepsilon_a = \frac{\Delta\varepsilon}{2} = \frac{\sigma'_f}{E}\left(2N_f\right)^b + \varepsilon'_f\left(2N_f\right)^c \tag{12.17}$$

1. *ASM Handbook,* Vol. 19, pp. 963–79.

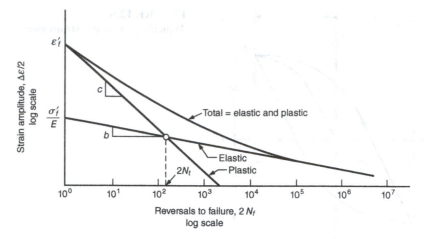

FIGURE 12.7
Typical strain-life curve for mild steel.

The exponent b ranges from about -0.06 to -0.14 for different materials, and a typical value is -0.1. The exponent c ranges from about -0.5 to -0.7, and -0.6 is a representative value. The term ε'_f, called the *fatigue ductility coefficient*, is approximately equal to the true fracture strain measured in the tension test. Likewise, the *fatigue strength coefficient* σ'_f is approximated by the true fracture stress.

An important use of the low-cycle fatigue approach is to predict the life to crack initiation at notches in machine parts where the nominal stresses are elastic but the local stresses and strain at the notch root are plastic. When there is plastic deformation, both a strain concentration K_ε and a stress concentration K_σ must be considered. Neuber's rule relates these by

$$K_f = \left(K_\sigma K_\varepsilon\right)^{1/2} \tag{12.18}$$

where K_f is the fatigue notch factor.

The situation is described in Fig. 12.8, where ΔS and Δe, as shown in Fig. 12.8b, are the elastic stress and strain increments at a location remote from the notch and $\Delta \sigma$ and $\Delta \varepsilon$ are the local stress and strain at the root of the notch.

$$K_\sigma = \frac{\Delta \sigma}{\Delta S} \quad \text{and} \quad K_\varepsilon = \frac{\Delta \varepsilon}{\Delta e}$$

$$K_f = \left(\frac{\Delta \sigma \, \Delta \varepsilon}{\Delta S \, \Delta e}\right)^{1/2} = \left(\frac{\Delta \sigma \, \Delta \varepsilon E}{\Delta S \, \Delta \varepsilon E}\right)^{1/2}$$

and

$$K_f \left(\Delta S \Delta e E\right)^{1/2} = \left(\Delta \sigma \Delta \varepsilon E\right)^{1/2} \tag{12.19}$$

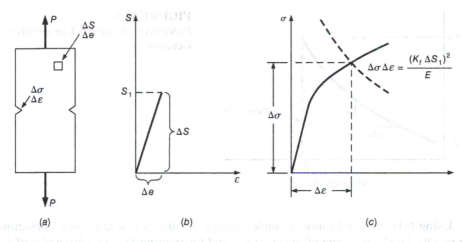

FIGURE 12.8
Stress analysis at a notch using Neuber's analysis.

For nominally elastic loading, $\Delta S = \Delta e E$ and

$$K_f \Delta S = (\Delta \sigma \Delta \varepsilon E)^{1/2} \qquad (12.20)$$

Thus, Eq. (12.20) allows stresses remotely measured from the notch to be used to predict stress and strain at the notch. Rearranging Eq. (12.20) gives

$$\Delta \sigma \Delta \varepsilon = \frac{\left(K_f \Delta S\right)^2}{E} = \text{constant} \qquad (12.21)$$

which is the equation of a rectangular hyperbola (Fig. 12.8c). If a nominal stress S_1 is applied to the notched specimen (Fig. 12.8b), then the right side of Eq. (12.21) is known provided K_f is known. The cyclic stress-strain curve also is plotted in Fig. 12.8c (solid curve), and its intersection with Eq. (12.21) gives the local stress and strain at the notch root.

The use of Neuber's analysis to find the stress and strain at a notch in conjunction with the strain-life curve (Fig. 12.7) is a widely accepted approach to predicting the number of fatigue cycles to the initiation of a crack. The basic assumption is that the local fatigue response at the critical region, usually at a flaw or geometric notch, is similar to the fatigue response of a small, smooth laboratory specimen subjected to the same cyclic strains and stresses as the critical region.

To use the prediction model it is necessary to

- Compute with FEA the local stresses and strains, including mean stress and stress range.
- Count the cycles and corresponding mean and range values of stress and strain.
- Nonzero mean cycles must be converted to equivalent completely reversed cycles.
- Have appropriate material properties to compute fatigue damage during each cycle.
- Sum the damage to give the prediction of the number of cycles to crack initiation.

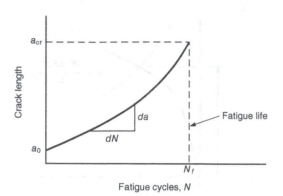

FIGURE 12.9
Process of fatigue crack propagation (schematic).

Using this method in more complex design situations would require correcting the results for the presence of mean stress, and for treating less regular cycles of fatigue stress.[1]

Strain-life analysis has found broad acceptance in the automotive and aerospace industries for component design in the early stages of embodiment design. It allows reasonable prediction of fatigue life before components are available for actual fatigue testing. This ability to predict results in a reduction in the number of design iterations and a speed-up in the PDP cycle. A number of computer programs are available to assist in this type of design analysis. See www.efatigue.com. See Example for Strain-Life Design under Chap.12 at www.mhhe.com/dieter.

12.3.5 Damage-Tolerant Design Strategy

Damage-tolerant design starts with the premise that the part contains a fatigue crack of known dimensions and geometry and predicts how many cycles of service are available before the crack will propagate to a catastrophic size that will cause failure, N_p. Thus, the emphasis is on fatigue crack growth. Figure 12.9 shows the process of crack propagation from an initial crack of length a_0 to a crack of critical flaw size a_{cr}. The crack growth rate da/dN increases with the cycles of repeated load.

An important advance in fatigue design was the realization that the fatigue crack growth rate da/dN can be related to the *range of the stress intensity factor;* $\Delta K = K_{max} - K_{min}$, for the fatigue cycle. Since the stress intensity factor $K = \alpha\sigma\sqrt{\pi a}$ is undefined in compression, K_{min} is taken as zero if σ_{min} is compression in the fatigue cycle.

Figure 12.10 shows a typical plot of rate of crack growth versus ΔK. The typical curve is sigmoidal in shape with three distinct regions. Region I contains the threshold value ΔK_{th} below which there is no observable crack growth. Below ΔK_{th}, fatigue cracks behave as nonpropagating cracks. The threshold starts at crack propagation rates of around 10^{-8} in./cycle and at very low values of K, for example 8 ksi$\sqrt{\text{in}}$.

---

1. Y-L Lee, J. Pan., R. Hathaway, and M. Barkey, *Fatigue Testing and Analysis,* Chap. 5, Elsevier, Boston, 2005.

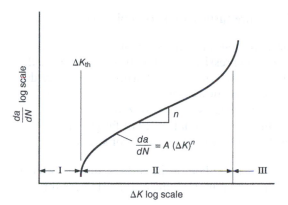

FIGURE 12.10
Schematic fatigue crack growth vs. ΔK curve.

for stress. Region II exhibits essentially a linear relation between log da/dN and log K, which results in

$$\frac{da}{dN} = A(\Delta K)^n \qquad (12.22)$$

where n = slope of the curve in region II

A = coefficient found by extending the line to $\Delta K = 1 \text{ksi} \sqrt{\text{in.}}$

Region III is a region of rapid crack growth that is soon followed by fracture.

The relation between fatigue crack growth ΔK and expressed by Eq. (12.22) ties together fatigue design[1] and linear elastic fracture mechanics (LEFM). The elastic stress intensity factor is applicable to fatigue crack growth even in low-strength, high-ductility materials because the K values needed to cause fatigue crack growth are very low and the plastic zone sizes at the tip are small enough to permit an LEFM approach. By correlating crack growth with stress intensity factor, it is possible to use data generated under constant-amplitude conditions with simple specimens for a broad range of design situations in which K can be calculated. When K is known for the component under relevant loading conditions, the fatigue crack growth life of the component can be obtained by integrating Eq. (12.22) between the limits of initial crack size and final crack size. See Example for Damage-Tolerant Design under Chap.12 at www.mhhe.com/dieter.

12.3.6 Further Issues in Fatigue Life Prediction

The three preceding sections have illustrated the major approaches for dealing with fatigue failure in design. The presentation has been necessarily brief and basic. Using

---

1. R. P. Wei, *Trans. ASME,* Ser. H., *J. Eng. Materials Tech.,* vol. 100, pp. 113–20, 1978; A. F. Liu, *Structural Life Assessment Methods,* ASM International, Materials Park, OH, 1998.

these design strategies most likely will require further study of topics such as:

- Accounting for geometric stress concentrations and stress gradients
- Making allowance for the presence of compressive residual stresses that are either purposely introduced to improve fatigue performance or harmful tensile residual stresses that are introduced by manufacturing processes
- Further consideration of the case when the mean stress is not zero
- Calculation for situations of two- or three-dimensional states of fatigue stresses
- Accounting for nonregular stress cycles, and for random cycles of stress.

The many references given in this section will be helpful in gaining this knowledge, while computer design software for fatigue will remove some of the drudgery inherent in these kinds of calculations.

Much progress has been made in designing for fatigue, especially through the merger of fracture mechanics and fatigue. Nevertheless, the interaction of many variables, which is typical of real fatigue situations, makes it inadvisable to depend on a design based solely on analysis. Full-scale prototype testing, often called *simulated service testing,* should be part of all critical designs for fatigue. The points of failure not recognized in design will be detected by these tests. Simulating the actual service loads requires great skill and experience. Often it is necessary to accelerate the test, but doing so may produce misleading results. For example, when time is compressed in that way, the full influence of corrosion or fretting is not measured, or the overload stress may appreciably alter the residual stresses. Also, it is common practice to eliminate many small load cycles from the load spectrum, but they may have an important influence in fatigue crack propagation.

12.4
DESIGN FOR CORROSION RESISTANCE

Failure of metal components by corrosion is as common as failure due to mechanical causes, such as brittle fracture and fatigue. The National Institute of Standards and Technology estimates that corrosion annually costs the United States $70 billion, of which at least $10 billion could be prevented by better selection of materials and design procedures. Although corrosion failures are minimized by proper materials selection and careful attention to control of metallurgical structure through heat treatment and processing, many corrosion-related failures can be minimized by proper understanding of the interrelation of the fundamental causes of corrosion and design details.[1]

12.4.1 Basic Forms of Corrosion

Corrosion of metals is driven by the basic thermodynamic force of a metal to return to the oxide or sulfide form, but it is more related to the electrochemistry of the reactions of a metal in an electrolytic solution (electrolyte). There are eight basic forms of corrosion.[2]

1. V. P. Pludek, *Design and Corrosion Control,* John Wiley & Sons, New York, 1977.
2. P. R. Roberge, *Corrosion Engineering: Principles and Practice,* McGraw-Hill, 2008.

TABLE 12.2
A Brief Galvanic Series for Commercial
Metals and Alloys

Noble (cathodic)	Platinum
	Gold
	Titanium
	Silver
	316 stainless steel
	304 stainless steel
	410 stainless steel
	Nickel
	Monel
	Cupronickel
	Cu-Sn bronze
	Copper
	Cast iron
	Steel
Active (anodic)	Aluminum
	Zinc
	Magnesium

Uniform attack. The most common form of corrosion is uniform attack. It is characterized by a chemical or electrochemical reaction of the material with its environment that proceeds uniformly over the entire exposed surface area. The metal becomes thinner, leading to eventual failure.

Galvanic corrosion. Galvanic corrosion occurs when two dissimilar metals are immersed in a corrosive or conductive solution creating an electrical potential difference between them. The less-resistant (anodic) metal is corroded relative to the cathodic metal. Table 12.2 gives a brief galvanic series for some commercial alloys immersed in seawater. In this table, for any two metals or alloys in contact in seawater, the metal that is more anodic (lower in the series) will be corroded. Note that the relative position in a galvanic series depends on the electrolytic environment as well as the metal's surface chemistry (presence of passive surface films).

Use pairs of metals that are close together in the galvanic series to minimize galvanic corrosion and avoid situations in which a small anodic area of metal is connected to a larger surface area of more noble metal. If two metals far apart in the series must be used in contact, they should be electrically insulated from each other. Do not coat the anodic surface to protect it, because most coatings are susceptible to pinholes. The coated anode surface would corrode rapidly in contact with a large cathodic area. When a galvanic couple is unavoidable, consider utilizing a third metal that is anodic and sacrificial to both of the other metals.

Crevice corrosion. An intense localized corrosion frequently occurs within crevices and other shielded areas on metal surfaces exposed to corrosive attack. This type of corrosion usually is associated with small volumes of stagnant liquid trapped in design features such as holes, gasket surfaces, lap joints, and crevices under bolt and rivet heads.

Pitting. Pitting is a form of extremely localized corrosive attack that produces holes in the metal. It is an especially insidious form of corrosion because it causes equipment to fail after exposure to only a small percentage of the designed-for weight loss.

Intergranular corrosion. Localized attack along the grain boundaries with only slight attack at the grain faces is called intergranular corrosion. It is especially common in austenitic stainless steel that has been sensitized by heating to the range 950 to 1450°F. It can occur either during heat treatment for stress relief or during welding. When it occurs during welding it is known as *weld decay.*

Selective leaching. The removal of one element from a solid-solution alloy by corrosion processes is called selective leaching. The most common example of it is the selective removal of zinc from brass (dezincification), but aluminum, iron, cobalt, and chromium also can be removed. When selective leaching occurs, the alloy is left in a weakened, porous condition.

Erosion-corrosion. Deterioration at an accelerated rate is caused by relative movement between a corrosive fluid and a metal surface; it is called *erosion corrosion.* Generally the fluid velocity is high and mechanical wear and abrasion may be involved, especially when the fluid contains suspended solids. Erosion destroys protective surface films and exacerbates chemical attack. Design plays an important role in erosion control by altering the form and features to reduce fluid velocity, eliminating situations in which direct impingement occurs, and minimizing abrupt changes in the direction of flow. Some erosion situations are so aggressive that neither selection of a suitable material nor design can ameliorate the problem. Here the role of design is to provide for easy detection of damage and for quick replacement of damaged components.

A special kind of erosion-corrosion is *cavitation,* which arises from the formation and collapse of vapor bubbles near the metal surface. Rapid bubble collapse can produce shock waves that cause local deformation of the metal surface.

Another special form of erosion-corrosion is *fretting corrosion.* It occurs between two surfaces under load that are subjected to cycles of relative motion. Fretting produces breakdown of the surface into an oxide debris and results in surface pits and cracks that usually lead to fatigue cracks.

Stress-corrosion cracking. Cracking caused by the simultaneous action of a tensile stress and contact with a *specific* corrosive medium is called stress-corrosion cracking (SCC). The stress may be a result of applied loads or "locked-in" residual stress. Only specific combinations of alloys and chemical environments lead to stress-corrosion cracking. However, many occur commonly, such as aluminum alloys and seawater, copper alloys and ammonia, mild steel and caustic soda, and austenitic steel and salt water.[1] Over 80 combinations of alloys and corrosive environments are known to cause stress-corrosion cracking. Design against stress-corrosion cracking involves selecting an alloy that is not susceptible to cracking in the service environment; but if that is not possible, then the stress level should be kept low. The concepts of fracture mechanics have been applied to designing against SCC.

12.4.2 Corrosion Prevention

Material Selection. Selecting a material with a low rate of corrosion in the environment of concern is the obvious first step to preventing corrosion. In general, the more noble the metal the slower it will corrode (see Table 12.2). The brief discussion of the corrosion mechanisms in Sec. 12.4.1 provides some insight, but there is a vast literature on corrosion, much of it compiled in handbook form[2] and in databases.[3] Metals are the most susceptible to corrosion. Plastics in general have much better corrosion resistance. Some polymers absorb moisture, which causes swelling and degradation of mechanical properties.

While some aspects of material selection for corrosion are straightforward—for example, avoiding materials that are attacked in the corrosive environment of interest or are subject to SCC in the environment—other aspects of corrosion can be quite subtle. Microscopic galvanic cells can be created in metallic alloys due to such things as microsegregation of alloying elements (especially at grain boundaries), local cold-worked regions, or differences in galvanic potential between phases in multiphase alloys. The behavior of a material in a corrosive environment can be significantly changed by seemingly small changes in the corrosive environment. Such change factors are temperature, the amount of dissolved oxygen, or impurities in the liquid.

Cathodic Protection. Cathodic protection reduces galvanic corrosion by supplying electrons to the anodic metal that needs to be protected. This can be done by connecting the anodic metal to a sacrificial anode of an even more anodic potential, such as Mg or Zn. The sacrificial anode must be in close proximity to the protected metal. It will be gradually corroded away, so it must be replaced periodically. Alternatively, cathodic protection can be achieved by applying a DC voltage to the corrosion site that will oppose the one caused by the electrochemical reaction of galvanic corrosion.

---

1. G. H. Koch, *ASM Handbook,* Vol. 19, pp. 483–506, ASM International, Materials Park, OH, 1996.
2. *ASM Handbooks,* Vol. 13: Vol. 13A, *Corrosion Fundamentals, Testing, and Protection;* Vol. 13B, *Corrosion: Materials* (provides extensive data on corrosion performance of metals and nonmetals); Vol. 13C, *Corrosion: Environments and Industries* (focuses on corrosion in specific environments and in specific industries). The Cambridge Engineering Selector software is a good place to do a first screen for corrosion properties.
3. For an excellent compilation of sources, see *ASM Handbook,* Vol. 13A, pp. 999–1001.

Corrosion Inhibitors. Specific chemical compounds can be added to the corrosive solution to reduce the diffusion of ions to the metal-electrolyte interface. In many cases the inhibitor forms an impervious, insulating film covering either the anode or cathode. The chromate salts added as inhibitors to radiator antifreeze are good examples. Other inhibitors act as scavengers to reduce the amount of dissolved oxygen in the electrolyte.

Protective Coatings. A common way to minimize corrosion is to provide a protective coating to the metal to provide a barrier to the corrosive environment. Common examples are porcelain enamel, paint, and polymer coating. Electroplated metal coatings such as chromium are used both for corrosion protection and for decorative purposes. Grease, oil, and wax are used as temporary coatings during shipment or storage. The most common problem with coatings is incomplete coverage of the anode, usually due to "pinholes" in the coating, or subsequent damage to the coating in service. This results in an unfavorable ratio of anode to cathode area, which then causes more rapid corrosion at the penetrated sites in the coating.

A special form of protective coating is the passive layer that forms on some metals in certain environments. Al_2O_3 forms on aluminum alloys subjected to the weather, and this can be improved by subjecting the aluminum to a high corrosion potential to form a thick and tough anodized layer. The excellent corrosion resistance of stainless steel is believed to be due to the formation of very thin but protective oxide layers.

Corrosion Prevention by Design

Some of the more obvious design rules for preventing corrosion failure have been discussed previously. The essential strategy is to prevent a corrosive solution from coming in contact with a vulnerable surface. Tanks and containers should be designed for easy draining and easy cleaning. Welded rather than riveted tanks will provide less opportunity for crevice corrosion. When possible, design to exclude air; if oxygen is eliminated, corrosion can often be reduced or prevented. Exceptions to that rule are titanium and stainless steel, which are more resistant to acids that contain oxidizers than to those that do not.

- An important factor in the design of metal parts to resist corrosion is the ratio of the area of the anode to the cathode. To minimize corrosion of the anode, its surface area should be as large as possible in comparison with the surface area of the cathode. This will result in lower current density at the anode, and a lower corrosion rate. For example, in galvanized steel the zinc coating (anode−) on the steel (cathode) will give protection to the steel, even if the zinc is scratched to the bare steel, because the area of the anode remains much larger than the cathode. However, if the steel is protected by a copper cladding, according to Table 12.2 the copper would be cathodic to the steel, and the scratch would lead to a high current density in the steel, with accelerated corrosion.
- Great attention needs to be given to minimizing the ways that the electrolyte can come in contact with the metal. This can occur by direct immersion, by exposure to spray or mist, by alternate periods of wetting and drying, as when it rains, by

contact with moist earth, or by humidity in the atmosphere. It is important that design details minimize crevices, and that pipes drain out all liquids so there is no residual liquid to cause corrosion. Unavoidable crevices should be sealed with an elastomeric material. Surfaces that are smooth collect less fluid and corrode less than rough surfaces.

- Allow provisions in the design to clean equipment exposed to such things as mud, dirt, corrosive atmospheres, and salt spray. Combinations of corrosives with dirt or mud can cause galvanic action.
- The severity of corrosion increases exponentially with temperature. Steep temperature gradients and high fluid velocities can also increase corrosion severity.
- Designing with a *corrosion allowance* is another strategy. In many situations, as in chemical plant design, it is more economical to select a material with poorer corrosion resistance than a more expensive material with better corrosion resistance and to make the part with larger dimensions (corrosion allowance). The part is kept in service until a critical dimension, such as the wall thickness, has corroded to a predetermined limit, and then it is replaced. This design approach requires that the design provides for easy replacement, and that a rigorous inspection and maintenance program is in effect.

Other examples of design details for corrosion prevention are given by Pludek[1] and Elliott.[2]

12.5
DESIGN AGAINST WEAR

Wear is mechanically induced surface damage that results in progressive removal of material. Wear usually results in a loss of weight and change of dimensions over time. In severe situations wear can result in fracture, usually from a surface-originated fatigue crack. Wear typically occurs when two solid materials are in close contact with each other, either in a sliding or rotational motion. The wear of a material is closely associated with the friction of the sliding surfaces, and its degree of damage is strongly influenced by the presence of a lubricant. For example, the wear rate for an unlubricated bearing can be 10^5 times greater than that of a properly lubricated bearing. The scientific study of friction, wear, and lubrication is known as *tribology*, and the mechanics analysis of wear problems is called *contact mechanics*. Wear, corrosion, and fatigue are the largest contributors to the failure of machine components.

Wear is not a true material property. It is a characteristic of a tribological system consisting of the contacting materials, their geometrical parameters (shape, size, surface roughness), their relative motion and the magnitude of the applied load during motion, the type of lubrication, and the environment.

1. V.R. Pludek, op. cit.
2. P. Elliott, "Design Details to Minimize Corrosion," in *Metals Handbook,* Vol. 13A, *Corrosion,* pp. 910–28, ASM International, Metals Park, OH, 2003.

12.5.1 Types of Wear

As in corrosion, there are many types of wear. In designing for wear it is first important to identify the main type of wear that is operative in a particular design. Often a given wear mechanism gives way to a different mechanism as wear progresses, or several mechanisms act together. There are four predominant situations where wear occurs:

- *Adhesive wear* occurs when two solid bodies are in contact and move relative to each other. The motion can be either sliding, rolling, or by impact.
- *Abrasive wear* occurs when hard particles slide or roll across a surface under pressure.
- *Erosion* is the loss of material from a solid surface due to the interaction of that surface with a fluid. The fluid may be a multicomponent fluid like steam, or a stream of solid particles.[1]
- *Surface fatigue* is a form of damage in which particles of metal are detached from a surface under cyclic stresses, causing pitting or spalling. The most common occurrences of surface fatigue are in rolling-contact systems[2], as in gear teeth and bearings, and fretting fatigue where there is small-amplitude oscillating motion between the surfaces.[3]

In this section we limit the discussion to only adhesive and abrasive wear, the two most common types of wear.

Adhesive Wear

Adhesive wear is localized bonding between contacting solids leading to material transfer or loss. Figure 12.11 shows the two contacting surfaces at high magnification. The surfaces meet at asperities, leading to high contact stresses and local bonding. As the surfaces slide across one another a bonded junction in Fig. 12.11*a* is torn away (*b*) by the shearing and forms a *wear debris particle*, Fig. 12.11*c*. This particle is lost from the surface, or alternatively, it may be transferred to the opposing surface. Either way, the integrity of the surface has been damaged.

The degree of wear damage, as determined by visual or microscopic examination of the wear surface, is usually described as mild or severe. Mild adhesive wear would be characterized by microscopic scratches aligned with the direction of motion. Severe wear is usually called *scoring* or *scuffing,* where the surface shows definite roughening and patches of solid-phase welding. This is an indication of an unacceptable wear rate. Extreme adhesive wear is called *galling.* The material actually flows up from the surface, eventually stopping the relative motion by *seizure.* When severe wear is observed it calls for immediate action, such as changing the material or modifying design parameters such as load or shape to achieve mild wear behavior.

---

1. For more detail see *ASM Handbook,* Vol. 18, *Friction, Lubrication, and Wear Technology,* ASM, 2002, pp. 199–235.
2. *ASM Handbook,* Vol. 18, pp. 257–62.
3. *ASM Handbook,* Vol. 18, pp. 242–56.

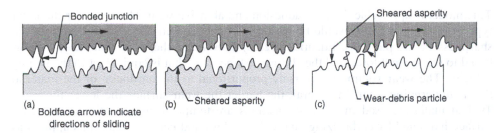

(a) Boldface arrows indicate directions of sliding

(b) Sheared asperity

(c) Wear-debris particle

FIGURE 12.11
Schematic description of the mechanism of adhesive wear.

Note that the transition from mild to severe wear behavior is often sharp as load, speed, or temperature is increased.

Abrasive Wear

Abrasive wear is caused by a hard, sharp particle imposed on and moving on a softer surface. Some abrasive wear is intentional. An example is the wear between an abrasive grinding wheel and a steel workpiece to produce a smooth, high-tolerance surface. This is an example of *two-body abrasion*. The most common form of abrasive wear is *three-body abrasion*, where hard and sharp particles, comprising the third body, are trapped between the contact surfaces.

Abrasive wear is usually divided into low-stress and high-stress abrasive wear. In low-stress wear, the particles plow wear scars like shallow furrows or scratches, but they do not fracture off chips. In high-stress abrasive wear the stress is sufficient to cause the abrasive particles to fracture or crush, producing many sharp edges that remove material by plowing the surface into deep scratches. Generally, abrasive wear is accompanied by adhesive wear, but the presence of abrasive particles results in higher wear rates (volume of material removed per unit time).

12.5.2 Wear Models

Many models have been proposed for wear processes. The most general relationship expresses wear by the volume of wear debris created, V.

$$V = k \frac{FS}{H} \tag{12.23}$$

where V is the wear volume, mm³
  k is a dimensionless proportionality constant, called the wear coefficient
  F is the compressive normal force, N
  S is the sliding distance, mm
  H is the hardness[1] of the softer member in contact, kg/mm²

---

1. Brinell, Vickers, and Knoop (microhardness) measurements are in kg/mm². To convert to MPa multiply by 9.81.

This model is applicable for both adhesion and abrasive wear. However, for the latter type of wear, Eq. (12.23) would be multiplied by a geometric term to account for the sharpness of the particles. Equation (12.23) shows that the wear volume is proportional to the normal force and the sliding distance, or what is equivalent to the sliding velocity. The wear volume is inversely proportional to the hardness[1] of the material that is undergoing wear. In general, the harder the wear surface the lower the wear. Typical materials used in wear applications are tempered-martensitic steels, steels surface hardened by carburizing, and cobalt alloys and ceramic materials applied as surface layers. Wear models are useful in designing against wear.[2] However, the level of detail involved precludes further examples here.

12.5.3 Wear Prevention

Design guidelines used to minimize wear can be divided into analysis methods, design details of the product, the use of lubricants, and appropriate materials selection.

Analysis Methods
- In identifying of the type of wear that can be expected, look carefully at the surface damage from wear failures from similar situations.[3] Examine the wear surface with a scanning electron microscope. Microscopic examination of the wear debris, and spectrographic analysis of lubricants, which may contain wear debris, can shed light on the nature of the wear processes that are taking place.
- Modeling can provide much insight in design for wear mitigation.

Design Details
- The overall aim in designing against wear should be to minimize contact stresses. One way to achieve this is to add details that help maintain good alignment between contact surfaces.
- Rolling contact is preferred to sliding contact. Avoid designs that lead to fretting motions.
- When satisfactory wear life cannot be achieved by other means, use a sacrificial design where one contact element is softer than the other and is intended to be replaced periodically.
- It is important to minimize the chance of buildup of abrasive particles in machines by giving proper attention to the design of oil filters, air cleaners, dust covers, and seals.

Lubrication
- The most general solution to excessive wear is lubrication. Lubrication provides a barrier between the contact surfaces that reduces both friction and wear. Lubricants are usually liquids, sometimes polymer solids, and rarely gases.

---

1. Hardness for metals is proportional to the yield strength, $H \approx 3\sigma_y$.
2. R.G. Bayer, "Design for Wear Resistance," *ASM Handbook,* Vol. 20, 1997, pp. 604–14; R.G. Bayer, *Engineering Design for Wear,* 2nd ed., Marcel Dekker, New York, 2004.
3. "Surface Damage," *ASM Handbook,* Vol. 18, pp. 176–183.

- There are several regimes of lubrication: the most common are boundary lubrication (each surface is covered by a chemically bonded fluid that may not give continuous coverage) and elastohydrodynamic lubrication, in which the friction and film thickness are determined by the elastic properties of the moving surfaces and the viscous properties of the lubricant.
- When a design depends on lubrication to control wear, a lubricant failure can be disastrous. Lubricant failure can occur because of chemical breakdown or contamination, change in properties due to excessive heat, or loss of lubricant.

Materials

- The higher the ratio of a material's surface hardness to the elastic modulus, the greater its resistance to adhesive wear. Avoid unlubricated sliding between similar materials, especially metals.
- Sliding between a hard metal surface and a softer metal surface will produce more wear on both members than if both surfaces were hard. A hard steel (BHN 650) coupled with a soft steel (BHN 250) will not protect the hard member from wear.
- The lowest metal-to-metal adhesive wear and resistance to galling is achieved with two hard surfaces (BHN > 650).
- Hard materials usually have low fracture toughness. An effective and economical approach is to provide a high hardness layer on the surface of a lower-hardness material. Depending on the base material, this is achieved by diffusion treatments, surface hardening (in steels), hard facing, and thermal spray coatings.[1]
- Diffusion treatments are usually applied to steels. The surface of low-carbon steel can be made hard and wear resistant by diffusing carbon atoms (*carburizing*) or nitrogen atoms (*nitriding*) into the surface of a part. These surface treatment processes have been widely adopted in the automotive industry.
- Diffusion treatments require hours at high temperature for the diffusion of atoms to produce a case depth of 0.010–0.020 in. The change in surface composition leads to a minor change in dimensions. Steels can also be surface hardened by *selective hardening* in which only the outer surface of the part is heated into the austenitic range for hardening, and then rapidly quenched to produce a hard martensitic layer on a soft, tough core. For large parts, heating is accomplished by heating with a gas torch (*flame hardening*), while heating with an induction coil or laser beam is used for smaller parts and greater precision in control of the depth of the surface layer.
- *Hardfacing* is the application of surface coatings using welding techniques. Surface layers of 1/8 in. are common. Typical materials applied by hardfacing are tool steels, iron chromium alloys for resistance against high-stress abrasive wear, and cobalt-based alloys for applications involving galling.

12

---

1. *ASM Handbook*, Vol. 5, *Surface Engineering*, 1994.

- *Thermal spraying* builds up a surface layer by melting the material into droplets and depositing them on the surface at high velocity. The droplets cool very rapidly and form an interlocking layer of *splats*. Typical processes are flame spraying and plasma arc spraying. These can deposit all wear-resistant metallic materials, and the higher-velocity spray processes can deposit ceramic materials such as chromium and aluminum oxide and tungsten carbide. Thermal spray processes can also be used to build up and repair worn parts.

12.6
DESIGN WITH PLASTICS

Most mechanical design is taught with the implicit assumption that the part will be made from a metal. However, plastics are increasingly finding their way into design applications because of their light weight, attractive appearance, relative freedom from corrosion, and the ease with which many parts may be manufactured from polymers.

Polymers are sufficiently different from metals to require special attention in design.[1] With respect to mechanical properties, steel has about 100 times the Young's modulus of a polymer and about 10 times the yield strength. Also, the strength properties of polymers are time-dependent at or near room temperature, which imposes a different way of looking at allowable strength limits. Polymers are 1/7th as dense as metals, but their thermal conductivity is 1/200th that of steel and their thermal expansion is seven times greater. These last two properties influence their processing (see Sec. 13.18). Therefore, these differences in properties must be allowed for in design with plastics.

12.6.1 Classification of Plastics and Their Properties

The majority of plastics are synthetic materials characterized by having a carbon-carbon backbone modified by other organic side groups. Plastics are made up of tens of thousands of small molecular units (mers) that are polymerized into long-chain macromolecules, hence the scientific term *polymers*. Depending on composition and processing, polymer chains can take many configurations (coiled, cross-linked, crystalline) to change the polymer properties.[2]

Polymers are divided into two general classes: *thermoplastics* (TP) and *thermosets* (TS). The difference between these two classes of plastics lies in the nature of

---

1. "Engineering Plastics," *Engineered Materials Handbook,* Vol. 2, ASM International, Materials Park, OH, 1988; M. L. Berins, ed., *Plastics Engineering Handbook of the Society of Plastics Industry,* 5th ed., Van Nostrand Reinhold, New York, 1991; E. A. Muccio, *Plastic Part Technology,* ASM International, Materials Park, OH, 1991; Dominic V. Rosato, Donald V. Rosato, and Marles G. Rosato, *Plastics Design Handbook*, Kluwer Academic Publishers, Boston, 2001; G. Erhard, *Designing with Plastics,* Hanser Gardner Publications, Cincinnati, 2006.

2. A. B. Strong, *Plastics: Materials and Processing,* Prentice Hall, Englewood Cliffs, NJ, 1996; G. Gruenwald, *Plastics: How Structure Determines Properties,* Hanser Publishers, New York, 1993; N. Mills, *Plastics: Microstructure and Engineering Applications,* 3rd ed., Butterworth-Heinemann, Woburn, MA, 2005.

bonding and their response to increases in temperature. When a TP is heated to a sufficiently high temperature it will become viscous and pliable. In this condition it can be formed into useful shapes and will retain these shapes when cooled to room temperature. If reheated to the same temperature, it again becomes viscous and then can be reshaped and retains the shape when cooled. When a TS polymer is heated, or a catalyst is added, covalent bonding occurs between the polymer chains, resulting in a rigid, cross-linked structure. This structure is "set in place," so that if the TS is reheated from a cooled state it will not return to the fluid viscous state but instead degrades and chars on continued heating.

Few polymers are used in their pure form. *Copolymers* are made by polymerizing two or more different mers so that the repeating unit in the polymer chain is not just a single mer. Think of these as polymer alloys. Styrene-acrylonitrile copolymer (SAN) and acrylonitrile-butadiene-styrene copolymer (ABS) are common examples. *Blends* are combinations of polymers that are mechanically mixed. Blends do not depend on chemical bonding, like copolymers; instead they need chemical compatibilizer additions to keep the components of the blend from separating. Copolymers and blends are developed to improve upon one or more property of a polymer, like impact resistance, without degrading its other properties.

About three-quarters of the plastics sold are thermoplastics. Therefore, we will concentrate on this category of plastics. For commercial purposes TP plastics can be divided into *commodity plastics* and *engineering plastics.* Commodity plastics are generally used in low-load-bearing applications like packaging, house siding, water pipes, toys, and furniture. Polyethylene (PE), polystyrene (PS), polyvinyl chloride (PVC), and polypropylene (PP) are good examples. These plastics generally compete with glass, paper, and wood. Engineering TPs compete with metals, since they can be designed to carry significant loads for a long period of time. Examples are polyoxymethylene or acetal (POM), polyamides or nylon (PA), polyamide-imide (PAl), polycarbonates (PC), polyethylene terephthalate (PET), and polyetheretherketones (PEEK).

Figure 12.12 compares the tensile stress-strain curve for polycarbonate with a soft mild steel and a high-carbon steel. We note that the level of strength is much lower and that yielding and fracture occur at much larger strains. The level of the stress-strain curve is strongly dependent on the strain rate (rate of loading). Increasing the strain rate raises the curve and decreases the ductility. Table 12.3 lists some short-time mechanical properties of metals and plastics at room temperature. Because many polymers do not have a truly linear initial portion of the stress-strain curve, it may be difficult to specify a yield strength, so the tensile strength usually is reported. Also, because of the initial curvature of the stress-strain curve, the modulus of elasticity is usually determined by the *secant modulus.*[1] Some polymers are brittle at room temperature because their glass-transition temperature is above room temperature. Brittleness is measured by the impact test. Note the marked improvement when glass fibers are introduced. The data in Table 12.3 are aimed at illustrating the difference in properties between metals and polymers. They should not be used for

12

---

1. The secant modulus is the slope of a line drawn from the origin of the stress-strain curve to a point on the curve determined by a designated strain, e.g., 2 percent.

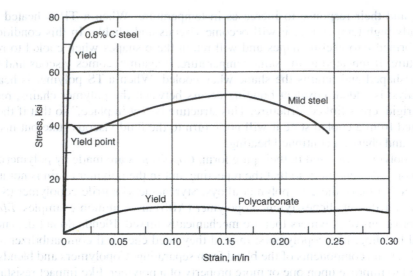

FIGURE 12.12
Comparison of engineering stress-strain curves for a thermoplastic polymer (polycarbonate) with 0.2% carbon mild steel and 0.8% carbon steel.

TABLE 12.3
Comparative Properties of Some Polymers and Metals

Material	Young's Modulus, psi × 10⁻⁶	Tensile Strength, ksi	Impact ft-lb/in.	Specific Gravity, g/cm³
Aluminum alloys	10	20–60		2.7
Steel	30	40–200		7.9
Polyethylene	0.08–0.15	3–6	1–12	0.94
Polystyrene	0.35–0.60	5–9	0.2–0.5	1.1
Polycarbonate	0.31–0.35	8–10	12–16	1.2
Polyacetal	0.40–0.45	9–10	1.2–1.8	1.4
Polyester-glass reinforced	1.5–2.5	20–30	10–20	1.7

design purposes. Moreover, it needs to be recognized that the mechanical properties of plastics are more subject than metals to variations due to blending and processing.

Tests other than the tensile and impact tests are frequently made on plastics and reported in the producer's literature as an aid in selecting polymers. One is a *flexure test*, in which the plastic is bent as a beam until it fractures.

A polymer's heat resistance is measured by the heat-deflection temperature (HDT). A plastic bar is bent under a low constant stress of 264 psi as the temperature of a surrounding oil bath is raised slowly from room temperature. The HDT is the temperature at which the sample shows a deflection of 0.010 in. While useful in ranking materials, such a test is of no value in predicting the structural performance of a plastic at a given temperature and stress.

Plastics do not corrode like metals because they are insulators, but they are susceptible to various types of environmental degradation. Some plastics are attacked by organic solvents and gasoline. Some are highly susceptible to the absorption of water vapor, which degrades mechanical properties and causes swelling. Many plastics are affected by sunlight, which causes cracking, fading of the color, or loss of transparency.

Many additives are compounded with the polymer to improve its properties. Fillers like wood flour, silica flour, or fine clay are sometimes added as *extenders* to decrease the volume of plastic resin used, and therefore the cost, without severely degrading properties. Chopped glass fiber is commonly added to increase the stiffness and strength of plastics. *Plasticizers* are used to enhance flexibility and melt flow during processing. *Flame retardants* are added to reduce the flammability of plastics, but some of these are proving to be toxic. *Colorants* such as pigments and dyes are used to impart color. Ultraviolet light absorbers are used to stabilize the color and lengthen the life of the product when exposed to sunlight. Antistatic agents are used to reduce the buildup of electrostatic charges on an insulating plastic surface.

A category of plastics that is growing in importance as a material for engineered structures is plastic matrix *composite materials*. In these materials high-modulus, strong, but often brittle fibers are embedded in a matrix of thermosetting plastic.[1] The fibers most often used are graphite or glass. Because of the high cost of materials and the complexity of fabrication, composite structures are used where a high premium is placed on lightweight, strong structures. Composite structures represent the ultimate in materials design in that the structure is laid up layer upon layer. Because the strength and stiffness reside in the fibers, and these are highly directional, great attention must be paid to directional properties in the design.

12.6.2 Design for Stiffness

Young's modulus is much lower for plastics than metals, so resistance to deflection (stiffness) is often a concern when using plastics. The stiffness of a structure is dependent upon the elastic modulus of the material and the part geometry. For example, the maximum deflection of a cantilever beam of length L, loaded with a concentrated load P at its end, is

$$\delta_{max} = \frac{PL^3}{3EI} \text{ where } I = \frac{bh^3}{12} \text{ and } b \text{ is the beam width and } h \text{ is its thickness} \quad (12.24)$$

Equation (12.24) shows that increasing the stiffness (i.e., reducing the deflection) can be achieved by either increasing the elastic modulus E or increasing the moment of inertia I, or both. Table 12.3 shows that the addition of short glass fibers to a nylon

---

1. *ASM Handbook,* Vol. 21, *Composites,* ASM International, Materials Park, OH, 2001; D. Hull, *An Introduction to Composite Materials,* Cambridge University Press, Cambridge, 1981; R. J. Diefendorf, "Design with Composites," *ASM Handbook,* Vol. 20, pp. 648–65, 1997.

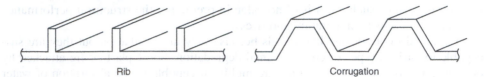

FIGURE 12.13
Examples of structural stiffening elements.

molding polymer can increase the elastic modulus. The addition of long glass fibers to an epoxy polymer matrix to make a composite material can increase the modulus much more, but at the complication of introducing extreme directionality (anisotropy) in properties.

The second way to increase the stiffness of the cantilever beam is to increase the value of the moment of inertia, I. An obvious way to do this is to increase the value of the thickness h that appears in Eq. (12.24) raised to the third power. If a polymer has $E = 300,000$ psi and a metal $E = 10,000,000$ psi, the polymer beam will have to be more than three times thicker to have the same flexural rigidity as the aluminum beam. This introduces problems in addition to the extra cost of the additional material that must be used. Polymers have low thermal conductivity, so there are issues in cooling thick sections fast enough in production forming machines like injection molding. In fact, the limit on wall thickness is around 4 to 5 mm.

However, significant increases in I can be achieved through modification of the shape of the cross section to move material far away from the neutral axis. This is expressed by the *shape factor*. The structural elements usually employed for this purpose are ribs or corrugation, Fig. 12.13. While corrugated surfaces provide greater stiffness, for equal weight and thickness, by a factor of about 1.8, they usually are avoided for aesthetic reasons and for the fact that they result in higher mold costs than do ribs.[1]

In their simplest form, reinforcing ribs are beam-like structures.[2] Generally they are placed inside of the part attached to the load-bearing surface and are oriented to run in the direction of maximum stress and deflection. As shown in Fig. 12.14, they can run the full length of the part or stop part way. When they do not intersect a vertical wall they should be tapered down gradually to aid in polymer melt flow and to avoid stress concentration. A short rib used to support a wall is often called a *gusset plate*. The ribs should be tapered slightly (draft) to aid in ejection from the mold, and they should be designed with a generous radius where they attach to the plate to avoid stress concentration. See www.dsm.com/en_US/html/dep/ribsandprofiledstructures.htm for information on design of ribs and corrugations. For greater detail see http://plastics.dupont.com/plastics/pdflit/americas/general/H76838.pdf.

_____

1. G. Erhard, *Designing with Plastics*, Chap. 10, Hanser Gardner Publications, Cincinnati, OH, 2006.
2. For a methodology with which to determine the value of *I,* see R. A. Malloy, *Plastic Part Design for Injection Molding,* Hanser Gardner Publications, Cincinnati, OH, 1994, pp. 213–30. For more discussion of design issues in plastic parts (ribs, bosses, shrinkage, and warpage) see Design Guide at DSM Engineering Plastics-Design.

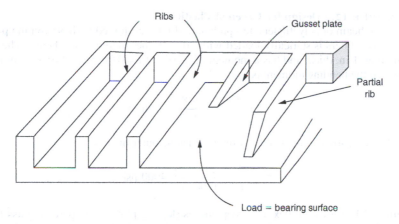

FIGURE 12.14
Typical rib designs used to increase the stiffness of an injection molded plate.

12.6.3 Time-Dependent Part Performance

The mechanical properties of plastics are *viscoelastic.*[1] This means that they vary with time under load, the rate of loading, and the temperature. This material behavior shows itself most prominently in the phenomena of creep and stress relaxation. *Creep* is the permanent deformation of a material over time, under constant load, and at constant temperature. *Stress relaxation* measures the decreasing stress required to cause a constant strain at a constant temperature. For example, the loosening of a snap fit with time, or the loosening of a self-tapping screw, is a result of stress relaxation.

The calculation of stress and strain from viscoelastic equations that describe the creep of plastics is an advanced topic. Nevertheless, design engineers have developed useful and simpler methods for designing with plastics. The procedure is as follows:

1. Determine the maximum service temperature and the time for which a constant load will be applied.
2. Calculate the maximum stress in the design using normal mechanics of material equations.
3. From a creep curve of strain versus time, find the strain from the appropriate temperature and stress.
4. Divide these values of strain into stress to determine an *apparent creep modulus.*
5. This creep modulus can be used in mechanics of materials equations to determine deformation or deflection.

---

1. J.G. Williams, *Stress Analysis of Polymers,* 2nd ed., John Wiley-Halsted Press, New York, 1980; A.S. Wineman and K.R. Rajagopal, *Mechanical Response of Polymers,* Cambridge University Press, New York, 2000.

EXAMPLE 12.2 Design for Creep of Plastics

A simple beam of polyethylene terephthalate (PET) reinforced with 30 volume percent of short glass fibers is statically loaded with 10 lb in the middle of the beam. The beam is 8 in. long, 1 in. thick, and has a moment of inertia of $I = 0.0025$ in.[4] For a beam loaded at midspan the bending moment is

$$M = \frac{PL}{4} = \frac{10 \times 8}{4} = 20 \text{ in-lb}$$

The bending stress at the center of the beam at its outer fiber is

$$\sigma = \frac{Mc}{I} = \frac{20 \times 0.5}{0.0025} = 4000 \text{ psi}$$

Figure 12.15 shows the flexural creep curves (log-log plot) for 30 percent glass-filled GF PET. Since we are interested in long-term behavior, we look for the creep strain at 1000 h with the environmental conditions of 4000 psi (27.6 MPa) and room temperature 73°F (23°C). This gives a strain of 0.7 percent or 0.007 in./in. Then the *apparent modulus* is

$$E_a = \frac{\sigma}{\varepsilon} = \frac{4000}{0.007} = 5.71 \times 10^5 \text{ psi}$$

For a simply supported beam the maximum deflection under the load at midspan is given by

$$\delta = \frac{PL^3}{48 E_a I} = \frac{10 \times 8^3}{48(5.71 \times 10^5)(0.0025)} = 0.075 \text{ in.}$$

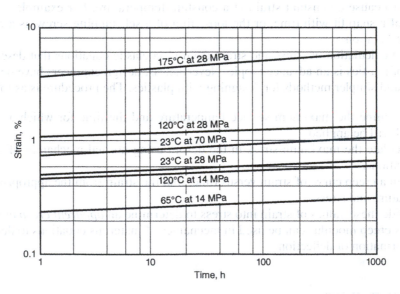

FIGURE 12.15
Creep curve in flexure for 30 percent GF PET. (*Engineered Materials Handbook,* vol. 2, p. 173, ASM International, Materials Park, OH, 1988. Used with permission.)

If the operating temperature is raised to 250°F (120°C) the creep strain at 1000 h is 1.2 percent or 0.012 in./in. E_a changes to 3.33×10^5 and the deflection under the load is 0.13 in. This shows the strong sensitivity of creep in plastics to temperature.

The static elastic modulus obtained from a tension test for this material at room temperature is 1.3×10^6 psi. If this value is used in the deflection equation a value of 0.033 is obtained. Thus, ignoring the viscoelastic nature of polymers results in a severe underestimation of deflection.

12.7
SUMMARY

This chapter provides an introduction to types of failure of materials not usually covered in any depth in courses in fundamentals of materials and mechanics of materials. The approach is at a rather elementary level and does not consider failure in situations of combined stress. Ample references are given to sources where students can extend their knowledge. The excellent text by Norman Dowling should be particularly helpful, as will be the various volumes of the ASM Handbook that are referenced.

NEW TERMS AND CONCEPTS

Abrasive wear	Endurance limit	Notch sensitivity factor
Adhesive wear	Failsafe design	Plane-strain fracture toughness
Apparent modulus	Fatigue failure	Stress corrosion fracture
Creep failure	Fracture mechanics	Viscoelastic behavior
Damage tolerant design	Low-cycle fatigue	Wear

BIBLIOGRAPHY

ASM Handbook, Vol. 19, *Fatigue and Fracture,* ASM International, Materials Park, OH, 1996.

ASM Handbook, Vol. 20, *Materials Selection and Design,* ASM International, Materials Park, OH, 1997.

Budynas, R. G., and J. K. Nisbett: *Shigley's Mechanical Engineering Design*, 9th ed., McGrawHill, New York, 2011.

Derby, B., D. A. Hills, and C. Ruiz: *Materials for Engineering: A Fundamental Design Approach*, John Wiley & Sons, New York, 1992.

Dieter, G. E.: *Mechanical Metallurgy,* 3rd ed., McGraw-Hill, New York, 1986.

Dowling, N. E.: *Mechanical Behavior of Materials,* 3rd ed., Prentice Hall, Englewood Cliffs, NJ, 2006.

Jones, D. R. H.: *Engineering Materials 3: Materials Failure Analysis*, Pergamon Press, Oxford, 1993.

PROBLEMS AND EXERCISES

12.1 Compare steels A and B for the construction of a pressure vessel measuring 30 in. inside diameter and 12 ft long. The pressure vessel must withstand an internal pressure of 5000 psi. Use a safety factor of 2 on the yield strength. For each steel, determine (a) critical flaw size and (b) flaw size for a leak-before-break condition. This is the situation where the crack penetrates the wall thickness before brittle fracture occurs.

Steel	Yield Strength, ksi	K_{Ic} ksi$\sqrt{\text{in.}}$
A	260	80
B	110	170

12.2 A high-strength steel has a yield strength of 100 ksi and a fracture toughness 150 ksi(in)$^{1/2}$. By the use of a certain nondestructive evaluation technique, the smallest-size flaw that can be detected routinely is 0.3 in. Assume that the most dangerous crack geometry in the structure is a single-edge notch so that $K = 1.12\sigma\sqrt{\pi a}$. The structure is subjected to cyclic fatigue loading in which $\sigma_{max} = 45$ ksi and $\sigma_{min} = 25$ ksi. The crack growth rate for the steel is given by $da/dN = 0.66 \times 10^{-8}(\Delta K)^{2.25}$. Estimate the number of cycles of fatigue stress that the structure can withstand.

12.3 A 3-inch-diameter steel shaft is rotating under a steady bending load. The shaft is reduced in diameter at each end to fit in the bearings, which gives a theoretical stress concentration factor at this radius of 1.7. The fatigue limit of the steel in a rotating beam test is 48,000 psi. Its fatigue notch sensitivity factor is 0.8. Since the fatigue limit is close to a 50 percent survivability value, we want to design this beam to a 99 percent reliability. What point in the beam will be the likely point for a failure, and what is the maximum allowable stress at that point?

12.4 A 1.50-inch-diameter steel bar is subjected to a cyclical axial load of $P_{max} = 75,000$ lb in tension and 25,000 lb in compression. The fatigue limit (endurance limit) for a stress cycle with $R = -1.0$ is 75 ksi. The ultimate tensile strength is 158 ksi and the yield strength is 147 ksi.

(a) Equation (12.13) expresses the relationship between alternating stress and the mean stress in fatigue known as the Goodman line. Plot Eq. (12.13) on coordinates of σ_a(y-axis) versus σ_m(x-axis) and label the critical points on this graph.

(b) To what value would the endurance limit be reduced by the stress cycle given here?

12.5 A high-carbon steel is used to make a leaf spring for a truck. The service loads the truck will experience can be approximated by an alternating stress from a maximum load to a zero load in tension ($R = 0$). In the heat-treated condition the leaf spring has a fatigue limit of 380 MPa based on laboratory tests made under completely reversed fatigue loading. The surface finish of the leaf spring is expected to reduce this value by 20 percent. The ultimate tensile strength of the steel is 1450 MPa. Before assembly the leaf spring is shotpeened to introduce a surface compressive residual stress that is estimated to be 450 MPa. Find the maximum stress amplitude that the surface of the spring could be expected to withstand for an infinite number of fatigue cycles.

12.6   A plastic beam (6 in. diameter) used to haul small loads into a loft extends 8 ft in the horizontal plane from the building. If the beam is made from glass-reinforced PET, how much would the beam deflect on a 23°C day when hauling up a 900 lb load?

12.7   Stress-corrosion failures that occur in the 304 and 316 stainless-steel recirculation piping have been a major problem with boiling-water nuclear reactors (BWR). What are the three conditions necessary for stress-corrosion cracking? Suggest remedies for the cracking.

12.8   What are the chief advantages and disadvantages of plastic gears? Discuss how material structure and processing are utilized to improve performance.

12

13

DESIGN FOR MANUFACTURING

13.1
ROLE OF MANUFACTURING IN DESIGN

Producing the design is a critical link in the chain of events that starts with a creative idea and ends with a successful product in the marketplace. With modern technology the function of production no longer is a mundane activity. Rather, design, materials selection, and processing are inseparable, as shown in Fig. 11.1.

There is confusion of terminology concerning the engineering function called *manufacturing*. Materials engineers use the term *materials processing* to refer to the conversion of semifinished products, like steel blooms or billets, into finished products, like cold-rolled sheet or hot-rolled bar. A mechanical, industrial, or manufacturing engineer is more likely to refer to the conversion of the sheet into an automotive body panel as *manufacturing*. Processing is the more generic term, but manufacturing is the more common term. Production engineering is a term used in Europe to describe what we call manufacturing in the United States. We will use the term *manufacturing* in this text to refer to converting a design into a finished product.

The first half of the 20th century saw the maturation of manufacturing operations in the western world. Increases in the scale and speed of operations brought about increases in productivity, and manufacturing costs dropped while wages and the standard of living rose. There was a great proliferation of available materials as basic substances were tailor-made to have selectively improved properties. One of the major achievements of this era was the development of the production line for mass-producing automobiles, appliances, and other consumer goods. Because of the preeminence in manufacturing that arose in the United States, there has been a recent tendency to take the manufacturing function for granted. Manufacturing has been downplayed in the education of engineers. Manufacturing positions in industry have been considered routine and not challenging, and as a result they have not attracted their share of the most talented engineering graduates. Fortunately, this situation is improving as the rapid rise of manufacturing in Asia has threatened jobs for a large segment of the workforce in the western world. The nature and perception of manufacturing is being changed by increasing automation and computer-aided manufacturing.

A serious problem facing manufacturing companies has been the tendency to separate the design and manufacturing functions into different organizational units. Barriers between design and manufacturing decision making can inhibit the close interaction that the two engineering functions should have, as discussed previously under concurrent engineering (Sec. 2.4.4). When technology is sophisticated and fast-changing, a close partnership between the people in research, design, and manufacturing is very necessary.

That has been demonstrated best in the area of solid-state electronic devices. As semiconductor devices replaced vacuum tubes, it became apparent that design and processing could no longer be independent and separable functions. Design using vacuum-tube technology was essentially a linear process in which specialists in materials passed on their input to specialists in components who passed on their input to circuit designers who, in turn, communicated with system designers. With the advent of transistors, the materials, device construction, and circuit design functions became closely coupled. Then, with the microelectronics revolution of large-scale integrated circuits, the entire operation from materials to system design became interwoven, and manufacturing became inseparable from design. The result was a situation of rapid technical advance requiring engineers of great creativity, flexibility, and breadth. The payoff in making the personal computer and workstations a reality has been huge. Never has productivity been enhanced as rapidly as during the microelectronics revolution. This should serve as a model of the great payoff that can be achieved by closer integration of research, design, and manufacturing.

The need to break down barriers between design and manufacturing is widely recognized today and is accomplished by the use of concurrent engineering and the involvement of manufacturing engineers in product design and development teams. Also, focus on improving the link between manufacturing and design has increased emphasis on codifying a set of practices that designers should follow to make their designs easier to manufacture. This topic, *design for manufacture* (DFM), is an emphasis of this chapter.

13.2
MANUFACTURING FUNCTIONS

More conventional manufacturing is divided into the following functions: (1) process engineering, (2) tool engineering, (3) work standards, (4) plant engineering, and (5) administration and control. *Process engineering* is the development of a step-by-step sequence of operations for production. The overall product is subdivided into its components and subassemblies, and the steps required to produce each component are arranged in logical sequence. An important part of process engineering is to specify the needed tooling. Vital parameters in process engineering are the rate of production and the cost of manufacturing a component. *Tool engineering* is concerned with the design of tools, jigs, fixtures, and gages to produce the part. *Jigs* both hold the part

13

and guide the tool during manufacture, while *fixtures* hold a part to be joined, assembled, or machined. *Tools* do the machining or forming; *gages* determine whether the dimensions of the part are within specification. *Work standards* are time values associated with each manufacturing operation that are used to determine standard costs to make the part. Other standards that need to be developed in manufacturing are tool standards and materials standards. *Plant engineering* is concerned with providing the plant facilities (space, utilities, transportation, storage, etc.) needed to carry out the manufacturing process. *Administration and control* deals with production planning, scheduling, and supervising to assure that materials, machines, tools, and people are available at the right time and in the quantities needed to produce the part.

Computer-automated machine tool systems, which include industrial robots and computer software for scheduling and inventory control, have demonstrated the ability to increase machine utilization time from an average of 5 percent to as much as 90 percent. The introduction of computer-controlled machining centers that can perform many operations in a single machine greatly increases the productivity of the

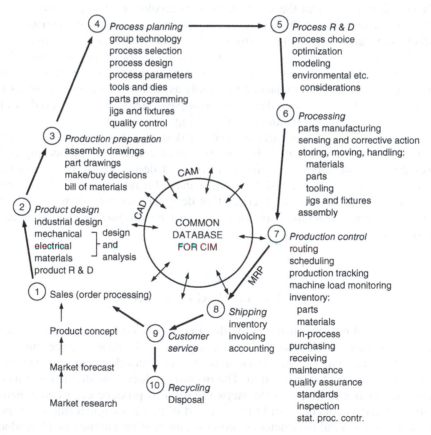

FIGURE 13.1
Spectrum of activities that are encompassed by manufacturing.

machine tool. The computer-automated factory carries this one step further. All steps in parts manufacture are optimized with software systems and at least half of the machine tools will have the capability for multiple machining operations with automatic parts handling between workstations. This automated factory differs from the stereotypical automobile assembly line in that it is a flexible manufacturing system capable of producing a wide variety of parts under computer control. This broad-based effort throughout industry to link computers into all aspects of manufacturing is called *computer-integrated manufacturing* (CIM).

Figure 13.1 shows the broad spectrum of activities that are encompassed by manufacturing. It begins in step 4, when design engineering turns the complete information for the design over to the process planners. As mentioned earlier, many tasks of process planning are done concurrently with the detail design phase. Process selection and design of tooling are major functions in this step. Step 5 involves fine-tuning a process, often by computer modeling or optimization processes, to improve throughput or improve yield (reduce defects) or decrease cost. Actual part manufacturing, step 6, involves production team training and motivation. In many instances a considerable amount of materials handling is required. The many issues involved with step 7 are vital for an effective manufacturing operation. Finally, in step 8, the product is shipped and sold to the customer. Customer service, step 9, handles warranty and repair issues, and eventually the product is retired from service, hopefully by recycling. The information gathered from customer service operations is fed back into the design of new products, step 2; the cycle is completed.

13.3
CLASSIFICATION OF MANUFACTURING PROCESSES

It is not an easy task to classify the tremendous variety of manufacturing processes. We start with the hierarchical classification of business and industry shown in Fig. 13.2. The service industries consist of enterprises, such as education, banking,

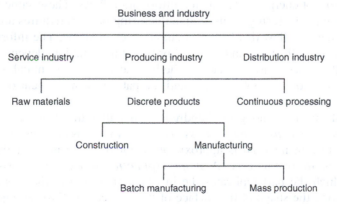

FIGURE 13.2
A simple hierarchical classification of business and industry.

insurance, communication, and health care that provide important services to modern society but do not create wealth by converting raw materials. The producing industries acquire raw materials (minerals, natural products, or fossil fuels) and process them, through the use of energy, machinery, and knowledge, into products that serve the needs of society. The distribution industries, such as merchandising and transportation, make those products available to the general public.

A characteristic of modern industrialized society is that an increasingly smaller percentage of the population produces the wealth that makes our affluent society possible. Just as the 20th century saw the United States change from a predominantly agrarian society to a nation in which only 3 percent of the population works in agriculture, so we have become a nation in which an ever-decreasing percentage of the workforce is engaged in manufacturing. In 1947 about 30 percent of the workforce was in manufacturing; in 1980 it was about 22 percent. By the year 2004 about 15 percent of U.S. workers were engaged in manufacturing and today the number is under 10 percent.

The producing industries can be divided conveniently into raw materials producers (mining, petroleum, agriculture), producers of discrete products (autos, consumer electronics, etc.), and industries engaged in continuous processing (gasoline, paper, steel, chemicals, etc.). Two major divisions of discrete products are construction (buildings, roads, bridges, etc.) and manufacturing. Under manufacturing we recognize batch (low-volume) manufacturing and mass production as categories.

13.3.1 Types of Manufacturing Processes

A manufacturing process converts a material into a finished part or product. The changes that occur with respect to part geometry can also affect the internal microstructure and therefore the properties of the material. For example, a sheet of brass that is being drawn into the cylindrical shape of a cartridge case is also being hardened and reduced in ductility by the process of dislocation glide on slip planes.

Recall from Chap. 6 that the functional decomposition of a design was described initially in terms of energy, material, and information flows. These same three factors are present in manufacturing. Thus, a manufacturing process requires an energy flow to cause the material flow that brings about changes in shape. The information flow, which consists of both shape and material property information, depends on the type of material, the process used—that is, whether mechanical, chemical, or thermal—the characteristics of the tooling used, and the pattern of movement of the material relative to the tooling.

A natural division among the hundreds of manufacturing processes is whether the process is *mass conserving* or *mass reducing.* In a mass conserving process the mass of the starting material is approximately equal to the mass of the final part. Most processes are of this type. A *shape replication* process is a mass conserving process in which the part replicates the information stored in the tooling by being forced to assume the shape of the surface of the tool cavity. Casting, injection molding, and closed-die forging are examples. In a *mass reducing* process, the mass of the starting material is greater than the mass of the final part. Such processes are *shape-generation* processes because the part shape is produced by the relative motion between the tool and the workpiece. Material removal is caused by controlled fracture, melting, or chemical reaction. A machining process, such as milling or drilling is an example of controlled fracture.

A different way of dividing manufacturing processes is to classify them into three broad families: (1) primary processes, (2) secondary processes, and (3) finishing processes.

- **Primary processes** take raw materials and create a shape. The chief categories are casting processes, polymer processing or molding processes, deformation processes, and powder processes.
- **Secondary processes** modify shape by adding features such as keyways, screw threads, and grooves. Machining processes are the main type of secondary processes. Other important categories are joining processes that fasten parts together, and heat treatment to change mechanical properties.
- **Finishing processes** produce the final appearance and feel of a product by processes such as coating, painting, or polishing.

The taxonomy structure used to classify materials in Sec. 11.2.1 can be applied to manufacturing processes. For example, the **Family** of Shaping Processes can be divided into the **Classes** of Casting, Polymer Molding, Deformation, and Powder processes. The class Deformation Processes can, in turn, be broken into many **Member** processes such as rolling, drawing, cold forming, swaging, sheet metal forming, and spinning. Then, for each process we would need to determine **Attributes** or *process characteristics* (PC) such as its applicability to certain ranges of part size, the minimum thickness that can be consistently produced by the process, the typical tolerance on dimensions and surface roughness produced by the process, and its economical batch size.

13.3.2 Brief Description of the Classes of Manufacturing Processes

This section provides further understanding of the major classes of manufacturing processes

1. Casting (solidification) processes: Molten liquid is poured into a mold and solidified into a shape defined by the contours of the mold. The liquid fills the mold by flowing under its own weight or with a modest pressure. Cast shapes are designed so the liquid flows to all parts of the mold cavity, and solidification occurs progressively so there are no trapped liquid pockets in a solidified shell. This requires a low-viscosity liquid, so casting is usually done with metals and their alloys. The various casting processes, and their costs, differ chiefly according to the expense and care used to prepare the mold. Great progress has been made using computer models to predict and control the flow and solidification of the liquid material, thereby minimizing casting defects.

2. Polymer processing (molding): The wide use of polymers has brought about the development of processes tailored to their high viscosity. In most of these processes a hot viscous polymer is either compressed or injected into a mold. The distinction between casting and molding is the viscosity of the material being worked. Molding can take such extreme forms as compression molding plastic pellets in a hot mold, or blowing a plastic tube into the shape of a milk bottle against a mold wall.

3. Deformation processes: A material, usually metal, is plastically deformed (hot or cold) to give it improved properties and change its shape. Deformation processes are also called metal-forming processes. Typical processes of this type are forging, rolling, extrusion, and wire drawing. Sheet-metal forming is a special category in which the deformation occurs in a two-dimensional stress state instead of three dimensions.

4. Powder processing: This rapidly developing manufacturing area involves the consolidation of particles of metal, ceramics, or polymers by pressing and sintering, hot compaction, or plastic deformation. It also includes the processing of composite materials. Powder metallurgy is used to make small parts with precision dimensions that require no machining or finishing. Powder processing is the best route for materials that cannot be cast or deformed, such as very high melting point metals and ceramics.

5. Material removal or cutting (machining) processes: Material is removed from a workpiece with a hard, sharp tool by a variety of methods such as turning, milling, grinding, and shaving. Material removal occurs by controlled fracture, producing chips. Machining is one of the oldest manufacturing processes, dating back to the invention of the power lathe early in the Industrial Revolution. Essentially any shape can be produced by a series of machining operations. Because a machining operation starts with a manufactured shape, such as bar stock, casting, or forging, it is classified as a secondary process.

6. Joining processing: Included in joining processing are all categories of welding, brazing, soldering, diffusion bonding, riveting, bolting, and adhesive bonding. These operations attach the parts to one another. Fastening occurs in the assembly step of manufacturing.

7. Heat treatment and surface treatment: This category includes the improvement of mechanical properties by thermal heat treatment processes as well as the improvement of surface properties by diffusion processes like carburizing and nitriding or by alternative means such as sprayed or hot-dip coatings, electroplating, and painting. The category also includes the cleaning of surfaces preparatory to surface treatment. This class of processes can be either secondary or finishing processes.

8. Assembly processes: In this, usually the final step in manufacturing, a number of parts are brought together and combined into a subassembly or finished product.

13.3.3 Sources of Information on Manufacturing Processes

In this book we cannot describe the many processes used in modern manufacturing in detail. Table 13.1 lists several readily available texts that describe the behavior of the material, the machinery, and the tooling to present a good understanding of how each process works.

TABLE 13.1
Basic Texts on Manufacturing Processes

J. T. Black and R. Kohser, *DeGarmo's Materials and Processes in Manufacturing,* 10th ed., John Wiley & Sons, Hoboken, NJ, 2008.

M. P. Groover, *Fundamentals of Modern Manufacturing,* 4th ed., John Wiley & Sons, New York, 2010.

S. Kalpakjian and S. R. Schmid, *Manufacturing Processes for Engineering Materials,* 5th ed., Pearson Prentice Hall, Upper Saddle River, NJ, 2008.

J. A. Schey, *Introduction to Manufacturing Processes,* 3d ed., McGraw-Hill, New York, 2000.

Also, Section 7, Manufacturing Aspects of Design, in *ASM Handbook* Vol. 20 gives an overview of each major process from the viewpoint of the design engineer.

The most important reference sources giving information on industrial practices are *Tool and Manufacturing Engineers Handbook,* 4th ed., published in nine volumes by the Society of Manufacturing Engineers, and various volumes of *ASM Handbook* published by ASM International devoted to specific manufacturing processes, see Table 13.5. In general, the *ASM Handbooks* have been updated more recently than the *Manufacturing Engineers Handbooks.* More books dealing with each of the eight classes of manufacturing processes are listed below.

Casting Processes

M. Blair and T. L. Stevens, eds., *Steel Castings Handbook,* 6th ed., ASM International, Materials Park, OH, 1995.

J. Campbell, *Casting,* 2nd ed., Butterworth-Heinemann, Oxford, UK, 2004.

H. Fredriksson and U. Åkerlind, *Material Processing During Casting,* John Wiley & Sons, Chichester, UK, 2006.

Casting, ASM Handbook, Vol. 15, ASM International, Materials Park, OH, 2008.

Polymer Processing

E. A. Muccio, *Plastics Processing Technology,* ASM International, Materials Park, OH, 1994.

A. B. Strong, *Plastics: Materials and Processing,* 3d ed., Prentice Hall, Upper Saddle River, NJ, 2006.

Plastics Parts Manufacturing, Tool and Manufacturing Engineers Handbook, Vol. 8, 4th ed., Society of Manufacturing Engineers, Dearborn, MI, 1995.

J. F. Agassant, P. Avenas, J. Sergent, and P. J. Carreau, *Polymer Processing: Principles and Modeling,* Hanser Gardner Publications, Cincinnati, OH 1991.

Z. Tadmor and C. G. Gogas, *Principles of Polymer Processing,* 2nd ed., Wiley-Interscience, Hoboken, NJ, 2006.

Deformation Processes

W. A. Backofen, *Deformation Processing,* Addison-Wesley, Reading, MA, 1972.

W. F. Hosfortd and R. M. Caddell, *Metal Forming: Mechanics and Metallurgy,* 2nd ed., Prentice Hall, Upper Saddle River, NJ, 1993.

E. Mielnik, *Metalworking Science and Engineering,* McGraw-Hill, New York, 1991.

R. H. Wagoner and J-L Chenot, *Metal Forming Analysis,* Cambridge University Press, Cambridge, UK, 2001.

K. Lange, ed., *Handbook of Metal Forming,* Society of Manufacturing Engineers, Dearborn, MI, 1985.

R. Pearce, *Sheet Metal Forming,* Adam Hilger, Bristol, UK, 1991.

Metalworking: Bulk Forming, *ASM Handbook,* Vol. 14A, ASM International, Materials Park, OH, 2005.

Metalworking: Sheet Forming. *ASM Handbook,* Vol. 14B, ASM International, Materials Park, OH, 2006.

Z. Marciniak and J. L. Duncan, *The Mechanics of Sheet Metal Forming,* Edward Arnold, London, 1992.

(Continued)

13

TABLE 13.1
(continued)

Powder Processing

R. M. German, *Powder Metallurgy Science,* Metal Powder Industries Federation, Princeton, NJ, 1985.

R. M. German, *Powder Metallurgy of Iron and Steel,* John Wiley & Sons, New York, 1998.

J. S. Reed, *Introduction to the Principles of Powder Processing,* 2nd ed., John Wiley & Sons, Hoboken, NJ, 1995.

ASM Handbook, Vol. 7, *Powder Metal Technologies and Applications,* ASM International, Materials Park, OH, 1998.

Powder Metallurgy Design Manual, 2nd ed., Metal Powder Industries Federation, Princeton, NJ, 1995.

Material Removal Processes

G. Boothroyd and W. W. Knight, *Fundamentals of Machining and Machine Tools,* 3d ed., Taylor & Francis, Boca Raton, FL, 2006.

E. M. Trent and P. K. Wright, *Metal Cutting,* 4th ed., Butterworth-Heinemann, Boston, 2000.

H. El-Hofy, *Fundamentals of Machining Processes: Conventional and Nonconventional Processes,* Taylor & Francis, Boca Raton, FL, 2007.

S. Malkin, *Grinding Technology: Theory and Applications,* Ellis Horwood, New York, 1989.

M. C. Shaw, *Metal Cutting Principles,* 2nd ed., Oxford University Press, New York, 2004.

Machining, Tool and Manufacturing Engineers Handbook, Vol. 1, 4th ed., Society of Manufacturing Engineers, Dearborn, MI, 1983.

ASM Handbook, Vol. 16, *Machining,* ASM International, Materials Park, OH, 1989.

Joining Processes

S. Kuo, *Welding Metallurgy,* John Wiley & Sons, New York, 1987.

R. W. Messler, *Joining of Materials and Structures,* Butterworth-Heinemann, Boston, 2004.

Engineered Materials Handbook, Vol. 3, *Adhesives and Sealants,* ASM International, Materials Park, OH, 1990.

R. O. Parmley, ed., *Standard Handbook for Fastening and Joining,* 3d ed., McGraw-Hill, New York, 1997.

ASM Handbook, Vol. 6A, *Welding Fundamentals and Processes,* ASM International, Materials Park, OH, 2011.

Welding Handbook, 9th ed., American Welding Society, Miami, FL, 2001.

Heat Treatment and Surface Treatment

Heat Treating, *ASM Handbook,* Vol. 4, ASM International, Materials Park, OH, 1991.

ASM Handbook, Vol. 5, *Surface Engineering,* ASM International, Materials Park, OH, 1994.

Tool and Manufacturing Engineers Handbook, Vol. 3, *Materials, Finishing, and Coating,* 4th ed., Society of Manufacturing Engineers, Dearborn, MI, 1985.

Assembly Processes

G. Boothroyd, *Assembly Automation and Product Design,* Marcel Dekker, New York, 1992.

P. H. Joshi, *Jigs and Fixtures Design Manual,* McGraw-Hill, New York, 2003.

A. H. Redford and J. Chal, *Design for Assembly,* McGraw-Hill, New York, 1994.

Fundamentals of Tool Design, 5th ed., Society of Manufacturing Engineers, Dearborn, MI, 2003.

Tool and Manufacturing Engineers Handbook, Vol. 9, *Assembly Processes,* 4th ed., Society of Manufacturing Engineers, Dearborn, MI, 1998.

13

TABLE 13.2
Characteristics of Production Systems

Characteristic	Job Shop	Batch Flow	Assembly Line	Continuous Flow
Equipment and Physical Layout				
Batch size	Low (1–100 units)	Moderate (100–10,000 units)	Large (10,000–millions/year)	Large. Measured in tons, gals., etc.
Process flow	Few dominant flow patterns	Some flow patterns	Rigid flow patterns	Well defined and inflexible
Equipment	General-purpose	Mixed	Specialized	Specialized
Setups	Frequent	Occasional	Few and costly	Rare and expensive
Process changes for new products	Incremental	Often incremental	Varies	Often radical
Information and Control				
Production information requirements	High	Varies	Moderate	Low
Raw material inventory	Small	Moderate	Varies; frequent deliveries	Large
Work-in-process	Large	Moderate	Small	Very small

13.3.4 Types of Manufacturing Systems

There are four general types of manufacturing systems: job shop, batch, assembly line, and continuous flow.[1] The characteristics of these production systems are listed in Table 13.2. The *job shop* is characterized by small batches of a large number of different part types every year. There is no regular work flow, so work-in-process must often wait in a queue for its turn on the machine. Hence, it is difficult to specify job shop capacity because it is highly dependent on the product mix. *Batch flow,* or decoupled flow line, is used when the product design is relatively stable and produced in periodic batches, but the volume for an individual product is not sufficient to warrant the cost of specialized, dedicated equipment. Examples are the production of heavy equipment or ready-to-wear clothing. With *assembly-line production,* the equipment is laid out in the sequence of usage. The large number of assembly tasks is divided into small subsets to be performed at successive workstations. Examples are the production of automobiles or consumer appliances. Finally, a *continuous-flow process* is the most specialized type. The equipment is highly specialized, laid out in a circuit, and usually automated. The material flows continuously from input to output. Examples are a gasoline refinery or a paper mill.

A process is said to be *mechanized* when it is being carried out by powered machinery and not by hand. Nearly all manufacturing processes in developed countries

---

1. G. Chryssolouris, *Manufacturing Systems,* 2nd ed., Springer, New York, 2006.

13

are mechanized. A process is *automated* when the steps in the process, along with the movement of material and inspection of the parts, are automatically performed or controlled by self-operating devices. Automation involves mechanization plus sensing and controlling capabilities (programmable logic controllers and PCs). Hard automation is hard-linked and hard-wired, while flexible automation includes the added capability of being reprogrammed to meet changing conditions.

13.4
MANUFACTURING PROCESS SELECTION

The factors that influence the selection of a process to make a part are:

- Quantity of parts required
- Complexity—shape, size, features
- Material
- Quality of part
- Cost to manufacture
- Availability, lead time, and delivery schedule

As emphasized in Chap. 11, there is a close interdependance between material selection and process selection.

The steps in selecting a manufacturing process are:

- Based on the part specification, identify the material class, the required number of parts, and the size, shape, minimum thickness, surface finish, and tolerance on critical dimensions of the part. These constitute constraints on the selection of the process.
- Decide what the objective of the process selection process is. Generally, the objective is to minimize the cost of the manufactured part. However, it might be to maximize the quality of the part, or to minimize the time to make it.
- Using the identified constraints, screen a large number of processes to eliminate the processes incapable of meeting them. This can be done using the information sources given in this chapter, or the screening charts found in M.F. Ashby, *Materials Selection in Mechanical Design,* 4th ed., Butterworth-Heinemann, Oxford, UK, 2011. The Cambridge Engineering Selector software from Granta Design Ltd., Cambridge, UK, 2010 greatly facilitates this process. It links material selection with possible processes and provides extensive data about each process. Figure 13.3 shows an example of the information provided about a process.
- Having narrowed the possible processes to a smaller number, rank them based on manufacturing cost. A quick ranking can be based on the economic batch size (Sec. 13.4.1), but a cost model is needed (Sec. 13.4.6) for making the final decision. However, before making this decision it is important to seek supporting information from among the references given in Table 13.1 and elsewhere in this chapter. Look for case studies and examples of industry practice that will lend credibility and support your decision.

Each factor affecting the selection of a manufacturing process for a particular part is discussed in the following sections.

13

INJECTION MOLDING of thermoplastics is the equivalent of pressure die casting of metals. Molten polymer is injected under high pressure into a cold steel mold. The polymer solidifies under pressure and the molding is then ejected.

Various types of injection molding machines exist, but the most common in use today is the reciprocating screw machine (shown schematically). Capital and tooling costs are very high. Production rate can be high, particularly for small moldings. Multicavity molds are often used. The process is used almost exclusively for large-volume production. Prototype moldings can be made using cheaper single-cavity molds of cheaper materials. Quality can be high but may be traded off against production rate. The process may also be used with thermosets and rubbers. Some modifications are required—this is dealt with separately. Complex shapes are possible, though some features (e.g., undercuts, screw threads, inserts) may result in increased tooling costs.

Process Schematic

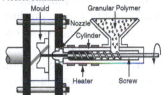

Physical Attributes

Adjacent section ratio	1	–	2	
Aspect ratio	1	–	250	
Mass range	0.02205	–	55.12	lb
Minimum hole diameter	0.02362	–		in
Minimum corner radius	0.05906	–		in
Range of section thickness	0.01575	–	0.248	in.
Roughness	7.874e-3	–	0.06299	mil
Quality factor (1–10)	1	–	6	
Tolerance	3.937e-3	–	0.03937	in.

Economic Attributes

Economic batch size (mass)	1.102e4	–	1.102e6	lb
Economic batch size (units)	1e4	–	1e6	

Cost Modelling

Relative cost index (per unit)	18.16	–	113.3	

Parameters: Material Cost = 4.309USD/lb, component Mass = 2.205lb, Batch size = 1000,

Capital cost	3.77e4	–	8.483e5	USD
Lead time	4	–	6	week(s)
Material utilization fraction	0.6	–	0.9	
Production rate (mass)	66.14	–	2205	lb/hr
Production rate (units)	60	–	3000	/hr
Tool life (mass)	1.102e4	–	1.102e6	lb
Tool life (units)	1e4	–	1e6	

Supporting Information

Design guidelines

Complex shapes are possible. Thick sections or large changes in section are not recommended. Small reentrant angles are possible.

Technical nodes

Most thermoplastics can be injection moulded. Some high melting point polymers (e.g., PTFE) are not suitable. Thermoplastic based composites (short fibre and particulate filled) are also processed.
Injection-moulded parts are generally thin-walled.

Typical uses

Extremely varied. Housings, containers, covers, knobs, tool handles, plumbing fittings, lenses, etc.

The economics

Tooling cost range covers small, simple to large, complex moulds. Production rate depends on complexity of component and number of mould cavities.

The environment

Thermoplastic sprues can be recycled. Extraction may be required for volatile fumes. Significant dust exposures may occur in the formulation of the resins. Thermostatic controller malfunctions can be extremely hazardous.

FIGURE 13.3
Typical process data sheet from CES EduPack, 2006 Granta Design Limited, Cambridge, UK, 2006.

13.4.1 Quantity of Parts Required

Two important factors in the choice of processes are the total number of parts to be produced and the rate of production, in units per time period. All manufacturing processes have a minimum number of pieces (volume) that must be made to justify their use. Some processes, like an automatic screw machine, are inherently high-volume processes, in that the setup time is long relative to the time needed to produce a single part. Others, like the hand layup of a fiberglass plastic boat, are low-volume processes. Here the setup time is minimal but the time to make a part is much longer.

The total volume of production often is insufficient to keep a production machine continuously occupied. As a result, production occurs in *batches* or *lots* representing a fraction of the number of parts needed for a year of product production. The batch size is influenced by the cost and inconvenience of setting up for a new production run on a particular machine, and by the cost of maintaining parts in inventory in a warehouse between production runs.

Figure 13.4 compares the cost of making an aluminum connecting rod by sand casting and die casting to illustrate the interplay between tooling and setup cost and quantity on process cost per part. Sand casting uses cheaper equipment and tooling, but it is more labor intensive to build the sand molds. Pressure die casting uses more costly equipment and expensive metal molds, but it is less labor intensive. The cost of material is the same in both processes. For a small number of parts the unit cost is higher for die casting, chiefly because of the more expensive tooling. However, as these costs are shared with a larger number of parts, the unit cost is decreased, and

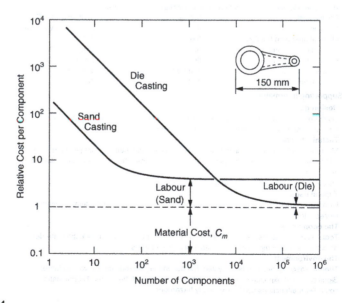

FIGURE 13.4

The relative cost of casting a part versus the number of parts produced using the sand casting and die casting processes. (From M. F. Ashby, *Materials Selection in Mechanical Design*, 2nd ed., p. 278. Copyright Elsevier, 1999. Used with permission.)

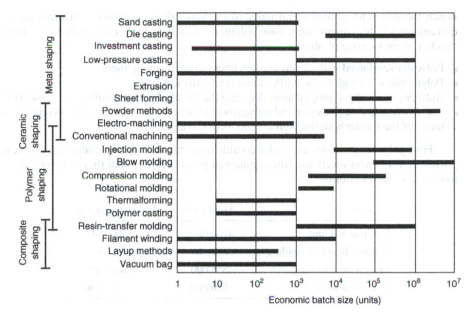

FIGURE 13.5
Range of economic batch size for typical manufacturing processes. (From M. F. Ashby, *Materials Selection in Mechanical Design,* 3d ed., p. 205. Copyright Elsevier, 2005. Used with permission.)

at about 3000 parts the die casting process has a lower unit cost. Note that the sand casting process leveled out at about 100 parts, maintaining a constant unit cost that is determined by the material cost plus the labor cost. The same thing happens for the die casting process, only here the labor cost is very low relative to the material cost.

The number of parts at which the unit cost of one process becomes lower than that of its competitors is called the *economic batch size.* The economic batch size for sand casting in this example is from 1 to 3600 parts, while that for die casting is 3600 and beyond. The economic batch size is a good rough guide to the cost structure of a process. It is a useful screening parameter for differentiating among candidate processes, as shown by Fig. 13.5. A more detailed cost model (Sec. 13.4.6) is then used to refine the ranking of the most promising process candidates.

The *flexibility* of the process is related to the economic batch size. Flexibility in manufacturing is the ease with which a process can be adapted to produce different products or variations of the same product. It is greatly influenced by the time needed to change and set up tooling. At a time when product customization is increasingly important, this process attribute has gained importance.

EXAMPLE 13.1
With the drive to reduce the weight of automobiles, there is strong interest in plastic bumpers. Such a bumper must have good rigidity to maintain dimensional limits, low-temperature impact resistance (for crashworthiness), and dimensional stability over the operating range of temperature.[1] In addition, it must have the ability to be finished to

1. L. Edwards and M. Endean, eds., *Manufacturing with Materials,* Butterworth, Boston, 1990.

13

match the adjoining painted metal parts. With these critical-to-quality performance requirements of chief importance, four polymeric materials were chosen from the large number of engineering plastics.

- Polyester reinforced with chopped-glass fiber to improve toughness
- Polyurethane with glass-flake filler to increase stiffness
- Rubber-modified polypropylene to decrease the ductile-brittle transition to below 30°C
- A polymer blend of polyester and polycarbonate to combine the excellent solvent resistance of the former with the high toughness of the latter.

Four polymer processes are under consideration for making the bumpers from these polymers. Each works well with the engineered plastics chosen, but they vary greatly in tooling costs and flexibility.

Process	Mold Cost	Labor Input/Unit
Injection molding	$450,000	3 min = $1
Reaction injection molding	$90,000	6 min = $2
Compression molding	$55,000	6 min = $2
Contact molding	$20,000	1 h = $20

Then, the part cost is the sum of the mold cost per part plus the labor input, neglecting the material cost, which is roughly the same for each.

	Cost per Part			
Process	1000 Parts	10,000 Parts	100,000 Parts	1,000,000 Parts
Injection molding	$451	$46	$5.50	$1.45
Reaction injection molding	$92	$11	$2.90	$2.09
Compression molding	$57	$7.50	$2.55	$2.06
Contact molding	$40	$22	$20.20	$20.02

Note how the unit part cost varies greatly with the quantity of parts required. The hand layup process of contact molding is the least expensive for a low part volume, while the low-cycle-time injection molding process excels at the highest part volume. Assuming that the material cost for the bumper is $30 per part, we see how material cost represents the largest fraction of the total cost as the part volume increases.

13.4.2 Shape and Feature Complexity

The complexity of a part refers to its *shape* and type and number of *features* that it contains. One way of expressing the complexity of a component is through its information content I, expressed in number of digital bits of information.

$$I = n \log_2 \left(\frac{\bar{l}}{\Delta l} \right)$$

(13.1)

where n = number of dimensions of the component

$$\bar{l} = \left(l_1 \bullet l_2 \bullet l_3 \dots l_n\right)^{1/n} \text{ is the geometric mean dimension}$$

$$\overline{\Delta l} = \left(\Delta l_1 \bullet \Delta l_2 \bullet \Delta l_3 \dots \Delta l_n\right)^{1/n} \text{ is the geometric mean of the tolerance}$$

$$\log_2(x) = \frac{\log_{10}(x)}{\log_{10}(2)}$$

Simple shapes contain only a few bits of information. Complex shapes, like integrated circuits, contain very many. A cast engine block might have 10^3 bits of information, but after machining the various features the complexity increases by both adding new dimensions (n) and improving their precision (reducing $\overline{\Delta l}$).

Most mechanical parts have a three-dimensional shape, although sheet metal fabrications are basically two-dimensional. Figure 13.6 shows a useful shape classification system. In this schema a shape of uniform cross section is given a complexity rating of 0.

The shape complexity increases from left to right in Fig. 13.6 with the addition of greater geometric complexity and added features, that is, greater information content.

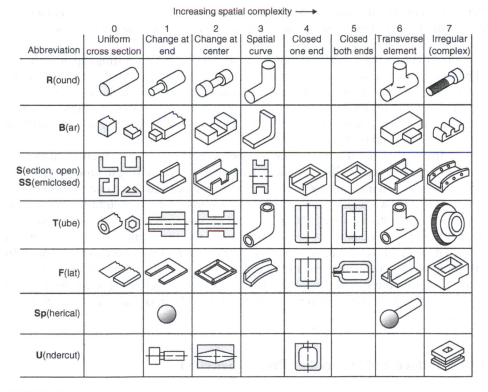

FIGURE 13.6
A classification system for basic shapes in design. (After J. A. Schey.)

Note that a small increase in information content can have major significance in process selection for making a part. Moving from the solid shape R0 (shape in column 0 of the Round row) to the hollow shape T0 (shape in column 0 of the Tube row) adds only one additional dimension (the hole diameter), but the change excludes some processes as the best choice for making the part or adds an additional operation step in other processes.

Manufacturing processes vary in their limitations for producing complex shapes. For example, there are many processes that do not allow the making of *undercuts*, shown in the bottom row in Fig. 13.6. Undercuts make it impossible to extract the part from the mold without complicated and expensive tooling. Other processes have limitations on how thin the wall thickness can be, or require the part to have uniform wall thickness. Extrusion processes require a part that is axially symmetric. Powder metallurgy cannot make parts with sharp corners or acute angles because the unsintered powder will crumble when transferring from the die. Lathe turning requires a part with cylindrical symmetry. Table 13.3 associates the shapes defined in Fig. 13.6 with the ability of various manufacturing processes to create them.

13.4.3 Size

Parts vary considerably in size. Because of the nature of the equipment used in a manufacturing process, each process has a range of part sizes for which it is economical to use that process. Figure 13.7 shows this.

Note that machining processes (i.e., removal of metal by cutting) span the complete range of sizes, and that machining, casting, and forging are able to produce the largest mass objects. However only a limited number of plants in the world can make very large parts. Therefore, to make very large products like aircraft, ships, and pressure vessels, it is necessary to assemble them from many parts using joining methods such as welding and riveting.

A limiting geometric factor in process selection often is section thickness. Figure 13.8 displays capabilities for achieving thickness according to process. Gravity-fed castings have a minimum wall thickness that they can produce due to surface tension and heat flow considerations. Thin sections may solidify before the rest of the casting, leaving internal voids. Minimum thickness can be extended by using pressure die casting. The availability of press tonnage and the occurrence of friction in metal deformation processes create a similar restriction on minimum section thickness. In injection molding there must be sufficient time for the polymer to harden before it can be ejected from the molding machine. Because high production rates are desired, the slow rate of heat transfer of polymers severely limits the maximum thickness that can be obtained.

13.4.4 Influence of Material on Process Selection

Just as shape requirements limit the available selection of processes, the selection of a material also places certain restrictions on the available manufacturing processes. The

TABLE 13.3
Ability of Manufacturing Processes to Produce Shapes in Fig. 13.6

Process	Capability for Producing Shapes
Casting processes	
Sand casting	Can make all shapes
Plaster casting	Can make all shapes
Investment casting	Can make all shapes
Permanent mold	Can make all shapes except T3, T5; F5; U2, U4, U7
Die casting	Same as permanent mold casting
Deformation processes	
Open-die forging	Best for R0 to R3; all B shapes; T1; F0; Sp6
Hot impression die forging	Best for all R, B, and S shapes; T1, T2; Sp
Hot extrusion	All 0 shapes
Cold forging/cold extrusion	Same as hot die forging or extrusion
Shape drawing	All 0 shapes
Shape rolling	All 0 shapes
Sheet-metal working processes	
Blanking	F0 to F2; T7
Bending	R3; B3; S0, S3, S7; T3; F3, F6,
Stretching	F4; S7
Deep drawing	T4; F4, F7
Spinning	T1, T2, T4, T6; F4, F5
Polymer processes	
Extrusion	All 0 shapes
Injection molding	Can make all shapes with proper coring
Compression molding	All shapes except T3, T5, T6, F5, U4
Sheet thermoforming	T4, F4, F7, S5
Powder metallurgy processes	
Cold press and sinter	All shapes except S3, T2, T3, T5, T6, F3, F5, all U shapes
Hot isostatic pressing	All shapes except T5 and F5
Powder injection molding	All shapes except T5, F5, U1, U4
PM forging	Same shape restrictions as cold press and sinter
Machining processes	
Lathe turning	R0, R1, R2, R7; T0, T1, T2; Sp1, Sp6; U1, U2
Drilling	T0, T6
Milling	All B, S, SS shapes; F0 to F4; F6, F7, U7
Grinding	Same as turning and milling
Honing, lapping	R0 to R2; B0 to B2; B7; T0 to T2, T4 to T7; F0 to F2; Sp

Based on data from J.A. Schey, *Introduction to Manufacturing Processes.*

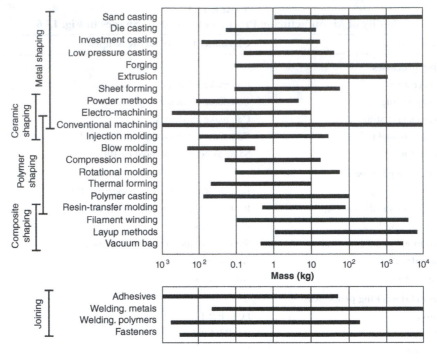

FIGURE 13.7

Process selection chart. Process versus range of size (mass). (From M. F. Ashby, *Materials Selection in Mechanical Design,* 3d ed., p. 199. Copyright Elsevier, 2005. Used with permission.)

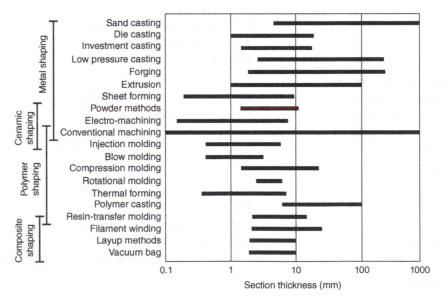

FIGURE 13.8

Range of available section thickness provided by different processes. (From M. F. Ashby, *Materials Selection in Mechanical Design,* 3d ed., p. 200. Copyright Elsevier, 2005. Used with permission.)

melting point of the material and its level of deformation resistance and ductility are the chief factors. The melting point of the material determines the casting processes that can be employed. Low-melting-point metals can be used with a wide number of casting processes, but as the melting point rises, problems with mold reaction and atmosphere contamination limit the available processes. Some materials, like ceramics, are too brittle for shape creation by deformation processes, while others are too reactive to have good weldability.

Figure 13.9 shows a matrix laying out the manufacturing processes generally used with the most common classes of engineering materials. The table is further divided with respect to the quantity of parts needed for economical production. Use this matrix as a way to narrow down the possibilities to a manageable few processes for final evaluation and selection. This table is part of the PRocess Information MAps (PRIMA) methodology for manufacturing process selection.[1]

Steels, aluminum alloys, and other metallic alloys can be purchased in a variety of metallurgical conditions other than the annealed (soft) state. Examples are quenched and tempered steel bars, solution-treated and cold-worked and aged aluminum alloys, or cold-drawn and stress-relieved brass rods. It may be more economical to have the metallurgical strengthening produced in the workpiece by the material supplier than to heat-treat each part separately after it has been manufactured.

When parts have very simple geometric shapes, as straight shafts and bolts have, the form in which the material is obtained and the method of manufacture are readily apparent. However, as the part becomes more complex in shape, it becomes possible to make it from several forms of material and by a variety of manufacturing methods. For example, a small gear may be machined from bar stock or, more economically, from a precision-forged gear blank. The selection of one of several alternatives is based on overall cost of a finished part (see Chap. 16 for details of cost evaluation). Generally, the production quantity is an important factor in cost comparisons, as was shown in Fig. 13.4. There will be a break-even point beyond which it is more economical to invest in precision-forged preforms in order to produce a gear with a lower unit cost than to machine it from bar stock. As the production quantity increases, it becomes easier economically to justify a larger initial investment in tooling or special machinery to lower the unit cost.

13.4.5 Required Quality of the Part

The quality of the part is defined by three related sets of characteristics: (1) freedom from external and internal defects, (2) surface finish, and (3) dimensional accuracy and tolerance. To a high degree, the achievement of high quality in these areas is influenced by the workability or formability of the material.[2] While different materials exhibit different workability in a given process, the same material may show different workability in different processes. For example, in deformation processing,

---

1. K.G. Swift and J.D. Booker, *Process Selection,* 2nd ed., Butterworth-Heinemann, Oxford, UK, 2003.
2. G.E. Dieter, H.A. Kuhn, and S.L. Semiatin, eds., *Handbook of Workability and Process Design,* ASM International, Materials Park, OH, 2003.

QUANTITY \ MATERIAL	IRONS	STEEL (carbon)	STEEL (tool, alloy)	STAINLESS STEEL	COPPER & ALLOYS	ALUMINIUM & ALLOYS	MAGNESIUM & ALLOYS	ZINC & ALLOYS
VERY LOW 1 TO 100	[1.5] [1.6] [1.7] [4.M]	[1.5] [1.7] [3.10] [4.M] [5.1] [5.5] [5.6]	[1.1] [1.5] [1.7] [3.10] [4.M] [5.1] [5.5] [5.6] [5.7]	[1.6] [1.7] [3.7] [3.10] [4.M] [5.1] [5.5] [5.6]	[1.5] [1.7] [3.6] [4.M] [5.1]	[1.5] [1.7] [3.7] [3.10] [4.M] [5.5]	[1.6] [1.7] [3.10] [4.M] [5.1][5.5]	[1.1] [1.7] [3.10] [4.M] [5.5]
LOW 100 TO 1,000	[1.2] [1.5] [1.6] [1.7] [4.M] [5.3] [5.4]	[1.2] [1.6] [1.7] [1.10] [4.M] [5.1] [5.3] [5.4] [5.5]	[1.1] [1.2] [1.7] [4.M] [5.1] [5.3] [5.4] [5.5] [5.6] [5.7]	[1.2] [1.7 [3.7] [3.10] [4.M] [5.1] [5.3] [5.4] [5.5]	[1.2] [1.3] [1.5] [1.7] [1.8] [3.3] [3.6] [4.M] [5.1] [5.3] [5.4]	[1.2] [1.5] [1.7] [1.8] [3.7] [3.10] [4.M] [5.3] [5.4] [5.5]	[1.6] [1.7] [1.8] [3.10] [4.M] [5.5]	[1.1] [1.7] [1.8] [3.10] [4.M] [5.5]
LOW TO MEDIUM 1,000 TO 10,000	[1.2] [1.3] [1.5] [1.6] [1.7] [3.11] [4.A] [5.2]	[1.2] [1.3] [1.5] [1.7] [3.1] [3.3] [3.10] [3.11] [4.A] [5.2] [5.3] [5.4] [5.5]	[1.2] [1.5] [1.7] [3.1] [3.4] [3.11] [4.A] [5.2] [5.3] [5.4] [5.5]	[1.2] [1.5] [1.7] [3.1] [3.2] [3.7] [3.10] [3.11] [4.A.] [5.2] [5.3] [5.4] [6.0]	[1.2] [1.3] [1.5] [1.8] [3.1] [3.2] [3.10] [3.11] [4.A] [5.2] [5.3] [5.4]	[1.2] [1.3] [1.4] [1.5] [3.1] [3.3] [3.7] [3.10] [3.11] [4.A] [5.3] [5.4] [5.5]	[1.3] [1.5] [1.8] [3.1] [3.3] [3.4] [3.10] [4.A] [5.5]	[1.3] [1.8] [3.3] [3.10] [4.A] [5.5]
MEDIUM TO HIGH 10,000 TO 100,000	[1.2] [1.3] [3.11] [4.A]	[1.0] [2.1] [3.3] [3.4] [3.5] [3.11] [3.12] [4.A] [5.2] [5.5]	[3.1] [3.4] [3.5] [3.11] [3.12] [4.A] [5.2]	[1.0] [3.1] [3.3] [3.4] [3.5] [3.11] [3.12] [4.A]	[1.2] [1.4] [1.8] [3.1] [3.3] [3.4] [3.5] [3.11] [3.12] [4.A]	[1.2] [1.3] [1.4] [1.8] [3.1] [3.3] [3.4] [3.5] [3.11] [3.12] [4.A] [5.5]	[1.3] [1.4] [3.1] [3.3] [3.4] [3.5] [3.12] 4.A]	[1.3] [1.4] [3.3] [3.4] [3.5] [3.12] [4.A]
HIGH 100,000+	[1.2] [1.3] [3.11] [4.A]	[1.9] [3.1] [3.2] [3.3] [3.4] [3.5] [3.12] [4.A]	[4.A]	[1.9] [3.2] [3.3] [4.A]	[1.2] [1.M] [3.1] [3.2] [3.3] [3.4] [3.5] [3.7] [3.8] [3.11] [3.12] [4.A]	[1.2] [1.3] [1.4] [1.6] [3.1] [3.2] [3.3] [3.4] [3.5] [3.8] [3.12] [4.A]	[1.3] [1.4] [3.1] [3.3] [3.4] [3.8] [3.12] [4.A]	[1.4] [3.2] [3.3] [3.4] [3.5] [4.A]
ALL QUANTITIES	[1.1]	[1.1] [1.5] [3.0] [3.0] [3.0]	[1.6] [3.6]	[1.1] [1.6] [3.6] [3.8] [3.9]	[1.1] [1.6] [3.6] [3.8] [3.9] 5.5]	[1.1] [1.6] [3.4] [3.5] [3.6]	[1.1] [3.5] [3.6] [3.8]	[3.6] [3.8] [3.9]

QUANTITY \ MATERIAL	TIN & ALLOYS	LEAD & ALLOYS	NICKEL & ALLOYS	TITANIUM & ALLOYS	THERMOPLASTICS	THERMOSETS	FR COMPOSITES	CERAMICS	REFRACTORY METALS	PRECIOUS METALS
VERY LOW 1 TO 100	[1.1] [1.7] [3.10] [4.M] [5.5]	[1.1] [3.10] [4.M] [5.5]	[1.5] [1.7] [3.10] [4.M] [5.1] [5.5] [5.6]	[1.1] [1.6] [3.7] [3.10] [4.M] [5.1] [5.5] [5.6] [5.7]	[2.5] [2.7]	[2.5] [3.7]	[2.2] [2.6] [5.7]	[1.5] [5.1] [5.5] [5.6] [5.7]	[1.1] [5.7]	[5.5]
LOW 100 TO 1,000	[1.1] [1.7] [1.8] [3.10] [4.M] [5.5]	[1.1] [1.8] [3.10] [4.M] [5.5]	[1.2] [1.5] [1.7] [3.10] [4.M] [5.1] [5.3] [5.4] [5.5] [5.6]	[1.1] [1.6] [3.7] [3.10] [4.M] [5.1] [5.3] [5.4] [5.5] [5.6] [5.7]	[2.3] [2.5] [2.7]	[2.2] [2.3]	[2.2] [2.3] [2.6] [5.7]	[5.1] [5.3] [5.5] [5.6] [5.7]	[5.7]	[5.5]
LOW TO MEDIUM 1,000 TO 10,000	[1.3] [1.5] [3.2] [3.10]	[1.3] [1.5] [3.3] [3.10]	[1.2] [1.3] [1.5] [1.7] [3.1] [3.2] [3.11] [4.A] [5.1] [5.3] [5.4] [5.5] [3.10]	[3.1] [3.7] [3.10] [3.11] [4.A] [5.2] [5.3] [5.4] [5.5]	[2.3] [2.5] [2.6] [2.7]	[2.2] [2.3] [2.4]	[2.1] [2.2] [2.3]	[5.2] [5.3] [5.4] [5.5]		[5.5]
MEDIUM TO HIGH 10,000 TO 100,000	[1.3] [1.4] [3.3] [3.4] [3.12]	[1.3] [1.4] [3.3] [3.4] [3.8] [3.12] [4.A]	[3.1] [3.3] [3.5] [3.4] [3.11] [3.12] [4.A] [5.2] [5.5]	[3.1] [3.4] [3.11] [3.12] [4.A] [5.2] [5.5]	[2.1] [2.3] [2.6] [2.7] [2.8]	[2.1] [2.3] [2.9]	[2.1] [2.3]	[3.11]	[3.12]	[3.5]
HIGH 100,000+	[1.4] [3.3] [3.4] [4.A]	[1.4] [3.2] [3.3] [3.4] [4.A]	[3.2] [3.3] [4.A]	[4.A]	[2.1] [2.6] [2.8]	[2.1] [2.3] [2.4] [2.9]		[3.7] [3.11]		[3.5]
ALL QUANTITIES		[3.5]	[1.1] [1.6] [3.4] [3.5] [3.8]	[3.8] [3.9]				[5.5]	[1.5]	[1.6]

KEY TO MANUFACTURING PROCESS PRIMA SELECTION MATRIX:

CASTING PROCESSES
[1.1] SAND CASTING
[1.2] SHELL MOULDING
[1.3] GRAVITY DIE CASTING
[1.4] PRESSURE DIE CASTING
[1.5] CENTRIFUGAL CASTING
[1.6] INVESTMENT CASTING
[1.7] CERAMIC MOULD CASTING
[1.8] PLASTER MOULD CASTING
[1.9] SQUEEZE CASTING

PLASTIC & COMPOSITE PROCESSING
[2.1] INJECTION MOULDING
[2.2] REACTION INJECTION MOULDING
[2.3] COMPRESSION MOULDING
[2.4] TRANSFER MOULDING
[2.5] VACUUM MOULDING
[2.6] BLOW MOULDING
[2.7] ROTATIONAL MOULDING
[2.8] CONTACT MOULDING
[2.9] CONTINUOUS EXTRUSION (PLASTICS)

FORMING PROCESSES
[3.1] CLOSED DIE FORGING
[3.2] ROLLING
[3.3] DRAWING
[3.4] COLD FORMING
[3.5] COLD HEADING
[3.6] SWAGING
[3.7] SUPERPLASTIC FORMING
[3.8] SHEET-METAL SHEARING
[3.9] SHEET-METAL FORMING
[3.10] SPINNING
[3.11] POWDER METALLURGY
[3.12] CONTINUOUS EXTRUSION (METALS)

MACHINING PROCESSES
[4.A] AUTOMATIC MACHINING
[4.M] MANUAL MACHINING

(THE ABOVE HEADINGS COVER A BROAD RANGE OF MACHINING PROCESSES AND LEVELS OF CONTROL TECHNOLOGY. FOR MORE DETAIL, THE READER IS REFERRED TO THE INDIVIDUAL PROCESSES.)

NTM PROCESSES
[5.1] ELECTRICAL DISCHARGE MACHINING (EDM)
[5.2] ELECTROCHEMICAL MACHINING (ECM)
[5.3] ELECTRON BEAM MACHINING (EBM)
[5.4] LASER BEAM MACHINING (LBM)
[5.5] CHEMICAL MACHINING (CM)
[5.6] ULTRASONIC MACHINING (USM)
[5.7] ABRASIVE JET MACHINING (AJM)

FIGURE 13.9
PRIMA selection matrix showing which materials and processes are usually used together, based on common practice. (From K. G. Swift and J. D. Booker, *Process Selection*, 2nd ed., p. 23. Copyright Elsevier, 2003. Used with permission.)

the workability increases with the extent that the process provides a condition of hydrostatic compression. Thus, steel has greater workability in extrusion than in forging, and even less in drawing, because the hydrostatic component of the stress state decreases in the order of the processes listed.

Defects

Defects may be internal to the part or concentrated mainly at the surface. Internal defects are such things as voids, porosity, cracks, or regions of different chemical-composition (segregation). Surface defects can be surface cracks, rolled-in oxide, extreme roughness, or surface discoloration or corrosion. The amount of material used to make the part should be just enough larger than the final part to allow for removal of surface defects by machining or another surface conditioning method. Thus, extra material in a casting may be needed to permit machining the surface to a specified finish, or a heat-treated steel part may be made oversized to allow for the removal of a decarburized layer.[1]

Often the manufacturing process dictates the use of extra material, such as sprues and risers in castings and flash in forgings and moldings. At other times extra material must be provided for purposes of handling, positioning, or testing the part. Even though extra material removal is costly, it usually is cheaper to purchase a slightly larger workpiece than to pay for a scrapped part.

Computer-based process modeling is being used effectively to investigate the design of tooling and the flow of material to minimize defect formation. Also, improved nondestructive inspection methods make more certain the detection of defects before a part is placed into service. Defects such as voids can often be eliminated by subjecting the part to a high hydrostatic pressure, such as 15,000 psi, at elevated temperature, in a process called hot-isostatic pressing (HIP).[2] HIPing has been used effectively with investment casting to replace parts previously made by forging.

Surface Finish

The surface finish of a part determines its appearance, affects the assembly of the part with other parts, and may influence its resistance to corrosion and wear. The surface roughness of a part must be specified and controlled because of its influence on fatigue failure, friction and wear, and assembly with other parts.

No surface is smooth and flat like the straight line we make on an engineering drawing. When viewed on a highly magnified scale every surface is rough, as sketched in Fig. 13.10. Surface roughness is measured with a profilometer, a precision instrument that traverses a line (typically a travel of 1 mm) with a very fine-tipped stylus. Several parameters are used to describe the state of surface roughness.[3]

---

1. For photographs and discussion of the formation of defects in deformation processing, see *ASM Handbook,* Vol. 11, *Failure Analysis and Prevention,* pp. 81–102, ASM International, Materials Park, OH, 2002.
2. H. V. Atkinson and B. A. Rickinson, *Hot Isostatic Pressing,* Adam Huger, Bristol, UK, 1991.
3. See Surface texture, ANSI Standard B46.1, ASME, 1985.

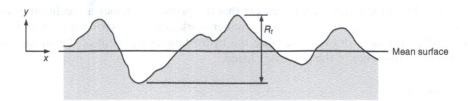

FIGURE 13.10
Cross-sectional profile of surface roughness with vertical direction magnified.

R_t is the height measured from maximum peak to the deepest trough. It is not
the most commonly used measure of surface roughness, but it is an important
value when roughness needs to be removed by polishing.

R_a is the arithmetic average based on the absolute value of the deviations from
the mean surface line. The mean surface is drawn such that the area under the
peaks and valleys is equal. This measure of roughness is also called the centerline average.

$$R_a = \frac{y_1 + y_2 + y_3 + \cdots + y_n}{n} \tag{13.2}$$

This measure of surface roughness is commonly used in industry. However, it
is not particularly useful for evaluating bearing surfaces.[1]

R_q is the root-mean square of the deviations from the mean surface.

$$R_q = \left(\frac{y_1^2 + y_2^2 + y_3^2 + \cdots + y_n^2}{n} \right)^{1/2} \tag{13.3}$$

R_q is sometimes given as an alternative to R_a because it gives more weight to the
higher peaks in the surface roughness. As an approximation, $R_q / R_a \approx 1.1$.

Surface roughness is usually expressed in units of μm (micrometer or micron) or μin
(microinch). 1μm $= 40\mu$in and 1μin $= 0.025\mu$m $= 25$nm.

There are other important characteristics of a surface besides the roughness. Surfaces usually exhibit a directionality of scratches characteristic of the finishing process. This is called *surface lay*. Surfaces may have a random lay, or an angular or
circular pattern of marks. Another characteristic of the surface is its *waviness*, which
occurs over a longer distance than the peaks and valleys of roughness. Allowable limits on these surface characteristics are specified on the engineering drawing by the
scheme shown in Fig. 13.11. The roughness cutoff length is used to separate the waviness from the roughness variations. The cutoff length is a specified length over which
measurements are made of the surface roughness. A sampling length of 0.030 in. will
generally filter out wavinesss from the surface roughness.

1. N. Judge, *Manufacturing Engineering,* Oct. 2002, pp. 60–68.

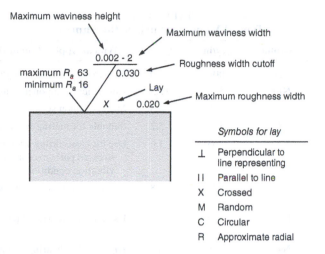

FIGURE 13.11
Symbols used to specify finish characteristics on an engineering drawing. Roughness given in microinches.

It is important to realize that specifying a surface by average roughness height is not an ideal approach. Two surfaces can have the same value of R_a and vary considerably in the details of surface profile.

Surface texture does not completely describe a surface. For example, there is an altered layer just below the surface texture layer. This layer is characteristic of the nature and amount of energy that has been put into creating the surface. It can contain small cracks, residual stresses, hardness differences, and other alterations. Control of the surface and subsurface layer as it is influenced by processing is called *surface integrity*.[1]

Table 13.4 gives a description of the various classes of surface finish, and gives some examples of different types of machine elements where each would be specified. The surfaces are defined in words and by the preferred values, N, given by the ISO surface roughness standard.

Control of surface roughness is important in many areas of engineering design.

1. Precision is required in many types of mating surfaces such as gaskets, seals, tools and dies.
2. Rough surfaces serve as notches and reduce fatigue life.
3. Roughness plays an important role in the tribological issues of friction, wear, and lubrication.
4. Surface roughness increases electrical and thermal contact resistance.
5. A rough surface will entrap corrosive fluids.

---

1. A. R. Marder, "Effects of Surface Treatments on Materials Performance," *ASM Handbook*, Vol. 20, pp. 470–90, 1997; E. W. Brooman, "Design for Surface Finishing," *ASM Handbook,* Vol. 20, pp. 820–27, ASM International, Materials Park, OH, 1997.

TABLE 13.4
Typical Values for Surface Roughness

Description	N-value	R_a, μin	R_a, μm	Typical Application in Design
Very rough	N11	1000	25.0	Nonstressed surface; rough cast surface
Rough	N10	500	12.5	Noncritical components; machined
Medium	N9	250	6.3	Most common surface for components
Average smooth	N8	125	3.2	Suitable for mating surfaces without motion
Better than avg.	N7	63	1.6	Use for close-fitting sliding surfaces and stressed parts except for shafts and vibration conditions
Fine	N6	32	0.8	Use where stress concentration is high: gears, etc.
Very fine	N5	16	0.4	Use for fatigue-loaded parts; precision shafts
Extremely fine	N4	8	0.2	High-quality bearings; requires honing/polishing
Superfinish	N3	4	0.1	For highest precision parts; requires lapping

6. The appearance of a product is influenced by the surface roughness which can vary from shiny to dull.
7. The adherence of surface coatings such as paint or plating is strongly influenced by roughness.

Dimensional Accuracy and Tolerances

Processes differ in their ability to meet close tolerances. Inability to hold close tolerances leads to problems with performance and interchangeability of parts. Generally, materials with good workability can be held to closer tolerances. Achieving dimensional accuracy depends on both the nature of the material and the process. Solidification processes must allow for the shrinkage that occurs when a molten metal solidifies. Polymer processes must allow for the much higher thermal expansion of polymers than metals, and hot working processes for metals must allow for oxidation of the surface.

Each manufacturing process has the capability of producing a part to a certain surface finish and tolerance range without incurring extra cost. Figure 13.12 shows this general relationship. The tolerances apply to a 1-inch dimension and are not necessarily scalable to larger and smaller dimensions for all processes. For economical design, the loosest possible tolerances and coarsest surface finish that will fulfill the function of the design should be specified. As Fig. 13.13 shows, processing cost increases nearly exponentially as the requirements for tolerances and surface finish are made more stringent.

13.4.6 Cost to Manufacture

The final decision on a manufacturing process is usually made on the basis of the cost to make a part, called the *unit cost*. Now that we have discussed the main factors that

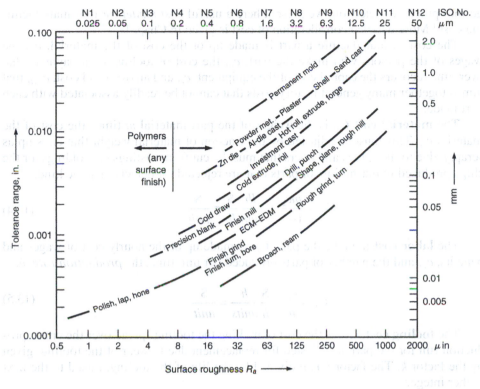

FIGURE 13.12

Approximate values of surface roughness and tolerance on dimensions typically obtained with different manufacturing processes. (J. A. Schey, *Introduction to Manufacturing Processes,* 3d ed. McGraw-Hill, 2000)

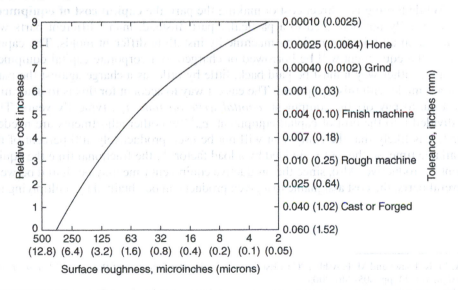

FIGURE 13.13

Influence of surface roughness and tolerance on processing costs (schematic).

go into processes selection, we present here a useful *cost model* for unit manufacturing cost.[1] More detailed consideration of cost is given in Chap. 17.

The cost to manufacture a part is made up of the cost of the material, c_m, the wages of the persons who make the part, c_w, the cost of tooling, c_t, a payment that over time recovers the capital cost of the equipment, c_e, and an overhead cost, c_{OH}, that lumps together many general factory costs that cannot be readily associated with each part made.

The **material cost** C_M is the weight of the part material m times the cost of the material c_m. This must be adjusted by the fraction of material weight that ends up as scrap, f, due to the sprues and risers that must be cut from castings or moldings, or the chips produced in machining, or parts that are rejected for defects of some kind.

$$C_M = \frac{mc_m}{1-f} \quad \frac{lb}{unit}\frac{\$}{lb} = \frac{\$}{unit} \tag{13.4}$$

The **labor cost** to make the part, C_L, is made up of the hourly cost of wages and benefits, c_w, and the number of parts produced per unit time, the *production rate*, $\dot{n}$.

$$C_L = \frac{c_w}{\dot{n}} \quad \frac{\$}{h}\frac{h}{units} = \frac{\$}{unit} \tag{13.5}$$

The **tooling cost**, C_T, is the cost of making the tooling spread over the entire production run for the part, n, adjusted for replacement due to wear of the tooling, given by the factor k. The factor k is n divided by the life of the tooling, raised to the next higher integer.

$$C_T = \frac{c_t k}{n} \quad \frac{\$}{units} \times (integer) \tag{13.6}$$

While tooling is a direct cost of making the part, the **capital cost of equipment,** C_E, is usually not dedicated to a particular part. Instead, many different parts will be made on an injection molding machine by installing different molds. The capital cost of the equipment will be borrowed or charged to a corporate capital equipment account. Either way it must be paid back, little by little, as a charge against the parts that are made with this equipment. The easiest way to account for this is to determine the time to pay off the equipment, *capital write-off time, t_{wo}*, typically years. This is divided into the cost of capital equipment, c_e.[2] Two other adjustments are needed. First, it is likely that the equipment will not be used productively 100 percent of the available time, so the cost is divided by a load factor, L, the fractional time the equipment is productive. Also, since the productive equipment time may be shared between several parts, the cost assignable to a given product can be obtained by multiplying the

---

[1] A. M. K. Esawi and M. F. Ashby, "Cost Estimates to Guide Pre-Selection of Processes," *Materials and Design,* vol. 24, pp. 605–616, 2003.
[2] This approach does not consider time value of money. For further details see Chap.16.

total cost by the appropriate fraction q. Finally, the cost in \$/hr is converted to \$/unit by dividing by the production rate $\dot{n}$.

$$C_E = \frac{1}{\dot{n}}\left(\frac{c_e}{Lt_{wo}}\right)q \quad \frac{h}{units}\frac{\$}{h} = \frac{\$}{unit} \tag{13.7}$$

Overhead costs are used due to the many costs in manufacturing a product that cannot be charged directly to each part or product because breaking these costs down to this level is too laborious. Examples are factory maintenance, tool crib operation, general supervision, or process R&D. These *indirect costs* are added up and then distributed to each part or product as an overhead charge. Often this is done in a fairly arbitrary way, as a cost per production time multiplied by the number of hours or seconds required to make the part. Thus, the total overhead pool is accumulated and then divided by the number of hours of production to give the hourly overhead rate, c_{OH}, \$/hr. Once again, we divide by the production rate, to find the unit overhead C_{OH}.

$$C_{OH} = \frac{c_{OH}}{\dot{n}} \quad \frac{\$}{h}\frac{h}{units} = \frac{\$}{unit} \tag{13.8}$$

Thus, the **unit cost of a part** is the sum of these five component costs: $C_U = C_M + C_L + C_T + C_E + C_{OH}$,

$$C_U = \frac{mc_m}{1-f} + \frac{c_w}{\dot{n}} + \frac{c_t k}{n} + \frac{1}{\dot{n}}\left(\frac{c_e}{Lt_{wo}}\right)q + \frac{c_{OH}}{\dot{n}} \quad \frac{\$}{unit} \tag{13.9}$$

This equation shows that the total unit cost of a part will depend on:

- Material cost, independent of the number of parts, but strongly dependent on its mass
- Tooling cost that varies inversely with the number of parts
- Labor cost, capital equipment cost, and overhead cost, which vary inversely with the rate of production

These dependencies lead to the concept of economic batch size shown in Sec. 13.4.1.

13.4.7 Availability, Lead Time, and Delivery

Next to cost, a critical business factor in selecting a manufacturing process is the availability of the production equipment, the lead time to make tooling, and the reliability of the expected delivery date for parts made by outside suppliers. Large structural parts, such as rotors for electrical generators, or the main structural forgings for military aircraft, can be made in only a few factories in the world because of equipment requirements. Careful scheduling with the design cycle may be needed to mesh with the production schedule. Complex forging dies and plastic injection molding dies can have lead times of a year. These kinds of issues clearly affect the choice of the manufacturing process and demand attention during the embodiment design phase.

13.4.8 Steps for Process Selection

The book by Schey[1] and the handbook chapter by the same author[2] are particularly helpful in the way they compare a wide spectrum of manufacturing processes. A comparison of manufacturing processes is given in Table 13.5. This is based on a series of data cards published by the Open University.[3]

This table is useful in two ways. First, it gives a quick way to screen for some broad process characteristics.

- Shape—the nature of the shapes that can be produced by the process
- Cycle time—time for a machine cycle to produce one part ($1/\dot{n}$)
- Flexibility—time to change tooling to make a different part
- Material utilization—percent of input material that ends up in finished part
- Quality—level of freedom from defects and ability to hold dimensions to drawing
- Equipment/tooling costs—level of equipment charges and tooling costs.

The rating scale for ranking processes according to these factors is in Table 13.6. (Another rating system using a more detailed listing of process characteristics is given by Schey.[4])

A second useful feature of Table 13.5 is the references to the extensive series of ASM Handbooks (AHB) and Engineered Materials Handbooks (EMH), which give many practical details on the processes.

The Manufacturing Process Information Maps (PRIMA) give much information that is useful for an initial selection of process.[5] The PRIMA selection matrix (Fig. 13.9) gives a set of 5 to 10 possible processes for different combinations of material and quantity of parts. Each PRIMA then gives the following information, which is a good summary of the information needed to make an intelligent decision on the manufacturing process:

- Process description
- Materials: materials typically used with the process
- Process variations: common variants of the basic process
- Economic factors: cycle time, minimum production quantity, material utilization, tooling costs, labor costs, lead times, energy costs, equipment costs
- Typical applications: examples of parts commonly made with this process
- Design aspects: general information on shape complexity, size range, minimum thickness, draft angles, undercuts, and limitations on other features
- Quality issues: describes defects to watch out for, expected range of surface finish, and process capability charts showing dimensional tolerances as a function of dimension

1. J. A. Schey, *Introduction to Manufacturing Processes,* 3d ed., McGraw-Hill, New York, 2000.
2. J. A. Schey, "Manufacturing Processes and Their Selection," *ASM Handbook,* Vol. 20, pp. 687–704, ASM International, Materials Park, OH, 1997.
3. Data cards to accompany L. Edwards and M. Endean, eds., *Manufacturing with Materials,* Butterworth, Boston, 1990.
4. J. A. Schey, "Manufacturing Processes and Their Selection," *ASM Handbook,* Vol. 20, pp. 687–704, ASM International, Materials Park, OH, 1997.
5. K. G. Swift and J. D. Booker, *Process Selection,* 2nd ed., Butterworth-Heinemann, Oxford, UK, 2003.

13

TABLE 13.5
Rating of Characteristics of Common Manufacturing Processes

Process	Shape	Cycle Time	Flexibility	Material Utilization	Quality	Equipment Tooling Costs	Handbook Reference
Casting							
Sand casting	3-D	2	5	2	2	1	AHB, vol. 15, p. 523
Evaporative foam	3-D	1	5	2	2	4	AHB, vol. 15, p. 637
Investment casting	3-D	2	4	4	4	3	AHB, vol. 15, p. 646
Permanent mold casting	3-D	4	2	2	3	2	AHB, vol. 15, p. 687
Pressure die casting	3-D solid	5	1	4	2	1	AHB, vol. 15, p. 713
Squeeze casting	3-D	3	1	5	4	1	AHB, vol. 15, p. 727
Centrifugal casting	3-D hollow	2	3	5	3	3	AHB, vol. 15, p. 665
Injection molding	3-D	4	1	4	3	1	EMH, vol. 2, p. 308
Reaction injection molding (RIM)	3-D	3	2	4	2	2	EMH, vol. 2, p. 344
Compression molding	3-D	3	4	4	2	3	EMH, vol. 2, p. 324
Rotational molding	3-D hollow	2	4	5	2	4	EMH, vol. 2, p. 360
Monomer casting contact molding	3-D	1	4	4	2	4	EMH, vol. 2, p. 338
Forming							
Forging, open die	3-D solid	2	4	3	2	2	AHB, vol. 14A, p. 99
Forging, hot closed die	3-D solid	4	1	3	3	2	AHB, vol. 14A, p. 111, 193
Sheet metal forming	3-D	3	1	3	4	1	AHB, vol. 14B, p. 293
Rolling	2-D	5	3	4	3	2	AHB, vol. 14A, p. 459
Extrusion	2-D	5	3	4	3	2	AHB, vol. 14A, p. 421
Superplastic forming	3-D	1	1	5	4	1	AHB, vol. 14B, p. 350
Thermoforming	3-D	3	2	3	2	3	EMH, vol. 2, p. 399
Blow molding	3-D hollow	4	2	4	4	2	EMH, vol. 2, p. 352

(Continued)

TABLE 13.5
(continued)

Process	Shape	Cycle Time	Flexibility	Material Utilization	Quality	Equipment Tooling Costs	Handbook Reference
Pressing and sintering	3-D solid	2	2	5	2	2	AHB, vol. 7, p. 326
Isostatic pressing	3-D	1	3	5	2	1	AHB, vol. 7, p. 605
Slip casting	3-D	1	5	5	2	4	EMH, vol. 14, p. 153
Machining							
Single-point cutting	3-D	2	5	1	5	5	AHB, vol. 16
Multiple-point cutting	3-D	3	5	1	5	4	AHB, vol. 16
Grinding	3-D	2	5	1	5	4	AHB, vol. 16, p. 421
Electrical discharge machining	3-D	1	4	1	5	1	AHB, vol. 16, p. 557
Joining							
Fusion welding	All	2	5	5	2	4	AHB, vol. 6, p. 175
Brazing/ soldering	All	2	5	5	3	4	AHB, vol. 6, p. 328, 349
Adhesive bonding	All	2	5	5	3	5	EMH, vol. 3
Fasteners	3-D	4	5	4	4	5	. . .
Surface treatment							
Shot peening	All	2	5	5	4	5	AHB, vol. 5, p. 126
Surface hardening	All	2	4	5	4	4	AHB, vol. 5, p. 257
CVD/PVD	All	1	5	5	4	3	AHB, vol. 5, p. 510

Rating scheme: 1, poorest; 5, best. From *ASM Handbook,* Vol. 20, p. 299, ASM International. Used with permission.

The book *Process Selection* is an excellent resource for process selection if the Cambridge Selection software is not available.

EXAMPLE 13.2

The selection of materials for an automobile fan, Example 11.2, was done with the assumption that the manufacturing costs for each material would be approximately equal since they were either casting or molding processes. The top-ranked materials were (1) an aluminum casting alloy, (2) a magnesium casting alloy, and (3) nylon 6/6 with 30 percent chopped glass fiber to increase the fracture toughness of the material. Casting or molding

TABLE 13.6
Rating Scale for Ranking Manufacturing Processes

Rating	Cycle Time	Flexibility	Material Utilization	Quality	Equipment Tooling Costs
1	>15 min	Changeover very difficult	Waste >100% of finished part	Poor quality	High machine and tooling costs
2	5 to 15 min	Slow changeover	Waste 50 to 100%	Average quality	Tooling and machines costly
3	1 to 5 min	Avg changeover and setup time	Waste 10 to 50%	Average to good quality	Tooling and machines relatively inexpensive
4	20 s to 1 min	Fast changeover	Waste < 10% finished part	Good to excellent	Tooling costs low
5	<20 s	No setup time	No appreciable waste	Excellent quality	Equip. and tooling very low

Rating scale: 1, poorest; 5, best

were given high consideration since we expect to be able to manufacture the component with the fan blades integrally attached to the fan hub.

Now we need to think more broadly about possible processes for making 500,000 parts per year. Figure 13.9 and Table 13.5 are used to perform a preliminary screening for potential processes before making a final decision based on costs calculated from Eq. (13.9). Table 13.7 shows the processes suggested in Fig. 13.9 for an aluminum alloy, a magnesium alloy, and the thermoplastic nylon 6/6.

In interpreting Table 13.7, the first consideration was whether Fig. 13.9 indicated that the process was suitable for one of the materials. The matrix of possible processes versus materials shows the greatest number of potential processes for an aluminum alloy, and the fewest for nylon 6/6. The first round of screening is made on the basis of the predominant shapes produced by each process. Thus, blow molding was eliminated because it produces thin, hollow shapes, extrusion and drawing because they produce straight shapes with high length-to-diameter ratios and because the blades must have a slight degree of twist. Sheet metal processes were eliminated because they create only 2-D shapes. In addition, Table 13.3 was consulted to see if any of the remaining processes were excluded based on shape. The bladed-hub is most similar to shape T7 in Fig. 13.6. None of the candidate processes were excluded, although information on some of the processes was missing. Machining was declared too costly by management edict. The preliminary screening left the following processes for further consideration:

Aluminum Alloy	Magnesium Alloy	Nylon 6/6
Shell molding	Gravity die casting	Injection molding
Gravity die casting	Pressure die casting	
Pressure die casting	Closed die forging	
Squeeze casting	Squeeze casting	
Closed die forging		

It is clear that injection molding is the only feasible process for the thermoplastic nylon 6/6. The available processes for aluminum or magnesium alloy come down to

13

TABLE 13.7
Initial Screening of Candidate Processes

Possible Process	Aluminum Alloy Yes or No?	Reject?	Magnesium Alloy Yes or No?	Reject?	Nylon 6/6 Yes or No?	Reject?	Reason for Elimination
1.2 Shell molding	Y		N		N		
1.3 Gravity die casting	Y		Y		N		
1.4 Pressure die casting	Y		Y		N		
1.9 Squeeze casting	Y		Y		N		
2.1 Injection molding	N		N		Y		
2.6 Blow molding	N		N		Y	R	Used for 3-D hollow shapes
2.9 Plastic extrusion	N		N		Y	R	Need to twist the blades
3.1 Closed die forging	Y		Y		N		
3.2 Rolling	Y		N		N		2-D process for making sheet
3.3 Drawing	Y	R	Y	R	N		Makes shapes with high L/D
3.4 Cold forming	Y	R	Y	R	N		Used for hollow 3-D shapes
3.5 Cold heading	Y	R	N	R	N		Used for making bolts
3.8 Sheet shearing	Y	R	Y	R	N		2-D forming process
3.12 Metal extrusion	Y	R	Y	R	N		Need to twist the blades
4A Automatic machining	Y	R	Y	R	N		Machining is ruled out by edict

several casting processes and closed die forging. These remaining processes are compared using the selection criteria given in Table 13.5. Investment casting is added as an additional process because it is known to make high-quality castings. Data for shell molding is not listed in Table 13.5, but its entry in Table 13.8 was constructed from data given in *Process Selection*. The gravity die casting process is most commonly found under the name of permanent mold casting, and the data for permanent mold casting from Table 13.5 was used in Table 13.8. The rating for each criterion is totaled for each process, as seen in Table 13.8.

The results of this process ranking are not very discriminating. All casting processes rank 13 or 14, except investment casting. The ranking for hot forging is slightly lower at 12. Moreover, designing a forging die to produce a part with 12 blades integrally attached to the fan hub is more difficult than designing a casting mold for the same shape. For this application there appears to be no advantage of forging over casting.

The next step (Example 13.3) in deciding on the manufacturing process is to compare the estimated cost to manufacture a part using Eq. (13.9). The following processes will be compared: injection molding for nylon 6/6, and low-pressure permanent mold casting, investment casting, and squeeze casting for metal alloys. Squeeze casting is

TABLE 13.8
Second Screening of Possible Manufacturing Processes

Process	Cycle Time	Process Flexibility	Material Utilization	Quality	Equipment & Tooling Costs	Total
Shell molding	5	1	4	3	1	14
Low pressure permanent mold	4	2	2	3	2	13
Pressure die casting	5	1	4	2	1	13
Squeeze casting	3	1	5	4	1	14
Investment casting	2	4	4	4	3	17
Hot closed die forging	4	1	3	3	1	12

included because it has the potential to produce low-porosity, fine detail castings when compared to shell molding and pressure die casting.

EXAMPLE 13.3

Now we use Eq. (13.9) to determine the estimated cost for making 500,000 units of the fan. By using either casting or molding we expect to be able to manufacture a component with the blades cast integral with the hub. This will eliminate assembling the blades into the hub, although there may be a requirement for a balancing step.

The radius of the bladed hub will be 9 in; see Fig. 11.9. The hub is 0.5 in. thick and has a diameter of 4 in. There are 12 blades cast into the hub, each of which is 1 in. wide at the root and 2.3 in. wide at the tip. Each blade is 0.4 in. thick, narrowing down somewhat toward the tip. About 0.7 of the volume envelope is hub and blades. Therefore, the volume of the casting is about 89 cu. in., and if cast in aluminum it would weigh 8.6 lb (3.9 kg).

Only casting or molding processes are considered, since we are interested in an integral hub and blade process. Low-pressure permanent mold casting (gravity die casting) is a variant of die casting in which the molten metal is forced upward into the die by applying low pressure on the liquid metal. Because the die cavity is filled slowly upwards, there is no entrapped air, and the casting has fewer defects. Squeeze casting is a combination of die casting and forming in which metal is introduced into the bottom half of the die and during solidification the top of the die applies high pressure to compress the semisolid material into the final shape.

The surface finish on the blades must be at least N8 (Table 13.3) to minimize fatigue failure. The tolerance on blade width and thickness should be ± 0.020 in. (0.50 mm). Figure 13.12 indicates that these quality conditions can be met by several metal casting processes, including die casting and investment casting. In addition, injection molding is the process of choice for 3-D thermoplastics, and squeeze casting was added as an innovative casting process that produces high-quality castings with high definition of details.

The requirements of the automotive fan are compared with the capabilities of four likely manufacturing processes in Table 13.9. The data for the first three processes were taken from the CES software. The data for squeeze casting was taken from Swift and Booker.[1] Note that data for investment casting has not been included because the Economic Batch Size for it is below one or two thousand parts, and we are planning for 500,000 parts annual production.

1. K. G. Swift, and J. D. Booker, *Process Selection,* 2nd ed., Butterworth-Heinemann, Oxford, UK, 2003.

13

TABLE 13.9
Comparison of Characteristics of Each Process with Requirements of the Fan

Process Requirements	Fan Design	Low-Pressure Permanent Mold Casting	Investment Casting	Injection Molding	Squeeze Casting
Size range, max mass (kg) (Fig. 13.7)	3.9	80		30	4.5
Section thickness, max (mm) (Fig. 13.8)	13	120		8	200
Section thickness, min (mm) (Fig. 13.8)	7.5	3		0.6	6
Tolerance (±mm)	0.50	0.5		0.1	0.3
Surface roughness (μm) R_a	3.2	4		0.2	1.6
Economic batch size, units (Fig. 13.5)	5×10^5	$>10^3$	$<10^3$	$>10^5$	$>10^4$

Each of the candidate processes is capable of producing symmetrical 3-D shapes. The screening parameter examined first was the economic batch size. Since it is expected that 500,000 units will be produced per year, investment casting was eliminated as a possibility because the economic batch size is less than 1000 units. Several of the other processes have borderline issues with respect to process capability, but they do not disqualify them from further analysis. For example, it may not be possible to obtain the maximum thickness of 13 mm with injection molding of nylon. This deficiency could be overcome by a different design of the hub using thinner sections and stiffening ribs; see Sec. 12.6.2. There is also a possibility that low-pressure permanent mold casting may not be able to achieve the required tolerance on critical dimensions. Experiments with process variables such as melt temperature and cooling rate will determine whether this proves to be a problem.

Now that we have narrowed the selection of a manufacturing process down to three alternatives, the final selection is based on the estimate of the cost to make one unit of the integral hub–blade fan using the cost model described in Sec. 13.4.6.

Calculations show that two machines operating three shifts for 50 weeks per year will be required to produce 500,000 units per year. This is reflected in the tooling and capital costs. Labor cost is based on one operator per machine. For the permanent mold casting and squeeze casting processes the material is A357 aluminum alloy. For injection molding the material used is nylon 6/6 reinforced with 30 percent chopped glass fibers.

It is clear from Table 13.10 that the cost of the material is the major cost category. It varies from 54 percent to 69 percent of total unit cost for the three processes studied. The production rate is also an important process parameter. It accounts for the higher cost of squeeze casting over permanent mold casting in the categories of labor cost and overhead. Process engineering studies using some of the TQM methods discussed in Chap. 4 might be able to increase the rate of production. However, there are physical limits to increasing this rate very greatly since all three processes are limited by the heat transfer rate that determines the time required to solidify the part sufficiently so that it can be ejected from the mold.

Low-pressure permanent mold casting is the obvious choice for producing the fan hub and blades. The only reason for rejecting this process would be if it was not possible to maintain required dimensions or tolerance, or if the castings contained porosity. Squeeze casting would be an attractive alternative, since the addition of mechanically induced

TABLE 13.10
Determination of Unit Cost for Three Processes Based on Cost Model in Sec. 13.4.6

Cost Element	Low-Pressure Permanent Mold	Injection Molding	Squeeze Casting
Material cost, c_m ($/lb)	0.60	1.80	0.60
Fraction of process that is scrap, f	0.1	0.05	0.1
Mass of part, m (lb)	8.6	4.1	8.6
C_M see Eq. (13.4) unit cost of material	$5.73	$7.77	$5.73
Labor cost, c_w ($/h)	25.00	25.00	25.00
Production rate, $\dot{n}$, (units/h)	38	45	30
C_L see Eq. (13.5) unit cost of labor	$0.66	$0.55	$0.83
Tooling cost, c_t ($/set)	80,000	70,000	80,000
Total production run, n (units)	500,000	500,000	500,000
Tooling life, n_t (units)	100,000	200,000	100,000
Sets of tooling required, k	5×2	3×2	5×2
C_T see Eq. (13.6) unit cost of tooling	$1.60	$0.84	$1.60
Capital cost, c_e ($)	$100,000 \times 2$	$500,000 \times 2$	200,000
Capital write-off time, t_{wo} (yrs)	5	5	5
Load fraction, L (fraction)	1	1	1
Load sharing fraction, q	1	1	1
C_E see Eq. (13.7) unit cost of capital equipment	$0.17	$0.74	$0.44
Factory overhead, c_{OH} ($/h)	60	60	60
Production rate, $\dot{n}$ (units/h)	38	45	30
C_{OH} see Eq. (13.8) unit cost of factory overhead	$1.58	$1.33	$2.00
Total unit cost $= C_M + C_L + C_T + C_E + C_{OH}$	$9.74	$11.23	$10.60

compressive stresses would result in less distortion of the metal on cooling, and the ability to hold tighter tolerances for a relatively small increase in unit cost. Injection molding of nylon 6/6 is the least attractive alternative due to the higher cost of the polymer compound.

The process selection shown in Examples 13.2 and 13.3 can be done more efficiently and with consideration of many more initial alternatives using a computer database. The CES EduPack 2010 contains datasheets on hundreds of processes similar to the one shown in Fig. 13.3.

13.5
DESIGN FOR MANUFACTURE (DFM)

For the past 20 years engineers have seen a large amount of effort devoted to the integration of design and manufacture, with the goals of reducing manufacturing cost and improving product quality. The processes and procedures that have been developed have become known as *design for manufacture* or design for manufacturability

(DFM). Associated with this is the closely related area of *design for assembly* (DFA). The field is often simply described by the abbreviation DFM/DFA or DFMA. DFMA methods should be applied during the embodiment stage of design.

Design for manufacture represents an awareness of the importance of design as the time for thoughtful consideration of all steps of production. To best achieve the goals of DFM requires a concurrent engineering team approach (Sec. 2.4.4) in which appropriate representatives from manufacturing, including outside suppliers, are members of the design team from the start.

13.5.1 DFM Guidelines

DFM guidelines are statements of good design practice that have been empirically derived from years of experience.[1] Using these guidelines helps narrow the range of possibilities so that the mass of detail that must be considered is within the capability of the designer.

1. **Minimize total number of parts:** Eliminating parts results in great savings. A part that is eliminated costs nothing to make, assemble, move, store, clean, inspect, rework, or service. A part is a good candidate for elimination if there is no need for relative motion, no need for subsequent adjustment between parts, and no need for materials to be different. However, part reduction should not go so far that it adds cost because the remaining parts become too heavy or complex.

    The best way to eliminate parts is to make minimum part count a requirement of the design at the conceptual stage of design. Combining two or more parts into an integral design architecture is another approach. Plastic parts are particularly well suited for integral design.[2] Fasteners are often prime targets for part reduction. Another advantage of making parts from plastics is the opportunity to use snap-fits instead of screws, Fig. 13.14*a*.[3]

2. **Standardize components:** Costs are minimized and quality is enhanced when standard commercially available components are used in design. The benefits also occur when a company standardizes on a minimum number of part designs (sizes, materials, processes) that are produced internally in its factories. The life and reliability of standard components may have already been established, so cost reduction comes through quantity discounts, elimination of design effort, avoidance of equipment and tooling costs, and better inventory control.

3. **Use common parts across product lines:** It is good business sense to use parts in more than one product. Specify the same materials, parts, and subassemblies in each product as much as possible. This provides economies of scale that drive down unit cost and simplify operator training and process control. Product data management (PDM) systems can be used to facilitate retrieval of similar designs.

---

1. H. W. Stoll, *Appl. Mech. Rev,* vol. 39, no. 9, pp. 1356–64, 1986; J. R. Bralla, *Design for Manufacturability Handbook,* 2nd ed., McGraw-Hill, New York, 1999; D. M. Anderson, *Design for Manufacturability,* 2nd ed., CIM Press, Cambria, CA, 2001.
2. W. Chow, *Cost Reduction in Product Design,* chap. 5, Van Nostrand Reinhold, New York, 1978.
3. P. R. Bonnenberger, *The First Snap-Fit Handbook,* 2nd ed., Hanser Gardener Publications, Cincinnati, OH, 2005.

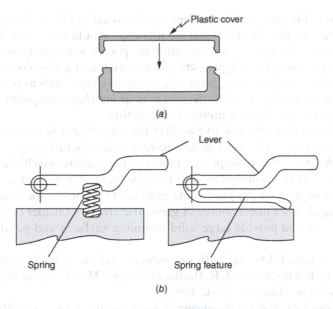

FIGURE 13.14
Some examples of applying DFM. (a) This product utilizes snap-fit principles to attach the cover, eliminating the need for screw fasteners. Since the cover is molded from plastic material and because of the taper of the snap-fit elements, it also illustrates *compliance*. (b) This illustrates a multifunctional part. By incorporating a spring function in the lever, the need for a separate coil spring is eliminated.

4. **Standardize design features:** Standardizing on design features like drilled hole sizes, screw thread types, and bend radii minimizes the number of tools that must be maintained in the tool room. This reduces manufacturing overhead cost. An exception is high-volume production where special tooling may be more cost effective.

    Space holes in machined, cast, molded, or stamped parts, so they can be made in one operation without tooling weakness. There is a limit on how close holes can be spaced due to strength in the thin section between holes.

5. **Aim to keep designs functional and simple:** Achieving functionality is paramount, but don't specify more performance than is needed. It is not good engineering to specify a heat-treated alloy steel when a plain carbon steel will achieve the performance with a little bit more careful analysis. When adding features to the design of a component, have a firm reason for the need. The product with the fewest parts, the least intricate shapes, the fewer precision adjustments, and the lowest number of manufacturing steps will be the least costly to manufacture. Also, the simplest design will usually be the most reliable and the easiest to maintain.

6. **Design parts to be multifunctional:** A good way to minimize part count is to design such that parts can fulfill more than one function, leading to integral architecture. For example, a part might serve as both a structural member and a spring, Fig. 13.14*b*. The part might be designed to provide a guiding, aligning, or self-fixturing feature in assembly. This rule can cancel out guideline 5 and break guideline 7 if it is carried too far.

7. **Design parts for ease of fabrication:** As discussed in Chap. 11, the least costly material that satisfies the functional requirements should be chosen. It is often the case that materials with higher strength have poorer workability or fabricability. Thus, one pays more for a higher-strength material, and it also costs more to process it into the required shape. Since machining to shape tends to be costly, manufacturing processes that produce the part to *near net shape* are preferred whenever possible so as to eliminate or minimize machining.

It is important to be able to visualize the steps that a machine operator will use to make a part so that you can minimize the manufacturing operations needed to make the part. For example, clamping a part before machining is a time-consuming activity, so design to minimize the number of times the operator will be required to reorient the part in the machine to complete the machining task. Reclamping also is a major source of geometric errors. Consider the needs for the use of fixtures and provide large solid mounting surfaces and parallel clamping surfaces.

Use generous fillets and radii on castings, and on molded, formed, and machined parts. For details see J. R. Bralla, *Design for Manufacturability Handbook,* 2nd ed., McGraw-Hill, New York, 1999.

8. **Avoid excessively tight tolerances:** Tolerances must be set with great care. Specifying tolerances that are tighter than needed results in increased cost; recall Fig. 13.13. These come about from the need for secondary finishing operations like grinding, honing, and lapping, from the cost of building extra precision into the tooling, from longer operating cycles because the operator is taking finer cuts, and from the need for more skilled workers. Before selecting a manufacturing process, be sure that it is capable of producing the needed tolerance and surface finish.

As a designer, it is important to maintain your credibility with manufacturing concerning tolerances. If in doubt that a tolerance can be achieved in production, always communicate with manufacturing experts. Never give a verbal agreement to manufacturing that they can loosen a tolerance without documentation and making the change on the part drawing. Also, be careful about how the statement for blanket tolerances on the drawing is worded and might be misinterpreted by manufacturing.

9. **Minimize secondary and finishing operations:** Minimize secondary operations such as heat treatment, machining, and joining and avoid finishing operations such as deburring, painting, plating, and polishing. Use only when there is a functional or safety reason for doing so. Machine a surface only when the functionality requires it or if it is needed for aesthetic purposes.

10. **Utilize the special characteristics of processes:** Be alert to the special design features that many processes provide. For example, molded polymers can be provided with "built-in" color, as opposed to metals that need to be painted or plated. Aluminum extrusions can be made in intricate cross sections that can then be cut to short lengths to provide parts. Powder-metal parts can be made with controlled porosity that provides self-lubricating bearings.

These rules are becoming the norm in every engineering design course and in engineering practice.

13.6
DESIGN FOR ASSEMBLY (DFA)

Once parts are manufactured, they need to be assembled into subassemblies and products. The assembly process consists of two operations, *handling*, which involves grasping, orienting, and positioning, followed by *insertion and fastening*. There are three types of assembly, classified by the level of automation. In *manual assembly* a human operator at a workstation reaches and grasps a part from a tray, and then moves, orients, and pre-positions the part for insertion. The operator then places the parts together and fastens them, often with a power tool. In *automatic assembly,* handling is accomplished with a parts feeder, like a vibratory bowl, that feeds the correctly oriented parts for insertion to an automatic workhead, which in turn inserts the part.[1] In *robotic assembly,* the handling and insertion of the part is done by a robot arm under computer control.

The cost of assembly is determined by the number of parts in the assembly and the ease with which the parts can be handled, inserted, and fastened. Design can have a strong influence in both areas. Reduction in the number of parts can be achieved by elimination of parts (e.g., replacing screws and washers with snap or press fits, and by combining several parts into a single component). Ease of handling and insertion is achieved by designing so that the parts cannot become tangled or nested in each other, and by designing with symmetry in mind. Parts that do not require end-to-end orientation prior to insertion, as a screw does, should be used if possible. Parts with complete rotational symmetry around the axis of insertion, like a washer, are best. When using automatic handling it is better to make a part highly asymmetric if it cannot be made symmetrical.

For ease of insertion, a part should be made with chamfers or recesses for ease of alignment, and clearances should be generous to reduce the resistance to assembly. Self-locating features are important, as is providing unobstructed vision and room for hand access. Figure 13.15 illustrates some of these points.

13.6.1 DFA Guidelines

The guidelines for design for assembly can be grouped into three classes: general, handling, and insertion.

General Guidelines

1. **Minimize the total number of parts:** A part that is not required by the design is a part that does not need to be assembled. Go through the list of parts in the assembly and identify those parts that are essential for the proper functioning of

1. G. Boothroyd, *Assembly Automation and Product Design,* 2nd ed., CRC Press, Boca Raton, FL, 2005; "Quality Control and Assembly," *Tool and Manufacturing Engineers Handbook,* vol. 4, Society of Manufacturing Engineers, Dearborn, MI, 1987.

13

Poor Assembly   **Improved Assembly**

Straight slot will tangle   Crank slot will not tangle

Difficult to orientate small chamfer on chip with mechanical tooling

Non-functional longitudinal feature simplifies orientation

Provide guide on surfaces to aid component placing

Component does not have a stable orientation

Flats on the sides make it easy to orientate with respect to small holes

FIGURE 13.15
Some design features that improve assembly.

the product. All others are candidates for elimination. The criteria for an *essential part,* also called a theoretical part, are:

- The part must exhibit motion relative to another part that is declared essential.
- There is a fundamental reason that the part be made from a material different from all other parts.
- It would not be possible to assemble or disassemble the other parts unless this part is separate, that is, it is an essential connection between parts.
- Maintenance of the product may require disassembly and replacement of a part.
- Parts used only for fastening or connecting other parts are prime candidates for elimination.

Designs can be evaluated for efficiency of assembly with Eq. (13.10), where the time taken to assemble a "theoretical" part is taken as 3 seconds.[1]

$$\text{Design assembly efficiency} = \frac{3 \times \text{"theoretical" minimum number of part}}{\text{total assembly time for all parts}} \qquad (13.10)$$

A theoretical part is one that cannot be eliminated from the design because it is needed for functionality. Typical first designs have assembly efficiencies of 5 to 10 percent, while after DFA analysis it is typically around 20 to 30 percent.

2. **Minimize the assembly surfaces:** Simplify the design so that fewer surfaces need to be prepared in assembly, and all work on one surface is completed before moving to the next one.

---

1. For small parts such as those found in household and electronic products, the assembly time runs from 2 to 10 seconds. On an automobile assembly line, times of 45 to 60 seconds are more typical.

3. **Use subassemblies:** Subassemblies can provide economies in assembly since there are fewer interfaces in final assembly. Subassemblies require connected parts that can be reoriented without falling apart and connect easily with other assembled components. Subassemblies can be built and tested elsewhere and brought to the final assembly area. When subassemblies are outsourced they should be delivered fully assembled and tested. Products made from subassemblies are easier to repair by replacing the defective subassembly.

4. **Mistake-proof the design and assembly:** An important goal in design for assembly is to ensure that the assembly process is unambiguous so that the operators cannot make mistakes in assembling the components. Components should be designed so that they can only be assembled one way. The way to orient the part in grasping it should be obvious. It should not be capable of being assembled in the reverse direction. Orientation notches, asymmetrical holes, and stops in assembly fixtures are common ways to mistake-proof the assembly process. For more on mistake-proofing, see Sec. 13.8.

Guidelines for Handling

5. **Avoid separate fasteners or minimize fastener costs:** Fasteners may amount to only 5 percent of the material cost of a product, but the labor they require for proper handling in assembly can reach 75 percent of the assembly costs. The use of screws in assembly is expensive. Snap fits should be used whenever possible. When the design permits, use fewer large fasteners rather than several small ones. Costs associated with fasteners can be minimized by standardizing on a few types and sizes of fasteners, fastener tools, and fastener torque settings. When a product is assembled with a single type of screw fastener it is possible to use auto-feed power screwdrivers.

6. **Minimize handling in assembly:** Parts should be designed to make the required position for insertion or joining obvious and easy to achieve. Orientation can be assisted by design features that help to guide and locate parts in the proper position. Parts that are to be handled by robots should have a flat, smooth top surface for vacuum grippers, or an inner hole for spearing, or a cylindrical outer surface for gripper pickup.

Guidelines for Insertion and Fastening

7. **Minimize assembly directions:** All products should be designed so that they can be assembled from one direction. Rotation of an assembly requires extra time and motion and may require additional transfer stations and fixtures. The best situation in assembly is when parts are added in a top-down manner to create a z-axis stack.

8. **Provide unobstructed access for parts and tools:** Not only must the part be designed to fit in its prescribed location, but there must be an adequate assembly path for the part to be moved to this location. This also includes room for the operator's arm and tools, which in addition to screwdrivers, could include wrenches or welding torches. If a worker has to go through contortions to perform an assembly operation, productivity and possibly product quality will suffer after a few hours of work.

9. **Maximize compliance in assembly:** Excessive assembly force may be required when parts are not identical or perfectly made. Allowance for this should be made

in the product design. Designed-in compliance features include the use of generous tapers, chamfers, and radii. If possible, one of the components of the product can be designed as the part to which other parts are added (part base) and as the assembly fixture. This may require design features that are not needed for the product function.

13.6.2 DFA Analysis

The most widely used design for assembly methodology is the Boothroyd-Dewhurst DFA method.[1] The method uses a step-by-step application of the DFA guidelines, to reduce the cost of manual assembly. The method is divided into an analysis phase and a redesign phase. In the first phase, the time required to handle and insert each part in the assembly is found from data tables based on time and motion study experiments. These values are derived from a part's size, weight, and geometric characteristics. If the part requires reorienting after being handled, that time is also included. Also, each part is identified as being essential or "theoretical," (whether it is a candidate for elimination in a redesign phase). The decision on the minimum number of theoretical parts is determined by applying the criteria listed under Guideline 1 in Sec. 13.6.1. Then the estimated total minutes to put together the assembly is determined. With this information the Design Assembly Efficiency can be determined using Eq. (13.10). This gives the designer an indication of how easily the design can be assembled, and how far the redesign phase should progress to increase assembly efficiency.

EXAMPLE 13.4
A design is needed for a motor-drive assembly that moves vertically on two steel guide rails.[2] The motor must be fully enclosed and have a removable cover for access to the position sensor. The chief functional requirement is that there be a rigid base that supports the motor and the sensor and moves up and down on the rails. The motor must be fully enclosed and have a removable cover so the position detection sensor can be adjusted.

Figure 13.16 shows the initial design of the motor-drive assembly. The rigid base is designed to slide up and down the steel guide rails (not shown). It also supports the linear motor and the position sensor. Two brass bushings are pressed into the base to provide suitable friction and wear characteristics for sliding on the steel rails. The end plate is fitted with a plastic grommet through which pass the connecting wires to the motor and the sensor. The box-shaped cover slides over the whole assembly from below the base and is held in place by four cover screws, two attached to the base and two passing into the end plate. In addition there are two stand-off rods that support the end plate and assorted screws to make a total of eight main parts and nine screws, for a total of 17 parts. The motor and sensor are outsourced subassemblies. The two guide rails are made from 0.5 in. diameter cold drawn steel bar stock. Because they are clearly

1. G. Boothroyd, P. Dewhurst and W. Knight, *Product Design for Manufacture and Assembly,* 2nd ed., Marcel Dekker, New York, 2002. DFA and DFM software is available from Boothroyd-Dewhurst, Inc. www.dfma.com.
2. G. Boothroyd, "Design for Manufacture and Assembly," *ASM Handbook,* Vol. 20, p. 676, ASM International, Materials Park, OH, 1997.

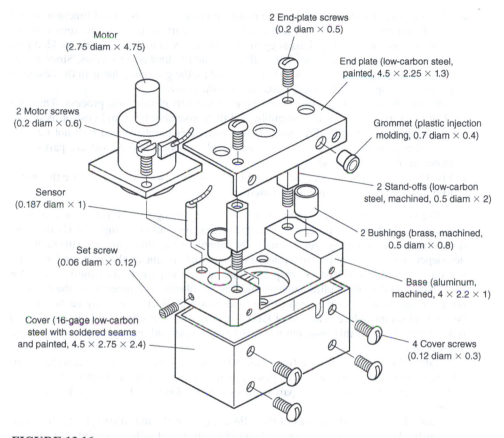

2 End-plate screws
(0.2 diam × 0.5)

Motor
(2.75 diam × 4.75)

End plate (low-carbon steel,
painted, 4.5 × 2.25 × 1.3)

2 Motor screws
(0.2 diam × 0.6)

Grommet (plastic injection
molding, 0.7 diam × 0.4)

Sensor
(0.187 diam × 1)

2 Stand-offs (low-carbon
steel, machined, 0.5 diam × 2)

2 Bushings (brass, machined,
0.5 diam × 0.8)

Set screw
(0.06 diam × 0.12)

Base (aluminum,
machined, 4 × 2.2 × 1)

Cover (16-gage low-carbon
steel with soldered seams
and painted, 4.5 × 2.75 × 2.4)

4 Cover screws
(0.12 diam × 0.3)

FIGURE 13.16
Initial design of the motor-drive assembly. (*ASM Handbook,* Vol. 20, p. 680, ASM International, Materials Park, OH, 1997. Used with permission.)

essential components of the design, and there is no apparent substitute, they are not involved in the analysis.

We now use the DFA criteria to identify the theoretical parts, those that cannot be eliminated, and the parts that are candidates for replacement, Sec. 13.6.1.

- The base is clearly an essential part. It must move along the guide rails, which is a "given" for any redesign. However, by changing the material for the base from aluminum to some other material there could be a savings in part count. Aluminum sliding on steel is not a good combination. The bushings are part of the base and are included in the design to provide the function of low sliding friction. However, it is known that nylon (a thermoplastic polymer) has a much lower sliding coefficient of friction against steel than aluminum. Using nylon for the base would permit the elimination of the two brass bushings.
- Now we consider the stand-off rods. We ask the question, Are they only there to connect two parts? Since the answer is yes, they are candidates for elimination. However, if eliminated the end plate would have to be redesigned.

- The end plate functions to protect the motor and sensor. This is a vital function, so the redesigned end plate is a cover and is a theoretical part. It must also be removable to allow access for servicing. This suggests that the cover could be a plastic molded part that would snap onto the base. This will eliminate the four cover screws. Since it will be made from a plastic, there is no longer a need for the grommet that is in the design to prevent fraying of the electrical leads entering the cover.
- Both the motor and the sensor are outside of the part elimination process. They are clearly essential parts of the assembly, and their assembly time and cost of assembly will be included in the DFA analysis. However, their purchase cost will not be considered because they are purchased from outside vendors. These costs are part of the material costs for the product.
- Finally, the set screw to hold the sensor in place and the two screws to secure the motor to the base are not theoretically required.

The time for manual assembly is determined by using lookup tables or charts[1] to estimate (1) the handling time, which includes grasping and orienting, and (2) the time for insertion and fastening. For example, the tables for handling time list different values depending on the symmetry, thickness, size, and weight of the part, and whether it requires one hand or two to grasp and manipulate the part. Extra time is added for parts with handling difficulties such as tangling, flexibility, or slipperiness, the need for optical magnification, or the need to use tools. For a product with many parts this can be a laborious procedure. The use of DFA software can be a substantial aid not only in reducing the time for this task, but in providing prompts and questions that assist in the decision process.

Tables for insertion time differentiate whether the part is secured immediately or whether other operations must take place before it can be secured. In the latter case it differentiates whether or not the part requires holding down, and how easy it is to align the part.

Table 13.11 shows the results of the DFA analysis of the initial design. As discussed previously, the base, motor, sensor, and end plate are found to be essential parts, so the theoretical part count is 4 out of a total of 19 parts. Therefore, according to Eq. (13.10), the design efficiency for the assembly is quite low, 7.5 percent, indicating that there should be ample opportunity for part elimination.

In Table 13.11 the cost of assembly is determined by multiplying the total assembly time by the hourly cost of assembly. In this example it is $30/h.

The results of the DFA analysis for the redesigned motor-drive assembly, Fig. 13.17, are given in Table 13.12. Note that the part count has been reduced from 19 to 7, with an increase in the assembly efficiency from 7.5% to 26%. There is a commensurate reduction in the cost of assembly from $1.33 to $0.384. The three nonessential parts are all screws that theoretically could be eliminated but have been retained for reliability and quality reasons. The next step is to do a design for manufacture analysis to determine whether the changes made in material and design have carried over to reduced part costs.

Example 13.4 shows the importance of DFA in design. Even though assembly follows part manufacturing, the DFA analysis contributes much more than reducing the cost of assembly, which rarely exceeds 20 percent of the product cost. A major

---

1. G. Boothroyd, et. al., op. cit., Chap. 3.

TABLE 13.11
Results of DFA Analysis for the Motor-Drive Assembly (Initial Design)

Part	No.	Theoretical Part Count	Assembly Time, s	Assembly Cost, ¢
Base	1	1	3.5	2.9
Bushing	2	0	12.3	10.2
Motor subassembly	1	1	9.5	7.9
Motor screw	2	0	21.0	17.5
Sensor subassembly	1	1	8.5	7.1
Setscrew	1	0	10.6	8.8
Stand-off	2	0	16.0	13.3
End plate	1	1	8.4	7.0
End-plate screw	2	0	16.6	13.8
Plastic bushing	1	0	3.5	2.9
Thread leads	...	...	5.0	4.2
Reorient	...	...	4.5	3.8
Cover	1	0	9.4	7.9
Cover screw	4	0	31.2	26.0
Total	19	4	160.0	133.0

Design efficiency for assembly = $(4 \times 3)/160 = 7.5\%$

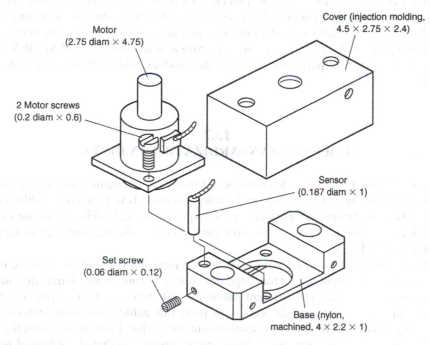

FIGURE 13.17
Redesign of motor-drive assembly based on DFA analysis. (*ASM Handbook,* Vol. 20, p. 68, ASM International, Materials Park, OH, 1997. Used with permission.)

TABLE 13.12
Results of DFA Analysis for Motor-Drive Assembly After Redesign

Part	No.	Theoretical Part Count	Assembly Time, s	Assembly Cost, ¢
Base	1	1	3.5	2.9
Motor subassembly	1	1	4.5	3.8
Motor screw	2	0	12.0	10.0
Sensor subassembly	1	1	8.5	7.1
Setscrew	1	0	8.5	7.1
Thread leads	. . .	. . .	5.0	4.2
Plastic Cover	1	1	4.0	3.3
Total	7	4	46.0	38.0

Design efficiency for assembly = (4 × 3)/46 = 26%

contribution of DFA is that it forces the design team to think critically about part elimination through redesign. A part eliminated is a part that does not require manufacturing.

DFM guidelines as "rules of thumb" have existed for centuries. The design for assembly methodology (DFA) was developed prior to the DFM software that will be described Sec. 13.9.1. DFA usually precedes DFM in time sequence since the emphasis on reducing part count in DFA serves as a driver for design for manufacture (DFM). Current thinking considers the two methodologies as complementary parts of a single methodology, *design for manufacture and assembly* (DFMA). DFA looks at the entire system of parts that make up the product, while DFM focuses on each individual part.

13.7
ROLE OF STANDARDIZATION IN DFMA

In Section 1.7 the important role of codes and standards in engineering design was introduced. There the emphasis was on the role of standards in protecting public safety and assisting the designer in performing high-quality work. In this section we extend these ideas about standardization to show the important role that part standardization can play in DFMA.

Part proliferation is an endemic problem in manufacturing unless steps are taken to prevent it from happening. One large automotive manufacturer found that in one model line alone it used 110 different radiators, 1200 types of floor carpet, and 5000 different fasteners. Reducing the variety of parts that achieve the same function can have many benefits to the product development enterprise. Firm numbers on the cost of part proliferation are difficult to obtain, but estimates are that about half of manufacturing overhead costs are related to managing too many part numbers.

13.7.1 Benefits of Standardization

The benefits of standardization occur in four areas: cost reduction, quality improvement, production flexibility, and manufacturing responsiveness.[1] The specifics of benefits in each area are outlined here.

Cost Reduction

- **Purchasing costs.** Standardization of parts and the subsequent reduction in part numbers[2] will result in large savings in procurement costs in outsourcing because parts will be bought in larger quantities. This allows for quantity discounts, flexible delivery schedules, and less work for the purchasing department.
- **Reduce costs through raw material standardization.** Cost for in-house production of parts can be reduced if raw materials can be standardized to a single size of bar stock, tubing, and sheet metal. Also, metal casting and plastic molding operations can each be limited to a single material. These standardization efforts allow for increased use of automated equipment with a minimum of cost for tool and fixture changing and setup.
- **Feature standardization.** Part features such as drilled, reamed, or threaded holes and bend radii in sheet metal all require special tools. Unless there is a dedicated machine for each size, the tools need to be changed for different dimensions, with the corresponding setup charge. Designers often specify an arbitrary hole size, when a standard size would do just as well. If the specification of radii in lathe turning or milling is not standardized it can cause a requirement for the shop to maintain a large inventory of cutting tools.
- **Reduction of inventory and floor space requirements.** The preceding cost reduction tactics assist in decreasing inventory costs either as incoming parts inventory, or the work-in-progress inventory, through fewer machine setups. Standardization makes building-on-demand more of a possibility, which will greatly decrease finished goods inventory. Reducing inventory has the advantage of reducing the required factory floor space. All of these issues, reduction of inventory and floor space, tooling costs, and purchasing and other administrative costs result in a decrease in overhead costs.

Quality Improvement

- **Product quality.** Having fewer parts of a given type greatly reduces the chance of using the wrong part in an assembly.
- **Prequalification of parts.** The use of standard parts means that there is much greater cumulative experience with using the particular part. This means that standard parts can be prequalified for use in a new product without the requirement for extensive testing.

---

1. D. M. Anderson, *Design for Manufacturability,* 2nd ed., Chap. 5, CIM Press, Cambria, CA, 2001.
2. A part number is the identification for a part (often a drawing number) and is not to be confused with the number of parts.

- **Supplier reduction means improved quality.** Standardization of parts means there will be fewer outside suppliers of parts. Those suppliers remaining should be those with a record of producing quality parts. Giving more business to fewer suppliers will be an incentive for developing stronger supplier relationships.

Production Flexibility

- **Material logistics.** The flow of parts within the plant will be easier with fewer parts to order, receive, stock, issue, assemble, test, and reorder.
- **Reliable delivery of standard low-cost parts.** These parts can be restocked directly to points of use in the plant by parts suppliers using long-term purchase agreements, much as food is delivered to a supermarket. This reduces overhead costs for purchasing and materials handling.
- **Flexible manufacturing.** Eliminating setup operations allows products to be made in any batch size. This allows the products to be made to order or to *mass customize* the product. This eliminates finished goods inventory and lets the plant make only the products for which it has an order.

Manufacturing Responsiveness

- **Parts availability.** Fewer part types used in greater volume will mean less chance of running out of parts and delaying production.
- **Quicker supplier deliveries.** Standardization of parts and materials should speed up deliveries. Suppliers will have the standard tools and materials in their inventory.
- **Financially stronger suppliers.** Part suppliers to OEMs have seen their profit margins narrow, and many have gone out of business. With larger volume orders and fewer part types to make, they can rationalize their business model, simplify their supply chain management, and reduce overhead costs. This will give them the resources to improve the quality and efficiency of their operations.

While the benefits from standardization seem very compelling, it may not always be the best course of action. For example, the compromises required by standardization may restrict the design and marketing options in undesirable ways. Stoll[1] presents advantages and disadvantages about part standardization.

13.7.2 Achieving Part Standardization

Many engineers do not realize that regardless of the cost of a part, there is real cost in ordering, shipping, receiving, inspecting, warehousing, and delivering the part to where it will be used on the assembly line. Thus, it is just as important to be concerned with standardization of inexpensive parts like fasteners, washers, and resistors as it is with more intricate molded parts.

A common misconception is that the way to achieve a minimum-cost design is to create a minimum-weight design. Certainly this may be true in aircraft and spacecraft design where weight is very important, but for most product design this design

---

1. H. W. Stoll, *Product Design Methods and Practices,* Chaps. 9 and 10, Marcel Dekker, New York, 1999.

philosophy should not be followed if it means using nonstandard parts. The most economical approach is to select the next larger *standard size* of motor, pump, or angle iron to achieve adequate strength or functionality. Special sizes are justified only in very special situations.

A common reason for the existence of part duplication is that the designer is not aware of the existence of an identical part. Even if she knows of its existence, it may be more difficult to find the part number and part drawing than it is to create a new part. This issue is discussed in Sec. 13.7.3.

13.7.3 Group Technology

Group technology (GT) is a methodology in which similar parts are grouped together in order to take advantage of their common characteristics. Parts are grouped into *part families* in terms of commonality of design features (see Fig. 13.6), as well as manufacturing processes and processing steps. Table 13.13 lists typical design and manufacturing characteristics that would be considered.

Benefits of Group Technology

- GT makes possible standardization of part design and elimination of part duplication. Since only about 20 percent of design is original design, new designs can be developed using previous similar designs, with a great saving in cost and time.
- By being able to access the previous work of the designer and the process planner, new and less experienced engineers can quickly benefit from that experience.
- Process plans for making families of parts can be standardized and retained for future use. Therefore, setup times are reduced and more consistent quality is obtained. Also, since the tools and fixtures are often shared in making a family of parts, unit costs are reduced.
- With production data aggregated in this way, cost estimates based on past experience can be made more easily, and with greater precision.

Another advantage of group technology addresses the trend among consumers for greater variety in products. This has pushed many consumer products from being mass produced products to batch production. Batch manufacturing facilities are typically organized in a *functional layout*, in which processing machines are arranged by

13

TABLE 13.13
Design and Manufacturing Characteristics That Are Typically Considered in GT Classification

Design Characteristics of Part		Manufacturing Characteristics of Part	
External shape	Part function	External shape	Annual production
Internal shape	Type of material	Major dimensions	Tooling and fixtures used
Major dimensions	Tolerances	Length/diameter ratio	Sequence of operations
Length/diameter ratio	Surface finish	Primary process used	Tolerances
Shape of raw material	Heat treatment	Secondary processes	Surface finish

common type; that is, lathes are arranged together in a common area, as are milling machines, grinders, and so on. Parts are moved from area to area as the sequence of machining operations dictates. The result is delays because of the need for tooling changes as part types change, or the machine stands idle waiting for a new batch of parts to be delivered. A functional layout is hardly a satisfactory arrangement for batch production.

A much better arrangement is using a *manufacturing cell layout.* This arrangement exploits the similarities provided by a part family. All the equipment necessary to produce a family of parts is grouped into a cell. For example, a cell could be a lineup of a lathe, milling machine, drill press, and cylindrical grinder, or it could be a CNC machining center that is equipped to do all of these machining operations, in turn, on a single computer-controlled machine. Using a cell layout, the part is transferred with minimum movement and delay from one unit of the cell to another. The machines are kept busy because GT analysis has insured that the part mix among the products made in the factory provides an adequate volume of work to make the cell layout economically viable.

Part Classification

Group technology depends on the ability to classify parts into families. At a superficial level this appears relatively easy to do, but to gain the real benefits of GT requires much experience and hard work. Classification of parts can be approached on four levels.

1. **Experience-based judgment.** The easiest approach is to assemble a team of experienced design engineers and process planners to classify parts into families based on part shape and knowledge of the sequence of processing steps used to make the part. This approach is limited in its search capabilities, and it may not assure an optimum processing sequence.
2. **Production flow analysis (PFA).** Production flow analysis uses the sequence of operations to make a part, as obtained from factory routing sheets or computer-aided process planning. Parts that are made by identical operations form a family. This is done by creating a matrix of part numbers (rows) versus machine numbers/ operation numbers. The rows and columns are rearranged, often with computer assistance, until parts that use the same process operations are identified by being grouped together in the matrix. These parts are then candidates for being incorporated into a manufacturing cell.

    The PFA method quickly ends up with very large, unwieldy matrices. A practical upper limit is several hundred parts and 20 different machines. Also, the method has difficulty if past process routing has not been done consistently.
3. **Classification and coding.** The previous two methods are chiefly aimed at improving manufacturing operations. Classification and coding is a more formal activity that is aimed at DFMA. The designer assigns a *part code* that includes such factors as basic shape, like in Fig. 13.6, external shape features, internal features, flat surfaces, holes, gear teeth, material, surface properties, manufacturing process, and operation sequences. As of yet, there is no universally applicable or accepted coding system. Some GT systems employ a code of up to 30 digits.

13

4. **Engineering database.** With the advent of large relational databases, many companies are building their own GT systems directly applicable to their own line of products. All information found on an engineering drawing plus processing information can be archived.

Software on the market does this in one of three ways:

- The designer sketches the shape of the part on the computer screen and the computer searches for all part drawings that resemble this shape.
- The software provides the capability to rapidly browse the library of hundreds of drawings, and the designer flags those that look interesting.
- The designer annotates the part drawing with text descriptors such as the part characteristics shown in Table 13.13. Then the computer can be asked, for example, to retrieve all part drawings with an L/D ratio between certain limits, or retrieve a combination of descriptors.

Determining part classification is an active area of research, stimulated by the widespread use of CAD. The power of computational algorithms combined with the capabilities of CAD systems assure that there will be continual improvement in the automation of part classification.

13.8
MISTAKE-PROOFING

An important element of DFMA is to anticipate and avoid simple human errors that occur in the manufacturing process by taking preventive action early in the product design process. Shigeo Shingo, a Japanese manufacturing engineer, developed this idea in 1961 and called it poka-yoke.[1] In English this is usually referred to as *mistake-proofing* or *error proofing.* A basic tenet of mistake-proofing is that human errors in manufacturing processes should not be blamed on individual operators but should be considered to be system errors due to incomplete engineering design. Mistake-proofing aims at reaching a state of *zero defects,* where a defect is defined as any variation from design or manufacturing specification.

Common mistakes in manufacturing operations are:

- Mistakes setting up workpieces and tools in machines or in fixtures
- Incorrect or missing parts in assemblies
- Processing the wrong workpiece
- Improper operations or adjustment of machines

Note that mistakes can occur not only in manufacturing but in design and purchasing as well. An infamous design mistake occurred with the 1999 orbiter to Mars, when it crashed on entering the Martian atmosphere. The contractor to NASA used conventional U.S. units instead of the specified SI units in designing and building the control rockets, and the error was never detected by those who designed the control system in SI units.

1. Pronounced POH-kah YOH-kay.

13.8.1 Using Inspection to Find Mistakes

A natural response to eliminating mistakes is to increase the degree of inspection of parts by machine operators and of products by assembly line workers. However, as shown by Example 13.5, even the most rigorous inspection of the process output cannot eliminate all defects caused by mistakes.

EXAMPLE 13.5 Screening with Self-Checks and Successive Checks
Assume a part is being made with a low average defect rate of 0.25% (0.0025). In an attempt to reduce defects even further, 100 percent inspection is employed. Each operator self-checks each part, and then the operator next in line checks the work of the previous operator.

A defect rate of 0.25% represents 2500 defects in each million parts produced (2500 ppm). If an operator has a 3% error rate in self inspection, and two operators inspect each part in succession, then the number of defective parts that pass through two successive inspections is $2500(0.03)(0.03) = 2.25$ ppm. This is a very low level of defective parts. In fact it is below the magic percentage of defects of 3.4 ppm for achieving the Six Sigma level of quality (see Sec. 15.4).

However, the product is an assemblage of many parts. If each product consists of 100 parts, and each part is 999,998 ppm defect free, then a product of 100 parts has $(0.999998)^{100}$ or 999,800 ppm that are defect free. This leaves 200 ppm of assembled products that are defective. If the product has 1000 parts there would be 1999 defective products out of a million made. However, if the product has only 50 parts the defective products would decrease to 100 ppm.

The prior example shows that even with extreme and expensive 100 percent inspection, it is difficult to achieve high levels of defect-free products, even when the product is not very complex. Example 13.5 also shows that decreasing product complexity (part count) is a major factor in reducing product defects. As Shingo showed,[1] a different approach from inspection is needed to achieve low levels of defects.

13.8.2 Frequent Mistakes

There are four categories of mistakes in part production. They are design mistakes, defective material mistakes, manufacturing mistakes, and human mistakes.

The following are mistakes attributable to the design process:

- Providing ambiguous information on engineering drawings or specifications: Failure to properly use GD&T dimensions and tolerances.
- Incorrect information: Mistake in conversion of units or just plain wrong calculations.
- A poorly developed design concept that does not fully provide the needed functionality. Hastily made design decisions that result in poorly performing products with low reliability, or with dangers to the safety of humans or hazards for the environment.

1. S. Shingo, *Zero Quality Control: Source Inspection and the Poka-yoke System,* Productivity Press, Portland, OR, 1986.

TABLE 13.14
Causes of Human Mistakes and Suggested Safeguards

Human Mistakes	Safeguard
Inattentiveness	Discipline; work standardization; work instructions
Forgetfulness	Checking at regular intervals
Inexperience	Skill enhancement; work standardization
Misunderstanding	Training; checking in advance; standard work practices
Poor identification	Training; attentiveness; vigilance

Defective material is another category of mistakes. These mistakes include:

- Material that is poorly chosen because not all performance requirements have been considered in the selection. Most commonly these involve long-term properties such as corrosion or wear.
- Material that does not meet specifications but gets into production, or purchased components that are not up to quality standards.
- Parts with hard-to-detect flaws such as internal porosity or fine surface cracks because of poorly designed dies or molds, or improper processing conditions (e.g., temperature, rate of deformation, poor lubrication) for the material that is being processed.

The most common mistakes in manufacturing parts or their assembly are listed below, in decreasing order of frequency.[1]

- Omitted operations: Failure to perform a required step in the process plan.
- Omitted part: Forgetting to install a screw, gasket, or washer.
- Wrong orientation of part: A part is inserted in the proper location but in the wrong orientation.
- Misaligned part: Alignment is not sufficiently accurate to give proper fit or function.
- Wrong location of part: Part is oriented properly but in wrong location. Example: The short bolt is put in the location for the long bolt.
- Selection of wrong parts: Many parts look very much alike. Example: A 1-inch bolt is used instead of 1¼-inch bolt.
- Misadjustments: An operation is incorrectly adjusted.
- Commit a prohibited action: Often this is an accident, like dropping a wrench, or a safety violation, like failure to lock-out a power panel before hooking up a motor.
- Added material or parts: Failure to remove materials., e.g., leaving on protective cover, or cores in a casting. Adding extra parts, e.g., dropping a screw into the assembly.
- Misread, mismeasure, or misinterpret: Error in reading instruments, measuring dimensions, or understanding correct information.

Some generic human mistakes, and safeguards that can be used against committing these mistakes, are given in Table 13.14.

Constructive checking and correction, along with training and work standardization, are the best ways to limit human mistakes. However, the ultimate way to eliminate

---

1. C. M. Hinckley, *Make No Mistake,* Productivity Press, Portland, OR, 2000.

mistakes is to engineer them out of the system through improved product design and manufacturing. This process is outlined in Sec. 13.8.3.

13.8.3 Mistake-Proofing Process

The steps in a mistake-proofing process follow a general problem-solving process:

- **Identify the problem.** The nature of the mistake is not always obvious. There is a natural human tendency to conceal mistakes. Work hard to develop a culture of openness and quality consciousness. Normal inspection by sampling will not give sufficient sample size of defects in a short time to identify the parts and processes causing the problem. Instead, use 100 percent inspection when looking for the cause of an error.
- **Prioritize.** Once the sources of mistakes have been identified, classify them with a Pareto chart to find the issues with the highest frequency of occurrence and which have the greatest impact on company profits.
- **Use cause finding methods.** To identify the root cause of the mistake use the TQM tools of cause-and-effect diagram, why-why chart, and interrelationship digraph (presented in Sec. 4.6) to identify the root cause of the mistake.
- **Identify and implement solutions.** General approaches for generating mistake-proofing solutions are discussed in Sec. 13.8.4. Many solutions will reduce the defect rate in manufacturing parts and reduce the mistake rate in assembling the parts. However, the greatest impact will occur in the initial design of the part if DFMA guidelines are rigorously followed during embodiment design.
- **Evaluate.** Determine if the problem has been solved. If the solution is ineffective, revisit the mistake-proofing process.

13.8.4 Mistake-Proofing Solutions

In the broadest sense, mistake-proofing is about introducing controls to prevent mistakes, detect mistakes, or detect defects arising from mistakes. Clearly it is better to prevent mistakes through appropriate design and operational controls than to only take action once a mistake has occurred.

Mistake-proofing operates in three areas of control.

- **Control of variability,** as when a part diameter varies from piece to piece as parts are made in a manufacturing process. Control of variability is vital to making a quality product. This topic is covered in some detail in Chap. 15 under the topic of robust design.
- **Control of complexity** is addressed chiefly through DFMA guidelines and can often be traced back to issues arising with product architecture decisions in embodiment design.
- **Control of mistakes** is implemented chiefly through the design and use of mistake-proofing devices[1] as were first suggested by the *poka-yoke* methodology.

---

1. 200 examples of mistake-proofing devices are described in the Appendix A to C.M. Hinckley, op. cit.

Mistake-proofing devices can be grouped into five broad classifications:

1. **Checklists.** These are written or computer-based lists of process steps or tasks that need to be done for completeness of operation. The checklist that a commercial aircraft pilot goes through before take-off is a good example. To catch errors in operations, *duplication of actions,* as when you enter a computer password twice, is sometimes used. In manual assembly processes, instructions must be accompanied by clear pictures.
2. **Guide pins, guide ways, and slots.** These design features are used in assembly to ensure that parts are located and oriented properly. It is important that guides should align parts before critical features mate.
3. **Specialized fixtures and jigs.** These devices deal with a broader case of geometries and orientation issues. They typically are intended to catch any errors between steps in the manufacturing process.
4. **Limit switches.** Limit switches or other sensors detect mistakes in location, or the absence of a problem. These sensors trigger warnings, shut down the process, or enable it to continue. Sensors typically are interlocked with other processing equipment.
5. **Counters.** Counters, either mechanical, electrical, or optical are used to verify that the proper number of machine operations or parts have been carried out. Timers are used to verify the duration of a task.

Although the methods and examples of mistake-proofing have been given in the context of manufacturing processes, the methods can be implemented in areas such as sales, order entry, and purchasing, where the cost of mistakes is probably higher than the cost of errors that occur in manufacturing. A very similar, but more formalized process called Failure Modes and Effects Analysis (FMEA) is used to identify and improve upon potential failure modes in design; see Sec. 14.5.

13.9
EARLY ESTIMATION OF MANUFACTURING COST

The decisions about materials, shape, features, and tolerances that are made in the conceptual design and embodiment design phases determine the manufacturing cost of the product. It is not often possible to get large cost reductions once production has begun because of the high cost of change at this stage of the product development process. Therefore, we need a way of identifying costly designs as early as possible in the design process.

One way to achieve this goal is to include knowledgeable manufacturing personnel on the product design team. The importance of this is unassailable, but it is not always possible from a practical standpoint due to conflicts in time commitments, or even because the design and manufacturing personnel may not be in the same location.

The method presented in Sec. 13.4.6 is useful for selecting between alternative possible processes on the basis of estimated unit part cost. While considerable information is used, the level of detail is not sufficient to do much better than give a relative ranking of competing manufacturing processes.

A system that is useful for cost estimation early in the design process was developed at the University of Hull.[1] It is based on data obtained from British automotive, aerospace, and light manufacturing companies. It allows for the reasonable calculation of part cost as changes are made in design details or for changes in part cost as different processes are used to manufacture the part. An important extension of the method in Sec. 13.4.6 is that the factor of part shape complexity is considered.

While DFM and DFA methods can be done manually on paper, the use of computerized methods greatly aids the designer by providing prompts and help screens, providing access to data that is often scattered in the literature, and making it easy to quickly see the effect of design changes. The use of DFMA software also teaches good design practice. Whatever the method, a major benefit from performing a DFMA analysis is that the rigor of using a formal analysis scheme invariably leads to asking better questions, and therefore to better solutions.

13.9.1 Concurrent Costing

The Design for Manufacture Concurrent Costing software developed by Boothroyd Dewhurst Inc. (www.dfma.com) allows real-time cost estimation of parts using much more detail than the methods discussed in Sec. 13.4.6. Typically the program starts by downloading a CAD file for the part that is being designed. If the design is not yet at a stage where a CAD drawing has been made, it is possible to input a shape envelope with dimensions of the part. We show the power of the software with a brief example.

EXAMPLE 13.6
We will describe the use of the software in the costing and design of a plastic cover. The material and process are selected from drop-down menus. Generally this starts with a menu of materials and processes, and selection of a class of materials gives the designer the option of selecting a specific material. Selecting the material greatly limits the choice of processes. Injection molding is the obvious choice for the hollow rectangular shell made from thermoplastic polypropylene.

Figure 13.18 shows the computer screen after the material and process have been selected. The values are determined by the part geometry that is entered as a drawing, and default values for the injection molding process. Because this is a molding process, much of the cost is determined by the cost of the mold. The DFM input will be concerned chiefly with how decisions on design details are reflected in the cost to make the tooling.

Following down the list of design parameters we come to part complexity. Part complexity is measured by the number of *surface patches* needed to describe both the inner and outer surface of the part in a 3-D CAD model. The inner surface of the cover is ribbed and contains bosses used for making screw connections, so the patches add up to a considerable value. The next two sections ask for input that further defines the mold cost.

- *Tolerances:* The tightness of the tolerance determines the care needed in machining the mold cavity.
- *Appearance:* If the part is transparent, the mold surface will need to be polished to a high degree.

1. K. G. Swift and J. D. Booker, *Process Selection,* 2nd ed., Butterworth-Heinemann, Oxford, UK, 2003; also A. J. Allen and K. G. Swift, *Proc. Instn. Mech. Engrs.,* Vol. 204, pp. 143–48, 1990.

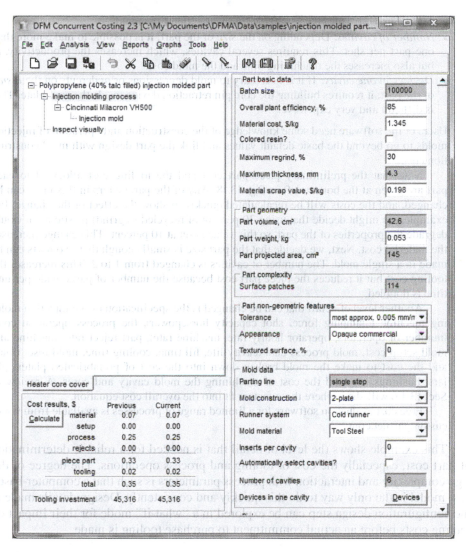

FIGURE 13.18

Design parameters used to determine cost of an injection-molded cover. (Used with permission of Boothroyd-Dewhurst, Inc.)

- *Texture:* If the surface needs to have a grain or leather appearance, this will require fine engraving of the mold surface.
- *Parting line:* The parting line refers to the shape of the surface across which the mold separates to eject the part. A straight parting plane is the least costly, but if the part design requires a stepped or curved surface, the mold cost will be significantly increased.
- *Mold construction:* A two-plate mold, a stationary cavity plate, and a moving core plate is the least complex mold system.
- *Runner system:* For high rates of production a heated runner is designed into the fixed plate. This also eliminates the need to separate the runners and sprues from the part. This mold feature adds cost to the mold.

- *Mold material:* This choice will determine the life of the mold.
- *Number of cavities:* Depending on the size of the part, it is possible to make more than one part per shot. This requires several cavities, which increases the productivity, $\dot{n}$, but also increases the machining time and the mold cost.
- *Devices in one cavity:* If it is necessary to mold depressions or undercuts on the *inside* of the part, it requires building the core pin retraction device inside the core plate. This is difficult and very expensive.

Users of the software need some knowledge of the construction and operation of injection molds to go beyond the basic default values and link the part design with mold construction costs.[1]

Note that the preliminary piece part cost and the tooling cost allocated to each part are given at the bottom left in Fig. 13.18. Any of the parameters in this table can be changed, and the costs will be recalculated quickly to show the effect of the change. For example, we might decide that using 30 percent of recycled (regrind) plastic resin would degrade the properties of the part, so this value is set at 10 percent. This change increases the material cost. Next, we decide that the part size is small enough that two parts can be made in a single mold. The number of cavities is changed from 1 to 2. This increases the tooling cost but it reduces the piece part cost because the number of parts made per unit time is doubled.

Another level of detail that can be changed is the specification of the injection molding machine (clamping force, shot capacity horsepower), the process operation costs (number of operators, operator hourly rate, machine rate), part reject rate, machine and mold setup cost, mold process data (cavity life, fill time, cooling time, mold reset time), and the cost to make the mold broken down into the cost of prefabricated plates, pillars, bushings, etc. and the cost of machining the mold cavity and cores. A review of Sec. 13.4.6 will show where these factors fit into the overall cost equation.

Free cost evaluation software for a limited range of processes is available from www.custompartnet.com.

This example shows the level of detail that is needed for a reliable determination of part cost, especially the cost of tooling and process operations. The degree of design complexity and interaction with process parameters is such that a computer-based cost model is the only way to do this quickly and consistently. Design details made at the configuration design step can be explored in a "what-if" mode for their impact on tooling costs before an actual commitment to purchase tooling is made.

13.9.2 Process Modeling and Simulation

Advances in computer technology and finite element analysis have led to industry's widespread adoption of computer manufacturing process models. Just as finite element and finite difference analyses and CFD have made possible refined design for performance of components that have reduced the cost of prototype testing, so have computer process models[2] reduced the development time and cost of tooling. The greatest application of

1. H. Rees and B. Catoen, *Selecting Injection Molds,* Hanser Gardner Publications, Cincinnati, OH, 2005.
2. ASM Handbook, Vol. 22A, *Fundamentals of Modeling for Metals Processing,* 2009; ASM Handbook, Vol. 22B, *Metals Process Simulation,* 2010, ASM International, Materials Park, OH.

process models has been with casting, injection molding, closed-die forging, and sheet metal forming processes.

Since most manufacturing processes use large equipment and expensive tooling, it is costly and time consuming to do process improvement development. A typical type of problem is making refinements to the mold to achieve complete material flow in all regions of a component made by casting or injection molding. In deformation processes like forging or extrusion, a typical problem is to modify the dies to prevent cracking in regions of high stress in the part. Today, these types of problems and many others can be solved quickly using commercially available simulation software. The results of the analysis can be seen as a series of color maps of a process parameter, such as temperature. Animations showing the actual solidification of the metal over time are commonplace. Modeling of microstructure and defects developed during casting and deformation processing have reached acceptable levels of reality. References to common process modeling software will be given as we describe the DFM guidelines for several processes in subsequent sections of this chapter.

13.10
PROCESS SPECIFIC DFMA GUIDELINES

Section 13.5 discussed general guidelines for design for manufacture (DFM), and Section 13.6 did the same for design for assembly (DFA). We have also seen how DFA can have an important impact on DFM by achieving reduction in part count. As emphasized by Boothroyd,[1] these are really complementary processes and it makes sense to consider them as a single unified process, *design for manufacturing and assembly,* DFMA.

The remaining sections in this chapter will be concerned with DFMA issues specific to the main classes of manufacturing processes. Many of these are aspects of shape that can minimize certain types of manufacturing defects, or issues with material behavior under processing conditions of which the designer needs to be aware.

Page restrictions preclude using much space to describe the process equipment and tooling. Fortunately, excellent descriptions can be found on the Internet. Two that we recommend are (1) www.custompartnet.com, which has excellent 3-D models showing the equipment and tooling, along with a detailed word descriptions about how parts are made using the process; and (2) http://aluminium.matter.org.uk, which has excellent animations that help you understand how the main process variables influence part manufacture. For access to (2) click on Processing at the upper right of the diagram, which should lead you to the process you are looking for. Another useful website (www.engineersedge.com/manufacturing_design.shtml) considers more processes but with more variability in presentation. Yet another (www.npd-solutions.com) leads to a listing of DFM rules for 15 common processes.

13

---

1. G. Boothroyd, P. Dewhurst, and W. Knight, *Product Design for Manufacture and Assembly,* 3rd ed., Taylor & Francis, Boca Raton, FL, 2010

A valuable resource for details on manufacturing processes is the big green ASM Handbooks available in most engineering libraries. They are also available in electronic form, and through knovel.com. Table 13.5 lists the most pertinent volumes in this handbook series.

13.11
DESIGN OF CASTINGS

One of the shortest routes from raw material to finished part is a casting process. In casting, a molten metal is poured into a mold or cavity that approximates the shape of the finished part (see www.custompartnet.com). Also, www.diecastingdesign.org gives useful details on this important process. Heat is extracted through the mold (in this case a sand mold), and the molten metal solidifies into the final solid shape. The chief design issues for the mold are to provide an entry for the molten metal into the mold that creates continuous laminar flow through the sprue and runner, and to provide a source of molten metal, suitably located in the mold so that it stays molten until all of the casting has been solidified. *Cores* are placed to provide hollow features for the part.

This seemingly simple process can be quite complex metallurgically, since the metal undergoes a complete transition from the superheated molten state to the solid state.[1] Liquid metal shrinks on solidification. Thus, the casting and mold must be designed so that a supply of molten metal is available to compensate for the shrinkage. The supply is furnished by introducing feeder heads (risers) that supply molten metal but must be removed from the final casting. Allowance for shrinkage and thermal contraction after the metal has solidified must also be provided in the design. Since the solubility of dissolved gases in the liquid decreases suddenly as the metal solidifies, castings are subject to the formation of trapped gas bubbles that result in porosity.

The mechanical properties of a casting are determined during solidification and subsequent heat treatment. The grain structure of the casting, and thus its properties, are determined by how fast each part of the casting freezes. The solidification time is roughly proportional to the square of the ratio of volume to surface area. Thus, bulky castings freeze much more slowly than thin section castings and have lower properties. A sphere of a given volume will freeze more slowly than a thin plate of the same volume because the plate has much more surface area to transfer heat into the mold.

The casting must be designed so that the flow of molten metal is not impeded by solidified metal before the entire mold cavity fills with molten metal. The casting should freeze progressively, with the region farthest from the source of molten metal freezing first so that the risers can supply liquid metal to feed shrinkage that occurs during solidification. Designing the needed solidification pattern can be achieved with finite element modeling to construct temperature distributions as a function of time.[2]

13

---

1. H. Fredriksson and U. Åkerlind, *Materials Processing During Casting,* John Wiley & Sons, Ltd., Chichester, UK, 2006.
2. Magmasoft, MAVIS-FLOW, ProCAST, and SOLIDcast are among the several commercial software programs.

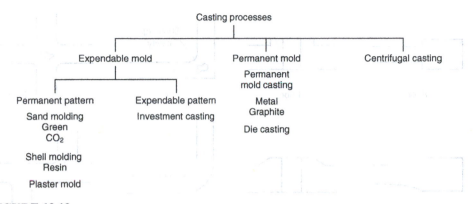

FIGURE 13.19
Classification of casting processes.

The FEA can predict shrinkage regions due to lack of feeding and grain size (property) distribution in the casting.

There are a large number of casting processes, which can be classified best with respect to the type of mold that is employed (Fig. 13.19). The two broad categories are *expendable mold* casting, in which the mold is destroyed after making each part, and *permanent mold casting*, for which many parts are made in each mold. For more details see Custompartnet.com.

13.11.1 Guidelines for the Design of Castings

Proper attention to design details can minimize casting problems and lead to lower costs.[1] Therefore, close collaboration between the designer and the foundry engineer is important. The use of computer-based solidification modeling in this design collaboration is recommended.

The chief consideration is that the shape of the casting should allow for orderly solidification by which the solidification front progresses from the remotest parts toward the points where molten metal is fed in. Whenever possible, section thickness should be uniform. Large masses of metal lead to hot spots, where freezing is delayed, and a *shrinkage cavity* is produced when the surrounding metal freezes first.

Figure 13.20 illustrates some design features that can eliminate the shrinkage cavity problem. A transition between two sections of different thicknesses should be made gradually (*a*). As a rule of thumb, the difference in thickness of adjoining sections should not exceed 2 to 1. Wedge-shaped changes in wall thickness should not have a taper exceeding 1 to 4. The thickness of a boss or pad (*b*) should be less than the thickness of the section the boss adjoins, and the transition should be gradual. The local

---

1. *Casting Design and Performance*, American Society for Metals, 2009; *ASM Handbook*, Vol. 15, American Society for Metals, Materials Park, OH, 2008; T. S. Piwonka, "Design for Casting," *ASM Handbook*, Vol. 20, pp. 723–29, ASM International, Materials Park, OH, 1997.

13

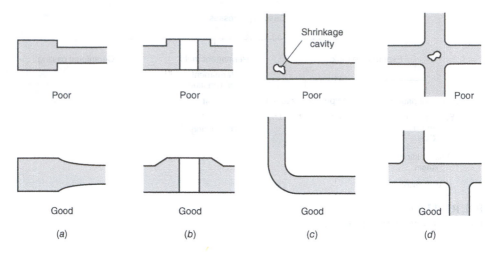

Poor   Poor   Poor   Poor

Good   Good   Good   Good

(a)   (b)   (c)   (d)

FIGURE 13.20
Some design details to minimize shrinkage cavity formation.

heavy section caused by omitting the outer radius at a corner (c) should be eliminated. The radius for good shrinkage control should be from one-half to one-third of the section thickness. A strong hot spot is produced when two ribs cross each other (d). These areas solidify after the thinner sections surrounding the junction so that the shrinkage cannot be fed with liquid metal, resulting in a shrinkage cavity. This problem can be eliminated by offsetting the ribs as shown in (d). A good way to evaluate where hot spots brought about by a large mass of molten metal occur is to inscribe a circle in the cross section of the part. The larger the diameter of the circle, the greater the thermal mass effect, and the more the concern with shrinkage cavity formation.

Castings must be designed to ensure that the pattern can be removed from the mold and the casting from a permanent mold. A *draft,* or taper, of from 6 to 3 degrees is required on vertical surfaces so the pattern can be removed from the mold. Projecting details or undercuts should be avoided, as these require extra cores. Molds made with extensive use of cores cost more money, so castings should be designed to minimize the use of cores. Also, provisions must be made for placing cores in the mold cavity and holding them in place when the metal flows into the mold.

Solidification stresses can occur when different sections of the casting solidify at different times and rates. If this happens while the alloy is cooling through the temperature range where both liquid and solids coexist, it can result in internal fracture called *hot tearing.* Uneven cooling as the temperature continues to drop can result in severe distortion or warping of the casting.

Some casting processes like die casting, permanent mold casting, and investment casting produce parts with excellent dimensional accuracy and smooth surfaces. Parts made with these processes are *net shape parts* that require no machining before using. Sand-cast parts always require machining after casting in order to attain the required dimensions and surface finish. Therefore, it is necessary to provide extra material in the casting as a *machining allowance.*

13.11.2 Producing Quality Castings

Casting offers exceptional design flexibility of shape at reasonable cost. It is an ancient metalworking process that has not always enjoyed a reputation for producing high-quality parts. A point not always understood by designers is that in a part made by casting, its mechanical properties depend on the design of the part because the properties depend on the size and shape of the grains. This depends on the solidification rate, which in turn depends on the thickness of different sections of the part. Furthermore, most casting processes are carried out in the surrounding atmosphere so that hot liquid metal can react with air to form oxide films and inclusions. Inclusions are nonmetallic particles formed by interaction with the mold or from chemical reactions among the constituents of the molten metal. These can serve as sites for crack initiation, and oxide films themselves can act like cracks.

Porosity of various origins can be present in a cast part. We have already discussed the macroporosity[1] produced by inadequate feeding of liquid while the casting is cooling. This is solved by modifications to the part geometry combined with better placement of risers in the casting. The use of solidification computer models is highly effective in identifying possible sites for macroporosity. The second type of porosity, microporosity, is more difficult to eliminate. As the metal solidifies it tends to form small, interlocking, tree-like structures called dendrites. When these approach 40 percent solid material by volume, the passages for further fluid inflow become blocked, leaving a fine porosity network. A second mechanism for pore formation arises from the fact that the solubility of gases in liquid metal decreases strongly with falling temperature. Thus, the dissolved gases are expelled from the liquid metal, and they form bubbles that grow into sizeable pores.

Microporosity can be minimized by the choice of alloy, but it is not something the designer can affect by the part design. It can be nearly eliminated by melting and pouring in the absence of air (vacuum melting), as is done for aircraft turbine blades, but this is very expensive. Another possibility is to use hot isostatic pressing (HIP) to close up any residual porosity. This process consists of enclosing the parts in a pressure vessel and subjecting them to a hydrostatic pressure of argon at 15 to 25 ksi at an elevated temperature for several hours. This will eliminate nearly all vestiges of microporosity, but again it is an expensive secondary process.

The complexity of successfully casting parts with high-quality metallurgical properties requires the designer to be able to predict the solidification of the part and the metallurgical structure of the final part. At a minimum this means being able to determine how the part will solidify before a casting is poured, and to make design alterations until a casting is obtained without macroporosity. It would be highly desirable to be able to map the temperature-time curve at critical points in the casting. To do this requires teaming up with a progressive foundry with an experienced foundry engineer who is skilled in using solidification software. The most advanced software is capable of predicting grain size and structure, and therefore being able to

---

1. Macroporosity is large enough to be seen with the unaided eye on radiographic inspection. Microporosity refers to pores that are not visible without magnification.

infer mechanical properties. This software is also capable of predicting the distortions that occur during casting.

Obtaining quality castings requires working with a foundry that is up-to-date on the latest casting technology.[1] Much of this newer technology deals with minimizing defects in castings. A high-tech foundry will be knowledgeable about and practicing such things as:

- The proper way to prepare the melt to minimize the level of inclusions and level of dissolved gas
- Design of sprue, runner, and ingate so as to minimize the distance the liquid metal falls or prevent molten metal from spilling into unfilled regions of the casting
- Mold design to keep the liquid metal front moving at all times
- Design to eliminate bubbles of entrained air in the liquid metal
- Design to avoid the need to feed metal uphill against gravity
- Finding out whether feeding requirements are established by calculation or guesswork
- Design to prevent convection problems by ensuring that thermal gradients act with rather than against gravity
- The use of filters to reduce inclusions in the casting

This knowledge area is not that of the part designer, but he should at least be aware of the issues so as to be able to evaluate the technical capability of his casting supplier.

Using casting simulation software to design the part and mold, and using a part supplier that employs modern casting operations concepts, will produce parts that perform admirably in many applications. Reducing the level of defects will result in parts with more reproducible mechanical properties than are usually expected from castings.

13.12
DESIGN OF FORGINGS

Forging processes are among the most important means of producing parts for high-performance applications. Forging is typical of a group of bulk deformation processes in which a solid billet is forced under high pressure by the use of a press to undergo extensive plastic deformation into a final near-to-finished shape.[2] Other examples of deformation processes are *extrusion,* in which a long object with a high L/D ratio is produced by pushing a metal billet through a die, *drawing,* in which a billet is pulled through a die, and *rolling,* in which a slab is passed through rolls to make a thin sheet. Figure 13.21 shows the range of bulk deformation processes that are used to convert cast material into semi-finished or final products.

1. J. Campbell, "Casting Practice-Guidelines for Effective Production of Reliable Castings," *ASM Handbook,* Vol. 15, *Casting,* pp. 497–512. J. Campbell, *Castings Practice: The Ten Rules of Castings,* 2nd ed., Butterworth-Heinemann, Oxford, UK, 2004.
2. B. L. Ferguson, "Design for Deformation Processes," *ASM Handbook,* Vol. 20, pp. 730–44, ASM International, Materials Park, OH, 1997.

13

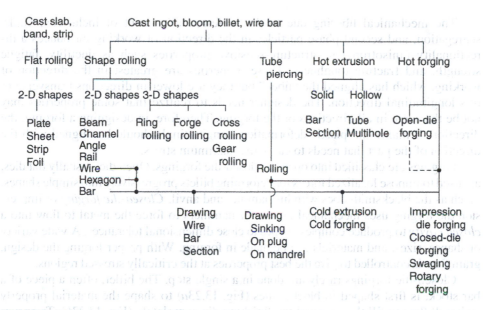

FIGURE 13.21
Bulk deformation processes that convert cast metals into semifabricated or finished products. (*After J.A. Schey*)

Forging usually is carried out on a hot workpiece, but other deformation processes such as cold extrusion or impact extrusion may be conducted cold, depending upon the material. Because of the extensive plastic deformation that occurs in forging, the metal undergoes metallurgical changes. Any porosity is closed up, and the grain structure and second phases are deformed and elongated in the principal directions of working, creating a "fiber structure." The forging billet has an axial fiber structure due to hot working, but this is redistributed depending upon the geometry of the forging (Fig. 13.22).

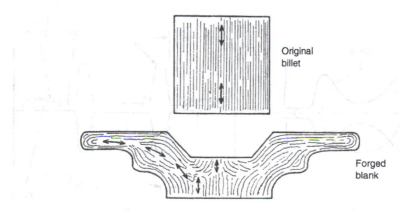

Original billet

Forged blank

FIGURE 13.22
The redistribution of the fiber structure direction during the forging of a part.

13

The mechanical fibering due to the preferred alignment of inclusions, voids, segregation, and second-phase particles in the direction of working introduces a directionality (anisotropy) to structure-sensitive properties such as ductility, fatigue strength, and fracture toughness. These properties are greatest in the direction of working, which has aligned the "fiber," but they are lower in directions transverse to this longitudinal direction. The designer needs to realize that some properties may not be the same in all directions of the forging. Therefore, in designing a forging, the direction of maximum plastic deformation (longitudinal) should be aligned with the direction of the part that needs to carry the maximum stress.

Forgings are classified into open- or closed-die forgings. Open dies, usually flat dies, are used to impose localized forces for deforming billets progressively into simple shapes, much as the blacksmith does with his hammer and anvil. *Closed-die forging* or impression die forging uses mechanical presses or hammers to force the metal to flow into a *closed cavity* to produce complex shapes to close dimensional tolerances. A wide variety of shapes, sizes, and materials can be made in forging. With proper forging die design, grain flow is controlled to give the best properties at the critically stressed regions.

Closed-die forgings rarely are done in a single step. The billet, often a piece of a bar stock, is first shaped in blocker dies (Fig. 13.23a) to shape the material properly so it will flow to fill the cavity of the finishing die completely (Fig. 13.23b). To ensure complete filling of the die cavity, a slight excess of material is used. It escapes into the flash surrounding the part, where it is "trapped" from further deformation. This causes the pressure over the rest of the workpiece to build up, forcing the workpiece material into the farthest recesses of the die. In this way, the details of the forging are achieved. Then the flash is trimmed off from the finished forging and recycled.

13.12.1 DFM Guidelines for Closed-Die Forging

Forging is essentially a molding process like casting, only now the material is a plastically deforming solid instead of a very low-viscosity fluid. Thus the DFM guidelines

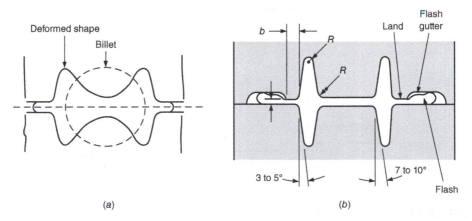

(a)   (b)

FIGURE 13.23
Schematic of closed-die forging. (a) Blocker die; (b) finishing die. (After J.A. Schey.)

for forging are very similar to casting.[1] Detailed rules for designing forgings are given in *ASM Handbook,* Vol. 14A, pp. 701–823. The Forging Industry Association (FIA) has an excellent website oriented toward showing engineering students how to design and use forgings: www.forging.org/engineer/Design_engineer.cfm. Click on Product Design Guide (electronic version) and go to Section 3.5.4.

13.12.2 Computer-Aided Forging Design

To predict the sequence of shapes to go from a piece of bar stock to a complex, defect-free forged shape requires great skill on the part of the die designer. This complex engineering task has been greatly aided by 30 years of research in applying FEA to the analysis of deformation processes. Currently, software is available for the desktop computer[2] that allows the designer to accurately determine not only press loads and die stresses but also such significant parameters as stress, strain, and temperature distribution throughout the deforming workpiece and free surface profile. An important feature of the software is the ability to visualize the geometrical changes in the workpiece as the dies close in each step of the process. The designer can make changes in the tooling design and observe on subsequent simulations whether these led to improvement in the material flow and eliminated flow defects like laps or incomplete die fill. The savings in the cost of reworking dies and trying out reworked dies have led to broad industry adoption of deformation processing simulation software. Complementary software models the change in grain size as the part undergoes forging at elevated temperature.

13.13
DESIGN FOR SHEET-METAL FORMING

Sheet metal is widely used for industrial and consumer parts because of its capacity for being bent and formed into intricate shapes. Sheet-metal parts comprise a large fraction of automotive, agricultural, and aircraft components. Successful sheet-metal forming depends on the selection of a material with adequate formability, the proper design of the part and the tooling, the surface condition of the sheet, the selection and application of lubricants, and the speed of the forming press.

13.13.1 Sheet Metal Stamping and Bending

The cold stamping of a strip or sheet of metal with dies can be classified as either a cutting or a forming operation.[3] Cutting operations are designed to *punch* holes in sheets or to separate entire parts from sheets by *blanking.* Cutting is a shearing operation which

1. J. G. Bralla, *Handbook of Product Design for Manufacturing,* Sec. 3.13, McGraw-Hill, New York, 1986.
2. DEFORM® from Scientific Forming Technologies, Columbus, OH and FORGE from Transvalor, Inc.
3. *ASM Handbook,* Vol.14B, *Metalworking: Sheet Forming,* ASM International, Materials Park, OH, I. Suchy, *Handbook of Die Design,* 2nd ed., McGraw-Hill, 2006; J. A. Schey, *Introduction to Manufacturing Processes,* 3d ed., Chap. 10, McGraw-Hill, New York, 2000.

culminates in controlled fracture. A blanked shape may be either a finished part or the first stage in a forming operation in which the final shape is created by plastic deformation, often a *bending* operation. Individual sheet metal parts are often joined into a structure by spot or laser welding. See custompartnet.com for descriptions of stamping and bending operations as well as discussion of bend allowance and springback allowance. The great variety of sheet forming operations is shown in Fig.13.24.

When holes are punched in metal sheet, only part of the metal thickness is sheared cleanly; that is, a hole with partially tapered sides is created. If the hole is to be used as a bearing surface, then a subsequent operation will be required to obtain parallel walls. Diameters of punched holes should not be less than the thickness of the sheet or a minimum of 0.025 in. Smaller holes result in excessive punch breakage and should be drilled. The minimum distance between holes, or between a hole and the edge of the sheet, should be at least equal to the sheet thickness. If holes are to be threaded, the sheet thickness must be at least one-half the thread diameter.

The ability to bend a metal without cracking at the bend improves when the bend is made across the "metal grain" (i.e., the line of the bend is perpendicular to the rolling direction of the sheet). The largest possible bend radius should be used in design to prevent cracking, and the bend radius should not be less than the sheet thickness t. The formability of sheet in bending is expressed in multiples of the sheet thickness; thus a $2t$ material has a greater formability than a sheet metal whose minimum bend radius is $4t$.

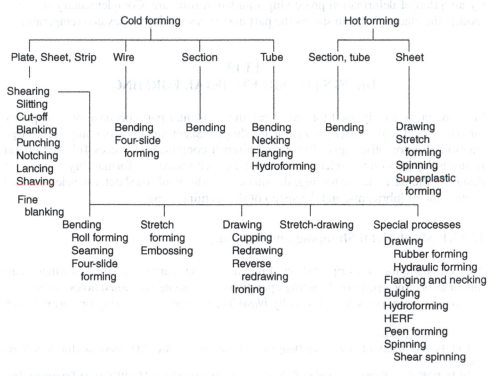

FIGURE 13.24
Classification of sheet-metalworking processes. (*After J.A. Schey*)

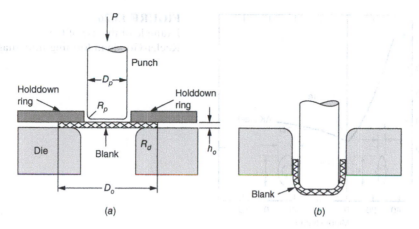

FIGURE 13.25
Deep drawing of a cylindrical cup (a) before drawing; (b) after drawing.

13.13.2 Stretching and Deep Drawing

Metal sheets are often formed into large contoured shapes such as the roof or fender of an automobile. To form such shapes requires a combination of stretching and deep drawing. In *stretching,* the sheet is clamped around its periphery and subjected to tension forces that elongate it and thin the sheet at the same time. The limit of deformation is the formation of a localized region of thinning (necking) in the sheet. This behavior is governed by the uniform elongation of the material in a tension test. The greater the capacity of the material to undergo strain hardening, the greater its resistance to necking in stretching.

The classic example of sheet drawing is *deep drawing,* as in the formation of a cup.[1] In deep drawing, the blank is "drawn" with a punch into a die, Fig. 13.25. In deep drawing the circumference of the blank is decreased when the blank is forced to conform to the smaller diameter of the punch. The resulting circumferential compressive stresses cause the blank to thicken and also to wrinkle at its outer circumference unless sufficient pressure is provided by the holddown ring or binder. However, as the metal is drawn into the die over the die radius, it is bent and then straightened while being subjected to tension. That results in substantial thinning of the sheet in the region between the punch and the die wall. The deformation conditions in deep drawing are substantially different from those in stretching. Success in deep drawing is enhanced by factors that restrict sheet thinning: a die radius about 10 times the sheet thickness, a liberal punch radius, R_p, and adequate clearance between the punch and die. Of considerable importance is the crystallographic texture or orientation of the grains relative to the sheet rolling direction. If the texture is such that the slip mechanisms favor deformation in the width direction over slip in the thickness direction of

13

---

1. See animation of deep drawing at aluminium.matter.com under Sheet Forming.

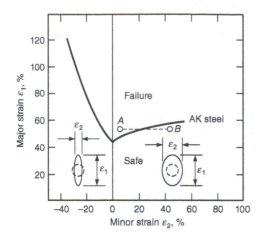

FIGURE 13.26
Example of the use of the
Keeler-Goodwin forming limit diagram.

the sheet, then deep drawing is facilitated. This property of the material can be measured with a tension test on the sheet from the *plastic strain ratio r.*

$$r = \frac{\text{strain in width direction of tension specimen}}{\text{strain in thickness direction}} = \frac{\varepsilon_w}{\varepsilon_t} \qquad (13.11)$$

The best deep-drawing sheet steels have an *r* of about 2.0.

An important tool in developing sheet-forming operations is the Keeler-Goodman forming limit diagram (Fig. 13.26). It is experimentally determined for each sheet material by placing a grid of circles on the sheet before deformation. When the sheet is deformed, the circles distort into ellipses. The major and minor axes of an ellipse represent the two principal strain directions in the stamping. Strains at points where the sheet just begins to crack are measured. The largest strain, ε_1, is plotted on the y-axis and the smaller strain, ε_2, is plotted along the x-axis. The strains are measured at points of failure for different stampings with different geometries to fill out the diagram. Strain states above the curve cause failure, and those below do not cause failure. The tension-tension sector is essentially stretching, whereas the tension-compression sector is closer to deep drawing. As an example of how to use the diagram, suppose point *A* represents the critical strains in a particular sheet metal stamping. This failure could be eliminated by changing the metal flow by either design changes to the die or the part to move the strain state to *B*. Alternatively, a material of greater formability in which the forming limit diagram was at higher values could be substituted.

13.13.3 Computer-Aided Sheet Metal Design

Several computer-aided design tools for designing dies for parts to be made by sheet metal forming[1] are used extensively by the automotive industry.[2] Another software,[3]

---

1. C-Y. Sa, "Computer-Aided Engineering in Sheet Metal Forming," *ASM Handbook,* Vol. 14B, pp. 766–90.
2. www.autoform.com; www.dynaform.com
3. www.csi-group.com

PAM-STAMP 2G, provides a completely integrated sheet metal forming simulation for a wide range of applications. The CAD model for the part is imported into the Diemaker Module where a parametric geometric model of a die is created in a matter of minutes. Next the Quikstamp Module takes the geometric model, and using elastic-plastic models of different steel and aluminum sheet materials it determines the feasibility of the formability of the design. This is done using forming limit diagrams similar to Fig. 13.26. After possibly several die design changes or changes of sheet material, the ability to make the part is verified. Then the design passes to the Autostamp Module, where a virtual die tryout is conducted in detail. The simulation can show the location of defects like splits and wrinkles and shows where *drawbeads* should be placed to alter metal deformation flow. Operating parameters such as the die hold down force and sheet lubrication can be changed to observe their effects on formability. Built-in springback prediction enables the designer to make changes in tooling geometry before any expensive tooling has been built. This software, and the others mentioned previously, allow the development of tooling in a few days, whereas with conventional "cut and try" methods it may take several months.

13.14
DESIGN FOR MACHINING

Machining operations represent the most versatile of manufacturing processes. Practically every part undergoes some kind of machining operation in its final stages of manufacture. Parts that are machined may have started out as castings or forgings and require only a few drilled holes and finishing, or they may be machined completely from bar stock or plate when only a small number of parts are needed.

There is a wide variety of machining processes with which the design engineer should be familiar.[1] Machining processes can be categorized by whether the tool translates or rotates or is stationary while the workpiece rotates. The classification of machining processes based on this system is shown in Fig. 13.27.

All machining operations produce a shape by cutting a succession of small chips from the workpiece with a hard, sharp cutting tool. There are many ways of removing material by chip formation. Some processes use a tool with a single cutting edge (e.g., lathe, shaper, planer), but most use a multipoint tool (milling, drilling, sawing, grinding). Two very different approaches to machining are forming and generating. A shape is *formed* when a cutting tool possessing the finished contour of the shape is fed (plunged) directly into the workpiece. The workpiece may be moving or stationary, as in drilling a hole.

Most machining processes *generate* a shape by relative motion between the tool and the workpiece. The *primary motion* moves the cutting tool into the workpiece, and the *feed motion* moves the point of engagement of the tool along the workpiece.

---

1. For examples see *ASM Handbook,* Vol. 16, ASM International, Materials Park, OH, 1989; J. T. Black and R. Kohser, *DeGarmo's Materials and Processes in Manufacturing,* 10th ed., John Wiley & Sons, Hoboken, NJ, 2008.

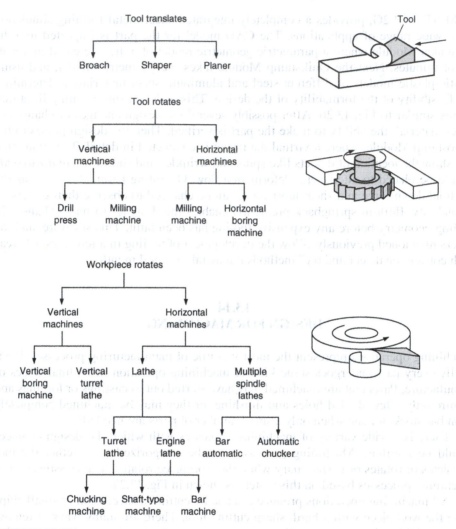

FIGURE 13.27
Classification of metal-cutting processes.

Fig. 13.28 shows some examples. See custompartnet.com for turning, milling, and hole-making machining operations. Pay special attention to equipment, process cycle, possible defects, and design rules. See aluminium.matter and follow the sequence Processing-Machining-Fund of Machining-Cutting. We suggest you look at topics in the order 1, 2, 7, 8, and 10. To gain deeper insight about individual processes, go to Cutting Processes and look at Milling, Turning, Drilling, Tapping, and Reaming as desired.

13.14.1 Machinability

Most metals and plastics can be machined, but they vary a great deal in the ease with which they can be machined, that is, their *machinability*. Machinability is a complex

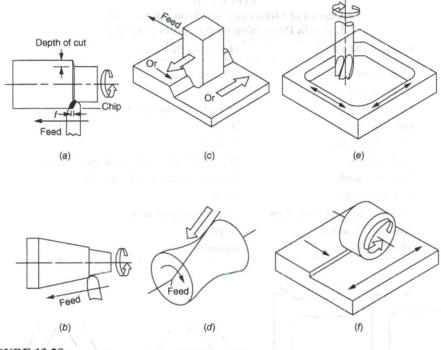

FIGURE 13.28
Programmed tool motion (feed) is necessary in generating a shape: (a) turning a cylinder and (b) a cone; (c) shaping (planing) a flat and (d) a hyperboloid; (e) milling a pocket; and (f) grinding a flat (principal motions are marked with hollow arrows, feed motions with solid arrows).

technological property that is difficult to define precisely. The machinability of a material is usually measured relative to a standard material in a particular machining process. A material has good machinability if the tool wear is low, the cutting forces are low, the chips break into small pieces instead of forming long snarls, and the surface finish is acceptable.

Machinability is a system property that depends on the workpiece material, the cutting tool material and its geometry, the type of machining operation, and its operating conditions.[1] Table 13.15 lists metallic alloys by decreasing order of machinability. The right column in this table lists various machining processes in decreasing order of machinability. For example, for any material, grinding is generally possible when other machining processes give poor results, and milling is easier to accomplish than generation of gear teeth.

Nothing has greater impact on machining costs and quality of machined parts than the machinability of the work material. Therefore, choose the material of highest machinability for the machining process you need to make the part. The one

1. D. A. Stephenson, "Design for Machining," *ASM Handbook,* Vol. 20, *Materials Selection and Design,* pp. 754–761, 1997; I. S. Jawahir, "Design for Machining: Machinability and Machining Performance Considerations," *Handbook of Metallurgical Process Design,* Chap. 22, pp. 919–959, Marcel Dekker, New York, 2004.

TABLE 13.15
Classes of Metals and Machining Processes,
Listed in Decreasing Order of Machinability

Classes of Metals	Machining Processes
Magnesium alloys	Grinding
Aluminum alloys	Sawing
Copper alloys	Turning with single-point tools
Gray cast iron	Drilling
Nodular cast iron	Milling
Carbon steels	High-speed, light feed, screw machine work
Low-alloy steels	Screw machining with form tools
Stainless steels	Boring
Hardened and high-alloy steels	Generation of gear teeth
Nickel-base superalloys	Tapping
Titanium alloys	Broaching

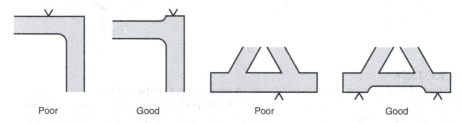

Poor   Good   Poor   Good

FIGURE 13.29
Examples of design details in castings that minimize the area of the machined surface.

generalization that can be applied to machinability is that the higher the hardness of the workpiece material, the poorer the machinability. Therefore, steel parts are usually machined in the annealed condition and then heat-treated and finished by grinding. It is necessary to leave a grinding allowance to remove any distortion from heat treatment.

13.14.2 DFM Guidelines for Machining

The following are general guidelines for designing parts that will be made by machining.[1]

- An important factor for economy in machining is to specify a machined surface only when it is needed for the functioning of the part. Two design examples for reducing the amount of machined area are shown in Fig. 13.29. A related issue is to specify the size of the workpiece so it requires only a minimum of material removal.

1. G. Boothroyd, and W. A. Knight, *Fundamentals of Machining and Machine Tools,* 2nd ed., Chap. 13, Marcel Dekker, New York, 1989; "Simplifying Machining in the Design Stage," *Tool and Manufacturing Engineers Handbook,* Vol. 6, *Design for Manufacturability* SME, Dearborn, MI, 1992.

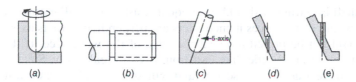

FIGURE 13.30
Some design details that affect machining operations.

- In designing a part, the sequence by which the part would be machined must be kept in mind so the design details that make machining easy are incorporated.[1] Computer modeling software for machining verifies numerical control tool paths, checks final part shape and dimensions, checks for interference between the cutting tool and the machine tool, and calculates material removal rate. Some machining software also optimizes cutting speed and feed for minimum cutting time. Most of the main CAD software vendors offer machining software.[2]
- The workpiece must have a reference surface that is suitable for holding it on the machine tool or in a fixture. A surface with three-point support is better than a large, flat surface because the workpiece is less likely to rock. Sometimes a supporting foot or tab must be added to a rough casting for support purposes. It will be removed from the final machined part.
- When possible, the design should permit all the machining to be done without reclamping the workpiece. If the part needs to be clamped in a second, different position, one of the already machined surfaces should be used as the reference surface.
- Whenever possible, the design should be such that existing tools can be used in production. When possible, the radius of the feature should be the same as the radius of the cutting tool, Fig. 13.30*a*.
- Design parts so that machining is not needed on the unexposed surfaces of the workpiece when the part is gripped in the work-holding device.
- Make sure that when the part is machined, the tool, tool holder, work-piece, and work-holding device do not interfere with one another.
- Remember that a cutting tool often requires a runout space because the tool cannot be retracted instantaneously, see Fig. 13.30*b*.
- Adjust the cutting conditions to minimize the formation of sharp burrs. A burr is a small projection of metal that adheres at the edges of the cut workpiece. If thicker than 0.4 mm, burrs cannot be removed by blast grit or tumbling methods and must be machined away.

The following guidelines pertain to drilling holes:

- The cost of a hole increases proportionately with depth, but when the depth exceeds three times the diameter, the cost increases more rapidly.

---

1. http//techtv.mit.edu/genres/24-how-to-videos/142-machine-shop-1. Also look at Fundamentals of Machining at aluminium.matters.
2. C. E. Fischer, "Modeling and Simulation of Machining," *ASM Handbook,* Vol. 22B, *Metals Process Simulation,* ASM International, 2010, pp. 361–371.

13

- When a drill is cutting, it should meet equal resistance on all cutting edges. It will if the entry and exit surfaces it encounters are perpendicular to its axis.
- Holes should not be placed too near the edge of the workpiece. If the workpiece material is weak and brittle, like cast iron, it will break away. Steel, on the other hand, will deflect at the thin section and will spring back afterward to produce a hole that is out of round.
- When there is a choice, design a through hole rather than a blind hole.

The following guidelines pertain to turning or milling operations.

- To avoid tool changing, radii should be designed to be the same as the edge of a milling cutter (Fig. 13.30a) or the nose radius of a lathe cutting tool. Of course, this rule should not supersede the need to have the appropriate radius for stress concentration considerations.
- The deflection of tools when boring or milling internal holes sets limits on the depth-to-diameter ratio.
- Undercuts can be machined if they are not too deep. It is essential to use an undercut if the design requires either external or internal threads.
- Designing features at an angle to the main tool movement direction call for special machines or attachments, Fig. 13.30c. They will be costly to make because of the need to interrupt operations and transfer to another machine.
- Placing features at an angle to the workplace surface will deflect the tool and prevent it from holding close tolerances, Fig. 13.30d. Fig. 13.30e shows an appropriate design to avoid this problem.

13.15
DESIGN OF WELDING

Welding is the most prominent process for joining large components into complex assemblies or structures. It is an important area of the wider topic of joining parts into assemblies. While welding is the process by which parts are permanently joined together, the assembly of parts is often called a *weldment*.

13.15.1 Joining Processes

Technology has created a myriad of joining processes, Fig. 13.31. They can be conveniently divided into permanent and nonpermanent joints. Nonpermanent joints are used when the assembly must be taken apart for maintenance, repair, or recycling.

Bolts and screws[1] and snap fits[2] (especially in plastic parts) are most common. Other nonpermanent joining methods are shrink and press fits, snap rings, pins, and various types of mechanical quick-release mechanisms like clamps and clips.

---

1. The design of bolts and screws is typically covered in machine design texts; see R. G. Budynas and J. K. Nisbet *Shigley's Mechanical Engineering Design,* 8th ed., Chap. 8, McGraw-Hill, New York, 2008.
2. P. R. Bonenberger, *The First Snap-Fit Handbook,* 2nd ed., Hanser Gardner Publications, Cincinnati, OH, 2005.

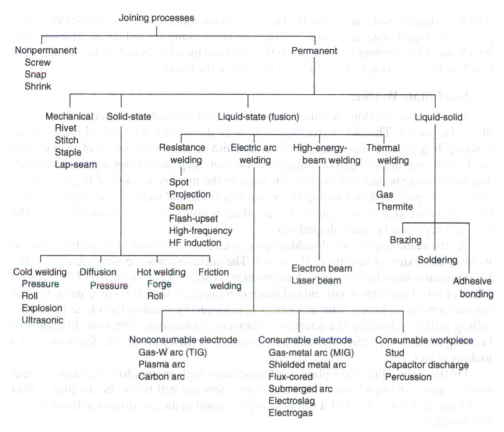

FIGURE 13.31
Classification of joining processes.

Permanent mechanical joining methods include riveting, stitching, and stapling of thin materials, and seams produced in sheet metal by making tight bends. Sometimes a sealer such as polymer or solder is used to make the seam impermeable. The majority of processes for making permanent joints involve melting, either the melting (fusion) of two metals at a joint (welding) or the addition of a molten material at a temperature where the metals at the joint have not melted (brazing, soldering, and adhesive bonding).

An extensive PRIMA selection matrix and data sheets have been developed for joining processes.[1]

13.15.2 Welding Processes

Figure 13.31 demonstrates that welding processes predominate among permanent joining processes.[2] Welding processes can be divided into solid-state processes, in

1. K. G. Swift and J. D. Booker, *Process Selection*, 2nd ed., pp. 31–34 and pp. 190–239, Butterworth-Heinemann, Oxford, UK, 2003.
2. *ASM Handbook, Welding Fundamentals and Processes,* ASM International, Materials Park, OH, 2011.

which neither the weld metal nor the base metals are melted, and two classes of fusion processes, liquid state, and liquid-solid. The latter comprise soldering and brazing, which use a low melting weld metal so that the base metal does not melt, and adhesive bonding that uses polymeric adhesives to achieve the bond.

Solid-State Welding

In solid-state welding, joining is carried out without melting either of the materials to be joined. The oldest welding process is the solid-state method called *forge welding*. It is the technique used by the blacksmith in which two pieces of steel or iron are heated and forged together under point contact. Slag and oxides are squeezed out, and interatomic bonding of the metal results. In the modern version of forge welding, steel pipe is produced by forming sheet into a cylinder and welding the edges together by forge-seam welding in which either the sheet is pulled through a conical die or the hot strip is passed between shaped rolls.

As the name implies, cold-welding processes are carried out at room temperature without any external heating of the metal. The surfaces must be very clean, and the local pressure must be high enough to produce substantial cold-working. The harmful effect of interface films is minimized when there is considerable relative movement of the surfaces to be joined. The movement is achieved by passing the metal through a rolling mill or subjecting the interface to tangential ultrasonic vibration. In explosive bonding there is very high pressure and extensive vorticity at the interfaces to form a locking effect.

Diffusion bonding takes place at a temperature high enough for diffusion to occur readily across the bond zone. Yet the temperatures are still below the melting points of the metals being joined. Hot roll bonding is a combination of diffusion bonding and roll bonding.

Friction welding (inertia welding) utilizes the frictional heat generated when two bodies slide over each other. In the usual way of doing friction welding, one part is held fixed and the other part (usually a shaft or cylinder) is rotated rapidly and, at the same time, forced axially against the stationary part. The friction quickly heats the abutting surfaces and, as soon as the proper temperature is reached, the rotation is stopped and the pressure is maintained until the weld is complete. The impurities are squeezed out into a flash, but essentially no melting takes place. The heated zone is very narrow, and therefore dissimilar metals are easily joined.

Liquid-State Welding (Fusion Welding)

In the majority of welding processes a bond between the two materials is produced by melting, usually with the addition of a filler metal in the joint between the base plates. In welding, the workpiece materials and the filler material in the joint have similar compositions and melting points. By contrast, in soldering and brazing, the filler material has a much different composition that is selected to have a lower melting point than the workpiece materials. Liquid-state welding processes in effect produce miniature castings in a mold made by the space between the plates to be welded. Thus they are subject to the defects found in castings (porosity, hot tearing) but with geometry and environmental constraints that are much more challenging. Cooling rates are rapid and much construction welding is carried out in the open air.

13

A welding process needs a controlled and often portable energy source to generate heat to melt the base metal and some way to protect the molten metal from oxidation by the environment. The large number of welding processes listed in Fig. 13.31 is the result of ongoing innovation in new processes. The first energy source was the hot gas flame provided by the combustion of acetylene and oxygen. A major improvement came with shielded metal arc process (SMAW) in which the energy is created by an electric arc struck between a consumable metal electrode and the base metal. The electrode is coated with a flux material that burns to give a protective gas around the weld and also forms a slag cover over the molten weld pool. Improvements in this process have used an arc submerged in a bed of flux and a metal arc surrounded by a cloud of inert argon gas (MIG). Another version uses a nonconsumable tungsten electrode and a metal filler rod (TIG). Major advances came with the development of high-energy beams, first electron beams and today high-energy laser beams. These make possible deep, narrow weld zones that are placed with precision in all kinds of materials. Not to be forgotten is resistance welding, which uses the heat produced when two metals are forced in contact under a high current flow. Spot welding and its variants seam and projection welding have found great application in the automotive and consumer appliance industries.

Excellent descriptions of welding processes and how they work can be found on the Internet. We suggest first exploring www.weldingengineer.com for descriptions of the fusion welding methods mentioned in the previous paragraph. Then go to www.twi.co.uk where a click on Technologies—Welding and Joining Technology will bring more information on arc welding, friction welding, resistance welding, electron beam processes, and welding of plastics. Wikipedia has an extensive article on welding (http://en.wikipedia.org/wiki/Welding) which leads to even more welding processes than those mentioned so far.

13.15.3 Welding Design

To design a weldment, consideration must be given to the selection of materials, the joint design, the selection of the welding process, and the stresses that must be resisted by the design. In liquid-state welding the process subjects the workpiece at the joint to a temperature that exceeds the melting point of the material. Heat is applied locally and rapidly to create a miniature casting in the weld pool. Often successive weld passes are laid down. The base metal next to the weld bead, the *heat-affected zone* (HAZ), is subjected to rapid heating and cooling, so the original microstructure and properties of the base metal are changed, Fig. 13.32. The figure shows coarse columnar grains characteristic of a casting in the weld joint. Into the base metal the elongated cold-worked grains have recrystallized and formed a large grain size near the original joint boundary, falling off in grain size throughout the region of the HAZ because of the difference in temperature and time that they have seen. Considerable opportunity for defects exists unless the welding process is properly carried out.

Material Behavior and Selection

Since fusion welding is a melting process, controls appropriate to producing quality castings must be applied. Reactions with the atmosphere are prevented by sealing

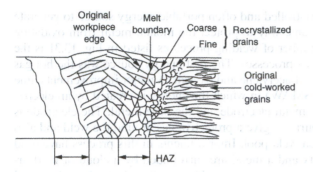

FIGURE 13.32
Sketch showing the grain structure in a section through an electric arc weld in two rolled metal plates.

off the molten pool with an inert gas or a slag or by carrying out the welding in a vacuum chamber. The surfaces of the weld joint should be cleaned of scale or grease before welding is undertaken. The thermal expansion of the weld structure upon heating, followed by solidification shrinkage, can lead to high internal tensile stresses that can produce cracking and/or distortion. Rapid cooling of alloy steels in welding can result in brittle martensite formation and consequent crack problems. Since the hardness of martensite increases with carbon content, it is common to limit welding to carbon steels with less than 0.3 percent carbon or to alloy steels in which the carbon equivalent[1] is less than 0.3 percent carbon. When steels with 0.3 to 0.6 percent carbon must be used because their high strength is required, welding without martensite cracking can be performed if the weld joint is preheated before welding and postheated after the weld bead has been deposited. These thermal treatments decrease the rate of cooling of the weld and heat-affected zone, and they reduce the likelihood of martensite formation.

Material selection for welding involves choosing a material with high weldability. Weldability, like machinability, is a complex technological property that combines many more basic properties. The melting point of the material, together with the specific heat and latent heat of fusion, will determine the heat input necessary to produce fusion. A high thermal conductivity allows the heat to dissipate and therefore requires a higher rate of heat input. Metals with higher thermal conductivity result in more rapid cooling and more problems with weld cracking. Greater distortion results from a high thermal expansion, with higher residual stresses and greater danger of weld cracking. There is no absolute rating of weldability of metals because different welding processes impose a variety of conditions that can affect the way a material responds.

Weld Joint Design

The basic types of welded joints are shown in Fig. 13.33. Many variations of these basic designs are possible, depending on the type of edge preparation that is used.

---

1. $C_{eq} = C + \dfrac{Mn}{6} + \dfrac{Cr + Mo + V}{5} + \dfrac{Ni + Cu}{15}$

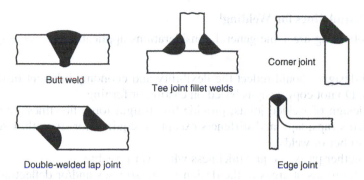

FIGURE 13.33
Basic types of welded joints.

A square-edged butt joint requires a minimum of edge preparation. However, an important parameter in controlling weld cracking is the ratio of the width of the weld bead to the depth of the weld. It should be close to unity. Since narrow joints with deep weld pools are susceptible to cracking, the most economical solution is to spend money shaping the edges of the plate to produce a joint design with a more acceptable width-to-depth ratio. Ideally, a butt weld should be a full-penetration weld that fills the joint completely throughout its depth. When the gap in a butt joint is wide, a backing strip is used at the bottom of the joint. Fillet welds (center, top row in Fig. 13.33) are the welds most commonly used in structural design. They are inherently weaker than full-penetration butt welds. A fillet weld fails in shear. The design of welded structures calls for specialized expertise that is discussed in machine design texts and books on welding design.[1]

Distortion in Welding

Distortion is ever-present in welding since it involves the rapid application of heat to a localized area, followed by the rapid removal of the heat. One of the best ways to eliminate welding distortion is to design the welding sequence with thermal distortion in mind. If, because of the geometry, distortion cannot be avoided, then the forces that produce the shrinkage distortion should be balanced with other forces provided by fixtures and clamps. Shrinkage forces can also be removed after welding by postwelding annealing and stress-relief operations. Distortion can be minimized by specifying in the design only the amount of weld metal that is absolutely required. Overwelding adds not only to the shrinkage forces but also to the costs.

13

---

1. R.G. Budynas and J.K. Nisbet, op. cit., Chap. 9; *Design of Weldments,* O. W. Blodgett, *Design of Welded Structures,* The James F. Lincoln Arc Welding Foundation, Cleveland, OH, 1963; T. G. F. Gray and J. Spencer, *Rational Welding Design,* 2nd ed., Butterworths, London, 1982; www.engineersedge. com/weld_design_menu.shtml.

DFM Guidelines for Welding[1]

The following are some general considerations applicable in designing a welded part.

- Welded designs should reflect the flexibility and economy inherent in the welding process. Do not copy designs based on casting or forging.
- In the design of welded joints, provide for straight force flowlines. Avoid the use of welded straps, laps, and stiffeners except as required for strength. Use the minimum number of welds.
- Weld together parts of equal thickness whenever possible.
- Locate the welds at areas in the design where stresses and/or deflections are least critical.
- Carefully consider the sequence with which parts should be welded together and include that information as part of the engineering drawing.
- Make sure that the welder or welding machine (for automatic welding) has unobstructed access to the joint so that a quality weld can be produced. Whenever possible, the design should provide for welding in the flat or horizontal position, not overhead.

13.15.4 Cost of Joining

We can adapt the cost of manufacture model presented in Sec. 13.4.6 to cover the cost of joining parts. The cost of joining *per unit* part, C_{unit}, is given by[2]

$$C_{unit} = \sum_{1}^{n_p} \left\{ c_{com} + (c_w \times t_{process}) + \frac{c_t}{n} + \left(\frac{c_e}{Lt_{wo}} + c_{OH} \right) \left(t_{process} + \frac{t_{setup}}{n_{batch}} \right) \right\} \quad (13.12)$$

where c_{com} is the cost of consumable materials (weld rod, fluxes, gases, adhesives, or fasteners) for one joing, \$/joint

c_t is the cost of dedicated jigs and fixtures, \$

n_p is the number of joints required to make the part (or one unit)

n is the number of joints to be made in the entire production run of the part or weldment

n_{batch} is the number of joints made per batch. For every batch there is one setup.

$t_{process}$ is the hours to make a single weld, adhesive joint, or to insert and torque one fastener

t_{setup} is the time to complete a tooling setup, h

c_w, c_e, c_{OH}, L, and t_{wo} have the same meaning as given in Sec. 13.4.6

---

1. Welding simulation software has been developed to aid in design for welding. It aids in selection of filler materials, determines the temperature distribution for given welding conditions, and estimates the properties of the HAZ. An important feature is the ability to determine residual stresses produced by the process and the resulting distortion. The most widely used is Sysweld (www.esi-group.com).
2. A. M. K. Esawi and M. F. Ashby, *Materials and Design,* Vol. 24, pp. 605–16, 2003.

The term in brackets has units of \$/joint. When summed over the n_p joints in the part it has the dimension of \$/unit. Note that c_{com} and $t_{process}$ could be different for each joint in a part.

13.16
RESIDUAL STRESSES IN DESIGN

Residual stresses are the system of stresses that can exist in a part when the part is free from external forces. They are sometimes referred to as internal stresses or locked-in stresses.[1] They arise from nonuniform plastic deformation of a solid, chiefly as a result of inhomogeneous changes in volume or shape that occur during manufacturing processes.

13.16.1 Origin of Residual Stresses

Residual stresses are developed due to nonuniform deformation. To understand how, consider an assembly that is made by joining a core and a tight-fitting tube[2] (Fig. 13.34a). Both components are made from the same material and have equal cross-sectional areas. The core was longer than the tube, so before they were joined by welding the core was compressed by a fixture to the same length as the tube, (Fig. 13.34b). After making the weld, the fixture compressing the core was removed and the assembly assumed a new length somewhere between the original lengths of the two components. Now the core wants to expand to its original length, and the tube wants to return to its original length, but they are now joined as a single assembly and the individual components cannot move. The tube has extended relative to its original length, so it is subjected to tensile residual stresses. The core has been compressed relative to its original length, so it is subjected to compressive residual stresses. Even though there is no external load on the assembly, it has a tensile stress at its surface and a compressive stress in its core (Fig. 13.34c). Because the areas of tube and core are equal, the stresses are equal and uniform in each region. The residual stress system existing in the assembly after reaching its final state (c) must be in static equilibrium. Thus, the total force acting on any plane through the body and the total moment of forces on any plane must be zero. For the longitudinal stress pattern in Fig. 13.34 this means that the area subjected to compressive residual stresses must balance the area subjected to tensile residual stresses.

The situation regarding residual stress generation is not quite so simple as is pictured in Fig. 13.34. Often residual stresses in deformation processes arise from having regions of heavy plastic deformation contiguous to regions of light deformation. The boundaries between these regions are not as simple nor are the volumes of the regions the same as in the previous example, but the results are the same. Regions of the part

---

1. W. B. Young, ed., *Residual Stresses in Design, Process and Materials Selection,* ASM International, Materials Park, OH, 1987; U. Chandra, "Control of Residual Stresses," *ASM Handbook,* Vol. 20, pp. 811–19, ASM International, Materials Park, OH, 1997.
2. J. A. Schey, *Introduction to Manufacturing Processes,* 3d ed., pp. 105–6, McGraw-Hill, New York, 2000.

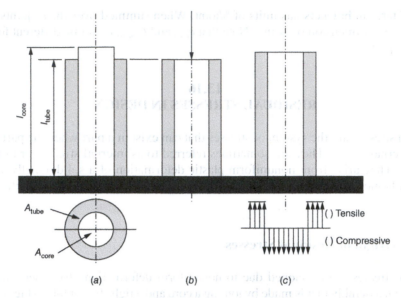

FIGURE 13.34
Example of the formation of residual stresses due to inhomogeneous deformation.

that have been required to deform in tension will upon unloading develop compressive residual stresses, and vice versa. In some cases the residual stresses acting in the three principal directions need to be known. The state of residual stress at any point is a combined stress derived from the residual stresses in the three principal directions. Frequently, because of symmetry, only the residual stress in one direction need be considered. A complete determination of the state of residual stress in three dimensions is a considerable undertaking.

Residual stresses cannot exceed the value of the yield stress of the material. A stress in excess of that value, with no external force to oppose it, will relieve itself by plastic deformation until it reaches the value of the yield stress. Residual stress and stresses from applied forces add algebraically, so long as their sum does not exceed the yield stress of the material. For example, if the maximum applied stress due to applied loads is 60,000 psi tension and the part already contains a tensile residual stress of 40,000 psi, the total stress at the critically stressed region is 100,000 psi. However if the residual stress is a compressive 40,000 psi produced by shot peening, then the actual stress is 20,000 psi.

In Fig. 13.34, if the weld holding together the two components was machined away, each component would be free to assume its original length. The assembly would undergo distortion from its intended shape. The same thing happens in parts with more complex residual stress distributions. If they are machined, removing material alters the residual stress distribution and the new balance of forces may cause distortion of the part.

Any process, whether mechanical, thermal, or chemical, that produces a permanent nonuniform change in shape or volume creates a residual stress pattern. Practically, all cold-working operations develop residual stresses because of nonuniform plastic flow.

In surface-working operations, such as shot peening, surface rolling, or polishing, the penetration of the deformation is very shallow. The distended surface layer is held in compression by the less-worked interior.

A surface compressive residual stress pattern is highly effective in reducing the incidence of fatigue failure.

Residual stresses arising from thermal processes may be classified as those due to a thermal gradient alone or to a thermal gradient in conjunction with a phase transformation, as in heat-treating steel. These stresses arise most frequently in quenching during heat treatment, or in heating and cooling experienced in casting and welding.

The control of residual stresses starts with understanding the fundamental source of the stress and identifying the parameters in the manufacturing process that influence the stress. Then, experiments are performed in varying the process parameters to produce the desired level of stress. FEA modeling has been used effectively in predicting how residual stresses can be reduced.[1]

13.16.2 Residual Stress Created by Quenching

The case of greatest practical interest involves the residual stresses developed during the quenching of steel for hardening. The residual stress pattern created is due to thermal volume changes plus volume changes resulting from the transformation of austenite to martensite. The simpler situation, in which the stresses are due only to thermal volume changes, will be considered first. This is the situation encountered in the quenching of a metal that does not undergo a phase change on cooling, like copper. This condition is also present when steel is quenched from a tempering temperature below the A_1 critical temperature.

The distribution of residual stress over the diameter of a quenched bar in the longitudinal, tangential, and radial directions is shown in Fig. 13.35a for the most common situation of a metal that contracts on cooling. Figure 13.35c shows that the opposite residual stress distribution is obtained if the metal expands on cooling (this occurs for only a few materials). The sequence of events producing the stress pattern shown in Fig. 13.35a is as follows: The relatively cool surface of the bar tends to contract into a ring that is both shorter and smaller in diameter than it was originally. This tends to extrude the hotter, more plastic center into a cylinder that is longer and thinner than it was originally. If the inner core were free to change shape independently of the outer region of the bar, it would change dimensions to a shorter and thinner cylinder on cooling. Mechanics of materials principles require that continuity must be maintained throughout the bar. Thus, the outer ring is drawn in (compressed) in the longitudinal, tangential, and radial directions at the same time the inner core is extended in the same directions. The stress pattern shown in Fig. 13.35a results.

The magnitude of the residual stresses produced by quenching depends on the stress-strain relationship for the metal and the degree of strain mismatch produced by the quenching operation. For a given strain mismatch, the higher the elastic modulus of the metal the higher the residual stress. Further, since the residual stress

---

1. U. Chandra, op. cit.

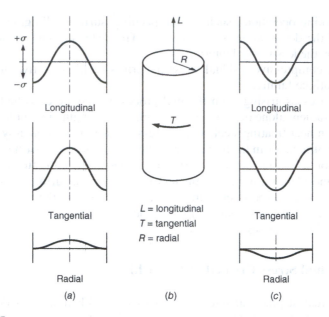

FIGURE 13.35
Residual stress patterns found in quenched bars and due to thermal strains (schematic).
(a) For metal that contracts on cooling; (b) orientation of directions; (c) for metal that expands on cooling.

cannot exceed the yield stress, the higher the yield stress the higher the possible residual stress. The yield stress–temperature curve for the metal also is important. If the yield stress decreases rapidly with increasing temperature, the strain mismatch will be small at high temperatures because the metal can accommodate to thermally produced volume changes by plastic flow. Metals that have a high yield strength at elevated temperatures, like nickel base superalloys, will develop strain mismatches in quenching, leading to high residual stresses.

The following physical properties will lead to high mismatch strains on quenching:

- Low thermal conductivity, k
- High specific heat, c
- High density, ρ
- High coefficient of thermal expansion, α

The first three factors can be combined into the thermal diffusivity $D = k/\rho c$. Low values of thermal diffusivity lead to high strain mismatch. Other process conditions that produce an increase in the temperature difference between the surface and center of the bar promote high quenching stresses. They are (1) a large diameter of the cylinder, (2) a large temperature difference between the initial temperature and the temperature of the quenching bath, and (3) rapid heat transfer at the metal-liquid interface.

In the quenching of steels, austenite (the high-temperature form of steel) begins to transform to martensite whenever the local temperature of the bar reaches the M_s temperature. Since an increase in volume accompanies the phase transformation, the metal expands as the martensite reaction proceeds on cooling from the M_s

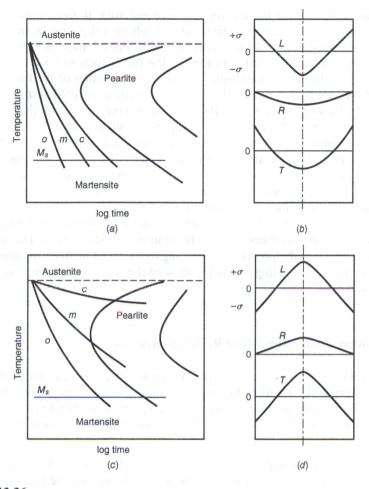

FIGURE 13.36
Transformation characteristics of a steel (a and c), and resulting residual stress distributions (b and d).

to M_f temperature.[1] This produces a residual stress distribution of the type shown in Fig. 13.35c. The residual stress distribution in a quenched steel bar is the resultant of the competing processes of thermal contraction and volume expansion due to martensite formation. The resulting stress pattern depends upon the transformation characteristics of the steel, as determined chiefly by composition and hardenability,[2] the heat-transfer characteristics of the system, and the severity of the quench.

Figure 13.36 illustrates some of the possible residual stress patterns that can be produced by quenching steel bars. On the left side of the figure is an isothermal

1. M_s and M_f are the temperatures at which martensite starts to form and finishes forming, respectively, on quenching.
2. Hardenability is an engineering measurement of the depth of hardening in a steel as a result of quenching into a specific medium.

transformation diagram for the decomposition of austenite. To form martensite the bar must cool fast enough to avoid entering the area where soft pearlite forms. The cooling rates of the outside, midradius, and center of the bar are indicated on the diagram by the curves marked o, m, and c. In Fig. 13.36a the quenching rate is rapid enough to convert the entire bar to martensite. By the time the center of the bar reaches the M_s temperature, the transformation has been essentially completed at the surface. The surface layers try to contract against the expanding central core, and the result is tensile residual stresses at the surface and compressive stresses at the center of the bar (Fig. 13.36b). However, if the bar diameter is rather small and the bar has been drastically quenched in brine so that the surface and center transform at about the same time, the surface will arrive at room temperature with compressive residual stresses. If the bar is slack-quenched so that the outside transforms to martensite while the middle and center transform to pearlite (Fig. 13.36c), there is little restraint offered by the hot, soft core during the time when martensite is forming on the surface, and the core readily accommodates to the expansion of the outer layers. The middle and center pearlite regions then contract on cooling in the usual manner and produce a residual stress pattern consisting of compression on the surface and tension at the center (Fig. 13.36d).

13.16.3 Other Issues Regarding Residual Stresses

The two previous sections were concerned with residual stresses produced by inhomogeneous plastic deformation due to mechanical forces or constraints, due to thermal expansion, or due to volume changes as a result of solid-state transformations. In this section we briefly discuss several additional important issues about residual stresses.

- The residual stresses in castings are often modeled by a quenched cylinder. However, the situation in castings is made more complicated by the fact that the mold acts as a mechanical restraint to the shrinking casting. Moreover, the casting design may produce greatly different cooling rates at different locations that are due to variations in section size and the introduction of chills. Chills are metal plates added to a sand mold to produce an artificially rapid cooling rate.
- Appreciable residual stresses are developed in welding, even in the absence of a phase transformation. As the weld metal and heat-affected zone shrink on cooling, they are restricted by the cooler surrounding plate. The result is that the weld region contains tensile residual stresses, which are balanced by compressive stresses in a region of the plate outside of the heat-affected zone. Because thermal gradients tend to be high in welding, the residual stress gradients also tend to be high. Residual stresses in welds are often involved in weld cracking.
- Chemical processes such as oxidation, corrosion, and electroplating can generate large surface residual stresses if the new surface material retains coherency with the underlying metal surface. Other surface chemical treatments such as carburizing and nitriding cause local volume changes by the diffusion of an atomic species into the surface, and this can result in residual stresses.

Residual stresses are measured by either destructive or nondestructive testing methods.[1] The destructive methods relax the locked-in stress by removing a layer from the body. The stress existing before the cut was made is calculated from the deformation produced by relaxing the stress. The nondestructive method depends on the fact that the spacing of atomic planes in a crystalline material is altered by stress. This change can be measured very precisely with a diffracted x-ray beam. The x-ray method is nondestructive, but it gives only the value of residual surface stress.

13.16.4 Relief of Residual Stresses

The removal or reduction in the intensity of residual stress is known as *stress relief.* Stress relief may be accomplished either by heating or by mechanical working operations. Although residual stresses will disappear slowly at room temperature, the process is very greatly accelerated by heating to an elevated temperature. The stress relief that comes from a stress-relief anneal is due to two effects. First, since the residual stress cannot exceed the yield stress, plastic flow will reduce the residual stress to the value of the yield stress at the stress-relief temperature. Only the residual stress in excess of the yield stress at the stress-relief temperature can be eliminated by immediate plastic flow. Generally, most of the residual stress will be relieved by time-dependent creep or stress relaxation. Since the process is extremely temperature-dependent, the time for nearly complete elimination of stress can be greatly reduced by increasing the temperature. Often a compromise must be made between the use of a temperature high enough for the relief of stress in a reasonable length of time and annealing the effects of cold working.

The differential strains that produce high residual stresses also can be eliminated by plastic deformation at room temperature. Products such as sheet, plate, and extrusions are often stretched several percent beyond their yield stress to relieve differential strains by yielding. In other cases the residual stress distribution that is characteristic of a particular working operation may be superimposed on the residual stress pattern initially present in the material. A surface that contains tensile residual stresses may have the stress distribution converted into beneficial compressive stresses by a surface-working process like rolling or shot peening. However, it is important in using this method of stress relief to select surface-working conditions that will completely cancel the initial stress distribution. If only very light surface rolling were used on a surface that initially contained tensile stresses, only the tensile stresses at the surface would be reduced. Dangerously high tensile stresses could still exist below the surface.

13.17
DESIGN FOR HEAT TREATMENT

Heat treatment is widely used to change the metallurgical structure of a part and by this process improve its mechanical properties. Common heat treatment processes are described here.

---

1. A. A. Denton, *Met. Rev.,* vol. 11, pp. 1–22, 1966; C. O. Ruud, *J. Metals,* pp. 35–40, July 1981.

- Annealing is heating a metal or alloy to an elevated temperature, holding at temperature for enough time to allow a desired metallurgical change to occur, and then cooling slowly to room temperature. It is used to relieve residual stresses, to homogenize a cast structure so that chemical segregation is minimized, or to remove the hardening effects of cold working and through the generation of new strain-free grains (recrystallization) create a structure that has adequate ductility to allow additional working.
- Quenching of steels to produce hard but brittle martensite has been described in the previous section. Quenching must be followed by heating below the transformation temperature (A_1) to temper the martensite into a precipitation of fine carbides. Quenched and tempered (Q&T) steels are the engineering materials with the best combination of readily achievable strength and toughness.
- Many nonferrous alloys, especially aluminum alloys, are strengthened by first heating to solution treatment temperature to put the alloying elements into solid solution, and then cooling rapidly to an aging temperature. The alloy is held (aged) for a sufficient time to allow the formation of a fine precipitate that hardens the alloy.

13.17.1 Issues with Heat Treatment

Processing by heat treatment requires energy. It also requires a protective atmosphere or surface coating on the metal to prevent the part from oxidizing or otherwise reacting with the furnace atmosphere. During long exposure at elevated temperatures, metal parts soften, creep, and eventually sag. Therefore, parts may require special fixtures to support them during heat treatment. Since heat treatment is a secondary processing step, it would be advantageous to eliminate the need whenever possible. Sometimes a part made from a cold-worked sheet or bar can be substituted for a heat-treated part to achieve the needed strength properties. Usually the flexibility and/or superior properties that result from heat treatment make it the preferred choice in manufacturing.

The best combination of high strength and high toughness is produced in a steel by first heating within the austenite temperature region (1400 to 1650°F) and then quenching rapidly enough that hard and brittle martensite is formed (see Fig. 13.36). The part is then reheated below the austenite region to allow the martensite to break down (temper) into a fine precipitation of carbides in a soft ferrite (iron) matrix. Achieving a proper quenched and tempered microstructure depends on cooling fast enough that pearlite or other nonmartensitic phases are not formed. This requires a balance between the heat transfer from the part (as determined chiefly by geometry), the cooling power of the quenching medium (brine, water, oil, or air), and the transformation kinetics of the steel (as controlled by the alloy chemistry). These factors are interrelated by the property called hardenability.[1]

In heating for austenitization, care should be taken to subject the parts to uniform temperature in the furnace. Long, thin parts are especially prone to distortion from nonuniform temperature. Parts containing residual stress from previous processing operations may distort on heating as the residual stresses are partially relieved.

---

1. *ASM Handbook,* Vol. 4, *Heat Treating,* ASM International, Materials Park, OH, 1991; C. A. Siebert, D. V. Doane, and D. H. Breen, *The Hardenability of Steels,* ASM International, Materials Park, OH, 1977.

Quenching is a severe treatment to impose upon a piece of steel.[1] In quenching, the part is suddenly cooled at the surface. The part must shrink rapidly because of thermal contraction (steel is at least 0.125 in. per ft larger before quenching from the austenitizing temperature), but it also undergoes a volume increase when it transforms to martensite at a comparatively low temperature. As discussed in Sec. 13.16.2 and shown in Fig. 13.36, this heat treatment can produce high residual stresses on the surface. Locally concentrated tensile stresses may be high enough to produce fractures called *quench cracks*. Local plastic deformation can occur in quenching even if cracks do not form, and that causes warping and distortion.

Problems with quench cracks and distortion are chiefly caused by nonuniformity of temperature distribution that result from part geometry as influenced by the design. Thus, many heat-treatment problems can be prevented by proper design. The most important guideline is to make the cross sections of the part as uniform as possible. In the ideal design for heat treatment, all sections should have equal ability to absorb or give up heat. Unfortunately, designing for uniform thickness or sectional area usually interferes with the functions of the design.

13.17.2 DFM for Heat Treatment

The following are good suggestions for avoiding problems in heat treatment.

- Design details that minimize stress concentrations to prevent fatigue also are good design features to minimize quench cracking.
- Distortion in heat treatment is minimized by designs that are symmetrical. A single keyway in a shaft is a particularly difficult design feature to deal with in quenching.
- A part with a severe distortion problem may have to be quenched in a special fixture that restrains it from distorting beyond the tolerance limits. Gears are often hardened this way.
- Design the part so that the quenching fluid has access to all critical regions that must be hardened. Since the quenching fluid produces a vapor blanket when it hits the hot steel surface, it may be necessary to add venting or access holes to the design of the part.

Sysweld is a design simulation software for heat treatment and welding.[2] This software assists with the steel selection and choice of the quenching media. It uses hardenability calculations to determine what the hardness distribution will be in the part. The software will also determine whether the risk of cracking is acceptable and whether distortion is within acceptable limits. It also evaluates whether compressive residual stresses are high enough and properly located in the part. DICTRA (www.thermcaic.com) is a simulation software for diffusion processes in alloy systems, such as occur in heat treating. It can be used to simulate such things as homogenization of an alloy, carburization, or coarsening of a precipitate phase.

13

---

1. A. J. Fletcher, *Thermal Stress and Strain Generation in Heat Treatment,* Elsevier Applied Science, New York, 1989.
2. See esi-group.com.

13.18
DESIGN FOR PLASTICS PROCESSING

The manufacturing processes used with plastics must accommodate to the unique flow properties of polymers. Compared to metals, the flow stress of a plastic part is much lower and highly strain rate dependent, the viscosity is much higher than liquid metal, and the formability is much greater. See Sec. 12.6 for a discussion of how the properties of plastics affect their use in design. Plastics divide broadly into (1) thermoplastic polymers (TP) that soften and melt on heating and harden when cooled. A TP can be reshaped by heating the part. (2) Thermosetting polymers (TS) that set or cross-link upon heating. On reheating they char instead of melting, so they cannot be reshaped. (3) Polymer composites that have either a TS or TP matrix reinforced with fibers of glass or graphite. TP polymers are polymerized in their primary manufacturing step and enter plastics processing as a granule or pellet resin. TS polymers are polymerized during the processing step, usually by the addition of a catalyst or simply by the addition of heat.

The plastic manufacturing processes considered in this section are:[1]

- Injection molding (mostly TP)
- Extrusion (TP)
- Blow molding (TP)
- Rotational molding (TP)
- Thermoforming (TP)
- Compression molding (mostly TS)
- Casting (mostly TS)
- Composite processing (mostly TS)

Plastic manufacturing processes excel in producing parts with good surface finish and fine detail.[2] By adding dyes and colorants, the part can be given a color that eliminates a secondary painting operation. However, the cycle time is usually longer than for metal-working processes. Depending on the plastic process, the cycle time can vary from 10 s to 10 h. Generally plastics manufacturing is the preferred method for producing small- to medium-sized parts for consumer and electronic products where mechanical stresses in parts are not too high. Thermoplastic resins are often tailored for best performance in specific processing operations. For example, a resin may be labelled "extrusion grade" because it has high molecular weight and works well in extrusion.

---

1. For a set of colorful and informative diagrams of polymer processing equipment see www.me.gatech.edu/jonathan.colton/me4210/polymer.pdf

2. *Tool and Manufacturing Engineers Handbook,* Vol. 8, *Plastic Part Manufacturing,* Society of Manufacturing Engineers, Dearborn, MI, 1996; "Engineering Plastics," *Engineered Materials Handbook* Vol. 2, Sec. 3, ASM International, Materials Park, OH, 1988; E. A. Muccio, "Design for Plastics Processing," *ASM Handbook,* Vol. 20, pp. 793–803, 1997.

13.18.1 Injection Molding

Injection molding is a process in which plastic granules are heated and forced under pressure into a die cavity (see Fig. 13.3). It is a fast process (5 to 60 s cycle time) that is economical for production runs in excess of 10,000 parts. It is well suited for producing three-dimensional shapes that require fine details like holes, snaps, and surface details. It is the plastics analog to pressure die casting in metals. There are more injection molding machines in use than any other type of plastic molding equipment. See custompartnet.com for details of equipment and mold design and also for an excellent discussion of DFM guidelines.

Design of the gating and feed system for the die is crucial to ensure complete die fill.[1] As in design for casting, it is important to design the molding so that solidification does not prevent complete mold fill. The design and location of the gates for entry of polymer into the die is a crucial design detail. In large parts there may need to be more than one gate through which resin will flow in two or more streams into the mold. These will meet inside of the mold to create a weld line. This may be a source of weakness or a surface blemish.

The mold must be designed so that the solid part can be ejected without distortion. Thus, the direction of mold closure, the parting surface between the two halves of the mold, and the part design must be considered concurrently. By proper consideration of part orientation in the mold it may be possible to avoid expensive mold costs like side cores. If at all possible, design the part so that it can be ejected in the direction of mold closure.

In addition to the economics of the process, the main DFM concerns involve the ability to achieve the required dimensional tolerances.[2] Mostly this deals with shrinkage, which is much larger in plastics than in metals. As the polymer cools from a plastic melt to a solid, the volume decreases (the density increases). Different plastics show different amounts of shrinkage. To minimize shrinkage, fillers, like glass fiber, wood flour, or natural fibers, are added during molding. Shrinkage can also be influenced by the rate and direction of injecting the melt into the mold. It is best to have any shrinkage occur while the part is confined by the mold.

13.18.2 Extrusion

Extrusion involves forcing a polymer through a die to produce long, axisymmetric shapes like rods, pipe, fiber, and sheet. Extrusion is commonly used for melting polymer resin and mixing in (compounding) fillers, colorants, and other additives like antioxidants. The homogeneous mixture is then extruded into thin rods and chopped into pellets. For information on equipment see Plastic Extrusion in Wikipedia. The chief

---

1. Software to aid in mold design and provide practical advice on manufacturing constraints is available as an add-on module with most 3-D CAD software. The most common software is Moldflow (www.moldflow.com).
2. R. A. Malloy, *Plastic Part Design for Injection Molding,* Hanser Publishers, New York, 1994.

DFM issues with the process are die swell and molecular orientation. In die swell the extrudate swells to a size greater than the die from which it just exited. Thus, the design must compensate for the swell. During extrusion, polymer molecules become highly oriented in one or two directions as a result of the strongly oriented flow inherent in the extrusion process. When molecular orientation is controlled, it can improve the properties of the material.

13.18.3 Blow Molding

Blow molding produces hollow products. A heated thermoplastic tube (called a parison) is held inside a mold and is expanded under air pressure to match the inner contour of the mold. The part cools, hardens, and is ejected from the mold. The process produces a part that is dimensionally defined on its external dimensions, but the interior surfaces are not controlled. Examples are milk bottles and automotive fuel tanks. The process does not lend itself to incorporating design details such as holes, sharp corners, or narrow ribs. For more detail see custompartnet.com.

13.18.4 Rotational Molding

Like blow molding, rotational molding produces a hollow part. Rotational molding uses a fine TP powder that is placed inside a hollow, heated metal mold. The mold is slowly rotated about two perpendicular axes. Gravity rather than centrifugal force causes even coating of the mold surface. While still rotating, the mold is cooled and the part solidifies and hardens. Rotational molding can produce large parts, like tanks up to 500 gal capacity. Since it is a low-pressure process and the plastic is not forced through narrow channels, rotational molding does not induce a significant amount of residual stress. Therefore, parts made by rotational molding exhibit a high degree of dimensional stability. For details on the process see rotational molding in Wikipedia.

13.18.5 Thermoforming

Thermoforming, or vacuum forming, is a sheet forming process in which a TP sheet is clamped to a mold and heated to soften it, and a vacuum is applied to draw the sheet into the contour of the mold. When the sheet cools, it will retain the shape of the mold. Traditionally, thermoforming is done with only a single mold, but for more precise control of dimensions, two matching mold halves are used, as is done in sheet metal forming. For more details see custompartnet.com.

13.18.6 Compression Molding

The oldest plastics process is compression molding. It is similar to powder metallurgy. A preform of polymer, usually TS, is placed in a heated mold cavity and a plunger applies pressure to force the polymer to fill the mold cavity. The plastic is allowed to cure and is then ejected from the mold. Because the amount of flow is much less than in injection molding or extrusion, the level of residual stress in the part is low.

A variation of compression molding is *transfer molding*. In this process the plastic is preheated in a transfer mold and then "shot" into the mold as a viscous liquid with a transfer ram. The ram holds the plastic under pressure until it begins to cure. Then the ram retracts, and the part completes its cure cycle and is ejected. Compression molding has a cycle time of 40 to 300 s, while for transfer molding the cycle time is 30 to 150 s. Also, because a liquid plastic enters the mold in transfer molding it is possible to mold in inserts or to encapsulate parts. However, parts made this way have sprues and runners which must be trimmed and which result in lower yield of material. For schematics of these processes see www.substech.com, Compression molding of polymers and Transfer molding of polymers.

13.18.7 Casting

Plastics are cast much less frequently than metals. The oldest applications are the casting of sheets and rods of acrylics and the "potting" of electrical components in epoxy. The development of a wider range of casting resins has led to consideration of casting as a way to make prototypes and low-volume production parts. Casting produces parts with low residual stress and a high degree of dimensional stability. Because of the high viscosity (low fluidity) of polymers it is difficult to fill molds by gravity alone and get fine detail without applying pressure, as in injection molding.

13.18.8 Composite Processing

The most common composite materials are plastics reinforced with glass, metal, or carbon fibers.[1] The reinforcement may be in the form of long, continuous filaments, short fibers, or flakes. TS polymers are the most common matrix materials. Except for filament winding, as in making a rocket motor case, the fiber and the matrix are combined in some preliminary form prior to processing. Molding compounds consist of TS resin with short, randomly dispersed fibers. Sheet molding compound (SMC) is a combination of TS resin and chopped fibers rolled into a sheet about 1/4 in. thick. Bulk molding compound (BMC) consists of the same ingredients made in billet form instead of sheet. SMC is used in the lay up of large structures. BMC is used in compression molding. Prepreg consists of long fibers in partially cured TS resin. Prepregs are available as tape or cross-plied sheets or fabrics.

Composites are made by either open-mold or closed-mold processes. In hand layups, successive layers of resin and fiber are applied to the mold by hand, with the resin being rolled into the fiber. An alternative is an open-mold process in which the

---

1. *ASM Handbook,* Vol. 21, *Composites,* ASM International, Materials Park, OH, 2001.

liquid resin and chopped glass fibers are sprayed into the surface of the mold. In bag molding, a plastic sheet or elastomer bag is clamped over the mold and pressure is applied either by drawing a vacuum or with compressed air.

Closed-mold composite processing closely follows the compression molding process. Variations have evolved to better place and orient the fibers in the composite. In resin transfer molding (RTM) a glass preform or mat is placed in the mold and a TS resin is transferred into the cavity under moderate pressure to impregnate the mat.

13.18.9 DFM Guidelines for Plastics Processing

The issues of designing with plastics have been discussed in Sec. 12.6. These issues chiefly result from the lower strength and stiffness of polymers compared with metals. This limits plastic parts to applications where stresses are low, and to parts designed with many internal stiffening features. In considering DFM it must be recognized that (1) polymers have much higher coefficients of thermal expansion and (2) much lower thermal conductivity than metals. The first issue means that molds must be carefully designed to achieve tight tolerances. The second issue means that because of slower heat conduction, the time for the part to cool from the melt to a solid object that can be ejected from the mold is too long to result in a desirably short cycle time. This drives the design for many plastic parts to have thin walls, usually less than 5 mm.

Since many of the design for manufacturing guidelines are common to all plastics processes, we have consolidated them here.

- The wall is the most important design feature of the plastic part. The wall thickness should be uniform. This is the most important design rule of all. The nominal wall thickness will vary from about 4 to 30 mm depending on the process and the plastic. The rate of change of the thickness of the nominal wall should be gradual to ensure mold filling. Avoid thick walls. They require more plastic, but more importantly, they reduce the cycle time by requiring longer time until the part is rigid enough to be ejected from the mold.
- The typical projections from the inside surface of a molded wall are ribs, webs, and bosses. Ribs and webs are used to increase stiffness rather than increasing wall thickness. A *rib* is a piece of reinforcing material between two other features that are more or less perpendicular. A *web* is a piece of bracing material between two features that are more or less parallel. A *boss* is a short block of material protruding from a wall which is used to drive a screw through or to support something in the design. Ribs should be made slightly thinner than the walls they reinforce in order to avoid sink marks (depressions) on the outside wall.
- It is important to design into a part as many features (e.g., holes, countersinks to receive fasteners, snap fits, and living hinges) as are needed rather than adding them as secondary operations. A big part of the attractiveness of plastic manufacture is that it minimizes the need for secondary operations.
- Part design and process selection affect the residual stresses formed in the part. These stresses arise from inhomogeneous flow as the polymer molecules flow through the passages of the mold. Generous radii, higher melt temperatures (which

result in longer cycle time), and processes which minimize polymer flow like rotational molding result in lower residual stresses. Lower residual stresses lead to better dimensional stability.

- Plastic parts are often used in consumer products where appearance is of great importance. An attractive feature of plastics is that they can be colored by adding color concentrates when compounding the polymer resin. The surface roughness of a molded part will reproduce the surface finish of the mold. By etching the surface of the mold, letters or logos that protrude about 0.01 mm above the surface of the part can be produced. It is much more expensive to mold depressed letters on the part, and this should be avoided if possible.
- As in forging and casting, the parting line should be chosen to avoid unnecessary complexity of the mold. Perfect mating of the parting surfaces is difficult to achieve when they are not flat. This results in a small flash all of the way around the perimeter of the part. If the flash is small due to good placement of the parting line it may be removed by tumbling, rather than a more expensive machining process. Avoid undercuts since they will require expensive movable inserts and cores.
- Tight radii can be molded but generous radii allow for better polymer flow, longer mold life, and lower stress concentrations. The minimum radii should be 1 to 1.5 mm. However, large radii lead to hot spots and sinks.
- Molded parts require a taper (draft angle) to remove the parts. The draft on exterior surfaces is a modest 0.5 to 2 degrees, with larger allowance on ribs and bosses.
- Metal inserts are often molded in plastic parts to provide functions such as screw attachments or electrical binding posts. The flow in these regions of the mold must be carefully designed to prevent weld lines. A *weld line* is formed when the fronts of two or more melt streams meet each other and fail to achieve complete intermolecular penetration. Often there is air trapped by the meeting of the flow streams. This reduces the mechanical properties in the region and affects the appearance of the surface of the part. Since there is no adhesion between the surface of metals and plastics, it is important to provide mechanical locking features, like knurling, for metal inserts in plastics.

The following sources from the Internet provide useful information and visuals about design for manufacture of plastics.

- http://engr.bd.psu.edu/pkoch/plasticdesign/e_frames.htm: Provides details of design for manufacture of plastic parts.
- http://www.protomold.com/DesignGuidelines_UniformWallThickness.aspx: Provides DFM guidelines for rapid prototyping service for injection molding.

13.19
SUMMARY

This chapter completes the core theme of the book that design, materials selection, and processing are inseparable. Decisions concerning the manufacturing of parts should be made as early as possible in the design process—certainly in embodiment design. We recognize that there is a great deal of information that the designer needs to intelligently make these decisions. To aid in this the chapter provides:

- An overview of the most commonly used manufacturing processes, with emphasis on the factors that need to be considered in design for manufacture
- References to a carefully selected set of books and handbooks that will provide both in-depth understanding of how the processes work and detailed data needed for design. Also, carefully selected websites that give clear illustrations of how the process works and that provide in-depth DFM guidelines.
- An introduction to a simple methodology for ranking manufacturing processes on a unit cost basis that can be used early in the design process
- Reference to some of the most widely used computer simulation tools for design for assembly and design for manufacturing

A material and a process for making a part must be chosen at the same time. The overall factor in deciding on the material and the manufacturing process is the cost to make a quality part. When making a decision on the material, the following factors must be considered:

- Material composition: grade of alloy or plastic
- Cost of material
- Form of material: bar, tube, wire, strip, plate, pellet, powder, etc.
- Size: dimensions and tolerance
- Heat-treated condition
- Directionality of mechanical properties (anisotropy)
- Quality level: control of impurities, inclusions, cracks, microstructure, etc.
- Ease of manufacture: workability, weldability, machinability, etc.
- Ease of recycling

The decision on the manufacturing process will be based on the following factors:

- Unit cost of manufacture
- Life cycle cost per unit
- Quantity of parts required
- Complexity of the part, with respect to shape, features, and size
- Compatibility of the process for use with candidate materials
- Ability to consistently make a defect-free part
- Economically achievable surface finish
- Economically achievable dimensional accuracy and tolerances
- Availability of equipment
- Lead time for delivery of tooling
- Make-buy decision. Should we make the part in-house or purchase from a supplier?

Design can decisively influence manufacturing cost. That is why we must adopt methods to bring manufacturing knowledge into the embodiment design. An integrated product design team that contains experienced manufacturing people is a very good way of doing this. Design for manufacture guidelines is another way. Some general DFM guidelines are:

- Minimize total number of parts in a design.
- Standardize components.
- Use common parts across product lines.
- Design parts to be multifunctional.
- Design parts for ease of manufacture.
- Avoid too-tight tolerances.
- Avoid secondary manufacturing and finishing operations.
- Utilize the special characteristics of a process.

Experience has shown that a good way to proceed with DFM is to first do a rigorous design for assembly (DFA) analysis in an attempt to reduce part count. This will trigger a process of critical examination that can be followed up by what-if exercises on critical parts to drive down manufacturing cost. Use manufacturing simulation software to guide part design in improving parts for ease of manufacture and reducing tooling costs.

NEW TERMS AND CONCEPTS

Batch flow process
Blanking
Continuous flow process
Deep drawing
Design for assembly (DFA)
Design for manufacturing (DFM)
Economic batch size
Feed motion in machining
Finishing process

Group technology
Heat affected zone (HAZ)
Job shop
Machinability
Mistake-proofing
Near net shape
Parting surface
Primary manufacturing process

Process cycle time
Process flexibility
Secondary manufacturing process
Shielded metal arc welding
Solidification
Tooling
Undercut

BIBLIOGRAPHY

Manufacturing Processes (see Table 13.1)

Benhabib, B.: *Manufacturing: Design, Production, Automation, and Integration,* Marcel Dekker, New York, 2003.

Creese, R.C.: *Introduction to Manufacuring Processes and Materials,* Marcel Dekker, New York, 1999.

Koshal, D.: *Manufacturing Engineer's Reference Book,* Butterworth-Heinemann, Oxford, UK, 1993.

Kutz, M., ed.: *Environmentally Conscious Manufacturing,* John Wiley & Sons, Hoboken, NJ, 2007.

Design for Manufacture (DFM)

Anderson, D. M.: *Design for Manufacturability and Concurrent Engineering,* CIM Press, Cambria, CA, 2010.

13

Boothroyd, G., P. Dewhurst, and W. Knight: *Product Design for Manufacture and Assembly,* 3d ed., Taylor & Francis, Boca Raton, FL, 2010.

Bralla, J. G., ed.: *Design for Manufacturability Handbook,* 2nd ed., McGraw-Hill, New York, 1999.

"Design for Manufacturability," *Tool and Manufacturing Engineers Handbook,* Vol. 6, Society of Manufacturing Engineers, Dearborn, MI, 1992.

Dieter, G. E., ed.: *ASM Handbook,* Vol. 20, *Materials Selection and Design,* ASM International, Materials Park, OH, 1997.

Poli, C.: *Design for Manufacturing,* Butterworth-Heinemann, Boston, 2001.

The following websites will connect you with DFM guidelines for many processes: www.engineersedge.com/manufacturing_design.shtml and www.npd-solutions.com.

PROBLEMS AND EXERCISES

13.1 Classify the following manufacturing processes as to whether they are shape-replication or shape-generative:

(a) honing the bore of a cylinder,

(b) powder metallurgy gear,

(c) rough turning a cast roll,

(d) extrusion of vinyl house siding.

13.2 A small hardware fitting is made from free-machining brass. For simplicity, consider that the production cost is the sum of three terms: (1) material cost, (2) labor costs, and (3) overhead costs. Assume that the fitting is made in production lots of 500, 50,000, and 5×10^6 pieces by using, respectively, an engine lathe, a tracer lathe, and an automatic screw machine. Schematically plot the relative distribution of the cost due to materials, labor, and overhead for each of the production quantities.

13.3 Product cycle time is the time it takes for raw materials to be transformed into a finished product. A firm makes 1000 products per day. Before it is sold, each product represents $200 in materials and labor.

(a) If the cycle time is 12 days, how many dollars are tied up with in-process inventory? If the company's internal interest rate is 10 percent, what is the annual cost due to in-process inventory?

(b) If the cycle time is reduced to 8 days as a result of process improvement, what is the annual cost saving?

13.4 You are the designer of a crankshaft for an automotive engine. You have decided to make this part from nodular cast iron using a casting process. During design you consult frequently with an experienced manufacturing engineer from the foundry where the part will be made. What design factors determine the manufacturing cost? Which of the costs are chiefly determined by the foundry and which by the designer?

13.5 Determine the shape complexity for a part with shape R0 in Fig. 13.6, and compare with shape R2. For shape R0 the diameter is 10 mm and the length is 30 mm. For shape R2 the overall length is 30 mm and the length of each shoulder is 10 mm. The large diameter is 10 mm and the small diameter is 6 mm. Use Eq. (13.1) to determine shape complexity. The tolerance on each dimension is ±0.4 mm.

13.6 Give four metrics that could be used to measure the complexity of an assembly operation.

13.7 Examine the processes in Example 13.2. One of the processes that was rejected in the second round of decision making has great potential for making the integral bladed hub for the fan from an aluminum alloy. This process selection would have required a creative design for the die that might have required considerable development time and cost. Identify the process, and briefly describe what technical issues prevented its selection.

Another approach is to abandon the concept that the hub and blades should be made as an integral piece. Instead, think about making the part as separate pieces to be assembled. What manufacturing processes does this open up for consideration?

13.8 Make a brief literature study of the hot isostatic process (HIP). Discuss the mechanics of the process, its advantages, and its disadvantages. Think broadly about how HIPing can improve more conventional processes, and how it can impact design.

13.9 The limiting draw ratio, the ratio of the diameter of the blank to the diameter of the deep drawn cup, is generally less than 2 for metal sheets. How then is a two-piece soft drink can made? A two-piece can is one that does not have a soldered longitudinal seam. The two parts of the can are the cylindrical can body and the top.

13.10 A manufacturing process to make a product consists of 10 separate processes. A mistake occurs in each process on average of once every 10,000 part produced. What is the product defect rate, expressed in parts per million (ppm)?

13.11 What kind of mistake-proofing device or assembly method would you suggest using the following situations?

(a) A check that the required number of bolts are available for assembling a product.

(b) A count that the proper number of holes has been drilled in a plate.

(c) Insurance that three wires are connected to the proper terminals.

(d) A simple method to ensure that a product identification label has not been glued upside down.

(e) A simple method to ensure that a plug is inserted in the proper orientation in an electrical plug.

13.12 As a team project, create a table comparing the plastic processing processes listed in Sec. 13.18. Develop your own list of process characteristics that are generic to all plastic processing processes. Include a clear schematic drawing of the equipment and tooling, along with a description of how the process works.

13

14

RISK, RELIABILITY, AND SAFETY

14.1
INTRODUCTION

We start this chapter by defining terms that are often confused in the public mind but actually have precise technical meanings. A *hazard* is a condition that has the potential for human, property, or environmental damage. A cracked steering linkage, a leaking fuel line, or a loose step all represent hazards. Another term for a hazard is an *unsafe condition*. This is a condition which, if not corrected, can reasonably be expected to result in failure and/or injury.

A *risk* is the likelihood, expressed either as a probability or as a frequency, of a hazard's potential for harm being realized. Risk exists only when a hazard exists and something of value is exposed to the hazard. It is part of our individual existence and that of society as a whole. As young children we were taught about risks. "Don't touch the stove." "Don't chase the ball into the street." As adults we are made aware of the risks of society in our everyday newspaper and newscast. Thus, depending upon the particular week, the news makes us concerned about the risk of all-out nuclear war, a terrorist attack, or an airplane crash. The list of risks in our highly complex technological society is endless. Risk is expressed as the product of the frequency of an event times the magnitude (consequence) of the event. The result is the probability of the event occurring over a specified time period, usually a year. An event can be an accident, death, or loss of property.

$$\text{Risk}\left(\frac{\text{consequence}}{\text{unit time}}\right) = \text{frequency}\left(\frac{\text{events}}{\text{unit time}}\right) \times \text{magnitude}\left(\frac{\text{consequence}}{\text{event}}\right) \quad (14.1)$$

For example, if there are 15 million automobile accidents in the United States per year, and on average 1 out of 300 accidents results in a fatality, the annual fatality risk is:

$$\text{Risk}\left(\frac{\text{fatality}}{\text{year}}\right) = 15 \times 10^6 \frac{\text{accidents}}{\text{year}} \times \frac{1 \text{ fatality}}{300 \text{ accidents}} = 50,000 \frac{\text{fatalities}}{\text{year}}$$

TABLE 14.1
Classification of Societal Hazards

Category of Hazard	Examples
1. Infections and degenerative diseases	Influenza, heart disease, AIDS
2. Natural disasters	Earthquakes, floods, hurricanes
3. Failure of large technological systems	Failure of dams, power plants, aircraft, ships, buildings
4. Discrete small-scale accidents	Automotive accidents, power tools, consumer and sport goods
5. Low-level, delayed-effect hazards	Asbestos, PCB, microwave radiation, noise
6. Sociopolitical disruption	Terrorism, nuclear weapons proliferation, oil embargo

From W. W. Lawrance, in R. C. Schwing and W. A. Albus (eds.), *Social Risk Assessment,* Plenum Press, New York, 1980.

Table 14.1 lists the six classes of hazards to which society is subject. We can see that categories 3 and 4 are directly within the responsibility of the engineer and categories 2, 5, and possibly 6 provide design constraints in many situations.

Risk assessment has become increasingly important in engineering design as the complexity of engineering systems has increased. The risks associated with engineering systems do not always arise because risk avoidance procedures were ignored. One category of risks arises from external factors that were considered acceptable at the time of design but subsequent research has revealed to be a health or safety hazard. A good example is the extensive use of sprayed asbestos coating as an insulation and fire barrier before the toxicity of asbestos fibers was known.[1]

A second category of risks comes from abnormal conditions that are not a part of the basic design concept in its normal mode of operation. Usually these abnormal events stop the operation of the system without harming the general public, although there may be danger to the operators. Other systems, such as passenger aircraft or a nuclear power plant, pose a potential risk and cost to the larger public. Risks in engineering systems are often associated with operator error. Although these should be eliminated by using mistake-proofing methods, Sec. 13.8, it is difficult to anticipate all possible future events. This topic is discussed in Secs. 14.4 and 14.5. Finally, there are the risks associated with poor decisions, design errors and accidents. Clearly, these should be eliminated, but since design is a human activity, errors and accidents will occur.[2]

Most reasonable people will agree that life is not risk-free and cannot be made so.[3] However, an individual's reaction to risk depends upon three main factors: (1) whether the person feels in control of the risk or whether the risk is imposed by some outside group, (2) whether the risk involves one big event (like an airplane crash) or many small, separate occurrences (like auto collisions), and (3) whether the hazard is familiar or is some strange, puzzling risk like a nuclear reactor. Through the medium of mass communication the general public has become better informed about

1. M. Modarres, *Risk Analysis in Engineering,* Taylor & Francis, New York, 2006.
2. T. Kletz, *An Engineer's View of Human Error,* 3d ed., Taylor & Francis, New York, 2006.
3. E. Wenk, *Tradeoffs: Imperatives of Choice in a High-Tech World,* Johns Hopkins University Press, Baltimore, 1986.

the existence of risks in society, but they have not been educated concerning the need to accept some level of risk and to balance risk avoidance against cost. It is inevitable that there will be conflict between various special-interest groups when trying to decide on what constitutes an acceptable risk.

Reliability is a measure of the capability of a part or a system to operate without failure in the service environment for a given period of time. It is always expressed as a probability; for example, a reliability of 0.999 implies that there is probability of failure of 1 part in every 1000. The mathematics of reliability is introduced in Sec. 14.3.

Safety is relative protection from exposure to hazards. A thing is safe if its risks are judged to be acceptable.[1] Therefore, two different activities are involved in determining how safe a design is: (1) a risk assessment, which is a probabilistic activity, and (2) a judgment of the acceptability of that risk, which is a societal value judgment.

14.1.1 Regulation as a Result of Risk

In a democracy, when the public perception of a risk reaches sufficient intensity, legislation is enacted to control the risk. That usually means the formation of a regulatory commission that is charged with overseeing the regulatory act. In the United States the first regulatory commission was the Interstate Commerce Commission (ICC). The following federal organizations have a major role to play in regulating technical risk:

> Consumer Product Safety Commission (CPSC)
> Environmental Protection Agency (EPA)
> Federal Aviation Agency (FAA)
> Federal Highway Administration (FHA)
> Federal Railway Administration (FRA)
> Nuclear Regulatory Commission (NRC)
> Occupational Safety and Health Administration (OSHA)

Some of the federal laws concerning product safety are listed in Table 14.2. The rapid acceleration of interest in consumer safety legislation is shown by the dates of enactment of these regulatory laws. Once a federal regulation becomes official it has the force of law. In the United States, some 60 federal agencies issue more than 1800 regulations a year; the Code of Federal Regulations (CFR), contains more than 130,000 pages.[2]

Legislation has the important result that it charges all producers of a product with the cost of complying with the product safety regulations. Thus, we are not faced with the situation in which the majority of producers spend money to make their product safe but an unscrupulous minority cuts corners on safety to save on cost. However, in complex engineering systems it may be very difficult to write regulations that do not conflict with each other and work at cross purposes. The automobile is a good example.[3] Here, separate agencies have promulgated regulations to influence fuel

---

1. W. W. Lawrance, *Of Acceptable Risk,* William Kaufman, Inc., Los Altos, CA, 1976.
2. *The Economist,* Aug. 2, 1997, p. 2; for CFR see www.gpo.gov/nara/cfr/index.html.
3. L. B. Lave, *Science,* vol. 212, pp. 893–99, May 22, 1981.

TABLE 14.2
A Sample of Federal Laws Concerning
Product Safety

Year	Legislation
1893	Railroad Appliance Safety Act
1938	Food, Drug, and Cosmetic Act
1953	Flammable Fabrics Act
1960	Federal Hazardous Substance Act
1966	National Traffic and Motor Vehicle Safety Act
1968	Fire Research and Safety Act
1969	Child Protection and Toy Safety Act
1970	Lead-Based Paint Poison Prevention Act
1970	Occupational Safety and Health Act
1972	Consumer Product Safety Act
1982	Nuclear Waste Policy Act
1990	Oil Pollution Act
1996	Mercury-Containing and Rechargeable Battery Management Act

economy, exhaust emissions, and crash safety. The law to control emissions also reduces fuel efficiency by 7.5 percent, but the fuel efficiency law has forced the building of smaller cars that have increased crash fatalities each year, until the widespread use of safety air bags. The need for strong technical input into the regulatory process should be apparent from this example.

A common criticism of the regulatory approach is that decisions are often made arbitrarily. That is understandable when we consider that a regulatory agency often has a congressional mandate to protect the public from "unreasonable risk." Since there usually are no widely agreed-on definitions of unreasonable risk, the regulators are accused of being hostile to or soft on the regulated industry, depending upon the individual's point of view. Sometimes the regulating agency specifies the technology for meeting the target level of risk. This removes the incentive for innovation in developing more effective methods of controlling the risk.

14.1.2 Standards

Design standards were first considered in Sec. 1.7. There we discussed the difference between a code and a standard, the different kinds of standards, and the types of organizations that develop standards. In Sec. 5.8 standards were discussed for their value as sources of information. In this section we consider standards and codes more broadly from the viewpoint of the role they play in minimizing risk. Standards are among the most important ways in which the engineering profession makes sure that society receives a minimum level of safety and performance.

14

The role that standards play in protecting public safety was first shown in the United States in the middle of the 19th century. This was a time of rapid adoption of steam power on railroads and in ships. The explosion of steam boilers was an all-too-frequent occurrence, until the ASME developed the Boiler and Pressure Vessel Code that prescribed detailed standards for materials, design, and construction. The ASME Boiler Code was quickly adopted as law by the individual states. Other examples of *public safety standards* are fire safety and structural codes for buildings and codes for the design, construction, maintenance, and inspection of elevators.

Other standards protect the general health and welfare. Examples are emission standards for cars and power plants to protect public health by minimizing air pollution, and standards on the discharge of effluents into rivers and streams.

Mandatory Versus Voluntary Standards

Standards may be mandatory or voluntary. Mandatory standards are issued by governmental agencies, and violations are treated like criminal acts for which fines and/or imprisonment may be imposed. Voluntary standards are prepared by a committee of interested parties (industry suppliers and users, government, and the general public), usually under the sponsorship of a technical society or a trade association. Approval of a new standard generally requires agreement by nearly all participants in the committee. Therefore, voluntary standards are consensus standards. They usually specify only the lowest performance level acceptable to all members of the standards committee. Thus, a voluntary standard indicates the lowest safety level that an industry intends to provide in the product it manufactures. In contrast, a mandatory standard indicates the lowest safety level the government will accept. Because mandatory standards frequently set more stringent requirements than voluntary standards do, mandatory standards force manufacturers to innovate and advance the state of the art. This is often at increased cost to the consumer.

Regulatory agencies often adopt an existing voluntary standard. They may do this by citing the voluntary standard by reference in the regulation. They may also choose to modify the existing voluntary standard before adopting it, or they may decide to ignore the voluntary standard and write their own standard.

14.1.3 Risk Assessment

The assessment of risk is an imprecise process involving judgment and intuition. However, triggered by the consumer safety movement and the public concern over nuclear energy, a growing literature has evolved.[1] The level of risk, as perceived by an individual or the public, can be classified as tolerable, acceptable, or unacceptable.[2]

---

1. C. Starr, *Science,* vol. 165, pp. 1232–38, Sept. 19, 1969; N. Rasmussen, et al., *Reactor Safety Study,* WASH-1400, U.S. Nuclear Regulatory Commission, 1975; W. D. Rowe, *An Anatomy of Risk,* John Wiley & Sons, New York, 1977; J. D. Graham, L. C. Green, and M. J. Roberts, *In Search of Safety,* Harvard University Press, Cambridge, 1988; M. Modarres, *Risk Analysis in Engineering,* CRC Press, Boca Baton, FL, 2006.
2. D. J. Smith, *Reliability, Maintainability, and Risk,* 5th ed., Butterworth-Heinemann, Oxford, 1997.

TABLE 14.3
Fatality Rate

Cause of Fatality	Fatality per Person per Year
Smoking (20 per day)	5×10^{-3}
Cancer, in general	3×10^{-3}
Race car driving	1×10^{-3}
Motor vehicle driving	3×10^{-4}
Fires	4×10^{-5}
Poison	2×10^{-5}
Industrial machinery	1×10^{-5}
Air travel	9×10^{-6}
Railway travel	4×10^{-6}
California earthquake	2×10^{-6}
Lightning	5×10^{-7}

Tolerable risk: Indicates that people are prepared to live with the level of risk but want to continue to review its causes and seek ways of reducing the risk.

Acceptable risk: Indicates that people accept the level of risk as reasonable and would not seek to expend much in resources to reduce it further. An acceptable risk is one that satisfies the general public. This is often influenced by the decisions of relevant government regulating agencies.

Unacceptable risk: Indicates that people do not accept this level of risk and would not participate in the activity or permit others to participate.

Many regulations are based on the principle of making the risk "as low as reasonably practicable" (ALARP). This means that all reasonable measures will be taken to reduce risks that lie in the tolerable region until the cost to achieve further risk reduction becomes greatly disproportionate to the benefit.

Data on risk are subject to considerable uncertainty and variability. In general, three classes of statistics are available: (1) financial losses (chiefly from the insurance industry), (2) health information, and (3) accident statistics. Usually the data are differentiated between fatalities and injuries. Risk is usually expressed as the probability of the risk of a fatality or accident per person per year. A risk that exceeds 10^{-3} fatalities per person per year (or 1 in 1,000) is generally considered unacceptable, while a rate that is less than 10^{-5} is not of concern to the average person.[1] The range 10^{-3} to 10^{-5} is the tolerable range. However, an individual's perception of risk depends upon the circumstances. If the risk is voluntarily assumed, like smoking or driving a car, then there is a greater acceptance of the risk than if the risk was assumed involuntarily, as with traveling in a train or breathing secondhand smoke. There is a large difference between individual risk and societal risk. Table 14.3 gives some generally accepted fatality rates for a variety of risks.

---

1. D. J. Smith, *Reliability, Maintainability, and Risk,* 5th ed., Butterworth-Heinemann, Oxford, 1997.

14.2
PROBABILISTIC APPROACH TO DESIGN

Conventional engineering design uses a deterministic approach. It disregards the fact that material properties, the dimensions of the parts, and the externally applied loads vary statistically. In conventional design these uncertainties are handled by applying a factor of safety. In critical design situations such as aircraft, space, and nuclear applications, it is often necessary to use a probabilistic approach to better quantify uncertainty and thereby increase reliability.[1]

14.2.1 Basic Probability Using the Normal Distribution

Many physical measurements follow the symmetrical, bell-shaped curve of the normal, or Gaussian frequency distribution. The distributions of yield strength, tensile strength, and reduction of area from the tension test follow the normal curve to a suitable degree of approximation. The equation of the normal curve is

$$f(x) = \frac{1}{\sigma\sqrt{2\pi}} \exp\left[-\frac{1}{2}\left(\frac{x-\mu}{\sigma}\right)^2\right] \tag{14.2}$$

where $f(x)$ is the height of the frequency curve corresponding to an assigned value x, μ is the mean of the population, and σ is the standard deviation of the population. The normal distribution extends from $x = -\infty$ to $x = +\infty$ and is symmetrical about the population mean μ. The existence of negative values and long "tails" makes the normal distribution a poor model for certain engineering problems.

In order to place all normal distributions on a common basis in a standardized way, the normal curve frequently is expressed in terms of the *standard normal variable* or the z variable.

$$z = \frac{x-\mu}{\sigma} \tag{14.3}$$

Now, the equation of the standard normal curve becomes

$$f(z) = \frac{1}{\sqrt{2\pi}} \exp\left(-\frac{z^2}{2}\right) \tag{14.4}$$

For the standardized normal curve $\mu = 0$ and $\sigma = 1$. The standardized normal curve was presented previously in Chap. 8, Fig. 8.23. The total area under the curve is unity. The probability of a value of z falling between $z = -\infty$ and a specified value of z is given by the area under the curve. Probability is the numerical measure of likelihood

1. E. B. Haugen, *Probabilistic Mechanical Design,* Wiley-Interscience, Hoboken, NJ, 1980; J. N. Siddal, *Probabilistic Engineering Design,* Marcel Dekker, New York, 1983.

TABLE 14.4
Areas Under Standardized Normal Frequency Curve

$z = \dfrac{x-\mu}{\sigma}$	Area	z	Area
−3.0	0.0013	−3.090	0.001
−2.0	0.0228	−2.576	0.005
−1.0	0.1587	−2.326	0.010
−0.5	0.3085	−1.960	0.025
0.0	0.5000	−1.645	0.050
+0.5	0.6915	1.645	0.950
+1.0	0.8413	1.960	0.975
+2.0	0.9772	2.326	0.990
+3.0	0.9987	2.576	0.995
		3.090	0.999

of an event. The probability, P, is bounded between $P = 0$ (an impossible event) and $P = 1$ (a certain event).

The area under the curve from $-\infty$ to $z = -1.0$ is 0.1587, so the probability of a value falling into that interval is $P = 0.1587$, or 15.87 percent. Since the curve is symmetric, the probability of a value falling into the interval $z = -1$ to $z = 1$ or $\mu \pm \sigma$ is $1.0000 - 2(0.1587) = 0.6826$. In a similar way it can be shown that $\mu \pm 3\sigma$ encompasses 99.73 percent of all values.

Some typical values for the area under the z curve are listed in Table 14.4. More complete values will be found in Appendix A. For example, if $z = -3.0$ the probability of a value being less than z is 0.0013 or 0.13 percent. The percentage of values greater than this z is $100 - 0.13 = 99.87$ percent. The fraction of values less than z is $1/0.0013 = 1$ in 769. Table 14.4 also shows that if we wanted to exclude the lowest 5 percent of the population values we would set z at −1.645.

EXAMPLE 14.1
A highly automated factory is producing ball bearings. The average ball diameter is 0.2152 in. and the standard deviation is 0.0125 in. These dimensions are normally distributed.

(a) What percentage of the parts can be expected to have a diameter less than 0.2500 in.? Note that up until now we have used μ and σ to represent the mean and standard deviation of the population. The *sample values* of the mean and standard deviation are given by $\bar{x}$ and s. In this example, where we are sampling literally millions of balls, these values are nearly identical.

Determining the standard normal variable

$$z = \frac{x-\mu}{\sigma} \approx \frac{x-\bar{x}}{s} = \frac{0.2500-0.2512}{0.0125} = \frac{-0.0012}{0.0125} = -0.096$$

From Appendix A, $P(z < -0.09) = 0.4641$ and $P(z < -0.10) = 0.4602$. Interpolating, the area under the z distribution curve at $z = -0.096$ is 0.4618. Therefore, 46.18 percent of the ball bearings are below 0.2500 in. diameter.

14

(b) What percentage of the balls are between 0.2574 and 0.2512 in.?

$$z = \frac{0.2512 - 0.2512}{0.0125} = 0.0$$ Area under curve from $-\infty$ to $z = 0$ is 0.5000.

$$z = \frac{0.2574 - 0.2512}{0.0125} = \frac{0.0062}{0.0125} = +0.50$$ Area under curve from $-\infty$ to $z = 0.5$ is 0.6915

Therefore, percentage of ball diameters in interval 0.2512 to 0.2574 is $0.6915 - 0.5000 = 0.1915$ or 19.15 percent.

14.2.2 Sources of Statistical Tables

All statistical texts contain tables for the z distribution, the confidence limits of the mean, and the t and F distributions, but tables of more esoteric statistics often needed in engineering may be more elusive. Here we mention two convenient sources of statistical tables and information.

The Microsoft spreadsheet program Excel provides access to many special mathematical and statistical functions. To display this menu of functions, click on the Insert Function button, f_x, on the formula bar at the top of a spreadsheet and search in the function box for the special function you might need. Click on the name of the function for a description and example of how to use it. Table 14.5 is a short listing of some of the more useful statistical functions.

The NIST/SEMATECH e-Handbook of Statistical Methods is the modern version of Experimental Statistics, edited by Mary Natrella and published in 1963 by the National Bureau of Standards as Handbook 91. It is available online at www.itl.nist .gov/div898/handbook.

TABLE 14.5
Some Statistical Functions Available in Excel

Function	Description of Excel Function
NORMDIST	Returns the $f(x)$ in Eq. (14.2) for given x, μ and σ
NORMINV	Returns x in Eq. (14.2) for given $f(x)$, μ and σ
NORMSDIST	Returns the area under Eq. (14.4) for a given z (probability)
NORMSINV	Returns the std normal variable z, given the probability (area under $f(z)$)
LOGNORMDIST	Returns the $f(x)$ for a distribution where $\ln x$ is normally distributed
EXPONDIST	Returns the exponential distribution
GAMMADIST	Returns values of the Gamma distribution (useful in Weibull distribution)
WEIBULL	Returns the Weibull distribution for values of x, and shape and scale parameters
ZTEST	Returns the two-tailed probability in a z test
TDIST	Returns values of the t distribution
FDIST	Returns values of the F distribution
FINV	Returns the inverse of the F probability distribution

TABLE 14.6
Typical Values of Coefficient of Variation

Variable x	Typical δ
Modulus of elasticity of metals	0.05
Tensile strength of metals	0.05
Yield strength of metals	0.07
Buckling strength of columns	0.15
Fracture toughness of metals	0.15
Cycles to failure in fatigue	0.50
Design load in mechanical components	0.05–0.15
Design load in structural systems	0.15–0.25

H. R. Millwater and P. H. Wirsching, "Analysis Methods for Probabilistic Life Assessment," *ASM Handbook,* Vol. 17, p. 251, ASM International, Materials Park, OH, 2002.

14.2.3 Variability in Material Properties

The mechanical properties of engineering materials exhibit variability. Fracture and fatigue properties show greater variability than the static tensile properties of yield strength and tensile strength (see Table 14.6). Most published mechanical property data do not give mean values and standard deviations. Haugen[1] has presented much of the published statistical data. *MMPDS-02 Handbook* presents extensive statistical data for materials used in aircraft.[2] Much other statistical data resides in the files of companies and government agencies.

Published mechanical property data without statistical attribution is usually taken to represent a mean value. If a range of values is given, the lower value is often taken to represent a conservative value for design. Although certainly not all mechanical properties are normally distributed, a normal distribution is a good first approximation that usually results in a conservative design. When statistical data are not available we can estimate the standard deviation by assuming that the upper x_U and lower x_L values of a sample are ± three standard deviations from the mean. Thus,

$$x_U - x_L = 6\sigma \quad \text{and} \quad s \approx \sigma = \frac{x_U - x_L}{6} \tag{14.5}$$

When the range of property values is not given, it is still possible to approximate the standard deviation by using the *coefficient of variation,* δ, which is a measure of the uncertainty of the value of the mean.

$$\delta = \frac{s}{\bar{x}} \tag{14.6}$$

14

---

1. E. B. Haugen, op. cit., Chap. 8 and App. 10A and 10B.
2. *Metallic Materials Properties Development and Standardization Handbook,* 5 volumes, 2005. This is the successor to MIL-HDBK-5 formerly published by DOD. The last (2003) version of this handbook can be viewed at www.mmpds.org.

The coefficient of variation is different for each mechanical property, but it tends to be relatively constant over a range of mean values. Thus, it is a way of estimating the standard deviation. Table 14.6 gives some values of coefficient of variation.

EXAMPLE 14.2

The yield strength of a sample of 50 tensile specimens from an alloy steel is $\bar{x} = 130.1$ ksi. The range of yield strength values is from 115 to 145 ksi. The estimate of standard deviation, which measures the variability in the strength values, is $s = \dfrac{x_U - x_L}{6} = \dfrac{145 - 115}{6} = 5$ ksi.

Assuming that a normal distribution applies, estimate the value of yield strength which 99 percent of the yield strengths will exceed. From Table 14.4, $z_{1\%} = -2.326$, and from Eq. (14.3)

$$-2.326 = \frac{x_{1\%} - 130.1}{5} \quad \text{and} \quad x_{1\%} = 118.5 \text{ ksi}$$

Note that if the range of yield strength had not been known, we could estimate the standard deviation from Table 14.6 and Eq. (14.6).

$$s = \bar{x}\delta = 130.1 \times 0.07 = 9.1 \text{ ksi. This results in } x_{1\%} = 108.9 \text{ ksi}$$

In Example 14.2, sample values of mean and standard deviation were used to determine the probability limits. This is inaccurate unless the sample size n is very large, possibly approaching $n = 1000$. This is because the sample values x and s are only estimates of the true population values μ and σ. The error in using sample values to estimate population values can be corrected if we used *tolerance limits*. Because we generally are interested in finding the lower limit of the property, we use the one-sided tolerance limit.

$$x_L = \bar{x} - (k_{R,C})s \tag{14.7}$$

To find $k_{R,C}$ statistical tables[1] we first need to decide on the confidence level, c. This is usually taken as 95 percent, indicating that we have a 95 percent confidence that the method will produce a true lower limit on the property. R is our expectation that the value of x_L will be exceeded R percent of the time. Usually R is taken at 90, 95 or 99 percent. Table 14.7 gives some values of $k_{R,C}$ for different values of sample size n.

EXAMPLE 14.3

Now we redo Example 14.2 using the one-sided tolerance limit. The sample size is $n = 50$, so $k_{R,C} = 2.86$ at a 95 percent confidence level and with $R = 0.99$. Then, $x_L = 130.1 - 2.86(5) = 115.8$ ksi. Note that x_L has been decreased from 118.5 to 115.8 ksi when we corrected for using sample statistics instead of population statistics. If n consisted of only 10 specimens, x_L would be 110.2 ksi.

---

1. J. Devore and N. Farnum, Table V, *Applied Statistics for Engineers and Scientists*, Duxbury Press, Pacific Grove, CA.

TABLE 14.7
One-Sided Tolerance Limit Factors
for 95% Confidence Level

n	$k_{90,95}$	$k_{99,95}$
5	3.41	5.74
10	2.35	3.98
20	1.93	3.30
50	1.65	2.86
100	1.53	2.68
500	1.39	2.48
∞	1.28	2.37

Note: These apply only to normally distributed variables.

14.2.4 Safety Factor

An important concept in risk and reliability analysis is that hazards are controlled, mitigated, or removed by *barriers*. Barriers can be physical objects like pipes, walls, or containment vessels, or active barriers such as human operators and computer controlled systems. On a more abstract level, the property of a material that is used to build a component can be considered a barrier. This situation is considered in a class of problems called Stress-Strength Models, Sec. 14.2.5. This model assumes that the barrier fails if the *stress* (mechanical, thermal, electrical, etc.) exceeds the resistance of the material to the stress, measured in terms of some material property like yield strength.

The use of a *safety factor* is the oldest and simplest stress-strength model. We will define the safety factor, SF, as the ratio of the strength, S, divided by the stress σ. Another way to view the safety factor is that it is the ratio of the capacity of the system to its load.

$$\text{SF} = \frac{S}{\sigma} = \frac{\text{strength}}{\text{stress}} = \frac{\text{capacity}}{\text{load}} \qquad (14.8)$$

The concept of safety factor is sometimes expressed by the *margin of safety*, MS.

$$\text{MS} = \text{capacity} - \text{load} \qquad (14.9)$$

The margin of safety indicates the amount by which the design capacity exceeds the load. If you have information on the mean values of strength and stress, then using Eq. (14.8) is advisable. However, this information is often unavailable.

Deciding on a safety factor requires experience. Often design standards or codes prescribe what SF to use. In the absence of this advice, the following is a rational way to arrive at a factor of safety.[1] Rather than using Eq. (14.8), break the safety factor into five components which measure how well you understand the capacity vs. load issues for the design of the part. Estimate how well you know the material properties, the loads and stress state, the manufacturing tolerances, the degree to which the design is based on a well-validated

---

1. D. G. Ullman, *The Mechanical Design Process*, 4th ed, pp. 405–406, McGraw-Hill, New York, 2010.

theory of failure, and finally, the level of reliability the application requires. Each of these factors is evaluated separately, and then multiplied to arrive at the overall SF.

$$SF = SF_{material} \times SF_{stress} \times SF_{tolerances} \times SF_{failure\ theory} \times SF_{reliability} \qquad (14.10)$$

Each component SF should be estimated from the following listing.

Estimating the Contribution from the Material

$SF_{material} = 1.0$	The properties of the material are well known, or they have been obtained from tests on the same material used for the design of the part.
$SF_{material} = 1.1$	The material properties are known from a handbook or from manufacturer's values.
$SF_{material} = 1.2–1.4$	The material properties are not well known.

Estimating the Contribution from the Load or Stress

$SF_{stress} = 1.0$	The load is well defined as static or fluctuating. There are no expected overloads or shock loads. An accurate method of analyzing stress has been used.
$SF_{stress} = 1.2–1.3$	Average overloads of 20–50%. The stress analysis method may result in errors less than 50%.
$SF_{stress} = 1.4–1.7$	The load is not well known or the stress analysis method is of doubtful accuracy.

Estimating the Contribution from Tolerances (Geometry)

$SF_{tolerances} = 1.0$	The manufacturing tolerances are tight and well held.
$SF_{tolerances} = 1.0$	The manufacturing tolerances are average.
$SF_{tolerances} = 1.1–1.2$	The dimensions are not closely held.

Estimating the Contribution from Failure Analysis

$SF_{failure\ theory} = 1.0–1.1$	The failure analysis used is based on static uniaxial or multiaxial state of stress, or fully reversed uniaxial fatigue stresses.
$SF_{failure\ theory} = 1.2$	Same as above, but now includes multiaxial fully reversed fatigue stresses or uniaxial nonzero mean fatigue stresses.
$SF_{failure\ theory} = 1.3–1.5$	Failure analysis not well developed, as with cumulative fatigue damage.

Estimating the Contribution from Reliability

$SF_{reliability} = 1.1$	The reliability of the part does not need to be high; less than 90%.
$SF_{reliability} = 1.2–1.3$	The reliability is on average 92–98%.
$SF_{reliability} = 1.4–1.6$	The reliability must be 99% or higher.

14

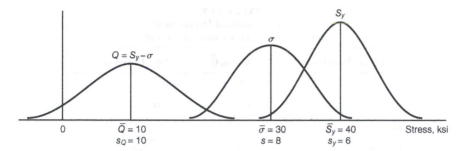

FIGURE 14.1
Distributions of yield strength S_y and stress.

The following section shows how the safety factor can be expressed in terms of probability.

14.2.5 Reliability-Based Safety Factor

Consider a structural member subjected to a static load that develops a stress σ. The variation in load or sectional area results in the distribution of stress shown in Fig. 14.1, where the mean is $\bar{\sigma}$ and the standard deviation[1] of the sample of stress values is s. The yield strength of the material S_y has a distribution of values given by $\bar{S}_y$ and s_y. However, the two frequency distributions overlap, and it is possible for $\sigma > S_y$, which is the condition for failure. The probability of failure is given by

$$P_f = P(\sigma > S_y) \qquad (14.11)$$

The reliability R is defined as

$$R = 1 - P_f \qquad (14.12)$$

If we subtract the stress distribution from the strength distribution, we get the distribution $Q = S_y - \sigma$ shown at the left in Fig. 14.1.

We now need to be able to determine the mean and standard deviation of the distribution Q constructed by performing algebraic operations on two independent random variables x and y, that is, $Q = x \pm y$. Without going into statistical details,[2] the results are as given in Table 14.8. Referring now to Fig. 14.1, and using the results in Table 14.8, we see that the distribution $Q = S_y - \sigma$ has a mean value and $\bar{Q} = 40 - 30 = 10$ and $\sigma_Q = \sqrt{6^2 + 8^2} = 10$. The part of the distribution to the left of $Q = 0$

14

---

1. Note that probabilistic design is at the intersection of two engineering disciplines: mechanical design and engineering statistics. Thus, confusion in notation is a problem.
2. E. B. Haugen, op. cit., pp. 26–56.

TABLE 14.8
Mean and Standard Deviation of
Independent Random Variables x and y

Algebraic Functions	Mean, $\bar{Q}$	Std. Deviation
$Q = C$	C	0
$Q = Cx$	$C\bar{x}$	$C\sigma_x$
$Q = x + C$	$\bar{x} + C$	σ_x
$Q = x \pm y$	$\bar{x} \pm \bar{y}$	$\sqrt{\sigma_x^2 + \sigma_y^2}$
$Q = xy$	$\bar{x}\bar{y}$	$\sqrt{\bar{x}^2\sigma_y^2 + \bar{y}^2\sigma_x^2}$
$Q = x/y$	$\bar{x}/\bar{y}$	$\left(\bar{x}^2\sigma_y^2 + \bar{y}^2\sigma_x^2\right)^{1/2}/\bar{y}^2$
$Q = 1/x$	$1/\bar{x}$	$\sigma_x/\bar{x}^2$

TABLE 14.9
Value of z to Give Different Levels
of Probability of Failure

Probability of Failure P_f	$z = (x - \mu)/\sigma$
10^{-1}	-1.28
10^{-2}	-2.33
10^{-3}	-3.09
10^{-4}	-3.72
10^{-5}	-4.26
10^{-6}	-4.75

represents the area for which $S_y - \sigma$ is a negative number; that is, $\sigma > S_y$, and failure occurs. If we transform to the standard normal variable, $z = (x - \mu)/\sigma$, we get, at $Q = 0$,

$$z = \frac{0 - \bar{Q}}{\sigma_Q} = -\frac{10}{10} = -1.0$$

From Table 14.4 we find that 0.16 of the area falls between $-\infty$ and $z = -1.0$. Thus, the probability of failure is $P_f = 0.16$, and the reliability is $R = 1 - 0.16 = 0.84$. Clearly, this is not a particularly satisfactory situation. If we select a stronger material with $\bar{S}_y = 50$ ksi, $\bar{Q} = 20$ and $z = 2.0$. The probability of failure now is about 0.02. Values of z corresponding to various values of failure probabilities are given in Table 14.9.

14.3
RELIABILITY THEORY

Reliability is the probability that a system, component, or device will perform without failure for a specified period of time under specified operating conditions. The discipline of reliability engineering basically is a study of the causes, distribution, and prediction of failure. If $R(t)$ is the reliability with respect to time t, then $F(t)$ is the unreliability (probability of failure) in the same time t. Since failure and nonfailure are mutually exclusive events,

$$R(t) + F(t) = 1 \qquad (14.13)$$

If N_0 components are put on test, the number surviving to or at time t is $N_s(t)$, and the number that failed between $t = 0$ and $t = t$ is $N_f(t)$.

$$N_s(t) + N_f(t) = N_0 \qquad (14.14)$$

From the definition of reliability

$$R(t) = \frac{N_s(t)}{N_0} = 1 - \frac{N_f(t)}{N_0} \qquad (14.15)$$

Taking the derivative with respect to time

$$\frac{dR(t)}{dt} = -\frac{1}{N_0} \frac{d(N_f)}{dt} \qquad (14.16)$$

or

$$\frac{dN_f}{dt} = -N_0 \frac{dR}{dt} \qquad (14.17)$$

We could find the failure rate from Eq. (14.17), but this would not be a valid metric since the numbers would depend on the sample size N_0. The larger of two samples of the same components under test will have more items failing per unit time. A much more meaningful measure of failure rate is the hazard rate or the instantaneous failure rate, $h(t)$.

$$h(t) = \frac{dN_f}{dt} \frac{1}{N_s(t)} = \frac{f(t)}{1 - F(t)} = \frac{f(t)}{R(t)} \qquad (14.18)$$

The last part of Eq. (14.18) uses statistical terminology to define $h(t)$. It is expressed as the probability density function of time to failure divided by the cumulative distribution function of nonfailures. It is the probability that a given test item will fail between t_1 and $t_1 + dt_1$ when it has already survived to t_1.

If we divide Eq. (14.17) by N_s and combine with Eq. (14.18)

$$h(t) = \frac{dN_f}{dt} \frac{1}{N_s} = -\frac{N_0}{N_s} \frac{dR}{dt} \quad \text{and since} \quad R(t) = \frac{N_s}{N_0} \quad \text{this simplifies to}$$

$$h(t) = -\frac{1}{R} \frac{dR}{dt} \quad \text{and} \quad h(t)dt = -\frac{dR}{R}$$

Integrating gives $\int_0^t h(t)dt = -\int_1^R \frac{dR}{R}$ where the reliability (probability of survival) at $t = 0$ is one.

$$-\int_0^t h(t)dt = \ln R \quad \text{or} \quad R(t) = \exp\left[-\int_0^t h(t)\,dt\right] \qquad (14.19)$$

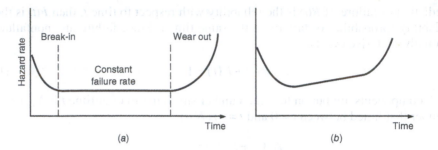

FIGURE 14.2
Forms of the failure curve: (a) three-stage (bath tube) curve typical of electronic equipment;
(b) failure curve more typical of mechanical equipment.

Equation (14.19) can be used to drive the reliability function from a known hazard rate function. Making a good estimate of the reliability depends on using an appropriate model for the hazard rate function. In this chapter we consider the constant failure rate model and the Weibull model.

The hazard rate or failure rate is given in terms like 1 percent per 1000 h or 10^{-5} per hour. Components in the range of failure rates of 10^{-5} to 10^{-7} per hour exhibit a good commercial level of reliability.

The general failure curve shown in Fig. 14.2 is the summation of three competing processes: (1) an early failure process, (2) a random failure process, and (3) a wearout process. The three-stage curve shown in Fig. 14.2a is typical of electronic components. At short lifetimes there is a high failure rate due to "infant mortality" arising from design errors, manufacturing defects, or installation defects. This is a period of shakedown, or debugging, of failures. These early failures can be minimized by improving production quality control, subjecting the parts to a proof test before service, or "running in" the equipment before sending it out of the plant. As these early failures leave the system, failure will occur less and less frequently until eventually the failure rate will reach a constant value. The time period of constant failure rate is a period in which failures can be considered to occur at random from random overloads or random flaws. These failures follow no predictable pattern. Finally, after what is hopefully a long time, materials and components begin to age and wear rapidly and the wearout period of accelerating failure rate begins. Mechanical components (Fig. 14.2b) do not exhibit a region of constant failure rate. After an initial break-in period, wear mechanisms operate continuously until failure occurs.

14.3.1 Definitions

Following are some definitions that are important in understanding reliability.

Cumulative time to failure (T): When N_0 components are run for a time t without replacing or repairing failed components,

$$T = [t_1 + t_2 + t_3 + \ldots t_k + (N_0 - k)t] \tag{14.20}$$

TABLE 14.10
**Average Failure Rates for a Variety of
Components and Systems**

Component	Failure Rate: Number of Failures per 1000 h
Bolts, shafts	2×10^{-7}
Gaskets	5×10^{-4}
Pipe joints	5×10^{-4}
Plastic hoses	4×10^{-2}
Valves, leaking	2×10^{-3}
Systems:	
Centrifugal compressor	1.5×10^{-1}
Diesel-driven generator	1.2–5
Refrigerator, household	4–6×10^{-2}
Mainframe computer	4–8
Personal computer	2–5×10^{-2}
Printed circuit board	7–10×10^{-5}

where t_1 is the occurrence of the first failure, etc., and k is the number of failed components.

Mean life: The average life of the N_0 components put on test or in service, measured over the entire life curve out to wearout.

Mean time to failure (MTTF): The sum of the survival time for all of the components divided by the number of failures. This can be applied to any period in the life of the component. MTTF is used for parts that are not repaired, like light bulbs, transistors, and bearings, or for systems containing many parts, like a printed circuit board or a spacecraft. When a part fails in a nonrepairable system, the system fails; therefore, system reliability is a function of the first part failure.

Mean time between failures (MTBF): The mean time between two successive component failures. MTBF is similar to MTTF, but it is applied to components or systems that are repaired.

Table 14.10 gives some rough ideas of average failure rates for different engineering components and systems.

14.3.2 Constant Failure Rate

For the special case of a constant failure rate, $h(t) = \lambda$, and Eq. (14.19) can be written

$$R(t) = \exp\left(-\int_0^t \lambda \, dt\right) = e^{-\lambda t} \tag{14.21}$$

The probability distribution of reliability, for this case, is a negative exponential distribution.

$$\lambda = \frac{\text{number of failures}}{\text{number of time units during which all items were exposed to failure}}$$

The reciprocal of λ, $\bar{T} = 1/\lambda$, is the mean time between failures (MTBF).

$$\bar{T} = \frac{1}{\lambda} = \frac{\text{number of time units during which all items were exposed to failure}}{\text{number of failures}}$$

so
$$R(t) = e^{-t/\bar{T}} \qquad (14.22)$$

Note that if a component is operated for a period equal to MTBF, the probability of survival is $1/e = 0.37$.

Although an individual component may not have an exponential reliability distribution, in a complex system with many components the overall reliability may appear as a series of random events, and the system will follow an exponential reliability distribution.

> **EXAMPLE 14.4.** If a device has a failure rate of 2×10^{-6} failures/h, what is its reliability for an operating period of 500 h? If there are 2000 items in the test, how many failures are expected in 500 h? Assume that strict quality control has eliminated premature failures so we can assume a constant failure rate.
>
> $$R(500) = \exp(-2 \times 10^{-6} \times 500) = e^{-0.001} = 0.999$$
> $$N_s = N_0 R(t) = 2000(0.999) = 1998$$
> $$N_f = N_0 - N_s = 2 \text{ failures expected}$$
>
> If the MTBF for the device is 100,000 h, what is the reliability if the operating time equals 100,000 h?
>
> $$t = \bar{T} = 1/\lambda$$
> $$R(t) = e^{-t/\bar{T}} = e^{-100,000/100,000} = e^{-1} = 0.37$$
>
> We note that a device has only a 37 percent chance of surviving as long as the MTBF.
>
> If the length of the constant failure rate period is 50,000 h, what is the reliability for operating for that length of time?
>
> $$R(50,000) = \exp(-2 \times 10^{-6} \times 5 \times 10^{4}) = e^{-0.1} = 0.905$$
>
> If the part has just entered the useful life period, what is the probability it will survive 100 h?
>
> $$R(100) = \exp(-2 \times 10^{-6} \times 10^{2}) = e^{-0.0002} = 0.9998$$
>
> If the part has survived for 49,900 h, what is the probability it will survive for the next 100 h?
>
> $$R(100) = \exp(-2 \times 10^{-6} \times 10^{2}) = e^{-0.0002} = 0.9998$$
>
> We note that the reliability of the device is the same for an equal period of operating time so long as it is in the constant-failure-rate (useful-life) region.

14.3.3 Weibull Frequency Distribution

The normal frequency distribution is an unbounded symmetrical distribution with long tails extending from $-\infty$ to $+\infty$. However, many random variables follow a bounded, nonsymmetrical distribution. The Weibull distribution describes the life of a component for which all values are positive (there are no negative lives) and for which there are occasional long-lived results.[1] It is useful for describing the probability of fracture in brittle materials, and also for describing fatigue life at a given stress level.

The two-parameter Weibull distribution function is described by[2]

$$f(x) = \frac{m}{\theta}\left(\frac{x}{\theta}\right)^{m-1}\exp\left[-\left(\frac{x}{\theta}\right)^m\right] \quad x > 0 \tag{14.23}$$

where $f(x)$ = frequency distribution of the random variable x
$\quad m$ = *shape parameter,* which is sometimes referred to as the Weibull modulus
$\quad \theta$ = *scale parameter,* sometimes called the characteristic value

The change in the Weibull distribution for various values of shape parameter is shown in Fig. 14.3, illustrating its flexibility for describing a wide range of situations. The probability of x being less than a value q for a Weibull distribution of given m and θ is given by

$$P(x \le q) = \int_o^q f(x)\,dx = 1 - e^{-(q/\theta)^m} \tag{14.24}$$

The mean of a Weibull distribution can be found from

$$\bar{x} = \theta - \Gamma\left(1 + \frac{1}{m}\right) \tag{14.25}$$

where Γ is the gamma function. Tables of the gamma function are available in many statistical texts or in Excel. The variance of a Weibull distribution is given by

$$\sigma^2 = \theta^2\left\{\Gamma\left(1+\frac{2}{m}\right) - \left[\Gamma\left(1+\frac{1}{m}\right)\right]^2\right\} \tag{14.26}$$

The cumulative frequency distribution of a Weibull distribution is given by

$$F(x) = 1 - \exp\left[-\left(\frac{x}{\theta}\right)^m\right] \tag{14.27}$$

---

1. W. Weibull, *J. Appl. Mech.,* vol. 18, pp. 293–97, 1951; *Materials Research and Stds.,* pp. 405–11, May 1962; C.R. Mischke, *Jnl. Mech. Design,* vol. 114, pp. 29–34, 1992.
2. Writers use different symbols for the Weibull parameter. Other uses are α or β for the shape parameter and β or η for the scale parameter.

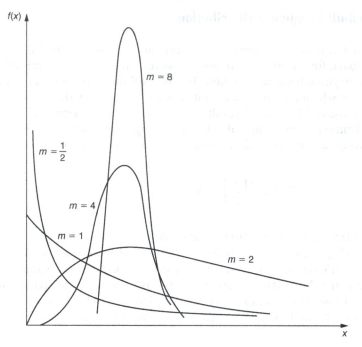

FIGURE 14.3
The Weibull distribution for $\theta = 1$ and different values of m.

Rewriting Eq. (14.27) as

$$\frac{1}{1 - F(x)} = \exp\left(\frac{x}{\theta}\right)^m$$

$$\ln\frac{1}{1 - F(x)} = \left(\frac{x}{\theta}\right)^m$$

$$\ln\left(\ln\frac{1}{1 - F(x)}\right) = m\ln x - m\ln\theta = m(\ln x - \ln\theta) \qquad (14.28)$$

This is a straight line of the form $y = mx + c$. Special Weibull probability paper is available to assist in the analysis according to Eq. (14.28). When the cumulative probability of failure is plotted against x (life) on Weibull paper, a straight line is obtained (Fig. 14.4). The slope is the Weibull modulus m. The greater the slope, the smaller the scatter in the random variable x.

θ is called the *characteristic value* of the Weibull distribution. If $x = \theta$, then

$$F(x) = 1 - \exp\left[-\frac{\theta}{\theta}\right] = 1 - e^{-1} = 1 - \frac{1}{2.718} = 0.632$$

For any Weibull distribution, the probability of being less than or equal to the characteristic value is 0.632. The value of x at a probability of 63 percent on the Weibull plot is the value of θ.

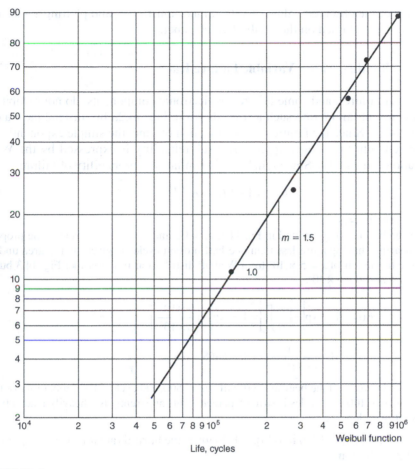

FIGURE 14.4
Weibull plot for life of ball bearings. (From C. Lipman and N. J. Sheth, *Statistical Design and Analysis of Engineering Experiments,* p. 41, 1974. Used with the permission of McGraw-Hill, New York.)

If the data do not plot as a straight line on Weibull graph paper, then either the sample was not taken from a population with a Weibull distribution or it may be that the Weibull distribution has a minimum value x_0 that is greater than $x_0 = 0$. This leads to the *three parameter Weibull distribution.*

$$F(x) = 1 - \exp\left[-\left(\frac{x - x_0}{\theta - x_0}\right)^m\right] \tag{14.29}$$

For example, in the distribution of fatigue life at a constant stress, it is unrealistic to expect a minimum life of zero. The easiest procedure for finding x_0 is to use the Weibull probability plot. First, plot the data as in the two-parameter case where $x_0 = 0$. Then, pick a value of x_0 between 0 and the lowest observed value of x and subtract it from each of the observed values of x. Continue adjusting x_0 and plotting $x - x_0$ until a straight line is obtained on the Weibull graph paper.

14.3.4 Reliability with a Variable Failure Rate

Mechanical failures, and some failures of electronic components, do not exhibit a period of constant failure rate such as that shown in Fig. 14.2a but instead have a curve like Fig. 14.2b. Since the failure rate is a function of time, the simple exponential relation for reliability no longer applies. Instead, reliability is expressed by the Weibull distribution, Eq. (14.27). Since reliability is 1 minus the probability of failure,

$$R(t) = 1 - F(t) = e^{-(t/\theta)^m} \tag{14.30}$$

EXAMPLE 14.5
For the ball bearings plotted in Fig 14.4, $m = 1.5$ and $\theta = 6 \times 10^5$ cycles. The proportion of bearings having a life less than one-half million cycles is given by the area under the curve to the left of $x = 5 \times 10^5$ for a Weibull distribution function like Fig. 14.3 but with $m = 1.5$ and $\theta = 6 \times 10^5$.

$$F(t) = 1 - \exp\left[-\left(\frac{t}{\theta}\right)^m\right] = 1 - \exp\left[-\left(\frac{5 \times 10^5}{6 \times 10^5}\right)^{1.5}\right] = 1 - e^{-0.760}$$

$$= 1 - \frac{1}{(2.718^{0.760})} = 1 - 0.468 = 0.532$$

Thus, 53 percent of the bearings will fail before 500,000 cycles. The probability of failure in less than 100,000 cycles is still 8.5 percent. This apparently is a heavily loaded bearing operating at low speed.

Substituting Eq. (14.29) into Eq. (14.18) gives the hazard rate for the three-parameter Weibull distribution.

$$h(t) = \frac{m}{\theta}\left(\frac{t - t_0}{\theta}\right)^{m-1} \tag{14.31}$$

For the special case $t_0 = 0$ and $m = 1$, Eq. (14.31) reduces to the exponential distribution with $\theta = \text{MTBF}$. When $m = 1$, the hazard rate is constant. When $m < 1$, $h(t)$ decreases as t increases, as in the break-in period of a three-stage failure curve. When $1 < m < 2$, $h(t)$ increases with time. When $m = 3.2$, the Weibull distribution becomes a good approximation of the normal distribution.

EXAMPLE 14.6
Ninety components, N, are tested for a total time of 3830 hours. At various times the tests are stopped and the number of failed components, n, is recorded. Instead of just plotting percentage failure versus time, we use the mean rank to estimate $F(t) = n/(N + 1)$.[1]

---

1. An alternative plotting metric is the median rank, $M = (n - 0.3)/(N + 0.4)$. See C.R. Mischke, "Fitting Weibull Strength Data and Applying It to Stochastic Mechanical Design," *Jnl of Mech. Design,* vol. 114, pp. 35–41, 1992. For a database of Weibull failure statistics, as well as many documents on reliability analysis, see http://www.barringer1.com.

TABLE 14.11

Time $t \times 10^2$ h	Cumulative Total Number of Failures, n	Cumulative Probability of Failure $F(t) = n/(90 + 1)$	Reliability $R(t) = 1 - F(t)$
0	0	0.000	1.000
0.72	2	0.022	0.978
0.83	3	0.033	0.967
1.0	4	0.044	0.957
1.4	5	0.055	0.945
1.5	6	0.066	0.934
2.1	7	0.077	0.923
2.3	9	0.099	0.901
3.2	13	0.143	0.857
5.0	18	0.198	0.802
6.3	27	0.297	0.703
7.9	33	0.362	0.638
11.2	52	0.571	0.429
16.1	56	0.615	0.385
19.0	69	0.758	0.242
38.3	83	0.912	0.088

(a) Plot the data in Table 14.11 and evaluate the parameters for the Weibull reliability, Eq. (14.29).

(b) Find the probability of survival for 700 h.

(c) Determine the instantaneous hazard rate from Eq. (14.31).

(a) $F(t)$ is plotted against time on Weibull probability paper to give the plot shown in Fig. 14.5. A straight line drawn through the data shows that the data follow a Weibull distribution. From Table 14.11, $t = 0 = t_0$. Thus, $R(t) = \exp[-(t/\theta)^m]$. When $t = \theta$, $R(t) = e^{-1} = 0.368$ and $F(t) = 1 - 0.368 = 0.632$. Thus, we can find the scale parameter θ from the value of t where a horizontal line $F(t) = 0.632$ intersects the line through the data points. From Fig. 14.5, $\theta = 1.7 \times 10^3$ h. To find the shape parameter m we need to find the slope of the line. The line has the equation $\ln\ln[1/1 - F(t)] = m\ln(t - t_0) - m\ln\theta$. The line passes through the points (100, 0.04) and (2000, 0.75). Then, its gradient is given by

$$m = \frac{\ln\left(\ln\dfrac{1}{1 - 0.75}\right) - \ln\left(\ln\dfrac{1}{1 - 0.04}\right)}{\ln(2000) - \ln(100)}$$

$$m = \frac{\ln(\ln 4.00) - \ln(\ln 1.0417)}{7.601 - 4.605}$$

$$m = \frac{0.327 - (-3.198)}{2.996} = \frac{3.525}{2.996} = 1.17$$

$$R(t) = \exp\left[-\left(\frac{t}{1700}\right)^{1.17}\right]$$

14

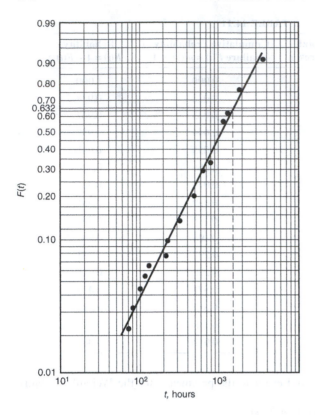

FIGURE 14.5
Plot of $F(t)$ vs. time on Weibull probability paper, for the data in Table 14.11.

$$(b) \qquad R(700) = \exp\left[-\left(\frac{700}{1700}\right)^{1.17}\right] = \exp\left[-(0.412)^{1.17}\right]$$

$$= \exp[-(0.354)] = 0.702 = 70.2\%$$

$$(c) \qquad h(t) = \frac{m}{\theta}\left(\frac{t - t_0}{\theta}\right)^{m-1} = \frac{1.17}{1.7 \times 10^3}\left(\frac{t - 0}{1.7 \times 10^3}\right)^{1.17-1}$$

$$= 6.88 \times 10^{-4}\left(\frac{t}{1700}\right)^{0.17}$$

The failure rate is slowly increasing with time.

14.3.5 System Reliability

Most mechanical and electronic systems comprise a collection of components. The overall reliability of the system depends on how the individual components with their individual failure rates are arranged.

If the components are arranged so that the failure of any component causes the system to fail, it is said to be arranged in series.

$$R_{\text{system}} = R_A \times R_B \times \cdots \times R_n \qquad (14.32)$$

It is obvious that if there are many components exhibiting series reliability, the system reliability quickly becomes very low. For example, if there are 20 components each with $R = 0.99$, the system reliability is $0.99^{20} = 0.818$. Most consumer products exhibit series reliability.

If we are dealing with a constant-failure-rate system,

$$R_{system} = R_A \times R_B = e^{-\lambda_A t} \times e^{-\lambda_B t} = e^{-(\lambda_A + \lambda_B)t}$$

and the value of λ for the system is the sum of the values of λ for each component.

A much better arrangement of components is one in which it is necessary for all components in the system to fail in order for the system to fail. This is called parallel reliability.

$$R_{system} = 1 - (1 - R_A)(1 - R_B) \cdots (1 - R_n) \tag{14.33}$$

If we have a constant-failure-rate system,

$$R_{system} = 1 - (1 - R_A)(1 - R_B) = 1 - \left(1 - e^{-\lambda_A t}\right)\left(1 - e^{-\lambda_B t}\right)$$
$$= e^{-\lambda_A t} + e^{-\lambda_B t} - e^{-(\lambda_A + \lambda_B)t}$$

Since this is not in the form e^{-const}, the parallel system has a variable failure rate.

A system in which the components are arranged to give parallel reliability is said to be redundant; there is more than one mechanism for the system functions to be carried out. In a system with full active redundancy, all but one component may fail before the system fails.

Other systems have partial active redundancy, in which certain components can fail without causing system failure, but more than one component must remain operating to keep the system operating. A simple example would be a four-engine aircraft that can fly on two engines but would lose stability and control if only one engine were operating. This type of situation is known as an n-out-of-m unit network. At least n units must function normally for the system to succeed rather than only one unit in the parallel case and all units in the series case. The reliability of an n-out-of-m system is given by a binomial distribution, on the assumption that each of the m units is independent and identical.

$$R_{n|m} = \sum_{i=n}^{m} \binom{m}{i} R^i (1 - R)^{m-i} \tag{14.34}$$

where

$$\binom{m}{i} = \frac{m!}{i!(m-i)!}$$

EXAMPLE 14.7

A complex engineering design can be described by a reliability block diagram as shown in Fig. 14.6. In subsystem A, two components must operate for the subsystem to function successfully. Subsystem C has true parallel reliability. Calculate the reliability of each subsystem and the overall system reliability.

14

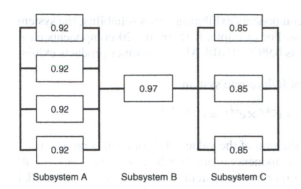

FIGURE 14.6
Reliability block diagram depicting complex design network.

Subsystem A is an n-out-of-m model for which $n = 2$ and $m = 4$. Using Eq. (14.34),

$$R_A = \sum_{i=2}^{4}\binom{4}{i}R^i(1-R)^{4-i}$$

$$\binom{4}{2}R^2(1-R)^2 + \binom{4}{3}R^3(1-R) + \binom{4}{4}R^4$$

$$6R^2(1-2R+R^2) + 4R^3(1-R) + (1)R^4$$

$$3R^4 - 8R^3 + 6R^2 = 3(0.92)^4 - 8(0.92)^3 + 6(0.92)^2 = 0.998$$

Since subsystem B is a single component, $R_B = 0.97$.
Subsystem C is a parallel system. Using Eq. (14.33),

$$R_C = 1 - (1-R_1)(1-R_2)(1-R_3) = 1 - (1-R)^3$$

$$= 1 - (1-0.85)^3 = 1 - (0.15)^3 = 1 - 3.375 \times 10^{-3} = 0.9966$$

The total system reliability can be calculated by visualizing the system reduced to three subsystems in series, of value $R_A = 0.998$, $R_B = 0.970$, and $R_C = 0.997$. From Eq. (14.33),

$$R_{Syst.} = R_A \times R_B \times R_C = (0.998)(0.970)(0.997) = 0.965$$

Another approach to redundancy is to employ a standby system, which is activated only when it is needed. An emergency diesel generating unit in a hospital is a common example. In the analysis of the standby redundant system,[1] the Poisson distribution is used. The reliability of a system of two components, one of which is on standby, is

$$R(t) = e^{-\lambda t}(1 + \lambda t) \tag{14.35}$$

If the units are not identical, but have failure rates λ_1 and λ_2, the reliability of the systems is given by

$$R(t) = \frac{\lambda_1}{\lambda_2 - \lambda_1}(e^{-\lambda_1 t} - e^{-\lambda_2 t}) + e^{-\lambda_1 t} \tag{14.36}$$

---

1. C. O. Smith, *Introduction to Reliability in Design*, pp. 50–59, McGraw-Hill, New York, 1976.

On a theoretical basis, the use of standby redundancy results in higher reliability than active redundancy. However, the feasibility of standby redundancy depends completely on the reliability of the sensing and switching unit that activates the standby unit. When this key factor is considered, the reliability of a standby system is little better than that of an active redundant system.

14.3.6 Maintenance and Repair

An important category of reliability problems deals with maintenance and repair of systems. If a failed component can be repaired while a redundant component has replaced it in service, then the overall reliability of the system is improved. If components subject to wear can be replaced before they have failed, then the system reliability will be improved.

Preventive maintenance is aimed at minimizing system failure. Routine maintenance, such as lubricating, cleaning, and adjusting, generally does not have a major positive effect on reliability, although the absence of routine maintenance can lead to premature failure. Replacement before wearout is based on knowledge of the statistical distribution of failure time; components are replaced sooner than they would normally fail. Here a small part of the useful life is traded off for increased reliability. This approach is greatly facilitated if it is possible to monitor some property of the component that indicates degradation toward an unacceptable performance.

Repairing a failed component in a series system will not improve the reliability, since the system is not operating. However, decreasing the repair time will shorten the period during which the system is out of service, and thus the maintainability and availability will be improved.

A redundant system continues to operate when a component has failed, but it may become vulnerable to shutdown unless the component is repaired and placed back in service. To consider this fact we define some additional terms.

$$\text{MTBF} = \text{MTTF} + \text{MTTR} \tag{14.37}$$

where MTBF = mean time between failures = $1/\lambda$ for constant failure rate
   MTTF = mean time to fail
   MTTR = mean time to repair
If the repair rate $r = 1/\text{MTTR}$, then for an active redundant system,

$$\text{MTTF} = \frac{3\lambda + r}{2\lambda^2} \tag{14.38}$$

As an example of the importance of repair, let $r = 6 \text{ h}^{-1}$ and $\lambda = 10^{-5}$ per h. With repair, the MTTF $= 3 \times 10^{10}$ h, but without repair it is 1.5×10^5 h.

Maintainability is the probability that a component or system that has failed will be restored to service within a given time. The MTTF and failure rate are measures of reliability, but the MTTR and repair rate are measures of maintainability.

$$M(t) = 1 - e^{-rt} = 1 - e^{-t/\text{MTTR}} \tag{14.39}$$

where $M(t)$ = maintainability
   r = repair rate
   t = permissible time to carry out the required repair

It is important to try to predict maintainability during the design of an engineering system.[1] The components of maintainability include (1) the time required to determine that a failure has occurred and to diagnose the necessary repair action, (2) the time to carry out the necessary repair action, and (3) the time required to check out the unit to establish that the repair has been effective and the system is operational. An important design decision is to establish what constitutes the least repairable assembly, that is, the unit of the equipment beyond which diagnosis is not continued but the assembly simply is replaced. An important design trade-off is between MTTR and cost. If MTTR is set too short for the labor hours to carry out the repair, then a large maintenance crew will be required at an increased cost.

Availability is the concept that combines both reliability and maintainability; it is the proportion of time the system is working "on line" to the total time, when that is determined over a long working period.

$$
\begin{aligned}
\text{Availability} &= \frac{\text{total on-line time}}{\text{total on-line time} + \text{total downtime}} \\
&= \frac{\text{total on-line time}}{\text{total on-line time} + (\text{no. of failures} \times \text{MTTR})} \\
&= \frac{\text{total on-line time}}{\text{total on-line time} + (\lambda \times \text{total on-line time} \times \text{MTTR})} \\
&= \frac{1}{1 + \lambda \text{MTTR}}
\end{aligned}
\tag{14.40}
$$

If $\text{MTTF} = 1/\lambda$, then

$$
\text{Availability} = \frac{\text{MTTF}}{\text{MTTF} + \text{MTTR}}
\tag{14.41}
$$

14.3.7 Further Topics

In Sec. 14.3 we have mainly covered situations dealing with continuous variables that can take on any value over a considerable range of data. Also, we have not covered in any detail how to examine a mass of data to determine its characteristics of central tendency and variability, or how to determine what frequency distribution best describes the data. Nor have we discussed how to determine with confidence whether two experimentally determined values are statistically significant from each other. Moreover, we have given scant attention to the statistics of discrete variables other than to mention the binomial and Poisson distributions in passing. For these and other topics of engineering we suggest reading beginning texts[2] or a formal course in statistics.

---

1. B. S. Blanchard, *Logistics Engineering and Management,* 2d ed., Prentice Hall, Englewood Cliffs, NJ, 1981; C. E. Cunningham and W. Cox, *Applied Maintainability Engineering,* John Wiley & Sons, New York, 1972; A. K. S. Jardine, *Maintenance, Replacement and Reliability,* John Wiley & Sons, New York, 1973.
2. W. Navidi, *Statistics for Engineers and Scientists,* 2d ed., McGraw-Hill, New York, 2008.

We have just begun to scratch the surface of such a dynamic and rich subject as reliability engineering. Models for realistic situations where reliability varies with time are important. Also, there are models for censored or truncated frequency distributions in which a life test is terminated after a specified number of failures have occurred. Another extension is models for the case where each element in a system fails by two mutually exclusive failure modes, and the further refinement where failure is by a common cause. A *common cause failure* is one where a single event can lead to multiple failures. Often such failure results from the location of a component, as when an aircraft turbine disk shatters and the flying pieces break several independent hydraulic control systems of the airplane. Additional complexity yet realism arises when the repairability condition is added to the above reliability models. These topics are beyond the scope of this book. The interested reader is referred to the references in the Bibliography at the end of this chapter.

14.4
DESIGN FOR RELIABILITY

The design strategy used to ensure reliability can fall between two broad extremes. The *fail-safe approach* is to identify the weak spot in the system or component and provide some way to monitor that weakness. When the weak link fails, it is replaced, just as the fuse in a household electrical system is replaced. At the other extreme is what can be termed "the one-horse shay" approach. The objective is to design all components to have equal life so the system will fall apart at the end of its useful lifetime just as the legendary one-horse shay did. Frequently an *absolute worst-case approach* is used; in it the worst combination of parameters is identified and the design is based on the premise that all can go wrong at the same time. This is a very conservative approach, and it often leads to overdesign.

Two major areas of engineering activity determine the reliability of an engineering system. First, provision for reliability must be established during the design concept stage, carried through the detailed design development, and maintained during the many steps in manufacture. Second, once the system becomes operational, it is imperative that provision be made for its continued maintenance during its service.[1]

The steps in building reliability into a design are shown in Fig. 14.7. The process starts at the beginning of conceptual design by clearly laying out the criteria for the success of the design, estimating the required reliability, the duty cycle, and carefully considering all of the factors that make up the service environment. In the configuration step of embodiment design the physical arrangement of components can critically

1. H. P. Bloch and F. K. Gleitner, *An Introduction to Machinery Reliability Assessment*, 2d ed., Gulf Publishing Co., Houston, TX, 1994; H. P. Bloch, *Improving Machinery Reliability*, 3d ed., Gulf Publishing Co., 1998; H. P. Bloch, and F. K. Geitner, *Machinery Failure Analysis and Troubleshooting*, Gulf Publishing Co., Houston, TX, 1997; C. Hales and C. Pattin, *ASM Handbook*, Vol. 11, *Design Review for Failure Analysis and Prevention*, pp. 40–49, ASM International, 2003.

Design Stage	Design Activity
Conceptual design	Problem definition:
	Estimate reliability requirement
	Determine likely service environment
Embodiment design	Configuration design:
	Investigate redundancy
	Provide accessibility for maintenance
	Parametric design:
	Select highly reliable components
	Build and test physical and computer prototypes
	Full environment tests
	Establish failure modes/FMEA
	Estimate MTBF
	User trials/modification
Detail design	Produce & test preproduction prototype
	Final estimate of reliablity
Production	Production models:
	Further environmental tests
	Establish quality assurance program
Service	Deliver to customer:
	Feedback field failures and MTBFs to designers
	Repair and replace
	Retirement from service

FIGURE 14.7
Reliability activities throughout design, production, and service.

affect reliability. In laying out functional block diagrams, consider those areas that strongly influence reliability, and prepare a list of parts in each block. This is the place to consider various redundancies and to be sure that physical arrangement allows good access for maintenance. In the parametric step of embodiment design, select components with high reliability. Build and test both computer and physical prototypes. These should be subjected to the widest range of environmental conditions. Establish failure modes and estimate the system and subsystem MTBF. Detail design is the place for the final revision of specifications, for building and testing the preproduction prototype, and the preparation of the final production drawings. Once the design is released to the production organization the design organization is not finished with it. Production models are given further environmental tests, and these help establish the quality assurance program (see Sec. 15.2) and the maintenance schedules. When the product is put into service with customers, there is a steady feedback concerning field failures and MTBFs that helps in redesign efforts and follow-on products.

14.4.1 Causes of Unreliability

The malfunctions that an engineering system can experience can be classified into five general categories.[1]

1. *Design mistakes:* Among the common design errors are failure to include all important operating factors, incomplete information on loads and environmental conditions, erroneous calculations, and poor selection of materials.
2. *Manufacturing defects:* Although the design may be free from error, defects introduced at some stage in manufacturing may degrade it. Some common examples are (1) poor surface finish or sharp edges (burrs) that lead to fatigue cracks and (2) decarburization or quench cracks in heat-treated steel. Elimination of defects in manufacturing is a key responsibility of the manufacturing engineering staff, but a strong relationship with the R&D function may be required to achieve it. Manufacturing errors produced by the production work force are due to such factors as lack of proper instructions or specifications, insufficient supervision, poor working environment, unrealistic production quota, inadequate training, and poor motivation.
3. *Maintenance:* Most engineering systems are designed on the assumption they will receive adequate maintenance at specified periods. When maintenance is neglected or is improperly performed, service life will suffer. Since many consumer products do not receive proper maintenance by their owners, a good design strategy is to design products that do not require maintenance.
4. *Exceeding design limits:* If the operator exceeds the limits of temperature, speed, or another variable for which it was designed, the equipment is likely to fail.
5. *Environmental factors:* Subjecting equipment to environmental conditions for which it was not designed, such as rain, high humidity, and ice, usually greatly shortens its service life.

14.4.2 Minimizing Failure

A variety of methods are used in engineering design practice to improve reliability. We generally aim at a probability of failure of $P_f < 10^{-6}$ for structural applications and $10^{-4} < P_f < 10^{-3}$ for unstressed applications.

Margin of Safety

We saw in Sec. 14.2.4 that variability in the strength properties of materials and in loading conditions (stress) leads to a situation in which the overlapping statistical distributions can result in failures. The variability in strength of materials has a major impact on the probability of failure, so failure can be reduced with no change in the mean value if the variability of the strength can be reduced.

---

1. W. Hammer, *Product Safety Management and Engineering,* Chap. 8, Prentice Hall, Englewood Cliffs, NJ, 1980.

Derating

The analogy to using a factor of safety in structural design is derating electrical, electronic, and mechanical equipment. The reliability of such equipment is increased if the maximum operating conditions (power, temperature, etc.) are derated below their nameplate values. As the load factor of equipment is reduced, so is the failure rate. Conversely, when equipment is operated in excess of rated conditions, failure will ensue rapidly.

Redundancy

One of the most effective ways to increase reliability is with redundancy. In parallel redundant designs the same system functions are performed at the same time by two or more components even though the combined outputs are not required. The existence of parallel paths may result in load sharing so that each component is derated and has its life increased by a longer-than-normal time.

Another method of increasing redundancy is to have inoperative or idling standby units that cut in and take over when an operating unit fails. The standby unit wears out much more slowly than the operating unit does. Therefore, the operating strategy often is to alternate units between full-load and standby service. The standby unit must be provided with sensors to detect the failure and switching gear to place it in service. The sensor and/or switching units frequently are the weak link in a standby redundant system.

Durability

The material selection and design details should be performed with the objective of producing a system that is resistant to degradation from such factors as corrosion, erosion, foreign object damage, fatigue, and wear. This usually requires the decision to spend more money on high-performance materials so as to increase service life and reduce maintenance costs. Life cycle costing is the technique used to justify this type of decision.

Damage Tolerance

Crack detection and propagation have taken on great importance since the development of the fracture mechanics approach to design (Sec. 12.2). A damage-tolerant material or structure is one in which a crack, when it occurs, will be detected soon enough after its occurrence so that the probability of encountering loads in excess of the residual strength is very remote. Figure 14.8 illustrates some of the concepts of damage tolerance. The initial population of very small flaws inherent in the material is shown at the far left. These are small cracks, inclusions, porosity, surface pits, and scratches. If they are less than a_1, they will not grow appreciably in service. Additional defects will be introduced by manufacturing processes. Those larger than a_2 will be detected by inspection and eliminated as scrapped parts. However, some cracks will be present in the components put into service, and they will grow to a size a_3 that can be detected by the nondestructive evaluation (NDE) techniques that can be used in service. The allowable design stresses must be so selected that the number of flaws of size a_3 or greater will be small. Moreover, the material should be damage-tolerant so that propagation to the critical crack size a_{cr} is slow.

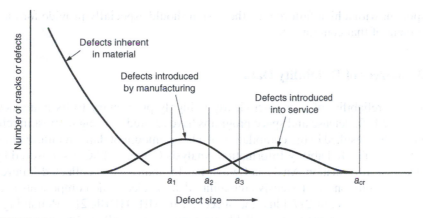

FIGURE 14.8
Distribution of defects in engineering components.

In conventional fracture mechanics analysis (Sec. 12.2), the critical crack size is set at the largest crack size that might be undetected by the NDE technique used in service. The value of fracture toughness of the material is taken as the minimum reasonable value. This is a safe but overly conservative approach. These worst-case assumptions can be relaxed and the analysis based on more realistic conditions by using probabilistic fracture mechanics (PFM).[1]

Ease of Inspection

The importance of detecting cracks should be apparent from Fig. 14.8. Ideally it should be possible to employ visual methods of crack detection, but special design features may have to be provided in order to do so. In critically stressed structures, special features to permit reliable NDE by ultrasonics or eddy current techniques may be required. If the structure is not capable of ready inspection, then the stress level must be lowered until the initial crack cannot grow to a critical size during the life of the structure. For that situation the inspection costs will be low but the structure will carry a weight penalty because of the low stress level.

Specificity

The greater the degree of specificity, the greater the inherent reliability of design. Whenever possible, be specific with regard to material characteristics, sources of supply, tolerances and characteristics of the manufacturing process, tests required for qualification of materials and components, and procedures for installation, maintenance, and use. Specifying standard items increases reliability. It usually means that the materials and components have a history of use so that their reliability is known. Also, replacement items will be readily available. When it is necessary to use

1. H. R. Millurter and P. H. Wirsching, "Analysis Methods for Probabilistic Life Assessment," *ASM Handbook,* Vol. 11, pp. 250–68, ASM International, Materials Park, OH, 2002.

14

a component with a high failure rate, the design should especially provide for the easy replacement of that component.

14.4.3 Sources of Reliability Data

Data on the reliability of a product clearly is highly proprietary to its manufacturer. However, the U.S. defense and space programs have created a strong interest in reliability, and this has resulted in the compilation of a large amount of data on failure rates and failure modes. The Reliability Information Analysis Center (RIAC),[1] sponsored by the DOD Defense Information Analysis Center, has for many years collected failure data on electronic components. Extensive reliability data on electronic components is available online, for a fee, in 217 Plus, the successor to MIL-HDBK-217. Reliability data on nonelectronic components is available on compact disk NPRD-95. Information on European sources of reliability data can be found in the book by Moss.[2] Appendix G of this book contains 20 pages of tables on failure rates. Data and failure rate λ for a wide selection of mechanical components is given by Fisher and Fisher.[3]

14.5
FAILURE MODE AND EFFECTS ANALYSIS (FMEA)

Failure mode and effects analysis (FMEA) is a team-based methodology for identifying potential problems with new or existing designs.[4] It was first used to identify and correct safety hazards. FMEA identifies the mode of failure of every component in a system and determines the effect on the system of each potential failure. By failure we mean inability to meet a customer's requirements as opposed to actual catastrophic material breakage or failure.

Thus, a failure mode is any way that a part could fail to perform its required function. For example, a cable used to lift I-beams could fray from wear, kink from misuse, or actually fracture from excessive load. Note that either fraying or kinking could lead to fracture, but fracture might occur without these events if a design error incorrectly estimated either the strength of the cable or the load it needed to support. Failure modes are discussed in more detail in Sec. 14.7.

There are many variations in detailed FMEA methodology, but they are all aimed at accomplishing three things: (1) predicting what failures could occur; (2) predicting the effect of the failure on the functioning of the system; and (3) establishing

1. http://quanterion.com/RIAC/index.asp
2. T. R. Moss, *The Reliability Data Handbook,* ASME Press, New York, 2005.
3. F. E. Fisher and J. R. Fisher, *Probabilistic Applications in Mechanical Design,* Appendix D, Marcel Dekker, New York, 2000.
4. R. E. McDermott, R. J. Mikulak, and M. R. Beauregard, *The Basics of FMEA,* 2d ed., CRC Press, New York, 2009. D. H. Stamatis, *Failure Mode and Effects Analysis: FMEA from Theory to Execution,* ASQ Quality Press, Milwaukee, WI, 1995; *ASM Handbook,* Vol. 11, *Failure Analysis and Prevention,* pp. 50–59, ASM International, 2003. MIL-STD-1629.

14

TABLE 14.12
Rating for Severity of Failure

Rating	Severity Description
1	The effect is not noticed by the customer
2	Very slight effect noticed by customer; does not annoy or inconvenience customer
3	Slight effect that causes customers annoyance, but they do not seek service
4	Slight effect, customer may return product for service
5	Moderate effect, customer requires immediate service
6	Significant effect, causes customer dissatisfaction; may violate a regulation or design code
7	Major effect, system may not be operable; elicits customer complaint; may cause injury
8	Extreme effect, system is inoperable and a safety problem; may cause severe injury
9	Critical effect, complete system shutdown; safety risk
10	Hazardous; failure occurs without warning; life-threatening

steps that might be taken to prevent the failure, or its effect on the function. FMEA is useful in identifying critical areas of the design that need redundant components and improved reliability. FMEA is a bottom-up process that starts with the required functions, identifies the components to provide the functions, and for each component, lists all possible modes of failure.

Three factors are considered in developing a FMEA.

1. The severity of a failure. Table 14.12 gives the scale for rating severity. Many organizations require that potential failures with a 9 or 10 rating require immediate redesign.
2. The probability of occurrence of the failure. Table 14.13 gives a scale for probability of occurrence. The probabilities given are very approximate and depend on the nature of the failure, the robustness of the design, and the level of quality developed in manufacturing.

TABLE 14.13
Rating for Occurrence of Failure

Rating	Approx. Probability of Failure	Description of Occurrence
1	$\leq 1 \times 10^{-6}$	Extremely remote
2	1×10^{-5}	Remote, very unlikely
3	1×10^{-5}	Very slight chance of occurrence
4	4×10^{-4}	Slight chance of occurrence
5	2×10^{-3}	Occasional occurrence
6	1×10^{-2}	Moderate occurrence
7	4×10^{-2}	Frequent occurrence
8	0.20	High occurrence
9	0.33	Very high occurrence
10	≥ 0.50	Extremely high occurrence

TABLE 14.14
Rating for Detection of Failure

Rating	Description of Detection
1	Almost certain to detect
2	Very high chance of detection
3	High chance of detection
4	Moderately high chance of detection
5	Medium chance of detection
6	Low chance of detection
7	Slight chance of detection
8	Remote chance of detection
9	Very remote chance of detection
10	No chance of detection; no inspection

3. The likelihood of detecting the failure in either design or manufacturing, before the product is used by the customer. Table 14.14 gives the scale for detection. Clearly, the rating for this factor depends on the quality review systems in place in the organization.

Usual practice is to combine the rating for the three factors into a *risk priority number* (RPN).

$$RPN = (\text{severity of failure}) \times (\text{occurrence of failure}) \times (\text{detection rating}) \quad (14.42)$$

Values of RPN can vary from a maximum of 1000, the greatest risk, to a minimum of 1. Numbers derived from Eq. (14.42) are often used to select the "vital few" problems to work on. This can be done by setting a threshold limit, for example, RPN = 200, and working on all potential failures above this limit. Another approach is to arrange the RPN values in a Pareto plot and give attention to those potential failures with the highest ratings. The next paragraph suggests an alternative approach.

Decisions on how to use the information provided from the FMEA should not be blindly based on the RPN values. Consider the results of a FMEA analysis shown in Table 14.15.

Compare failure modes A and B. A has nearly four times the RPN of B, yet B has a severity of failure that would cause safety risk and complete system shutdown. Failure by A would cause only a slight effect on product performance. It achieves its high

TABLE 14.15
Results of a FMEA Analysis

Failure Mode	Severity	Occurrence	Detection	RPN
A	3	4	10	120
B	9	4	1	36
C	3	9	3	81

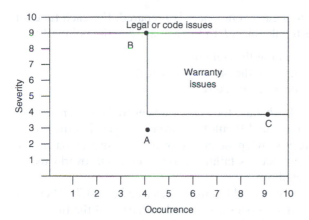

FIGURE 14.9

A rational way to interpret FMEA results.

RPN value because it is not possible to detect the defect that is causing the failure. Certainly failure B is more critical than A and should be given prompt attention for design of the product. Failure mode C has over 2 times the RPN of B, but because the severity of the failure is low it should be given lower priority than B even though the occurrence of failure is high.

A rational way to interpret the results of FMEA analysis has been given by Harpster,[1] Fig. 14.9. Often product specifications include a requirement that action should be taken if the RPN value exceeds some number like 100 or 200. It may not be rational to require a design change if the reason for the high RPN is due to a very hard-to-detect defect or if detectability scores high because no inspection process is in use. Using a plot such as Fig. 14.9 gives better guidance on which design details (failure modes) require remedial action than simply basing all decisions on the RPN value.

14.5.1 Creating a FMEA Chart

The development of a FMEA is best done as a team effort that employs many of the problem-solving tools presented in Sec. 4.7. FMEA can be done on a design, a manufacturing process, or a service. While there is no well-defined format, as there is for a House of Quality (HOQ), a FMEA is usually developed in a spreadsheet format.[2] First you must clearly identify the system or subassembly that you are investigating. Then the following steps are performed and the results recorded in the spreadsheet, Example 14.8.

1. The design is reviewed to determine the interrelations of assemblies and the interrelations of the components of each subassembly. Identify how each component might fail to perform its required function. A complete list of the components in

1. R. A. Harpster. *Quality Digest,* pp. 40–42, June 1999.

2. FMEA software is available to aid in the process. Two examples are FailMode®, from Item Software, and FMEAplus®, developed by Ford Motor Co. and available from the Society of Automotive Engineers.

each assembly and the function of each component is prepared. For each function ask, What if this function fails to take place? To further sharpen the point ask:

- What if the function fails to occur at the right time?
- What if the function fails to occur in the proper sequence?
- What if this function fails to occur completely?

2. Now look more broadly, and ask what are the consequences to the system of each failure identified in step 1. It may be difficult to answer this question in systems for which the subsystems are not independent. A frequent cause of hazardous failures is that an apparently innocuous failure in one subsystem overloads another subsystem in an unexpected way.
3. For each of the functions, list the potential failure modes (see Sec. 14.7). There are likely to be several potential failure modes associated with each of the functions.
4. For each of the failure modes identified in step 3, describe the consequences or effects of the failure. First list the local effect on the particular component; then extend the effects analysis to the entire subassembly and to the total system.
5. Using the severity of failure table (Table 14.12), enter the numerical value. This is best done as a team using consensus decision methods.
6. Identify the possible causes of the failure mode. Try to determine the root cause by using a why-why diagram and interrelationship digraph. (see Sec. 4.7).
7. Using the occurrence of failure table (Table 14.13), enter a value for the occurrence of the cause of each failure.
8. Determine how the potential failure will be detected. This might be a design checklist, a specific design calculation, a visual quality inspection, or a nondestructive inspection.
9. Using Table 14.14, enter a rating that reflects the ability to detect the cause of failure identified in step 8.
10. Calculate the risk priority number (RPN) from Eq. (14.42). Those potential failures with the highest RPN values will be given priority action. In making decisions about where to deploy the resources, also consider Fig. 14.9.
11. For each potential failure, determine the corrective action to remove a potential design, manufacturing, or operational failure. One of the actions might be "no action required." Assign ownership for the removal of each potential failure.

EXAMPLE 14.8

Rifle bolts are made by a powder forging process in which steel preforms of the rifle bolt are made by cold pressing and sintering, followed by hot forging to the required shape and dimensions. The completed chart for the FMEA is given on the following page. Note that the analysis rates the part design and process as it performs in service, and then recommends design or process changes that are expected to improve the RPN of the design.

When the bolt in a rifle fractures, it is the most severe type of failure since the product no longer functions, but of more importance, someone's life is in great danger. The corrective action is to scan all finished parts with 3-D x-ray tomography, the most precise nondestructive inspection method, to reject any parts with fine cracks in the interior of the metal part. This is an expensive step taken while the powder forging process is studied in detail to identify the source of the fine cracks. If these cannot be eliminated, then another manufacturing process would be used to make the part. Note that the severity of the event

Failure Modes and Effect Analysis   **Prepared by:**   **Sheet No. _____ of _____**

Product name: **Part name: Rifle bolt**   **Primary design responsibility:**

Product code: **Part no.:**   **Design deadline:**

1	2	3	4	5	6	7	8	9	10	11	12	13	14
Function	Failure mode	Effects of failure	Causes of failure	Detection	S	O	D	RPN	Recommended corrective action	S	O	D	RPN
1. Chambers and fires the round.	Brittle fracture	Destroys rifle. Injures people.	Internal fine cracks	Dye penetrant	10	4	8	320	Scan all parts with x-ray tomography.	10	1	2	20
2. Seals against gas blowback.													
3. Extracts cartridge case.	Jamming after firing 4 clips in succession.	Rifle will not fire another round.	CTQ dimensions out of spec.	Dimensions checked with gages.	8	6	3	144	Rework tolerances to incl. thermal expansion. Start SPC.	3	4	2	24

is not changed by the corrective action, but the occurrence is set at one chance in a million because of the consequences of failure of this part.

The second failure found in the rifle is jamming of the bolt in the chamber. This makes the rifle inoperable, but is less life threatening than a failure by fracture. Examination of the design notebooks showed that thermal expansion of the bolt due to heating produced by extensive rapid fire was not taken into account when setting the tolerances for the original design. When this is done, and statistical process control (SPC) is initiated for the critical-to-quality dimensions, it is expected that this will eliminate failure by jamming.

FMEA is a powerful design tool, but it can be tedious and time consuming. It requires top-level corporate support to make sure it is used routinely. However, FMEA reduces total life cycle cost by avoiding cost due to warranty problems, service calls, customer dissatisfaction, product recalls, and damaged reputation.

14.6
FAULT TREE ANALYSIS

Fault tree analysis (FTA) is a systematic method to identify undesired events (faults) in a system. A fault is when a system does something it is not supposed to do or does not do something it is supposed to do. Often these faults are reliability or safety issues.

Fault tree analysis starts with the top undesired event and develops in a tree-like fashion all potential causes for that event. It considers component failure, subassembly failures (modules), operating conditions, and human error. An important attribute of FTA is the ability to identify combinations of events that can affect the top undesired event. FTA is in contrast to FMEA which is a much more bottom-up approach that identifies all of the failure modes that can affect the components of a system in isolation. Building a FTA starts with the function structure established in conceptual design.

Basically, a fault tree is a logic diagram in which logic gates are used to determine the relations between input events and output events.[1] A fully quantitative FTA uses Boolean algebra in the logical analysis, and probabilities of failure are computed for each event. However, considerable insight can be gained from the graphical relations portrayed by the fault tree. Our discussion of PTA will be restricted to this qualitative level.

Each fault tree deals with a specific event, for example, failure of a lawn mower engine to start (see Fig. 14.10). FTA is a "top-down approach" that starts with the top event and then determines the contributory events that would lead to the top event. Most top events would be suggested from a preliminary hazard analysis. They can be either hardware failures or human errors.

Different kinds of events are identified with specific symbols on the fault tree diagram.

An output event ⬜ is an event that should be developed or analyzed further to determine how it can occur. It is the only symbol on the fault tree diagram that will have a logic gate and input events below it. Except when it represents the undesired top event, the rectangle also serves as an input event to another output event.

An independent event ○ is an event that does not depend on other components within the system for its occurrence. A common example is a failure of a system component.

A normal event ⬠ is an event that is expected to occur during system operation. It always occurs unless failure occurs.

An undeveloped event ◇ is an event that has not been developed further because of the lack of information or because it is not of sufficient consequence.

A transfer symbol △ is a connection to another part of the fault tree within the same branch.

The AND gate ⌓ is used to represent a logic condition in which all inputs below the gate must be present for the output (at the top of the gate) to occur.

The OR gate ⌂ is used to represent the situation in which any of the input events will lead to the output.

These symbols are employed in Fig. 14.10, which is a fault tree for the failure of a lawn mower engine. The bottom branch of the tree consists of failures or initiating events that are errors. They point out events where FMEA should be done. Starting with the top event and moving down the branches, we enter a hierarchy of causes. Each event branches into other events. In constructing the tree, it is important to proceed slowly and deliberately down from the top event and list every direct immediate cause before going on to consider the next level of causes. The description used for each event must be carefully chosen to indicate the cause.

---

1. J. B. Fassell, *Nuc. Sci. Eng.,* vol. 2, pp. 433–438, 1973; G. J. Powers and F. C. Tompkins, Jr., *AIChE J.,* vol. 20, pp. 376–387, 1974; W. E. Vessey *et al.,* NASA Fault Tree Handbook with Aerospace Applications, NASA, Washington DC, 2002; W. Vesely, *et al., Fault Tree Handbook with Aerospace Applications,* see www.hq.nasa.gov/office/codeq/doctree/fthb.pdf.

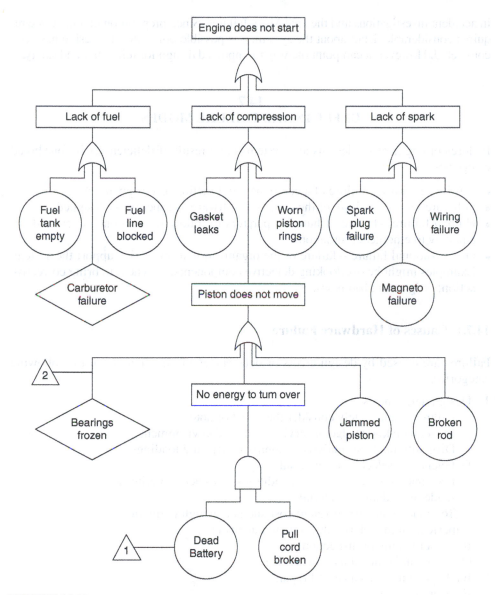

FIGURE 14.10
Fault tree for the failure of a lawn mower engine to start.

The fault tree clearly indicates where corrective action should be taken. For example, the need for disassembling the piston and/or bearings is indicated. Also, the need for preventive maintenance and inspection of the battery or pull cord is called for.

In summary, FTA points out the critical areas in a complex system where further study in failure mode analysis and reliability engineering is required. It also can be used as a tool in troubleshooting a piece of equipment that will not operate. FTA is very useful in accident investigations and the analysis of failures. Since preparation of a fault tree requires considerable detail about the system, it is possible only after the design has been completed. However, it can point the way to improved design for reliability and safety.

14.7
DEFECTS AND FAILURE MODES

Failures of engineering designs and systems are a result of deficiencies in four broad categories.

- Hardware failure—failure of a component to function as designed.
- Software failure—failure of the computer software to function as designed.
- Human failure—failure of human operators to follow instructions or respond adequately to emergency situations.
- Organizational failure—failure of the organization to properly support the system. Examples might be overlooking defective components, slowness to bring corrective action, or ignoring bad news.

14.7.1 Causes of Hardware Failure

Failures are caused by design errors or deficiencies in one or more of the following categories:

1. Design deficiencies
   Failure to adequately consider the effect of notches
   Inadequate knowledge of service loads and environment
   Difficulty of stress analysis in complex parts and loadings
2. Deficiency in selection of material
   Poor match between service conditions and selection criteria
   Inadequate data on material
   Too much emphasis given to cost and not enough to quality
3. Imperfection in material due to manufacturing
4. Improper testing or inspection
5. Overload and other abuses in service
6. Inadequate maintenance and repair
7. Environmental factors
   Conditions beyond those allowed for in design
   Deterioration of properties with time of exposure to environment

Deficiencies in the design process, or defects in the material or its processing, can be classified in the following ways: At the lowest level is a lack of conformance to a stated specification. An example would be a dimension "out of spec" or a strength property below specification. Next in severity is a lack of satisfaction by the customer or user. This may be caused by a critical performance characteristic set at an improper value, or it may be a system problem caused by rapid deterioration. The ultimate defect is one that causes failure of the product. Failure may be an actual fracture or disruption of physical continuity of the part, or failure may be inability of the part to function properly.

14.7.2 Failure Modes

The specific modes of failure of engineering components can usually be grouped into four general classes:

1. Excessive elastic deformation
2. Excessive plastic deformation
3. Fracture
4. Loss of required part geometry through corrosion or wear

The most common failure modes are listed in Table 14.16. Some of these failure modes are directly related to a standard mechanical property test, but most are more

TABLE 14.16
Failure Modes for Mechanical Components

1. Elastic deformation	d. Surface fatigue wear
2. Yielding	e. Deformation wear
3. Brinelling	f. Impact wear
4. Ductile failure	g. Fretting wear
5. Brittle fracture	9. Impact
6. Fatigue	a. Impact fracture
a. High-cycle fatigue	b. Impact deformation
b. Low-cycle fatigue	c. Impact wear
c. Thermal fatigue	d. Impact fretting
d. Surface fatigue	e. Impact fatigue
e. Impact fatigue	10. Fretting
f. Corrosion fatigue	a. Fretting fatigue
g. Fretting fatigue	b. Fretting wear
7. Corrosion	c. Fretting corrosion
a. Direct chemical attack	11. Galling and seizure
b. Galvanic corrosion	12. Scoring
c. Crevice corrosion	13. Creep
d. Pitting corrosion	14. Stress rupture
e. Intergranular corrosion	15. Thermal shock
f. Selective leaching	16. Thermal relaxation
g. Erosion-corrosion	17. Combined creep and fatigue
h. Cavitation	18. Buckling
i. Hydrogen damage	19. Creep buckling
j. Biological corrosion	20. Oxidation
k. Stress corrosion	21. Radiation damage
8. Wear	22. Bonding failure
a. Adhesive wear	23. Delamination
b. Abrasive wear	24. Erosion
c. Corrosive wear	

TABLE 14.17
Examples of Failure Modes in Components

Component	Failure Mode	Possible Cause
Battery	Discharged	Extended use
Check valve	Stuck closed	Corrosion
Piping	Sagging pipes	Designed inadequate supports
Valve	Leaks	Faulty packing
Lubricant	No flow	Clogged by debris/no filter
Bolt	Threads stripped	Excessive tightening torque

complex, and failure prediction requires using a combination of two or more properties. However, not all failures are related to material behavior. Table 14.17 gives some failures modes for common engineering components.

14.7.3 Importance of Failure

It is a human tendency to be reluctant to talk about failure or to publish much information about failures. Spectacular system failures, like the Tacoma Narrows bridge or the O-ring seal on the space shuttle *Challenger* solid rocket booster, have caught the public's attention, but most failures go undocumented.[1] This is a shame, because much learning in engineering occurs by studying failures. Simulated service testing and proof-testing of preproduction prototypes are important steps at arriving at a successful product. While the literature on engineering failures is not extensive, there are several useful books on the subject.[2] For information on conducting failure analysis[3] see Techniques for Failure Analysis at www.mhhe.com/dieter.

14.8
DESIGN FOR SAFETY

Safety may well be the paramount issue in product design.[4] Normally we take safety for granted, but the recall of an unsafe product can be very costly in terms of product liability suits, replaced product, or tarnished reputation. The product must be safe to

14

---

1. An assortment of failures of aircraft, bridges, engineering machines and strictures, and software are described in Wikipedia under Engineering Failures.
2. *Case Histories in Failure Analysis,* ASM International, Materials Park, OH, 1979; H. Petroski, *Success through Failure: the Paradox of Design,* Princeton University Press, Princeton, NJ, 2006; V. Ramachandran, et al., *Failure Analysis of Engineering Structures: Methodology and Case Histories,* ASM International, Materials Park, OH, 2005; *Microelectronics Failure Analysis Desk Reference,* 5th ed., ASM International, Materials Park, OH, 2004. A. Sofronas, *Analytical Troubleshooting of Process Machinery and Pressure Vessels,* John Wiley & Sons, Hoboken, NJ, 2006.
3. Extensive information on conducting a failure analysis can be found in *ASM Handbook,* Vol 11: *Failure Analysis and Prevention,* 2002, pp. 315–556.
4. C. O. Smith, "Safety in Design," *ASM Handbook,* Vol. 20, pp. 139–45, ASM International, Materials Park, OH, 1997.

manufacture, to use, and to dispose of after use.[1] Also, a serious accident in which a life is lost can be very traumatic to the person responsible, and possibly career ending to the responsible engineer.

A safe product is one that does not cause injury or property loss. Also included under safety is injury to the environment. Achieving safety is no accident. It comes from a conscious focus on safety during design, and in knowing and following some basic rules. There are three aspects to design for safety:

1. Make the product safe; that is, design all hazards out of the product.
2. If it is not possible to make the product inherently safe, then design in protective devices like guards, automatic cutoff switches, and pressure-relief valves, to mitigate the hazard.
3. If step 2 cannot remove all hazards, then warn the user of the product with appropriate warnings like labels, flashing lights, and loud sounds.
4. Provide training and protective clothing or devices (glasses, ear mufflers) to the user or operator of the equipment.

A *fail-safe design* seeks to ensure that a failure will either not affect the product or change it to a state in which no injury or damage will occur. There are three variants of fail-safe designs.

- Fail-passive design. When a failure occurs, the system is reduced to its lowest-energy state, and the product will not operate until corrective action is taken. A circuit breaker is an example of a fail-passive device.
- Fail-active design. When failure occurs, the system remains energized and in a safe operating mode. A redundant system kept on standby is an example.
- Fail-operational design. The design is such that the device continues to provide its critical function even though a part has failed. A valve that is designed so that it will remain in the open position if it fails is an example.

14.8.1 Potential Dangers

We list here some of the general categories of safety hazards that need to be considered in design.

- Acceleration/deceleration—falling objects, whiplash, impact damage
- Chemical contamination—human exposure or material degradation
- Electrical—shock, burns, surges, electromagnetic radiation, power outage
- Environment—fog, humidity, lightning, sleet, temperature extremes, wind
- Ergonomic—fatigue, faulty labeling, inaccessibility, inadequate controls
- Explosions—dust, explosive liquids, gases, vapors, finely powdered materials
- Fire—combustible material, fuel and oxidizer under pressure, ignition source
- Human factors—failure to follow instructions, operator error
- Leaks or spills

---

1. For a comprehensive safety website see http://www.safetyline.net.

- Life cycle factors—frequent startup and shutdown, poor maintenance
- Materials—corrosion, weathering, breakdown of lubrication, wear
- Mechanical—fracture, misalignment, sharp edges, stability, vibrations
- Physiological—carcinogens, human fatigue, irritants, noise, pathogens
- Pressure/vacuum—dynamic loading, implosion, vessel rupture, pipe whip
- Radiation—ionizing (alpha, beta, gamma, x-ray), laser, microwave, thermal
- Structural—aerodynamic or acoustic loads, cracks, stress concentrations
- Temperature—changes in material properties, burns, flammability, volatility

Product hazards are often controlled by government regulation. The U.S. Consumer Product Safety Commission is charged with this responsibility.[1] Products designed for use by children are held to much higher safety standards than products intended to be used by adults. The designer must also be cognizant that in addition to providing a safe product for the customer, it must be safe to manufacture, sell, install, and service.

In our society, products that cause harm invariably result in lawsuits for damages under the product liability laws. Design engineers must understand the consequences of these laws and how they must practice to minimize safety issues and the threat of litigation. This topic is covered in Chap. 18, which is available at www.mhhe.com/dieter.

14.8.2 Guidelines for Design for Safety[2]

1. Recognize and identify the actual or potential hazards, and then design the product so they will not affect its functioning.
2. Thoroughly test prototypes of the product to reveal any hazards overlooked in the initial design.
3. Design the product so it is easier to use safely than unsafely.
4. If field experience turns up a safety problem, determine the root cause and redesign to eliminate the hazard.
5. Realize that humans will do foolish things, and allow for it in your design. More product safety problems arise from improper product use than from product defects. A user-friendly product is usually a safe product.
6. There is a close correspondence between good ergonomic design and a safe design. For example:
   - Arrange the controls so that the operator does not have to move to manipulate them.
   - Make sure that fingers cannot be pinched by levers or other features.
   - Avoid sharp edges and corners.
   - Point-of-operation guards should not interfere with the operator's movement.
   - Products that require heavy or prolonged use should be designed to avoid cumulative trauma disorders like carpal tunnel syndrome. This means avoiding awkward positions of the hand, wrist, and arm and avoiding repetitive motions and vibration.

---

1. See the CPSC website, www.cpsc.gov.
2. C. O. Smith, op. cit.; J. G. Bralla, *Design for Excellence,* Chap. 17, McGraw-Hill, New York, 1996.

7. Minimize the use of flammable materials, including packaging materials.
8. Paint and other surface finishing materials should be chosen to comply with EPA and OSHA regulations for toxicity to the user and for safety when they are burned, recycled, or discarded.
9. Think about the need for repair, service, or maintenance. Provide adequate access without pinch or puncture hazards to the repairer.
10. Electrical products should be properly grounded to prevent shock. Provide electrical interlocks so that high-voltage circuits will not be energized unless a guard is in the proper position.

14.8.3 Warning Labels

With rapidly escalating costs of product liability, manufacturers have responded by plastering their products with warning labels. Warnings should supplement the safety-related design features by indicating how to avoid injury or damage from the hazards that could not be feasibly designed out of the product without seriously compromising its performance. The purpose of the warning label is to alert the user to a hazard and tell how to avoid injury from it.

For a warning label to be effective, the user must receive the message, understand it, and act on it. The engineer must properly design the label with respect to the first two issues to achieve the third. The label must be prominently located on the product. Most warning labels are printed in two colors on a tough, wear-resistant material, and fastened to the product with an adhesive. Attention is achieved by printing *Danger, Warning,* or *Caution,* depending on the degree of the hazard. The message to be communicated by the warning must be carefully composed to convey the nature of the hazard and the action to be taken. It should be written at the sixth-grade level, with no long words or technical terms. For products that will be used in different countries, the warning label must be in the local language.

14.9
SUMMARY

Modern society places strong emphasis on avoiding risk, while insisting on products that last longer and require less service or repair. This requires greater attention to risk assessment in the concept of a design, in using methods for deciding on potential modes of failure, and in adopting design techniques that increase the reliability of engineered systems.

A *hazard* is the potential for damage. *Risk* is the likelihood of a hazard materializing. *Danger* is the unacceptable combination of hazard and risk. *Safety* is freedom from danger. Thus, we see that the engineer must be able to identify hazards to the design, evaluate the risk in adopting a technology or course of action, and understand when conditions constitute a danger. Design methods that mitigate a danger lead to safe design. One of the common ways this is achieved is by designing with respect to accepted codes and standards.

14

Reliability is the probability that a system or component will perform without failure for a specified time. Most systems follow a three-stage failure curve: (1) an early burn-in or break-in period, in which the failure rate decreases rapidly with time, (2) a long period of nearly constant failure rate (useful life), and (3) a final wearout period of rapidly increasing failure rate. The failure rate is usually expressed as the number of failures per 1000 h, or by its reciprocal, the mean time between failures (MTBF). System reliability is determined by the arrangement of its components, that is, in series or parallel.

System reliability is heavily influenced by design. The product design specification should contain a reliability requirement. The configuration of the design determines the degree of redundancy. The design details determine the level of defects. Early estimation of potential failure modes by FMEA lead to more reliable designs. Other methods to increase the reliability of the design are use of highly durable materials and components, derating of components, reduction in part count and simplicity of the design, and adoption of a damage-tolerant design coupled with ready inspection. Extensive testing of preproduction prototypes to "work the bugs out" is a method that works well. Methods for carrying out a root cause analysis of the reasons for a failure are an important means of improving the reliability of designs.

A safe design is one that instills confidence in the customer. It is a design that will not incur product liability costs. In developing a safe design, the primary objective should be to identify potential hazards and then produce a design that is free from the hazards. If this cannot been done without compromising the functionality of the design, the next best approach is to provide protective devices that prevent the person from coming in contact with the hazard. Finally, if this cannot be done, then warning labels, lights, or buzzers must be used.

NEW TERMS AND CONCEPTS

Availability	Failure mode and effects analysis	Reliability
Break-in period	Hazard	Risk
Common cause failure	Hazard rate	Root cause analysis
Derating	Maintainability	Safety
Design redundancy	Mandatory standard	Safety factor
Fail-safe design	Mean time between failure	Wearout period
Failure mode	Mean time to failure	Weibull destribution

BIBLIOGRAPHY

Risk Assessment

Haimes, Y. Y.: *Risk Modeling, Assessment, and Management,* 2d ed., Wilex-Interscience, Hoboken, NJ, 2004.
Michaels, J. V.: *Technical Risk Management,* Prentice Hall, Upper Saddle River, NJ, 1996.
Schwing, R. C. and W. A. Alpers, Jr. (eds.): *Societal Risk Assessment: How Safe Is Enough?* Plenum Publishing Co., New York, 1980.

14

Failures and Failure Prevention

Booker, J. D., M. Raines and K.G. Swift, *Designing Capable and Reliable Products,* Butterworth-Heinemann, Boston, 2001.

Evan, W. M. and M. Manion: *Minding Machines: Preventing Technological Disasters,* Prentice Hall, Upper Saddle River, NJ, 2003.

Evans, J. W. and J. Y. Evans (eds.): *Product Integrity and Reliability in Design,* Springer-Verlag, London, 2000.

Petroski, H.: *Success through Failure: The Paradox on Design,* Princeton University Press, Princeton, NJ, 2006.

Witherell, C. E.: *Mechanical Failure Avoidance: Strategies and Techniques,* McGraw-Hill, New York, 1994.

Reliability Engineering

Bentley, J. P.: *An Introduction to Reliability and Quality,* John Wiley & Sons, New York, 1993.

Ebeling, C. E.: *Reliability and Maintainability Engineering,* McGraw-Hill, New York, 1997.

Ireson, W. G. (ed.): *Handbook of Reliability Engineering and Management,* 2d ed., McGraw-Hill, New York, 1996.

O'Connor, P. D. T.: *Practical Reliability Engineering,* 4th ed., John Wiley & Sons, New York, 2002.

Rao, S. S., *Reliability-Based Design,* McGraw-Hill, New York, 1992.

Smith, D. J.: *Reliability, Maintainability, and Risk,* 7th ed., Butterworth-Heinemann, Oxford, 2005.

Safety Engineering

Brauer, R. L. and R. Brauer, *Safety and Health for Engineers,* 2d ed., John Wiley & Sons, New York, 2005.

Covan, J.: *Safety Engineering,* John Wiley & Sons, New York, 1995.

Hunter, T. A.: *Engineering Design for Safety,* McGraw-Hill, New York, 1992.

Wong, W.: *How Did That Happen?: Engineering Safety and Reliability,* Professional Engineering Publishing Ltd., London, 2002.

PROBLEMS AND EXERCISES

14.1 Assume you are part of a federal commission established in 1910 to consider the risk to society of the expected widespread use of the motor car powered with highly flammable gasoline. Without the benefit of hindsight, what potential dangers can you contemplate? Use a worst-case scenario. Now, taking advantage of hindsight, what lesson can you draw about evaluating the hazards of future technologies? Do this as a team exercise.

14.2 Give some examples of voluntary standards that have been adopted by the cooperating industry, and others that industries have not adopted until forced to by competitive pressures.

14

14.3 A steel tensile link has a mean yield strength of $\overline{S}_y = 27,000$ psi and a standard deviation on strength of $S_y = 4000$ psi. The variable applied stress has a mean value of $\overline{\sigma} = 13,000$ psi and a standard deviation $s = 3000$ psi.

   (a) What is the probability of failure taking place? Show the situation with carefully drawn frequency distributions.
   (b) The factor of safety is the ratio of the mean material strength divided by the mean applied stress. What factor of safety is required if the allowable failure rate is 5 percent?
   (c) If absolutely no failures can be tolerated, what is the lowest value of the factor of safety?

14.4 A machine component has average life of 120 h. Assuming an exponential failure distribution, what is the probability of the component operating for at least 200 h before failing?

14.5 A nonreplacement test was carried out on 100 electronic components with a known constant failure rate. The history of failures was as follows:

1st failure after	93 h
2nd failure after	1,010 h
3rd failure after	5,000 h
4th failure after	28,000 h
5th failure after	63,000 h

The testing was discontinued after the fifth failure. If we can assume that the test gives an accurate estimate of the failure rate, determine the probability that one of the components would last for (a) 10^5 h and (b) 10^6 h.

14.6 The failure of a group of mechanical components follows a Weibull distribution, where $\theta = 10^5$ h, $m = 4$, and $t_0 = 0$. What is the probability that one of these components will have a life of 2×10^4 h?

14.7 A complex system consists of 550 components in a series configuration. Tests on a sample of 100 components showed that 2 failures occurred after 1000 h. If the failure rate can be assumed to be constant, what is the reliability of the system to operate for 1000 h? If an overall system reliability of 0.98 in 1000 h is required, what would the failure rate of each component have to be?

14.8 A system has a unit with MTBF $= 30,000$ h and a standby unit (MTBF $= 20,000$ h). If the system must operate for 10,000 h, what would be the MTBF of a single unit (constant failure rate) that, without standby, would have the same reliability as the standby system?

14.9 A reliability block diagram for an engineering system is given in Fig. 14.11 (on the following page). Determine the overall system reliability.

14.10 An electronic component has a constant failure rate of $\lambda = 100 \times 10^{-6}$ per h.

   (a) Calculate the MTBF measured in years.
   (b) Calculate the reliability of the component after one year of service.
   (c) The unavailability of the system, $\overline{A}$, equals 1 minus A. Calculate the unavailability if the mean downtime, MDT, is 10 h. MDT $\approx$ MTTR.
   (d) What is the effect on the unavailability of doubling the MTTR?

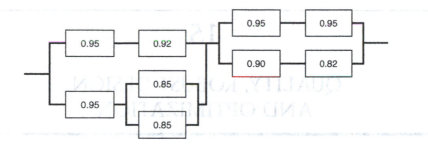

FIGURE 14.11
Reliability block diagram for Problem 14.9.

14.11 Make a failure modes and effects analysis for a ballpoint pen.

14.12 List a number of reasons why the determination of product life is important in engineering design.

14.13 Using the principles of mechanics of materials, what would a torsion failure look like in a ductile material and a brittle material?

14.14 Read one of the following detailed accounts of a failure analysis:

    (*a*) C. O. Smith, "Failure of a Twistdrill," *Trans. ASME, J. Eng. Materials Tech.,* vol. 96, pp. 88–90, April 1974.
    (*b*) C. O. Smith, "Failure of a Welded Blower Fan Assembly," ibid., vol. 99, pp. 83–85, January 1977.
    (*c*) R. F. Wagner and D. R. McIntyre, "Brittle Fracture of a Steel Heat Exchanger Shell," ibid., vol. 102, pp. 384–87, October 1980.

14.15 Consult the home page of the Consumer Product Safety Commission to determine what products have recently received rulings. Divide the work up between teams, and together, prepare a set of detailed design guidelines for safe product design.

14.16 Discuss the practice of using consumer complaints to establish that a product is hazardous and should be recalled.

14

15

QUALITY, ROBUST DESIGN, AND OPTIMIZATION

15.1
THE CONCEPT OF TOTAL QUALITY

In the 1980s many manufacturers in the United States and Western Europe became threatened by the high quality of products produced by Japan. Not only were these products of high quality but they were competitively priced. The threat forced a frantic search for the "magic bullet" that enabled Japanese manufacturers to capture market share. However, what the investigators found was a system of continuous quality improvement, *kaizen*, using simple statistical tools, emphasizing working in teams, and focusing on delighting the customer. We have introduced many of these concepts throughout this text, starting with quality function deployment (QFD) in Chap. 3 and team methods and most of the quality problem-solving tools in Chap. 4. The concepts learned from the Japanese became known as total quality management (TQM) in the western world. More recently, the ideas of TQM have been extended using a more rigorous statistical approach and strong focus on increasing the revenue from new products in a quality methodology called Six Sigma.

An important lesson learned from Japan is that the best way to achieve high quality in a product is to design it into the product from the beginning, and then to assure that it is maintained throughout the manufacturing stage. A further lesson, advanced by Dr. Genichi Taguchi, is that the enemy of quality is variability in the performance of a product and in its manufacture. A *robust design* is one that has been created with a system of design tools that reduces product or process variability, while simultaneously guiding the performance toward a near-optimal setting. A product that is robustly designed will provide customer satisfaction even when subjected to extreme conditions in the service environment.

15.1.1 Definition of Quality

Quality is a concept that has many meanings depending upon your perspective. Quality implies the ability of a product or service to satisfy a stated or implied need.

680

Additionally, a quality product or service is one that is free from defects or deficiencies. In Sec. 3.3.1 we discussed Garvin's[1] eight basic dimensions of quality for a manufactured product. These serve as a general specification of a quality product.

In another foundational paper, Garvin[2] identified the five distinct approaches toward the achievement of quality.

- *The transcendent approach:* This is a philosophical approach that holds that quality is some absolute and uncompromising high standard that we learn to recognize only through experience.
- *Product-based approach:* This is completely opposite from the transcendent approach and views quality as a precise and measurable parameter. A typical parameter of quality might be the number of product features, or its expected life.
- *Manufacturing-based approach:* In this view quality is defined by conformance to requirements or specifications. High quality is equated with "doing it right the first time."
- *Value-based approach:* In this view quality is defined in terms of costs and prices. A quality product is one that provides performance at an acceptable price. This approach equates quality (excellence) with value (worth).
- *User-based approach:* This approach views quality as "being in the eyes of the beholder." Each individual is considered to have a highly personal and subjective view of quality.

The phrase "total quality" denotes a broader concept of quality[3] than simply checking the parts for defects as they come off the production line. The idea of preventing defects by improved design, manufacturing, and process control plays a big role in total quality. We refer to the first aspect as off-line quality control, while the latter is on-line quality control. In order for total quality to be achieved it must be made the number one priority of the organization. In a study in which companies were ranked by an index of perceived quality, the firms in the top third showed an average return on assets of 30 percent, while the firms in the bottom third showed an average return of 5 percent.

Quality is meeting customer requirements consistently. To do this we must know who our customers are and what they require. This attitude should not be limited to external customers. Those we interact with are our customers. This means that a manufacturing unit providing parts to another unit for further processing should be just as concerned about defects as if the parts were shipped directly to the customer.

Total quality is achieved by the use of facts and data to guide decision making. Thus, data should be used to identify problems and to help determine when and if action should be taken. Because of the complex nature of the work environment, this requires considerable skill in data acquisition and analysis with statistical methods.

---

1. D. A. Garvin, *Harvard Business Review*, November–December 1987, pp. 101–9.
2. D. A. Garvin, "What Does Product Quality Really Mean?" *Sloan Management Review*, Fall 1984, pp. 25–44.
3. A. V. Feigenbaum, *Total Quality Control*, 3rd ed., McGraw-Hill, New York, 1983.

15.1.2 Deming's 14 Points

Work by Walter Shewhart, W. Edwards Deming, and Joseph Juran in the 1920s and 1930s pioneered the use of statistics for the control of quality in production. These quality control methods were mandated by the War Department in World War II for all ordnance production in the United States and were found to be very effective. After the war, with a pent-up demand for civilian goods and relatively cheap labor and materials costs, these statistical quality control (SQC) methods were largely abandoned as unnecessary and an added expense.

It was a different story in Japan, whose industry had been largely destroyed by aerial bombing. The Japanese Union of Scientists and Engineers invited Dr. W. Edwards Deming to Japan in 1950 to teach them SQC. His message was enthusiastically received, and SQC became an integral part of the rebuilding of Japanese industry. An important difference between how Americans and Japanese were introduced to SQC is that in Japan the first people converted were top management, while in America it was largely engineers who adopted it. The Japanese have continued to be strong advocates of SQC methods and have extended it and developed new adaptations. Today Japanese products are viewed as having quality. In Japan, the national award for industrial quality, a very prestigious award, is called the Deming Prize.

Dr. Deming viewed quality as one principle in a broader philosophy of management,[1] as expressed by his fourteen points.

1. Create a constancy and consistency of purpose toward improvement of product and service. Aim to become competitive and to stay in business and to provide jobs.
2. Adopt the philosophy that we are in a new economic age. Western management must awaken to the challenge, must learn their responsibilities, and take on the leadership of change.
3. Stop depending on inspection to achieve quality. Eliminate the need for production line inspection by building quality into the product's design.
4. Stop the practice of awarding business only on the basis of price. The goal should be to minimize total cost, not just acquisition cost. Move toward a single supplier for any one item. Create a relationship of loyalty and trust with your suppliers.
5. Search continually for problems in the system and seek ways to improve it.
6. Institute modern methods of training on the job. Management and workers alike should know statistics.
7. The aim of supervision should be to help people and machines to do a better job. Provide the tools and techniques for people to have pride of workmanship.
8. Eliminate fear, so that everyone may work effectively for the company. Encourage two-way communication.
9. Break down barriers between departments. Research, design, sales, and production must work as a team.

---

1. W. E. Deming, *Out of Crisis,* MIT Center for Advanced Engineering Study, Cambridge, MA, 1986; M. Tribus, *Mechanical Engineering,* January 1988, pp. 26–30.

10. Eliminate the use of numerical goals, slogans, and posters for the workforce. Eighty to 85 percent of the causes of low quality and low productivity are the fault of the system, 15 to 20 percent are because of the workers.
11. Eliminate work standards (quotas) on the factory floor and substitute leadership. Eliminate management by objective, management by numbers, and substitute leadership.
12. Remove barriers to the pride of workmanship.
13. Institute a vigorous program of education and training to keep people abreast of new developments in materials, methods, and technology.
14. Put everyone in the company working to accomplish this transformation. This is not just a management responsibility—it is everybody's job.

15.2
QUALITY CONTROL AND ASSURANCE

Quality control[1] refers to the actions taken throughout the engineering and manufacturing of a product to prevent and detect product deficiencies and product safety hazards. The American Society for Quality (ASQ) defines quality as the totality of features and characteristics of a product or service that bear on the ability to satisfy a given need. In a narrower sense, *quality control* (QC) refers to the statistical techniques employed in sampling production and monitoring the variability of the product. *Quality assurance* refers to those systematic actions that are vital to providing satisfactory confidence that an item or service will fulfil defined requirements.

Quality control received its initial impetus in the United States during World War II when war production was facilitated and controlled with QC methods. The traditional role of quality control has been to monitor the quality of raw materials, control the dimensions of parts during production, eliminate imperfect parts from the production line, and assure functional performance of the product. With increased emphasis on tighter tolerance levels, slimmer profit margins, and stricter interpretation of liability laws by the courts, there has been even greater emphasis on quality control. The heavy competition for U.S. markets from overseas producers who have emphasized quality has placed even more emphasis on QC by U.S. producers.

15.2.1 Fitness for Use

An appropriate engineering definition of quality is to consider that it means fitness for use. The consumer may confuse quality with luxury, but in an engineering context quality has to do with how well a product meets its design and performance specifications.

---

1. J. A. Defeo, ed., *Juran's Quality Handbook,* 6th ed., McGraw-Hill, New York, 2010; F. M. Gryna and R. C. H. Chura, *Juran's Quality Planning and Analysis for Enterprise Quality,* 5th ed., McGraw-Hill, New York, 2007.

The majority of product failures can be traced back to the design process. It has been found that 75 percent of defects originate in product development and planning, and that 80 percent of these remain undetected until the final product test or in service.[1]

The particular technology used in manufacturing has an important influence on quality. We saw in Chap. 13 that each manufacturing process has an inherent capability for maintaining tolerances, generating a shape, and producing a surface finish. This has been codified into a methodology called conformability analysis.[2] This technique aims, to identify the potential process capability problems in component manufacture and assembly and to estimate the level of potential failure costs for a given design.

As computer-aided applications pervade manufacturing, there is a growing trend toward automated inspection. This permits a higher volume of part inspection and removes human variability from the inspection process. An important aspect of QC for both manual and automated inspection is the design of inspection fixtures and gaging.[3]

The skill and attitude of production workers can have a great deal to do with quality. Where there is pride in the quality of the product, there is greater concern for quality on the production floor. A technique used successfully in Japan and meeting with growing acceptance in the United States is the *quality circle,* in which small groups of production workers meet regularly to suggest quality improvements in the production process.

Management must be solidly behind total quality or it will not be achieved. There is an inherent conflict between achieving quality and wanting to meet production schedules at minimum cost. This is another manifestation of the perennial conflict between short- and long-term goals. There is general agreement that the greater the autonomy of the quality function in the management structure, the higher the level of quality in the product. Most often the quality control and manufacturing departments are separate, and both the QC manager and the production manager report to the plant manager.

Field service comprises all the services provided by the manufacturer after the product has been delivered to the customer: equipment installation, operator training, repair service, warranty service, and claim adjustment. The level of field service is an important factor in establishing the value of the product to the customer, so that it is a real part of the fitness-for-use concept of quality control. Customer contact by field service engineers is one of the major sources of input about the quality level of the product. Information from the field "closes the loop" of quality assurance and provides needed information for redesign of the product.

15.2.2 Quality-Control Concepts

A basic tenet of quality control is that variability is inherent in any manufactured product. There exists an economic balance between reducing the variability and the

---

1. K. G. Swift and A. J. Allen, "Product Variability, Risks, and Robust Design," *Proc. Instn. Mech. Engrs.,* vol. 208, pp. 9–19, 1994.
2. K. G. Swift, M. Raines, and I. D. Booker, "Design Capability and the Costs of Failure," *Proc. Instn. Mech. Engrs.,* vol. 211, Part B, pp. 409–23, 1997.
3. C. W. Kennedy, S. D. Bond, and E. G. Hoffman, *Inspection and Gaging,* 6th ed., Industrial Press, Inc., New York, 1987.

cost of manufacture.[1] Statistical quality control regards part of the variability as inherent in the materials and process, and it can be changed only by changing those factors. The remainder of the variability is due to assignable causes that can be reduced or eliminated if they can be identified.

There are four basic questions in establishing a QC policy for a part: (1) What do we inspect? (2) How do we inspect? (3) When do we inspect? (4) Where do we inspect?

What to Inspect

The objective of inspection is to focus on a few critical characteristics of the product that are good indicators of performance. These are the critical-to-quality parameters. This is chiefly a technically based decision. Another decision is whether to emphasize nondestructive or destructive inspection. Obviously, the chief value of an NDI technique is that it allows the manufacturer to inspect a part that will actually be sold. Also, the customer can inspect the same part before it is used. Destructive tests, like tensile tests, are done with the assumption that the results derived from the test are typical of the population from which the test samples were taken. Often it is necessary to use destructive tests to verify that the nondestructive test is measuring the desired characteristic.

How to Inspect

The basic decision is whether the characteristic of the product to be monitored will be measured on a continuous scale (inspection by variables) or whether the part passes or fails some go/no-go test. The latter situation is known as measurement by attributes. Inspection by variables uses the normal, lognormal, or some similar frequency distribution. Inspection by attributes uses the binomial and Poisson distributions.

When to Inspect

The decision on when to inspect determines the QC method that will be employed. Inspection can occur either while the process is running (process control) or after it has been completed (acceptance sampling). A process control approach is used when the inspection can be done nondestructively at low unit cost. An important benefit of process control is that the manufacturing conditions can be continuously adjusted on the basis of the inspection data to reduce the percent defectives. Acceptance sampling often involves destructive inspection at a high unit cost. Since not all parts are inspected, it must be expected that a small percentage of defective parts will be passed by the inspection process. The development of sampling plans[2] for various acceptance sampling schemes is an important aspect of statistical quality control.

Where to Inspect

This decision has to do with the number and location of the inspection steps in the manufacturing process. There is an economic balance between the cost of inspection and the cost of passing defective parts to the later stages of the production sequence or

---

1. I. L. Plunkett and B. G. Dale, *Int. J. Prod. Res.,* vol. 26, pp. 1713–26, 1988.
2. See MIL-STD-105E and MIL-STD-414. See http://www.sqconline.com/acceptance-sampling-plans.html.

to the customer. The number of inspection stations will be optimal when the marginal cost of another inspection exceeds the marginal cost of passing on some defective parts. Inspection operations should be conducted before production operations that are irreversible, that is, operations that are very costly or where rework is impossible. Inspection of incoming raw material to a production process is one such place. Steps in the process that are most likely to generate flaws should be followed by an inspection. In a new process, inspection operations might take place after every process step; but as experience is gathered, the inspection would be maintained only after steps that have been shown to be critical.

15.2.3 Newer Approaches to Quality Control

The success of the Japanese in designing and producing quality products has led to new ideas about quality control. Rather than flooding the receiving dock with inspectors who establish the quality of incoming raw material and parts, it is cheaper and faster to require the supplier to provide statistical documentation that the incoming material meets quality standards. This can only work where the buyer and seller work environment in an of cooperation and trust.

In traditional QC an inspector makes the rounds every hour, picks up a few parts, takes them back to the inspection area, and checks them out. By the time the results of the inspection are available it is possible that bad parts have been manufactured and it is likely that these parts have either made their way into the production stream or have been placed in a bin along with good parts. If the latter happens, the QC staff will have to perform a 100 percent inspection to separate good parts from bad. We end up with four grades of product—first quality, second quality, rework, and scrap. To achieve close to real-time control, inspection must be an integral part of the manufacturing process. Ideally, those responsible for making the parts should also be responsible for acquiring the process performance data so that they can make appropriate adjustments. This has resulted in using electronic data collectors to eliminate human error and to speed up analysis of data.

15.2.4 Quality Assurance

Quality assurance is concerned with all corporate activities that affect customer satisfaction with the quality of the product. There must be a quality assurance department with sufficient independence from manufacturing to act to maintain quality. This group is responsible for interpreting national and international codes and standards in terms of each purchase order and for developing written rules of operating practice. Emphasis should be on clear and concise written procedures. A purchase order will generate a great amount of in-plant documentation, which must be accurate and be delivered promptly to each workstation. Much of this paper flow has been computerized, but there must be a system by which it gets on time to the people who need it. There must also be procedures for maintaining the *identity and traceability of materials* and semifinished parts while in the various stages of processing. Definite policies and procedures for dealing with defective material and parts must be in place. There must be

TABLE 15.1
ISO 9000 Standards

Standard	Subject
ISO 9000	Guidelines for Selection and Use
ISO 9001	Quality Assurance in Design, Production, Installation, and Servicing
ISO 9002	Quality Assurance in Production, Installation, and Servicing
ISO 9003	Quality Assurance in Final Inspection
ISO 9004	Guidelines for Implementation

a way to decide when parts should be scrapped, reworked, or downgraded to a lower quality level. A quality assurance system must identify which records should be kept and must establish procedures for accessing those records as required.

Quality control is not something that can be put in place and then forgotten. There must be procedures for training, qualifying, and certifying inspectors and other QC personnel. Funds must be available for updating inspection and laboratory equipment and for the frequent calibration of instruments and gauges.

15.2.5 ISO 9000

An important aspect of quality assurance is the audit of an organization's quality system against written standards.[1] The most prevalent quality standard is ISO 9000, and its companion standards, which are issued by the International Organization for Standards (ISO). ISO 9000 is required by companies doing business in the European Union, and since it is a worldwide marketplace, companies around the world have become ISO 9000 certified. Certification to ISO 9000 is accomplished by submitting to an audit by an accredited ISO registrar.

The system of standards that make up ISO 9000 is listed in Table 15.1. ISO 9001 is the most complete since it extends from design to field service.[2] Clause 4.4, Design Control, lays out many of the issues discussed in this text, as seen from the outline in Table 15.2.

15.3
STATISTICAL PROCESS CONTROL

Collecting manufacturing performance data and keeping charts on this data is common practice in industrial plants. Walter Shewhart[3] showed that such data could be interpreted and made useful through a simple but statistically sound method called a control chart.

---

1. D. Hoyle, *ISO 9000: Quality System Assessment Handbook,* 5th ed., Butterworth-Heinemann, Oxford, 2006.

2. F. P. Dobb, *ISO 9001:2000 Registration Step-by-Step,* Elsevier Butterworth-Heinemann, Boston, 2004.

3. W. A. Shewhart, *Economic Control of Quality in Manufactured Product,* Van Nostrand Reinhold Co., New York, 1931.

15

TABLE 15.2
Topics Covered in ISO 9001,
Clause 4.4, Design Control

Subclause	Topic
4.4.1	General
4.4.2	Design and development planning
4.4.3	Organizational and technical interfaces
4.4.4	Design input
4.4.5	Design output
4.4.6	Design review
4.4.7	Design verification
4.4.8	Design validation
4.4.9	Design changes

15.3.1 Control Charts

The use of the control chart is based on the viewpoint that every manufacturing process is subject to two sources of variation: (1) *chance variation*, also called *common causes of variation,* and (2) *assignable variation*, or that due to *special causes*. Chance variation arises from numerous factors in the operation of the process that are individually of small importance. These can be considered the "noise" in the process. They are an expected but uncontrollable variability. An assignable variation is a variation that can be detected and controlled. It is due to a special cause like poorly trained operators or worn production tooling. The *control chart* is an important quality control[1] tool for detecting the existence of assignable causes.

In constructing a control chart, a process is sampled at regular time intervals and a variable appropriate to the product is measured on each sample. Generally the sample size n is small, between 3 and 10. The number of samples, k, is typically over 20. The theory behind the control chart is that the samples should be chosen such that all of the variability in the samples should likely be due to common causes and none should be due to special causes. Thus, when a sample shows atypical behavior, it can be assumed to be due to a special cause. There are two approaches generally used in selecting samples: (1) all items in each sample represent parts made near the time of sampling, or (2) the sample is representative of all the parts made since the last sample was taken. The choice between these two selection methods is based on the engineer's opinion of which would be more likely to detect the special cause of variation that is highest on the "suspect list."

EXAMPLE 15.1
Consider a commercial heat-treating operation in which bearing races are being quenched and tempered in a conveyor-type furnace on a continuous 24-hour basis. Every hour the

15

1. D. Montgomery, *Introduction to Statistical Quality Control,* 6th ed., John Wiley & Sons, New York, 2009.

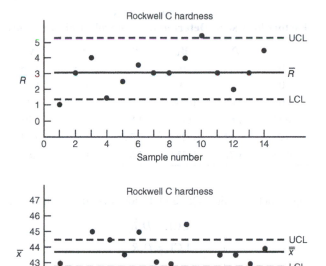

FIGURE 15.1

Control charts for R (top) and $\bar{x}$ (bottom).

Rockwell hardness[1] is measured on 10 bearing races to determine whether the product conforms to the specifications. The mean of the sample, $\bar{x}$, approximates the process mean μ. The range of sample values, $R = x_{max} - x_{min}$, typically is used to approximate the process standard deviation, σ. The variable hardness is assumed to follow a normal frequency distribution.

If the process is in statistical control, the values of mean and range will not vary much from sample to sample, but if the process is out of control then they will vary significantly. Control limits need to be drawn to establish how much variation constitutes out-of-control behavior indicative of the presence of an assignable cause.

Usually the control chart for R is drawn first to make certain that the variation from sample to sample is not too great. If some points on the R are out of the control limits, then the control limits on the $\bar{x}$ chart will be inflated. Figure 15.1 shows the control chart based on range. The centerline of the R chart is $\bar{R}$ and is calculated by averaging the ranges of the k samples.

$$\bar{R} = \frac{1}{k}\sum_{i=1}^{k} R_i \tag{15.1}$$

---

1. The Rockwell hardness test measures the depth of penetration, on an arbitrary scale, of a hard indenter into a metal surface.

TABLE 15.3
Factors for Use in Determining Control Limit for Control Charts

Sample size, n	D_3	D_4	B_3	B_4	A_2	A_3	d_2	c_4
2	0	3.27	0	3.27	1.88	2.66	1.13	0.798
4	0	2.28	0	2.27	0.73	1.63	2.06	0.921
6	0	2.00	0.030	1.97	0.48	1.29	2.53	0.952
8	0.14	1.86	0.185	1.82	0.37	1.10	2.70	0.965
10	0.22	1.78	0.284	1.72	0.27	0.98	2.97	0.973
12	0.28	1.71	0.354	1.65	0.22	0.89	3.08	0.978

The upper control limit, UCL, and the lower control limit, LCL, are determined by

$$UCL = D_4 \bar{R}$$
$$LCL = D_3 \bar{R} \tag{15.2}$$

The constants D_3 and D_4 can be found in Table 15.3. These can be used only if the process variable is normally distributed. Examination of the range control chart shows that two points are outside of the control limits. Based on the assumption of a normal distribution, 0.27 percent of the observations would be expected to fall outside of these $\pm 3\sigma$ limits if these were due to common causes. Therefore, we must examine these points to determine if there are assignable causes for them. Sample 1 was done first thing on Monday morning, and a strip chart was found that determined that the furnace had not reached its proper temperature. This was an operator error, and these data were dropped for assignable cause. No reason could be found for sample 10 being beyond the UCL. This casts some doubt on the results, but this set of data was also dropped when calculating the control chart based on mean values.

The centerline of the $\bar{x}$ control chart is "x double bar," the grand average of the k sample means.

$$\bar{\bar{x}} = \frac{1}{k} \sum_{i=1}^{k} \bar{x}_i \tag{15.3}$$

Again, the UCL and LCL are set at $\pm 3\sigma$ about the mean. If we knew the population mean and standard deviation, this would be given by UCL $= \mu + 3(\sigma \sqrt{n})$, where the term in parentheses is the standard error of the mean. Since we do not know these parameters, the approximations for the control limits is

$$UCL = \bar{\bar{x}} + A_2 \bar{R}$$
$$LCL = \bar{\bar{x}} - A_2 \bar{R} \tag{15.4}$$

Note that the upper and lower control limits depend not only on the grand mean but also on the sample size, through A_2, and the mean of the range.

The $\bar{x}$ control chart in Fig. 15.1 shows many excursions of the mean outside of the control limits even after the control limits have been recalculated to eliminate the two out-of-control samples from the range chart. It is concluded that this particular batch of steel does not have sufficient homogeneity of alloy content to respond consistently to heat treatment within such narrow specification limits. If this is unexpected, then the process should be investigated to see if there was some special cause for the lack of quality control.

15.3.2 Other Types of Control Charts

The $\bar{R}$ and $\bar{x}$ charts were the first types used for quality control. The range was chosen to measure variability because of its ease of calculation in a period before electronic calculators made it quick and easy to calculate standard deviations. Also, for small sample sizes the range is a more efficient statistic than the standard deviation.

Today it is much more convenient to use standard deviation in control charts. The average standard deviation $\bar{s}$ of k samples is given by

$$\bar{s} = \frac{1}{k} \sum_{i=1}^{k} s_i \tag{15.5}$$

Equation (15.5) represents the centerline of the s chart. The upper and lower control limits are set at the ± 3-sigma limits for the sample standard deviation according to Equation (15.6).

$$\text{UCL} = B_4 \bar{s} \quad \text{and} \quad \text{LCL} = B_3 \bar{s} \tag{15.6}$$

A control chart often is used to detect a shift in the process mean during a production run. A succession of 6 to 10 points above or below the centerline of the chart is an indication of a shift in the mean. The sensitivity in detecting a mean shift can be increased by taking the deviation of each sample from the centerline and adding these in a cumulative way for each succeeding sample to form a CUSUM chart.[1]

The preceding discussion of control charts was based on a variable measured on a *continuous quantitative scale*. Often in inspection it is quicker and cheaper to check the product on a *go/no-go basis*. The part is either "not defective" or "defective" based on a gage or predetermined specification. In this type of *attribute testing,* we deal with the fraction or proportion of defects in a sample. The *p chart,* based on the binomial distribution, deals with the fraction of defective parts in a sample over a succession of samples. The *c chart*, based on the Poisson distribution, monitors the number of defects per sample. Other important issues in statistical quality control are the design of sampling plans and the intricacies of sampling parts on the production line.[2]

15.3.3 Determining Process Statistics from Control Charts

Because control charts are commonly established for manufacturing processes, they are a useful source of process statistics for determining the process capability index, Sec. 15.5. The grand average $\bar{\bar{x}}$ of the means of k samples, Eq. (15.3), is the best estimate, $\hat{\mu}$ for the true process mean, μ.

1. W. Navidi, *Statistics for Engineers and Scientists,* 2d ed., pp. 782–84, McGraw-Hill, New York, 2008.
2. D. H. Besterfield, *Quality Control,* 5th ed., Prentice Hall, Upper Saddle River, NJ, 1998; A. Mitra, *Fundamentals of Quality Control and Improvement,* 2d ed., Prentice Hall, Upper Saddle River, NJ, 1998.

The estimate of the process standard deviation is given by Eq. (15.7), depending on whether the R chart or s chart has been used to measure the variability in the process.

$$\hat{\sigma} = \frac{\bar{R}}{d_2} \quad \text{or} \quad \hat{\sigma} = \frac{\bar{s}}{c_4} \tag{15.7}$$

All of the equations for determining the process parameters are based on the assumption that they follow a normal distribution.

15.4
QUALITY IMPROVEMENT

Four basic costs are associated with quality.

- *Prevention*—those costs incurred in planning, implementing, and maintaining a quality system. Included are the extra expense in design and manufacturing to ensure the highest-quality product.
- *Appraisal*—costs incurred in determining the degree of conformance to the quality requirements. The cost of inspection is the major contributor.
- *Internal failure*—costs incurred when materials, parts, and components fail to meet the quality requirements for shipping to the customer. These parts are either scrapped or reworked.
- *External failure*—costs incurred when products fail to meet customer expectations. These result in warranty claims, loss of future business, or product liability suits.

Simply collecting statistics on defective parts and weeding them out of the assembly line is not sufficient for quality improvement and cost reduction. A proactive effort must be made to determine the root causes of the problem so that permanent corrections can be made. Among the problem-solving tools described in Sec. 4.6, the Pareto chart and cause-and-effect diagram are most commonly used in cause finding.

15.4.1 Pareto Chart

In 1897 an Italian economist, Vilfredo Pareto, studied the distribution of land ownership in Italy and found that 80 percent of the land was owned by 20 percent of the population. This was published and became known as Pareto's law. Shortly after World War II, inventory control analysts observed that about 20 percent of the items in the inventory accounted for about 80 percent of the dollar value. In 1954 Joseph Juran generalized Pareto's law as the "80/20 rule"; that is, 80 percent of sales are generated by 20 percent of the customers, 80 percent of the product defects are caused by 20 percent of the parts, and so on. While there is no widespread validation of the 80/20 rule, it is widely quoted as a useful axiom. Certainly Juran's admonition *"to concentrate on the vital few and not the trivial many"* is excellent advice in quality improvement, as in other aspects of life.

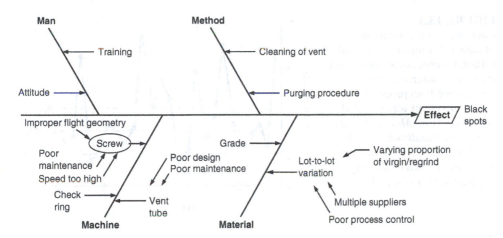

FIGURE 15.2
Cause-and-effect (Ishikawa) diagram for black spot defects on automobile grille. (From *Tool and Manufacturing Engineers Handbook,* 4th ed., vol. 4, p. 2–23, 1987, courtesy of Society of Manufacturing Engineers, Dearborn, MI.)

15.4.2 Cause-and-Effect Diagram

Cause-and-effect analysis uses the "fishbone diagram" or Ishikawa diagram,[1] Fig. 15.2, to identify possible causes of a problem. Poor quality is associated with four categories of causes: (operator) man, machine, method, and material. The likely causes of the problem are listed on the diagram under these four main categories. Suggested causes of the problem are generated by the manufacturing engineers, technicians, and production workers meeting to discuss the problem. The use of the cause-and-effect diagram provides a graphical display of the possible causes of the problem.

EXAMPLE 15.2
A manufacturing plant was producing injection-molded automobile grilles.[2] The process was newly installed, and the parts produced had a number of defects. Therefore, a quality improvement team consisting of operators, setup people, manufacturing engineers, production supervisors, quality control staff, and statisticians was assembled to improve the situation. The first task was to agree on what the defects were and how to specify them. Then a sampling of 25 grilles was examined for defects. Figure 15.3a shows the control chart (see Sec. 15.3.1 for more details on control charts) for the grilles produced by the process. It shows a mean of 4.5 defects per part. The pattern is typical of a process out of control.

A Pareto diagram was prepared to show the relative frequency of the various types of defects, Fig. 15.4. This was based on the data in Fig. 15.3a. It shows that black spots (degraded polymer patches on the surface) are the most prevalent type of defect. Therefore, it was decided to focus attention on this defect.

Focusing on the causes of the black spots resulted in the "fishbone" diagram shown in Fig. 15.2. The causes are grouped under the four Ms of manufacturing. Note that for

1. K. Ishikawa, *Guide to Quality Control,* 2d ed., UNIPUB, New York, 1982.
2. This example is based on *Tool and Manufacturing Engineer's Handbook,* 4th ed., vol. 4, pp. 2–20 to 2–24, Society of Manufacturing Engineers, Dearborn, MI, 1987.

FIGURE 15.3
Control chart for the number
of defects for injection-molded
grilles: (a) process out of control;
(b) process after injection screw
was changed; (c) process after
new vent system was installed.
(From *Tool and Manufacturing
Engineers Handbook,* 4th ed.,
vol. 4, p. 2–22, 1987, courtesy
of Society of Manufacturing
Engineers, Dearborn, MI.)

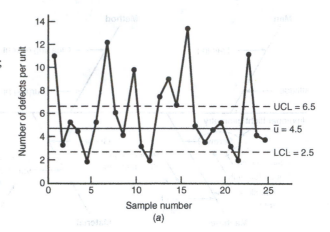

(a)

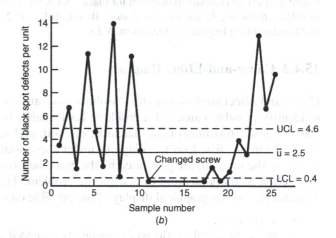

(b)

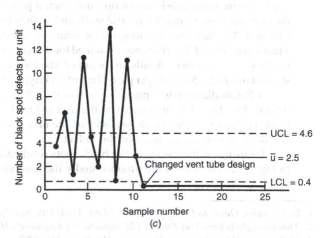

(c)

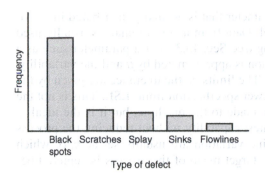

FIGURE 15.4
Pareto diagram for defects in automotive grille. (From *Tool and Manufacturing Engineers Handbook,* 4th ed., vol. 4, p. 2–22, 1987, courtesy of Society of Manufacturing Engineers, Dearborn, MI.)

some items, like the injector screw, the level of detail is greater. The team decided that the screw had been worn through too much use and needed to be replaced.

When the screw was changed, the black spots completely disappeared (see control chart in Fig. 15.3*b*). Then after a few days the black spots reappeared at about the same level of intensity as before. Thus, it must be concluded that the root cause of black spots had not been identified. The quality team continued to meet to discuss the black spot problem. It was noted that the design of the vent tube on the barrel of the injection molding machine was subject to clogging and was difficult to clean. It was hypothesized that polymer either accumulated in the vent tube port, became overheated and periodically broke free and continued down the barrel, or it was pushed back into the barrel during cleaning. A new vent tube design that minimized these possibilities was designed and constructed, and when installed the black spots disappeared, Fig. 15.3*c*.

Having solved the most prevalent defect problem the team turned its attention to scratches, the defect with the second-highest frequency of occurrence. A machine operator proposed that the scratches were caused by the hot plastic parts falling on the metal lacings of the conveyor belt. He proposed using a continuous belt without metal lacings. However, this type of belt cost twice as much. Therefore, an experiment was proposed in which the metal lacings were covered with a soft latex coating. When this was done the scratches disappeared, but after time they reappeared as the latex coating wore away. With the evidence from this experiment, the belt with metal lacings was replaced by a continuous vulcanized belt, not only on the machine under study but for all the machines in the shop.

15.5
PROCESS CAPABILITY

In Section 13.4.5 we discussed how important it is to select a manufacturing process that is able to make a part within the required tolerance range. Not only is knowledge about process capability important when setting tolerances, but it is important information to have when deciding which outside supplier should get the contract to make the part. In this section we show how statistical information about the parts produced by a machine or process can be used to determine the percentage of parts that fall outside of a specified tolerance band.

Process capability is measured by the *process capability index, C_p.*

$$C_p = \frac{\text{Acceptable part variation}}{\text{Machine or process variation}} = \frac{\text{Tolerance}}{\pm 3\hat{\sigma}} = \frac{\text{USL} - \text{LSL}}{3\hat{\sigma} - (-3\hat{\sigma})} = \frac{\text{USL} - \text{LSL}}{6\hat{\sigma}} \quad (15.8)$$

Equation 15.8 applies to a design parameter that is normally distributed in a process that is in a state of statistical control. Data from a control chart is usually used to describe how the process is performing (see Sec. 15.3). For a parameter such as a CTQ dimension, the mean of the population is approximated by $\hat{\mu}$ and the variability, measured by the standard deviation, by $\hat{\sigma}$. The limits on the tolerance are given by the upper specification limit, USL, and the lower specification limit, LSL. This is not the usual case unless careful adjustments are made to the machine, but it is the ideal to be achieved because it results in the greatest capability without reducing the process standard deviation. The limits on machine variation are usually set at $\pm 3\sigma$, which gives 0.27% defects when $C_p = 1$ and the target mean of the process is centered between the LSL and the USL.

Figure 15.5 shows three situations of the distributions of the design variable of the part produced by the process compared with the upper and lower limits of the tolerance. Figure 15.5a shows the situation where the process variability (spread) is greater than the acceptable part variation (tolerance range). According to Eq. 15.8, $C_p \leq 1$, and the process is not capable. To make it capable the variability in the process will have to be reduced, or the tolerance will have to be loosened. Figure 15.5b is the case where the tolerance range and the process variability just match, so $C_p = 1$. This is a tenuous situation, for any shift of the process mean, for example, to the right, will increase the number of defective parts. Finally, in Fig. 15.5c, the process variability is much less

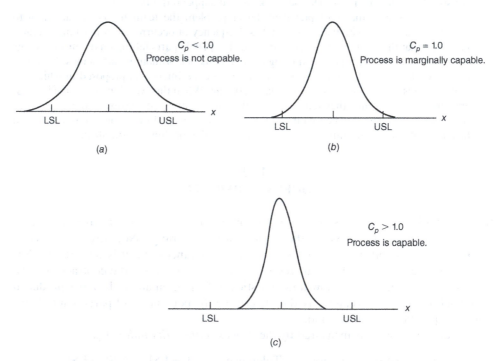

FIGURE 15.5
Examples for different process capability situations.

than the tolerance range. This provides a considerable margin of safety because the process mean could move quite a bit before the distribution reaches the USL or LSL. For mass production, where the percentage of defects is critical, the acceptable level of C_p is required to exceed 1.33.

EXAMPLE 15.3

(a) A machine spindle has a specification (tolerance) on its diameter of 1.500 ± 0.009 inches. If $C_p = 1.0$, what is the standard deviation of the spindles being produced by the cylindrical grinder?

$$C_p = 1.0 = \frac{1.509 - 1.491}{6\hat{\sigma}} \qquad \hat{\sigma} = \frac{(0.018)}{6(1.0)} = 0.003 \text{ inches}$$

(b) What would the standard deviation have to be to achieve a process capability index of 1.33?

$$1.33 = \frac{0.018}{6\hat{\sigma}} \qquad \hat{\sigma} = \frac{0.018}{7.98} = 0.00226$$

With a C_p value of 1.33, the process mean is four standard deviations from each specification limit. This is considered good manufacturing practice.

EXAMPLE 15.4

If $C_p = 1.33$ and the process mean is centered within the tolerance range, how many oversized parts would be expected in grinding the spindle described in Example 15.3b? (Note: this is the same type of problem discussed in Example 14.1.)

We can visualize this problem with the help of Fig.15.5c. Using the standard normal variable, z,

$$z = \frac{x - \mu}{\sigma} \approx \frac{USL - \hat{\mu}}{\hat{\sigma}} = \frac{1.509 - 1.500}{0.00226} = 3.982$$

The z value is far out on the right end of the z distribution. Most tables stop at about $z = 3.9$, but using the NORMDIST function in Excel gives 0.999966. This is the area under the curve from $-\infty$ to 3.982. Therefore, the area under the very small piece of the right tail is $1 - 0.999966 = 0.000034$ or 0.0034 percent or 34 ppm (parts per million).

The problem asked for the percentage of oversized parts, but there also will be parts with undersized diameters. Since the z distribution is symmetrical, the total percentage of defects (over and undersized) is 0.0068 or 68 defective parts for every million parts produced.

In the previous examples the process mean was centered midway between the upper and lower specification limits. This is not easy to achieve and maintain in practice. If the process starts out centered, there is a tendency for the mean to move off center with time due to tool wear and process changes. The midpoint of the tolerance range $(USL + LSL)/2$ equals m (the target for the process mean). The distance between the actual process mean, $\hat{\mu}$, and the midpoint is $\hat{\mu} - m$, where $m \le \hat{\mu} \le USL$ or $LSL \le \hat{\mu} \le m$.

15

TABLE 15.4
Effect of Shift in Process Mean on Defect Rate

Tolerance range*	C_p	Process Centered		Process Mean 1.5 Sigma from Center	
		Percent good parts	Defective parts ppm	Percent good parts	Defective parts ppm
± 3 sigma	1.00	99.73	2,700	93.32	697,700
± 4 sigma	1.33	99.9932	68	99.605	3,950
± 6 sigma	2.00	99.9999998	0.002	99.99966	3.4

*Indicates the number of times the process sigma fits within the tolerance range (specification limits). ppm is parts per million. 10,000 ppm = 1 percent.

The parameter k is the ratio of the deviation of the actual process mean from m to one-half of the tolerance range. The value of k varies from 0 to 1.

$$k = \frac{|m - \hat{\mu}|}{(USL - LSL)/2} = \frac{|(USL + LSL)/2 - \hat{\mu}|}{(USL - LSL)/2} \tag{15.9}$$

The process capability index when the mean is not centered should be calculated by C_{pk}.

$$C_{pk} = \text{minimum}\left[\frac{USL - \hat{\mu}}{3\hat{\sigma}}, \frac{\hat{\mu} - LSL}{3\hat{\sigma}}\right] \tag{15.10}$$

C_{pk} defines the process capability by the lesser of the ranges from the mean to the specification limit. C_p and C_{pk} are related through the equation

$$C_{pk} = (1 - k)C_p \tag{15.11}$$

When k equals zero, the mean is centered and $C_{pk} = C_p$.

Table 15.4 shows how the percentage of good parts and defective parts varies with the number of process standard deviations, "sigmas," that can be accommodated within the tolerance range. It also shows the dramatic increase in defective parts that results from a 1.5 sigma shift of the process mean. A shift of the process mean by this amount is considered to be typical of the average manufacturing process.

EXAMPLE 15.5
The process mean has moved 1.5 $\hat{\sigma}$ from the center of the tolerance range. From Example 15.3, $\hat{\sigma} = 0.00226$ inches. The shift $k = 1.5(0.00226) = 0.003$ inches toward the USL.
Now $\hat{\mu} = 1.500 + 0.003 = 1.503$. From Eq. (15.10):

$$C_{pk1} = \frac{USL - \hat{\mu}}{3\hat{\sigma}} = \frac{1.509 - 1.503}{3(0.00226)} = 2.655$$

$$C_{pk2} = \frac{\hat{\mu} - LSL}{3\hat{\sigma}} = \frac{1.503 - 1.491}{3(0.00226)} = 1.770$$

The calculation shows that $C_{pk1} \neq C_{pk2}$, so the process mean is not centered. However, the process capability index of 1.77 shows that the process is capable. To determine the percentage of expected defective parts, we use the standard normal variable z.

$$z_{USL} = \frac{USL - \hat{\mu}}{\hat{\sigma}} = \frac{1.509 - 1.503}{0.00226} = 2.655 \text{ and } z_{LSL} = \frac{LSL - \hat{\mu}}{\hat{\sigma}} = \frac{1.491 - 1.503}{0.00226} = -5.31$$

The probability of parts falling outside the tolerance range is given by

$$P(z \leq -5.31) + P(z \geq 2.655) = 1 - (0 + 0.99605) = 0.0039$$

Thus, the probability is approximately 0.0039 or 0.39% or 3950 ppm. While the defect rate still is relatively low, it has increased from 68 ppm when the process was centered in the middle of the tolerance range, Example 15.4.

15.5.1 Six Sigma Quality Program

Table 15.4 shows that the percentage of good parts is exceedingly high if the process variability is so low that ±6 standard deviations (a width of 12 $\hat{\sigma}$) will fit within the specification limits, Fig. 15.5c. This is the origin of the name of the quality program called *Six Sigma* that has been pursued vigorously by many world-class corporations. It is generally recognized that achieving the 2 parts per billion defect level that is shown in Table 15.4 is not realistic, since most processes show some mean shift. Therefore, the practical six sigma goal is usually stated to be the 3.4 ppm of defective parts that is given in Table 15.4. Even that goal is exceedingly difficult and rarely, if ever, attained.

Six sigma can be viewed as a major extension of the TQM process described in Chap. 4. Six Sigma incorporates the problem-solving tools of TQM with many others discussed in this text such as QFD, FMEA, reliability, and Design of Experiments, as well as extensive tools for statistical analysis.[1] Compared with TQM, Six Sigma has more of a financial focus than a customer focus, with emphasis on cutting cost and improving profit. Six Sigma has stronger emphasis on training of special teams, using a more structured approach, and setting stretch goals.[2] As seen above, the idea of Six Sigma came from the concept of process capability, so it is no surprise that a major focus is on reducing process defects by systematically reducing process variability. However, with the strong emphasis on cost reduction that has evolved, many of the most spectacular results of Six Sigma projects have come from process simplification and reducing non–value-added activities.

Six Sigma uses a disciplined five-stage process with the acronym DMAIC to guide improvement processes.

1. R. C. Perry and D. W. Bacon, *Commercializing Great Products with Design for Six Sigma*, Pearson Education, Upper Saddle River, NJ, 2007.
2. G. Wilson, *Six Sigma and the Development Cycle*, Elsevier Butterworth-Heinemann, Boston, 2005.

15

- *Define the Problem*: During this stage the team works to identify the customers involved and to determine their needs. It is necessary to determine that the problem is important and traceable to either customer needs or business goals. The team defines the scope of the project, its time frame, and the potential financial gains. These are recorded in a team charter.
- *Measure:* During the second stage the team develops metrics, which allow them to evaluate the performance of the process. This task requires accurate measurement of current process performance so it can be compared with the desired performance. At this stage it is important to begin to understand those process variables that cause significant variations in the process.
- *Analyze:* The team analyzes the data taken in the previous stage to determine the root causes of the problem and identify any non–value-added process steps. The team should determine which process variables actually affect the customer, and by how much. They should examine possible combinations of variables on the process and how changing each process variable affects process performance. Process modeling is often used to advantage in this phase.
- *Improve:* This phase pertains to solution generation and implementation. It involves selecting the solution that best addresses the root cause. Tools like cost/benefit analysis using financial tools such as net present value are employed. The development of a clear implementation plan and its communication to management are essential at this stage of the process.
- *Control:* This final stage institutionalizes the change and develops a monitoring system so that the gains of the improvement are maintained over time. Aim to mistake-proof the revised process. Part of the plan should be to translate the opportunities discovered by the project beyond the immediate organization to the corporation as a whole. The project should be documented thoroughly so that in the future another Six Sigma team may use the results to initiate another improvement project using the same process.

15.6
TAGUCHI METHOD

A systematized statistical approach to product and process improvement has developed in Japan under the leadership of Dr. Genichi Taguchi.[1] This took a total quality emphasis but developed quite unique approaches and terminology. It emphasizes moving the quality issue upstream to the design stage and focusing on prevention of defects by process improvement. Taguchi has placed great emphasis on the importance of minimizing variation as the primary means of improving quality. Special attention is given to the idea of designing products so that their performance is insensitive to changes in the environment in which the product functions, also called noise. The process of achieving this through the use of statistically designed experiments has been called robust design (see Sec. 15.7).

1. G. Taguchi, *Introduction to Quality Engineering,* Asian Productivity Organization, Tokyo, 1986, available from Kraus Int. Publ., White Plains, NY; G. Taguchi, *Taguchi on Robust Technology Development,* ASME Press, New York, 1993.

15.6.1 Quality Loss Function

Taguchi defines the quality level of a product to be the total loss incurred by society due to the failure of the product to deliver the expected performance and due to harmful side effects of the product, including its operating cost. This may seem a backward definition of quality because the word *quality* usually denotes desirability, while the word *loss* conveys the impression of undesirability. In the Taguchi concept some loss is inevitable because of the realities of the physical world from the time a product is shipped to the customer and the time it is put in use. Thus, all products will incur some quality loss. The smaller the loss, the more desirable the product.

It is important to be able to quantify this loss so that alternative product designs and manufacturing processes can be compared. This is done with a quadratic loss function (Fig. 15.6a):

$$L(y) = k(y-m)^2 \tag{15.12}$$

where $L(y)$ is the *quality loss* when the quality characteristic is y, m is the target value for y, and k is a constant, the quality loss coefficient.

Figure 15.6a shows the loss function for the common situation where the specification on a part is set at a target value, m, with a bilateral tolerance band $\pm\Delta$. The conventional approach to quality considers a part with all dimensions falling within the tolerance range to be a good part, while one with any dimension outside of the USL-LSL region is a defective part. The analogy can be made to the goalposts in football, where any kick that went through the uprights is a score, no matter how close it came to the upright. In football, no extra points are awarded for a kick that goes right between the middle of the goal posts.

Taguchi argues that this conventional approach is not realistic for defining quality. While it may be reasonable in football to award the same score so long as the ball falls in the interval 2Δ, for a quality engineering approach where variability is the enemy of quality, any deviation from the design target is undesirable and degrading to quality. Moreover, defining the quality loss function as a quadratic instead of a linear expression emphasizes the importance of being close to the target value.

It is evident from Fig. 15.6a that y exceeds the tolerance Δ when $L(y) = A$. A is the loss incurred when a product falls outside of the tolerance range and is rejected, or when a part in service needs to be repaired or replaced. When this occurs, $y = \text{USL} = m + \Delta$. Substituting into Eq. (15.12),

$$L(m+\Delta) = A = k[(m+\Delta)-m]^2 = k\Delta^2$$
$$k = A/\Delta^2$$

Substituting into Eq. (15.12) gives:

$$L(y) = \frac{A}{\Delta^2}(y-m)^2 \tag{15.13}$$

15

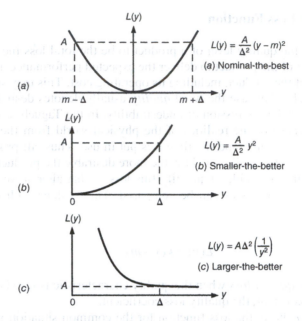

FIGURE 15.6
Plots of the loss function curve for three common situations.

This is the form of the quality loss equation that is most often used for the case where the highest quality (lowest loss) is achieved when the quality characteristic is as close as possible to the target value, and it is symmetrical about the target. Note that $L(y) = 0$ only when $y = m$. A CTQ dimension on a part is an example of a nominal-is-better design parameter.

Two other common situations are shown in Fig. 15.6, along with the appropriate equation for the loss function. Figure 15.6b illustrates the case where the ideal value is zero and the smallest deviation from this target produces the highest quality. An example would be if y represented pollution from an automobile exhaust. Figure 15.6c shows the opposite situation, where the largest deviation from zero would produce the lowest loss function. Design for the strength of a part would fall in this category.

EXAMPLE 15.6
A power supply for an electronic product must deliver a nominal output voltage of 115 volts. When the output voltage varies from the nominal by more than 20 volts, the customer will experience degraded performance or the product will be damaged and repairs will be necessary at an average cost of $100. What is the loss if the product is shipped with a power supply having an output of 110 volts? From this statement of the problem we may write:

$$m = 115 \text{ volts} \quad y = 110 \quad \Delta = 20 \text{ volts} \quad A = \$100 \quad k = A/\Delta^2 = 100 / (20)^2 = \$0.25/\text{volts}$$

$$L(110) = k(y - m)^2 = \$0.25(110 - 115)^2 = \$6.25$$

This is the customer's perceived quality loss when the power supply delivers 110 instead of 115 volts.

EXAMPLE 15.7

Suppose the manufacturer could recalibrate the power supply at the end of the production line to bring it closer to the target voltage. Whether this should be done, from an economic point of view, depends upon whether the cost of repair is less than the customer's perceived quality loss. In this case, let A = cost of rework = $3 per unit. How great should the deviation from target be before the manufacturer should rework the power supply? The loss to the customer is given in Example 15.6.

$$L(y) = 0.25(y - m)^2 \text{ and } y = m - \Delta$$

$$L(y) = \$3 \text{ at the decision point}$$

$$3 = 0.25(m - \Delta - m)^2 = 0.25\Delta^2 \quad \Delta = \sqrt{\frac{3}{0.25}} = \sqrt{12} = 3.46 \text{ volts}$$

Providing that the output voltage is within 3.5 volts of the target (115 v) the manufacturer should not spend $3 per unit to recalibrate the unit. This value is the manufacturer's economic tolerance limit. Beyond this point the customer's loss increases beyond acceptable limits.

The *average quality loss* of a sample of products, obtained by summing the individual losses and dividing by their number, is given by:[1]

$$\bar{L}(y) = k[\sigma^2 + (\bar{y} - m)^2] \tag{15.14}$$

where $\bar{L}(y)$ is the average quality loss

σ^2 is the population variance on y due to common causes in the process. It usually is approximated by the sample variance

$\bar{y}$ is the mean of all y_i in the sample, or $\hat{\mu}$

$(\bar{y} - m)^2$ is the square of the deviation of $\bar{y}$ from the target value m, due to assignable variation

Equation (15.14) is an important relationship because it breaks the quality loss into the component of the loss that is due to product or process variability and the amount that is due to the mean of the sample being displaced from the target value.

EXAMPLE 15.8

A manufacturing process has a standard deviation of 0.00226 inches and a mean of 1.503 inches (see Example 15.5). The specification for the CTQ dimension of the part is 1.500 ± 0.009 inches. The part can no longer be assembled into a subsystem if y exceeds 1.5009 and it is reworked at a cost of $16.

15

---

1. W. Y. Fowlkes and C. M. Creveling, *Engineering Methods for Robust Product Design*, Chap. 3, Addison-Wesley, Reading, MA, 1995.

(a) What is the average quality loss for parts made from this process?

First we need to find the quality loss coefficient, k, for the process.

$$k = A/\Delta^2 = \$16/(0.009)^2 = 197{,}531 \ \$/in.^2$$

$$L(y) = k[\hat{\sigma}^2 + (\hat{\mu} - m)^2] = 197{,}531[(0.00226)^2 + (1.503 - 1.500)^2]$$

$$= 197{,}531[5.108 \times 10^{-6} + 9 \times 10^{-6}] = \$2.787$$

Note that the quality loss due to the shift of the mean is about twice that due to process variability.

(b) If the process mean is centered with the target mean for the part, what is the quality loss factor?

Now $(\hat{\mu} - m) = (1.500 - 1.500) = 0$ and the quality loss factor is due entirely due to variation of the process. $\bar{L}(y) = 197{,}531(5.108 \times 10^{-6}) = \$1.175$

As we will see in Sec. 15.7, the usual approach using the Taguchi method is to first search for choices of the design parameters that minimize the product's susceptibility to variation, and then having found the best combination, adjust the process conditions to bring the product mean and the process mean into coincidence.

15.6.2 Noise Factors

The input parameters that affect the quality of the product or process may be classified as design parameters and disturbance factors. The former are parameters that can be specified freely by the designer. It is the designer's responsibility to select the optimum levels of the design parameters. Disturbance factors are the parameters that are either inherently uncontrollable or impractical to control.

Taguchi uses the term *noise factors* to refer to those parameters that are either too difficult or too expensive to control when a product is in service or during manufacture of its components. The noise factors can be classified into four categories:

- *Variational noise* is the unit-to-unit variation that nominally identical products will exhibit due to the differences in their components or their assembly.
- *Inner noise* is the long-term change in product characteristics over time due to deterioration and wear.
- *Design noise* is the variability introduced into the product due to the design process. This consists mostly of the tolerance variability that practical design limitations impose on the design.
- *External noise*, also called outer noise, represents the disturbance factors that produce variations in the environment in which the product operates. Examples of external noise factors are temperature, humidity, dust, vibration, and the skill of the operator of the product.

The Taguchi method is unusual among methods of experimental investigation in that it places heavy emphasis on including noise factors in every experimental design.

Taguchi was the first to articulate the importance of considering external noise directly in design decisions.

15.6.3 Signal-to-Noise Ratio

Whenever a series of experiments is to be carried out, it is necessary to decide what *response* or output of the experiment will be measured. Often the nature of the experiment provides a natural response. For example, in the control chart in Fig. 15.1, which evaluated the effectiveness of a heat-treating process for hardening steel bearings, a natural response was the Rockwell hardness measurement. The Taguchi method uses a special response variable called the *signal-to-noise ratio, S/N*. The use of this response is somewhat controversial, but its use is justified on the basis that it encompasses both the mean (signal) and the variation (noise) in one parameter, just as the quality loss function does.[1]

Following are three forms of the *S/N* ratio corresponding to the three forms of the loss function curves shown in Fig. 15.6.

For the nominal-is-best type of problem,

$$S/N = 10 \log\left(\frac{\mu}{\sigma}\right)^2$$ (15.15)

where

$$\mu = \frac{1}{n}\sum_{i=1}^{n} y_i \quad \text{and} \quad \sigma^2 = \frac{1}{n-1}\sum_{i=1}^{n}(y_i - \mu)^2$$

and n is the number of external noise observation combinations used for each design parameter matrix (control factors) combination. For example, if four tests are made to allow for noise for each combination of the control parameters, then $n = 4$.

For the smaller-the-better type of problem,

$$S/N = -10 \log\left(\frac{1}{n}\sum y_i^2\right)$$ (15.16)

For the larger-the-better type of problem, the quality performance characteristic is continuous and nonnegative. We would like y to be as large as possible. To find the *S/N*, we turn this into a smaller-the-better problem by using the reciprocal of the performance characteristic.

$$S/N = -10 \log\left(\frac{1}{n}\sum \frac{1}{y_i^2}\right)$$ (15.17)

---

1. Dr. Taguchi was an electrical engineer with the national telephone system of Japan, so the concept of signal-to-noise ratio, the ratio of signal strength to unwanted interference in a communications circuit, was very familiar to him.

15.7
ROBUST DESIGN

Robust design is the systematic approach to finding optimum values of design factors that lead to economical designs with low variability. The Taguchi method achieves this goal by first performing parameter design, and then, if the outcomes still are not optimum, by performing tolerance design.

Parameter design[1] is the process of identifying the settings of the design parameters or process variables that reduce the sensitivity of the design to sources of variation. This is done in a two-step process. First, *control factors* are identified. These are design parameters that primarily affect the S/N ratio but not the mean. Using statistically planned experiments, we find the level of the control factors that minimize the variability of the response. Second, once the variance has been reduced, the mean response can be adjusted by using a suitable design parameter, known as the *signal factor*.

15.7.1 Parameter Design

Parameter design makes heavy use of planned experiments. The approach involves statistically designed experiments that are based on fractional factorial designs.[2] With factorial designs only a small fraction of the total number of experiments must be performed when compared with the conventional approach of varying one parameter at a time in an exhaustive testing program. The meaning of a fractional factorial design is shown in Fig. 15.7. Suppose we identify three control factors P_1, P_2, and P_3 that influence the performance of the design. We want to determine their influence on the design variable. The response is measured at two levels of the design parameters, one low (1) and one high (2). In the conventional approach of varying one factor at a time, this would require $2^3 = 8$ tests as illustrated in Fig. 15.7a. However, if we use a fractional factorial Design of Experiment (DoE), essentially the same information is obtained with half as many tests, as illustrated in Fig. 15.7b. All common fractional factorial designs are orthogonal arrays. These arrays have the balancing property that every setting of a design parameter occurs with every setting of all other design parameters the same number of times. They keep this balancing property while minimizing the number of test runs. Taguchi presented the orthogonal arrays in an easy-to-use form that uses only parts of the fractional factorial test plan. The trade-off is that the number of tests is minimized, but detailed information about interactions is lost.

Figure 15.8 shows two commonly used orthogonal arrays. The columns represent the control factors, A, B, C, and D, and the rows represent the setting of the parameters for each experimental run. The L4 array deals with three control factors at two levels, while the L9 array considers four factors each at three levels. Note that the L9 array reduces the full experiment of $3^4 = 81$ runs to only 9 experimental runs. This

---

1. The terminology is a bit tenuous. The process called *parameter design* is firmly established in the Taguchi method for robust design. This work is generally conducted in the *parametric design stage* of the *embodiment phase* of the design process.
2. W. Navidi, op. cit., pp. 735–38.

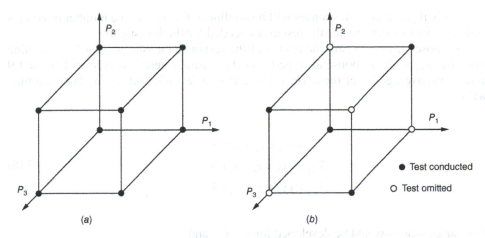

FIGURE 15.7
Designed experiment plan. Three factors P tested at two levels. (a) All test combinations considered. (b) Fractional factorial design.

	L4 Array		
Run No.	A	B	C
1	1	1	1
2	1	2	2
3	2	1	2
4	2	2	1

	L9 Array			
Run No.	A	B	C	D
1	1	1	1	1
2	1	2	2	2
3	1	3	3	3
4	2	1	2	3
5	2	2	3	1
6	2	3	1	2
7	3	1	3	2
8	3	2	1	3
9	3	3	2	1

FIGURE 15.8
Orthogonal arrays; left, the L4 array and right, the L9 array.

reduction is accomplished by confounding the interaction effects (AB, etc.) with the main effects (A, B, etc.). Note also the balance between the levels of the control factors. Each level for each control factor appears in the same number of runs. For example, level 1 of B appears in runs 1, 4, and 7; level 2 occurs in runs 2, 5, and 8; while level 3 occurs in runs 3, 6, and 9. This balance between control factor levels allows averages to be computed that isolate the effect of each factor.

The choice of which orthogonal array to use depends on the number of control factors and noise factors.[1] The decision of whether to use an array with factors at two or three levels depends on whether you are seeking more resolution in the results,

---

1. G. Taguchi, *System of Experimental Design: Engineering Methods to Optimize Quality and Minimize Cost*, 2 vols., Quality Resources, White Plains, NY, 1987; M. S. Phadke, *Quality Engineering Using Robust Design*, Prentice Hall, Upper Saddle River, NJ, 1989; W. Y. Fowlkes and C. M. Creveling, op. cit., Appendix C.

15

especially if you feel the responses will be nonlinear. Of course, the number of control and noise factors determines the resources needed for the investigation.

Suppose y_1, y_2 ... y_9 are the results of the response measured in each of the nine runs. Let $\bar{y}_{B1}$ be the response averaged over those runs where B is at level 1 in the L9 array; $\bar{y}_{B2}$ averaged over those runs where B is at level 2, and so on. Then we may write:

$$\bar{y}_{B1} = (y_1 + y_4 + y_7)/3$$
$$\bar{y}_{B2} = (y_2 + y_5 + y_8)/3 \qquad\qquad (15.18)$$
$$\bar{y}_{B3} = (y_3 + y_6 + y_9)/3$$

Similar equations would be developed for $\bar{y}_{Ai}$, $\bar{y}_{Ci}$, and $\bar{y}_{Di}$.

The Taguchi design of experiments usually consists of two parts. The first part is a design parameter matrix from which the effects of the control parameters are determined through the use of a suitable orthogonal array. The second part is the noise matrix, a smaller orthogonal array consisting of noise parameters. Often the first matrix is called the inner array and the noise matrix is termed the outer array. It is common to use an L9 array with nine runs for the inner array and an L4 array with four runs for the outer array. Thus, for run 1 in the L9 array [all factors at the low (1) level] there are four trials, one for each combination of factors in the noise matrix, the L4 array. For run 2 there are another four trials, and so on, so that a total of $9 \times 4 = 36$ test conditions will be evaluated. The responses are evaluated for each of the four trials in the first run and statistics like the mean and standard deviation are determined. This evaluation is performed for each of the nine runs for the design parameter matrix.

The creation of a robust design using the Taguchi Method proceeds in six steps:

1. Define the problem, including the selection of the parameter to be optimized and the objective function.
2. Select the design parameters—often called the control factors—and the noise factors. The control factors are parameters under the control of the designer that may be calculated or determined experimentally. The noise factors are those parameters that contribute to the variation caused by the environment.
3. Design the experiment by selecting the appropriate fractional factorial array (see Fig. 15.8), the number of levels to be used, and the range of the parameters that correspond to these levels.
4. Conduct the experiments according to the DoE. These may be actual physical experiments or computer simulations.
5. Analyze the experimental results by calculating the signal-to-noise ratio (S/N) as shown in Sec. 15.6.3. If the analysis does not give a clear optimum value, then repeat steps 1 through 4 with new values of the design levels, or perhaps, with a change in the control parameters.
6. When the method gives a set of optimal parameter values, perform a confirming experiment to validate the results.

EXAMPLE 15.9

In Example 4.1 in Sec. 4.6 we showed how to use the TQM tools to find the root cause in a design problem concerned with a failed indicator light in a prototype of a new game box. In the example we found that the root cause of poor solder joints was the use of improper solder paste, which consists of solder balls and flux. We decide to improve the situation by using the Taguchi method to establish the best conditions for making strong solder joints. We decide that four control parameters are important and that there are three main noise parameters. Thus, it is appropriate to employ the L9 orthogonal array for the parameter matrix and the L4 array for the noise matrix as shown in Fig. 15.8.

Selection of Control Factors and Range of Factors for the L9 Orthogonal Array

Control Factor	Level 1	Level 2	Level 3
A—solder ball size	30 micron	90 micron	150 micron
B—screen print diameter	0.10 mm	0.15 mm	0.20 mm
C—type flux	Low activity	Moderate activity	High activity
D—temperature	500°F	550°F	600°F

The control factors listed above fall into the category of variational noise factors. The objective of this study is to find the process conditions where the part-to-part variation in these factors is minimized.

Selection of Noise Factors for the L4 Orthogonal Array

Noise Factors	Level 1	Level 2
A—shelf life of paste	New can	Opened 1 yr ago
B—surface cleaning method	Water rinse	Chlorocarbon solvent
C—cleaning application	Horizontal spray	Immersion

The first noise factor is an inner noise factor, while the other two are outer noise factors.

We now conduct the experiments according to the experimental design. For example, run 2 in L9 is executed four times to include the noise matrix. In the first trial the conditions would be: 30 micron solder ball, 0.15 mm screen diameter, flux with moderate activity, 550°F temperature, a new can of paste, water rinse, and horizontal spray. The last three factors are from run 1 of the L4 (noise) array. In the fourth trial of run 2 the conditions for L9 would be identical, but the noise factors would change to using a can of paste opened one year ago, a chlorocarbon cleaning agent, and horizontal spray for cleaning. For each of the four trials of run 2, we measure a response that represents the objective function that we are attempting to optimize. In this case, the response is the shear strength of the solder joint measured at room temperature. For the four trials, we average the strength measurements and determine the standard deviation. For run 2, the results are:

$$\bar{y}_2 = (4.175 + 4.301 + 3.019 + 3.3134)/4 = 3.657 \text{ ksi}$$

$$\text{and } \sigma = \sqrt{\frac{\sum (y_{2i} - \bar{y}_2)^2}{n-1}} = 0.584$$

In robust design the appropriate response parameter is the signal-to-noise ratio. Because we are trying to find the conditions to maximize the shear strength of the solder joints, the larger-is-best form of the S/N is selected.

$$S/N = -10 \log\left(\frac{1}{n}\sum \frac{1}{y_i^2}\right)$$

For each of the runs in the L9 array we calculate a signal-to-noise ratio. For run 2,

$$(S/N)_{run2} = -10 \log\left\{\frac{1}{4}\left[\frac{1}{(4.175)^2} + \frac{1}{(4.301)^2} + \frac{1}{(3.019)^2} + \frac{1}{(3.134)^2}\right]\right\} = 10.09$$

The following table shows the results of similar calculations for all of the runs in the parameter matrix.[1]

Run No.	Control Matrix				S/N
	A	B	C	D	
1	1	1	1	1	9.89
2	1	2	2	2	10.09
3	1	3	3	3	11.34
4	2	1	2	3	9.04
5	2	2	3	1	9.08
6	2	3	1	2	9.01
7	3	1	3	2	8.07
8	3	2	1	3	9.42
9	3	3	2	1	8.89

Next, it is necessary to determine the average response for each of the four control parameters at each of its three levels. We have noted previously that this result is obtained by averaging over those runs where A is at level 1, or where C is at level 3, etc. From the preceding table, it is evident that the average S/N for factor B at level 2 is (10.09 + 9.08 + 9.42)/3 = 9.53. Performing this calculation for each of the four factors at the three levels creates the response table shown below.

Response Table

	Average S/N			
Level	A	B	C	D
1	10.44	9.00	9.44	9.29
2	9.04	9.53	9.34	9.05
3	8.79	9.75	9.49	9.93

---

1. Note that these numbers are to illustrate the design method. They should not be considered to be valid design data.

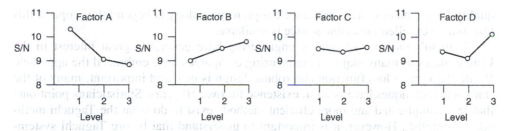

FIGURE 15.9
Linear graphs showing the S/N for the four control parameters.

The average S/N ratios are plotted against test level for each of the four control parameters as shown in Fig. 15.9. These linear graphs show that factor A, solder ball size, and factor B, diameter of the holes in the print screen, have the greatest influence on the shear strength of the solder joints. Also, factor C, activity of the flux, is not an important variable. As a result of these graphs, we conclude that the optimum settings for the control parameters are:

Control Parameter	Optimum Level	Parameter Setting
A—solder ball size	1	30 micron
B—screen print diameter	3	0.20 mm
C—type of flux	—	No strong trend Prefer moderate activity
D—temperature	3	600°F

Note that these experimental conditions are different from any of the nine runs in the control matrix. To verify this result we perform an additional set of four trials at the above test conditions. The validity of the optimization is confirmed when we calculate a S/N of 11.82, which is larger than any of the S/N values measured at the 36 test points.

Example 15.9 used a relatively small number of experiments to study a number of design variables (four control parameters and three noise factors) to provide a new set of control parameters that are closer to an optimum than an informed guess, and which are robust to the noise factors.

15.7.2 Tolerance Design

Often, as in Example 15.9, the parameter design results in a design optimized for robustness and with a low variability. However, there are situations when the variability is too large and it becomes necessary to reduce tolerances to decrease variability. Typically, analysis of variance (ANOVA) is used to determine the relative contribution of each control parameter so as to identify those factors that should be considered for tolerance tightening, substituting an improved material, or some other means of improving quality. Since these methods often incur additional cost, the Taguchi method of tolerance design provides careful methods for balancing increased quality (lower

15

quality loss) with cost. This tolerance design methodology is beyond the scope of this text, but an excellent, readable source is available.[1]

Taguchi's methods of quality engineering have generated great interest in the United States as many major manufacturing companies have embraced the approach. While the idea of loss function and robust design is new and important, many of the statistical techniques have been in existence for over 50 years. Statisticians point out[2] that less complicated and more efficient methods exist to do what the Taguchi methods accomplish. However, it is important to understand that before Taguchi systematized and extended these ideas into an engineering context, they were largely unused by much of industry. The growing acceptance of the Taguchi method comes from its applicability to a wide variety of industrial problems with a methodology that does not require a high level of mathematical skills to achieve useful results.

15.8
OPTIMIZATION METHODS

The example described in the previous section is a search for the best combination of design parameters using a statistically designed set of experiments when the desired outcome is clear. There is more than one solution to a design problem, and the first solution is not necessarily the best. Thus, the need for optimization is inherent in the design process. A mathematical theory of optimization has become highly developed and is being applied to design where design functions can be expressed mathematically. The applicability of the mathematical methods usually depends on the existence of a continuously differentiable objective function. Where differentiable equations cannot be developed, numerical methods, aided by computer-based computation, are used to carry out optimization. These optimization methods require considerable depth of knowledge and mathematical skill to select the appropriate optimization technique and work it through to a solution.

Optimization has always been a goal of engineering design, but designers have not had the computational capability to perform true optimization in the mathematical sense until the last 10 years, when methods for finding near-optimal solutions were developed. These methods fall into the following broad categories.[3]

- *Evolution:* There is a close parallel between technological evolution and biological evolution. Most designs in the past have been optimized by an attempt to improve upon existing similar designs. Survival of the resulting variations depends on the natural selection of user acceptance.
- *Intuition:* The art of engineering is the ability to make good decisions without having exact mathematical justification. Intuition is knowing what to do without knowing exactly why one does it. The gift of intuition seems to be closely related to the

1. C. M. Creveling, *Tolerance Design: A Handbook for Developing Optimal Specifications,* Addison-Wesley Longman, Reading, MA, 1997.
2. R. N. Kackar, *Jnl of Quality Tech.,* vol. 17, no. 4, pp. 176–209, 1985.
3. J. N. Siddall, *Trans. ASME, J. Mech. Design,* vol. 101, pp. 674–81, 1979.

unconscious mind. The history of technology is full of examples of engineers who used intuition to make major advances. Although the knowledge and tools available today are so much more powerful, there is no question that intuition continues to play an important role in the development of good designs. This intuition is often in the form of remembering what worked in the past.

- *Trial-and-error modeling:* This refers to the usual situation in engineering design where it is recognized that the first feasible design is not necessarily the best. Therefore, the design model is exercised for a few iterations in the hope of finding an improved design. This works best when the designer has sufficient experience to make an informed choice of initial design values. The parametric design of a spring in Sec. 8.6.2 is an example of this approach. However, this mode of operation is not true optimization. Some refer to this approach as *satisficing*, as opposed to optimizing, to mean a technically acceptable job done rapidly and presumably economically. Such a design should not be called an optimal design.
- *Numerical algorithm:* This approach to optimization, in which mathematically based strategies are used to search for an optimum, has been enabled by the ready availability of fast, powerful digital computation. It is currently an area of active engineering research.

By the term *optimal design* we mean the best of all feasible designs. Optimization is the process of maximizing a desired quantity or minimizing an undesired one. Optimization theory is the body of mathematics that deals with the properties of maxima and minima and how to find maxima and minima numerically. In the typical design optimization situation, the designer has defined a general configuration for which the numerical values of the independent variables have not been fixed. An *objective function*[1] that defines the overall value of the design in terms of the n design variables, expressed as a vector $\mathbf{x}$, is established.

$$f(\mathbf{x}) = f(x_1, x_2, \ldots x_n) \tag{15.19}$$

Typical objective functions can be expressed in terms of cost, weight, reliability, and material performance index or a combination of these. By convention, objective functions are usually written to minimize their value. However, maximizing a function $f(\mathbf{x})$ is the same as minimizing $-f(\mathbf{x})$.

Generally when we are selecting values for a design we do not have the freedom to select arbitrary points within the design space. Most likely the objective function is subject to certain constraints that arise from physical laws and limitations or from compatibility conditions on the individual variables. *Equality constraints* specify relations that must exist between the variables.

$$h_j(\mathbf{x}) = h_j(x_1, x_2, \ldots x_n) = 0; j = 1 \text{ to } p \tag{15.20}$$

---

1. Also called the criterion function, the payoff function, or cost function.

For example, if we were optimizing the volume of a rectangular storage tank, where $x_1 = l_1$, $x_2 = l_2$, and $x_3 = l_3$, then the equality constraint would be volume $V = l_1, l_2, l_3$. The number of equality constraints must be no more than the number of design variables, $p \leq n$.

Inequality constraints, also called regional constraints, are imposed by specific details of the problem.

$$g_i(\mathbf{x}) = g_i(x_1, x_2, \ldots x_n) \leq 0; \, i = 1 \text{ to } m \tag{15.21}$$

There is no restriction on the number of inequality constraints.[1] A type of inequality constraint that arises naturally in design situations is based on specifications. *Specifications* define points of interaction with other parts of the system. Often a specification results from a decision to carry out a suboptimization of the system by establishing a fixed value for one of the design variables.

A common problem in design optimization is that there often is more than one design characteristic that is of value to the user. One way to handle this case in formulating the optimization problem is to choose one predominant characteristic as the objective function and to reduce the other characteristics to the status of constraints. Frequently they show up as rather "hard" or severely defined specifications. In reality, such specifications are usually subject to negotiation (soft specifications) and should be considered to be target values until the design progresses to such a point that it is possible to determine the penalty that is being paid in trade-offs to achieve the specifications. Siddal[2] has shown how this may be accomplished in design optimization through the use of an interaction curve.

EXAMPLE 15.10
The example helps to clarify the definitions just presented. We wish to design a cylindrical tank to store a fixed volume of liquid V. The tank will be constructed by forming and welding thin steel plate. Therefore, the cost will depend directly on the area of plate that is used.

The design variables are the tank diameter D and its height h. Since the tank has a cover, the surface area of the tank is given by

$$A = 2(\pi D^2 / 4) + \pi Dh$$

We choose the objective function $f(x)$ to be the cost of the material for constructing the tank.

$$f(x) = C_m A = C_m(\pi D^2/2 + \pi Dh), \text{ where } C_m \text{ is the cost per unit area of steel plate.}$$

An equality constraint is introduced by the requirement that the tank must hold a specified volume:

$$V = \pi D^2 h / 4$$

---

1. It is conventional to write Eq. (15.21) as ≤ 0. If the constraint is of the type ≥ 0, convert to this form by multiplying through by -1.
2. J. N. Siddall and W. K. Michael, *Trans. ASME, J. Mech. Design*, vol. 102, pp. 510–16, 1980.

Inequality constraints are introduced by the requirement for the tank to fit in a specified location or to not have unusual dimensions.

$$D_{min} \leq D \leq D_{max} \quad h_{min} \leq h \leq h_{max}$$

There are no universal optimization methods for engineering design. If the problem can be formulated by analytical mathematical expressions, then using the approach of calculus is the most direct path. However, most design problems are too complex to use this method, and a variety of optimization methods have been developed. Table 15.5 lists most of these methods. The task of the designer is to understand whether the problem is linear or nonlinear, unconstrained or constrained, and to select the method most applicable to the problem. Brief descriptions of various approaches to design optimization are given in the rest of this section. For more depth of understanding about optimization theory, consult the various references given in Table 15.5.

Linear programming is the most widely applied optimization technique when constraints are known, especially in business and manufacturing production situations. However, most design problems in mechanical design are nonlinear; see Example 15.10.

TABLE 15.5
Listing of Numerical Methods Used in Optimization Problems

Type of Algorithm	Example	Reference (see footnotes)
Linear programming	Simplex method	1
Nonlinear programming	Davison-Fletcher-Powell	2
Geometric programming		3
Dynamic programming		4
Variational methods	Ritz	5
Differential calculus	Newton-Raphson	6
Simultaneous mode design	Structual optimization	7
Analytical-graphical methods	Johnson's MOD	8
Monotonicity analysis		9
Genetic algorithms		10
Simulated annealing		11

1. W. W. Garvin, *Introduction to Linear Programming*, McGraw-Hill, New York, 1960.
2. L. T. Biegler, *Nonlinear Programming*, Society of Industrial and Applied Mathematics, Philadelphia, 2010.
3. C. S. Beightler and D. T. Philips: *Applied Geometric Programming*, John Wiley & Sons, New York, 1976.
4. S. E. Dreyfus and A. M. Law, *The Art and Theory of Dynamic Programming*, Academic Press, New York, 1977.
5. M. H. Denn, *Optimization by Variational Methods*, McGraw-Hill, New York, 1969.
6. F. B. Hildebrand, *Introduction to Numerical Analysis*, McGraw-Hill, 1956.
7. L. A. Schmit (ed.), *Structural Optimization Symposium*, ASME, New York, 1974.
8. R. C. Johnson, *Optimum Design of Mechanical Elements*, 2d ed., John Wiley & Sons, New York, 1980.
9. P. Y. Papalambros and D. J. Wilde, *Principles of Optimal Design*, 2d ed., Cambridge University Press, New York, 2000.
10. D. E. Goldberg, *Genetic Algorithm*, Addison-Wesley, Reading, MA, 1989.
11. S. Kirkpatrick, C. D. Gelatt, and M. P. Vecchi, "Optimization by Simulated Annealing," *Science*, vol. 220, pp. 671–79, 1983.

15

FIGURE 15.10
Different types of extrema
in the objective function
curve.

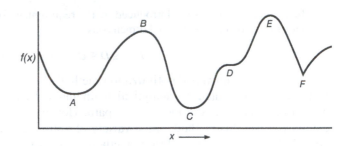

15.8.1 Optimization by Differential Calculus

We are all familiar with the use of the calculus to determine the maximum or minimum values of a mathematical function. Figure 15.10 illustrates various types of extrema that can occur. A characteristic property of an extremum is that the derivative of the function is 0 at that point. $f(x)$ is momentarily stationary at the point. The familiar condition for a stationary point is

$$\frac{df(x)}{dx} = 0 \qquad (15.22)$$

If the curvature is negative, then the stationary point is a maximum. The point is a minimum if the curvature is positive.

$$\frac{d^2 f(x)}{dx^2} \leq 0 \text{ indicates a local maximum} \qquad (15.23)$$

$$\frac{d^2 f(x)}{dx^2} \geq 0 \text{ indicates a local minimum} \qquad (15.24)$$

Both point B and point E are mathematical maxima. Point B, which is the smaller of the two maxima, is called a local maximum. Point E is the global maximum and Point C is the global minimum. Point D is a point of inflection. At an inflection point, the slope is zero and the curve is horizontal, but the second derivative is zero. When $d^2f(x)/dx^2 = 0$, higher-order derivatives must be used to find a derivative that becomes nonzero. If the zero-valued derivative's order is odd (e.g., 3rd or 5th derivative), the point is an inflection point, but if the derivative's order is even it is a local optimum. Point F is not a minimum point because at point F the objective function is not continuous; the point F is only a cusp in the objective function. Using the derivative of the function to infer maxima or minima only works with a continuous function.

We can apply this simple optimization technique to the tank problem described in Example 15.10. The objective function, expressed in terms of the equality constraint $V = \pi D^2 h/4$, is

$$f(\mathbf{x}) = C_m \pi \frac{D^2}{2} + C_m \pi Dh = \frac{C_m \pi D^2}{2} + C_m \pi D \left(\frac{4}{\pi} V D^{-2} \right) \qquad (15.25)$$

$$\frac{df(x)}{dD} = 0 = C_m \pi D - \frac{4C_m V}{D^2} \tag{15.26}$$

$$D = \left(\frac{4V}{\pi}\right)^{1/3} = 1.084 V^{1/3} \tag{15.27}$$

The value of diameter established by Eq. (15.27) results in minimum cost because the second derivative of Eq. (15.26) is positive. Note that while some problems yield to analytical expressions in which the objective function is a single variable, most engineering problems involve objective functions with more than one design variable.

Lagrange Multiplier Method

The Lagrange multipliers provide a powerful method for finding optima in multi-variable problems involving equality constraints. We have the original objective function $f(\mathbf{x}) = f(x, y, z)$ subject to the equality constraints $h_1 = h_1(x, y, z)$ and $h_2 = h_2(x, y, z)$. We establish a new objective function, the Lagrange expression (LE)

$$LE = f(x, y, z) + \lambda_1 h_1(x, y, z) + \lambda_2 h_2(x, y, z) \tag{15.28}$$

where λ_1 and λ_2 are the Lagrange multipliers. The following conditions must be satisfied at the optimum point.

$$\frac{\partial LE}{\partial x} = 0 \qquad \frac{\partial LE}{\partial y} = 0 \qquad \frac{\partial LE}{\partial z} = 0 \qquad \frac{\partial LE}{\partial \lambda_1} = 0 \qquad \frac{\partial LE}{\partial \lambda_2} = 0 \tag{15.29}$$

EXAMPLE 15.11

This example illustrates the determination of the Lagrange multipliers for use in optimization.[1] A total of 300 linear feet of tubes must be installed in a heat exchanger in order to provide the necessary heat-transfer surface area. The total dollar cost of the installation includes: (1) the cost of the tubes, $700; (2) the cost of the shell $25D^{2.5}L$; (3) the cost of the floor space occupied by the heat exchanger = $20DL$. The spacing of the tubes is such that 20 tubes must fit in a cross-sectional area of 1 ft² inside the heat exchanger tube shell.

The purchase cost C is taken as the objective function. The optimization should determine the diameter D and the length of the *heat exchanger L* to minimize the purchase cost. The objective function is the sum of three costs.

$$C = 700 + 25D^{2.5}L + 20DL \tag{15.30}$$

The optimization of C is subject to the *equality constraint* based on total length and cross-sectional area of the tube shell.

---

1. W. F. Stoecker, *Design of Thermal Systems*, 2d ed., McGraw-Hill, New York, 1980.

15

Total ft^3 of tubes $\times$ 20 tubes/ft^2 = total length (ft).

$$\frac{\pi D^2}{4} L \times 20 = 300$$

$$5\pi D^2 L = 300 \qquad \lambda = L - \frac{300}{5\pi D^2}$$

The Lagrange equation is: $LE = 700 + 25D^{2.5}L + 20DL + \lambda\left(L - \frac{300}{5\pi D^2}\right)$

$$\frac{\partial LE}{\partial D} = 2.5(25)D^{1.5}L + 20L + 2\lambda\frac{60}{\pi D^3} = 0 \qquad\qquad (15.31)$$

$$\frac{\partial LE}{\partial L} = 25D^{2.5} + 20D + \lambda = 0 \qquad\qquad (15.32)$$

$$\frac{\partial LE}{\partial \lambda} = L - \frac{300}{5\pi D^2} = 0 \qquad\qquad (15.33)$$

From Eq. (15.33), $L = \frac{60}{\pi D^2}$; From Eq. (15.32) $\lambda = -25D^{2.5} - 20D$

Substituting into Eq. (15.31):

$$62.5D^{1.5}\left(\frac{60}{\pi D^2}\right) + 20\left(\frac{60}{\pi D^2}\right) + 2(-25D^{2.5} - 20D)\left(\frac{60}{\pi D^3}\right) = 0$$

$$12.5D^{1.5} = 20 \quad D = (1.6^{0.666}) = 1.37\text{ft}$$

Substituting into the functional constraint between D and L gives $L = 10.2$ ft. Substituting the optimum values for D and L into the equation for the objective function, Eq. (15.30), gives the optimum cost as $1538.

This is an example of a closed form optimization for a single objective function with two design variables, D and L, and a single equality constraint.

By their nature, design problems tend to have many variables, many constraints limiting the acceptable values of some variables, and many objective functions to describe the desired outcomes of a design. A feasible design is any set of variables that simultaneously satisfies all the design constraints and fulfills the minimum requirements for functionality. An engineering design problem is usually underconstrained, meaning that there are not enough relevant constraints to set the value of each variable. Instead, there are many feasible values for each constraint. That means there are many feasible design solutions. As pointed out in the discussion of morphological methods (see Sec. 6.6), the number of feasible solutions grows exponentially as the number of variables with multiple possible values increases.

15.8.2 Search Methods

When it becomes clear that there are many feasible solutions to a design problem, it is necessary to use some method of searching through the design space to find the best one. Finding the globally optimal solution (the absolute best solution) to a design problem can be difficult. There is always the option of using brute calculation power to identify all design solutions and evaluate them. Unfortunately, design options reach into the thousands, and design performance evaluation can require multiple, complicated objective functions. Together, these logistical factors make an exhaustive search of the problem space impossible. There are also design problems that do not have one single best solution. Instead they may have a number of sets of design variable values that produce the same overall performance by combining different levels of the performance of one embedded objective function. In this case, we seek a set of best solutions. This set is called a Pareto Set.

We can identify several classes of search problems. A *deterministic search* is one in which there is little variability so all problem parameters are known. In a *stochastic search,* there is a degree of randomness in the search process that can lead to different solutions. We can have a search involving only a single variable or the more complicated and more realistic situation involving a search over multiple variables. We can have a *simultaneous search,* in which the conditions for every experiment are specified and all the observations are completed before any judgment regarding the location of the optima is made, or a *sequential search,* in which future experiments are based on past outcomes. Many search problems involve *constrained optimization,* in which certain combinations of variables are forbidden. Linear programming and dynamic programming are techniques that deal well with situations of this nature.

Golden Section Search

The *golden section search* is an efficient search method for a single variable with the advantage that it does not require an advance decision on the number of trials. The search method is based on the fact that the ratio of two successive Fibonacci numbers $F_{n-1}/F_n = 0.618$ for all values of $n > 8$. A Fibonacci series, named after a 13th century mathematician, is given by $F_n = F_{n-2} + F_{n-1}$ where $F_0 = 1$ and $F_1 = 1$.

n	0	1	2	3	4	5	6	7	8	9	...
F_n	1	1	2	3	5	8	13	21	34	55	

This same ratio was discovered by Euclid, who called it the *golden mean*. He defined it as a length divided into two unequal segments such that the ratio of the length of the whole to the larger segment is equal to the ratio of the length of the larger segment to the smaller segment. The ancient Greeks felt 0.618 was the most pleasing ratio of width to length of a rectangle, and they used it in the design of many of their buildings.

In using the golden section search, the first two trials are located at $0.618L$ from either end of the range of x that needs to be explored, Fig. 15.11. The goal is to find the *minimum* value of the function or response. In the first trial, $x_1 = 0.618L = 6.18$ and $x_2 = (1 - 0.618)L = 3.82$. If $y_2 > y_1$, the region to the left of x_2 is eliminated since we are searching for a minimum value of x and the assumption is that the function is unimodal.

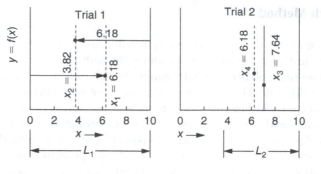

FIGURE 15.11
Example of use of the golden section search.

For the second trial, the search interval L_2 is from $x = 3.82$ to 10, a distance of 6.18 units. The values of the two points are $x_3 = 0.618(6.18) + 3.82 = 7.64$ (from 0 to the right) and $x_4 = 10 - 6.18(0.618) = 10 - 3.82 = 6.18$ from $x = 0$. Note that $x_4 = x_1$, so only one new data point is required. Once again, if $y_4 > y_3$, we can eliminate the region to the left of y_4. The new search interval is 3.82 units wide. The process is continued, placing a search point at 0.618 times the search interval, from both ends of the interval, until we reach as close to the minimum as is desired. Note that the golden section search cannot deal with functions that have multiple extrema between their limits. If this is suspected to occur, then start the search at one end of the domain and proceed in equal intervals across the limits.

Multivariable Search Methods

When the objective function depends on two or more variables, the geometric representation is a response surface (Fig. 15.12a). It usually is convenient to work with

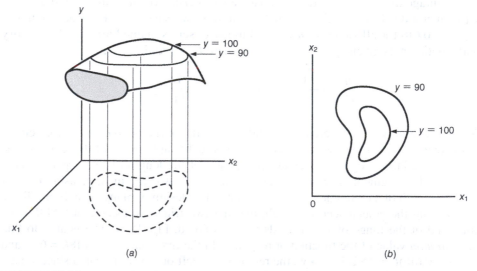

FIGURE 15.12
(a) Contour lines on surface created by $x_1 x_2$; (b) contour lines projected onto $x_1 x_2$ plane.

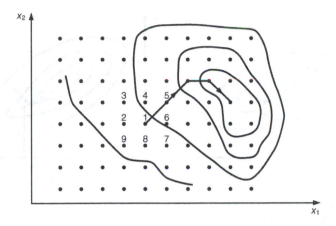

FIGURE 15.13
Procedure for a lattice search.

contour lines produced by the intersection of planes of constant y with the response surface and projected on the x_1x_2 plane (Fig. 15.12b).

Lattice Search

In the lattice search, which is an analog to the single-variable search, a two-dimensional grid lattice is superimposed over the projections of the contours (Fig. 15.13). In the absence of special knowledge about the location of the maximum, the starting point is selected near the center of the region, at point 1. The objective function is evaluated for points 1 through 9 surrounding point 1. If point 5 turns out to be the largest value, it becomes the central point for the next search. The procedure continues until the location reached is one at which the central point is greater than any of the other eight points. Frequently, a coarse grid is used initially and a finer grid is used after the maximum is approached.

Univariate Search

The univariate search is a one-variable-at-a-time method. All of the variables are kept constant except one, and it is varied to obtain an optimum in the objective function. That optimal value is then substituted into the function, and the function is optimized with respect to another variable. The objective function is optimized with respect to each variable in sequence, and an optimal value of a variable is substituted into the function for the optimization of the succeeding variables. This requires independence between the variables.

Figure 15.14a shows the univariate search procedure. Starting at point 0 we move along x_2 = constant to a maximum at point 1 by using a single-variable search technique. Then we move along x_1 = constant to a maximum at point 2 and along x_2 = constant to a maximum at 3. We repeat the procedure until two successive moves are less than some specified value. If the response surface contains a ridge, as in Fig. 15.14b, then the univariate search can fail to find an optimum. If the initial value is at point 1,

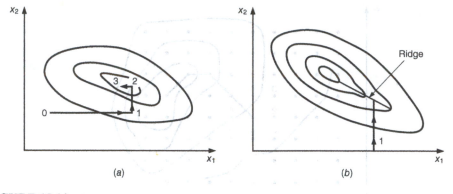

FIGURE 15.14
Univariate search procedures.

it will reach a maximum at $x_1 = $ constant at the ridge, and that will also be a maximum for $x_2 = $ constant. A false maximum is obtained.

An alternating single-variable search, as shown in Fig. 15.14, is sometimes used with the aid of a spreadsheet on a computer[1] when there are several design variables. The search procedure is to cycle through the design variables one at a time, selecting one variable for adjustment while holding the other variables constant. The objective function for variable 1 is first optimized using the golden section search, then variable 2, then 3, and so on. The cycle of variable searches will need to be repeated several times. The optimum is detected when running through a cycle of changes in design variables produces very little improvement in the value of the objective function.

Gradient Methods

A common local search method is to follow the steepest ascent (hill climbing) up the response surface. Imagine that we are walking at night up a hill. In the dim moonlight we can see far enough ahead to follow the local steepest slope. Thus, we would tend to climb in a direction normal to the contour lines in short segments and adjust the direction of climb as the terrain comes progressively into view. The gradient method does this with mathematics. We change the direction of the search to the direction of maximum slope, but we must do this in finite straight segments.

The gradient method starts with a best guess location and determines the direction with the gradient vector which by definition is normal to the local contour line. The gradient vector is expressed in terms of partial derivatives of the function describing the surface and the unit vectors $i, j,$ and k.

$$\nabla f(x, y, z) = \frac{\partial f}{\partial x} i + \frac{\partial f}{\partial y} j + \frac{\partial f}{\partial z} k \tag{15.34}$$

1. J. R. Dixon and C. Poli, *Engineering Design and Design for Manufacture,* pp. 18-13 to 18-14, Fieldstone Publishers, Conway, MA, 1995.

If the objective function is in analytical form, the partial derivatives can be obtained by calculus. If not, a numerical procedure such as the finite-difference method must be used. An important consideration is the choice of the step length. Too short a step makes the process very slow, while too large a step makes a zigzag path because it overshoots the changes in the direction of the gradient vector. The relative simplicity of hill climbing makes it a frequent choice when the time available to search is limited. The chief disadvantage is that steepest ascent will only find a local maximum. The method is also dependent upon the starting point of the search. Gradient descent uses the same approach to find a local minimum by using steps proportional to the negative of the gradient vector.

15.8.3 Nonlinear Optimization Methods

The methods discussed previously are not practical optimization techniques for engineering design problems with a large number of design variables and constraints. Numerical methods are needed to find solutions. The solution process starts with the best estimate of the optimum design. The objective function and the constraint functions, as well as their derivatives, are evaluated at that point. Then the design is moved to a new point, and to another, and so on, until optimality conditions or some other stopping criteria are met.

Mutivariable Optimization

Multivariable optimization of nonlinear problems has been a field of great activity, and many computer-based methods are available. Space permits mention of only a few of the more useful methods. Because an in-depth understanding requires considerable mathematics for which we do not have space, only a brief word description can be given. The interested student is referred to the text by Arora.[1]

Methods for unconstrained multivariable optimization are discussed first. Newton's method is an indirect technique that employs a second-order approximation of the function. This method has very good convergence properties, but it can be an inefficient method because it requires the calculation of $n(n + 1)/2$ second-order derivatives, where n is the number of design variables. Therefore, methods that require the computation of only first derivatives and use information from previous iterations to speed up convergence have been developed. The DFP (Davidon, Fletcher, and Powell) method is one of the most powerful methods.[2]

Optimization of nonlinear problems with constraints is a more difficult area. A common approach is to successively linearize the constraints and objective function of a nonlinear problem and solve using the technique of linear programming. The name of the method is sequential linear programming (SLP). A limitation of SLP is a lack of robustness. A robust computer algorithm is one that will converge to the same solution regardless of the starting point. The challenge of achieving robustness is

---

1. J. S. Arora, *Introduction to Optimum Design,* 2d ed., Elsevier Academic Press, San Diego, CA, 2004.
2. R. Fletcher and M. J. D. Powell, *Computer J.*, vol.6, pp.163–80, 1963; Arora, op. cit., pp. 324–327.

improved by using quadratic programming (QP) in determining the step size.[1] There is general agreement that the class of sequential quadratic programming (SQP) algorithms is the best overall choice for nonlinear multivariable optimization as they provide balance between efficiency (minimal CPU time) and robustness.

Many computer programs for doing multivariable optimization have been developed. A search of wikipedia under the heading of constrained nonlinear optimization found 80 entries.

- Because FEA is often used to search over a design space, many finite element software packages now come with optimization software. Vanderplaats Research and Development Inc. (www.vrand.com) was an early pioneer in the optimization of structures and provides optimization software linked with finite element analysis.
- iSIGHT, sold by Engineous Software (www.engenious.com), is popular in industry because of its broad capabilities and easy-to-use GUI interface.

Both Excel and MATLAB offer optimization tools. The Microsoft Excel Solver uses a generalized reduced gradient algorithm to find the maximum or minimum in nonlinear multivariable optimization problems.[2] For further information, go to www.office. microsoft.com and search for Excel Solver.

EXAMPLE 15.12 Excel Solver
The Excel Solver probably is the most-used optimization software program. In Section 8.6.2 we showed the iterative calculations used in a "cut and try" procedure to arrive at optimal dimensions for a helical compression spring. The time to carry out the needed design decisions and "run the numbers" with an electronic calculator was about 45 minutes. This example shows how a somewhat better result could be obtained in less than 15 minutes with the Solver function in Microsoft Excel.

In Sec. 8.6.2 the first design task was to determine the diameter of the spring wire, d, and the outside diameter of the coil spring, D, to withstand an axial force of 820 lb without yielding. The chief constraint was that the inside diameter, ID, of the spring must be at least 2.20 in. to permit the tie rod to pass easily through the spring.

Figure 15.15 on the left shows the Excel worksheet after completion of the optimization process. At the top is the equation expressing d in terms of D. This is in cell C6. An initial value of $D = 3.00$ was chosen for cell C5.

Turning now to the Solver dialog box, an absolute reference for cell C6 (C6) is placed in the Set Target Cell field. The Solver is set to minimize the value of the wire diameter, d, as the coil diameter, D, is changed in C5. All of this is subject to the constraint C9 > 2.20.

The optimum values of d and D are given in the spreadsheet. As discussed in Sec. 8.6.2, a figure of merit (f.o.m.) for this design is Dd^2N_p where a smaller f.o.m. is better. The optimized design gives a f.o.m. of 9.38, which is better than the best value of 11.25 in iteration 3, but not as good as the 8.5 found when using the stronger and more expensive Q&T wire (see Iteration 10, Table 8.5).

MATLAB has a number of optimization capabilities in its Optimization Toolbox, Table 15.6. For more information on these functions, enter MATLAB and at the command prompt and type help followed by the name of the function.

---

1. J. S. Arora, op. cit., Chaps. 8 and 10.
2. J. S. Arora, op. cit., pp. 369–73.

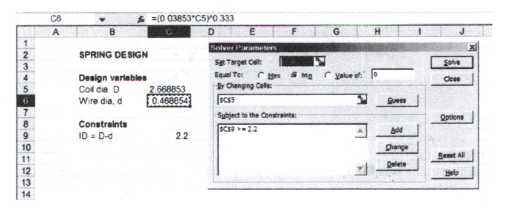

FIGURE 15.15
Excel worksheet on left, and Solver dialog box on right.

TABLE 15.6
Optimization Functions Provided by MATLAB

Class of Problem	MATLAB Function	Comments
Linear programming	linprog	
Nonlinear optimization		
Single-objective, unconstrained	fminuc	Can be set for steepest descent
Multiple variables	fminsearch	Uses Nelder-Mead Simplexsearch which does not require gradients
Single-objective, constrained		
Single variable	fminbnd	
Multiple variables	fmincon	Uses gradient based on finite diff.
Multiobjectives	fminimax	

For examples in the use of these functions, see Arora[1] and Magrab.[2]

Multiobjective Optimization

Multiobjective optimization refers to the solution of problems with more than one objective function. The design objectives in these problems are inherently in conflict. Consider a shaft loaded in torsion with the two design objectives of maximizing strength and minimizing weight (cost). As the diameter of the shaft is reduced to decrease weight, the stress is increased, and vice versa. This is the classical problem of *design trade-off*. During the optimization process the designer reaches a point where

1. J. S. Arora, op. cit., Chap. 12.
2. E. B. Magrab, et al. *An Engineer's Guide to MATLAB,* 2d ed., Chap. 13, S. Azarm, "Optimization," Prentice Hall, Upper Saddle River, NJ, 2005.

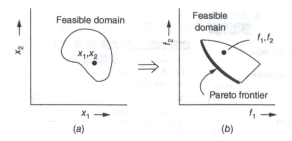

FIGURE 15.16
Feasible domains in (a) the design variable space and (b) the objective function space with its
Pareto frontier.

it is no longer possible to improve both design objectives. Such a point is referred to
as a Pareto point, and the locus of these points defines the Pareto frontier, Fig. 15.16*b*.

All points on a Pareto frontier have the same objective function value even though
the variable values are different. To solve such problems, the optimization method
finds the set of Pareto solutions. The actual decision maker can be queried for his or
her preferences, and the designer can rank order the preferences.

15.8.4 Other Optimization Methods

Monotonicity Analysis

Monotonicity analysis is an optimization technique that may be applied to design problems with monotonic properties, that is, where the change in objective
function and constraints steadily increases (or decreases) over the design space.
This is a situation that is very common in design problems. Engineering designs
tend to be strongly defined by physical constraints. When these specifications and
restrictions are monotonic in the design variables, then monotonicity analysis can
often show the designer which constraints are active at the optimum. An *active constraint* refers to a design requirement that has a direct impact on the location of the
optimum. This information can be used to identify the improvements that could be
achieved if the feasible domain were modified, which would point out directions for
technological improvement.

The ideas of monotonicity analysis were first presented by Wilde.[1] Subsequent
work by Wilde and Papalambros has applied the method to many engineering problems[2] and to the development of a computer-based method of solution.[3]

1. D. J. Wilde, *Trans. ASME, Jnl. of Engr for Industry,* vol. 94, pp. 1390–94, 1975.
2. P. Papalambros and D. J. Wilde, *Principles of Optimal Design,* 2d ed., Cambridge University Press,
New York, 2000.
3. S. Azarm and P. Papalambros, *Trans. ASME, Jnl. of Mechanisms, Transmissions, and Automation in
Design,* vol. 106, pp. 82–89, 1984.

15

Dynamic Programming

Dynamic programming is a mathematical technique that is well suited for the optimization of staged processes. The word *dynamic* in the name of this technique has no relationship to the usual use of the word to denote changes with respect to time. Dynamic programming is related to the calculus of variations and is not related to linear and nonlinear programming methods. The method is well suited for allocation problems, as when x units of a resource must be distributed among N activities in integer amounts. It has been broadly applied within chemical engineering to problems like the optimal design of chemical reactors. Dynamic programming converts a large, complicated optimization problem into a series of interconnected smaller problems, each containing only a few variables. This results in a series of partial optimizations that require a reduced effort to find the optimum. Dynamic programming was developed by Richard Bellmann[1] in the 1950s. It is a well-developed optimization method.[2]

Genetic Algorithms

Genetic algorithms (GA) are a form of computational design that uses simulated biological evolution as its search strategy. Genetic algorithms are stochastic in that there are probabilistic parameters that govern the GA's operation. GAs are also iterative because they involve many cycles of generating designs and checking for the best options.

Genetic algorithms mimic biological evolution. The basic idea of genetic algorithms is to transform the problem into one solved by evolution as defined in the natural sciences. Under evolution by natural selection, the fittest (i.e., best suited to thrive in the environment) members of a population survive and produce offspring. It's likely that the offspring inherit some of the characteristics that led to their parents' survival. Over time, the average fitness of a population increases as natural selection acts. The principles of genetics allow random mutation in a small percentage of the population. This is how some new characteristics arise over time.

The most unique contribution of genetic algorithms is the representation of each design as a string of binary computer code. The creation of new designs for a next generation is complex because several rules are used to mimic the action of genetic inheritance. Using binary computer code to represent designs enables computational shortcuts in manipulating designs to offset the complexity and allow iterations of tens of generations of populations of 100 designs each. Genetic algorithms are not widely used in mechanical design optimization, but their potential is so great that one expects them to increase in popularity. To find more information on all aspects of genetic algorithms (e.g., research papers, MATLAB codes), visit the site for the International Society for Genetic and Evolutionary Computation at www.isgec.org.

For a review of current design optimization methodologies and references, see A. Van der Velden, P. Koch, and S. Tiwari, *Design Optimization Methodologies, ASM Handbook*, Vol. 22B, pp. 614–624, ASM International, Materials Park, OH, 2010.

15

1. R. E. Bellman, *Dynamic Programming,* Princeton University Press, Princeton, NJ, 1957.
2. G. L. Nernhauser, *Introduction to Dynamic Programming,* John Wiley & Sons, New York, 1960; E. V. Denardo, *Dynamic Programming Models and Applications,* Prentice Hall, Englewood Cliffs, NJ, 1982.

Evaluation Considerations in Optimization

We have presented optimization chiefly as a collection of computer-based mathematical techniques. However, of more importance than knowing how to manipulate the optimization tools is knowing where to use them in the design process. In many designs a single design criterion drives the optimization. In consumer products it usually is cost, in aircraft it is weight, and in implantable medical devices it is power consumption. The strategy is to optimize these "bottleneck factors" first. Once the primary requirement has been met as well as possible, there may be time to improve other areas of the design, but if the first is not achieved, the design will fail. In some areas of design there may be no rigid specifications. An engineer who designs a talking, walking teddy bear can make almost any trade-off he or she wants between cost, power consumption, realism, and reliability. The designers and market experts will work together to decide the best combination of characteristics for the product, but in the end the four-year-old consumers will decide whether it is an optimal design.

15.9
DESIGN OPTIMIZATION

It has been a natural development to combine computer-aided-engineering (CAE) analysis and simulation tools with computer-based optimization algorithms.[1] Linking optimization with analysis tools creates CAE design tools by replacing traditional trial-and-error approaches with a systematic design-search approach. This extends the designer's capability from being able with FEA to quantify the performance of a particular design to adding information about how to modify the design to better achieve critical performance criteria.

Figure 15.17 shows a general framework for CAE-based optimal design. Starting with an initial design (size and shape parameters), a numerical analysis simulation, such as FEA, is performed on the design to compute the performance measures, such as von Mises stress, and the sensitivity of the performance measures with respect to the design parameters. Then an optimization algorithm computes new design parameters, and the process is continued until an optimum design is achieved. Often this is not a mathematical optimum but a set of design variables for which the objective function shows appreciable improvement.

Most FEA packages offer optimization routines that integrate design simulation, optimization, and design-sensitivity analysis into a comprehensive design environment. The user inputs preliminary design data and specifies acceptable variables and required constraints. The optimization algorithm generates successive models, in conjunction with remeshing routines, until it ultimately converges on an optimized design. For example, structural optimization of a turbine wheel design resulted in a 12 percent reduction in mass and a 35 percent reduction in stress.

---

1. D. E. Smith, "Design Optimization," *ASM Handbook,* vol. 20, pp. 209–18, ASM International, Materials Park, OH, 1997.

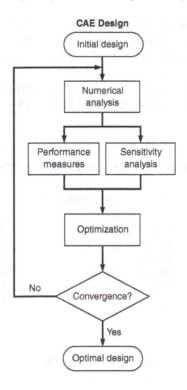

FIGURE 15.17
General framework for CAE-based design optimization. (From D. E. Smith, *ASM Handbook,*
Vol. 20, p. 211. Used with permission.)

15.10
SUMMARY

This chapter presents many of the modern views about design. The overarching concept is that quality is built into products during design. Manufacturing cannot compensate for errors in design. Second, we have emphasized that variability during manufacture and in service is the the challenge to a quality design. We aim for a robust design that is less sensitive to process variations and to extreme conditions in service.

Quality must be viewed as a total system from the perspective called total quality management (TQM). TQM places the customer at the center and solves problems with a data-driven approach using simple but powerful tools (see Sec. 4.6). It emphasizes continuous improvement where large changes are achieved by many small improvements made over time.

Statistics plays a significant role in achieving quality and robustness. A control chart shows whether the variability of a process is within reasonable bounds. The process capability index C_p, tells whether the selected tolerance range is easily achievable by a particular manufacturing process.

New ways of looking at quality have been introduced by Taguchi. The loss function provides a better way of looking at quality than the traditional upper and lower tolerance limits around a mean value. The signal-to-noise parameter (S/N) provides a powerful metric to search for design situations that minimize variability. Orthogonal experimental designs provide a useful and widely adopted methodology to find the design or process conditions that are most robust.

The search for optimum conditions has been a design goal for many years. A wide selection of optimization methods is described in Sec. 15.8.

NEW TERMS AND CONCEPTS

Design optimization	Multiobjective optimization	Robust design
Equality constraint	Noise factors	Signal-to-noise ratio
Genetic algorithm	Objective function	Six Sigma quality
Golden section search	Process capability index	Statistical process control
Inequality constraint	Quality	Steepest descent search
ISO 9000	Quality assurance	Taguchi method
Lattice search	Quality control	Univariate search
Loss function	Range	Upper control limit

BIBLIOGRAPHY

Quality

Besterfield, D. H.: *Total Quality Management,* 3rd ed., Prentice Hall, Upper Saddle River, NJ, 2003.
Gevirtz, C. D.: *Developing New Products with TQM,* McGraw-Hill, New York, 1994.
Kolarik, W. J.: *Creating Quality,* McGraw-Hill, New York, 1995.
Summers, D. C. S.: *Quality,* 5th ed., Prentice Hall, Upper Saddle River, NJ, 2009.

Robust Design

Ealey, L. A.: *Quality by Design,* 2nd ed., ASI Press, Dearborn, MI, 1984.
Fowlkes, W. Y., and C. M. Creveling: *Engineering Methods for Robust Product Design,* Addison-Wesley, Reading MA, 1995.
Roy, K. R.: *A Primer on the Taguchi Method,* 2nd ed., Society of Manufacturing Engineers, Dearborn, MI, 2010.
Wu, Y., and A. Wu: *Taguchi Methods for Robust Design,* ASME Press, New York, 2000.

Optimization

Arora, J. S.: *Introduction to Optimum Design,* 3rd ed., Elsevier Academic Press, San Diego, CA, 2011.
Papalambros, P. Y., and D. J. Wilde: *Principles of Optimal Design,* 2nd ed., Cambridge University Press, New York, 2000.
Park, G. J.: *Analytic Methods for Design Practice,* Spriner-Verlag, London, 2007.
Ravindran, A., Ragsdell, K. M., and G. V. Reklaitis: *Engineering Optimization,* 2nd ed., John Wiley & Sons, Hoboken, NJ, 2006.

PROBLEMS AND EXERCISES

15.1 Discuss as a class how Deming's 14 points could be applied to higher education.

15.2 Divide into teams and use the TQM problem-solving process introduced in Sec. 4.6 to decide how to improve the quality in several of your courses (one course per team).

15.3 Discuss the concept of quality circles. What would be involved in implementing a quality circle program in industry? How could the concept be applied to the classroom?

15.4 Use the concept of statistical hypothesis testing to identify and classify the errors that can occur in quality-control inspection.

15.5 Dig deeper into the subject of control charts and find some rules for identifying out-of-control processes.

15.6 For the control chart shown in Fig. 15.1, determine C_p. Note: Hardness is only recorded to the nearest 0.5 RC.

15.7 A product has specification limits of 120 ± 10 MN and a target value of 120 MN. The standard deviation of the products coming off the process line is 3 MN. The mean value of strength is initially 118 MN, but it shifts to 122 MN and then 125 MN without any change in variability. Determine C_p and C_{pk}.

15.8 The equations in Sec. 15.5 for process capability index are for parameters that have two-sided tolerances about the target value. What if your design parameter was fracture toughness, K_{Ic}. What would the equation for C_p be when you are only concerned with a one-sided tolerance below the target value?

15.9 A grinding machine is grinding the root of gas turbine blades where they attach to the disk. The critical dimension at the root must be 0.450 ± 0.006 in. Thus, a blade falls out of specs in the range 0.444 to 0.456 and has to be scrapped at a cost of $120.

(a) What is the Taguchi loss equation for this situation?

(b) Samples taken from the grinder had the following dimensions: 0.451; 0.446; 0.449; 0.456; 0.450; 0.452; 0.449; 0.447; 0.454; 0.453; 0.450; 0.451.

What is the average loss function for the parts made on the machine?

15.10 The weather strip that seals the door of an automobile has a specification on width of 20 ± 4 mm. Three suppliers of weather strip produced the results shown below.

Supplier	Mean Width	Variance s^2	C_{pk}
A	20.0	1.778	1.0
B	18.0	0.444	1.0
C	17.2	0.160	1.0

Field experience shows that when the width of the weather strip is 5 mm below the target, the seal begins to leak and about 50 percent of the customers will complain and insist that it be replaced at a cost of $60. When the strip width exceeds 25 mm, door closure becomes difficult and the customer will ask to have the weather strip replaced. Historically, the three suppliers had the following number of parts out of spec in deliveries of 250,000 parts: A: 0.27%; B: 0.135%; C: 0.135%.

(a) Compare the three suppliers on the basis of loss function.

(b) Compare the three suppliers on the basis of cost of defective units.

15.11 Part of the pollution control system of an automobile engine consists of a nylon tube inserted in a flexible elastomeric connector. The tubes had been coming loose, so an experimental program was undertaken to improve the robustness of the design. The effectiveness of the design was measured by the pounds of force needed to pull the nylon tube out of the connector. The control factors for this design were:

A—interference between the nylon tube and the elastomer connector
B—wall thickness of the elastomer connector
C—depth of insertion of the tube in the connector
D—the percent, by volume, of adhesive in the connector pre-dip

The environmental noise factors that conceivably could affect the strength of the bond had to do with the conditions of the pre-dip that the end of the connector was immersed in before the tube was inserted. There were three:

X—time the predip was in the pot 24 h and 120 h

Y—temperature of the predip 72°F and 150°F

Z—relative humidity 25% and 75%

(a) Set up the orthogonal arrays for the control factors (inner array) at three levels and the noise factors (outer array). How many runs will be required to complete the tests?

(b) The calculated S/N ratio for the pull-off force of the tube for the nine experimental conditions of the control matrix are, in order: (1) 24.02; (2) 25.52; (3) 25.33; (4) 25.90; (5) 26.90; (6) 25.32; (7) 25.71; (8) 24.83; (9) 26.15. What type of S/N ratio should be used? Determine the best settings for the design parameters.

15.12 Conduct a robust design experiment to determine the most robust design of paper airplanes. The control parameters and noise parameters are given in the following table.

Control Parameters			
Parameter	**Level 1**	**Level 2**	**Level 3**
Weight of paper (A)	One sheet	Two sheets	Three sheets
Configuration (B)	Design 1	Design 2	Design 3
Width of paper (C)	4 in	6 in	8 in
Length of paper (D)	6 in	8 in	10 in

Noise Parameter		
Parameter	**Level 1**	**Level 2**
Launch height (X)	Standing on ground	Standing on chair
Launch angle (Y)	Horizontal to ground	45° above horizontal
Ground surface	Concrete	Polished tile

All planes are launched by the same person in a closed room or hallway with no air currents. When launching a plane, the elbow must be touching the body and only the forearm, wrist, and hand are used to send the plane into flight. Planes are made from ordinary copy paper. The class should decide on the three designs, and once this is decided, the designs will not be varied throughout the experiment. The objective function to be optimized is the distance the plane flies and glides to a stop on the floor, measured to the nose of the plane.

15.13 We want to design a hot-water pipeline to carry a large quantity of hot water from the heater to the place where it will be used. The total cost is the sum of four items: (1) the cost of pumping the water, (2) the cost of heat lost from the pipe, (3) the cost of the pipe, (4) the cost of insulating the pipe.

(a) By using basic engineering principles, show that the system cost is

$$C = K_p \frac{1}{D^5} + K_h \frac{1}{\ln[(D+x)/D]} + K_m D + K_i x$$

where x is the thickness of insulation on a pipe of inside diameter D.

(b) If $K_p = 10.0$, $K_h = 2.0$, $K_m = 3.0$, and $K_i = 1.0$ and initial values of x and D are equal to 1.0, find the values of D and x that will minimize the system cost. Use an alternating single-variable search.

15.14 Find the maximum value of $y = 12x - x^2$ with the golden section search method for an original interval of uncertainty of $0 \le x \le 10$. Carry out the search until the difference between the two largest calculated values of y is 0.01 or less. Do the same problem using the Excel Solver.

15

16

ECONOMIC DECISION MAKING

16.1
INTRODUCTION

Throughout this book we have repeatedly emphasized that the engineer is a decision maker and that engineering design is a process of making a series of decisions over time. We also have emphasized from the beginning that engineering involves the application of science to real problems of society. In this authentic context, one cannot escape the fact that economics may play a role as big as, or bigger than, that of technical considerations in the decision making process of design.

The major engineering infrastructure that built this nation—the railroads, major dams, and waterways—required a methodology for predicting costs and balancing them against alternative courses of action. In any engineering project, costs and revenues will occur at various points of time in the future. The methodology for handling this class of problems is known as engineering economy or engineering economic analysis. Familiarity with the concepts and approach of engineering economy generally is considered to be part of the standard engineering toolkit. Indeed, an examination on the fundamentals of engineering economy is required for professional engineering registration in all disciplines in all states.

The chief concept in engineering economy is that *money has a time value.* Paying out $1.00 today is more costly than paying out $1.00 a year from now. A dollar invested today is worth a dollar plus interest a year from now. Engineering economy recognizes the fact that the *use of money* is a valuable asset. Money can be rented in the same way one can rent an apartment, but the charge for using it is called interest rather than rent. This time value of money makes it more profitable to push expenses into the future and bring revenues into the present as much as possible.

Before proceeding into the mathematics of engineering economy, it is important to understand where engineering economy sits with regard to related disciplines like economics and accounting. Economics generally deals with broader and more global issues than engineering economy, such as the forces that control the money supply and trade between nations. Engineering economy uses the interest rate established by the economic forces to solve more specific and detailed problems. However, it usually is a problem concerning alternative costs in the future. The accountant is more concerned with determining exactly, and often in great detail, what costs have been incurred in the past. One might say that the economist is an oracle, the engineering economist is a fortune teller, and the accountant is a historian.

16.2
MATHEMATICS OF TIME VALUE OF MONEY

If we borrow a present sum of money or principal P at a simple interest rate i, the annual cost of interest is $I = Pi$. If the loan is repaid in a lump sum F at the end of n years, the amount required is

$$F = P + nI = P + nPi = P(1 + ni) \tag{16.1}$$

where F = future worth

P = present worth

I = annual cost of interest

i = annual interest rate

n = number of years

If we borrow \$1000 for 6 years at 10 percent simple interest rate, we must repay at the end of 6 years:

$$F = P(1 + ni) = \$1000[1 + 6(0.10)] = \$1600$$

Therefore, we see that \$1000 available today is not equivalent to \$1000 available in 6 years. Actually, \$1000 in hand today is worth \$1600 available in only 6 years at 10 percent simple interest.

We can also see that the *present worth* of \$1600 available in 6 years and invested at 10 percent is \$1000.

$$P = \frac{F}{1 + ni} = \frac{\$1600}{1 + 0.6} = \$1000$$

In making this calculation we have discounted the future sum back to the present time. In engineering economy the term *discounted* refers to bringing dollar values *back in time* to the present.

16.2.1 Compound Interest

You are aware from your personal banking experiences that financial transactions usually use compound interest. In *compound interest,* the interest due at the end of a period is not paid out but is instead added to the principal. During the next period, interest is paid on the total sum.

First period: $\quad F_1 = P + Pi = P(1+i)$

Second period: $\quad F_2 = P(1+i) + iP(1+i) = [P(1+i)](1+i) = P(1+i)^2$

Third period: $\quad F_3 = P(1+i)^2 + iP(1+i)^2 = P[(1+i)^2](1+i) = P(1+i)^3$

nth period: $\quad F_n = P(1+i)^n$

(16.2)

We can write Eq. (16.2) in a short notation that is convenient to use when the engineering economy relationships become more complex.

$$F_n = P(1+i)^n = P(F/P, i, n)$$

(16.3)

In Eq. (16.3) the function $(F/P, i, n)$ has the meaning: Find the equivalent amount F given the amount P compounded at an interest rate i for n interest periods.

EXAMPLE 16.1 How long will it take money to double if it is compounded annually at a rate of 10 percent per year?

$\quad F = P(F/P, 10, n) \quad$ but $F = 2P$, (we want to find the doubling time)

$\quad 2P = P(F/P, 10, n)$

Therefore, the answer is found in a table of single-payment compound-amount factors at the year n for which $(F/P) = 2.0$. Examining the table in Appendix F2 at www.mhhe.com/dieter, we see that, for $n = 7$, $F/P = 1.949$ and, for $n = 8$, $(F/P) = 2.144$. Linear extrapolation gives us $(F/P, 10, 7.2) = 2.000$. We can generalize the result to establish the financial rule of thumb that the number of years to double an investment is 72 divided by the interest rate (expressed as an integer).

Usually in engineering economy, n is given in years and i is an annual interest rate. However, in banking circles the interest may be compounded at periods other than one year. Compounding at the end of shorter periods, such as daily, raises the effective interest rate. If we define r as the nominal annual interest rate and p as the number of interest periods per year, then the interest rate per interest period is $i = r/p$ and the number of interest periods in n years is pn. Using this notation, Eq. (16.2) becomes

$$F = P\left[\left(1 + \frac{r}{p}\right)^p\right]^n$$

(16.4)

Note that when $p = 1$, the above expression reduces to Eq. (16.2). Standard compound interest tables that are prepared for $p = 1$ can be used for other than annual periods. To do so, use the table for $i = r/p$ and for a number of years equal to $p \times n$. Alternatively, use the interest table corresponding to n years and an *effective rate* of yearly return equal to $(1 + r/p)^p - 1$.

TABLE 16.1
Influence of Compounding Period on Effective Rate of Return

Frequency of Compounding	No. Annual Interest Periods p	Interest Rate for Period, %	Effective Rate of Yearly Return, %
Annual	1	12.0	12.0
Semiannual	2	6.0	12.4
Quarterly	4	3.0	12.6
Monthly	12	1.0	12.7
Continuously	∞	0	12.75

Table 16.1 shows the influence of the number of interest periods per year on the effective rate of return.

16.2.2 Cash Flow Diagram

Engineering economy was developed to deal with financial transactions taking place at various times in the future. This can be best understood in terms of *cash flows*. Some of these will be cash inflows (receipts), like revenue from sale of products, reduction in operating cost, sale of used machinery, or tax savings. Others will be cash outflows (disbursements), such as the costs incurred in designing a product, the operating costs in making the product, and the periodic maintenance costs in keeping the factory running. The net cash flow is given by

$$\text{Net cash flow} = \text{cash inflows (receipts)} - \text{cash outflows (disbursements)} \qquad (16.5)$$

Cash flows occur frequently and take place at varying times within the time period of the problem. In the cash flow diagram, Fig. 16.1, the horizontal axis represents time and the vertical axis is cash flow. Cash inflows are positive and are represented by arrows above the x-axis. Cash outflows are negative and are below the x-axis.

It has been mentioned that engineering economy is chiefly concerned with assisting decision making about future financial decisions in an engineering project. Since

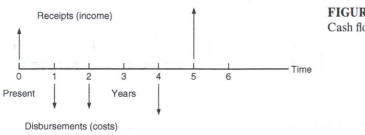

FIGURE 16.1
Cash flow diagram.

16

future prediction of cash flows is likely to be imprecise, it is not worth carefully locating each cash flow on the diagram in time. Instead, the *end-of-period convention* is used in which the cash flows within a period are assumed to occur at the end of the interest period.

16.2.3 Different Types of Annual Transactions

In many situations we are concerned with a uniform series of receipts or disbursements occurring equally at the end of each period. Examples are the payment of a debt on the installment plan, setting aside a sum that will be available at a future date for replacement of equipment, and a retirement annuity that consists of a series of equal payments instead of a lump sum payment. We will let A be the equal end-of-the-period payment that makes up the uniform annual series.

Figure 16.2 shows that if an annual sum A is invested at the end of each year for 3 years, the total sum F at the end of 3 years will be the sum of the compound amount of the individual investments A

$$F = A(1+i)^2 + A(1+i) + A$$

and for the general case of n years,

$$F = A(1+i)^{n-1} + A(1+i)^{n-2} + \cdots + A(1+i)^2 + A(1+i) + A \text{ which simplifies to}$$

$$F = A\frac{(1+i)^n - 1}{i} \tag{16.6}$$

Equation (16.6) gives the future sum of n uniform payments of A when the interest rate is i. This equation may also be written:

$$F_n = A(F/A, i, n) \tag{16.7}$$

where $(F/A, i, n)$ is the uniform-series compound amount factor that converts a series A to a future worth F.

By solving Eq. (16.6) for A, we have the uniform series of end-of-period payments factor, that, at compound interest i, provide a future sum F.

$$A = F\frac{i}{(1+i)^n - 1} \tag{16.8}$$

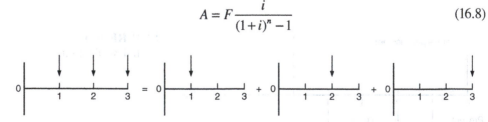

FIGURE 16.2
Equivalence of a uniform annual series.

This type of calculation often is used to set aside money in a *sinking fund* to provide funds for replacing worn-out equipment, or for investing money to send a child to college.

$$A = F(A/F, i, n) \tag{16.9}$$

where $(A/F, i, n)$ is the sinking fund factor. It sets up a future fund F by investing A each interest period n at a rate i.

By combining Eq. (16.2) with Eq. (16.6), we develop the relation for the present worth of a uniform series of payments A:

$$P = A\frac{(1+i)^n - 1}{i(1+i)^n} = A(P/A, i, n) \tag{16.10}$$

Solving Eq. (16.10) for A gives the important relation for *capital recovery*:

$$A = P\frac{i(1+i)^n}{(1+i)^n - 1} = P(A/P, i, n) \tag{16.11}$$

where $(A/P, i, m)$ is the capital recovery factor. The A in Eq. (16.11) is the annual payment needed to return the initial capital investment P *plus* interest on that investment at a rate i over n years.

Capital recovery is an important concept in engineering economy. It is important to understand the difference between capital recovery and sinking fund. Consider the following example:

EXAMPLE 16.2 What annual investment must be made at 10 percent to provide funds for replacing a $10,000 machine in 20 years?

$A = F(A/F, 10, 20) = \$10,000(0.01746) = \$174.60$ per year put into the sinking fund

What is the annual cost of capital recovery of $10,000 at 10 percent over 20 years?

$A = P(A/F, 10, 20) = \$10,000(0.011746) = \$1174.60$ per year for capital recovery

We see that $(A/P, i, n) = (A/F, i, n) + i$

$$0.11746 = 0.01746 + 0.10000$$

Annual cost of capital recovery = annual cost of sinking fund + annual interest cost
$$\$1174.60 = \$174.60 + 0.10\ (\$10,000)$$

With a sinking fund we put away each year a sum of money that, over n years, together with accumulated compound interest, equals the required future amount F. With capital recovery we put away enough money each year to provide for replacement in n years *plus we charge ourselves interest on the invested capital*. The use of capital recovery is a conservative but valid economic strategy. The amount of money invested in capital equipment ($10,000 in Example 16.2) represents an *opportunity cost*, since we are forgoing the revenue that the $10,000 could provide if invested in interest-bearing securities.

TABLE 16.2
Summary of Compound Interest Factors

Item	Conversion	Algebraic Relation	Factor	Factor Name	EXCEL Factor
1	P to F	$F = P(1+i)^n$	$(F/P, i, n)$	Single payment, compound amount factor	$= FV(i, Nper, 0, 1, 0)$
2	F to P	$P = \dfrac{F}{(1+i)^n}$	$(P/F, i, n)$	Single payment, present worth factor	$= PV(i, Nper, 0, 1, 0)$
3	A to P	$P = A\dfrac{(1+i)^n - 1}{i(1+i)^n}$	$(P/A, i, n)$	Uniform payment, present worth factor	$= PV(i, Nper, 1, 0, 0)$
4	P to A	$A = P\dfrac{i(1+i)^n}{(1+i)^n - 1}$	$(A/P, i, n)$	Capital recovery factor	$= PMT(i, Nper, 1, 0, 0)$
5	A to F	$F = A\dfrac{(1+i)^n - 1}{i}$	$(F/A, i, n)$	Uniform series, compound amount factor	$= FV(i, Nper, 1, 0, 0)$
6	F to A	$A = F\dfrac{i}{(1+i)^n - 1}$	$(A/F, i, n)$	Sinking fund factor	$= PMT(i, Nper, 0, 1, 0)$

A summary of the compound interest relationships among F, P, and A is given in Table 16.2. Table 16.2 gives relationships for a uniform series of payments or receipts. Two other series often used in engineering economy are a gradient series in which the cash flow increases (or decreases) by a fixed increment at each time period, and a geometric series in which the cash flow changes by a fixed percentage at each time period.

The last column in Table 16.2 shows how the compound interest factors can be obtained directly from Microsoft Excel, where i is the interest rate for the period expressed as a decimal, and $Nper$ is the number of periods. When i is the annual interest rate, then $Nper$ is the number of years.

Consider an arithmetic gradient which changes each year by a constant amount G. The magnitude of G is the difference between the value in year 2 and year 1; G starts in the second year, the base year. Equation (16.13) converts the gradient that occurs over n years to a present worth P located on the cash flow diagram at year zero.

$$P = \frac{G}{i}\left[\frac{(1+i)^n - 1}{i(1+i)^n} - \frac{n}{(1+i)^n}\right] \qquad (16.12)$$

Equation (16.13) converts an annual gradient G to an equivalent annual uniform series A. If the gradient starts at year 2 and goes for n years the uniform series will start at year 1 and spread over n years.

$$A = G\left[\frac{1}{i} - \frac{n}{(1+i)^n - 1}\right] \qquad (16.13)$$

16

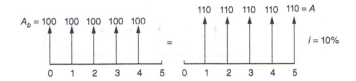

FIGURE 16.3
A uniform series paid at the beginning of the interest period, and the equivalent series paid at the end of the period.

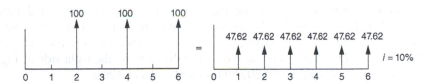

FIGURE 16.4
Conversion of payments every two years into annual payments.

Using symbolic notation, as shown in Table 16.2, simplifies writing the equations and aids in making calculations. For example, many compound interest tables do not contain a table for determining A (sinking fund factor) when F is known. However, using the symbolic factors this can be obtained by simply multiplying factors.

$$A = F(A/F) = F(P/F)(A/P) \qquad (16.14)$$

16.2.4 Irregular Cash Flows

Payment at the Beginning of the Interest Period

In working with a uniform series of payments or receipts, A, it is conventional practice to assume that A occurs at the end of each period. However, sometimes a series of payments begin immediately so that the payments are made at the beginning of each time period, A_b.

As Fig. 16.3 shows, this is equivalent to increasing each annual payment by the interest earned in one period of the accumulation of interest. Thus, Eq. (16.6) would be written as

$$F = A_b(1+i)\left[\frac{(1+i)^n - 1}{i}\right] \qquad (16.15)$$

Payments in Alternate Years

The left side of Fig. 16.4 shows uniform payments in alternate years with $i = 10\%$. One approach to finding the present value, P, would be to consider this as three future payments and determine P as follows:

$$P = 100(P/F, 10, 2) + 100(P/F, 10, 4) + P(P/F, 10, 6) = 82.64 + 68.30 + 56.45 = \$207.39.$$

FIGURE 16.5
Finding present values of a uniform series that does not extend to time zero.

An alternative approach is to consider the first annual payment to be a future pay-ment over two years and determine the annual payment (sinking fund factor) to produce $100. This would then be an annual payment paid over six years, since the payments are at the end of every two years, for six years total. $A = 100\,(A/F,\,10,\,2) = 100(0.4762) = \$47.62$ and is represented by the diagram on the right side of Fig. 16.4. The equivalent present value of $207.39 is calculated as follows.

$$P = 47.62(P/A,10,6) = 47.62(4.3553) = \$207.39$$

Uniform Payments Not Extending to Time Zero

Consider the uniform payments, A, extending from years 4 to 10 in Fig. 16.5 to find the present value, $P = A(P/A,\,i,\,7)$. This present value is located at the end of year 3, because the compound interest equations for the P/A factor assume that P will be determined one interest period prior to the first A in the series. Then to find the pres-ent value at time zero, P_3 must be discounted to the present. $P = F(P/F,\,i,\,3)$ where $F = P_3$.

16.3
COST COMPARISON

Having discussed the usual compound interest relations, we now are in a position to use them to make economic decisions. A typical decision is which of two courses of action is less expensive when the time value of money is considered. Generally the rate of interest to be used in these calculations is set by the *minimum attractive rate of return,* MARR. This is the lowest rate of return a company will accept for investing its money. The MARR is established by the corporate finance officer based on cur-rent market opportunities for investing money or on the importance of the project to advancing the company.

16.3.1 Present Worth Analysis

When the two alternatives have a common time period, a comparison on the basis of present worth is advantageous.

16

EXAMPLE 16.3 Two machines each have a useful life of 5 years. If money is worth 10 percent, which machine is more economical?

	A	B
Initial cost	$25,000	$15,000
Yearly maintenance cost	2,000	4,000
Rebuilding at end of third year	—	3,500
Salvage value	3,000	
Annual benefit from better quality production	500	

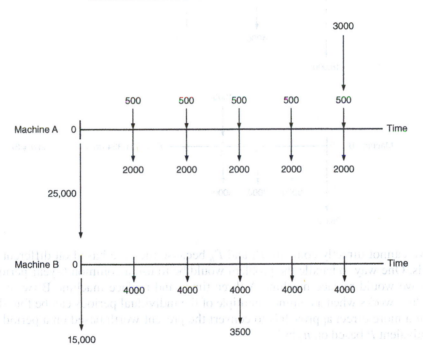

From the *cost* diagrams shown above we see that the cash flows definitely are different for the two alternatives. To place them on a common basis for comparison, we discount all costs back to the present time.

$$P_A = 25,000 + (2000 - 500)(P/A, 10, 5) - 3000(P/F, 10, 5)$$
$$= 25,000 + 1500(3.791) - 3000(0.621) = \$28,823$$
$$P_B = 15,000 + 4000(P/A, 10, 5) + 3500(P/F, 10, 3)$$
$$= 15,000 + 4000(3.791) + 3500(0.751) = \$32,793$$

Machine A is more economical because it has the lower cost on a present worth basis. In this example we considered both (1) costs plus benefits (savings) due to reduced scrap rate and (2) resale value at the end of the period of useful life. Thus, we really determined the *net present worth* for each alternative. We should also point out that present worth analysis is not limited to the comparison of only two alternatives. We could consider any number of alternatives and select the one with the smallest net present worth of costs.

In Example 16.3, both alternatives had the same life. Thus, the time period was the same and the present worth could be determined without ambiguity. Suppose we want to use present worth analysis for the following situation:

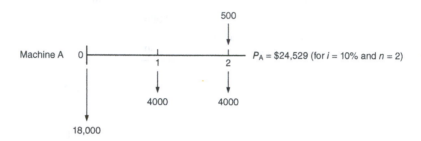

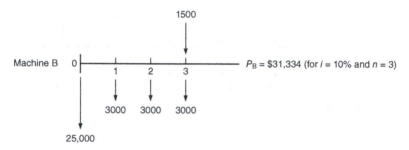

We cannot directly compare P_A and P_B because they are based on different time periods. One way to handle the problem would be to use a common 6-year period, in which we would replace machine A three times and replace machine B twice. This procedure works when a common multiple of the individual periods can be found easily, but a more direct approach is to convert the present worth based on a period n_1 to an equivalent P based on n_2 by[1]

$$P_{n_2} = P_{n_1} \frac{(A/P, i, n_1)}{(A/P, i, n_2)} \tag{16.16}$$

For our example, we convert P_B from a 3-year time period to a 2-year period.

$$P_{B_2} = P_{B_1} \frac{(A/P, i, n_1)}{(A/P, i, n_2)} = 31,334 \frac{(A/P, 10, 3)}{(A/P, 10, 2)} = 31,334 \left(\frac{0.40211}{0.57619}\right) = \$21,867$$

1. For a derivation of Eq. (16.16), see F. C. Jelen and J. H. Black, *Cost and Optimization Engineering*, 2nd ed., p. 28, McGraw-Hill, New York, 1983.

Since $P_A = \$24,529$, machine B is the more economical when compared on the basis of present worth *for equal time periods.*

16.3.2 Annual Cost Analysis

In the annual cost method, the cash flow over time is converted to an equivalent uniform annual cost or benefit. In this method no special procedures need be used if the time period is different for each alternative, because all comparisons are on an annual basis ($n = 1$).

Example	Machine A	Machine B
First cost	$10,000	$18,000
Estimated life	20 years	35 years
Estimated salvage	0	$3000
Annual cost of operation	$4000	$3000

$$A_A = 10,000(A/P, 10, 20) + 4000 = 10,000(0.1175) + 4000 = \$5175$$
$$A_B = (18,000 - 3000)(A/P, 10, 35) + 3000(0.10) + 3000 = \$4855$$

Machine B has the lower annual cost and is the more economical. Note that in calculating the annual cost of capital recovery for machine B we used the difference between the first cost and the salvage value; for it is only this amount of money that must be recovered. However, although the salvage value is returned to us, we are required to wait until the end of the useful life of the machine to recover it. Therefore, a charge for the annual cost of the interest on the investment tied up in the salvage value is made as part of the annual cost analysis.

Perhaps a more direct way to handle the case of machine B in the preceding example is to determine the equivalent annual cost based on the cash disbursements minus the annual benefit of the future resale value. The result is the same.

$$A_B = 18,000(A/P, 10, 35) + 3000 - 3000(A/F, 10, 35)$$
$$= 18,000(0.1037) + 3000 - 3000(0.0037) = \$4855$$

16.3.3 Capitalized Cost Analysis

Capitalized cost is a special case of present worth analysis. The capitalized cost of a project is the present value of providing for that project in perpetuity ($n = \infty$). The concept was originally developed for use with public works, such as dams and waterworks, that have long lives and provide services that must be maintained indefinitely. Capitalized cost subsequently has been used more broadly in economic decision making because it provides a method that is independent of the time period of the various alternatives.

Since capitalized cost is the present worth when $n = \infty$ we can use Eq. (16.16), and by setting $n_2 = \infty$ and $n_1 = n$, after simplification the capitalized cost K for a present value is

$$K = P\frac{(1+i)^n}{(1+i)^n - 1} = P(K/P, i, n) \qquad (16.17)$$

To find the capitalized cost for a series of uniform payments A, start with the relationship between P and A.

$$P = A\left[\frac{(1+i)^n - 1}{i(1+i)^n}\right] \text{ If we divide the numerator and denominator by } (1+i)^n$$

$$P = A\left[\frac{1 - \dfrac{1}{(1+i)^n}}{i}\right] \text{ and as } n \text{ approaches } \infty \text{ the equation reduces to } P = \frac{A}{i}.$$

Capitalized cost is P when $n = \infty$, so

$$K = A/i \qquad (16.18)$$

16.3.4 Using Excel Functions for Engineering Economy Calculation

The compound interest factors needed for engineering economy calculations can be determined on a calculator or looked up in the tables in all engineering economy textbooks.[1] Microsoft Excel provides an extensive menu of time value of money functions and other financial functions. When combined with the computational features of Excel and its "what if" capability, this makes an excellent general-purpose tool for engineering economic decision making. Table 16.3 gives a brief description of the most common functions for compound interest calculations. For details on using the functions, see the help pages in Excel or engineering economy texts.[2]

16

---

1. Several tables of F, P, and A and their combinations are given in the Appendix to this chapter.
2. L. T. Blank and A. J. Tarquin, op. cit., Appendix A.

TABLE 16.3
Useful Excel Functions for Compound Interest Calculations

Function	Description
FV(i, n, A, PV, type)	Calculates future value, FV, given int. rate per period, no. of periods, constant payment amount, A, present value PV, type = 0 end of period payment; type = 1, beginning of period payment
PV(i, n, A, FV, type)	Calculates present value PV, given i, n, periodic payments (–) or income (+) and future single payments or receipts
NPV(i, Incl, Inc2 . . .)	Calculates net present value, NPV, of a series of irregular future incomes (+) or expenses (–) at periodic interest i
PMT(i, n, PV, FV, type)	Calculates uniform payments A based on either a present value and/or a future value
RATE(n, A, PV, FV, type, g)	Calculates interest rate per period. g requires a guess for i, about 10%
NOMINAL(effect i, npery)	Calculates the nominal annual interest rate given the effective rate and number of compounding periods per year, npery
EFFECT(non i, npery)	Calculates the effective interest rate given the nominal interest rate and npery

16.4
DEPRECIATION

Capital equipment suffers a loss in value over time. This may occur by corrosion or wear, deterioration, or obsolescence, which is a loss of economic efficiency because of technological advances. Therefore, a company should lay aside enough money each year to accumulate a fund to replace the obsolete or worn-out equipment. This allowance for loss of value is called depreciation. Depreciation is an accounting expense on the income statement of the company. It is a noncash expense that is deducted from gross profits as a cost of doing business. In a capital-intensive business, depreciation can have a strong influence on the amount of taxes that must be paid.

$$\text{Taxable income} = \text{total income} - \text{allowable expenses} - \text{depreciation}$$

The basic questions to be answered about depreciation are: (1) what is the time period over which depreciation can be taken, and (2) how should the total depreciation charge be spread over the life of the asset? Obviously, the depreciation charge in any given year will be greater if the depreciation period is short (a rapid write-off).

The U.S. tax laws permit two methods for determining depreciation: straight-line depreciation and the MACRS method. The Economic Recovery Act of 1981 introduced the *accelerated cost recovery system* (ACRS) as the required depreciation method for tax purposes in the United States. This was modified in the 1986 Tax Reform Act for Modified Accelerated Cost Recovery System (MACRS). The

16

statute sets depreciation recovery periods based on the expected useful life. Some examples are:

- Special manufacturing devices; some motor vehicles: 3 years
- Computers; trucks; semiconductor manufacturing equipment: 5 years
- Office furniture; railroad track; agricultural buildings: 7 years
- Durable-goods manufacturing equipment; petroleum refining: 10 years
- Sewage treatment plants; telephone systems: 15 years

Residential rental property is recovered in 27.5 years and nonresidential rental property in 31.5 years. Land is a nondepreciable asset, since it is never used up.

16.4.1 Straight-Line Depreciation

In straight-line depreciation an equal amount of money is set aside yearly. The annual depreciation charge D is

$$D = \frac{\text{initial cost} - \text{salvage value}}{n} = \frac{C_i - C_s}{n} \tag{16.19}$$

where n is the recovery period in years.

The *book value* of the asset changes over time. The book value is the initial cost minus the sum of the depreciation charges that have been made. For straight-line depreciation, the book value B at the end of the jth year is

$$B_j = C_i - \frac{j}{n}(C_i - C_s) \tag{16.20}$$

where n is the recovery period.

16.4.2 Modified Accelerated Cost Recovery System (MACRS)

In MACRS the annual depreciation is computed using the relation

$$D = qC_i \tag{16.21}$$

where q is the recovery rate obtained from Table 16.4 and C_i is the initial cost. In MACRS the value of the asset is completely depreciated even though there may be a true salvage value. The recovery rates are based on starting out with a declining-balance method (see any engineering economy text) and switching to the straight-line method when it offers a faster write-off. MACRS uses a half-year convention that assumes that all property is placed in service at the midpoint of the initial year. Thus, only 50 percent of the first year depreciation applies for tax purposes, and a half year of depreciation must be taken in year $n + 1$.

Table 16.5 compares the annual depreciation charges for these two methods of calculation.

TABLE 16.4
Recovery Rates q Used in MACRS Method

	Recovery Rate, q, %				
Year	$n = 3$	$n = 5$	$n = 7$	$n = 10$	$n = 15$
1	33.3	20.0	14.3	10.0	5.0
2	44.5	32.0	24.5	18.0	9.5
3	14.8	19.2	17.5	14.4	8.6
4	7.4	11.5	12.5	11.5	7.7
5		11.5	8.9	9.2	6.9
6		5.8	8.9	7.4	6.2
7			8.9	6.6	5.9
8			4.5	6.6	5.9
9				6.5	5.9
10				6.5	5.9
11				3.3	5.9
12–15					5.9
16					3.0

n = recovery period, years

TABLE 16.5
Comparison of Depreciation Methods

	$C_i = \$6000$, $C_s = \$1000$, $n = 5$	
Year	Straight Line	MACRS
1	1000	1200
2	1000	1920
3	1000	1152
4	1000	690
5	1000	690
6	—	348

16.5
TAXES

Taxes are an important factor to be considered in engineering economic decisions. The chief types of taxes that are imposed on a business firm are:

1. *Property taxes:* Based on the value of the property owned by the corporation (land, buildings, equipment, inventory). These taxes do not vary with profits and usually are not too large.

16

2. *Sales taxes:* Imposed on sales of products. Sales taxes usually are paid by the retail purchaser, so they generally are not relevant to engineering economy studies of a business.
3. *Excise taxes:* Imposed on the manufacture of certain products like gasoline, tobacco, and alcohol. Also usually passed on to the consumer.
4. *Income taxes:* Imposed on corporate profits or personal income. Gains resulting from the sale of capital property also are subject to income tax.

Generally, federal income taxes have the most significant impact on engineering economic decisions. Although we cannot delve into the complexities of tax laws, it is important to incorporate the broad aspects of income taxes into our analysis.

The income tax rates are strongly influenced by politics and economic conditions. The United States as of 2011 has a corporate *graduated tax schedule* as follows:

Taxable Income	Tax Rate
$1–$50,000	0.15
$50,001–$75,000	0.25
$75,001–$100,000	0.34
$100,001–$335,000	0.39
$335,001–$10 M	0.34
$10M–$15M	0.35
$15M–$18.3M	0.38
Over $18.3 M	0.35

Most states and some cities and counties also have an income tax. For simplicity in economic studies a single effective tax rate is often used. This commonly varies from 35 to 50 percent. Since state taxes are deductible from federal taxes, the effective tax rate is given by

$$\text{Effective tax rate} = \text{state rate} + (1 - \text{state rate})(\text{federal rate}) \qquad (16.22)$$

The chief effect of corporate income taxes is to reduce the rate of return on a project or venture.

$$\text{After-tax rate of return} = \text{before-tax rate of return} \times (1 - \text{income tax rate})$$

$$r = i(1 - t) \qquad (16.23)$$

where t is the effective tax rate from Eq. (16.22).

Note that this relation is true only when there are no depreciable assets. For the usual case when we have depreciation, capital gains or losses, or investment tax credits, Eq. (16.23) is a rough approximation. The importance of depreciation in reducing taxes is shown in Fig. 16.6. The depreciation charge appreciably reduces the gross profit, and thereby the taxes. However, since depreciation is retained in the corporation, it is available for growing the enterprise.

Revenues	Operations expenses	
	Gross profit	Taxes
		Net profit
	Depreciation	

FIGURE 16.6
Distribution of corporate revenues.

EXAMPLE 16.4 High-Tech Pumps has a gross income in 1 year of $15 million. Operating expenses (salaries and wages, materials, etc.) are $10 million. Depreciation is $2.6 million. Also, this year there is a *depreciation recapture* of $800,000 because a specialized CNC machine tool that is no longer needed is sold for more than its book value. (*a*) Compute the company's federal income taxes. (*b*) What is the average federal tax rate? (*c*) If the state tax rate is 11 percent, what is the total income taxes paid?

(*a*) Taxable income (TI) = gross income − operating expenses − depreciation + depreciation recapture

$$TI = 15 - 10 - 2.6 + 0.8 = \$3.2M$$
$$\text{Taxes} = (\text{TI range})(\text{marginal rate})$$
$$= (50,000)0.15 + (25,000)0.25 + (25,000)0.34$$
$$+ (235,000)0.39 + (3.2M - 0.335M)0.34$$
$$= 7500 + 6250 + 8500 + 91,650 + 974,100 = \$1,088,000$$

(*b*) Average federal tax rate $= \dfrac{1,088,000}{3,200,000} = 0.34$

(*c*) From Eq. (16.22)

$$\text{Effective tax rate} = 0.11 + (1 - 0.11)(0.34) = 0.11 + 0.3026 = 0.4126$$

$$\text{Total income taxes} = 32,2000,000(0.4126) = \$1,320,320$$

Note that including state taxes makes a difference.

Consider a depreciable capital investment $C_d = C_i - C_s$. At the end of each year depreciation amounting to $D_f C_d$ is available to reduce the taxes by an amount $D_f C_d t$, where t is the tax rate.

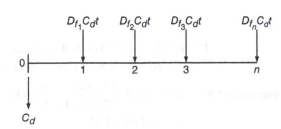

Note that the fractional depreciation charge each year D_f may vary from year to year depending on the method used to establish the depreciation schedule. See, for example, Table 16.5. The present value of this series of costs and benefits is

$$P = C_d - C_d t \left[\frac{D_{f_1}}{1+r} + \frac{D_{f_2}}{(1+r)^2} + \frac{D_{f_3}}{(1+r)^3} + \cdots + \frac{D_{f_n}}{(1+r)^n} \right] \qquad (16.24)$$

The exact evaluation of the term in brackets will depend on the depreciation method selected.

EXAMPLE 16.5 A manufacturing company of modest size is considering an investment in energy-efficient electric motors to reduce its large annual energy cost. The initial cost would be $12,000, and over a 10-year period it is estimated that the firm would save $2200 annually in electricity costs. The salvage value of the motors is estimated at $2000. Determine the after-tax rate of return.

Solution

First we will establish the before-tax rate of return. We need to determine the cash flow for each year. Cash flow, in this context, is the net profit or savings for each year. We shall use straight-line depreciation to determine the depreciation charge. Table 16.6 shows the cash flow results. The before-tax rate of return is the interest rate at which the before-tax cash flow savings just equals the purchase cost of the motors.

$$12,000 = 2200(P/A, i, 10) + 2000(P/F, i, 10)$$

We find the rate of return by trying different values of i in the compound interest tables. For $i = 14$ percent,

$$12,000 = 2200(5.2161) + 2000(0.2697)$$
$$= 11,475 + 539 = 12,014$$

Therefore, the before-tax rate of return is very slightly more than 14 percent. To find the after-tax rate of return, we use the after-tax cash flow in Table 16.6. From Eq. (16.23) we estimate the after-tax rate of return to be 7 percent.

$$12,000 = 1600(P/F, i, 10) + 2000(P/F, i, 10)$$

For $i = 6\%$:
$$12,000 = 1600(7.3601) + 2000(0.5584)$$
$$= 11,776 + 1117 = 12,893 \quad i \text{ too low}$$

For $i = 8\%$:
$$12,000 = 1600(6.7101) + 2000(0.4632)$$
$$= 10,736 + 926 = 11,662 \quad i \text{ too high}$$

Interpolating between 6 and 8 %: $i = 6\% + 2\% \dfrac{12,893 - 12,000}{12,893 - 11,662} = 6\% + 2\%(0.72)$

$$i = 6 + 1.44 = 7.44\%$$

TABLE 16.6
Cash Flow Calculations for Example 16.5

Year	Before-Tax Cash Flow	Depreciation	Taxable Income	50% Income Tax	After-Tax Cash Flow
0	−12,000				−12,000
1 to 9	2,200	1000	1200	−600	1,600
10	2,200	1000	1200	−600	1,600
Salvage value	2,000				2,000

Tax Considerations

For tax purposes the expenditures that a business incurs are divided into two broad categories. Those for facilities and production equipment with lives in excess of one year are called capital expenditures; they are said to be "capitalized" in the accounting records of the business. Other expenses for running the business, such as labor and material costs, direct and indirect costs, and facilities and equipment with a life of one year or less, are ordinary business expenses. Usually they total more than the capital expenses. In the accounting records, they are said to be "expensed." The ordinary expenses and depreciation are directly subtracted from the gross income to determine the taxable income.

When a capital asset is sold, a capital gain or loss is established by subtracting the book value of the asset from its selling price. If a capital asset is sold for more than its current book value, this establishes an income called *depreciation recapture.* Both capital gains and depreciation capture must be added to the gross income. Frequently in our modern history, capital gains have received special treatment by being taxed at a rate lower than for ordinary income.

Investment in capital is a vital step in the innovation process that leads to increased national wealth. Therefore, the federal government frequently uses the tax system to stimulate capital investment. This most often takes the form of a tax credit, usually 7 percent but varying with time from 4 to 10 percent. This means that 7 percent of the purchase price of qualifying equipment can be deducted from the taxes that the firm owes the U.S. government. Moreover, the depreciation charge for the equipment is based on its full cost.

16.6
PROFITABILITY OF INVESTMENTS

One of the principal uses for engineering economy is to determine the profitability of proposed projects or investments. The decision to invest in a project generally is based on three different sets of criteria.

Profitability: Determined by techniques of engineering economy to be discussed in this section. Profitability is an analysis that estimates how rewarding in monetary terms an investment will be.

Financial analysis: How to obtain the necessary funds and what it will cost. Funds for investment come from three broad sources: (1) retained earnings of the corporation, (2) long-term commercial borrowing from banks, insurance companies, and pension funds, and (3) the equity market through the sale of stock.

Analysis of intangibles: Legal, political, or social consideration or issues of a corporate image often outweigh financial considerations in deciding on which project to pursue. For example, a corporation may decide to invest in the modernization of an old plant because of its responsibility to continue employment for its employees when investment in a new plant 1000 miles away would be economically more attractive.

However, in our free-enterprise system a major goal of a business firm is to maximize profit. It does so by committing its funds to ventures that appear to be profitable. If investors do not receive a sufficiently attractive profit, they will find other uses for their money, and the growth—even the survival—of the firm will be threatened.

Four methods of evaluating profitability are commonly used. Accounting rate of return and payback period are simple techniques that are readily understood, but they do not take time value of money into consideration. Net present value and discounted cash flow are the most common profitability measures in which time value of money is considered. Before discussing them, however, we need to look a bit more closely at the concept of cash flow.

Cash flow measures the flow of funds into or out of a project. Funds flowing in constitute positive cash flow; funds flowing out are negative cash flow. The cash flow for a typical plant construction project is shown in Fig. 16.7.

Strictly speaking depreciation is not a cash flow component, since it is not a real flow of cash. In industries with billions of dollars of installed capital equipment, depreciation plays a major role in reducing the taxable investment. As we will see, depreciation is important in calculations of cash flow after taxes (CFAT). From the point of view of the tax collector, *taxable income* (TI) is given by

$$TI = \text{gross income} - \text{operating expenses} - \text{depreciation}$$
$$= GI - OE - D \tag{16.25}$$

From the view point of engineering economy analysis, depreciation enters only as it affects taxes. Therefore, *cash flow before taxes* (CFBT) is given by

$$CFBT = \text{gross income} - \text{operating expenses} - \text{capital investment} + \text{salvage value}$$
$$= GI - OE - P + S \tag{16.26}$$

Depreciation affects the *cash flow after taxes* (CFAT) through its role on taxable income. To determine CFAT Eq. (16.25), is first multiplied by the effective tax rate t_e, Eq. (16.22) to give the taxes that have been paid and then this is subtracted from CFBT to give CFAT.

$$CFAT = (GI - OE - P + S) - (GI - OE - D)t_e \tag{16.27}$$

Table 16.7 shows a step-by-step procedure for determining CFAT.

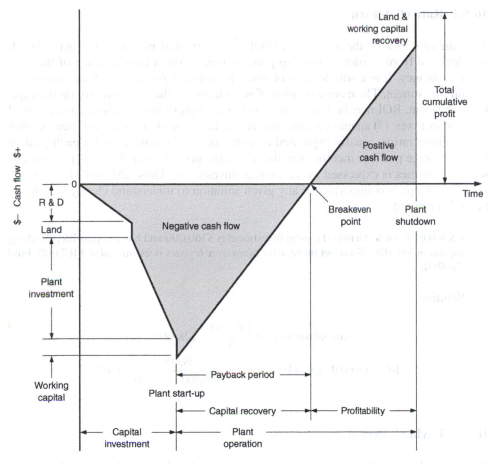

FIGURE 16.7
Typical costs in the cycle of a plant investment.

TABLE 16.7
Calculation of Cash Flow After Taxes (CFAT)

(1) Revenue (over 1-year period)	$500,000
(2) Operating costs	360,000
(3) (1) − (2) = gross margin	140,000
(4) Annual depreciation charge	60,000
(5) (3) − (4) = taxable income	80,000
(6) (5) × 0.35 = income tax	28,000
(7) (5) − (6) = net income after taxes	52,000
Net cash flow (after taxes)	
(7) + (4) = 52,000 + 60,000	112,000

The same result for cash flow after taxes will be found
using Eq.(16.27)

16.6.1 Rate of Return

The *rate of return* on the investment (ROI) is the simplest measure of profitability. It is calculated from a strict accounting point of view without consideration of the time value of money. It is a simple ratio of some measure of profit or cash income to the capital investment. There are a number of ways to assess the rate of return on the capital investment. ROI may be based on (1) net annual profit before taxes, (2) net annual profit after taxes, (3) annual cash income before taxes, or (4) annual cash income after taxes. These ratios, usually expressed as percents, can be computed for each year or on the average profit or income over the life of the project. In addition, capital investment sometimes is expressed as the average investment. Thus, although the ROI is a simple concept, it is important in any given situation to understand clearly how it has been determined.

> **EXAMPLE 16.6** An initial capital investment is $360,000 and has a 6-year life. Working capital is $40,000. Total net profit after taxes over 6 years is estimated at $167,000. Find the ROI.

Solution

$$\text{Annual net profit} = \frac{167,000}{6} = \$28,000$$

$$\text{ROI on initial capital investment} = \frac{28,000}{360,000 + 40,000} = 0.07$$

16.6.2 Payback Period

The payback period is the period of time necessary for the cash flow to fully recover the initial total capital investment (Fig. 16.7). Although the payback method uses cash flow, it does not include a consideration of the time value of money. Emphasis is on rapid recovery of the investment. Also, in using the method, no account is taken of cash flows or profits recovered after the payback period.

Consider the payback period example shown in Table 16.8. By the payback period criterion, project A is more desirable because it recovers the initial capital investment in 3 years. However, project B, which returns a cumulative cash flow of $110,000, obviously is more profitable overall.

16.6.3 Net Present Worth

In Sec. 16.3, as one of the techniques of cost comparison, we introduced the criterion of net present worth (NPW).

Net present worth = present worth of benefits − present worth of cost

16

TABLE 16.8
Payback Period Example

Year	Cash Flow	
	Project A	Project B
0	$ –100,000	$ –100,000
1	50,000	0
2	30,000	10,000
3	20,000	20,000
4	10,000	30,000
5	0	40,000
6	0	50,000
7	0	60,000
	$10,000	$110,000
Payback period	3 years	5 years

By this technique the expected cash flows (both + and –) through the life of the project are discounted to time zero at an interest rate representing the minimum acceptable return on capital, MARR. The project with the greatest positive value of NPW is preferred. NPW depends upon the project life, so strictly speaking the net present worths of two projects should not be compared if the projects have different service lives.

Obviously, the value of NPW will be dependent upon the interest rate used for the calculation. Low interest rates will tend to make NPW more positive, for a given set of cash flows, and large values of interest will push NPW in a negative direction. There will be some value of i for which the sum of the discounted cash flows equals zero; NPW = 0. This value of i is called the internal rate of return, IRR.

16.6.4 Internal Rate of Return

In the beginning of this chapter we considered calculation methods that determined what sum of money at the present time, when invested at a given interest rate, *is equivalent* to a larger sum at a future time. Now with the internal rate of return, we find what interest rate makes the present sum and the future sum *equivalent*. This value of interest rate is called the *internal rate of return,* IRR. This is the rate of return for which the net present value equals zero.

If, for example, the internal rate of return is 20 percent, it implies that 20 percent per year will be earned on the investment in the project, in addition to which the project will generate sufficient funds to repay the original investment. Depreciation is considered implicitly in NPW and IRR calculations through the definition of cash flow as discussed earlier in this chapter.

Because the decision on profitability is expressed as a percentage rate of return in the IRR method, it is more readily understood and accepted by engineers and business people than the NPW method, which produces a sum of money as an answer. In the

NPW method it is necessary to select an interest rate for use in the calculations, and that may be a difficult and controversial thing to do. But by using the IRR method, we compute a rate of return from the cash flows. One situation in which NPW has an advantage is that individual values of NPW for a series of subprojects may be added to give the NPW for the complete project. That cannot be done with the rate of return developed from IRR analysis.

EXAMPLE 16.7 A machine has a first cost of $10,000 and a salvage value of $2000 after a 5-year life. Annual benefits (savings) from its use are $5000, and the annual cost of operation is $1800. The tax rate is 50 percent. Find the IRR rate of return.

Solution

Using straight-line depreciation, the annual depreciation charge is

$$D = \frac{C_i - C_s}{n} = \frac{10,000 - 2,000}{5} = \$1600$$

The annual cash flow after taxes is the sum of the net receipts and depreciation.

$$(CF)_a = (5000 - 1800)(1 - 0.50) + 1600(0.50)$$
$$= 1600 + 800 = \$2400$$

Year	Cash Flow
0	−10,000
1	2,400
2	2,400
3	2,400
4	2,400
5	2,400 + 2,000 (C_s)

$$NPW = 0 = -10,000 + 2400(P/A, i, 5) + 2000(P/F, i, 5)$$

If $i = 10$ percent, NPW = +340; if $i = 12$ percent, NPW = −214. Thus, we have the IRR bracketed, and interpolating gives

$$i = 10\% + (12\% - 10\%)\frac{340}{340 + 214}$$

$$= 10 + 2\left(\frac{340}{564}\right) = 10 + 1.2 = 11.2\%$$

The IRR function in Microsoft Excel can be used to quickly determine the internal rate of return. To use IRR, the net benefits or costs (−) are entered in a column of cells, one for each period. Enter a 0 for any period where there is no cash flow. Finally, enter a guess as the starting point for the calculation. For example: =IRR(A2:A8,5)

It is an important rule of engineering economy that *each increment* of investment capital must be justified on the basis of earning the minimum required rate of return.

EXAMPLE 16.8 A company has the option of investing in one of the two machines described in the following table. Which investment is justified?

	Machine A	Machine B
Initial cost C_i	$10,000	$15,000
Useful life	5 years	10 years
Salvage value C_s	$2,000	0
Annual benefits	$5,000	$7,000
Annual costs	$1,800	$4,300

Solution

Assume a 50 percent tax rate and a minimum attractive rate of return of 6 percent. The conditions for machine A are identical with those in Example 16.7, for which $i = 11.2$ percent. Calculation of the IRR for machine B shows it is slightly in excess of the minimum rate of 6 percent. However, that is not the proper question. Rather, we should ask whether the *increment of investment* ($15,000 − $10,000) is justified. In addition, because machine B has twice the useful life of machine A, we should place them both on the same time basis (see Table 16.9).

TABLE 16.9
Cash Flow, Example 16.8

Year	Machine A	Machine B	Difference, B − A
0	−10,000	−15,000	−5,000
1	2,400	2,100	−300
2	2,400	2,100	−300
3	2,400	2,100	−300
4	2,400	2,100	−300
5	2,400 − 10,000 + 2,000	2,100	−300 + 8,000
6	2,400	2,100	−300
7	2,400	2,100	−300
8	2,400	2,100	−300
9	2,400	2,100	−300
10	2,400 + 2,000	2,100	−300 − 2,000

$$\text{NPW} = 0 = -5000 - 300(P/A, i, 10) + 8000(P/F, i, 5) - 2000(P/F, i, 10)$$

But, even at $i = \frac{1}{4}$ percent, NPW = −2009, and there is no way that the extra investment in machine B can be justified economically.

When only costs—not income (or savings)—are known, we can still use the IRR method for incremental investments, but not for a single project. We assume that the lowest capital investment is justified without being able to determine the internal rate of return, and we then determine whether the additional investment is justified.

EXAMPLE 16.9 On the basis of the data in the following table, determine which machine should be purchased.

	Machine A	Machine B
First cost	$3000	$4000
Useful life	6 years	9 years
Salvage value	$500	0
Annual operating cost	$2000	$1600

Solution

This solution will be based on cash flow before taxes. To place the machines on a common time frame, we use a common life of 18 years.

TABLE 16.10
Cash Flow for Example 16.9

Year	Machine A	Machine B	Difference, B – A
0	−3000	−4000	−1000
1 to 5	−2000	−1600	+400
6	−2000 − 2500	−1600	+400 + 2500
7 to 8	−2000	−1600	+400
9	−2000	−1600 − 4000	+400 − 4000
10, 11	−2000	−1600	+400
12	−2000 − 2500	−1600	+400 + 2500
13 to 17	−2000	−1600	+400
18	−2000 + 500	−1600	+400 − 500

$$\text{NPW} = 0 = -1000 + 400(P/A, i, 18) + 2500(P/F, i, 6) + 2500(P/F, i, 12)$$
$$- 4000(P/F, i, 9) - 500(P/F, i, 18)$$

Trial and error shows that $i \approx 47$ percent, which clearly justifies purchase of machine B.

We have presented information on the four most common techniques for evaluating the profitability of an investment. The rate-of-return method has the advantage of being simple and easy to use. However, it ignores the time value of money and the consideration of cash flow. The payback period also is a simple method, and it is particularly attractive for industries undergoing rapid technological change. Like the rate-of-return method, it ignores the time value of money, and it places an undue emphasis on projects that achieve a quick payoff. The net present worth method takes both cash flow and time value of money into account. However, it suffers from the problem of ambiguity in setting the required rate of return, and it may present problems when

16

projects with different service lives are compared. Internal rate of return has the advantage of producing an answer that is the real internal rate of return. The method readily permits comparison between alternatives, but it is assumed that all cash flows generated by the project can be reinvested to yield a comparable rate of return.

16.7
OTHER ASPECTS OF PROFITABILITY

Innumerable factors affect the profitability of a project in addition to the mathematical expressions discussed in Sec. 16.6. The purpose of this section is to round out our consideration of the crucial subject of profitability.

We need to realize that profit and profitability are not the same concept. Profit is measured by accountants, and its value in any one year can be manipulated in many ways. Profitability is inherently a long-term parameter of economic decision making. As such, it should not be influenced much by short-term variations in profits. In recent years there has been a strong trend toward undue emphasis on quick profits and short payoff periods that work to the detriment of long-term investment in engineering projects.

Estimation of profitability requires the prediction of future cash flows, which in turn requires reliable estimates of sales volume and sales price by the marketing staff, and of material price and availability. The quadrupling of crude oil prices in 2005 was a dramatic example of how changes in raw material costs can greatly influence profitability predictions. Similarly, trends in operating costs must be looked at carefully, especially with respect to whether it is more profitable to reduce operating costs through increased investment, as with automation.

A number of technical decisions are closely related to investment policy and profitability. At the design stage it may be possible to ensure a level of product superiority that is more than that needed by the current market. Later, when competitors enter the market, the superiority would prove useful, but it is not achieved without an initial cost to profitability. Economics generally favor building as large a production facility as the market can absorb. However, this increased profitability is achieved at some risk to maintaining continuity of production should the facility be down for repairs. Thus, there often is a trade-off between the increased reliability of having a number of smaller production facilities over which to spread the operating risk and a single large facility with somewhat higher profitability.

The profitability of a particular product line can be influenced by decisions of cost allocation. Such factors as overhead, utility costs, transfer prices between divisions of a large corporation, or scrap value often require arbitrary decisions for allocation between various products. Thus, the situation often favors certain products and discriminates against others because of cost allocation policies. Sometimes corporations take a position of milking an established product line with a limited future (a "cash cow") in order to stimulate the growth of a new and promising product line. Another profit decision is whether to charge a particular item as a current expense or capitalize it to make a future expense. In a period of inflation there is strong pressure to increase present profitability by deferring costs into the future by capitalizing them.

16

It is argued that a fixed dollar amount deferred into the future will have less consequence in terms of future dollars.

The role of the government in influencing profitability is very great. In the broader sense the government creates the general economic climate through its policies on money supply, taxation, and foreign affairs. It provides subsidies to stimulate selected parts of the economy. Government's regulatory powers have had an increasing influence on profitability in such areas as pollution control, occupational health and safety, consumer protection and product safety, use of federal lands, antitrust, minimum wages, and working hours.

16.8
INFLATION

Since engineering economy deals with decisions based on future flows of money, it is important to consider inflation in the total analysis. Ten dollars in purchases and services in 1980 would require $27 to buy the same items in 2011. While the *inflation rate, f,* in the United States for consumer goods has typically been below 5 percent in the past 20 years, in the early 1980s there were three years of double-digit inflation.

Inflation exists when prices of goods and services are increasing so that a given amount of money buys less and less as time goes by. Interest rates and inflation are directly related. The basic interest rate is about 2 to 3 percent higher than the inflation rate. Thus, in a period of high inflation, not only does the dollar purchase less each month but the cost of borrowing money also rises.

Price changes may or may not be considered in an economic analysis. For meaningful results, costs and benefits must be computed in comparable units. It would not be sensible to calculate costs in 2012 dollars and benefits in 2017 dollars.

Inflation is measured by the change in the Consumer Price Index (CPI), as determined by the U.S. Department of Labor, Bureau of Labor Statistics.[1] The CPI is reported monthly, based on a survey of the price of a "market basket" of goods and services purchased by consumers. In 1984 the CPI was re-centered at 100, and items for price volatile areas such as food and energy were removed, to create the Core CPI. The CPI in January 2011 was 220, compared with 100 in 1984. This means that it would take $220 in 2011 dollars to purchase the same goods and services as in 1984. For design purposes the CPI is less important than the Producer Price Index (PPI), which is discussed in Sec. 17.10.1.

> **EXAMPLE 16.10** The CPI in 1984 was 100.0, and in 2011 it was 220.2. Find the rate of inflation over this period of 27 years.
>
> $$F = P(F/P, i, n) \qquad 220.2 = 100(F/P, i, 27) \qquad (F/P, i, 27) = 2.202$$
>
> and on interpolation between $i = 0.02$ and 0.03 we find $i = 0.0291$.

---

1. www.bls.gov/cpi/cpifaq.htm

Another way to find the inflation rate is to use the equation for *annualized return*.

$$\text{Annualized return} = \left(\frac{\text{current value}}{\text{orginal value}}\right)^{\frac{1}{n}} - 1$$

$$= \left(\frac{220.2}{100.0}\right)^{\frac{1}{27}} = 1.0297 - 1.000 = 0.0297 = 3.0\%$$

(16.28)

Money in one time period t_1 can be brought to the same value as money in another time period t_2 by the equation

$$\text{Dollars in period } t_1 = \frac{\text{Dollars in period } t_2}{\text{Inflation rate between } t_1 \text{ and } t_2}$$

(16.29)

It is useful to define two situations: then-current money and constant-value money. Let the dollars in period t_1 be constant-value dollars. Constant-value represents equal purchasing power at any future time. Current money, in time period t_2, represents ordinary money units that decline in purchasing power with time. For example, if an item cost $10 in 1998 and inflation was 3 percent during the previous year, in constant 1997 dollars, the cost is equal to $10/1.03 = $9.71.

There are three different rates to be considered when dealing with inflation.

Ordinary or inflation-free interest rate i: This is the rate at which interest is earned when effects of inflation have been ignored. This is the interest rate we have used up until now in this chapter.

Market interest rate i_f: This is the interest rate that is quoted on the business news every day. It is a combination of the real interest rate i and the inflation rate f. This is also called the *inflated interest rate.*

Inflation rate f: This is a measure of the rate of change in the value of the currency.

Consider the equation for the present worth of a future sum F in current dollars. F must be first discounted for the real interest rate and then for the inflation rate.

$$P = \frac{F}{(1+i)^n} \frac{1}{(1+f)^n} = F\frac{1}{(1+i+f+if)^n} = \frac{F}{(1+i_f)^n}$$

(16.30)

where $i_f = i + f + if$ is the market interest rate, also called the inflated interest rate.

EXAMPLE 16.11 A project requires an investment of $10,000 and is expected to return, in future, or "then current," dollars, $2500 at the end of year 1, $3000 at the end of year 2, and $7000 at the end of year 3. The monetary (ordinary) interest rate is 10 percent, and the inflation rate is 6 percent per year. Find the net present worth of this investment opportunity.

16

Solution

The inflated interest rate is $0.10 + 0.06 + (0.10)(0.06) = 0.166$; for simplicity we shall use $i_f = 0.17$.

Current-Dollars Approach

Year	Cash Flow	(P/F, 17, n)	Present Worth
0	−10,000	1.00	−10,000
1	2,500	0.8547	2,137
2	3,000	0.7305	2,191
3	7,000	0.6244	4,971
			NPW = −711

Constant-Value-Dollars Approach

Year	Cash Flow*	(P/F, 10, n)	Present Worth
0	−10,000	1.00	−10,000
1	2,358	0.9091	2,144
2	2,670	0.8264	2,206
3	5,877	0.7513	4,415
			NPW = −1,235

*Adjusted by finding present worth in inflated dollars by dividing each cash flow in current dollars by $(1 + 0.06)^n$.

In the current-dollars approach the inflated interest rate is used to discount the cash flows to the present time. For the constant-dollars approach the cash flow is adjusted by [constant (real) $] = [current (actual) $] $(1 + f)^{-n}$.

The difference in the NPWs found by the two treatments is due to using an approximate combined discount rate instead of the more accurate value of $i_f = 0.166$. However, the approximation is justified in view of the uncertainty in predicting the rate of inflation. It should be noted that, for this example, the NPW is +$10 if inflation is ignored. That emphasizes the fact that neglecting the influence of inflation overemphasizes the profitability.

When profitability is measured by the internal rate of return i, the inclusion of the inflation rate f results in an effective rate of return i' based on constant-value money.[1]

$$1 + i = (1 + i')(1 + f)$$
$$i' = i - f - i'f \approx i - f \tag{16.31}$$

Equation (16.31) shows that as a first approximation, the internal rate of return is reduced by an amount equivalent to the average inflation rate.

16

---

1. F. A. Holland and F. A. Watson, *Chem. Eng.*, pp. 87–91, Feb. 14, 1977.

Interest rates are quoted to investors in current money i, but investors generally expect to cover any inflationary trends and still receive an acceptable return. In other words, investors hope to obtain a constant-value interest rate i'. If the calculation is to be made with constant-value money, discounting should be done with the normal interest rate i. If calculations are in terms of current money, then the discount rate should be $i_f = i + f$.

Note that tax allowance for depreciation has a reduced benefit when constant money is used for profitability evaluations. By law, depreciation is defined in terms of current money. Therefore, under high inflation when constant-money conditions are appropriate, a full tax credit for depreciation is not achieved.

Another effect of inflation is that it increases the cash flow because the prices received for goods and services rise as the value of money falls. Even when constant-value money is used, the yearly cash flows should display the current money situation.

16.9
SENSITIVITY AND BREAK-EVEN ANALYSIS

A *sensitivity analysis* determines the influence of each factor in a valuation on the final result. Therefore, it determines which factors are most critical in the economic decision. Since there is a considerable degree of uncertainty in predicting future events like sales volume, salvage value, and rate of inflation, it is important to see how much the economic analysis depends on the magnitude of the estimates. To perform a sensitivity analysis, one factor is varied over a reasonable range and the others are held at their mean (expected) value. The amount of computation involved in a sensitivity analysis of an engineering economy problem can be considerable, but the use of computers has made sensitivity analysis a much more practical endeavor.

A *break-even analysis* often is used when there is particular uncertainty about one of the factors in an economic study. The break-even point is the value for the factor at which the project is just marginally justified.

EXAMPLE 16.12 Consider a $20,000 investment with a 5-year life. The salvage value is $4000, and the minimal acceptable return is 8 percent. The investment produces annual benefits of $10,000 at an operating cost of $3000. Suppose there is considerable uncertainty as to whether the new machinery will survive 5 years of continuous use. Find the break-even point, in terms of life, at which the project just becomes economically viable.

Solution

Using the annual cost method,

$$\$10,000 - 3000 - (20,000 - 4000)(A/P, 8, n) - 4000(0.08) = 0$$

$$(A/P, 8, n) = \frac{6680}{16,000} = 0.417$$

and interpolating in the interest tables gives us $n = 2.8$ years. Thus, if the machine does not last 2.8 years, the investment cannot be justified.

Break-even analysis frequently is used in problems dealing with staged construction. The usual problem is to decide whether to invest more money initially in unused capacity or to add the needed capacity at a later date when needed, but at higher unit costs.

EXAMPLE 16.13 A new plant will cost $100 million for the first stage and $120 million for the second stage at n years in the future. If it is built to full capacity now, it will cost $140 million. All facilities are expected to last 40 years. Salvage value is neglected. Find the preferable course of action.

Solution

The annual cost of operation and maintenance is assumed to be the same for a two-stage construction and full-capacity construction. We shall use a present worth (PW) calculation with a 10 percent interest rate. For full-capacity construction now, PW = $140 million ($140M). For two-stage construction

$$PW = \$100M + \$120M(P/F, 10, n)$$
$$n = \;5 \text{ years: PW} = 100 + 120(0.6201) = \$174M$$
$$n = 10 \text{ years: PW} = 100 + 120(0.3855) = \$146M$$
$$n = 20 \text{ years: PW} = 100 + 120(0.1486) = \$118M$$
$$n = 30 \text{ years: PW} = 100 + 120(0.0573) = \$107M$$

These results are plotted in Fig. 16.8. The break-even point (12 years) is the point at which the two alternatives have equivalent cost. If the full capacity will be needed before 12 years, then full-capacity construction now would be the preferred course of action.

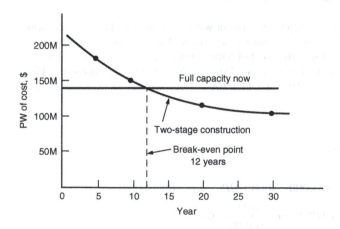

FIGURE 16.8
Break-even plot for Example 16.13.

16.10
UNCERTAINTY IN ECONOMIC ANALYSIS

In the preceding sections we discussed the fact that engineering economy deals chiefly with decisions based on future estimates of costs and benefits. Since none of us has a completely clear crystal ball, such estimates are likely to contain considerable uncertainty. In all of the examples presented so far in this chapter we have calculated best estimates of the future values without taking into account uncertainty.

Now that we are willing to recognize that estimates of the future may not be very precise, there are some ways by which we can guard against the imprecision. The simplest procedure is to supplement your estimated most likely value with an optimistic value and a pessimistic value. The three estimates are combined into a weighted mean value by

$$\text{Weighted mean value} = \frac{\text{optimistic value} + 4\,(\text{most likely value}) + \text{pessimistic value}}{6} \quad (16.32)$$

In Eq. (16.32) the distribution of values is assumed to be represented by a beta frequency distribution. The mean value determined from the equation is used in the economic analysis.

The next level of improved modeling would be to associate a probability with certain factors in the economic analysis. In a sense, by this approach we are transferring the uncertainty from the value itself to the selection of the probability.

> **EXAMPLE 16.14** The expected life of a piece of mining equipment is highly uncertain. The machine costs $40,000 and is expected to have $5000 salvage value. The new machine will save $10,000 per year, but it will cost $3000 annually for operations and maintenance. The service life is estimated to be:
>
> 3 years, with probability = 0.3
> 4 years, with probability = 0.4
> 5 years, with probability = 0.5
>
> **Solution**
>
> For 3-year life: (based on capital recovery)
>
> $$\text{Net annual cost} = (10{,}000 - 3000) - (40{,}000 - 5000)(A/P, 10, 3) - 5000(0.10)$$
> $$= 7000 - 35{,}000(0.4021) - 500 = -8573$$
>
> For 4-year life:
>
> $$\text{Net annual cost} = 7000 - 35{,}000(0.3155) - 500 = -4542$$
>
> For 5-year life:
>
> $$\text{Net annual cost} = 7000 - 35{,}000(0.2638) - 500 = -2733$$
>
> Expected value of net annual cost $= E(\text{AC}) = \sum \text{AC} \times P(\text{AC}) =$
>
> $$-8573(0.3) + [-4542(0.4)] + [-2733(0.3)] = -5207$$

16

16.11
BENEFIT-COST ANALYSIS

An important class of engineering decisions involves the selection of the preferred system design, material, purchased subsystem, and so on etc., when economic resources are constrained. The methods of making cost comparisons and profitability analysis described in Secs. 16.3 and 16.6 are important decision-making tools in this type of situation.

Frequently, comparisons are based on a *benefit-cost ratio,* which relates the desired benefits to the capital investment required to produce the benefits. This method of selecting alternatives is most commonly used by governmental agencies for determining the desirability of public works projects. A project is considered viable when the net benefits associated with its implementation exceed its associated costs. *Benefits* are advantages to the public (or owner), expressed in terms of dollars. If the project involves disadvantages to the owner, these *disbenefits* must be subtracted from the benefits. The costs to be considered include the expenditures for construction, operation, and maintenance, less salvage. Both benefits, disbenefits, and costs must be expressed in common monetary terms by using the present worth or annual cost concept.

$$\text{Benefit-cost ratio (BCR)} = \frac{\text{benefits} - \text{disbenefits}}{\text{costs}} \tag{16.33}$$

A design or project for which BCR < 1 does not cover the cost of capital to create the design. Generally, only projects for which BCR > 1 are acceptable. The benefits used in the BCR would be factors like improved component performance, increased payload through reduced weight, and increased availability of equipment. Benefits are defined as the advantages minus any disadvantages, that is, the net benefits. Likewise, the costs are the total costs minus any savings. The costs should represent the initial capital cost as well as costs of operation and maintenance.

In problems of choosing between several alternatives, the incremental or marginal benefits and costs associated with changes beyond a base level or reference design should be used. The alternatives are ranked with respect to cost, and the lowest-cost situation is taken as the initial reference. This is compared with the next higher-cost alternative by computing the incremental benefit (Benefits of Alternative 2 − Benefits of Alternative 1) and incremental cost (Costs of Alternative 2 − Costs of Alternative 1). If $\Delta B / \Delta C < 1$, then alternative 2 is rejected because the first alternative is superior. Alternative 1 now is compared with alternative 3. If $\Delta B / \Delta C > 1$, then alternative 1 is rejected and alternative 3 becomes the current best solution. Alternative 3 is compared with number 4, and if $\Delta B / \Delta C < 1$, then alternative 3 is the best choice. We should note that this may not be the alternative with the largest overall benefit-cost ratio.

EXAMPLE 16.15 You are asked to recommend a site for a small dam to generate hydroelectric power. The construction cost at various sites is given in the following table. These vary with topography and soil conditions. Each estimate includes $3M for the turbines and generators. The annual benefits from the sale of electricity vary between sites because of stream velocity.

We require an annual return of 10 percent. The life of the dam is infinite for purposes of calculation, so this is treated as a capitalized cost, Sec. 16.3.3. The hydroelectric machinery (H-E) has a 40-year life.

Site	Construction Cost, $M	Cost of Machinery, $M	Cost of Dam, $M	Annual Income, $M
A	9	3	6	1.0
B	8	3	5	0.9
C	12	3	9	1.25
D	6	3	3	0.5

Since the benefit (income) is on an annual basis, we have to convert the cost to an annual basis. Also, we are going to make our decision on an *incremental basis*. We construct Table 16.11, placing the alternatives in order of increasing annual cost of capital recovery (from left to right).

For example, the annual cost of capital recovery for site A is given by $A = P_D i + P_{H\text{-}E}(A/P, 10, 40)$ where P_D is the cost of the dam construction and $P_{H\text{-}E}$ is the cost of the hydroelectric equipment.

$$\text{Site A Annual cost of capital recovery} = 6,000,000(0.10) + 3,000,000(0.1023)$$
$$= 600,000 + 306,900 = \$907,000$$

We note that when compared to not building a dam (zero cost and benefit), $\Delta B/\Delta C$ for site D is less than 1.0. The $\Delta B/\Delta C$ for the next-lowest-cost dam site, B, is greater than 1.0, so it is selected in comparison to not building a dam. The benefit-cost ratio for sites A and C also is greater than 1.0, but now, having found a low-cost qualifying site (B), we need to determine whether the increment in benefits and costs is better than B. We see that on a $\Delta B/\Delta C$ basis, A and C are not better choices than B. Therefore, we select site B.

TABLE 16.11
Site Benefit-Cost Analysis, Example 16.15

	D	B	A	C
Annual cost of capital recovery ($1000)	607	807	907	1207
Annual benefits ($1000)	500	900	1000	1250
Comparison	D to do nothing	B to do nothing	A to B	C to B
Δ capital recovery	607	807	100	400
Δ annual benefits	500	900	100	350
$\Delta B/\Delta C$	0.82	1.11	1.00	0.87
Selection	Do nothing	B	B	B

16

When used in a strictly engineering context to aid in the selection of alternative materials, the benefit-cost ratio is a useful decision-making tool. However, it often is used with regard to public projects financed with tax monies and intended to serve the overall public good. There is a psychological advantage to the BCR concept over the internal rate of return in that it avoids the connotation that the government is profiting from public monies. Here questions that go beyond economic efficiency become part of the decision process. Many of the broader issues are difficult to quantify in monetary terms. Of even greater difficulty is the problem of relating monetary cost to the real values of society.

Consider the case of a hydroelectric facility. The dam produces electricity, but it also will provide flood control and an area for recreational boating. The value of each of the outputs should be included in the benefits. The costs include the expenditures for construction, operation, and maintenance. However, there may be social costs like the loss of virgin timberland or a scenic vista. Great controversy surrounds the assignment of costs to environmental and aesthetic issues and are included in the broad topic of Design for Sustainability (see Chap. 10).

Although benefit-cost analysis is a widely used methodology, it is not without problems. The assumption is that costs and benefits are relatively independent. Basically, it is a deterministic method that does not deal with uncertainty in a major way. As with most techniques, it is best not to try to push it too far. Although the quantitative ratios provided by Eq. (16.33) should be used to the greatest extent possible, they should not preempt the utilization of common sense and good judgment.

16.12
SUMMARY

Engineering economy is a methodology that promotes rational decision making about the allocation of amounts of money at various points in time and in various ways—for example, as a uniform series over time or a single payment in the future. As such, engineering economy accounts for the time value of money.

The basic engineering economy relationship is the compound interest formula that relates the future sum F to the present sum P over n years at an interest rate i.

$$F = P(1+i)^n$$

If P is solved for in this equation, we are discounting the future sum F back to the present time. If the money occurs as equal end-of-the-period amounts A, then

$$F = A\frac{(1+i)^n - 1}{i}$$

If this equation is solved for A, it gives the annual payment to provide a sinking fund to replace worn-out equipment. More important is the annual payment to return the initial capital investment *plus* paying interest on the principal P tied up in the investment, where CRF is the capital recovery factor.

$$A = P\frac{i(1+i)^n}{(1+i)^n - 1} = P(CRF)$$

Engineering economy allows rational decisions to be made about alternative courses of action involving money. To do this, each alternative must be placed on an equivalent basis. There are four common ways of doing this.

- *Present-worth analysis:* All costs or receipts are discounted to the present time to calculate the net present worth. This method works best when the alternatives have a common time period.
- *Annual cost analysis:* The cash flow over time is converted to an equivalent annual cost or benefit. This method works well when the alternatives have different time periods.
- *Capitalized cost analysis:* This is a special case of present worth analysis for a project that exists in perpetuity ($n = \infty$).
- *Benefit-cost ratio:* This method analyzes the costs and benefits of a project on one of the above three bases, and then decides to fund the project if the ratio of benefits to costs is greater than 1.0.

Realistic economic analysis requires consideration of *taxes*, chiefly federal income tax. Accurate determination of the taxable income requires allowance for *depreciation*, the reduction in value of owned assets due to wear and tear or obsolescence. Realistic economic analysis also requires allowance for *inflation*, the decrease in the value of currency over time.

An important use of engineering economy is in determining the profitability of proposed projects or investments. This usually starts with estimating the cash flow to be generated by the project.

$$\text{Cash flow} = \text{net annual cash income} + \text{depreciation}$$

Two common methods of estimating profitability are rate of *return on the investment* (ROI) and *payback period.*

$$\text{ROI} = \frac{\text{average annual net profit}}{\text{capital investment} + \text{working capital}}$$

Payback period is the period of time for the cumulative cash flow to fully recover the initial total capital investment. Both of these methods suffer from not considering the time value of money. A better method to measure profitability is *net present worth.*

$$\text{Net present worth} = \text{present worth of benefits} - \text{present worth of costs}$$

With this method the expected cash flows (both + and −) through the life of the project are discounted to time zero at an interest rate representing the minimum acceptable return on capital. The *internal rate of return* (IRR) is the interest rate for which the net present worth equals zero.

$$\text{Net PW} = \text{PW(benefits)} - \text{PW(costs)} = 0$$

16

Since there is considerable uncertainty in estimating future income streams and costs, engineering economic studies often estimate a range of values and utilize a mean value. Another approach is to place probabilities on the values and use an expected value in the analysis.

NEW TERMS AND CONCEPTS

Annual cost analysis	Discounting to the present	Net present worth
Benefit-cost analysis	Effective interest rate	Nominal interest rate
Capitalized cost	Future value	Payback period
Capital recovery factor	Inflation rate	Present value
Cash flow	Internal rate of return	Present worth analysis
Current dollars	MACRS	Sinking fund factor
Cash flow after taxes	Marginal incremental return	Time value of money
Depreciation	MARR	Uniform annual series

BIBLIOGRAPHY

Blank, L. T., and A. J. Tarquin: *Engineering Economy,* 7th ed., McGraw-Hill, New York, 2012.

Canada, J. R., W. G. Sullivan, D. J. Kulonda, and J. A. White: *Capital Investment Analysis for Engineering and Management,* 3rd ed., Prentice Hall, Englewood Cliffs, NJ, 2005.

Humphreys, K. K.: *Jelen's Cost and Optimization Engineering,* 3rd ed., McGraw-Hill, New York, 1990.

Park, C. S.: *Contemporary Engineering Economics,* 2nd ed., Addison-Wesley, Reading, MA, 1996.

White, J. A., K. E. Case, D. B. Pratt: *Principles of Engineering Economic Analysis,* 5th ed., John Wiley & Sons, New York, 2010.

PROBLEMS AND EXERCISES

The interest tables at www.mhhe.com/dieter are available to help you solve these problems. Also, note that computer spreadsheet software provides most of the financial functions discussed in this chapter. It is recommended that you use a spreadsheet to solve the problems.

16.1. (a) Calculate the amount realized at the end of 7 years through annual deposits of $1000 at 10 percent compound interest.

  (b) What would the amount be if interest were compounded semiannually?

16.2. A young woman purchases a used car. After down payment and allowances, the amount to be paid is $8000. If money is available at 10 percent, what is the monthly payment to pay off the loan in 4 years? What would it be at 4 percent interest?

16

16.3. A new machine tool costs $15,000 and has a $5000 salvage value at the end of 5 years. The interest rate is 10 percent. The annual cost of capital recovery is the annual depreciation charge (use straight-line depreciation) plus the equivalent annual interest charge. Work this out on a year-by-year basis and show that it equals the number obtained quickly by using the capital recovery factor.

16.4. A father desires to establish a fund for his new child's college education. He estimates that the current cost of a year of college education is $20,000 and that the cost will escalate at an annual rate of 4 percent.
  (a)   What amount is needed on the child's eighteenth, nineteenth, twentieth, and twenty-first birthdays to provide for a 4-year college education?
  (b)   If a rich aunt gives $10,000 on the day the child is born, how much in additional funds must be set aside at 4 percent on each of the first through seventeenth birthdays to build up the college fund?

16.5. A major industrialized nation manages its finances in such a way that it runs an annual trade deficit with other countries of $100 billion. If the cost of borrowing is 10 percent, how long will it be before the debt (accumulated deficit) is one trillion dollars ($1000B)? If nothing is done, how long will it take to accumulate the second $1000B debt?

16.6. Machine A costs $8500 and has annual operating costs of $4500. Machine B costs $7000 and has an annual operating cost of $4800. Each machine has an economic life of 10 years. If the minimum required rate of return is 10 percent, compare the advantages of machine A by
  (a)   present worth method,
  (b)   annual cost method, and
  (c)   rate of return on investment.

16.7. Make a cost comparison between two conveyor systems for transporting raw materials.

	System A	System B
Installed cost	$25,000	$15,000
Annual operating cost	6,000	11,000

The service life of each system is 5 years and the write-off period is 5 years. Use straight-line depreciation and assume no salvage value for either system. At what rate of return *after taxes* would B be more attractive than A?

16.8. A resurfaced floor costs $5000 and will last 2 years. If money is worth 10 percent, how long must a new floor costing $19,000 last to be economically justified? Use the capitalized cost method for your analysis.

16.9. You are concerned with the purchase of a heat-treating furnace for the gas carburizing of steel parts. Furnace A will cost $325,000 and will last 10 years; furnace B will cost $400,000 and will also last 10 years. However, furnace B will provide closer control on case depth, which means that the heat treater can shoot for the low side of the specification range on case depth. That will mean that the production rate for furnace B will be 2740 lb/h compared with 2300 lb/h for furnace A. Total yearly production is required to be 15,400,000 lb. The cycle time for furnace A is 16.5 h, and that for furnace B is 13.8 h. The hourly operating cost is $64.50 per h.

Justify the purchase of furnace B on the basis of
(a) payout time and
(b) discounted cash flow rate of return after taxes.
(c) Assume money is worth 10 percent and the tax rate is 50 percent.

16.10. The cost of capital has a strong influence on the willingness of management to invest in long-term projects. In the early 1990s the cost of capital in America was 10 percent and in Japan 4 percent. What must the return be after 2 years on a 2-year investment of $1 million for each of the situations to provide an acceptable return on the investment? Repeat the analysis for a 20-year period. Explain how your results support the opening sentence of this problem.

16.11. Find the present worth of a new experimental fuel cell project based on the following cost estimate.

Initial cost of equipment	$350,000
Development period	5 years
Fixed charges	20 percent of initial equipment cost each year
Variable charges	$40,000 first year, escalating at 6 percent each year with inflation starting at $t = 0$
MARR	$i = 10\%$

16.12. Determine the net present worth of the costs for a major construction project under the following set of conditions:
(a) Estimated cost $300 million over 3 years (baseline case).
(b) Project is delayed by 3 years with rate of inflation 4 percent and interest cost 10 percent.
(c) Project is delayed 6 years with rate of inflation 4 percent and interest costs 10 percent.

16.13. As a new professional employee you need to worry about your retirement many years in the future. Construct a table showing how much you need to have invested, at 4 percent, 8 percent, and 12 percent annual rate of return, to provide each $100 of monthly income. Assume that inflation will increase at 3 percent annually, so the numbers you calculate will be in inflation-adjusted dollars. Calculate the monthly amount needed for a retirement period of 25, 30, 35, and 40 years. Assume that the investments are made in tax-sheltered accounts.

16.14. At what annual mileage is it cheaper to provide your field representatives with cars than to pay them $0.55 per mile for the use of their own cars? The costs of furnishing a car are as follows:

Purchase price	$20,000
Life	4 years
Salvage	$3000
Storage	$400 per year
Maintenance	$0.15 per mile

(a) Assume $i = 10$ percent.
(b) Assume $i = 16$ percent.

16

16.15. To *levelize expenditures* means to create a uniform end-of-year payment that will have the same present worth as a series of irregular end-of-year payments. To illustrate, consider the estimated 5-year maintenance budget for a development lab. Develop a levelized cost assuming that $i = 0.10$ and the annual inflation escalation will be 5 percent.

Year	Maintenance Budget Estimate ($)
1	25,000
2	150,000
3	60,000
4	70,000
5	300,000

16.16. The marketing department made the following estimates about four different product designs. Use benefit-cost analysis to determine which design to pursue.

Design	Unit Manufacturing Cost	Sales Price	Est. Annual Sales
A	$12.50	$25.00	$250,000
B	22.00	40.00	200,000
C	15.00	25.00	250,000
D	15.00	20.00	300,000

16.17. You buy 100 shares of stock in QBC Corp. at $40 per share. It is a good buy, for 4 years, 3 months later you sell these shares of stock for $148 per share. What is the annual rate of return on this fortunate investment?

17

COST EVALUATION

17.1
INTRODUCTION

An engineering design is not complete until we have a good idea of the cost required to build the design or manufacture the product. Generally, among functionally equivalent alternatives, the lowest-cost design will be successful in a free marketplace. The fact that we have placed this chapter on cost evaluation toward the end of the text does not reflect the importance of the subject.

Understanding the elements that make up cost is vital because competition between companies and between nations is fiercer than ever. The world has become a single gigantic marketplace in which newly developing countries with very low labor costs are acquiring technology and competing successfully with the well-established industrialized nations. Maintaining markets requires a detailed knowledge of costs and an understanding of how new technology can lower costs.

Decisions made in the design process commit 70 to 80 percent of the cost of a product. It is in the conceptual and embodiment design stages that a majority of the costs are locked into the product. Thus, this chapter emphasizes how accurate cost estimates can be made early in the design process.

Cost estimates are used in the following ways:

1. To provide information to establish the selling price of a product or a quotation for a good or service.
2. To determine the most economical method, process, or material for manufacturing a product.
3. To become a basis for a cost-reduction program.
4. To determine standards of production performance that may be used to control costs.
5. To provide input concerning the profitability of a new product.

It can be appreciated that cost evaluation inevitably becomes a very detailed and "nitty-gritty" activity. Detailed information on cost analysis rarely is published in

776

the technical literature, partly because it does not make interesting reading but more important, because cost data are highly proprietary. Therefore, the emphasis in this chapter will be on the identification of the elements of costs and on some of the more generally accepted cost evaluation methods. Cost estimation within a particular industrial or governmental organization will follow highly specialized and standardized procedures particular to the organization. However, the general concepts of cost evaluation described here will still apply.

17.2
CATEGORIES OF COSTS

We can divide all costs into two broad categories: product costs and period costs. *Product costs* are those costs that vary with each unit of product made. Material cost and labor cost are good examples. *Period costs* derive their name from the fact that they occur over a period of time regardless of the amount (volume) of product that is made or sold. An example would be the insurance on the factory equipment or the expenses associated with selling the product. Another name for a product cost is *variable cost*, because the cost varies with the volume of product made. Another name for period cost is *fixed cost*, because the costs remain the same regardless of the volume of product made. Fixed costs cannot be readily allocated to any particular product or service that is produced.

Yet another way of categorizing costs is by direct cost and indirect cost. A *direct cost* is one that can be directly associated with a particular unit of product that is manufactured. In most cases, a direct cost is also a variable cost, like materials cost. Advertising for a product would be a direct cost when it is assignable to a specific product or product line, but it is not a variable cost because the cost does not vary with the quantity produced. An *indirect cost* cannot be identified with any particular product. Examples are rent on the factory building, cost of utilities, or wages of the shop floor supervisors. Often the line between direct costs and indirect costs is fuzzy. For example, equipment maintenance would be considered a direct cost if the machines are used exclusively for a single product line, but if many products were manufactured with the equipment, their maintenance would be considered an indirect cost.

Returning to the cost classifications of fixed and variable costs, examples are:

Fixed costs

1. Indirect plant cost
   (*a*) Investment costs
         Depreciation on capital investment
         Interest on capital investment and inventory
         Property taxes
         Insurance
   (*b*) Overhead costs (burden)
         Managers and supervisors not directly associated with a specific product or manufacturing process

17

       Utilities and telecommunications
       Nontechnical services (office personnel, security, etc.)
       General supplies
       Rental of equipment

2. Management and administrative expenses
   (*a*) Share of cost of corporate executive staff
   (*b*) Legal and auditing services
   (*c*) Share of corporate research and development staff
   (*d*) Marketing staff

3. Selling expenses
   (*a*) Sales force
   (*b*) Delivery and warehouse costs
   (*c*) Technical service staff

Variable costs

1. Materials
2. Direct labor (including fringe benefits)
3. Direct production supervision
4. Maintenance costs
5. Quality-control staff
6. Intellectual property licenses
7. Packaging and storage costs
8. Scrap losses and spoilage

Fixed costs such as marketing and sales costs, legal expenses, security costs, financial staff expense, and administrative costs are often lumped into an overall category known as *general and administrative expenses* (G&A expenses). The preceding list of fixed and variable costs is meant to be illustrative of the chief categories of costs, but it is not exhaustive.

The way the elements of cost build up to establish a selling price is shown in Fig. 17.1. The chief cost elements of direct material, direct labor, and any other direct expenses determine the *prime cost*. To it must be added indirect manufacturing costs such as light, power, maintenance, supplies, and factory indirect labor. This is the *factory cost*. The *manufacturing cost* is made up of the factory cost plus general fixed expenses such as depreciation, engineering, taxes, office staff, and purchasing. The *total cost* is the manufacturing cost plus the sales expense. Finally, the *selling price* is established by adding a profit to the total cost.

Another important cost category is *working capital*, the funds that must be provided in addition to fixed capital and land investment to get a project started and provide for subsequent obligations as they come due. It consists of raw material on hand, semifinished product in the process of manufacture, finished product in inventory, accounts receivable,[1] and cash needed for day-to-day operation. The working capital is

_____

1. Accounts receivable represents products that have been sold but for which your company has not yet been paid.

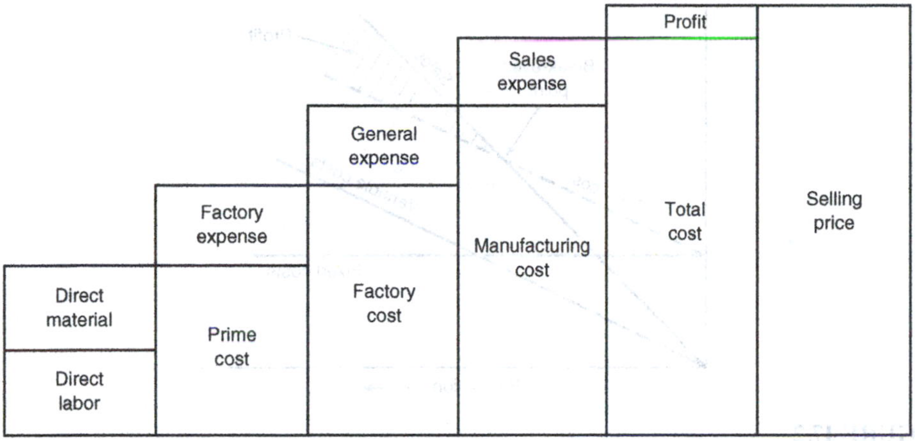

FIGURE 17.1
Elements of cost that establish the selling price.

tied up during the life of the plant, but it is considered to be fully recoverable at the end of the life of the project.

Break-Even Point

Separating costs into fixed and variable costs leads to the concept of the break-even point (BEP), Fig. 17.2. The break-even point is the sales or production volume at which sales and costs balance. Operating beyond the BEP results in profits; operating below the BEP results in losses. Let P be the unit sales price ($/unit), v be the variable cost ($/unit), and f be the fixed cost ($). Q is the number of production units, or the sales volume of products sold. The gross profit Z is given by [1]

$$Z = PQ - (Qv + f)$$

At the break-even point, $Q = Q_{BEP}$ and $Z = 0$

$$Q_{BEP}(P - v) = f \quad \text{Therefore,} \quad Q_{BEP} = \frac{f}{P - v} \tag{17.1}$$

EXAMPLE 17.1 A new product has the following cost structure over one month of operation. Determine the break-even point.

Labor cost 2.50 $/unit   Material cost 6.00 $/unit

G & A expenses $1200 Depreciation on equipment $5000

Factory expenses $800   Sales & distribution overhead $1000

Profit $1.70 $/unit

Total variable cost, v, = $2.50 + 6.00 = 8.50$ $/unit

---

1. Gross profit is the profit before subtracting general and administrative expenses and taxes.

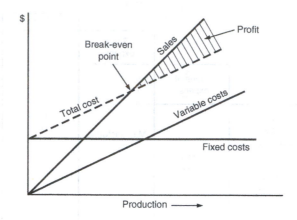

FIGURE 17.2
Break-even curve showing relation between fixed and variable costs and profit before taxes.

Total fixed cost, f, $= 1200 = 5000 + 800 + 1000 = \$8000$

Sales price, $P = 8.50 + 1.70 = \$10.20$

$$Q_{BEP} = \frac{f}{P - v} = \frac{8000}{10.20 - 8.50} = 4706 \text{ units}$$

What sales price would be needed for the product to break even at 1000 units?

$$P = \frac{f + Q_{BEP}v}{Q_{BEP}} = \frac{8000 + 1000(8.50)}{1000} = \frac{16,500}{1000} = 16.50 \$ / \text{units}$$

17.3
THE COST OF OWNERSHIP

Having discussed the various ways to categorize costs, in this section we address the basic contributions to the cost of a product from the viewpoint of the purchaser and owner of the product. In the next section we examine cost from the viewpoint of the manufacturer of the product.[1]

Purchase Price

The sales price to the purchaser S_p can be expressed by:

$$S_p = (nC_U + C_s + P_x)/n \tag{17.2}$$

where n is the total number of product units produced over the lifetime of the product.
C_U is the unit manufacturing cost, Eq. (13.9)
C_s is the total cost of selling the product (marketing, advertising, distribution, salaries of sales personnel, and rebates)

17

1. E. B. Magrab, S. K. Gupta, F. P. McCluskey, and P. A. Sandborn, *Integrated Product and Process Design and Development,* 2d ed., Chap. 3, CRC Press, Boca Raton, FL.

P_x is the sum of all the profits included in the distribution chain, starting with the manufacturer's profit and adding the markups by the distributor (wholesaler) and retailer

From the viewpoint of the product owner, the true cost is greater than the purchase price given in Eq. (17.2). Equation (17.3) lists the costs of ownership that need to be considered for a total cost of ownership C_T for n_p units purchased in a single transaction

$$C_T = n_p(S_p + C_x + C_o + C_{ps}) + C_{sp} + C_t + C_Q \qquad (17.3)$$

where S_p is the price on a per unit basis

C_x is product related taxes such as sales tax, import duty, or tariffs on a per unit basis

C_o is cost of operation on a per unit basis

C_{ps} is support (technical help, maintenance contract, etc.) per unit

C_{sp} is cost of spare parts to support n_p units

C_t is cost of operator training

C_Q is cost of certification or qualification (ISO 9000, UL approval, etc.)

Note that the purchase price of a product often depends on the number of units ordered. For large orders sellers are often willing to reduce their profit for the sake making the sale.

17.4
MANUFACTURING COST

This section extends the discussion of manufacturing cost found in Sec. 13.4.6. The total product cost from the viewpoint of the manufacturer, C_{TM}, is given by

$$C_{TM} = n(C_M + C_L + C_T + C_E + C_W + OH_f) + C_D + C_{WR} + C_Q + OH_c \qquad (17.4)$$

As in Eq. (17.2), n is the total number of units of product sold over its lifetime. The first four terms in parentheses are unit costs defined in Sec. 13.4.6 for materials, labor, tooling, and capital equipment. C_W is the unit cost for disposal, including recycling, for hazardous and nonhazardous waste generated in the manufacturing process, and OH_f is the factory overhead. These are all variable costs since they depend on the number of products manufactured.

The remaining terms are fixed costs. C_D is the one-time design and development cost including design through detail design, costs of reliability testing, software development, and protection of intellectual property. C_{WR} is manufacturer dependent life cycle costs, chiefly warranty costs. C_Q is defined in Sec. 17.2. OH_c is corporate overhead. *Corporate overhead* is based on the costs of running the company that are outside the manufacturing activities. Corporate overhead can include the salaries and fringe benefits of corporate executives, sales and marketing personnel, accounting and finance, legal staff, R&D, corporate engineering and design staff, and the operation of the corporate headquarters building. These costs are allocated to income producing units in the corporation.

Component (part) costs can be divided into two categories: cost of *custom parts* made according to the company's design from semi-finished materials (e.g., bar stock, sheet metal, or plastic pellets) and cost of *standard parts* that are purchased from suppliers. Custom parts are made in the company's own plants or outsourced to suppliers. Standard parts comprise standard components like bearings, motors, electronic chips, and screws, but they also include OEM subassemblies (parts made by suppliers for original equipment manufacturers) like diesel engines for trucks and seats and instrument panels for automobiles. No matter the origin of its manufacture, the cost of making a part includes the material cost, the cost of labor, the cost of tooling, and the cost of tool changing and setup. For outsourced parts, these costs are in the purchase price of the part along with a profit for the supplier.

The cost for manufacturing a product consists of (1) the costs of the parts, as defined by the parts drawings and the bill of materials for the product, (2) the cost for assembling the parts into the product, and (3) overhead costs. Assembly generally requires labor costs for assembly, and often special fixtures and other equipment. Overhead is the cost category that accounts for those costs of manufacture that cannot be directly attributed to each unit of production. This is discussed in Sec. 17.5.

The profit to the manufacturer is Profit = Selling Price − Cost of Product as described in Sec. 17.9. The profit percentage (margin) is determined by the acceptance and competition in the marketplace for the product. For unique products it may be 40 to 60 percent, but 10 to 30 percent is a more typical value.

17.5
OVERHEAD COST

Perhaps no aspect of cost evaluation creates more confusion and frustration in the young engineer than overhead cost. Many engineers consider overhead to be a tax on their creativity and efforts, rather than the necessary and legitimate cost it is. Overhead can be computed in a variety of ways. Therefore, you should know something about how accountants assign overhead charges.

An overhead cost[1] is any cost not specifically or directly associated with the production of identifiable goods or services. The two main categories of overhead costs are factory or plant overhead and corporate overhead. *Factory overhead* includes the costs of manufacturing that are not related to a particular product. *Corporate overhead* is based on the costs of running the company that are outside the manufacturing or production activities. Since many manufacturing companies operate more than one plant, it is important to be able to determine factory overhead for each plant and to lump the other overhead costs into corporate overhead.

One overhead rate may be assigned to an entire factory, but it is more common to designate different overhead rates to departments or cost centers. How the overhead is to be distributed is a management decision that is implemented by accountants.

$$\text{Overhead rate} = OH = \frac{\text{Overhead charges}}{\text{Basis}} \qquad (17.5)$$

1. The term "overhead" arose in early 20th century factories where the bosses were generally located in second-floor offices over the factory floor.

Historically, the most common basis for allocating overhead charges is direct labor dollars or hours. This was chosen in the beginning of cost accounting because most manufacturing was highly labor intensive, and labor represented the major fraction of the total cost. Other bases for distributing overhead charges are machine hours, materials cost, number of employees, and floor space.

EXAMPLE 17.2 A modest-sized corporation operates three plants with direct labor and factory overhead as follows:

Cost	Plant A	Plant B	Plant C	Total
Direct labor	$750,000	400,000	500,000	1,650,000
Factory overhead	900,000	600,000	850,000	2,350,000
Total	1,650,000	1,000,000	1,350,000	4,000,000

In addition, the cost of management, engineering, sales, accounting, etc., is $1,900,000. Find the corporate overhead rate based on direct labor.

$$\text{Corporate overhead rate} = \frac{1,900,000}{1,650,000} = 1.15 = 115\%$$

Then, the allocation of corporate overhead to Plant A would be $750,000(1.15) = $862,500

In the next example of overhead costs, we consider the use of factory overhead in determining the cost of performing a manufacturing operation.

EXAMPLE 17.3 A batch of 100 parts requires 0.75 h of direct labor each in the gearcutting operation. If the cost of direct labor is $20 per h and the factory overhead is 160 percent, determine the total cost of processing a batch.

The cost of processing a batch is: $(100 \text{ parts})(0.75 \text{ h/parts})($20.00 / \text{h}) = 1500

The factory overhead charge is: $\$1500(1.60) = \$2400$

The cost of gear cutting for a batch of 100 parts is processing cost + overhead charge = $1500 + 2400 = $3900. The unit cost is $39.00.

The overhead rate for a particular cost center or remanufacturing process is often expressed in dollars per direct labor hour ($/DLH). In Example 17.3, this is $2400/(100 × 0.75) = 32$/DLH. The allocation of overhead on the basis of DLH sometimes can cause confusion as to the real cost when process improvement results in an increase in manufacturing productivity.

EXAMPLE 17.4 A change from a high-speed steel-cutting tool to a new coated WC tool results in halving the time for a machining operation because the new carbide tool can cut at a much faster speed without "losing its edge." The data for the old tool and the new tool are shown in columns 1 and 2 of the following table. Because the cost of overhead is based on DLH, the cost of overhead apparently is reduced along with the cost of direct labor. The apparent savings per piece is 200 − 100 = $100. However, a little reflection will show that the cost elements that make up the overhead (supervision, tool room, maintenance, etc.) will not change because the DLH is reduced. Since the overhead is expressed

17

as $/DLH, the overhead will actually double if DLH is halved. This true cost is reflected in column (3). Thus, the actual savings per piece is $200 - 160 = \$40$. To take full advantage of the new technology it will be necessary to find creative ways to reduce the costs contributing to overhead or find a more realistic way to define overhead.

	(1) Old Tool	(2) New Tool (Apparent Cost)	(3) New Tool (True Cost)
Machining time, DLH	4	2	2
Direct labor rate, $/h	$20	$20	$20
Direct labor cost	$80	$40	$40
Overhead rate, $/DLH	$30	$30	$60
Cost of overhead	$120	$60	$120
Cost of direct labor and overhead	$200	$100	$160

In many manufacturing situations, overhead allocation based on something other than DLH may be appropriate. Consider a plant whose major cost centers are a machine shop, a paint line, and an assembly department. We see that it is reasonable for each cost center to have a different overhead rate in units appropriate to the function that is performed.

Cost center	Est. Factory Overhead	Est. Number of Units	Overhead Rate
Machine shop	$250,000	40,000 machine hours	$6.25 per machine hour
Paint line	80,000	15,000 gal of paint	$5.33 per gallon of paint
Assembly dept.	60,000	10,000 DLH	$6.00 per DLH

The preceding examples show that the allocation of overhead on the basis of DLH may not be the best way to do it. This is particularly true of automated production systems where overhead has become the dominant manufacturing cost. In such situations, overhead rates are often between 500 and 800 percent of the direct labor cost. In the limit, the overhead rate for an unmanned manufacturing operation would be infinity.

17.6
ACTIVITY-BASED COSTING

In a traditional cost accounting system, indirect costs are assigned to products using direct labor hours or some other unit-based measure to determine overhead cost. We have already seen (Example 17.4) where traditional cost accounting does not accurately represent cost when a large productivity gain has been made. Other types of distortion caused by the cost accounting system are concerned with timing; for example, the R&D costs of future products are charged to products currently being produced, and more complex products will require support costs in greater proportion to their

17

production volume. For these and other reasons a new way of assigning indirect costs called *activity-based costing* (ABC) has been developed.[1]

Rather than assigning costs to an arbitrary reference like direct labor hours or machine hours, ABC recognizes that products incur costs by the *activities* that are required for their design, manufacture, sale, delivery, and service. In turn, these activities create cost by consuming support services such as engineering design, production planning, machine setup, and product packing and shipping. To implement an ABC system you must identify the major activities undertaken by the support departments and identify a *cost driver* for each. Typical cost drivers might be hours of engineering design, hours of testing, number of orders shipped, or number of purchase orders written.

EXAMPLE 17.5 A company assembles electronic components for specialized test equipment. Two products A75 and B20 require 8 and 10.5 min, respectively, of direct labor, which costs $16 per hour. Product A75 consumes $35.24 of direct materials and product B20 consumes $51.20 of direct materials.

Using a traditional cost accounting system where all overhead costs are allocated to direct labor hours at a rate of $230 per DLH, the cost of a unit of product would be:

$$\text{Direct labor cost} + \text{direct material cost} + \text{overhead cost}$$

For product A75: $16(8/60) + \$35.24 + 230(8/60) = 2.13 + 35.24 + 30.59 = \$67.96$

For product B20: $16(10.5/60) + \$51.20 + \$230(10.5/60) = 2.80 + 51.20 + 40.25 = \$94.25$

In an attempt to get a more accurate estimate of costs, the company turns to the ABC approach. Six cost drivers are identified for this manufacturing system.[2]

Activity	Cost Driver	Rate
Engineering	Hours of engineering services	$60.00 per hour
Production setup	Number of setups	$100.00 per setup
Materials handling	Number of components	$0.15 per component
Automated assembly	Number of components	$0.50 per component
Inspection	Hours of testing	$40.00 per hour
Packing and shipping	Number of orders	$2.00 per order

The level of activity of each cost driver must be obtained from cost records.

	Product A75	Product B20
Number of components	36	12
Hours of engineering services	0.10	0.05
Production batch size	50	200
Hours of testing	0.05	0.02
Units per order	2	25

---

1. R.S. Kaplan and R.E. Cooper, *Cost and Effect: Using Integrated Cost Systems to Drive Profitability and Performance,* Harvard Business School Press, Boston, MA, 1998.
2. In a real ABC study there would be many more activities and cost drivers than are used in this example.

In building the cost comparison between products we start with direct labor and direct material costs, using the traditional cost accounting method. Then we turn to ABC in allocating the overhead costs. We apply the activity level of the cost drivers to the cost rate of the driver. For example, for Product A75,

Engineering services: 0.10 h/unit $\times$ \$60/h = \$6.00/unit

Production setups: $100 \dfrac{\$}{\text{setup}} \dfrac{1 \ \text{setup}}{50 \ \text{unit}} = 2.00 \dfrac{\$}{\text{unit}}$

Since number of units per setup equals batch size.

Materials handling: $36 \dfrac{components}{unit} \times 0.15 \dfrac{\$}{component} = 5.40 \dfrac{\$}{unit}$

Packing and shipping: $2.00 \dfrac{\$}{\text{order}} \dfrac{1 \ \text{order}}{2 \ \text{units}} = 1.00 \dfrac{\$}{unit}$

Comparison of the Two Products on Activity-Based Costing

	A75	B20
Direct labor	2.13	2.80
Direct materials	35.24	51.20
Engineering	6.00	3.00
Production setups	2.00	0.50
Materials handling	5.40	1.80
Assembly	18.00	6.00
Testing	2.00	0.80
Packing and shipping	1.00	0.80
	\$71.77	\$66.90

We see that by using ABC, we find that product B20 is less costly to produce. This shift has come entirely from changing the allocation of overhead costs from DLH to cost drivers based on the main activities in producing the product. B20 incurs lower overhead charges chiefly because it is a less complex product using fewer components and requiring less support for engineering, materials handling, assembly, and testing.

Using ABC leads to improved product-based decisions through more accurate cost data. This is especially important when manufacturing overhead accounts for a large fraction of manufacturing costs. By linking financial costs with activities, ABC provides cost information to complement nonfinancial indicators of performance like quality. The preceding data clearly show the need to reduce the number of components to lower the cost of materials handling and assembly. On the other hand, using only a single cost driver to represent an activity can be too simple. More complex factors can be developed, but at a considerable cost in the complexity of the ABC system.

ABC cost accounting is best used when there is diversity in the product mix of a company in terms of such factors as complexity, different maturity of products, production volume or batch sizes, and need for technical support. Computer-integrated manufacturing is a good example of a place where ABC can be applied because it has such high needs for technical support and such low direct labor costs.

17

There is more work in using ABC than traditional cost accounting, but this is partly compensated by the use of computer technology to accumulate the cost data. A big advantage of ABC is that when the system is in place it points to those areas of indirect cost where large savings could be made. Thus, ABC is an important component of a total quality management program aimed at process improvement and cost reduction.

17.7
METHODS OF DEVELOPING COST ESTIMATES

The methods to develop cost evaluations fall into three categories: (1) analogy, (2) parametric and factor methods, and (3) methods engineering.

17.7.1 Analogy

In cost estimation by analogy, the future costs of a project or design are based on past costs of a similar project or design, with due allowance for cost escalation and technical differences. The method therefore requires a database of experience or published cost data. This method of cost evaluation commonly is used for feasibility studies of chemical plants and process equipment.[1] When cost evaluation by analogy is used, future costs must be based on the same state of the art. For example, it would be valid to use cost data on a 777 jet transport aircraft to estimate costs for a larger 777, but it would not be correct to use the same data to predict the cost of the Boeing 787 because the main structures have changed from riveted aluminum construction to autoclave-bonded polymer-graphite fiber construction.

A concern with determining cost by analogy is to be sure that costs are being evaluated on the same basis. Equipment costs often are quoted FOB (free on board) the manufacturer's plant location, so delivery cost must be added to the cost estimate. Costs sometimes are given for the equipment not only delivered to the plant site but also installed in place, although it is more usual for costs to be given FOB some from the shipping point.

17.7.2 Parametric and Factor Methods

In the *parametric* or statistical approach to cost estimation, techniques such as regression analysis are used to establish relations between system cost and key parameters of the system, such as weight, speed, and power. This approach involves cost estimation at a high level of aggregation, so it is most helpful in the problem definition stage of conceptual design. For example, the cost of developing a turbofan aircraft engine might be given by

$$C = 0.13937 x_1^{0.7435} x_2^{0.0775}$$

where C is in millions of dollars, x_1 is maximum engine thrust, in pounds, and x_2 is the number of engines produced by the company. Cost data expressed in this empirical

---

1. M.S. Peters, K.D. Timmerhaus, and R.E. West, *Plant Design and Economics for Chemical Engineers,* 5th ed., McGraw-Hill, New York, 2003.

form can be useful in trade-off studies in the concept design phase. Parametric cost studies are often used in feasibility studies of large military systems. One must be careful not to use models of this type outside the range of data for which they apply.

Factor methods are related to parametric studies in that they use empirical relationships based on cost data to find useful predictive relationships. Equation (17.6) represents a factor method for determining the unit manufacturing cost of a part.[1]

$$C_u = VC_{mv} + P_c(C_{mp} \times C_c \times C_s \times C_{ft})$$ (17.6)

where C_u is the manufacturing cost to make one unit of a part

V is the volume of the part

C_{mv} is the material cost per unit volume

P_c is the basic cost to process an ideal shape by a particular process

C_{mp} is a cost factor that indicates the relative ease with which a material can be shaped in a particular process

C_c is a relative cost associated with shape complexity

C_s is a relative cost associated with achieving minimum section thickness

C_{ft} is the cost of achieving a specified surface finish or tolerance.

It is important to understand that equations based on cost factors are not just made up in a haphazard fashion. Basic physics and engineering logic are carried as far as possible before employing empirical analysis of data. Equation (17.6) is aimed at estimating the cost to make a part in the conceptual design phase when many of the details of the features of the part have not been established. Its goal is to use part cost as a way of selecting the best process to make the part by including more design details than are included in the model for manufacturing cost described in Sec. 13.4.6. Equation (17.6) recognizes that material cost is often the main cost driver in part cost, so it separates this factor from those associated with the process. Equation (17.6) introduces P_c, the basic cost to make an "ideal shape" by one of several common manufacturing processes. This factor aggregates all of the costs of production (labor, tooling, capital equipment, overhead) as a function of the production volume. Note that for a specific company, P_c could be decomposed into an equation representing its actual cost data. The factors in the parentheses are all factors that increase the cost over the ideal case.[2] Of these, shape complexity and tolerances (surface finish) have the greatest effect.

Models for developing cost for manufacturing use physics-based principles to determine such process parameters as the forces, flow rates, or temperatures involved. Eventually empirical cost factors are needed when dealing with process details. For example, the number of hours for machining a metal mold to be used in injection molding is given by[3] $M = 5.83(x_i + x_o)^{1.27}$ where x_i and x_o are contours of the inner and outer surfaces of the mold, respectively, and in turn, are given by x_i or $x_o = 0.1 N_{SP}$ where N_{SP} is the number of surface patches or sudden changes in slope or curvature of the surface.

1. K. G. Swift and J. D. Booker, *Process Selection*, 2d ed., Butterworth-Heinemann, Oxford, UK, 2003.
2. Building a model by starting with an ideal case and degrading it with individual factors is a common approach in engineering model building. In Sec. 12.3.4 we started with an ideal endurance limit and reduced its value by applying factors for stress concentration, diameter, and surface finish.
3. G. Boothroyd, P. Dewhurst, and W. Knight, *Product Design for Manufacture and Assembly,* 2d ed., pp. 362–64, Marcel Dekker, New York, 2002.

Factor methods of cost evaluation are used for estimating costs in the early stages of embodiment design and are employed in the concurrent costing software described in Sec. 13.9.1. For more details on parametric cost models, see the *Parametric Cost Estimation Handbook*, cost.jsc.nasa.gov/pcehhtml/pceh.htm.

17.7.3 Detailed Methods Costing

Once the detailed design is completed and the final detailed drawings of the parts and assemblies have been prepared, it is possible to prepare a cost evaluation to ±5 percent accuracy. This approach is sometimes called methods analysis, process flow method, or the industrial engineering approach. The cost evaluation requires a detailed analysis of every operation to produce the part and a good estimate of the time required to complete the operation. A similar method is used to determine the costs of buildings and civil engineering projects.[1]

At the outset of developing the cost estimate, the following information should be available:

- Total quantity of product to be produced
- Schedule for production
- Detailed drawings and/or CAD file
- Bill of materials (BOM)

In complicated products the bill of materials may be several hundred lines. This makes it important that a system be in place to keep track of all parts and make sure none are left out of the cost analysis.[2] The BOM should be arranged in layers, starting with the assembled product, then the first layer of subassemblies, then the subassemblies feeding into this layer, all the way down to the individual parts. The total number of a given part in an assembled product is the number used at the lowest level multiplied by the number used at each other level of assembly. The total number of each part to be made or purchased is the number per product unit times the total number of products to be produced.

Detailed methods costing analysis is usually prepared by a process planner or a cost engineer. Such a person must be very familiar with the machines, tooling, and processes used in the factory. The steps to determine cost to manufacture a part are:

1. *Determine the material costs.* Since the cost of material makes up 50 to 60 percent of the cost of many products, this is a good place to start. Usually the cost of material is measured on a mass basis, but sometimes it is based on volume, and in other instances, as when machining bar stock, it might be measured per foot. Issues concerning the cost of materials were discussed in Sec. 11.5 and Sec. 13.4.6.

    It is important to account for the cost of material that is lost in the form of scrap. Most manufacturing processes have an inherent loss of material. Sprues and

---

1. Historical cost data is published yearly by R. S. Means Co. and in the *Dodge Digest of Building Costs.* Also see P. F. Ostwald, *Construction Cost Analysis and Estimating,* Prentice Hall, Upper Saddle River, NJ, 2001.
2. P. F. Ostwald, *Engineering Cost Estimating,* 3d ed., pp. 295–97, Prentice Hall, Upper Saddle River, NJ, 1992.

risers that are used to introduce molten material into a mold must be removed from castings and moldings. Chip generation occurs in all machining processes, and metal stamping leaves unused sheet scrap. While most scrap materials can be recycled, there is an economic loss in all cases.

2. *Prepare the operations route sheet.* The route sheet is a sequenced list of all operations required to produce the part. An operation is the smallest category of work done on the workpiece while on one machine or in one holding device on the machine. Several different workpiece faces may be shaped in one operation. The term step is also used in place of operation. For example, an operation on an engine lathe might be to face the end of a bar, then rough turn the diameter to 0.610 in. and finish machine to 0.600. The *process* is the sequence of operations from the time the workpiece is taken from inventory until it is completed and placed in finished goods inventory. Part of developing the route sheet is to select the actual machine in the shop to perform the work. This is based on availability, the capacity to deliver the necessary force, depth of cut, or precision required by the part specification.

3. *Determine the time required to carry out each operation.* Whenever a new part is first made on a machine, there must be a *setup period* during which old tooling is taken out and new tooling is installed and adjusted. Depending on the process, this can be a period of minutes or several days, but two hours is a more typical setup time. Each process has a *cycle time,* which consists of loading the workpiece into the machine, carrying out the operation, and unloading the workpiece. The process cycle is repeated many times until the number of parts required for the batch size has been made. Often there is a downtime for shift change or for maintenance on the machine or tooling.

   Databases of standard times to perform small elements of typical operations are available.[1] Computer software with databases of operation times and cost calculation capability are available for most processes. If the needed information cannot be found in these sources, then carefully controlled time studies must be made.[2] A sampling of standard times for elements of operations is given in Table 17.1.

   An alternative to using standard times for operation elements is to calculate the time to complete an operation element with a physical model of the process. These models are well developed for machining processes[3] and for other manufacturing processes.[4] An example of the use of this method for metal cutting is given in Sec. 17.13.1.

4. *Convert time to cost.* The times for each element in each operation are added to find the total time to complete each operation of the process. Then the time is multiplied by the fully loaded wage rate ($/h) to give the cost of labor. A typical

1. P. F. Ostwald, *AM Cost Evaluator,* 4th ed., Penton Publishing Co., Cleveland, OH, 1988; W. Winchell, *Realistic Cost Estimating for Manufacturing,* 2d ed., Society of Manufacturing Engineers, Dearborn, MI, 1989.

2. B. Niebel and A. Freivalds, *Methods, Standards, and Work Design,* 11th ed., McGraw-Hill, New York, 2003.

3. G. Boothroyd and W. A. Knight, *Fundamentals of Machining and Machine Tools,* 2d ed., Chap. 6, Marcel Dekker, New York, 1989.

4. R. C. Creese, *Introduction to Manufacturing Processes and Materials,* Marcel Dekker, New York, 1999.

TABLE 17.1
A Sampling of Cycle-Time Elements

Operation Element	Minutes
Set up a lathe operation	78
Set up a drilling fixture	6
Brush away chips	0.14
Start or stop a machine tool	0.08
Change spindle speed	0.04
Index turret on turret lathe	0.03

product will require parts made by different processes, and some parts purchased rather than made in-house. Typically, different labor rates and overhead rates prevail in different cost centers of the factory.

EXAMPLE 17.6 A ductile cast iron V-belt pulley driven from a power shaft is made in a batch of 600 units. The material cost is $50.00 per unit. Table 17.2 gives estimates of labor hours, labor rates, and overhead charges. Determine the unit product cost.

TABLE 17.2
Process Plan for Ductile Iron Pulley (Batch Size 600 Units)

Cost Center	Operation	(1) Setup Time h/batch	(2) Cycle Time h/100 units	(3) Time to Finish Batch, h	(4) Wage Rate $/h	(5) Batch Labor Cost	(6) Batch Over-head	(7) Labor & Overhead Per Batch	(8) Unit Cost
Outsource	Purchase 600 units, rough castings, part no. 437837								$50.00
Machine Shop—lathe	Total costs for operation	2.7	35	212.7	$32.00	$6806	$7200	$14,006	$23.34
	1. Machine faces								
	2. Machine V-groove in OD								
	3. Rough machine hub								
	4. Finish machine ID of bore								
Machine Shop—drills	1. Drill and tap 2 holes for set screws	0.1	5	30.1	$28.00	$843	$1050	$1893	$3.15
Finishing Dept.	Total cost for operation	6.3	12.3	80.1	$18.50	$1482	$3020	$4502	$7.50
	1. Sand blast								
	2. Paint								
	3. Install 2 set screws								0.06
Totals		9.1	52.3	322.9		$9131	$11,616	$20,401	$84.05

17

The estimates of the standard costs for the elements of each operation give the cycle time per 100 units given in column (2). In a similar way the setup costs for a batch are estimated in column (1) for each cost center. Multiplying (2) by 6 (the batch size is 600) plus adding in the setup cost gives the time to produce a batch of 600 units. With this and the wage rate (4), we determine the batch labor cost, column (5). The overhead cost for each cost center, based on a batch of 600 units, is given in (6). Adding (5) and (6) gives all of the in-house costs for that batch. These costs are placed on a per-unit basis in (8). Note that the unit cost of $50.00 for the rough casting that was purchased from an outside foundry includes the overhead costs and profit for that company. The unit costs for the completed part developed in Table 17.2 do not include any profit, since that will be determined for the entire product for which the pulley is only one part.

Developing costs by an aggregated method is a lot of work, but computer databases and calculation aids make it much less of an onerous task than in the past. As already noted, this cost analysis requires a detailed process plan, which cannot be made until decisions on all of the design features, tolerances, and other parameters have been made. The chief drawback, then, is if a part cost turns out too high it may not be possible to make design changes to correct the problem. As a result, considerable effort is being given to cost methods that are capable of determining and controlling costs as the design process is being carried out. This topic, design to cost, is discussed in Sec. 17.12.

17.8
MAKE-BUY DECISION

One of the uses of a detailed cost evaluation method such as was described in Example 17.6 is to decide whether it is less costly to manufacture a part in-house than to purchase it from an outside supplier. In that example, where the rough casting was bought from an outside foundry, it was decided that the volume of cast parts that will be used by the manufacturer does not justify the cost of equipping a foundry and hiring the expertise to make quality castings.

The parts that go into a product fall into three categories related to whether they should be made in-house or purchased from suppliers.

- Parts for which there is no in-house process capability obviously need to be purchased from suppliers.
- Parts that are critical to the quality of the product, involve proprietary manufacturing methods or materials, or involve a core technical competency need to be made in-house.
- Parts other than those in the previous categories, the majority of parts, offer no compelling reason to either use in-house manufacture or purchase from a supplier. The decision is usually based on which approach is least costly to obtain quality parts. Today the make-buy decision is being made not just with respect to suppliers in the vicinity of the manufacturer's plant, but in locations anywhere in the world where low-cost labor and manufacturing skill exist. This phenomenon of *offshoring* is made possible by rapid communication via the Internet and cheap

water transportation with container ships. It has led to a boom in low-cost manufacturing of consumer goods in China and elsewhere in Asia.

Many factors other than cost enter into a make-buy decision.

Advantages of Outsourcing

- Lower cost of manufacture provides lower prime costs (materials and labor), especially with overseas suppliers.
- Suppliers can provide special expertise in design and manufacturing that the product developer may not have.
- Outsourcing provides increased manufacturing flexibility due to reduction in fixed costs. This lowers the breakeven point for a product.
- Manufacturing in a foreign country may result in access to a foreign market for the product.

Disadvantages of Outsourcing

- Outsourcing results in a loss of in-house design and manufacturing knowledge that is transferred to the supplier, and maybe to your competitors.
- It is more difficult to improve design for manufacture when in-house manufacturing capability is gone.
- Possible unsatisfactory quality.
- In offshoring the supply chain is much longer. There is always a danger of delays in supply due to delay in gaining entry into port, strikes on the docks, and severe weather in transit.
- Offshoring may present such issues as currency exchange, communication in a different language and business culture, and the added expense in coordinating with an external supplier.

17.9
PRODUCT PROFIT MODEL

The total cost for manufacturing n units a product was given in Eq. (17.4). Keeping Eq. (17.4) in mind, we can develop a simple cost model of product profitability.

(1) Net sales = (number of units sold) × (sales price)

(2) Cost of product sold = (number of units sold) × (unit cost*)

   *terms inside ( ) in Eq. (17.4).

(3) Gross margin = (1) − (2) = Net sales − Cost of product sold

(4) Operating expenses = Terms outside ( ) in Eq. (17.4)

(5) Operating income (profit) = (3) − (4) = gross margin − operating expenses

   Percentage profit = (profit/net sales) × 100

Unit cost will be arrived at from Eq. (17.4) and by the methods discussed in Sec. 17.7. The number of units sold will be estimated by the marketing staff. Other costs will be provided by cost accounting or historical corporate records.

Note that the profit determined by the profit model is not the "bottom line" net profit found on the income statement of the annual report of a company. The net profit is the aggregate profit of many product development projects. To get from the operating income of a company to its net profit, many additional deductions must be made, the chief of which are the interest on borrowed debt and federal and state tax payments.

It is convenient to build the profit model with a computer-based spreadsheet program. Figure 17.3 shows a typical cost projection for a consumer product. Note that the sales price is projected to decline slightly as other competitors come into the market, but the sales volume is expected to increase over most of the life of the product as it gains acceptance through use by customers and advertising. This results in a nearly constant gross margin over the life of the product.

The development cost is broken out as a separate item in Fig. 17.3. The product was developed in a two-year period spread over 2012 to 2014. After that a modest annual investment was made in small improvements to the product. It is encouraging to see that the product was an instant hit and recovered its development cost in 2014, the year it was introduced to the market. This is a strong indication that the product development team understood the needs of the customer and satisfied them with its new product.

Considerable marketing and sales activity began the year of product introduction and are planned to continue at a high level throughout the expected life of the product. This is a reflection of the competition in the marketplace and the recognition that a company must be aggressive in placing its products before the customer. The "other" category in the spreadsheet mostly comprises factory and corporate overhead charges.

Trade-Off Studies

The four key objectives associated with developing a new product are:

- Bringing the cost of the product under the agreed-upon target cost.
- Producing a quality product that exceeds the expectation of the customer.
- Conducting an efficient product development process that brings the product to market, on schedule.
- Completing the development process within the approved budget for the product.

A product development team must recognize that not everything will go smoothly during the development process. There may be delays in the delivery of tooling, costs for outsourced components may increase because of higher fuel costs, or several parts may not interface in assembly according to specification. Whatever the reason, when faced with issues such as these, it is helpful to be able to estimate the impact of your plan to fix the problem on the profitability of your product. This is done by creating trade-off decision rules using the spreadsheet cost model.

17

	Year								
	2012	2013	2014	2015	2016	2017	2018	2019	2020
Sales Price			$180.00	$178.00	$175.00	$173.00	$170.00	$168.00	$165.00
Unit Sales			100,000	110,000	120,000	130,000	130,000	120,000	110,000
Net Sales			$18,000,000	$19,580,000	$21,000,000	$22,490,000	$22,100,000	$20,160,000	$18,150,000
Unit Cost			$96.00	$95.00	$94.00	$93.00	$92.00	$92.00	$92.00
Cost of Product Sold			$9,600,000	$10,450,000	$11,280,000	$12,090,000	$11,960,000	$11,040,000	$10,120,000
Gross Margin ($)			$8,400,000	$9,130,000	$9,720,000	$10,400,000	$10,140,000	$9,120,000	$8,030,000
Gross Margin (%)			46.67%	46.63%	46.29%	46.24%	45.88%	45.24%	44.24%
Development Cost	$750,000	$1,500,000	$750,000	$350,000	$350,000	$250,000	5250,000	$250,000	$250,000
Marketing			$2,340,000	$2,545,400	$2,730,000	$2,923,700	$2,873,000	$2,620,800	$2,359,500
Other			$2,160,000	$2,349,600	$2,520,000	$2,698,800	$2,652,000	$2,419,200	$2,178,000
Total Operating Expense	$750,000	$1,500,000	$5,250,000	$5,245,000	$5,600,000	$5,872,500	$5,775,000	$5,290,000	$4,797,500
Operating Income (Profit)	($750,000)	($1,500,000)	$3,150,000	$3,885,000	$4,120,000	$4,527,500	$4,365,000	$3,830,000	$3,242,500
Op Income (%)			17.50%	19.84%	19.62%	20.13%	19.75%	19.00%	17.87%
Cumulative Op Income	($750,000)	($2,250,000)	$900,000	$4,785,000	$8,905,000	$13,432,500	$17,797,500	$21,627,500	$24,870,000

Cumulative Sales	$141,480,000
Cumulative Gross Margin	$64,940,000
Cumulative Op Income	$24,870,000
Average % Gross Margin	45.90%
Average % Op Income	17.58%

FIGURE 17.3
Cost projections for a consumer product.

795

TABLE 17.3
Trade-Off Decision Rules Based on Deviation from Baseline Conditions

Type of Shortfall	Baseline Oper. Income	Reduced Oper. Income	Cumulative Impact on Profit	Rule of Thumb
50% development cost overrun	$24,870,000	$23,370,000	−$1,500,000	$30,000 per %
5% overrun on product cost	$24,870,000	$21,043,000	−$3,827,000	$765,400 per %
10% reduction in sales due to performance issues	$24,870,000	$21,913,000	−$2,957,000	$295,700 per %
3-month delay in product introduction to market	$24,870,000	$23,895,000	−$957,000	$975,000 per %

Figure 17.3 represents the baseline profit model if everything goes according to plan. Other cost models can easily be determined for typical shortfalls from plan. For example:

- A 50% cost overrun in development cost.
- A 5% cost overrun in unit cost.
- A 10% reduction in sales due to poor performance and customer acceptance.
- A 3-month delay in introducing the product into the marketplace.

Table 17.3 shows the impact on the cumulative operating income as a result of these changes from the baseline condition.

The trade-off *rule of thumb* is based on the assumption that changes are linear and each shortfall is independent of the others. For example, if a 10 percent decrease in sales causes a $2,957,000 reduction in cumulative operating profit, then a 1 percent decrease in sales will decrease operating profit by $295,700. Note that the trade-off rules apply only to the particular case under study. They are not universal rules of thumb.

> **EXAMPLE 17.7** An engineer estimated that a savings of $1.50 per unit could be made by eliminating the balancing operation on the fan of the product for which data is given in Table 17.3. However, marketing estimated there would be a 5 percent loss in sales due to increased vibration and noise of the product. Use the trade-off rules to decide whether the cost saving is a good idea.
>
> Potential benefit: The unit cost is $96.00. The percentage saving is $1.50/96 = 0.0156 = 1.56\%$
>
> $1.56 \times \$765,400$(per 1% change in unit cost) = $1,194,000
>
> Potential cost: $5 \times \$295,700 = \$1,478,500$.
>
> Benefit/cost is close but says that the potential cost in lost sales outweighs the savings. On the other hand, the estimate of lost sales of 5 percent is just an educated guess. One strategy might be to ask the engineer to do the cost saving estimate in greater detail, and if the cost saving holds up, make a trial lot that are sold in a limited geographic area where complaints and returns could be closely monitored. However, before doing this the

17

product made without fan balancing needs to be carefully studied for noise and vibration with regard to OSHA requirements.

17.9.1 Profit Improvement

Three strategies commonly used to achieve increased profits are: (1) increased prices, (2) increased sales, (3) and reduced cost of product sold. Example 17.8 shows the impact of changes in these factors on the profit using the profit model described in the previous section.

EXAMPLE 17.8 *Case A* is the current distribution of cost elements for the product.
Case B shows what would happen if price competition would allow a 5 percent increase in price without loss in units sold. The increased income goes right to the bottom line.
Case C shows what would happen if sales were increased by 5 percent. There would be a 5 percent increase in the four cost elements, while unit cost remains the same. Costs and profits rise to the same degree and percentage profit remains the same.
Case D shows what happens with a 5 percent productivity improvement (5 percent decrease in direct labor) brought about by a process-improvement program. The small increase in overhead results from the new equipment that was installed to increase productivity. Note that the profit per unit has increased by 10 percent.
Case E shows what happens with a 5 percent decrease in the cost of materials or purchased components. About 65 percent of the cost content of this product is materials. This cost reduction could result from a design modification that allows the use of a less expensive material or eliminates a purchased component. In this case, barring a costly development program, all of the cost savings goes to the bottom line and results in a 55 percent increase in the unit profit.

	Case A	Case B	Case C	Case D	Case E
Sales price	$100	$105	$100	$100	$100
Units sold	100	100	105	100	100
Net sales	$10,000	$10,500	$10,500	$10,000	$10,000
Direct labor	$1,500	$1,500	$1,575	$1,425	$1,500
Materials	$5,500	$5,500	$5,775	$5,500	$1,225
Overhead	$1,500	$1,500	$1,575	$1,525	$1,500
Cost of product sold	$8,500	$8,500	$8,925	$8,450	$8,225
Gross margin	$1,500	$2,000	$1,575	$1,550	$1,775
Total operating expenses	$1,000	$1,000	$1,050	$1,000	$1,000
Pretax profit	$500	$1,000	$525	$550	$775
Percentage profit	5%	9.5%	5%	5.5%	7.75%

A fourth profit improvement strategy, not illustrated by the example, is to upgrade the mix of products made and sold by the company. With this approach, greater emphasis is given to products with higher profit margins while gradually phasing out the product lines with lower profit margins.

17

17.10
REFINEMENTS TO COST ANALYSIS METHODS

Several refinements to cost estimating methods have appeared over the years aimed at giving more accurate cost evaluations. In this section we discuss (1) adjustments for cost inflation, (2) relationships between product or part size and cost, and (3) reduction in manufacturing costs because of learning.

17.10.1 Cost Indexes

Because the purchasing power of money decreases with time, all published cost data are out of date. To compensate for this, cost indexes are used to convert past costs to current costs. The cost at time 2 is the cost at time 1 multiplied by the ratio of the cost indexes.

$$C_2 = C_1 \left(\frac{\text{Index @ time 2}}{\text{Index @ time 1}} \right) \qquad (17.7)$$

The most readily available cost indexes are:

- Consumer Price Index (CPI)—gives the price of consumer goods and services
- Producer Price Index (PPI)—measures the entire market output of U.S. producers of goods. The Finished Goods Price Index of the PPI is roughly split between durable goods (not in the CPI) and consumer goods. No services are measured by the PPI. Both the CPI and PPI are available at www.bls.gov.
- The *Engineering News Record* provides indexes on general construction costs.
- The Marshall and Swift Index, found in *Chemical Engineering* magazine, provides an index of industrial equipment costs. The same magazine publishes the Chemical Engineering Plant Equipment Index, which covers equipment such as heat exchangers, pumps, compressors, piping, and valves.

Many trade associations and consulting groups also maintain specialized cost indexes.

> **EXAMPLE 17.9** An oilfield diesel engine cost $5500 when it was purchased in 1982. What did it cost to replace the diesel engine in 1997?
>
> $$C_{1997} = C_{1982} \left(\frac{I_{1997}}{I_{1982}} \right) = 5500 \left(\frac{156.8}{121.8} \right) = 5500(1.29) = \$7095$$
>
> What did it cost to replace the engine in 2006 if the *finished goods price index* for oil and gas field machinery was 210.3?
>
> $$C_{2006} = C_{1997} \left(\frac{210.3}{156.8} \right) = 7095(1.34) = \$9516$$
>
> We see there was an average increase in price of 1.9 percent over the first 15 years, and a 3.8 percent yearly average over the last 9 years. This is a reflection of the rapid acceleration of oil and gas business in the recent past. Similar calculations for the automobile parts business would see hardly any price increase since 1997, an indication of the fierce competition in this relatively stagnant market.

You should be aware of some of the pitfalls inherent in using cost indexes. First, you need to be sure that the index you plan to use pertains to the problem you must solve. The cost indexes in the *Engineering News Record* index would not apply to estimating costs of computer parts. Also, the indexes are aggregate values, and do not generally pertain to a particular geographic area or labor market. Of more basic concern is the fact that the cost indexes reflect the costs of past technology and design procedures.

17.10.2 Cost-Size Relationships

The cost of most capital equipment is not directly proportional to the size or capacity of the equipment. For example, doubling the horsepower of a motor increases the cost by only about one-half. This *economy of scale* is an important factor in engineering design. The cost-capacity relation usually is expressed by

$$C_1 = C_0 \left(\frac{L_1}{L_0} \right)^x \tag{17.8}$$

where C_0 is the cost of equipment at size or capacity L_0. The exponent x varies from about 0.4 to 0.8, and it is approximately 0.6 for many items of process equipment. For that reason, the relation in Eq. (17.8) often is referred to as the "six-tenths rule." Values of x for different types of equipment are given in Table 17.4.

Logically, cost indexes can be combined with cost-size relationships to provide for cost inflation as well as economy of scale.

$$C_1 = C_0 \left(\frac{L_1}{L_0} \right)^x \left(\frac{I_1}{I_0} \right) \tag{17.9}$$

The six-tenths rule applies only to large process or factory-type equipment. It does not apply to individual machine parts or smaller kinds of mechanical systems

TABLE 17.4
Typical Values of Size Exponent for Equipment

Equipment	Size Range	Capacity Unit	Exponent x
Blower, single stage	1000–9000	ft³/min	0.64
Centrifugal pumps, S/S	15–40	hp	0.78
Dust collector, cyclone	2–7000	ft³/min	0.61
Heat exchanger, shell and tube, S/S	50–100	ft²	0.51
Motor, 440-V, fan-cooled	1–20	hp	0.59
Pressure vessel, unfired carbon steel	6000–30,000	lb	0.68
Tank, horizontal, carbon-steel	7000–16,000	lb	0.67
Transformer, 3-phase	9–45	kW	0.47

Source: R. H. Perry and C. H. Chilton, Chemical Engineers' Handbook, 5th ed., p. 25–18, McGraw-Hill, New York, 1973.

17

like transmissions. To a first approximation, the material cost of a part, MtC, is proportional to the volume of the part, which in turn is proportional to the cube of a characteristic dimension, L. Thus, the material cost increases as a power of its dimension.

$$MtC_1 = MtC_0 \left(\frac{L_1}{L_0} \right)^n \tag{17.10}$$

where n was found for steel gears to be 2.4 in the range of diameters from 50 to 200 mm and $n = 3$ for diameters from 600 to 1500 mm.[1]

In another example of a cost growth law, the production cost, PC, for machining, based on time to complete an operation, might be expected to vary with the surface area of the part, i.e., with L^2.

$$PC_1 = PC_0 \left(\frac{L_1}{L_0} \right)^p \tag{17.11}$$

Again, p depends on processing condition. The exponent is 2 for finish machining and grinding and 3 for rough machining, where the depth of cut is much deeper.

Information about how processing cost depends on part size and geometry is very scanty. This information is needed to find better ways to calculate part cost early in the design process as different features and part sizes are being explored.

17.10.3 Learning Curve

A common observation in a manufacturing situation is that as the workers gain experience in their jobs they can make or assemble more product in a given unit of time. That, of course, decreases costs. This learning is due to an increase in the worker's level of skill, to improved production methods that evolve with time, and to better management practices involving scheduling and other aspects of production planning. The extent and rate of improvement also depend on such factors as the nature of the production process, the standardization of the product design, the length of the production run, and the degree of harmony in worker-management relationships.

The improvement phenomenon usually is expressed by a *learning curve*, also called a product improvement curve. Figure 17.4 shows the characteristic features of an 80 percent learning curve. Each time the cumulative production doubles ($x_1 = 1$, $x_2 = 2$, $x_3 = 4$, $x_4 = 8$, etc.) the production time (or production cost) is 80 percent of what it was before the doubling occurred. For a 60 percent learning curve the production time would be 60 percent of the time before the doubling. Thus, there is a constant percentage reduction for every doubled[2] production. Such an obviously exponential

---

1. K. Erlenspiel et al., *Cost-Efficient Design*, p. 161, Springer, New York, 2007.
2. The learning curve could be constructed for a tripling curve of production or any other amount, but it is customary to base it on a doubling.

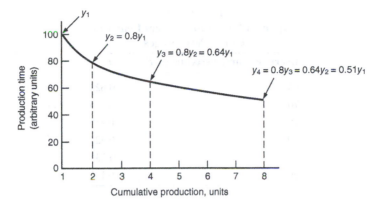

FIGURE 17.4
An 80 percent learning curve.

curve will become linear when plotted on loglog coordinates (Fig. 17.5). Note that a 60 percent learning curve gives a greater cost reduction than an 80 percent learning curve.

The learning curve is expressed by

$$y = kx^n \tag{17.12}$$

where y is the production effort, expressed either as h/unit or \$/unit
$\quad k$ is the effort to manufacture the first unit of production
$\quad x$ is the unit number, that is, $x = 5$ or $x = 45$
$\quad n$ is the negative slope of the learning curve, expressed as a decimal. Values for n are given in Table 17.5.
The value for n can be found as follows: For an 80 percent learning curve,

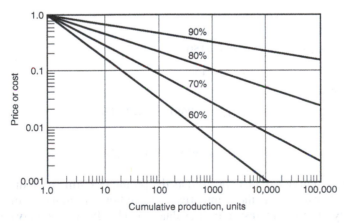

FIGURE 17.5
Standard learning curves.

17

TABLE 17.5
Exponent Values for Typical Learning Curve Percentages

Learning Curve Percentages, P	n
65	−0.624
70	−0.515
75	−0.415
80	−0.322
85	−0.234
90	−0.152

$y_2 = 0.8y_1$ for $x_2 = 2x_1$. Then,

$$\frac{y_2}{y_1} = \left(\frac{x_2}{x_1}\right)^n$$

$$\frac{0.8y_1}{y_1} = \left(\frac{2x_1}{x_1}\right)^n$$

$$n \log 2 = \log 0.8$$

$$n = \frac{-0.0969}{0.3010} = -0.322$$

Note that the learning curve percentage, expressed as a decimal, is $P = 2^n$.

EXAMPLE 17.10 The first of a group of 80 machines takes 150 h to build and assemble. If you expect a 75 percent learning curve, how much time would it take to complete the fortieth machine and the last machine?

$y = kx^n$

For $P = 75\%$, $n = -0.415$, and $k = 150$

$$y = 150(x^{-0.415})$$

For $x = 40$

$$y_{40} = 150(40^{-0.415}) = 32.4 \, \text{h}$$

For $x = 80$

$$y_{80} = 150(80^{-0.415}) = 24.3 \, \text{h}$$

The learning curve can be expressed as the production time in hours to produce a particular number unit or as the cumulative average hours to make N units. The latter term is usually of more interest in cost evaluation. The distinction between these two ways of expressing the output is shown in Table 17.6. Note that, for a given number of units of output, the cumulative average is greater than the unit values. However, the

TABLE 17.6
Based on an 80 Percent Learning Curve

x units	y, h/unit	Cumulative Total Hours	y, Cumulative Average h/unit
1	100.00	100.00	100.00
2	80.00	180.00	90.00
3	70.22	250.22	83.41
4	64.00	314.22	78.55
5	59.56	373.78	74.76
6	56.16	429.94	71.66
7	53.44	483.38	69.05
8	51.19	534.57	66.82

learning improvement percentage (80 percent) that applies to the unit values does not apply to the cumulative values. Similarly, if the unit values are derived from cumulative values, the constant percentage does not apply. In constructing learning curves from historical data we are more likely to find records of cumulative total hours than the hours to build each unit.

The total hours, T_c, required to manufacture a cumulative total of N units is given by

$$T_c = y_1 + y_2 + \ldots y_N = \sum_{i=1}^{N} y_i \qquad (17.13)$$

The average time to produce N parts, T_a, is

$$T_a = \frac{T_c}{N} \qquad (17.14)$$

An approximation for Eq. (17.14) when N is greater than 20 is

$$T_a \simeq \frac{1}{(1-n)} kN^n \qquad (17.15)$$

17.11
COST OF QUALITY

The four basic costs of quality were discussed in Section 15.4. In this section we give some relationships for determining three of these costs. The fourth cost, the prevention of defects by the use of quality enhancing methods in design and manufacture, does not lend itself to the kind of relationships presented here. Generally there is a cost in staff training to provide the knowledge and practice to become expert in the use of the quality methods (see Sec.15.1 to 15.7), but that should be small in comparison with the benefits in customer acceptance and high quality products that will result.

17

Internal Failures

These are the costs when materials, parts, and modules fail to meet the requirements for shipping to the customer. As a result the parts will be scrapped or reworked, and cost is incurred. Moreover, if there is a high incidence of defects, as often occurs with complex integrated circuit chips, it is necessary to overproduce the number of units to achieve the needed number of parts.

The number of acceptable defect-free parts N_a compared to the total number of parts N that enter a manufacturing operation is the yield Y_{op}.

$$Y_{op} = \frac{N_a}{N} \tag{17.16}$$

The quantity Y_{op} is the probability of achieving a part with zero defects from the particular manufacturing operation. If the manufacturing process has p independent operations, then the yield of the entire process is given by

$$Y_{process} = \prod_{i=1}^{p} Y_{op} \tag{17.17}$$

For example, if the process involves three operations, with yields of 90, 85, and 95%, the process yield will be $(0.90)(0.85)(0.95) = 0.73$. The unit cost to make a part then would be $C_U/0.73$.

Testing Costs

A slightly more detailed expression for the cost of internal failures is[1] Eq. (17.18). This allows for a process where there is a series of tests after various operations.

$$C_{out} = \frac{C_{in} + C_{test}}{Y_{in}} \tag{17.18}$$

where C_{out} is the unit cost of a part that has passed the test station
$\quad Y_{in}$ is the yield of parts as they enter the $p+1$ test station; this is found from Eq. (17.17)
$\quad C_{test}$ is the cost of making the test on one unit

External Failures

A warranty is a contract between the manufacturer and the buyer that the product will perform as advertised. The warranty cost needs to be estimated in order that the cost of servicing the warranty can be properly accounted for in the price or maintenance agreement for the product. For the simplest class of warranty, where the product is replaced or repaired free of cost to the customer, the warranty reserve amount C_{WR} for a time period t_w is given by[2]

$$C_{WR} = C_{fc} + nM(t_w)C_{rc} \tag{17.19}$$

1. P. Sandborn, *Course Notes on Manufacturing and Life Cycle Cost Analysis of Electronic Systems,* CALCE EPSC Press, College Park, MD, 2005.
2. E. B. Magrab, S. K. Gupta, F. P. McCluskey and P. A. Sandborn, *Integrated Product and Process Design and Development,* pp. 51–52, CRC Press, Boca Raton, FL, 2010.

where C_{fc} is fixed cost of maintaining the warranty system

n is number of products sold

C_{rc} is average recurring cost of replacing or repairing products, on a per unit basis

$M(t_w)$ is the number of replacement products expected to be needed in the interval 0 to t_w, where t_w is the time until actuating the warranty

$M(t_w) = \lambda t_w$ where λ is the constant failure rate; see Sec. 14.3.2

17.12
DESIGN TO COST

Design to cost, also called *target costing*, is the approach in which a target value, (sometimes called "should-cost" data), for the cost of a product is established at the beginning of a product development project. All design decisions are examined for their impact on keeping below the target cost. This is in contrast with the more usual practice of waiting for a complete cost analysis in the detail design phase. If this proves to be excessive, then the only practical recourse is to try to wring the excess cost out of the manufacturing process or to substitute a less expensive material, often at the expense of quality.

The steps in accomplishing design to cost are:[1]

- *Establish a realistic and reliable target cost.* The target cost is the difference between a realistic estimate of what the customer will pay for the product when developed minus the expected profit. This requires effective and realistic market analysis and an agile product development process that gets the product to market in minimum time.
- *Divide the target cost into subunits.* The basis for dividing the total cost can be (1) cost of subsystems and components in similar designs, (2) division according to competitors' component costs, as determined from dissection of competitor products[2], or (3) on the basis of estimates of what the customer is willing to pay for various functions and features of the product.
- *Oversight of compliance with cost targets.* A major difference in the design to cost approach is that the cost projections will be evaluated after each design phase as well as before going into production. For this to be effective there must be cost evaluation methods that can be applied at an earlier stage than detail design. There must also be a systematic way of quickly making cost comparisons.

17.12.1 Order of Magnitude Estimates

At the very early stage of product development where the market for a new product is being studied, comparison is usually made with similar products already on the market. This gives bounds on the expected selling price. Often the cost is estimated with

1. K. Ehrlenspiel et al., op. cit., pp. 44–63.
2. For details see K. T. Ulrich and S. Peterson, *Management Science*, vol. 44, no.3, pp. 352–369, 1998.

a single factor. Weight is most commonly used. For example,[1] products can be divided roughly into three categories:

1. Large functional products—automobile, front-end loader, tractor
2. Mechanical/electrical—small appliances and electrical equipment
3. Precision products—cameras, electronic test equipment

Products in each category cost roughly the same on a weight basis, but the cost between categories increases by a factor of approximately 10.

A slightly more sophisticated method is to estimate cost on the basis of the percentage of the share of the total cost that is due to materials cost.[2] For example, about 70 percent of the cost of an automobile is material cost, about 50 percent for a diesel engine, about 25 percent for electrical instruments, and about 7 percent for china dinnerware.

> **EXAMPLE 17.11** What is the total cost of a diesel engine that weighs 300 lb? The engine is made from ductile iron that costs \$2/lb. The material cost share for the engine is 0.5.
>
> $$Cost = (300 \times \$2)/0.5 = \$1200$$

Another rule of thumb is the one-three-nine rule.[3] This states the relative proportions of material cost to manufacturing cost to selling price are in the ratio of 1:3:9. In this rule the material cost is inflated by 20 percent to allow for scrap and tooling costs.

> **EXAMPLE 17.12** A 2 lb part is made from an aluminum alloy costing \$1.50/lb. What is the estimated material cost, part cost, and selling price?
>
> $$Material\ cost = 1.2 \times 1.50\ \$/lb \times 2\ lb = \$3.60$$
> $$Part\ cost = 3 \times material\ cost = 3 \times \$3.60 = \$10.80$$
> $$Selling\ price = 3 \times part\ cost = 3 \times \$10.80 = \$32.40\ or$$
> $$Selling\ price = 9 \times material\ cost = 9 \times \$3.60 = \$32.40$$

17.12.2 Costing in Conceptual Design

At the conceptual design stage, few details have been decided about the design. Costing methods are required that allow for direct comparison between different types of designs that would perform the same functions. An accuracy of ±20 percent is the goal.

Relative costs are often used for comparing the costs of different design configurations, standard components, and materials. The base cost is usually the cost of the lowest-cost or most commonly used item. An advantage of relative cost scales is that they change less with time than do absolute costs. Also, there are fewer problems with proprietary issues with relative costs. Companies are more likely to release relative cost data than they are absolute costs.

1. R. C. Creese, M. Adithan, and B. S. Pabla, *Estimating and Costing for the Metal Manufacturing Industries,* Marcel Dekker, New York, 1992, p. 101.
2. R. C. Creese et al., op. cit., pp. 102–5.
3. H. F. Rondeau, *Machine Design,* Aug. 21, 1975, pp. 50–53.

Parametric methods work well where designs tend to be variants of earlier designs. The costing information available at the conceptual design stage usually consists of historical cost for similar products. For example, cost equations for two-engine small airplanes have been developed,[1] and similar types of cost relationships exist for coal-fired power plants and many types of chemical plants. However, for mechanical products, where there is a wide diversity of products, few such relationships have been published. This information undoubtedly exists within most product manufacturing companies.

Cost calculations in conceptual design must be done quickly and without the amount of cost detail used in Example 17.6. One saving grace is that not all parts in a product will require cost analysis. Some parts may be identical to parts in other products, for which the cost is known. Other parts are standard components or are parts that will be outsourced, and the costs are known with a firm quotation. An additional group of parts will be similar parts that differ only by the addition or subtraction of some physical features. The cost of these parts will be the cost of the original part plus or minus the cost of the operations to create the features that are different.

For those parts that require a cost analysis, "quick cost calculations" are used. The development of quick cost methods is an ongoing activity, chiefly in Germany.[2] The methods are too extensive to detail here, other than to give an example of an equation for scaling unit manufacturing cost C_u from size L_0 to size L_1.

$$C_u = \frac{PCsu}{n}\left(\frac{L_1}{L_0}\right)^{0.5} + PCt_0\left(\frac{L_1}{L_0}\right)^2 + MtC_0\left(\frac{L_1}{L_0}\right)^3 \tag{17.20}$$

In the equation, $PCsu$ is the processing cost for tool setup, PCt_0 is the processing cost for making the original part based on total operation time, MtC_0 is the material cost for the original size L_0, and n is the batch size.

An intellectually satisfying approach to determining costs early in design is functional costing.[3] The idea behind this approach is that once the functions to be performed have been determined, the minimum cost of the design has been fixed. Since it is in conceptual design that we identify the needed functions and work with alternative ways of achieving them, linking functions to cost gives us a direct way of designing to cost. A start has been made with standard components like bearings, electric motors, and linear actuators, where the technology is relatively mature and costs have become rather competitive. Linking function with cost is the basic idea behind value analysis. This is discussed in the next section.

Probably the greatest progress in finding ways to determine cost early in the design process is with the use of special software. A number of software programs that

---

1. J. Roskam, *J. Aircraft,* vol. 23, pp. 554–560, 1986.
2. K. Ehrlenspiel, op. cit., pp. 430–456.
3. M. J. French, *Jnl. Engr. Design,* vol. 1, no. 1, pp. 47–53, 1990; M. J. French and M. B. Widden, *Design for Manufacturability 1993, DE,* vol. 52, pp. 85–90, ASME, New York, 1993.

17

incorporate quick design calculations, cost models of processes, and cost catalogs are available. Some sources where you can find additional information are:

- *SEER-MFG* by Galorath[1] uses advanced parametric modeling to estimate manufacturing costs early in the design process. The software is able to deal with the following processes: machining, casting, forging, molding, powder metals, heat treatment, coating, fabrication of sheet metal, composite materials, printed circuit boards, and assembly. *SEER-H* provides system-level cost analysis and management in product development from work breakdown structure to the cost of operation and maintenance.
- *DFM Concurrent Costing* by Boothroyd Dewhurst[2] was discussed in Sec. 13.10.2. This software requires minimum part detail to provide relative costs for process selection.
- *CustomPartNet*[3] is the only online source that provides free cost estimation tools for material and process selection. Processes considered are injection molding, sand and die casting, and machining. They also provide a collection of special calculators called "widgets" for common design and manufacturing problems.
- *Costimator* by MTI Systems[4] provides detailed cost estimates for parts made by machining. As one of the early suppliers in this field, its software contains extensive cost models, labor standards, and material cost data. It specializes in providing a fast, accurate, and consistent method that allows job shops to estimate cycle times and costs for preparing quotations.

17.13
VALUE ANALYSIS IN COSTING

Value analysis or value engineering is a problem-solving process to improve the value of a product for the customer.[5] Value is defined as the worth of a part, feature, or assembly related to its cost. Value analysis is often the first step in a redesign of a product, where the objective is to improve the functionality at fixed cost, or to reduce the cost keeping the functionality the same.

The value analysis methodology seeks to improve the design by finding answers to the following questions.

- Can we do without the part? (Use DFA analysis)
- Does the part do more than required?
- Does the part cost more than it is worth?
- Is there something that does the job better?
- Is there a less costly way to make the part?
- Can a standard item be used in place of the part?

---

1. www.galorath.com
2. www.dfma.com
3. www.custompartnet.com
4. www.mtisystems.com
5. T. C. Fowler, *Value Analysis in Design,* Van Nostrand Reinhold, New York, 1990.

TABLE 17.7
Cost Structure for a Centrifugal Pump

Cost Category	Part	Manufacturing Cost $	Manufacturing Cost %	Type of Cost, % Material	Type of Cost, % Production	Type of Cost, % Assembly
A	Housing	5500	45.0	65	25	10
A	Impeller	4500	36.8	55	35	10
B	Shaft	850	7.0	45	45	10
B	Bearings	600	4.9	Purchased	Purchased	Purchased
B	Seals	500	4.1	Purchased	Purchased	Purchased
B	Wear rings	180	1.5	35	45	20
C	Bolts	50	<1	Purchased	Purchased	Purchased
C	Oiler	20	<1	Purchased	Purchased	Purchased
C	Key	15	<1	30	50	20
C	Gasket	10	<1	Purchased	Purchased	Purchased

From M. S. Hundal, *Systematic Mechanical Design,* ASME Press, New York, 1997. Used with permission.

- Can an outside supplier provide the part at less cost without affecting quality or delivery schedule?

The first step in a value analysis study is to determine the costs of the parts and relate these to the functions which they provide. Example 17.13 shows how to do this. For more information on value analysis see the webpage of the Society of Value Engineers[1] and the online copy of the classic book by the originator of value analysis, Lawrence Miles.[2]

EXAMPLE 17.13 Table 17.7 shows the cost structure for a centrifugal pump.[3] In this table the components of the pump have been classified into three categories, A, B, and C, according to their manufacturing costs. Components in class A comprise 82 percent of the total cost. These "vital few" need to be given the greatest thought and attention.

We now focus attention on the functions provided by each component of the pump (Table 17.8). This table of functions is added to the cost structure table to create Table 17.9. Note that an estimate has been made of how much each component contributes to each function. For example, the shaft contributes 60 percent to transfer of energy (F2) and 40 percent to supporting the parts (F6). Multiplying the cost of each component by the fraction it serves to provide a given function gives the total cost for each function. For example, the function support parts (F6) is provided partly by the housing, shaft, and bearings.

$$\text{Cost of F6} = 0.5(5500) + 0.4(850) + 1.0(600) = \$3690$$

These calculations are summarized in Table 17.10. This table shows that the expensive functions of the pump are containing the liquid, converting the energy, and supporting the

1. www.value-eng.org/education_publications_function_monographs.php
2. http://wendt.library.wisc.edu/miles/milesbook.html
3. M. S. Hundal, *Systematic Mechanical Design,* ASME Press, New York, 1997, pp. 175, 193–96.

17

TABLE 17.8

Functions Provided by Each Component of the Centrifugal Pump

Function	Description	Components
F1	Contain liquid	Housing, seals, gasket
F2	Transfer energy	Impeller, shaft, key
F3	Convert energy	Impeller
F4	Connect parts	Bolts, key
F5	Increase life	Wear rings, oiler
F6	Support parts	Housing, shaft, bearings

From M. S. Hundal, *Systematic Mechanical Design,* ASME Press. Used with permission.

parts. Thus, we know where to focus attention in looking for creative solutions in reducing costs in the design and manufacture of the pump.

Table 17.7 shows the cost of the parts arranged in descending order, as in a Pareto chart. Thus, the housing and impeller would be logical places to look for cost reduction. The housing shares roughly equally in providing the functions of containing liquid (F1) and providing structural support (F6). These are, respectively, #2 and #1, and together constitute 57 percent of function cost. The housing would be the prime candidate for cost reduction since the impeller is the most critical part in making the pump. One might conceive that by using advanced casting methods like investment casting and FEA analysis a lighter and cheaper housing could be designed without any loss in structural rigidity of the pump.

TABLE 17.9

Cost Structure for Centrifugal Pump with Function Cost Allocation

Cost Class	Part	Manufacturing Cost $	Manufacturing Cost %	Type of Cost, % Material	Type of Cost, % Production	Type of Cost, % Assembly	Function Allocation, %			
A	Housing	5500	45.0	65	25	10	F1	50	F6	50
A	Impeller	4500	36.8	55	35	10	F2	30	F3	70
B	Shaft	850	7.0	45	45	10	F2	60	F6	40
B	Bearings	600	4.9	Purchased	Purchased	Purchased	F6	100		
B	Seals	500	4.1	Purchased	Purchased	Purchased	F1	100		
B	Wear rings	180	1.5	35	45	20	F5	100		
C	Bolts	50	<1	Purchased	Purchased	Purchased	F4	100		
C	Oiler	20	<1	Purchased	Purchased	Purchased	F5	100		
C	Key	15	<1	30	50	20	F2	80	F4	20
C	Gasket	10	<1	Purchased	Purchased	Purchased	F1	100		

From M. S. Hundal, *Systematic Mechanical Design,* ASME Press, New York, 1997. Used with permission.

TABLE 17.10
Calculation of Function Costs for Centrifugal Pump

Function	Part	% of Part Cost for Function	Part Cost, $	Function Cost of Individual Part, $	Total Function Cost $	Total Function Cost %
F1: Contain liquid	Housing	50	5500	2750		
	Seals	100	500	500		
	Gasket	100	10	10	3260	26.7
F2: Transfer energy	Impeller	30	4500	1350		
	Shaft	60	850	510		
	Key	80	15	12	1872	15.3
F3: Convert energy	Impeller	70	4500	3150	3150	25.8
F4: Connect parts	Key	20	15	3		
	Bolts	100	50	50	53	0.4
F5: Increase life	Wear rings	100	180	180		
	Oiler	100	20	20	200	1.6
F6: Support parts	Housing	50	5500	2750		
	Shaft	40	850	340		
	Bearings	100	600	600	3690	30.2

From M. S. Hundal, *Systematic Mechanical Design,* ASME Press, New York, 1997. Used with permission.

17.14
MANUFACTURING COST MODELS

The importance of modeling in the design process has been emphasized throughout this text. Modeling can show which elements of a design contribute most to the cost; that is, it can identify cost drivers. With a cost model it is possible to determine the conditions that minimize cost or maximize production (cost optimization).

17.14.1 Machining Cost Model

Extensive work has been done on cost models for metal removal processes.[1] For a background on machining, see Sec. 13.14. Broken down into its simplest cost elements, a machining process can be described by Fig. 17.6. The time designated A is

---

1. E. J. A. Armarego and R. H. Brown, *The Machining of Metals,* Chap. 9, Prentice Hall, Englewood Cliffs, NJ, 1969; G. Boothroyd and W. A. Knight, *Fundamentals of Machining and Machine Tools,* 3d ed., CRC Press, Boca Raton, FL, 2006.

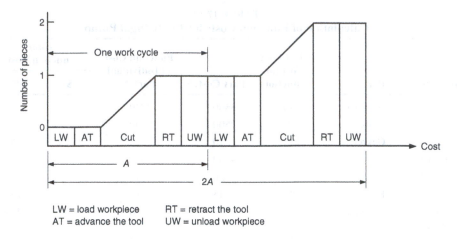

LW = load workpiece   RT = retract the tool
AT = advance the tool   UW = unload workpiece

FIGURE 17.6
Elements of a machining operation.

the machining plus work-handling costs per piece. If B is the tool cost, including the costs of tool changing and tool grinding, in dollars per tool, then

$$\text{Cost/piece} = \frac{nA+B}{n} = A + \frac{B}{n} \qquad (17.21)$$

where n is the number of pieces produced per tool.

We shall now consider a more detailed cost model for turning down a bar on a lathe (Fig. 17.7). The machining time for one cut, t_c, is

$$t_c = \frac{L}{V_{feed}} = \frac{L}{fN} \qquad (17.22)$$

where V_{feed} = feed velocity, in./min
$\qquad f$ = feed rate, in./rev
$\qquad N$ = rotational velocity, rev/min

Equation (17.22) provides detail only for the process of turning a cylindrical bar. For other geometries or other processes such as milling or drilling, different expressions would be used for L or V_{feed}.

The total cost of a machined part is the sum of the machining cost C_{mc}, the cost of the cutting tools, C_t, and the cost of the material C_m.

$$C_u = C_{mc} + C_t + C_m \qquad (17.23)$$

where C_u is the total unit (per piece) cost. The machining cost, C_{mc} ($/h), depends on the machining time t_{unit} and the costs of the machine, labor, and overhead.

17

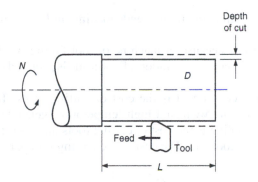

FIGURE 17.7
Details of lathe turning.

$$C_{mc} = [M(1+OH_m)+W(1+OH_{op})]t_{unit} \qquad (17.24)$$

where M is machine cost rate, \$/h
$\quad OH_m$ is machine overhead rate, decimal
$\quad W$ is labor rate for machine operator, \$/h
$\quad OH_{op}$ is operator overhead rate, decimal

The machine cost includes the cost of interest, depreciation, and maintenance. It is found with the methods of Chap. 16 by determining these costs on an annual basis and converting them to per-hour costs on the basis of the number of hours the machine is used in the year. The machine overhead cost includes the cost of power and other services and a proportional share of the building, taxes, insurance, and other such expenses.

The production time for a unit is the sum of the machining time t_m and the non-production or idle time t_i.

$$t_{unit} = t_m + t_i \qquad (17.25)$$

The machining time t_m is the machining time for one cut, t_c, multiplied by the number of cuts.

$$t_m = t_c \text{ (number of cuts)} \qquad (17.26)$$

The idle time is given by

$$t_i = t_{set} + t_{change} + t_{hand} + t_{down} \qquad (17.27)$$

where t_{set} = total time for job setup divided by number of parts in the batch
$\quad t_{change}$ = prorated time for changing the cutting tool

$$= \text{tool change time} \times \frac{t_m}{\text{tool life}}$$

t_{hand} = time the machine operator spends loading and unloading the work on the machine

t_{down} = downtime lost because of machine or tool failure, waiting for material or tools, or maintenance operations. Downtime is prorated per units production.

An important cost component is the cost of cutting tools. Tools lose their cutting edge from the extreme wear and high temperature generated at the tool-metal interface. The cost of tooling is the cost of cutting tools and a prorated cost of special fixtures used to hold the tool bits. The cost of the cutting tool per unit piece is

$$C_t = C_{tool} \frac{t_m}{T} \tag{17.28}$$

where C_{tool} is the cost of a cutting tool, $

t_m is the machining time (min), given by Eq. (17.26)

T is the tool life (min) given by Eq. (17.29)

Tool life usually is expressed by the Taylor tool life equation, which relates tool life T to surface (tangential) velocity v. For turning in a lathe, the tangential velocity (cutting speed) is $v = \pi DN$ where πD is the circumference, in./rev, and N is the rpm.

$$vT^p = K \tag{17.29}$$

A log-log plot of tool life (min) versus surface velocity (ft/min) will give a straight line. K is the surface velocity at $T = 1$ min and p is the reciprocal of the negative slope.

For a cutting tool that uses an insert in a tool holder,

$$C_{tool} = \frac{K_i}{n_i} + \frac{K_h}{n_h} - \tag{17.30}$$

where K_i is the cost of one tool insert, $

n_i is the number of cutting edges on a tool insert

K_h is the cost of a tool holder, $

n_h is the number of cutting edges in the life of a tool holder

Substituting the tool life T from Eq. (17.29) into Eq. (17.28) gives

$$C_t = C_{tool} t_m \left(\frac{v}{K} \right)^{1/p} \tag{17.31}$$

The time needed to change tools can be significant, so we separate it out as t_{tool} from the other times listed in Eq. (17.27) and express t_{change} with Eq. (17.32).

$$t_{change} = t_{tool} \left(\frac{t_m}{T} \right) \tag{17.32}$$

17

The other three terms in Eq. (17.27) are independent of tool life, and are designated by t_0. The expression for the time to machine one piece, Eq. (17.25), now can be written as

$$t_{unit} = t_m + t_i = t_m + t_{change} + t_0 = t_m + t_{tool}\frac{t_m}{T} + t_0 = t_m\left(1 + \frac{t_{tool}}{T}\right) + t_0 \qquad (17.33)$$

Substituting Eqs. (17.24), (17.33), and (17.28) into Eq. (17.23) gives

$$C_u = [M(1 + OH_m) + W(1 + OH_{op})]\left[t_m\left(1 + \frac{t_{tool}}{T}\right) + t_0\right] + C_t\frac{t_m}{T} + C_m \qquad (17.34)$$

This equation gives the cost of a unit machined piece. Both the machining time, t_m, and the tool life, T, depend on the cutting velocity through Eqs. (17.22), (17.26), and (17.29). If we plot unit cost versus cutting velocity (Fig. 17.8), there will be an optimum cutting velocity to minimize cost. That is so because machining time decreases with increasing velocity; but as velocity increases, tool wear and tool costs increase also. Thus, there is an optimum cutting velocity. An alternative strategy would be to operate at the cutting speed that results in maximum production rate. Still another alternative is to operate at the speed that maximizes profit. The three criteria do not result in the same operating point.

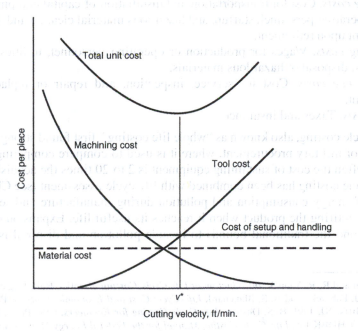

FIGURE 17.8
Variation of unit cost with cutting velocity, showing an optimum cutting velocity.

17

The machining cost model illustrates how a physical model of the process, along with standard times for elements of the operation, can be used to determine realistic part costs. Also, the problem shows how overhead costs can be allocated to both labor and material costs. Compare this with the approach given in Sec 17.5 where a single factory overhead cost was used.

The machining cost model is based chiefly on physical models. When a good physical model is not available the process still can be broken down into discrete steps, with times and costs for each step. The procedure for this can be found under Process Cost Modeling on the website for this text (www.mhhe.com/dieter).

17.15
LIFE CYCLE COSTING

Life cycle costing (LCC) is a methodology that attempts to capture all of the costs associated with a product throughout its life cycle.[1] A typical problem is whether it is more economical to spend more money in the initial purchase to obtain a product with lower operating and maintenance costs, or whether it is less costly to purchase a product with lower first costs but higher operating costs. Life cycle costing goes into the analysis in much detail in an attempt to evaluate all relevant costs, both present and future.

The costs that enter into life cycle costing can be divided into five categories.

- *First costs.* Purchase cost of equipment or plant.
- *One-time costs.* Cost for transportation and installation of capital equipment, training of operating personnel, startup, and hazardous material cleanup and disposal of equipment upon retirement.
- *Operating costs.* Wages for production or operating personnel, utilities, supplies, materials, disposal of hazardous materials.
- *Maintenance costs.* Cost for service, inspection, and repair or replacement of equipment.
- *Other costs.* Taxes and insurance.

Life cycle costing, also known as "whole life costing," first found strong advocates in the area of military procurement, where it is used to compare competing weapons systems.[2] Often the cost of sustaining equipment is 2 to 20 times the acquisition cost.

Life cycle costing has been combined with life cycle assessment (see Chap. 10) of the costs of energy consumption and pollution during manufacture and service, and the costs of retiring the product when it reaches its useful life. Expansion of the cost models beyond the traditional bounds to include pollution and disposal is an active

---

1. R. J. Brown and R. R. Yanuck, *Introduction of Life Cycle Costing,* Prentice Hall, Englewood Cliffs, NJ, 1985; W. J. Fabrycky and B. S. Blanchard, *Life-Cycle Cost and Economic Analysis,* Prentice Hall, Englewood Cliffs, NJ, 1991. B. S. Dhillon, *Life Cycle Costing for Engineers,* CRC Press, Boca Raton, FL, 2010; NIST-HDBK-135, *Life-Cycle Costing Manual for the Federal Energy Management Program,* February 1996, available online at www.barringer1.com, listed under Military Documents.
2. MIL-HDBK 259, Life Cycle Costs in Navy Acquisitions.

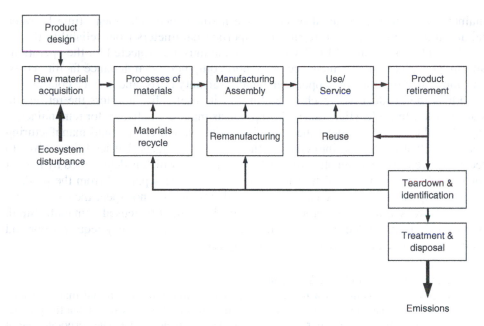

FIGURE 17.9
Total life cycle of a product.

area of research that will place the design engineer in a better position to make critical trade-off decisions.

Typical elements in the life cycle of a product are shown in Fig. 17.9. This figure emphasizes the overlooked impact on society costs (OISC) that are rarely quantified and incorporated into a product life cycle analysis.[1] Starting with design, the actual costs incurred here are a small part of the LCC, but the costs committed in design comprise about 75 percent of the avoidable costs within the life cycle of the product. Moreover, it is about 10 times less costly to make a change or correct an error in design than in manufacturing. Acquiring the raw materials, usually by mining or oil extraction, and processing the materials, can create large environmental costs. These areas also often have considerable inventory and transportation costs. We have concentrated in previous sections on the costs in manufacturing and assembly of products.

The cost of ownership of a product is the traditional aspect of LCC. Equation (17.3) lists the chief contributors to LCC. Useful life is commonly measured by cycles of operation, length of operation, or shelf life. In design we attempt to extend life for use and service by using durable materials and reliable components. Product obsolescence is dealt with through modular product architecture.

Maintenance costs, especially maintenance labor costs, usually dominate other use/service costs. Most analyses divide maintenance costs into scheduled or preventive

1. N. Nasr and E. A. Varel, "Total Product Life-Cycle Analysis and Costing," *Proceedings of the 1997 Total Life Cycle Conference,* P-310, pp. 9–15, Society of Automotive Engineers, Warrendale, PA, 1997.

maintenance and unscheduled or corrective maintenance. The mean time between failure and the mean time to repair are important parameters from reliability theory (see Sec. 14.3.6) that affect LCC. Other costs that must be projected for the operations and support phase are maintenance of support equipment, maintenance facility costs, pay and fringe benefits for support personnel, warranty costs, and service contracts.

Once the product has reached the limit of its useful life it enters the retirement stage of the life cycle. High-value-added products may be candidates for remanufacturing. By value-added we mean the cost of materials, labor, energy, and manufacturing operations that have gone into creating the product. Products that lend themselves to recycling are those with an attractive reclamation value, which is determined by market forces and the ease with which different materials can be separated from the product. Reuse components are subsystems from a product that have not spent their useful life and can be reused in another product. Materials that cannot be reused, remanufactured, or recycled are discarded in an environmentally safe way. This may require labor and tooling for disassembly or treatment before disposal.

EXAMPLE 17.14 Life Cycle Costing

The costs and income for a product development project to design and make a short-turning-radius lawnmower are given in the following chart. It is assumed that the product will be obsolete 10 years after the start of the development project. The corporate rate of return is 12 percent and its tax rate is 35 percent. Use the concepts of the time value of money presented in Chap. 16 to find the net present value (NPV) of the project and the average annual profit margin based on sales.

Category	Yr 1	Yr 2	Yr 3	Yr 4	Yr 5	Yr 6	Yr 7	Yr 8	Yr 9	Yr 10	Avg.
1. Development costs	0.8	1.90	0.4	0.4	0.4	0.4	0.4	0.2	0.2	0.2	
2. Cost of product sold			12.0	13.5	15.0	16.1	16.8	16.0	15.2	15.3	14.8
3. Sales & marketing			2.1	3.0	3.5	2.8	2.7	2.8	2.9	2.6	2.8
4. G&A plus overhead			0.8	1.5	2.0	2.0	2.0	2.0	2.0	2.0	1.7
5. Special production equipment, P		4.1									
6. Salvage value, S										0.5	
7. Depreciation on equip.	0.4	0.4	0.4	0.4	0.4	0.4	0.4	0.4	0.4	0.4	0.4
8. Environ. cleanup										1.1	
9. Net sales			28.2	31.3	36.2	39.8	40.0	39.1	38.0	35.0	35.95

All figures in millions of dollars.

Present Value of Costs

(1) PV of development costs $= 0.8(P/F,12,1) + 1.90\ (P/F,12,2) + 0.4(P/A,12,5)(P/F,12,2) + 0.2(P/A,12,3)(P/F,12,7) = \$3.47M$

(2) PV of cost of product sold $= 14.8(P/A,12,8)(P/F,12,2) = \$58.7M$

(3) PV of sales and marketing costs $= 2.8(P/A,12,8)(P/F,12,2) = \$11.17M$

(4) *PV* of G&A and overhead = 1.7(*P/A*,12,8)(*P/F*,12,2) = $6.73M
(5) Annual straight-line depreciation charge on (5), year 2 through 10 = (*P* − *S*)/*n* = (4.1 − 0.5)/9 = 0.40.
(7) *PV* of depreciation = 0.4(*P/A*,12,9)(*P/F*,12,1) = $1.90M
(8) *PV* of cost of environmental cleanup = 1.1(*P/F*,12,10) = $0.35M

Present value of total costs = 3.47 + 58.70 + 11.17 + 6.73 + 1.90 + 0.35 = $82.32

Present Value of Income or Savings

(9) Present value of net sales = 35.95(*P/A*,12,8)(*P/F*,12,2) = $130.8M
Present value of sale of equipment for salvage PV = 0.5(*P/F*,12,10) = $0.16M
Present value of tax reduction (0.35)(1.90) = $0.66M*
Present value of total income or savings =

Net present value = present value of income − present value of costs = 131.6 − 82.3 = $ 49.3M over 10 years, or an average of $ 4.93M per year

Annual profit margin = 4.93/35.95 = 13.7% per year

Note that an average of annual income and cost was used to simplify calculation. The use of a spreadsheet would have given more accurate numbers, but this is not warranted by the precision of the estimates.

Example 17.14 is typical of life cycle analysis for a product development project. Another common application is estimating the LCC costs for a major capital purchase. Since there is no income stream in this type of application, the selection would be based on the alternative that minimized the LCC. Using the cost of ownership model in Eq. (17.3) we divide the costs into *nonrecurring costs* (S_P, C_x, C_p, and C_Q) that only appear at year one and *recurring costs* (C_o, C_{ps}, and C_{sp}) that occur out into the future. The nonrecurring costs have been discussed in Sec. 17.11.

The cost of operating the equipment, C_o, depends on staffing levels as recommended by the supplier, the pay level of the operator, and the hours of operation.

The cost of support, C_{ps}, is chiefly the cost of maintenance, which depends greatly on the criticality of the operation and reliability of the equipment. For corrective maintenance, the number of maintenance events in a year can be estimated from the mean time between failure (MTBF). See Sec. 14.3.1 and 14.3.6 for discussion of MTBF.

Number of maintenance events = (scheduled operating hrs/year)/MTBF   (17.35)

The cost of *corrective maintenance* equals (number of maintenance events) × mean time to repair (MTTR) × (hourly labor cost). The cost of *preventive maintenance* is based on a monthly estimate of the labor cost.

The cost of spare parts, C_{sp}, is not an inconsequential cost in many situations. This involves the purchase of the spares, the cost of money tied up in their purchase,

---

*Item (7) gives the *PV* of 9 years of depreciation charges. These charges reduced the annual income on which taxes were paid at a 35% rate. This represents a savings of 0.35 × Item(7).

17

the cost of warehousing them in storage, and the cost of transporting them to the site of the repair. Often the cost of lost production from inoperable machines is the largest cost of all. Each of these costs represents a row that would be added to a present value calculation such as shown in Example 17.14.

17.16
SUMMARY

Cost is a primary factor of design that no engineer can afford to ignore. It is important to understand the basics of cost evaluation so that you can produce high-functioning, low-cost designs. Cost buildup begins in conceptual design and continues through embodiment and detail design.

To be cost literate you need to understand the meaning of such concepts as nonrecurring costs, recurring costs, fixed costs, variable costs, direct costs, indirect costs, overhead, and activity-based costing.

Cost estimates are developed by three general methods.

1. Cost estimation by analogy with previous products or projects. This method requires past experience or published cost data. Because this uses historical data, the estimates must be corrected for price inflation using cost indexes, and for differences of scale using cost-capacity indexes. This method is often used in the conceptual phase of design.
2. The parametric or factor approach uses regression analysis to correlate past costs with critical design parameters like weight, power, and speed, Software programs that use parametric relationships and cost databases are becoming increasingly useful for the calculation of costs in conceptual and embodiment design.
3. A detailed breakdown of all the steps required to manufacture a part with an associated cost of materials, labor, and overhead for each step for each operation is needed to determine the cost to produce the part. This method is generally used in the final cost estimates in the detail design stage.

Costs may sometimes be related to the functions performed by the design. This is a highly desired situation because it allows optimization of the design concept with respect to cost.

Manufacturing costs generally decrease with time as more experience is gained in making a product. This is known as a learning curve.

Computer cost models are gaining in use as a way to pinpoint the steps in a manufacturing process where cost savings must be achieved. Simple spreadsheet models are useful for determining product profitability and making trade-offs between aspects of the business situation.

Life cycle costing attempts to capture all the costs associated with a product throughout its life cycle, from design to retirement from service. Originally LCC focused only on the costs incurred in using a product, such as maintenance and repair, but more and more LCC is attempting to capture the costs that affect society from environmental issues and issues of energy use.

NEW TERMS AND CONCEPTS

Activity-based costing	General & administrative	Period costs
Break-even point	costs	Prime cost
Cost commitment	Indirect costs	Product costs
Cost index	Learning curve	Target costing
Design to cost	Life cycle costs	Value analysis
Fixed cost	Make-buy decision	
Functional costing	Overhead cost	

BIBLIOGRAPHY

Creese, R. C., M. Aditan, and B. S. Pabla: *Estimating and Costing for the Metals Manufacturing Industries,* Marcel Dekker, New York, 1992.

Ehrlenspiel, K, A. Kiewert, and U. Lindemann: *Cost-Efficient Design,* Springer, New York, 2007.

Malstrom, E. M. (ed.): *Manufacturing Cost Engineering Handbook,* Marcel Dekker, New York, 1984.

Michaels, J. V., and W. P. Wood: *Design to Cost,* John Wiley & Sons, New York, 1989.

Ostwald, P. F.: *Engineering Cost Estimating,* 3d ed., Prentice Hall, Englewood Cliffs, NJ, 1992.

Ostwald, P. F. and T. S. McLaren, *Cost Analysis and Estimating for Engineering and Management,* Prentice Hall, Upper Saddle River, NJ, 2004.

Winchell, W. (ed.): *Realistic Cost Estimating for Manufacturing,* 2d ed., Society of Manufacturing Engineers, Dearborn, MI, 1989.

PROBLEMS AND EXERCISES

17.1. In an environmental upgrade of a minimill making steel bar, it is found that a purchase must be made for a large cyclone dust collector. It is the time of the year for capital budget submissions, so there is no time for quotations from suppliers. The last unit of that type was purchased in 1985 for $35,000. It had a 100 ft³/min capacity. The new installation in 2012 will require 1000 ft³/min capacity. The cost escalation for this kind of equipment has been about 5 percent per year. For budget purposes, estimate what it will cost to purchase the dust collector.

17.2. Many consumer items today are designed in the United States and manufactured overseas where labor costs are much lower. A middle range athletic shoe from a name brand manufacturer sells for $70 in the U.S. The shoe company buys the shoe from an off-shore supplier for $20 and sells it to the retailer for $36. The profit margin for each unit in the chain is: supplier—9 percent; shoe company—17 percent; retailer—13 percent. Estimate the major categories of cost breakdown for each unit in the chain. Do this as a team problem and compare the results for the entire class.

17

17.3. The type of tooling to make for a manufacturing process depends on the expected total quantity of parts. Tooling made from standard components and less wear-resistant materials (soft tooling) can be made more quickly and cheaply than conventional tooling made from hardened steel (hard tooling). Use the concept of break-even point to determine the production quantity for which soft tooling can be justified. The following cost data applies:

	Soft Tooling	Hard Tooling
Tooling cost	C_S $600	C_H $7500
Setup cost	S_S $100	S_H $60
Unit part cost	C_{ps} $3.40	C_{pH} $0.80

The total production run is expected to be 5000 units. Parts are made in batches of 500.

17.4. A manufacturer of small hydraulic turbines has the annual cost data given here. Calculate the manufacturing cost and the selling price for a turbine.

Raw material and components costs	$2,150,000
Direct labor	950,000
Direct expenses	60,000
Plant manager and staff	180,000
Utilities for plant	70,000
Taxes and insurance	50,000
Plant and equipment depreciation	120,000
Warehouse expenses	60,000
Office utilities	10,000
Engineering salaries (plant)	90,000
Engineering expenses (plant)	30,000
Administrative staff salaries	120,000
Sales staff, salaries and commissions	100,000
Total annual sales: 60 units	
Profit margin: 15%	

17.5. A jewel case for a compact disc is made from polycarbonate ($2.20 per lb) by a thermoplastic molding process. Each CD case uses 20 grams of plastic. The parts will be made in a 10-cavity mold that makes 1400 parts per hour at an operating cost of $20 per hour. Manufacturing overhead is 40 percent. Since the parts are sold in large lots, the G&A expenses are a low 15 percent. Profit is 10 percent. What is the estimated selling price of each CD case?

17.6. Two competing processes for making high-quality vacuum melted steel are the vacuum arc refining process (VAR) and electroslag remelting (ESR). The estimated costs for operating each of the processes are:

Cost Component	VAR	ESR
Direct labor, one melter and one helper	$89,000	$89,000
Manufacturing overhead, 140% direct labor	$124,600	$124,600
Melting power	0.3 kWh/lb 1000 lb/h 10¢/kWh	0.5 kWh/lb 1250 lb/h 10¢/kWh
Cooling water (annual charge)	$5,500	$6,800
Slag	—	$42,000

The capital cost of a VAR system is $1.3M and for an ESR system it is $0.9M. Each melting system has a 10-year useful life. Each uses 1000 ft² of factory space, which costs $40 per ft². Assume both furnaces operate for 15 eight-hour shifts per week for 50 weeks in the year. Estimate the cost of melting a pound of high-grade steel for each process.

17.7. The accounting department established the costs given in the following table for producing two products, X and Z, over a given time period.

(a) Give an example of typical costs that would be put in each of the 10 cost categories listed.

(b) Determine the overhead and unit cost for each product in terms of direct labor cost.

(c) Determine the overhead and unit cost for each product on the basis of direct labor hours (DLH).

(d) Determine the overall overhead rate per DLH and use it to determine the unit cost of product X.

(e) Determine the overhead and unit cost for each product on the basis of the proportion of direct material costs.

Item	Product X	Product Z
Quantity	3000	5000
Machine hours	70	90
Direct labor hours (DLH)	400	600
Factory floor space	150	50

	Labor Rate $/h	Labor Amount, h	Material Cost $/unit	Material Amount, units	Cost $
Product X					
Direct labor	18.00	400			7,200
Direct material			6.50	3000	19,500
Product Z					
Direct labor	14.00	600			8,400
Direct material			7.50	5000	37,500

17

Cost Item	Product X	Product Z	Factory	Admin.	Sales	Total Cost, $
1. Direct labor	7,200	8,400				15,600
2. Indirect labor			3,000			3,000
3. Direct material	19,500	37,500				57,000
4. Indirect material			7,000			7,000
5. Direct engineering	900	2,500				3,400
6. Indirect engineering			1,500			1,500
7. Direct expense	1,000	700				1,700
8. Other factory burden			5,500			5,500
9. Admin. expense				11,000		11,000
10. Sales and distribution						
Direct	900	1,100				2,000
Indirect					8,000	8,000
	29,500	50,200	17,000	11,000	8,000	115,700

17.8. Determine the unit cost for making products X and Z in Prob.17.7 using activity-based costing. Use the cost drivers in Example 17.5, but omit automated assembly. The resources used on a per-batch basis are:

	Product X	Product Z
Number of components	18	30
Hours of engineering services	15	42
Production batch size	300	500
Hours of testing	3.1	5.2
Units per order	100	200

17.9. A manufacturer of high-performance pumps has the cost and profit data given in the following table. The company invests $1.2M in an aggressive two-year design and development program to reduce manufacturing costs by 20 percent. When this is completed, what will be the impact on profit? What business aspects need to be considered that are not covered by this analysis? What questions does it leave unanswered?

	Existing Design	Improved Design
Sales price	$500	$500
Units sold	20,000	20,000
Revenues	$10M	$10M
Direct labor	1.5M	
Materials	5.0M	
Overhead	2.0M	
Cost of product sold	8.5M	
Gross margin	1.5M	
Total operating expenses	1.0M	
Pretax profit	0.5M	
% Profit	5%	

17

17.10. A company has received an order for four sophisticated space widgets. The buyer will take delivery of one unit at the end of the first year and one unit at the end of each of the succeeding three years. He will pay for a unit immediately upon receipt and not before. However, the manufacturer can make the units ahead of time and store them at negligible cost for future delivery.

The chief component of cost of the space widget is labor at $25 per h. All units made in the same year can take advantage of an 80 percent learning curve. The first unit requires 100,000 h of labor. Learning occurs only in one year and is not carried over from year to year. If money is worth 16 percent after a 52 percent tax rate, decide whether it would be more economical to build four units the first year and store them, or build one unit in each of the four years.

17.11. Develop a cost model to compare the cost of drilling 1000 holes in steel plate with a standard high-speed steel drill and a TiN-coated H.S.S. drill. Each hole is 1 in. deep. The drill feed is 0.010 per rev. Machining time costs $10 per minute, and the cost of changing a tool is $5.

| | | Tool Life (No. of Holes) | |
	Price of a Drill	500 rpm	900 rpm
Std. H.S.S. drill	$12.00	750	80
TiN-coated H.S.S.	$36.00	1700	750

(a) Compare the costs at fixed conditions of 500 rpm.
(b) Compare the costs at a constant tool life of 750 holes.

17.12. Determine which system is more economical on a life cycle costing basis.

	System A	System B
Initial cost	$300,000	$240,000
Installation	23,000	20,000
Useful life	12 years	12 years
Operators needed	1	2
Operating hours	2100	2100
Operating wage rate	$20 per h	$20 per h
Parts and supplies cost (% of initial cost)	1%	2%
Power	8 kW at 10¢/kWh	9 kW at 10¢/kWh
Escalation of operating costs	6%	6%
Mean time between failures	600 h	450 h
Mean time to repair	35 h	45 h
Maintenance wage rate	$23 per h	$23 per h
Maintenance escalation rate	6%	6%
Desired rate of return	10%	10%
Tax rate	45%	45%

17.13. Discuss the automobile safety standards and air pollution standards in terms of the concept of life cycle costs.

Appendix A

AREA UNDER THE CUMULATIVE DISTRIBUTION FUNCTION FOR Z

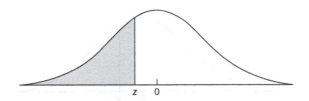

z	0.00	0.01	0.02	0.03	0.04	0.05	0.06	0.07	0.08	0.09
−3.6	.0002	.0002	.0001	.0001	.0001	.0001	.0001	.0001	.0001	.0001
−3.5	.0002	.0002	.0002	.0002	.0002	.0002	.0002	.0002	.0002	.0002
−3.4	.0003	.0003	.0003	.0003	.0003	.0003	.0003	.0003	.0003	.0002
−3.3	.0005	.0005	.0005	.0004	.0004	.0004	.0004	.0004	.0004	.0003
−3.2	.0007	.0007	.0006	.0006	.0006	.0006	.0006	.0005	.0005	.0005
−3.1	.0010	.0009	.0009	.0009	.0008	.0008	.0008	.0008	.0007	.0007
−3.0	.0013	.0013	.0013	.0012	.0012	.0011	.0011	.0011	.0010	.0010
−2.9	.0019	.0018	.0018	.0017	.0016	.0016	.0015	.0015	.0014	.0014
−2.8	.0026	.0025	.0024	.0023	.0023	.0022	.0021	.0021	.0020	.0019
−2.7	.0035	.0034	.0033	.0032	.0031	.0030	.0029	.0028	.0027	.0026
−2.6	.0047	.0045	.0044	.0043	.0041	.0040	.0039	.0038	.0037	.0036
−2.5	.0062	.0060	.0059	.0057	.0055	.0054	.0052	.0051	.0049	.0048
−2.4	.0082	.0080	.0078	.0075	.0073	.0071	.0069	.0068	.0066	.0064
−2.3	.0107	.0104	.0102	.0099	.0096	.0094	.0091	.0089	.0087	.0084
−2.2	.0139	.0136	.0132	.0129	.0125	.0122	.0119	.0116	.0113	.0110
−2.1	.0179	.0174	.0170	.0166	.0162	.0158	.0154	.0150	.0146	.0143
−2.0	.0228	.0222	.0217	.0212	.0207	.0202	.0197	.0192	.0188	.0183
−1.9	.0287	.0281	.0274	.0268	.0262	.0256	.0250	.0244	.0239	.0233
−1.8	.0359	.0351	.0344	.0336	.0329	.0322	.0314	.0307	.0301	.0294
−1.7	.0446	.0436	.0427	.0418	.0409	.0401	.0392	.0384	.0375	.0367
−1.6	.0548	.0537	.0526	.0516	.0505	.0495	.0485	.0475	.0465	.0455
−1.5	.0668	.0655	.0643	.0630	.0618	.0606	.0594	.0582	.0571	.0559
−1.4	.0808	.0793	.0778	.0764	.0749	.0735	.0721	.0708	.0694	.0681
−1.3	.0968	.0951	.0934	.0918	.0901	.0885	.0869	.0853	.0838	.0823
−1.2	.1151	.1131	.1112	.1093	.1075	.1056	.1038	.1020	.1003	.0985
−1.1	.1357	.1335	.1314	.1292	.1271	.1251	.1230	.1210	.1190	.1170
−1.0	.1587	.1562	.1539	.1515	.1492	.1469	.1446	.1423	.1401	.1379
−0.9	.1841	.1814	.1788	.1762	.1736	.1711	.1685	.1660	.1635	.1611
−0.8	.2119	.2090	.2061	.2033	.2005	.1977	.1949	.1922	.1894	.1867
−0.7	.2420	.2389	.2358	.2327	.2296	.2266	.2236	.2206	.2177	.2148
−0.6	.2743	.2709	.2676	.2643	.2611	.2578	.2546	.2514	.2483	.2451
−0.5	.3085	.3050	.3015	.2981	.2946	.2912	.2877	.2843	.2810	.2776
−0.4	.3446	.3409	.3372	.3336	.3300	.3264	.3228	.3192	.3156	.3121
−0.3	.3821	.3783	.3745	.3707	.3669	.3632	.3594	.3557	.3520	.3483
−0.2	.4207	.4168	.4129	.4090	.4052	.4013	.3974	.3936	.3897	.3859
−0.1	.4602	.4562	.4522	.4483	.4443	.4404	.4364	.4325	.4286	.4247
−0.0	.5000	.4960	.4920	.4880	.4840	.4801	.4761	.4721	.4681	.4641

Appendix A (Continued)

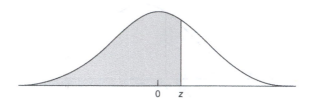

z	0.00	0.01	0.02	0.03	0.04	0.05	0.06	0.07	0.08	0.09
0.0	.5000	.5040	.5080	.5120	.5160	.5199	.5239	.5279	.5319	.5359
0.1	.5398	.5438	.5478	.5517	.5557	.5596	.5636	.5675	.5714	.5753
0.2	.5793	.5832	.5871	.5910	.5948	.5987	.6026	.6064	.6103	.6141
0.3	.6179	.6217	.6255	.6293	.6331	.6368	.6406	.6443	.6480	.6517
0.4	.6554	.6591	.6628	.6664	.6700	.6736	.6772	.6808	.6844	.6879
0.5	.6915	.6950	.6985	.7019	.7054	.7088	.7123	.7157	.7190	.7224
0.6	.7257	.7291	.7324	.7357	.7389	.7422	.7454	.7486	.7517	.7549
0.7	.7580	.7611	.7642	.7673	.7704	.7734	.7764	.7794	.7823	.7852
0.8	.7881	.7910	.7939	.7967	.7995	.8023	.8051	.8078	.8106	.8133
0.9	.8159	.8186	.8212	.8238	.8264	.8289	.8315	.8340	.8365	.8389
1.0	.8413	.8438	.8461	.8485	.8508	.8531	.8554	.8577	.8599	.8621
1.1	.8643	.8665	.8686	.8708	.8729	.8749	.8770	.8790	.8810	.8830
1.2	.8849	.8869	.8888	.8907	.8925	.8944	.8962	.8980	.8997	.9015
1.3	.9032	.9049	.9066	.9082	.9099	.9115	.9131	.9147	.9162	.9177
1.4	.9192	.9207	.9222	.9236	.9251	.9265	.9279	.9292	.9306	.9319
1.5	.9332	.9345	.9357	.9370	.9382	.9394	.9406	.9418	.9429	.9441
1.6	.9452	.9463	.9474	.9484	.9495	.9505	.9515	.9525	.9535	.9545
1.7	.9554	.9564	.9573	.9582	.9591	.9599	.9608	.9616	.9625	.9633
1.8	.9641	.9649	.9656	.9664	.9671	.9678	.9686	.9693	.9699	.9706
1.9	.9713	.9719	.9726	.9732	.9738	.9744	.9750	.9756	.9761	.9767
2.0	.9772	.9778	.9783	.9788	.9793	.9798	.9803	.9808	.9812	.9817
2.1	.9821	.9826	.9830	.9834	.9838	.9842	.9846	.9850	.9854	.9857
2.2	.9861	.9864	.9868	.9871	.9875	.9878	.9881	.9884	.9887	.9890
2.3	.9893	.9896	.9898	.9901	.9904	.9906	.9909	.9911	.9913	.9916
2.4	.9918	.9920	.9922	.9925	.9927	.9929	.9931	.9932	.9934	.9936
2.5	.9938	.9940	.9941	.9943	.9945	.9946	.9948	.9949	.9951	.9952
2.6	.9953	.9955	.9956	.9957	.9959	.9960	.9961	.9962	.9963	.9964
2.7	.9965	.9966	.9967	.9968	.9969	.9970	.9971	.9972	.9973	.9974
2.8	.9974	.9975	.9976	.9977	.9977	.9978	.9979	.9979	.9980	.9981
2.9	.9981	.9982	.9982	.9983	.9984	.9984	.9985	.9985	.9986	.9986
3.0	.9987	.9987	.9987	.9988	.9988	.9989	.9989	.9989	.9990	.9990
3.1	.9990	.9991	.9991	.9991	.9992	.9992	.9992	.9992	.9993	.9993
3.2	.9993	.9993	.9994	.9994	.9994	.9994	.9994	.9995	.9995	.9995
3.3	.9995	.9995	.9995	.9996	.9996	.9996	.9996	.9996	.9996	.9997
3.4	.9997	.9997	.9997	.9997	.9997	.9997	.9997	.9997	.9997	.9998
3.5	.9998	.9998	.9998	.9998	.9998	.9998	.9998	.9998	.9998	.9998
3.6	.9998	.9998	.9999	.9999	.9999	.9999	.9999	.9999	.9999	.9999

Appendix B

VALUES OF *t* STATISTIC

The t distribution

Given v, the table gives (a) the one-tail t_0 value with α of the area about it, that is, $P(t \geq t_0) = \alpha$, or (b) the two-tail $+t_0$ and $-t_0$ values with $\alpha/2$ in each tail, that is, $P(t \leq -t_0) + P(t \geq +t_0) = \alpha$

(a) One-tail α

(b) Two-tail α

	One-tail α							Two-tail α					
	0.10	0.05	0.02	0.01	0.005	0.001		0.10	0.05	0.025	0.01	0.005	0.001
	Two-tail α							Two-tail α					
v	0.20	0.10	0.05	0.025	0.01	0.002	v	0.20	0.10	0.05	0.02	0.01	0.002
1	3.078	6.314	12.706	31.821	63.657	318.300	19	1.328	1.729	2.093	2.539	2.861	3.579
2	1.886	2.920	4.303	6.965	9.925	22.327	20	1.325	1.725	2.086	2.528	2.845	3.552
3	1.638	2.353	3.182	4.541	5.841	10.214	21	1.323	1.721	2.080	2.518	2.831	3.527
4	1.533	2.132	2.776	3.747	4.604	7.173	22	1.321	1.717	2.074	2.508	2.819	3.505
5	1.476	2.015	2.571	3.305	4.032	5.893	23	1.319	1.714	2.069	2.500	2.807	3.485
6	1.440	1.943	2.447	3.143	3.707	5.208	24	1.318	1.711	2.064	2.492	2.797	3.467
7	1.415	1.895	2.365	2.998	3.499	4.785	25	1.316	1.708	2.060	2.485	2.787	3.450
8	1.397	1.860	2.306	2.896	3.355	4.501	26	1.315	1.706	2.056	2.479	2.779	3.435
9	1.383	1.833	2.262	2.821	3.250	4.297	27	1.314	1.703	2.052	2.473	2.771	3.421
10	1.372	1.812	2.228	2.764	3.169	4.144	28	1.313	1.701	2.048	2.467	2.763	3.408
11	1.363	1.796	2.201	2.718	3.106	4.025	29	1.311	1.699	2.045	2.462	2.756	3.396
12	1.356	1.782	2.179	2.681	3.055	3.930	30	1.310	1.697	2.042	2.457	2.750	3.385
13	1.350	1.771	2.160	2.650	3.012	3.852	40	1.303	1.684	2.021	2.423	2.704	3.307
14	1.345	1.761	2.145	2.624	2.977	3.787	60	1.296	1.671	2.000	2.390	2.660	3.232
15	1.341	1.753	2.131	2.602	2.947	3.733	80	1.292	1.664	1.990	2.374	2.639	3.195
16	1.337	1.746	2.120	2.583	2.921	3.686	100	1.290	1.660	1.984	2.365	2.626	3.174
17	1.333	1.740	2.110	2.567	2.898	3.646	∞	1.282	1.645	1.960	2.326	2.576	3.090
18	1.330	1.734	2.101	2.552	2.878	3.611							

Reprinted, with permission, from L. Blank, *Statistical Procedures for Engineering, Management and Science*, McGraw-Hill, New York, 1980.

Appendix C

MATERIALS COMMONLY USED IN ENGINEERING COMPONENTS

Metals are indicated by their SAE/AISI designation, e.g., 1040, or their ASTM Specification, e.g., A36. Plastics are indicated by their common name or abbreviation. The most commonly used material is given first in the list.

Component	Materials
Aircraft structural parts	Aluminum alloys 2024, 6061, 7075; Ti alloy 6-4; graphite-epoxy composites
Automotive engine block	Gray cast iron; A356 cast aluminum alloy
Automobile interior	ABS, polypropylene plastics
Automobile bodies	1005 steel; A619 drawing quality; A620 special killed, DQ steel
Automobile exhaust	409 stainless steel
Bearing	52100 high C-Cr steel; 440C stainless steel; bronze, nylon
Beverage container	1100 aluminum; 1005 steel; PET plastics
Biomedical devices	Ti-6Al-4V; 316L stainless steel; Co-Cr-Ni-Mo alloy; tantalum
Boat hulls (small)	6061 aluminum; fiberglass/epoxy composite
Bolts	1020, 1040, 4140 steel
Bridge structure	A36 steel
Cabinets and housings	1010 steel sheet; 356 die cast aluminum; polypropylene; polyethylene; epoxy
Chemical/food processing	304 stainless steel; CP titanium
Compact discs	Polycarbonate plastic
Computer case	ABS plastic; AZ81 magnesium alloy
Crankshaft	Forged 1040 steel; ductile cast iron
Cutting tool	High-speed steel (M2); cemented carbide (W-Co)
Dies for molding	O1 tool steel
Electrical contacts	Phosphor bronze; tungsten; palladium-silver-copper
Electrical wiring	OFHC copper; 1100 aluminum
Engine cylinder liners	Gray cast iron
Fixtures	O1 and A2 tool steel; filled epoxy; 6061 aluminum
Gaskets, O-rings	Neoprene; natural rubber; soft metal sheets
Gears	Carburized 4615 steel; flame-hardened 1045 steel; 4340 Q&T steel; ductile iron; powder metallurgy steel; nylon
Heat exchanger parts	316 stainless steel; CP titanium
Hoses	Neoprene; Buna A (NPR); nylon
Machine parts (general)	A36 steel; 1020 steel
Machine structural parts	A284 steel; 1020 steel
Machine tool base	Gray cast iron; ductile iron; 1020 steel
Nails and wire	1010 steel
Pressure vessels	4340 steel Q&T; carbon fiber/polymer composite
Shafts, light duty	1040 cold drawn bar; 1141 (free-mach. steel) plus surface hardening
Shafts, heavy duty	4140 or 4340 Q&T; 8620 plus carburized surface
Springs, coil	1080 steel (music wire); 9255 steel Q&T
Truck/railcar frames	A27 and A656 steel
Truck/railcar sides	6061 aluminum
Valve bodies	Ductile cast iron; cast stainless steel

NAME INDEX

Note: Page numbers in *italics* are references in web chapter.

SUBJECT INDEX